S0-CAS-642

JOURNAL OF CHROMATOGRAPHY LIBRARY - VOLUME 23A

# chromatography of alkaloids

## part A: thin-layer chromatography

JOURNAL OF CHROMATOGRAPHY LIBRARY


Volume 1 Chromatography of Antibiotics, by G.H. Wagman and M.J. Weinstein

Volume 2 Extraction Chromatography, edited by T. Braun and G. Ghersini

Volume 3 Liquid Column Chromatography. A Survey of Modern Techniques and Applications, edited by Z. Deyl, K. Macek and J. Janák

Volume 4 Detectors in Gas Chromatography, by J.Ševčík

Volume 5 Instrumental Liquid Chromatography. A Practical Manual on High-Performance Liquid Chromatographic Methods, by N.A. Parris

Volume 6 Isotachophoresis. Theory, Instrumentation and Applications, by F.M. Everaerts, J.L. Beckers and Th.P.E.M. Verheggen

Volume 7 Chemical Derivatization in Liquid Chromatography, by J.F. Lawrence and R.W. Frei

Volume 8 Chromatography of Steroids, by E. Heftmann

Volume 9 HPTLC — High Performance Thin-Layer Chromatography, edited by A. Zlatkis and R.E. Kaiser

Volume 10 Gas Chromatography of Polymers, by V.G. Berezkin, V.R. Alishoyev and I.B. Nemirovskaya

Volume 11 Liquid Chromatography Detectors, by R.P.W. Scott

Volume 12 Affinity Chromatography, by J. Turková

Volume 13 Instrumentation for High-Performance Liquid Chromatography, edited by J.F.K. Huber

Volume 14 Radiochromatography. The Chromatography and Electrophoresis of Radiolabelled Compounds, by T.R. Roberts

Volume 15 Antibiotics. Isolation, Separation and Purification, edited by M.J. Weinstein and G.H. Wagman

Volume 16 Porous Silica. Its Properties and Use as Support in Column Liquid Chromatography, by K.K. Unger

Volume 17 75 Years of Chromatography — A Historical Dialogue, edited by L.S. Ettre and A. Zlatkis

Volume 18A Electrophoresis. A survey of Techniques and Applications. Part A: Techniques, edited by Z. Deyl

Volume 18B Electrophoresis. A Survey of Techniques and Applications. Part B: Applications, edited by Z. Deyl

Volume 19 Chemical Derivatization in Gas Chromatography, by J. Drozd

Volume 20 Electron Capture. Theory and Practice in Chromatography, edited by A. Zlatkis and C.F. Poole

Volume 21 Environmental Problem Solving using Gas and Liquid Chromatography, by R.L. Grob and M.A. Kaiser

Volume 22A Chromatography. Fundamentals and Applications of Chromatographic and Electrophoretic Methods. Part A: Fundamentals, edited by E. Heftmann

Volume 22B Chromatography. Fundamentals and Applications of Chromatographic and Electrophoretic Methods. Part B: Applications, edited by E. Heftmann

Volume 23A Chromatography of alkaloids. Part A: Thin-Layer Chromatography, by A. Baerheim Svendsen and R. Verpoorte

JOURNAL OF CHROMATOGRAPHY LIBRARY - volume 23A

# chromatography of alkaloids

## part A: thin-layer chromatography

A. Baerheim Svendsen

and

R. Verpoorte

Department of Pharmacognosy, State University Leyden, Leyden, The Netherlands

ELSEVIER SCIENTIFIC PUBLISHING COMPANY
Amsterdam — Oxford — New York 1983

ELSEVIER SCIENTIFIC PUBLISHING COMPANY
Molenwerf 1
P.O. Box 211, 1000 AE Amsterdam, The Netherlands

*Distributors for the United States and Canada:*

ELSEVIER SCIENCE PUBLISHING COMPANY INC.
52, Vanderbilt Avenue
New York, NY 10017

**Library of Congress Cataloging in Publication Data**

Baerheim-Svendsen, A.
Chromatography of alkaloids.

(Journal of chromatography library ; v. 23A-
Bibliography: p.
Includes indexes.
Contents: pt. A. Thin-layer chromatography.
1. Alkaloids--Analysis. 2. Chromatographic analysis.
I. Verpoorte, R. II. Title. III. Series.
QD421.B125 1982 547.7'2046 82-20976
ISBN 0-444-42145-9 (U.S. : pt. A)

ISBN 0-444-42145-9 (Vol. 23A)
ISBN 0-444-41616-1 (Series)



Printed in The Netherlands

Dedicated to the memory of
Friedrich Wilhelm Sertürner
who was the first to isolate an alkaloid (morphine).
On the occasion of the 200th anniversary of his birth.

CONTENTS

Preface . . . . . . . . . . XI
Introduction . . . . . . . . . . XIII

I. GENERAL PART

Chapter 1. Adsorbents, solvent systems and TLC techniques . . . . . . . 3
Chapter 2. Detection of alkaloids in TLC . . . . . . . . . . 11
Chapter 3. TLC separation and identification of alkaloids in general . 19
Chapter 4. Isolation of alkaloids . . . . . . . . . . 51
4.1. Isolation methods . . . . . . . . . . 51
4.2. Solvents and artefact formation . . . . . . . . . . 53

II. SPECIAL PART

II.1. *Pyrrolidine, pyrrolizidine, pyridine, piperidine and quinolizidine alkaloids*
Chapter 5. Pyrrolidine, pyrrolizidine, pyridine, piperidine and quinolizidine alkaloids . . . . . . . . . . 61
Chapter 6. Tobacco alkaloids . . . . . . . . . . 79
II.2. *Tropane alkaloids*
Chapter 7. Tropine alkaloids . . . . . . . . . . 91
Chapter 8. Pseudotropine alkaloids . . . . . . . . . . 113
II.3. *Quinoline alkaloids*
Chapter 9. *Cinchona* alkaloids . . . . . . . . . . 131
II.4. *Phenylethylamine and isoquinoline alkaloids*
Chapter 10. Cactus alkaloids . . . . . . . . . . 159
Chapter 11. Isoquinoline alkaloids . . . . . . . . . . 171
11.1. Protoberberine and protopine alkaloids . . . . . . . . 172
11.2. Benzophenanthridine alkaloids . . . . . . . . . . 174
11.3. *Hydrastis* (phthalide) alkaloids and benzylisoquinoline alkaloids . . . . . . . . . . 175
11.4. Bisbenzylisoquinoline alkaloids; tertiary and quaternary alkaloids . . . . . . . . . . 177
11.5. Aporphine alkaloids . . . . . . . . . . 180
11.6. Rhoeadine alkaloids . . . . . . . . . . 181

11.7. *Ipecacuanha* alkaloids . . . . . . . . . . . . . . . . 182
11.8. Various isoquinoline alkaloids . . . . . . . . . . . . 184
Chapter 12. Opium alkaloids . . . . . . . . . . . . . . . . . . . . . 221
12.1. Opium alkaloids . . . . . . . . . . . . . . . . . . . . . 221
12.2. Drugs of abuse . . . . . . . . . . . . . . . . . . . . . 233
II.5. *Indole alkaloids*
Chapter 13. Terpenoid indole alkaloids and simple indole alkaloids . . 307
13.1. *Strychnos* alkaloids . . . . . . . . . . . . . . . . . 307
13.2. *Rauwolfia* alkaloids . . . . . . . . . . . . . . . . . 310
13.3. Heteroyohimbine and oxindole alkaloids . . . . . . . . 316
13.4. *Vinca* and *Catharanthus* alkaloids . . . . . . . . . . 318
13.5. Physostigmine . . . . . . . . . . . . . . . . . . . . . 321
13.6. Various indole alkaloids . . . . . . . . . . . . . . . 323
Chapter 14. Ergot alkaloids . . . . . . . . . . . . . . . . . . . . . 375
Chapter 15. *Psilocybe* alkaloids . . . . . . . . . . . . . . . . . . 409
II.6. *Steroidal alkaloids*
Chapter 16. Steroidal alkaloids . . . . . . . . . . . . . . . . . . . 415
II.7. *Miscellaneous alkaloids*
Chapter 17. Xanthine alkaloids . . . . . . . . . . . . . . . . . . . 435
Chapter 18. Diterpene alkaloids . . . . . . . . . . . . . . . . . . . 461
Chapter 19. Colchicine and related alkaloids . . . . . . . . . . . . 469
Chapter 20. Imidazole alkaloids . . . . . . . . . . . . . . . . . . . 477
Chapter 21. Quaternary ammonium compounds . . . . . . . . . . . . . . 483
*Appendix*
Detection methods and spray reagents . . . . . . . . . . . . . . . . 497
*Subject index* . . . . . . . . . . . . . . . . . . . . . . . . . . . . 513
*Compound index* . . . . . . . . . . . . . . . . . . . . . . . . . . . 517

## LIST OF ABBREVIATIONS USED FOR DESCRIPTION OF TLC SYSTEMS

| | |
|---|---|
| AcOH | acetic acid |
| AmOAc | amyl acetate |
| AmOH | amyl alcohol |
| $BuNH_2$ | butylamine |
| BuOH* | butanol |
| $Bu_2O$ | dibutyl ether |
| cell | cellulose |
| DEA | diethylamine |
| DMFA | dimethylformamide |
| EtOAc | ethyl acetate |
| EtOH | ethanol |
| $Et_2O$ | diethyl ether |
| FMA | formamide |
| imp | impregnated |
| isoAmOH | isoamyl alcohol |
| isoPrOH | isopropanol |
| $(isoPr)_2O$ | diisopropyl ether |
| $Me_2CO$ | acetone |
| MeEtCO | methyl ethyl ketone |
| MeOAc | methyl acetate |
| MeOH | methanol |
| MePrCO | methyl propyl ketone |
| lt. petr. | light petroleum |
| PrOH | propanol |
| $Pr_2O$ | dipropyl ether |
| THF | tetrahydrofuran |
| TrEA | triethylamine |
| two-dim | two-dimensional |

*In the Tables the prefix *n*, *sec* or *t* is added if it was mentioned in the original literature.

## PREFACE

In the analysis of alkaloids, thin-layer chromatography (TLC) has so far been applied more than any other chromatographic technique, as can be seen from the hundreds of papers published.

In this part of *Chromatography of Alkaloids* TLC is dealt with. The alkaloids have been classified in groups according to their chemical skeleton (indole, isoquinoline, piperidine, etc.). In some instances, however, alkaloids of the same botanical origin (opium alkaloids, ipecacuanha alkaloids) and alkaloids of importance from a special point of view (drugs of abuse) have been treated separately. The papers published have been examined chronologically and the references are listed at the end of each chapter in the same way to give the reader the opportunity of tracing the development that has taken place in the TLC analysis within each group of alkaloids.

In a number of published investigations the aim has been to develop a systematic approach to the identification of alkaloids by means of TLC. Other studies deal with newly developed TLC systems which are useful for various alkaloids. Publications of this kind have been dealt with separately.

Data collected in the literature on the separation of the many representatives of a group of alkaloids, *viz.*, solvent systems, adsorbents and the aim of the TLC analysis, have been listed in tables. A selection of such TLC separation systems which proved to be the best for solving a series of specific analytical problems has been made, and they are presented in other tables together with the $hR_F$ values obtained for the alkaloids analysed. Detection reagents for the various groups of alkaloids are described, together with the colours obtained with the various reagents. The compositions and the preparation of the most important TLC spray reagents for alkaloids are given in the Appendix.

Sample preparation is less important in TLC than in GLC and HPLC, as a TLC plate is used only once. In TLC it is not necessary to the same extent as in GLC and HPLC to remove non-alkaloidal compounds from the sample to be analysed prior to the analysis, provided that they do not interfere in the separation. However, for the analysis of extremely small amounts of alkaloids in mixtures with other compounds, such as in biological material, extensive purification and removal of other compounds from the sample may be necessary. Isolation and clean-up procedures for such analyses are given in Chapter 12.

Quantitative analysis in TLC can be performed either directly on the plate (*in situ*) or indirectly, after elution of the alkaloid from the spot on the plate. For the direct, densitometric determinations the necessary data, wavelengths, etc., are given. For indirect determinations all kinds of quantitative methods can be applied after the elution of the alkaloid to be analysed from the plate. Quantitative TLC is discussed in detail in papers by Shellard[1] and Touchstone and Sherma[2].

Special techniques involved in the sample preparation prior to or in connection with the application of the alkaloids on the TLC plate, e.g., the TAS technique and reaction chromatography, are described in a separate section for each group of alkaloids.

Lists of abbreviations used in the tables for alkaloids, etc., are given in each chapter. The abbreviations used for organic solvents are listed on page VII.

INTRODUCTION

"No other discovery has exerted as great influence and widened the field of investigation of the organic chemist as much as Tswett's chromatographic adsorption analysis. Research in the field of vitamins, hormones, carotinoids and numerous other natural compounds could never have progressed so rapidly and achieved such great results if it had not been for this new method, which also disclosed the enormous variety of closely related compounds in nature". (Karrer[3].)

The application of chromatographic techniques in the isolation, purification and fractionation of alkaloids has also created new possibilities in the important field of alkaloid research, which started with the isolation of morphine by Sertürner in 1806. The introduction of many groups of important alkaloids in therapy, such as the curare alkaloids, the *Rauwolfia* alkaloids and the *Catharanthus* alkaloids, was made possible to a great extent owing to the application of chromatographic separation methods.

In addition to the isolation and purification of alkaloids from plant material, investigations on alkaloids are also performed for several other reasons: for studies of the distribution of alkaloids in various parts of the plant as a function of the development of the plant, the plant growth conditions and time of harvesting, for the quantitative determination of total and individual alkaloids, for the control of crude drugs and pharmaceutical alkaloid-containing preparations, for international control of traffic in narcotics and other alkaloid-containing drugs, for toxicological investigations and for many other chemical, pharmaceutical, plant physiological, pharmacological and toxicological reasons.

Although Tswett[4,5] is commonly credited with the first successful application of chromatography, no great success was made until its use by Kuhn et al.[6] attracted attention. The application of chromatography in alkaloid research started when Karrer and Nielsen[7] separated quinine and cinchonine on a Floridin XXF column by adsorption chromatography. Späth et al.[8] isolated sanguinarine from *Sanguinaria canadensis* L. by adsorption on aluminium oxide, and Kondo et al.[9], using the same adsorbent, isolated methylisochondodendrine from *Cissampelos insularis* Makido. Stoll and Hofmann[10] also applied adsorption chromatography on aluminium oxide to separate ergometrine and ergometrinine. In their studies on Calebas curare alkaloids, Wieland et al.[11] separated the alkaloids on aluminium oxide as reineckates and perchlorates. Pease and Elderfield[12] isolated α- and β-earleine from *Astragalus earlei* by adsorption chromatography of their picrates.

The development of partition chromatography by Martin and Synge[13,14] created further new possibilities in the complex field of alkaloid research. Evans and

Partridge[15] introduced the method into alkaloid analysis by using a Kieselguhr column impregnated with phosphate buffer as stationary phase and diethyl ether as mobile phase for the separation of hyoscyamine and scopolamine.

The introduction of partition chromatography brought about a revolution in alkaloid research - especially when partition chromatography was applied in the form of paper chromatography. This technique was introduced by Consden et al.[16] and the first applications in alkaloid analysis were published by Munier and Macheboeuf[17] on various groups of alkaloids, and by Foster et al.[18] on ergot alkaloids.

In this work the available literature on the chromatography of alkaloids has been reviewed. However, because at the present time *thin-layer chromatography, gas-liquid chromatography and high-performance liquid chromatography* are the chromatographic techniques mostly applied in alkaloid research, we have dealt with these techniques only.

Review articles on the chromatography of alkaloids can be found in *Handbuch der Papierchromatographie* by I.M. Hais and K. Macek[19]; *Chromatography* by E. Lederer and M. Lederer[21], *Dünnschicht-Chromatographie* by E. Stahl[21] and *Chromatography* by E. Heftmann[22].

To facilitate the use of the information available in the literature on the application of the three above chromatographic techniques in the various fields of alkaloid research, and because of the great amount of information available, it was decided to publish this work in two volumes, one for thin-layer chromatography and one for gas-liquid and high-performance liquid chromatography.

In each volume the alkaloids have been classified in groups according to their chemical skeleton. However, often a classification into subgroups according to the botanical origin of the alkaloids has been used.

There is generally no *a priori* "best solution" to any analytical problem: a number of solutions can be tried and one will emerge as the most acceptable for a number of reasons. Different scientists will probably prefer different methods, depending upon personal choice and experience, the instrumentation available, and so on.

These volumes of *Chromatography of Alkaloids* should be used as a source of information, ideas and methods tested in order tackle a problem effectively.

REFERENCES

1 E.J. Shellard (Editor), *Quantitative Paper and Thin-layer Chromatography*, Academic Press, London, 1968, 140 pp.
2 J.C. Touchstone and J. Sherma, *Densitometry in Thin-layer Chromatography, Practice and Applications*, Wiley, Chicester, 1979, 747 pp.
3 P. Karrer, *Int. Congr. Pure Appl. Chem.*, London, 1947.
4 M. Tswett, *Ber. Deut. Bot. Ges.*, 24 (1906) 316.
5 M. Tswett, *Ber. Deut. Bot. Ges.*, 24 (1906) 384.
6 R. Kuhn, A. Wintersteiner and E. Lederer, *Z. Physiol. Chem.*, 197 (1931) 141.

7 P. Karrer and N. Nielsen, *Trennung von Substanzgemischen in Chromatogramm und Ultrachromatogramm, Zangger-Festschrift,* Rascher & Cp, Zürich, 1934, p. 954.
8 E. Späth, F. Schlemmer, G. Schenck and A. Gempp, *Ber. Deut. Chem. Ges.*, 70 (1937) 1677.
9 H. Kondo, M. Tomita and S. Uyeo, *Ber. Deut. Chem. Ges.*, 70 (1937) 1890.
10 A. Stoll and A. Hofmann, *Z. Physiol. Chem.*, 251 (1938) 155.
11 H. Wieland, W. Konz and R. Sonderhoff, *Justus Liebigs Ann. Chem.*, 525 (1937) 152.
12 D.C. Pease and R.C. Elderfield, *J. Org. Chem.*, 5 (1940) 192.
13 A.J.P. Martin and R.L.M. Synge, *Biochem. J.*, 35 (1941) 91.
14 A.J.P. Martin and R.L.M. Synge, *Biochem. J.*, 35 (1941) 1358.
15 W.C. Evans and M.W. Partridge, *Quart. J. Pharm. Pharmacol.*, 21 (1948) 126.
16 R. Consden, A.H. Gordon and A.J.P. Martin, *Biochem. J.*, 38 (1944) 224.
17 R. Munier and M. Macheboeuf, *Bull. Soc. Chim. Biol.*, 31 (1949) 114.
18 G.E. Foster, J. MacDonald and T.S.G. Jones, *J. Pharm. Pharmacol.*, 1 (1949) 802.
19 I.M. Hais and K. Macek, *Handbuch der Papierchromatographie,* VEB Gustav Fischer Verlag, Jena, 1958, p. 545.
20 E. Lederer and M. Lederer, *Chromatography*, Elsevier, Amsterdam, 1957, p. 209.
21 E. Stahl, *Dünnschicht-Chromatographie,* Springer Verlag, Berlin, 2nd ed., 1967, p. 405.
22 E. Heftmann, *Chromatography*, Van Nostrand Reinhold, New York, 3rd ed., 1975, p. 697.

# I. GENERAL PART

CONTENTS

## I. GENERAL PART

CHAPTER 1.

ADSORBENTS, SOLVENT SYSTEMS AND TLC TECHNIQUES....3
1.1. Adsorbents....3
1.1.1. Silica gel....3
1.1.2. Aluminium oxide....4
1.1.3. Miscellaneous....4
1.2. Solvent systems....5
1.3. Tank and development....5
1.4. Application of the sample....9
References....10

CHAPTER 2.

DETECTION OF ALKALOIDS IN TLC....11
2.1. Dragendorff's reagent....11
2.2. Potassium iodoplatinate....14
2.3. Other reagents....15
References....18

CHAPTER 3.

TLC SEPARATION AND IDENTIFICATION OF ALKALOIDS IN GENERAL....19
3.1. Identification....19
3.2. TLC of various alkaloids....38
References....48

CHAPTER 4.

ISOLATION OF ALKALOIDS....51
4.1. Isolation methods....51
4.2. Solvents and artefact formation....53
References....57

Chapter 1

# ADSORBENTS, SOLVENT SYSTEMS AND TLC TECHNIQUES

## 1.1. ADSORBENTS

### *1.1.1. Silica gel*

Silica gel is the most widely used adsorbent in the TLC analysis of alkaloids. With the commonly used 0.25 mm thick layer a reasonable linear adsorption isotherm is generally obtained for amounts of 5-50 μg of alkaloid. However, because of the weakly acidic properties of the silanol groups (pH = 4-5) chemisorption of basic compounds may occur. This is observed as pronounced tailing from the point of application of the alkaloid and its final spot after the chromatographic run. Chemisorption can be prevented by using basic mobile phases or by impregnating the silica gel with basic buffers or mineral bases*. As an alternative the alkaloids can be chromatographed as their salts by using polar, acidic mobile phases or ion-pair chromatography. A review of the latter technique was given by Tomlinson et al.[4].

Neutral mobile phases in combination with silica gel layers may lead to the formation of double spots if basic compounds are applied as their salts. This phenomenon has been described by Wesley-Hadzya[5] for fenfluramine and by Beckett and Choulis[6] for some sympathomimetic amines on cellulose layers. The formation of double spots is explained as a partial deprotonation of the amine, resulting in separation of the salt (ion pair) and the base in a neutral solvent system. A review of double spot formation in TLC was given by Keller and Giddings[7].

In the first period of TLC analysis, silica gel was always activated by heating at 110$^{o}$C for 30 min prior to its use, and the plates were stored over drying agents. The advantage of such activation is, however, doubtful. According to Geiss[8] activation is of little or no use, because during the application of the sample to be analysed the silica gel will adsorb moisture from the air; within 3 min an activated silica gel layer will then adsorb about 50% of its maximum water content. Therefore, more reproducible analytical results are obtained when the silica gel plates are stored in open air. In this way a relatively constant activity of the silica gel layers is obtained (under normal laboratory conditions).

---

*For more extensive discussions, see Refs. 1, 2 and 3.

References p. 10

Only if the application of the sample and the development of the TLC plates in the solvent system are performed under strictly controlled conditions can activation be of value.

The properties of silica gels from different manufacturers may differ widely as to active surface area (300-1000 $m^2/g$), particle size (10-40 μm), pore size (20-150 Å), binding agents (plaster of Paris, starch) and fluorescent indicator. Therefore, differences in separations may occur even if the same solvent system is used because of such differences in silica gel quality. Unger[9] has discussed the varying properties of silica gel.

### *1.1.2. Aluminium oxide*

Aluminium oxide is weakly basic (pH = 9). However, it also contains acidic adsorption sites and, by treatment with acids, neutral or acidic aluminium oxide can be obtained. Little is known about the adsorption mechanism, but hydroxyl groups are probably of minor importance, because upon heating at 200°C the activity increases, whereas the hydroxyl groups disappear.

Aluminium oxide exists in different modifications, the properties of which vary considerably. The surface area can be in the range 5-200 $m^2/g$. The activity, expressed on the scale proposed by Brockmann, varies from I to V, of which I has the highest activity, and the lowest water content.

Activated aluminium oxide layers of 0.2 mm thickness adsorb water from the air extremely rapidly. Upon exposure to open air for 4-12 min saturation will be achieved, the activity being about III. For normal laboratory use, activation of aluminium oxide layers is not necessary. Activation may even influence the reproducibility in a negative way.

Because of the weakly basic properties of aluminium oxide, neutral solvent systems are often used in the TLC analysis of alkaloids.

### *1.1.3. Miscellaneous*

Kieselguhr and particularly cellulose have found some application as supports in the TLC partition chromatography of alkaloids. Magnesium oxide and calcium carbonate have been used as sorbents in the analysis of alkaloids, because of their solubility in mineral acids. A few applications of the organic sorbent polyamide have also been reported.

Some applications of the adsorbents (stationary phases) mentioned in the TLC analysis af alkaloids are given in Chapter 3.

## 1.2. SOLVENT SYSTEMS

The solvent system is responsible for the great versatility of TLC and determines to a great extent the quality of the separations. In partition chromatography the solvent system must be immiscible with the stationary phase (usually water, formamide, etc.), but not in adsorption chromatography, where the choice of solvent system is free. All kind of combinations of solvents can be made to achieve the best possible solvent for a separation. However, increased complexity of the composition of a solvent system causes increased problems for the reproducibility of the analytical results. Therefore, for the sake of reproducibility, single-component solvent systems are to be preferred. However, only very seldom can such solvent systems be found that will give a satisfactory TLC separation of the alkaloids in question. The polarity of a solvent system can be varied by addition of non-polar or strongly polar components to medium polarity solvents. Often basic modifiers (ammonia, diethylamine and even triethylamine and triethanolamine) are added to reduce tailing due to chemisorption of alkaloids.

For the use of pure solvents in adsorption chromatography, eluotropic series have been described[2,8]. However, the solvent strength of mixtures of solvents, as usual in TLC, is difficult to derive from these data. Snyder[10,11,12] has described a classification of solvents in which not only the polarity of a solvent, but also its possibility to act as a proton acceptor ($x_a$), proton donor ($x_d$) or strong dipole interactor ($x_n$) are taken into account. These parameters can be used to predict the polarity of a mixture of solvents by addition of volume fractions of the polarity parameter ($P'$).

Solvent mixtures can be composed with similar $P'$ values, having different $x_e$, $x_d$ and $x_n$ values (see Table 1.1.) and thus different selectivities. A categorization of common solvents according to the parameters $x_e$, $x_d$ and $x_n$ as present by Snyder is given in Table 1.2. More details were given by Snyder[10,11,12].

## 1.3. TANK AND DEVELOPMENT

The ascending development technique with TLC plates in tanks in which the atmosphere is saturated with the vapour of the solvent system by lining the walls with filter-paper is mostly used. If solvents with a boiling point below 100°C are used, saturation is achieved within 5 min. If the walls are not lined with filter-paper, saturation may take several hours[8]. With a few exceptions all TLC separations described for alkaloids have been performed with this technique.

In the so-called sandwich chamber technique, the TLC plate and a glass plate are clamped together with a thin strip along three of the edges. The open end is

References p. 10

TABLE 1.1.

CLASSIFICATION OF SOLVENTS[11]

| No. | Solvent | $P'$[a] | Group | $x_e$ | $x_d$ | $x_n$ |
|---|---|---|---|---|---|---|
| 1 | carbon disulfide | 0.3 | -[b] | - | - | - |
| 2 | cyclohexane | 0.2 | -[b] | - | - | - |
| 3 | triethylamine | 1.9 | I[c] | 0.56 | 0.12 | 0.32 |
| 4 | ethyl ether | 2.8 | I | 0.53 | 0.13 | 0.34 |
| 5 | ethylbromide[d] | - | - | - | - | - |
| 6 | *n*-hexane | 0.1 | -[b] | - | - | - |
| 7 | iso-octane | 0.1 | -[b] | - | - | - |
| 8 | tetrahydrofuran | 4.0 | III | 0.38 | 0.20 | 0.42 |
| 9 | *i*-propylether | 2.4 | I | 0.48 | 0.14 | 0.38 |
| 10 | toluene | 2.4 | VII | 0.25 | 0.28 | 0.47 |
| 11 | benzene | 2.7 | VII | 0.23 | 0.32 | 0.45 |
| 12 | *p*-xylene | 2.5 | VII | 0.27 | 0.28 | 0.45 |
| 13 | chloroform | 4.1 | VIII[c] | 0.25 | 0.41 | 0.33 |
| 14 | carbon tetrachloride | 1.6 | -[b] | - | - | - |
| 15 | buthyl ether | 2.1 | I | 0.44 | 0.18 | 0.38 |
| 16 | dichloromethane | 3.1 | V | 0.29 | 0.18 | 0.53 |
| 17 | *n*-decane | 0.4 | -[b] | - | - | - |
| 18 | chlorobenzene | 2.7 | VII | 0.23 | 0.33 | 0.44 |
| 19 | bromobenzene | 2.7 | VII | 0.24 | 0.33 | 0.43 |
| 20 | fluorobenzene | 3.2 | VII | 0.24 | 0.32 | 0.45 |
| 21 | 2,6-lutidine | 4.5 | III | 0.45 | 0.20 | 0.36 |
| 22 | squalane | 1.2 | -[b] | - | - | - |
| 23 | hexafluorobenzene[d] | - | - | - | - | - |
| 24 | ethoxybenzene | 3.3 | VII | 0.28 | 0.28 | 0.44 |
| 25 | 2-picoline | 4.9 | III | 0.44 | 0.21 | 0.36 |
| 26 | dichloroethane[d] | 3.5 | V | 0.30 | 0.21 | 0.49 |
| 27 | ethyl acetate | 4.4 | VIa | 0.34 | 0.23 | 0.43 |
| 28 | iodobenzene | 2.8 | VII | 0.24 | 0.35 | 0.41 |
| 29 | methylethyl ketone | 4.7 | VIa | 0.35 | 0.22 | 0.43 |
| 30 | *bis*-(2-ethoxy ethyl)ether | 4.6 | VIa | 0.37 | 0.21 | 0.43 |
| 31 | anisole | 3.8 | VII | 0.27 | 0.29 | 0.43 |
| 32 | *n*-octanol | 3.4 | II | 0.56 | 0.18 | 0.25 |
| 33 | cyclohexanone | 4.7 | VIa | 0.36 | 0.22 | 0.42 |
| 34 | *t*-butanol | 4.1 | II | 0.56 | 0.20 | 0.24 |
| 35 | tetramethyl guanidine | 6.1 | I | 0.47 | 0.17 | 0.35 |
| 36 | *i*-pentanol | 3.7 | II | 0.56 | 0.19 | 0.26 |
| 37 | pyridine | 5.3 | III | 0.41 | 0.22 | 0.36 |
| 38 | dioxane[d] | 4.8 | VIa | 0.36 | 0.24 | 0.40 |
| 39 | *n*-butanol | 3.9 | II | 0.59 | 0.19 | 0.25 |
| 40 | *i*-propanol | 3.9 | II | 0.55 | 0.19 | 0.27 |
| 41 | *n*-propanol | 4.0 | II | 0.54 | 0.19 | 0.27 |
| 42 | phenyl ether | 3.4 | VII | 0.27 | 0.32 | 0.41 |
| 43 | acetone | 5.1 | VIa | 0.35 | 0.23 | 0.42 |
| 44 | benzonitrile | 4.8 | VIb | 0.31 | 0.27 | 0.42 |
| 45 | tetramethyl urea | 6.0 | III | 0.42 | 0.19 | 0.39 |
| 46 | benzyl ether | 4.1 | VII | 0.30 | 0.28 | 0.42 |
| 47 | acetophenone | 4.8 | VIa | 0.33 | 0.26 | 0.41 |
| 48 | hexamethyl phosphoric acid triamide | 7.4 | I | 0.47 | 0.17 | 0.37 |
| 49 | ethanol[d] | 4.3 | II | 0.52 | 0.19 | 0.29 |
| 50 | quinoline | 5.0 | III | 0.41 | 0.23 | 0.36 |
| 51 | nitrobenzene | 4.4 | VII | 0.26 | 0.30 | 0.44 |
| 52 | *m*-cresol | 7.4 | VIII | 0.38 | 0.37 | 0.25 |
| 53 | N,N-dimethyl acetamide | 6.5 | III | 0.41 | 0.20 | 0.39 |
| 54 | acetic acid | 6.0 | IV | 0.39 | 0.31 | 0.30 |
| 55 | nitroethane | 5.2 | VII | 0.28 | 0.29 | 0.43 |

Table 1.1. (*continued*)

| No. | Solvent | $P'$[a] | Group | $x_e$ | $x_d$ | $x_n$ |
|---|---|---|---|---|---|---|
| 56 | methanol | 5.1 | II | 0.48 | 0.22 | 0.31 |
| 57 | benzyl alcohol | 5.7 | IV | 0.40 | 0.30 | 0.30 |
| 58 | dimethyl formamide | 6.4 | III | 0.39 | 0.21 | 0.40 |
| 59 | tricresyl phosphate | 4.6 | VIa | 0.36 | 0.23 | 0.41 |
| 60 | methoxy ethanol | 5.5 | III | 0.38 | 0.24 | 0.38 |
| 61 | nonylphenol oxyethylate[d] | | III | 0.38 | 0.22 | 0.40 |
| 62 | N-methyl-2-pyrrolidone | 6.7 | III | 0.40 | 0.21 | 0.39 |
| 63 | acetonitrile | 5.8 | VIb | 0.31 | 0.27 | 0.42 |
| 64 | aniline | 6.3 | VIb | 0.32 | 0.32 | 0.36 |
| 65 | methyl formamide | 6.0 | III | 0.41 | 0.23 | 0.36 |
| 66 | cyano morpholine | 5.5 | VIa | 0.35 | 0.25 | 0.40 |
| 67 | butyrolactone | 6.5 | VIa | 0.34 | 0.26 | 0.40 |
| 68 | nitromethane | 6.0 | VII | 0.28 | 0.31 | 0.40 |
| 69 | dodecafluoroheptanol | 8.8 | VIII | 0.33 | 0.40 | 0.27 |
| 70 | formyl morpholine | 6.4 | VIa | 0.36 | 0.24 | 0.39 |
| 71 | propylene carbonate | 6.1 | VIb | 0.31 | 0.27 | 0.42 |
| 72 | dimethyl sulfoxide | 7.2 | III | 0.39 | 0.23 | 0.39 |
| 73 | tetrafluoropropanol | 8.6 | VIII | 0.34 | 0.36 | 0.30 |
| 74 | tetrahydrothiophene-1,1-dioxide | 6.9 | VIb | 0.33 | 0.28 | 0.39 |
| 75 | *tris*-cyanoethoxypropane | 6.6 | VIb | 0.32 | 0.27 | 0.41 |
| 76 | oxydipropionitrile | 6.8 | VIb | 0.31 | 0.29 | 0.40 |
| 77 | diethylene glycol | 5.2 | III | 0.44 | 0.23 | 0.33 |
| 78 | triethylene glycol | 5.6 | III | 0.42 | 0.24 | 0.34 |
| 79 | ethylene glycol[d] | 6.9[e] | IV | 0.43 | 0.29 | 0.28 |
| 80 | formamide[d] | 9.6[e] | IV | 0.36 | 0.33 | 0.30 |
| 81 | water[d] | 10.2 | VIII | 0.37 | 0.37 | 0.25 |

[a] values multiplied by 0.9, in order to preserve parity with preceding values.
[b] selectivity group irrelevant, because of low $P'$ value.
[c] these two solvents (triethyl amine and chloroform) fall close to groups I and VIII, respectively, but are not included within these groups.
[d] one or more $K$ values missing, so that $P'$ and selectivity parameters require estimated value; in some cases, no reasonable estimate is possible.
[e] approximate value, due to missing $K_o$ value.
(Published with permission of Preston Publ. Inc.).

TABLE 1.2.

CLASSIFICATION OF SOLVENT SELECTIVITY[11]

| Group | Solvent |
|---|---|
| I | aliphatic ethers, tetramethylguanidine, hexamethyl phosphoric acid amide, (trialkyl amines)[a] |
| II | aliphatic alcohols |
| III | pyridine derivatives, tetrahydrofuran, amides (except formamide), glycol ethers, sulfoxides |
| IV | glycols, benzyl alcohol, acetic acid, formamide |
| V | dichloromethane, dichloroethane |
| VI | (a)[b] tricresyl phosphate, aliphatic ketones and esters, poly-ethers ( 30, 61), dioxane<br>(b)[b] sulfones, nitriles (including 75, 76), propylene carbonate |
| VII | aromatic hydrocarbons, halo-substituted aromatic hydrocarbons, nitro compounds, aromatic ethers |
| VIII | fluoroalkanols, *m*-cresol, water, (chloroform)[c] |

[a] somewhat more basic than other group I solvents
[b] this group is rather broad and can be subdivided as indicated into groups VIa and VIb; however, normally there is no point to this in practical usage of the present scheme
[c] somewhat less basic than other group VIII solvents
(Published with permission of Preston Publ. Inc.).

placed in a trough with the solvent system for the development of the chromatogram. Several horizontal modifications of the sandwich chambers have also been described. They offer possibilities for equilibrating the TLC layer with the vapour of the solvent system or for adjusting the relative humidity (Vario KS chamber, VP chamber). A detailed description and discussion of the various methods for development of the TLC plates with the solvents were given by Geiss[8]. The results obtained with the "normal" development technique are usually better, because of less band spreading of the compounds in the sample (smaller spot size). It is also usually faster, and the $hR_F$ values mostly lower than with a sandwich chamber.

Special techniques, such as continuous flow TLC, radial TLC and vapour-programmed TLC, have only limited application in the analysis of alkaloids.

## 1.4. APPLICATION OF THE SAMPLE

A good TLC separation depends to a great extent on the application of the sample. The final spot size obtained after the development of the chromatogram is related to the size of the spot on application. This depends first on the care with which the sample is applied, but also on the volume of sample applied, the solvent used to dissolve the sample and the character of the sample (alkaloid base or salt). The smaller the volume of sample applied, the more easily are small spots obtained. The sample size applied at the starting point is usually in the range 0.1-10 μg, which gives a ratio of amount of sample to amount of adsorbent of 1:1000 to 1:10,000.

If the polarity of the sample solution is low, a spot with the highest concentration in the middle is obtained. If the polarity is high, a spot with the highest concentration at the edges is obtained. The best TLC results are obtained in the former instance.

The character of the compounds to be analysed must be taken into account before application. Alkaloids can be applied as a base or as a salt. When applied as salts, some problems can arise with neutral solvent systems, *viz.*, tailing or double spot formation. However, alkaloids applied as salts can be converted into free bases when treated with a drop of base prior to the development. With solvent systems of basic or acidic character no such problems will occur when the alkaloids are applied as salts.

The choice of an appropriate solvent for the sample to be applied in TLC is determined by the polarity of the alkaloid (salt or base) but the stability of the alkaloid must also be taken into account. Because of the proton donor character of chloroform, it is a suitable solvent for alkaloid bases. For more polar compounds small amounts of ethanol or methanol can be added. For alkaloid salts alcohols are preferred.

A special technique of application is the TAS technique. However, not all alka-

**References p. 10**

loids are sufficiently stable or volatile to be transferred from the sample to the TLC plate by this technique[13-15].

REFERENCES


1 E. Stahl (Editor), *Dünnschicht-Chromatographie, Ein Laboratoriumshandbuch,* Springer-Verlag, Berlin, 1967, 979 pp.
2 E. Heftmann (Editor), *Chromatography*, Van Nostrand Reinhold, New York, 3rd ed., 1975.
3 K. Macek (Editor), *Pharmaceutical Applications of Thin-layer and Paper Chromatography*, Elsevier, Amsterdam, 1972, 743 pp.
4 E. Tomlinson, T.M. Jefferies and C.M. Riley, *J. Chromatogr.*, 159 (1978) 315.
5 R. Wesley-Hadzya, *J. Chromatogr.*, 79 (1973) 243.
6 A.H. Beckett and N.H. Choulis, *J. Pharm. Pharmacol.*, 15 (1963) 236t.
7 R.A. Keller and J.C. Giddings, *J. Chromatogr.*, 3 (1960) 205.
8 F. Geiss, *Die Parameter der Dünnschichtschromatografie,* Friedr. Vieweg and Sohn, Braunschweig, 1972, 282 pp.
9 K.K. Unger, *Porous Silica, J. Chromatogr. Library,* Vol. 16, Elsevier, Amsterdam, 1979, 336 pp.
10 L.R. Snyder, *J. Chromatogr.*, 92 (1974) 223.
11 L.R. Snyder, *J. Chrom. Sci.*, 16 (1978) 223.
12 L.R. Snyder and J.J. Kirkland, *Introduction to Modern Liquid Chromatography*, Wiley, New York, 1974, pp. 215, 255 and 444.
13 E. Stahl, *J. Chromatogr.*, 37 (1969) 99.
14 G.H. Jolliffe and E.J. Shellard, *J. Chromatogr.*, 48 (1970) 125.
15 E. Stahl and W. Schmitt, *Arch. Pharm. (Weinheim),* 308 (1975) 570.

Chapter 2

# DETECTION OF ALKALOIDS IN TLC

For the detection of alkaloids many different methods have been described, ranging from non-selective to highly selective ones. Most of them were already in use in combination with paper chromatography.

The non-selective methods, which are used for the detection of all kind of organic compounds in TLC, such as quenching of UV light on fluorescent plates, iodine vapour or iodine spray reagents and concentrated sulphuric acid, are usually fairly sensitive, allowing a detection limit of less than 1 μg. However, because of a lack of specificity of these reagents other methods are usually preferred for the detection of alkaloids. Methods by means of which alkaloids can be selectively detected are particularly important. Selective and specific alkaloids reagents are various modifications of Dragendorff's reagent and potassium iodoplatinate. Both reagents react with tertiary and quaternary nitrogen atoms.

## 2.1. DRAGENDORFF'S REAGENT (REAGENT 39a-39g)

The modifications of this reagent were originally described for the detection of alkaloids on paper chromatograms. The first spray reagent described for this purpose were the modifications proposed by Munier and Macheboeuf[1] and Munier[2]. Both modifications are aqueous solutions, the difference being the acid used to dissolve the bismuth salt (acetic and tartaric acid, respectively). The sensitivity of the reagent for alkaloids is in the range 0.01-1 μg.

Thies and Reuther[3] developed a modification using a solution in acetic acid and ethyl acetate instead of water. Because the ethyl acetate evaporates faster than water, less diffusion of the spots will occur, resulting in sharper edges of the spots. However, the sensitivity is lower than that of the two already mentioned modifications[4].

Vāgūjfalvi[4] modified the method of Thies and Reuther by spraying the chromatogram with 0.05 *N* sulphuric acid after spraying with Dragendorff's reagent. The sensitivity increased ten-fold in paper chromatography (PC). The author assumed that the complex formation with the ethyl acetate-containing reagent is different from that with the aqueous reagent.

Later this method was modified for TLC by Vāgūjfalvi[5]: 10% sulphuric acid was used instead of 0.05 *N* sulphuric acid to increase the sensitivity of the reaction.

**References p. 18**

According to the same author, maximum intensity in colour is observed after 15-60 min; however, exposure to strong light for 15-30 min makes the colour decrease. Further, it was stated that a four-fold more dilute reagent than used in PC was more sensitive for alkaloid detection in TLC.

Trabert[6] described a Dragendorff reagent dissolved in diethyl ether-methanol. In this form the reagent is much more sensitive in detecting alkaloids in ethereal solutions and on paper chromatograms. Robles[7] stated that hydrochloric acid is more suitable than acetic acid in preparing Dragendorff's reagent, because too high a pH leads to faster fading of the spots. Dilution of the reagent should be done with 0.1 *N* hydrochloric acid instead of water, otherwise a cloudy solution is obtained. The same is observed if the amount of potassium iodide is too small.

If alkaloids are to be detected on a buffered cellulose layer, it is necessary to add a non-volatile acid. Citric acid can be used with good results. If bismuth nitrate is used instead of bismuth subnitrate, the colour of the alkaloid spots is more intense[7].

Primary and secondary amines are not (or only at high concentrations) detectable with Dragendorff's reagent. However, treatment of such compounds with dimethyl sulphate, giving quaternary nitrogen atoms, enables detection with Dragendorff's reagent to be achieved[8].

Roper et al.[9] reported on the sensitivity and the stability of Dragendorff's reagent prepared according to Munier and Macheboeuf. They found that neither the reagent concentrate nor the diluted spray reagent showed any decrease in alkaloid sensitivity under normal laboratory storage conditions for at least 6 months. There seemed to be no reason for storing either the reagent concentrate or the diluted spray reagent in amber-glass bottles or under refrigeration. However, the best results were obtained when the spray reagent was prepared from a reagent concentrate at least 2 days after its preparation, and further that spray reagents made from such concentrates should not be used before 6 days after dilution.

To increase the sensitivity of the reaction with Munier's or Munier and Macheboeuf's modification of Dragendorff's reagent, Fike[10], Puech et al.[11] and McLean and Jewers[12] sprayed afterwards with 10% sodium nitrite solution. By this treatment the background of the spots is decolorized from yellow to white, improving the contrast of the spots. For several groups of alkaloids the sensitivity is increased in this way, the sensitivity being in the range 0.01 - 0.1 μg[13].

Various non-alkaloidal compounds also react with Dragendorff's reagent[8,13-15]. In Table 2.1 examples of such compounds are summarized. Other compounds which are reported to precipitate with Dragendorff's reagent are proteins, including albuminous substances, peptones, ptomaines[15] and choline[16]. However, amino acids do not give any precipitate with Dragendorff's reagent[17].

Except for choline, the compounds mentioned have not been tested in TLC for a

TABLE 2.1

NON-ALKALOIDAL COMPOUNDS WITH POSITIVE DRAGENDORFF REACTIONS

Positive reaction: minimum detectable amount 10 μg or less[13-15]

| Type | Compounds with positive reactions |
|---|---|
| α-Pyrone | Kawain, yangonin, desmethoxyyangonin, dihydrokawain, methysticin, dihydromethysticin and some synthetic products |
| Benz-α-pyrones | Coumarin, scopoletin, bishydroxycoumarin, ethylbis(4-hydroxy-coumarinyl) acetate |
| Psoralens | bergapten, xanthotoxin, imperatorin, isopimpinellin |
| γ-Pyrones | Maltol, kojic acid, kojic acid methyl ether |
| Furochromones | Khellin |
| Chalcones | Chalcone, β-methoxychalcone, *cis*- and *trans*-1,4-diphenyl-2-butene-1,4-dione, cardenolides, steroids and triterpenes, digitoxin, ouabain, sitosterol, stigmasterol, cholesterol, quillaic acid, β-amyrin, lupenone, β-glycyrrhetinic acid, oleanolic acid. Friedelanol and cholestane are observed after spraying with sodium nitrite |
| Miscellaenous compounds | Acrolein, amylcinnamaldehyde, cinnamaldehyde, eugenol, hydroxy-citronellal, menadione, menthyl salicylate, ninhydrin, phenyl salicylate, salicylic acid, benzyl acetate, camphor, eucalyptol, methyl salicylate, piperonal, 2-phenylethanol, cinnamyl alcohol, choline, phosphatidylcholine. Geraniol, resorcinol, orcinol, anethol and thymol give a positive reaction on spraying with sodium nitrite. Further, acetylsalicylic acid, cinnamic acid, guaiacol, hydroquinone, phloroglucinol, quinone, terpineol and vanillin give a positive reaction after 24 h. |
| Polyethylene glycols | |

possible positive or negative Dragendorff reaction. Farnsworth[17] surveyed the false-positive and false-negative reactions of alkaloids. According to Farnsworth et al.[15], the minimum structural features for non-alkaloids to give a positive Dragendorff reaction are conjugated carbonyl (ketone or aldehyde) or lactone functions. Anderson et al.[13] observed that the sensitivity for such compounds increased if the plate was sprayed with sulphuric acid or sodium nitrite after spraying with Dragendorff's reagent. The authors concluded that the minimum structural requirement for a positive Dragendorff reaction is a hydroxyl group and an isolated double bond.

## 2.2. POTASSIUM IODOPLATINATE (REAGENTS 64a-64f)

This reagent also exists in various modifications. The sensitivity is about the same as for Dragendorff's reagent (0.01 - 0.1 μg), but potassium iodoplatinate reagent has several advantages.

It is non-destructive and the alkaloids can be recovered after application of the spray reagent. Holdstock and Stevens[18] used the following method for the recovery of the alkaloids. After drying the plate the spot material was scraped off the plate and transferred to a test-tube. A few drops of a saturated sodium sulphite solution were added, followed by 1 ml of 0.5 *N* sulphuric acid. If necessary the mixture was heated to decolorize the sample. The aqueous solution was then saturated with sodium chloride, followed by basification with strong ammonia. Subsequently the alkaloid bases could be extracted with diethyl ether, chloroform or butanol-chloroform (1:9). A recovery of 50-60% was found for tertiary amines and 70-80% for primary and secondary amines. For application in PC, see also Goldbaum and Kazyak[19].

Another advantage of potassium iodoplatinate reagent is that different colours are obtained with different alkaloids, the colours varying from brown through violet to blue. A blue-violet background colour due to starch present can be decolorized by spraying with sodium hydrogen sulphite solution.

Formamide impregnation of TLC plates can also interfere with the alkaloid detection, but by spraying the plate with a 0.25% solution of sodium nitrite in 0.5% hydrochloric acid the formamide will be converted into formic acid, which does not interfere with the alkaloid detection.

Diethylamine also interferes with alkaloid detection when using Dragendorff's reagent or iodoplatinate reagent. Even after drying the plate at elevated temperature the diethylamine does not disappear completely, resulting in a dark background after spraying with one of the reagents mentioned, with a consequent decrease in detection sensitivity[11].

## 2.3. OTHER REAGENTS

In addition to the two widely used reagents mentioned above, various other reagents have been described for the detection of alkaloids.

Marquis reagent (formaldehyde in sulphuric acid, no. 53)[20,30], ammonium vanadate in nitric or sulphuric acid (nos. 3, 91 and 103)[21,22,29], Fröhde's reagent (sulphomolybdic acid, no. 90)[21,29] and phosphomolybdic acid in nitric acid (no. 79)[21] have been used in the detection of alkaloids. Two reagents which may be mentioned separately because of their usefulness in the selective detection of several groups of alkaloids are iron(III) chloride in perchloric acid (nos. 46a-46c) and cerium(IV) sulphate in sulphuric acid or orthophosphoric acid (nos. 12 - 14c).

Rücker and Taha[23] described the use of π-acceptors for the detection of alkaloids. Several reagents were used. In Table 2.2 the colours observed with these reagents are summarized for some alkaloids. TCNQ reagent (no. 95) was found to be the most sensitive and to have the widest applicability. The sensitivity was in the range of 0.5-10 μg. Vinson and co-workers[24,25] described the use of TCBI reagent (no. 94) for the detection of drugs, including a number of alkaloids. Various colours are obtained for the test compounds (see Table 2.3), the detection limit being in the range 0.05-0.25 μg.

Grant[26] described the use of aqueous cobalt thiocyanate solution in the detection of various drugs, including a series of alkaloids. Several alkaloids give positive reactions. Sometimes derivatization may lead to a positive reaction, e.g., acetylation of morphine. Ranieri and McLaughlin[27] described the use of fluorescamine for the detection of primary and secondary amines. They proposed the following procedure. The TLC plate with the alkaloid extract is first sprayed with the fluorescamine solution (no. 50) (detection of primary and secondary amines), then with dansyl chloride reagent (no. 28) (detection of phenols, imidazoles and the fluorescamine conjugates of the secondary amines are converted into dansyl conjugates) and finally with iodoplatinate (detection of tertiary amines). Menn and McBain[28] described a detection method specifically for cholinesterase inhibitors. First the plate is sprayed with human plasma, and after 30 min with a mixture of 1 part of 0.6% bromothymol in 0.1 *N* sodium hydroxide solution and 15 parts of 1% aqueous acetylcholine chloride solution. The inhibiting substances show a blue colour against a yellow background. This method is applicable only on cellulose plates and does not work on silica gel or Florisil plates.

For each group of alkaloids the appropriate detection methods will be discussed in more detail. A list of spray reagents and their preparation are given in the Appendix.

**References p. 18**

TABLE 2.2

COLOURS OF TLC SPOTS OF ALKALOIDS WITH π-ACCEPTORS[23]

Colours: Bl = blue; Br = brown; Gn = green; Go = gold; Gy = grey; nc = no colour; Or = orange; R = red; V = violet; W = white; Y = yellow. White background unless otherwise stated.

| Alkaloid | TCNQ* (no.95) | TNF (no. 100b) | | TetNF (no.97b) after heating | DDQ** (no.29) | DNFB (no.38) |
|---|---|---|---|---|---|---|
| | | In the cold | After heating | | | |
| Atropine | Y-W | nc | V | Gn-Gy*** | Or | Y |
| Scopolamine | Gn-Y | nc | Gy | Gy*** | nc | Y |
| Homatropine | Go | nc | V | V-Gy*** | Or | Y |
| Homatropine·MeBr | Go | nc | R | V*** | nc | Y-W |
| Atropine·$MeNO_3$ | Y-Or | nc | R | V*** | nc | Y-W |
| Tropine | Go | nc | V-Gy | V*** | Gn | Y-W |
| Pilocarpine | Go | nc | Gy | V-Gy*** | V | Y |
| Ephedrine | Gn-Y | nc | Gy | Faint*** | nc | Y-Or |
| Cocaine | Go | nc | V-Gy | V*** | nc | Or |
| Morphine | Bl | Or-Br | Or | Or | Gn | Or |
| Codeine | Y-Go | Or-Br | Y | Gy | V-Br | Or |
| Papaverine | Y-Go | Or | Y | Br | V | Or |
| Quinine | Y-Go | Y | V-Br | V | V | Or |
| Brucine | Go | V | V | Gy | V | Or |
| Strychnine | Y | Y | Y | Or | Or | Or |
| Veratrine | Go | Or | Or | Gy | Or | Or |
| Reserpine | Gn | Gy | Gy | Or | Gn | Or-R |
| Ergotamine | Y-Gn | Gy-Br | Gy-Br | Gy | Br | Or-R |

*Colours given were on a pale bluish green background.
**Pale violet background.
***No colour in the cold.

TABLE 2.3

COLOURS OF VARIOUS DRUGS SPRAYED WITH TCBI (No.94)[24]

| Drug | Colour | Drug | Colour |
|---|---|---|---|
| Acetophenazine | Yellow-green | Mephenytoin | Blue |
| Amobarbital | Blue | Mescaline | Pink-brown |
| Amphetamine | Grey-purple | Methadone | Blue-green |
| Anileridine | Brown-green | Methamphetamine | Green |
| Apomorphine | Blue-green | Methapyrilene | Brown-green |
| Atropine | Blue-green | Methyldopa | Brown |
| Benzocaine | Orange-brown | Methylphenidate | Grey-green |
| Bufotenine | Brown-green | Nicotine | Grey-purple |
| Chlordiazepoxide | Green | Nortryptiline | Brown |
| Chloroquine | Green | Oxazepam | Blue |
| Chlorothiazide | Grey-brown | Pentazocine | Grey-green |
| Chlorpromazine | Grey-purple | Perphenazine | Grey-purple |
| Cocaine | Green | Phendimetrazine | Grey |
| Codeine | Brown-green | Phenmetrazine | Yellow-green |
| Dextromethorphan | Grey | Phenobarbital | Blue |
| Dicyclomine | Green | Phentermine | Green |
| Diethylpropion | Grey-brown | Phenylephrine | Green |
| Dimethyltryptamine | Brown-green | Procaine | Brown-green |
| Diphenylhydantoin | Blue-grey | Promazine | Purple |
| Diphenhydramine | Blue-green | Promethazine | Purple |
| Doxylamine | Brown | Quinine | Grey-green |
| Ephedrine | Grey-green | Quinidine | Green |
| Epinephrine | Brown | Reserpine | Brown |
| Ethinamate | Orange-brown | Secobarbital | Blue |
| Glutethimide | Grey-green | Strychnine | Green |
| Heptabarbital | Blue | Sulfamerazine | Grey |
| Heroin | Yellow-green | Sulfathiazole | Brown-blue |
| Hydrochlorothiazide | Grey-purple | Tetrahydrocannabinol ($\Delta^9$) | Red-brown |
| Ibogaine | Brown-green | Thiopropazate | Blue |
| Lidocaine | Green | Thioridazine | Grey-brown |
| Lysergic acid diethylamide | Grey-brown | Trifluoperazine | Blue-purple |
| Meperidine | Grey-green | Trimeprazine | Blue-grey |
| Mephentermine | Green | | |

REFERENCES

1 R. Munier and M. Macheboeuf, *Bull. Soc. Chim. Biol.*, 31 (1949) 1144.
2 R. Munier, *Bull. Soc. Chim. Biol.*, 35 (1953) 1225.
3 H. Thies and F.W. Reuther, *Naturwissenschaften*, 41 (1954) 230.
4 D. Vágújfalvi, *Planta Med.*, 8 (1960) 34.
5 D. Vágújfalvi, *Planta Med.*, 13 (1965) 79.
6 H. Trabert, *Pharm. Zentralhalle*, 93 (1954) 463.
7 M.A. Robles, *Pharm. Weekbl.*, 94 (1959) 178.
8 I.M. Hais and K. Macek, *Handbuch der Papierchromatographie*, VEB Gustav Fischer Verlag, Jena, 1958, pp. 547.
9 E.C. Roper, R.N. Blomster, N.R. Farnsworth and F.J. Draus, *Planta Med.*, 13 (1965) 98.
10 W.W. Fike, *Anal. Chem.*, 38 (1966) 1697.
11 A. Puech, M. Jacob and D. Gaudy, *J. Chromatogr.*, 68 (1972) 161.
12 W.F.H. McLean and K. Jewers, *J. Chromatogr.*, 74 (1972) 297.
13 L.A. Anderson, N.S. Doggett and M.S.F. Ross, *Planta Med.*, 32 (1977) 125.
14 M.E. Ginn, C.L. Church and J.C. Harris, *Anal. Chem.*, 33 (1961) 143.
15 N.R. Farnsworth, N.A. Pilewski and F.J. Draus, *Lloydia*, 25 (1962) 312.
16 U. Beiss, *J. Chromatogr.*, 13 (1964) 104.
17 N.R. Farnsworth, *J. Pharm. Sci.*, 55 (1966) 225.
18 T.M. Holdstock and H.M. Stevens, *Forensic Sci.*, 6 (1975) 187.
19 L.R. Goldbaum and L. Kazyak, *Anal. Chem.*, 28 (1956) 1289.
20 R. Paquin and M. Lapage, *J. Chromatogr.*, 12 (1963) 57.
21 W. Court and M.S. Habib, *J. Chromatogr.*, 89 (1973) 101.
22 M. Malaiyandi, J.P. Barette and M. Lanonette, *J. Chromatogr.*, 101 (1974) 155.
23 G. Rücker and A. Taha, *J. Chromatogr.*, 132 (1977) 165.
24 J.A. Vinson and J.E. Hooyman, *J. Chromatogr.*, 105 (1975) 415.
25 J.A. Vinson, J.E. Hooyman and C.E. Ward, *J. Forensic Sci.*, 20 (1975) 552.
26 F.W. Grant, *J. Chromatogr.*, 116 (1976) 230.
27 R.L. Ranieri and J.L. McLaughlin, *J. Chromatogr.*, 111 (1975) 234.
28 J.J. Menn and J.B. McBain, *Nature (London)*, 209 (1966) 1351.
29 L. Vignoli, J. Guillot, F. Goueze and J. Catalin, *Ann. Pharm. Fr.*, 24 (1966) 461.
30 F. Eiden and G. Kammash, *Pharm. Ztg.*, 117 (1972) 1994.

Chapter 3

# TLC SEPARATION AND IDENTIFICATION OF ALKALOIDS IN GENERAL

A number of publications on the analysis of alkaloids deal with the separation and identification of alkaloids belonging to various chemical groups. Some of the publications deal with the systematic identification of alkaloids and some with the testing of new chromatographic methods with a series of alkaloids. Because of these different aims, these two groups will be dealt with separately.

## 3.1. IDENTIFICATION

The TLC systems used in a systematic analysis for the identification of unknown compounds have to fulfil other demands than TLC systems used to separate a group of chemically related alkaloids. By means of a minimum number of TLC systems, a maximum number of alkaloids have to be characterized. Often the problem is to separate alkaloids of widely varying polarity, which requires special TLC systems. Waldi et al.[1] described the analysis of a series of alkaloids by means of eight TLC systems. The 54 alkaloids analysed were divided into two groups according to their $R_F$ values in solvents S2 and S3 (Table 3.1). Group I (alkaloids 1-37) had $R_F$ values lower than 0.9 in system S2 and lower than 0.3 in system S3. Group II (alkaloids 38-54) had $R_F$ values higher than 0.9 in S2 and higher than 0.3 in S3. The alkaloids of group I were eventually identified by means of the $R_F$ values in TLC systems S1 and S2 in combination with observations made under UV light (366 nm) and colours observed after spraying with iodoplatinate reagent. In case of doubt, systems S6-S8 might be used to confirm the identification. The alkaloids of group II were identified by their $R_F$ values in system S3 and eventually in systems S4 and S5 in combination with their appearance under UV light (366 nm) and their colour reaction with iodoplatinate spray reagent.

Drost and Reith[15] used the TLC systems as described by Waldi et al.[1] for a systematic toxicological analysis. For the identification of the alkaloids the spots were scraped off the plate and eluted with 1,2-dichloroethane and the UV spectra were recorded. Dickes[11] used solvent systems S2 and S3 of Waldi et al.[1] for the identification of various alkaloids. Vidic and Schütte[2] described a systematic analysis of basic compounds of toxicological and pharmaceutical interest by means of PC. Although TLC was mentioned, it was used only for preliminary investigations or for identification purposes. The compounds were divided in groups

References p. 48

TABLE 3.1

SYSTEMATIC ANALYSIS ALKALOIDS ON TLC PLATES[1]

TLC systems:

| | | |
|---|---|---|
| S1 | silica gel G, activated | Chloroform - acetone - diethylamine (5:4:1) |
| S2 | silica gel G, activated | Chloroform - diethylamine (9:1) |
| S3 | silica gel G, activated | Cyclohexane - chloroform - diethylamine (5:4:1) |
| S4 | silica gel G, activated | Cyclohexane - diethylamine (9:1) |
| S5 | silica gel G, activated | Benzene - ethyl acetate - diethylamine (7:2:1) |
| S6 | aluminium oxide G, activated | Chloroform |
| S7 | aluminium oxide G, activated | Cyclohexane - chloroform (3:7) + 0.05% diethylamine |
| S8 | silica gel G, impregnated with 0.1 mol/l sodium hydroxide, activated | Methanol |

| No. | Alkaloid | $hR_F$ values | | | | | | | | Fluorescence in UV light (366 nm) | Colour with iodoplatinate reagent (pink background) |
|---|---|---|---|---|---|---|---|---|---|---|---|
| | | S1 | S2 | S3 | S4 | S5 | S6 | S7 | S8 | | |
| 1 | Narceine | 3 | 0 | 0 | 0 | 0 | 0 | 0 | 0 | - | Deep blue |
| 2 | Cupreine | 3 | 0 | 0 | 0 | 0 | 0 | 0 | 46 | Brownish yellow | Red - brown |
| 3 | Sarpagine | 12 | 4 | 0 | 0 | 0 | 0 | 0 | 0 | - | Beige |
| 4 | Ergometrine | 14 | 6 | 0 | 0 | 2 | 3 | 0 | 64 | Violet-blue | White |
| 5 | Morphine | 10 | 8 | 0 | 0 | 3 | 3 | 0 | 34 | - | Deep blue |
| 6 | Dihydroergotamine | 21 | 12 | 0 | 0 | 3 | 7 | 0 | 61 | Violet-blue | Brownish |
| 7 | Serpentine | 24 | 15 | 0 | 0 | 4 | 0 | 0 | 0 | Dark brown | Red-brown |
| 8 | Ergotamine | 24 | 16 | 0 | 0 | 3 | 10 | 5 | 59 | Violet-blue | Pink |
| 9 | Boldine | 16 | 16 | 3 | 0 | 5 | 24 | 6 | 58 | Violet | Beige |
| 10 | Dihydromorphinone | 24 | 23 | 8 | 1 | 11 | 5 | 8 | 16 | - | Brownish yellow |
| 11 | Ergometrinine | 42 | 25 | 3 | 0 | 8 | 12 | 10 | 62 | Violet-blue | Violet-blue |
| 12 | Ephedrine | - | - | - | - | - | - | - | - | - | Light brown |
| 13 | Quinine | 19 | 26 | 7 | 0 | 17 | 9 | 18 | 43 | Blue | Yellow-white |
| 14 | Dihydroergocristine | 42 | 30 | 3 | 0 | 7 | 15 | 7 | 69 | Violet-blue | Brownish |
| 15 | Hordenine | 33 | 36 | 14 | 5 | 28 | 0 | 15 | 35 | - | White |
| 16 | Ergocristine | 51 | 38 | 14 | 5 | 13 | 46 | 15 | 70 | Violet-blue | Beige-light brown |
| 17 | Quinidine | 33 | 40 | 15 | 0 | 25 | 12 | 18 | 50 | Blue | Light yellow |
| 18 | Atropine | 38 | 40 | 16 | 5 | 12 | 0 | 10 | 17 | - | Violet-blue |
| 19 | Colchicine | 47 | 41 | 4 | 0 | 4 | 11 | 0 | 57 | - | Light grey |
| 20 | Ajmaline | 47 | 42 | 12 | 3 | 30 | 6 | 13 | 56 | Blue | Beige |
| 21 | Cinchonine | 38 | 44 | 17 | 7 | 27 | 0 | 22 | 40 | - | Beige-brown |
| 22 | Homatropine | 37 | 45 | 15 | 5 | 23 | 4 | 24 | 15 | - | Violet-blue |
| 23 | Ergotaminine | 24 | 51 | 0 | 0 | 14 | 42 | 15 | 68 | Violet-blue | Pink |
| 24 | Pilocarpine | 41 | 52 | 9 | 0 | 13 | 32 | 25 | 55 | - | Light brown |
| 25 | Codeine | 38 | 53 | 16 | 4 | 26 | 12 | 27 | 35 | - | Pink-violet |
| 26 | Dihydrocodeine | 38 | 54 | 18 | 6 | 28 | 10 | 30 | 25 | Blue | Violet-blue |
| 27 | Serpentinine | 53 | 56 | 8 | 0 | 10 | 0 | 3 | 12 | Yellow-green | Yellow-brown |
| 28 | Ergocristinine | 61 | 57 | 13 | 0 | 20 | 0 | 27 | 70 | Violet-blue | Light brown |
| 29 | Scopolamine | 56 | 60 | 19 | 3 | 34 | 30 | 0 | 52 | - | Violet |
| 30 | Yohimbine | 63 | 62 | 18 | 3 | 37 | 33 | 15 | 60 | Green-blue | Light yellow |
| 31 | Brucine | 42 | 63 | 18 | 0 | 19 | 50 | 54 | 12 | - | Violet-brown |
| 32 | Cephaeline | 56 | 63 | 19 | 2 | 23 | 25 | 17 | 37 | Violet-blue | White |
| 33 | Rauwolscine | 55 | 63 | 18 | 4 | 36 | 36 | 15 | 68 | Yellow-green | Faint beige |
| 34 | Dihydrocodeinone | 51 | 65 | 21 | 4 | 30 | 48 | 43 | 18 | - | Violet |
| 35 | Apoatropine | 54 | 67 | 40 | 20 | 26 | 15 | 40 | 16 | - | Violet-blue |
| 36 | Strychnine | 53 | 76 | 28 | 5 | 38 | 57 | 60 | 22 | - | Yellow |
| 37 | Reserpine | 72 | 80 | 20 | 0 | 46 | 63 | 35 | 69 | Green-yellow | White |
| 38 | Physostigmine | 65 | >90 | 32 | 4 | 44 | 59 | 50 | 46 | - | Pink |
| 39 | Aconitine | 68 | >90 | 35 | 3 | 49 | 36 | 60 | 65 | - | Red-brown |
| 40 | Bulbocapnine | 65 | >90 | 35 | 7 | 54 | 78 | 70 | 48 | Blue | White |

TABLE 3.1 (*continued*)

| No. | Alkaloid | $hR_F$ values | | | | | | | | Fluorescence in UV light (366 nm) | Colour with iodoplatinate reagent (pink background) |
|---|---|---|---|---|---|---|---|---|---|---|---|
| | | S1 | S2 | S3 | S4 | S5 | S6 | S7 | S8 | | |
| 41 | Emetine | 67 | 90 | 40 | 6 | 45 | 38 | 58 | 50 | Blue | Red-brown |
| 42 | Papaverine | 67 | 90 | 42 | 3 | 47 | 85 | 84 | 70 | Yellowish | Yellow |
| 43 | Cotarnine | 60 | 90 | 43 | 31 | 45 | 0 | 25 | 0 | Green-yellow | Violet |
| 44 | Scopoline | 60 | 90 | 44 | 20 | 44 | 46 | 50 | 37 | - | White |
| 45 | Lobeline | 68 | 90 | 48 | 14 | 48 | 55 | 60 | 55 | - | Red-brown |
| 46 | Noscapine | 72 | 90 | 51 | 10 | 57 | 81 | 79 | 72 | Blue | Light yellow |
| 47 | Thebaine | 65 | 90 | 51 | 16 | 50 | 71 | 76 | 40 | - | Red-brown |
| 48 | Aspidospermine | 65 | 90 | 54 | 20 | 49 | 50 | 60 | 65 | - | White |
| 49 | Tropacocaine | 65 | 90 | 56 | 34 | 45 | 58 | 78 | 35 | - | Violet |
| 50 | Arecoline | 66 | 90 | 56 | 34 | 48 | 0 | 0 | 0 | - | White |
| 51 | Hydrastinine | 66 | 90 | 58 | 41 | 50 | 0 | 25 | 0 | Steel blue | Violet-blue |
| 52 | Psicain new | 66 | 90 | 60 | 35 | 53 | 83 | 82 | 59 | - | Yellow |
| 53 | Cocaine | 73 | 90 | 65 | 36 | 58 | 84 | 77 | 62 | - | Violet |
| 54 | Sparteine | 70 | 90 | 68 | 68 | 55 | 0 | 55 | 5 | - | Violet |

based on their polarity and basicity. For this purpose a special extraction scheme was used. Based on the spreading of the compounds over four fractions, six groups of substances were distinguished. With the aid of PC the individual compounds of each group were identified.

Macek et al.[9] divided 161 drugs into groups by extraction with diethyl ether at pH 3-4 and pH 10-11 and finally by ion exchangers for the (polar) substances. Each group was further divided into subgroups by means of their $R_F$ values in PC. For the alkaline group, fractionated PC was used. Eventually the compounds of each subgroup were identified by means of PC, TLC and spray reagents specified for each compound. Sunshine et al.[12] reported a characterization scheme for 138 drugs, includings a series of alkaloids. The drugs were identified by their $hR_F$ values in seven different solvent systems. Fike[10] used the data mentioned in ref. 12 to study the structure versus $hR_F$ correlations.

Marozzi and Falzi[6] used $R_F$ values and relative $R_F$ values in the identification of non-volatile organic poisons, including some alkaloids. The TLC systems used were chloroform - acetone (9:1) and benzene - acetic acid (9:1), both in combination with silica gel plates.

Haywood and Moss[16] modified the Curry and Powel toxicological screening by using cellulose TLC plates instead of paper chromatography. The cellulose plates were impregnated with 5% sodium dihydrogen citrate and developed with butanol - 5% citric acid (9:1). El Gendi et al.[8] described the identification of a number of basic compounds in toxicology by means of TLC. They used silica gel plates in combination with the solvents chloroform - acetone (9:1), chloroform - ethanol (9:1 or 8:2) and methanol. Sixteen alkaloids were among the 58 compounds analysed.

Noirfalise and Mees[14] presented an identification scheme for 34 alkaloids based on their $R_F$ values. Seven solvent systems were described; two to five chromatograms were usually necessary to identify a compound. The $R_F$ values found are summarized in Table 3.2.

Simon and Lederer[19] presented an identification scheme for 16 alkaloids based on letter codes for $R_F$ values obtained with the same mobile phase on four different stationary phases. Each $R_F$ interval of 0.05 was given a letter of the alphabet. None of the alkaloids investigated gave the same letter combination (Table 3.3).

Egli[18,22] used an identification scheme for drugs, of which 11 were alkaloids, based on two $R_F$ values obtained in an acidic and in an alkaline solvent system and their colours in the presence of phenothiazine by exposure to UV light or bromine vapour, as well as their colours obtained with iodine-pyrrole reagent (nos. 62 and 78). The universal solvent systems were chloroform - methanol - 85% formic acid (85:10:5) and chloroform - methanol - concentrated ammonia (85:14:1) on activated silica gel plates. For basic substances chloroform - methanol - 100% formic acid (85:5:10) and chloroform - benzene (1:1) saturated with ammonia and ammonia in the atmosphere were found to be useful variants[18].

TABLE 3.2

TLC ANALYSIS OF SOME ALKALOIDS AND AMINES[14]

Silica gel DSF 5, activated.

Solvent systems:
S1 Chloroform - acetone (9:1)
S2 Chloroform - methanol (9:1)
S3 Acetone
S4 Acetone - 25% ammonia (99:1)
S5 Methanol
S6 Methanol - 25% ammonia (99:1)
S7 Chloroform - methanol (1:1)
S8 Chloroform - methanol - 25% ammonia (47.5:47.5:5)
S9 Chloroform - methanol - acetic acid (47.5:47.5:5)

| Substance | $hR_F$ value | | | | | | | | |
|---|---|---|---|---|---|---|---|---|---|
| | S1 | S2 | S3 | S4 | S5 | S6 | S7 | S8 | S9 |
| Acetyldihydrocodeinone | - | 33 | 6 | 17 | 16 | 28 | 37 | 81 | 37 |
| Amylocaine | 22 | 63 | 46 | 63 | 47 | 57 | 60 | 87 | 44 |
| Apomorphine | - | 25 | 25 | 36 | 43 | 54 | 50 | 79 | 35 |
| Atropine | - | 2 | - | 3 | 6 | 11 | 7 | 54 | 27 |
| Benzocaine | 41 | 57 | 55 | 60 | 59 | 61 | 61 | 79 | 81 |
| Brucine | - | 13 | - | 2 | 3 | 7 | 14 | 72 | 40 |
| Butylscopolamine | - | 2 | - | 3 | 5 | 2 | 18 | 45 | 36 |
| Caffeine | 7 | 48 | 24 | 42 | 40 | 49 | 55 | 79 | 68 |
| Cytisine | - | - | - | 3 | 11 | 22 | - | 60 | 13 |
| Cocaine | 4 | 40 | 30 | 72 | 29 | 47 | 43 | 84 | 21 |
| Codeine | - | 13 | - | 13 | 13 | 33 | 27 | 71 | 24 |
| Colchicine | - | 39 | 10 | 15 | 45 | 49 | 59 | 84 | 82 |
| Dextromoramide | - | 62 | 49 | 73 | 58 | 69 | 75 | 85 | 76 |
| Diacetylmorphine (heroin) | - | 31 | 9 | 30 | 23 | 49 | 42 | 83 | 34 |
| Dihydroergotamine | - | 33 | 23 | 34 | 40 | 50 | 56 | 86 | 72 |
| Dihydromorphinone | - | 7 | - | 4 | 8 | 25 | 13 | 53 | 24 |
| Ecgonine | - | - | - | - | 10 | 14 | 5 | 15 | 10 |
| Ephedrine | - | - | - | - | 12 | 18 | 8 | 47 | 41 |
| Physostigmine | - | 23 | 11 | 23 | 18 | 28 | 31 | 67 | 25 |
| (*dl*)-Methadone | - | 17 | 20 | 59 | 16 | 48 | 20 | 80 | 54 |
| Morphine | - | 4 | 3 | 8 | 15 | 33 | 24 | 58 | 23 |
| Narceine | - | 4 | - | - | 19 | 32 | 17 | 45 | 34 |
| Noscapine | 28 | 72 | 58 | 60 | 54 | 59 | 70 | 84 | 76 |
| Nicotine | - | 30 | 10 | 31 | 32 | 44 | 42 | 79 | 13 |
| N,N-Diethylnicotinamide | 9 | 49 | 29 | 40 | 48 | 54 | 60 | 78 | 71 |
| Papaverine | 22 | 68 | 38 | 48 | 51 | 56 | 70 | 82 | 78 |
| Procaine | - | 19 | 18 | 44 | 29 | 45 | 36 | 81 | 33 |
| Quinidine | - | 20 | 7 | 13 | 23 | 37 | 44 | 80 | 66 |
| Quinine | - | 9 | 3 | 10 | 26 | 35 | 31 | 77 | 60 |
| Scopolamine | - | 26 | 15 | 29 | 37 | 47 | 43 | 79 | 22 |
| Sparteine | - | - | - | 4 | 4 | 3 | 2 | 21 | 25 |
| Strychnine | - | 16 | - | 5 | 6 | 12 | 16 | 74 | 37 |
| Thebaine | 3 | 34 | 6 | 17 | 17 | 28 | 35 | 81 | 40 |
| Veratrine | - | 21 | 22 | 56 | 18 | 42 | 29 | 80 | 72 |

References p. 48

TABLE 3.3

$hR_F$ VALUES OF ALKALOIDS ON VARIOUS THIN LAYERS[19]

Solvent: *n*-butanol-acetic acid-water (45:3:12)

| Compound | Cellulose (Carlo Erba) | Acetylated cellulose AC-10 (MN) | Aluminium oxide (Carlo Erba) | Silica gel (Eastman) | $R_F$ "word" |
|---|---|---|---|---|---|
| Tropine | 38 | 57 | 53 | 17 | FLKD |
| Atropine | 68 | 83 | 60 | 35 | NQLG |
| Homatropine | 64 | 76 | 57 | 33 | MPLG |
| Belladonnine | 20 | 98 | 42 | 5 | DTIA |
| Cocaine | 69 | 90 | 69 | 30 | NSNF |
| Scopolamine·HCl | 46 | 76 | 46 | 28 | JPJF |
| Hyoscyamine | 68 | 90 | 68 | 33 | NTNG |
| Tropacocaine | 46 | 89 | 72 | 44 | JROJ |
| Narceine | 68 | 98 | 55 | 25 | NTKE |
| Morphine | 32 | 98 | 46 | 25 | GTJE |
| Papaverine | 83 | 94 | 83 | 63 | QSQH |
| Cotarnine | 43 | 66 | 52 | 22 | INKE |
| Noscapine | 84 | 97 | 84 | 60 | QTQL |
| Heroin | 60 | 89 | 60 | 35 | LRLG |
| Apomorphine | 56 | 90 | 47 | 44 | LREJ |
| Hydrastine | 78 | 93 | 71 | 48 | PSOF |

TABLE 3.4

TLC ANALYSIS OF SOME BASIC DRUGS[32] (from: Springer, Heidelberg)

TLC systems:

S1 Silica gel G — Benzene - chloroform - diethylamine (6:3:1);
S2 Silica gel G — Cyclohexane - diethylamine (9:1)
S3 Silica gel G — Methanol - 25% ammonia (100:1);
S4 Aluminium oxide (ready-made plates) — Chloroform - heptane - acetone - isopropanol - formic acid (90:30:30:25:20);
S5 Silica gel G — Chloroform - isopropanol - diethylamine (18:2:1).

| Substance | $hR_F$ values | | | | |
|---|---|---|---|---|---|
| | S1 | S2 | S3 | S4 | S5 |
| Quinine | 10 | 0 | 55 | 20 | 28 |
| Quinidine | 23 | 0 | 61 | 36 | 55 |
| Codeine | 39 | 8 | 36 | 27 | 61 |
| Dihydrocodeine | 41 | 13 | 25 | 30 | 55 |
| Scopolamine | 47 | 11 | 65 | 22 | 63 |
| Aminopyrine | 78 | 35 | 70 | 43 | 81 |
| Doxylamine | 78 | 56 | 43 | 23 | 74 |
| Hydrocodone | 48 | 4 | 24 | 36 | 65 |
| Pholedrine | 19 | 4 | 24 | 31 | 24 |
| Atropine | 29 | 10 | 16 | 36 | 44 |
| Ethylmorphine | 41 | 10 | 41 | 34 | 61 |
| Procaine | 51 | 4 | 67 | 31 | 66 |
| Chloroquine | 53 | 17 | 27 | 21 | 64 |
| Strychnine | 55 | 11 | 22 | 44 | 66 |
| Dimetindene | 77 | 47 | 40 | 24 | 72 |
| Prothipendyl·HCl | 80 | 58 | 44 | 53 | 70 |
| Ephedrine | 0 | 0 | 28 | 40 | 93 |
| Trimethoprim | 7 | 0 | 65 | 33 | 42 |
| Metoclopramide | 36 | 0 | 41 | 42 | 56 |
| Opipramol | 43 | 8 | 62 | 23 | 64 |
| Bamifylline | 57 | 5 | 76 | 40 | 75 |
| Phenmetrazine | 61 | 17 | 56 | 43 | 63 |
| Carbinoxamine | 62 | 35 | 46 | 17 | 70 |
| Thioproperazine | 63 | 8 | 44 | 25 | 72 |
| Diacetylmorphine (heroin) | 72 | 17 | 52 | 44 | 75 |
| Thebacone | 73 | 30 | 46 | 51 | 78 |
| Hexobendine | 78 | 10 | 48 | 34 | 88 |
| Mepyramine | 78 | 45 | 55 | 18 | 79 |
| Methapyrilene | 78 | 53 | 55 | 24 | 78 |
| Tripelenamine | 80 | 60 | 58 | 27 | 79 |
| Ketobemidone | 19 | 0 | 49 | 41 | 16 |
| Oxomemazine | 71 | 20 | 54 | 47 | 74 |
| Cocaine | 87 | 58 | 70 | 46 | 87 |

Schmidt[26] used the solvents as described by Egli[18,22], but 85% formic acid was replaced with 100% formic acid to avoid demixing. Several spray reagents were used for the detection and identification of 200 drugs.

Ahrend and Tiess[23,25] described the identification of 180 substances of toxicological interest by means of TLC using 19 different spray reagents. Thirty-nine alkaloids, both tertiary and quaternary, were studied. The mobile phases used were methanol, chloroform - acetone (9:1), chloroform - *n*-butanol - 25% ammonia (70:40:5) and *n*-butanol - acetic acid - water (4:1:5) (the thin-layer plate may only be immersed in the upper phase), all on silica gel. The compounds were arranged according to their $R_F$ values in first-mentioned TLC system (methanol/silica gel). The identification was based on the $R_F$ values in combination with the colours obtained with spray reagents. Two series of spray reagents were described, which can be used in succession in order to detect all 180 compounds. The first series consists of 1% aqueous mercuronitrate, iodoplatinate, 5% iron(III) chloride in 0.1 *M* hydrochloric acid, 1% iodine in chloroform and *p*-dimethylaminobenzaldehyde, and the second series is 1% aqueous mercuronitrate, 0.1% potassium permanganate and 5% iron(III) chloride in 0.1 *M* hydrochloric acid.

Vidic and Klug[32] made a systematic approach to the identification of 300 basic drugs by giving code numbers to the substances based on their relative retention behaviour in comparison with different test substances in one PC and five TLC systems (Table 3.4).

Moffat and co-workers[27,28,31] designed an identification scheme for basic substances, including a number of alkaloids, by determining the discriminating power of some TLC systems. The discriminating power was defined as the effectiveness of chromatographic systems in terms of separating two compounds selected at random from a specific drug population. Of 37 TLC systems investigated, the best results were obtained with cyclohexane - toluene - diethylamine (75:15:10), chloroform - methanol (90:10) and acetone, all on silica gel plates impregnated with 0.1 *N* sodium hydroxide. These systems might be used in combination because of their low correlation coefficients. Cellulose plates impregnated with 5% sodium dihydrogen citrate with the solvent system *n*-butanol - water - citric acid (87:13:0.48)[28,31] might also be useful, although it has the disadvantage of a long development time. For the identification of the compounds the $R_F$ values obtained were corrected graphically with the aid of $R_F$ values for four reference compounds. The corrected $R_F$ values were compared with the known values and the identification was thus made.

Armstrong[30] identified commonly used prescription drugs by means of TLC. For the separation of 76 non-acidic drugs, including about 17 alkaloids, he used silica gel plates in combination with the solvent systems methanol - ammonia (100:1.5), chloroform - diethylamine (9:1) and cyclohexane - diethylamine (8:1).

TABLE 3.5

TLC ANALYSIS OF SOME ALKALOIDS, PHENOTHIAZINES, AND ORGANIC BASES[20]

Equipment: standard thin-layer chromatography equipment, glass plates 20 x 20 cm. S-tank (S-chamber according to Stahl) and glass developing tank.

Thin layers: A = Silica gel G
B = Silica gel G + 0.1 *N* NaOH
C = Silica gel G + 0.5 *N* KOH
D = Silica gel G + 0.1 *N* $KHSO_4$
E = Aluminium oxide

Solvents: I = Methanol
II = Chloroform-acetone-diethylamine (50:40:10)
III = Cyclohexane-chloroform-diethylamine (50:40:10)
IV = Methanol-acetone-triethylamine (50:50:1.5)
V = Chloroform-ethanol (90:10)
VI = Cyclohexane-diethylamine (90:10)
VII = Cyclohexane-chloroform (30:70) + 0.05% diethylamine
VIII = 95% ethanol
IX = Acetone
X = Methyl acetate
XI = Cyclohexane-benzene-diethylamine (75:15:10)

| No. | Compound | $hR_F$ value | | | | | | | | | | | | |
|---|---|---|---|---|---|---|---|---|---|---|---|---|---|---|
| | | A/I | D/I | B/I | A/II | B/III | A/IV | A/V | A*/VI | E/VII | D/VIII | B/IX | B/X | B/XI |
| 1 | Aconitine | 50 | 90 | 85 | | 20 | 44 | | | 40 | | | | 17 |
| 2 | Atropine | 8 | 56 | 20 | 3 | 12**<br>30 | 30 | 5 | 34 | 72 | 31 | 3 | 1 | 7<br>37 |
| 3 | Berberine | 7 | 80 | 68 | 2 | 21 | 35 | 6 | 4 | 37 | 35 | 1 | 80 | 40 |
| 4 | Butaperazine | 27 | 65 | 60 | 56 | 54 | 54 | 67 | 29 | 9 | 3 | 14 | 61 | 43 |
| 5 | Chlordiazepoxide | 79 | 76 | 79 | 70 | 36<br>42 | 74 | 40 | 16 | 8 | 89 | 95 | 61 | 14 |
| 6 | Chlorpromazine | 33 | 50 | 50 | 60 | 65 | 63 | 50 | 81 | 68 | 25 | 20 | 47 | 53 |
| 7 | Cocaine | 25<br>47 | 65<br>34 | 30<br>77 | 90 | 29 | 43 | 48<br>66 | 41 | 19 | 23 | 16 | 7 | 5 |
| 8 | Codeine | 21 | | 41 | 90 | 49 | 40 | 40 | 17 | 7 | | 19 | 14 | 26 |
| 9 | Desipramine | 45 | | 31<br>80 | 85 | 37 | 82 | 24 | 71 | 10 | 87 | 13 | 2 | 37 |
| 10 | Diazepam | 93 | 81 | 72 | 98 | 42 | 88 | 48 | 39 | 9 | 82 | 99 | 92 | 30 |
| 11 | Dixyrazine | 52 | 35 | 60 | 75 | 15<br>40 | 75 | 53 | 6 | 48<br>19 | 21 | 40 | 60<br>40 | 15 |
| 12 | Ephedrine | 6 | | 5 | 100 | 43 | 35 | 71 | 74 | 43 | 42 | 2 | 0 | 43 |
| 13 | Ethylmorphine | 23 | 35 | 49<br>41 | 96 | | 31 | 49 | | 37 | 34 | 12 | 6 | 2 |
| 14 | Flufenazine | | 17 | 77 | 79 | 38 | 84 | 65 | 3 | 4 | 9<br>82 | 31 | | 11 |
| 15 | Heroin | 35 | 36 | 46 | 100 | 62 | 64<br>40 | 43 | 15 | 31<br>40 | 29 | 22 | 11 | 16 |
| 16 | Hydrocodone | 16 | 19 | 62<br>26 | 96 | 46 | 21 | 45 | 17 | 39 | 20 | 66 | 0 | 11 |
| 17 | Hydroxyzine | 67 | 67 | 79 | 81 | 41 | 78 | 50 | 19 | 7 | 48 | | 50 | 50 |
| 18 | Hyoscyamine | 9 | 55<br>7 | 19 | 2 | 32 | 34 | 23<br>4 | 0 | | 25 | 2 | 0 | 10 |
| 19 | Imipramine | 29 | 74 | 55 | 28 | 69 | 36 | 60 | 59 | 50 | 54 | 27 | 41 | 45 |
| 20 | Levomepromazine | 22 | 54 | 47<br>68 | 69 | 58 | 63 | 16 | 3 | 49 | 29<br>47 | 28<br>80 | 0 | 13 |
| 21 | Methadone | 27 | 62 | 50 | 100 | 62 | 39 | 42 | 99 | 40 | 60 | 66<br>100 | 35 | 36 |
| 22 | Methaqualone | 92 | 81 | | | | 91 | 54 | | 11 | | 98 | 90 | 40<br>60 |

TABLE 3.5 (*continued*)

| No. | Compound | A/I | D/I | B/I | A/II | B/III | A/IV | A/V | A*/VI | E/VII | D/VIII | B/IX | B/X | B/XI |
|---|---|---|---|---|---|---|---|---|---|---|---|---|---|---|
| 23 | Mogadon | 83 | 84 | 81 | 75 | 15 | | 54 | 2 | 0 | 89 | 99 | 98 | 29 |
| | | | | | | | | | | | | | | 3 |
| 24 | Morphine | 20 | 35 | 38 | 34 | 7 | 28 | 23 | 0 | 0 | 34 | 6 | 3 | 30 |
| 25 | Nicotine | 48 | 17 | 69 | 36 | 57 | 50 | 65 | 13 | 0 | 5 | 37 | 16 | 56 |
| 26 | Opipramol | 49 | 19 | 68 | 85 | 36 | 83 | 46 | 14 | 5 | 11 | 13 | 2 | 12 |
| 27 | Papaverine | 88 | 60 | 87 | 100 | 51 | 83 | 49 | 16 | 50 | 57 | 96 | 72 | 7 |
| | | | 25 | | | | | | | | | | 82 | |
| 28 | Periciazine | 58 | 16 | 70 | 58 | 37 | 79 | 10 | 2 | 34 | 19 | 30 | 48 | 10 |
| | | | | | | | | 37 | | | | | | |
| 29 | Perphenazine | 54 | 29 | 15 | 50 | 17 | 70 | 50 | 3 | 44 | 15 | 0 | 0 | 14 |
| | | | | | | 49 | | | | | | | | |
| 30 | Pethidine | 48 | 40 | 55 | 100 | | 43 | 45 | 92 | 49 | 27 | 30 | 19 | 35 |
| 31 | Procaine | 50 | 80 | 77 | 87 | 42 | 57 | 53 | 6 | 36 | 48 | 71 | 61 | 9 |
| 32 | Promazine | 18 | 36 | 33 | 63 | 31 | 60 | 41 | 4 | 49 | 30 | 29 | 51 | 12 |
| | | | | | | 58 | | | 46 | | | 80 | | 27 |
| 33 | Promethazine | 20 | 35 | 55 | 58 | 41 | 56 | 15 | 4 | 45 | 20 | 15 | 3 | 10 |
| | | | | | | 53 | | | | | 49 | | | |
| 34 | Promoton | 16 | 33 | 39 | 55 | 12 | 37 | 7 | 3 | 44 | 15 | 0 | 0 | 14 |
| | | | | | | 51 | | | | | | | | |
| 35 | Quinine | 28 | 67 | 64 | 20 | 55 | 57 | 5 | 4 | 52 | 32 | 90 | | 46 |
| | | 45 | | | | | | 65 | | | | | | |
| 36 | Scopolamine | 57 | 61 | 65 | 100 | 42 | 64 | 60 | 55 | 27 | 24 | 48 | 48 | 33 |
| 37 | Strychnine | 12 | 41 | 25 | 5 | 55 | 52 | 50 | 13 | 57 | 4 | 31 | 1 | 10 |
| | | | | | | | | 55 | | | 16 | | | |
| 38 | Tetracaine | 46 | 71 | 70 | 60 | 45 | 50 | 60 | 15 | 37 | 45 | | 59 | 28 |
| 39 | Thioproperazine | 11 | 50 | 30 | 58 | 50 | 56 | 55 | 4 | 59 | 16 | 10 | 61 | 19 |
| 40 | Thioridazine | 20 | 50 | 64 | 77 | 55 | 50 | 24 | 57 | 49 | 30 | 15 | 42 | 10 |
| | | | | | | | 69 | | | | | 35 | | |
| 41 | Trifluoperazine | 45 | 85 | 53 | 71 | 69 | 65 | 53 | 30 | 48 | 7 | 34 | 21 | 15 |
| | | | | | | | | | | 55 | | | | 47 |
| 42 | Triflupromazine | 44 | 23 | 30 | 66 | 65 | 76 | 20 | 69 | 47 | 54 | 0 | 63 | 41 |
| | | | | 18 | | | | | | 58 | 19 | | | 18 |
| 43 | Rhodamine B | 61 | 73 | 79 | 97 | | 34 | 42 | | 14 | 82 | 51 | 12 | 15 |

*The chromatogram was developed in a glass tank.
**Where two $hR_F$ values are mentioned, the lower one is the main one.

References p. 48

TABLE 3.6

TLC ANALYSIS OF SOME ALKALOIDS ON CELLULOSE IMPREGNATED WITH FORMAMIDE[24]

The stationary phase Cellulose Lucefol Quick is impregnated with a 20 or 40% solution of formamide (FMA) in ethanol (96%) to which either 2.4 g of ammonium formate per 100 ml or 1.0 g of trishydroxymethylaminomethane per 100 ml is added to obtain acidic (pH = 6.8-6.9) or basic plates (pH = 9.9), respectively.

TLC systems:

| System | Stationary phase | Mobile phase |
|---|---|---|
| S1 | Cellulose, impregnated with acidic 20% FMA | Chloroform |
| S2 | Cellulose, impregnated with acidic 40% FMA | Chloroform |
| S3 | Cellulose, impregnated with acidic 20% FMA | Benzene |
| S4 | Cellulose, impregnated with acidic 40% FMA | Benzene |
| S5 | Cellulose, impregnated with acidic 20% FMA | Cyclohexane |
| S6 | Cellulose, impregnated with acidic 40% FMA | Cyclohexane |
| S7 | Cellulose, impregnated with basic 20% FMA | Chloroform |
| S8 | Cellulose, impregnated with basic 40% FMA | Chloroform |
| S9 | Cellulose, impregnated with basic 20% FMA | Benzene |
| S10 | Cellulose, impregnated with basic 40% FMA | Benzene |
| S11 | Cellulose, impregnated with basic 20% FMA | Cyclohexane |
| S12 | Cellulose, impregnated with basic 40% FMA | Cyclohexane |
| S13 | Cellulose, impregnated with acidic 20% FMA | Chloroform - benzene (7:3) |
| S14 | Cellulose, impregnated with acidic 40% FMA | Chloroform - benzene (7:3) |
| S15 | Cellulose, impregnated with acidic 20% FMA | Chloroform - benzene (1:1) |
| S16 | Cellulose, impregnated with acidic 40% FMA | Chloroform - benzene (1:1) |
| S17 | Cellulose, impregnated with basic 20% FMA | Cyclohexane - chloroform (1:1) |
| S18 | Cellulose, impregnated with basic 40% FMA | Cyclohexane - chloroform (1:1) |
| S19 | Cellulose, impregnated with basic 20% FMA | Chloroform - benzene (7:3) |
| S20 | Cellulose, impregnated with basic 40% FMA | Chloroform - benzene (7:3) |

| Compound | $hR_F$ value | | | | | | | | | | | | | | | | | | | |
|---|---|---|---|---|---|---|---|---|---|---|---|---|---|---|---|---|---|---|---|---|
| | S1 | S2 | S3 | S4 | S5 | S6 | S7 | S8 | S9 | S10 | S11 | S12 | S13 | S14 | S15 | S16 | S17 | S18 | S19 | S20 |
| Aconitine | 58 | 41 | 32 | 20 | 0 | 0 | 100 | 100 | 86 | 58 | 15 | 15 | | | | | | | | |
| Atropine | 12 | 11 | 0 | 0 | 0 | 0 | 100 | 100 | 50 | 26 | 0 | 0 | | | | | | | | |
| Methylatropine | 0 | 0 | 0 | 0 | 0 | 0 | 0 | 0 | 0 | 0 | 0 | 0 | | | | | | | | |
| Brucine | 95 | 95 | 0 | 0 | 0 | 0 | 100 | 100 | 83 | 69 | 0 | 0 | | | 50 | 28 | | | | |
| Diacetylmorphine (heroin) | 95 | 95 | 29 | 19 | 0 | 0 | 100 | 100 | 100 | 93 | 22 | 12 | | | 78 | 60 | 90 | 79 | | |
| Dihydrocodeine | 32 | 26 | 0 | 0 | 0 | 0 | 100 | 100 | 70 | 49 | 9 | 7 | | | | | 50 | 24 | | |
| Emetine | 84 | 84 | 0 | 0 | 0 | 0 | 100 | 100 | 100 | 100 | 31 | 23 | 37 | 22 | | | | | | |
| Ethylmorphine | 65 | 52 | 0 | 0 | 0 | 0 | 100 | 100 | 88 | 73 | 19 | 11 | | | | | 68 | 45 | | |
| Physostigmine | 74 | 54 | 0 | 0 | 0 | 0 | 100 | 100 | 85 | 67 | 7 | 5 | 42 | 27 | | | 56 | 30 | | |
| Homatropine | 12 | 9 | 0 | 0 | 0 | 0 | 100 | 100 | 53 | 30 | 9 | 3 | | | | | | | | |
| Dihydrocodeinone | 80 | 67 | 0 | 0 | 0 | 0 | 100 | 100 | 93 | 81 | 12 | 8 | 44 | 29 | | | | | | |
| Dihydromorphinone | 6 | 5 | 0 | 0 | 0 | 0 | 90 | 72 | 9 | 6 | 0 | 0 | | | | | | | 71 | 46 |
| Quinidine | 100 | 95 | 0 | 0 | 0 | 0 | 100 | 100 | 78 | 55 | 7 | 0 | | | 54 | 29 | 44 | 22 | | |
| Quinine | 100 | 86 | 0 | 0 | 0 | 0 | 100 | 100 | 73 | 54 | 5 | 0 | | | 50 | 25 | 42 | 20 | | |
| Codeine | 38 | 31 | 0 | 0 | 0 | 0 | 100 | 100 | 71 | 53 | 7 | 6 | | | | | 47 | 23 | | |
| Cocaine | 100 | 95 | 49 | 35 | 0 | 0 | 100 | 100 | 100 | 87 | 95 | 87 | | | | | | | | |
| Lobeline | 81 | 81 | 11 | 6 | 0 | 0 | 100 | 100 | 100 | 100 | 84 | 67 | | | | | | | | |
| Morphine | 0 | 0 | 0 | 0 | 0 | 0 | 41 | 24 | 0 | 0 | 0 | 0 | | | | | | | 31 | 15 |
| Narceine | 50 | 41 | 0 | 0 | 0 | 0 | 28 | 21 | 0 | 0 | 0 | 0 | | | | | | | | |
| Noscapine | 100 | 100 | 100 | 100 | 22 | 14 | 100 | 100 | 100 | 100 | 52 | 25 | | | | | | | | |
| Nicotine | 80 | 71 | 19 | 14 | 0 | 0 | 100 | 90 | 0 | 0 | 0 | 0 | | | | | | | 9 | 5 |
| Oxycodone | 74 | 62 | 0 | 0 | 0 | 0 | 100 | 100 | 100 | 96 | 30 | 15 | 41 | 28 | | | | | | |
| Papaverine | 100 | 100 | 100 | 93 | 6 | 0 | 100 | 100 | 100 | 100 | 14 | 9 | | | | | 88 | 78 | | |
| Pilocarpine | 74 | 58 | 0 | 0 | 0 | 0 | 100 | 100 | 18 | 10 | 0 | 0 | 43 | 31 | | | | | | |
| Butylscopolamine | 0 | 0 | 0 | 0 | 0 | 0 | 0 | 0 | 0 | 0 | 0 | 0 | | | | | | | | |
| Scopolamine | 43 | 34 | 0 | 0 | 0 | 0 | 100 | 100 | 51 | 28 | 0 | 0 | | | | | 21 | 8 | | |
| Sparteine | 17 | 17 | 0 | 0 | 0 | 0 | 92 | 57 | 60 | 29 | 60 | 24 | | | | | | | | |
| Strychnine | 93 | 84 | 4 | 4 | 0 | 0 | 100 | 100 | 94 | 87 | 12 | 8 | | | | | 73 | 55 | | |
| Thebaine | 100 | 100 | 16 | 13 | 0 | 0 | 100 | 100 | 100 | 100 | 61 | 35 | | | 71 | 53 | | | | |
| Veratridine | 84 | 55 | 0 | 0 | 0 | 0 | 100 | 100 | 88 | 65 | 0 | 0 | 27 | 12 | | | 43 | 23 | | |
| Yohimbine | 40 | 27 | 0 | 0 | 0 | 0 | 100 | 100 | 62 | 38 | 0 | 0 | | | | | 22 | 11 | | |

TABLE 3.7

TLC ANALYSIS OF SOME ALKALOIDS ON POLYAMIDE PLATES[13]

Solvent systems:
- S1 Dioxane-cyclohexane-diethylamine (10:20:0.5)
- S2 Chloroform-cylohexane-diethylamine (10:20:0.5)
- S3 Methyl ethyl ketone-cyclohexane-diethylamine (20:30:0.5)
- S4 Chloroform-ethanol-acetic acid (200:20:0.5)
- S5 Water-ethanol-pyridine (10:0.5:0.3)
- S6 Cyclohexane-ethyl acetate-*n*-propanol-dimethylamine (30:2.5:0.9:0.1)
- S7 Water-ethanol-dimethylamine (88:12:0.1)

| Alkaloid (10 cm development) | $hR_F$ value | | | | | | | Color of spots with | | |
|---|---|---|---|---|---|---|---|---|---|---|
| | S1 | S2 | S3 | S4 | S5 | S6 | S7 | Iodoplatinate | Iodine | Ultraviolet |
| Atropine | 65 | 61 | 26 | 85 | 31 | 40 | 33 | Yellow | + | Blue |
| Brucine | 45 | 84 | 41 | 95 | 87 | 20 | 34 | Violet-blue | + | Blue |
| Caffeine | 90 | 65 | 39 | 95 | 85 | 13 | 80 | - | - | Blue |
| Cinchonine | 55 | 69 | 25 | 96 | 67 | 50 | 0 | Yellow | + | Yellow |
| Cocaine | 88 | 95 | 92 | 94 | 27 | 85 | 22 | Yellow | + | Blue |
| Codeine | 62 | 76 | 42 | 90 | 45 | 41 | 49 | Yellow | + | Blue |
| Colchicine | 3 | 21 | 20 | 97 | 54 | 2 | 48 | Yellow | + | Yellow |
| Cyclanoline chloride | 0 | 2 | 0 | 3 | 87 | 3 | 88 | Violet-red | + | Blue |
| Diacetylmorphine (heroin) | 77 | 85 | 65 | 97 | * | 69 | 45 | Yellow | + | Blue |
| Dihydrocodeine | 78 | 86 | 75 | 95 | 95 | 58 | 40 | Yellow | + | Blue |
| Ephedrine | 15 | 72 | 40 | 81 | 50 | 45 | 50 | Yellow | + | Blue |
| Ergometrine malate | 2 | 1 | 20 | 94 | 19 | 0 | 14 | Greyish violet | + | Blue |
| Ergotamine tartrate | 5 | 15 | 11 | 97 | 10 | 1 | 3 | Greyish violet | + | Blue |
| Ethylmorphine | 71 | 75 | 63 | 91 | 40 | 54 | 40 | Yellow | + | Blue |
| Homatropine | 60 | 68 | 30 | 85 | 41 | 43 | 39 | Yellow | + | Blue |
| Jatrorrhizine picrate | 21 | 0 | 0 | 50 | 20 | 1 | 34 | Violet | + | Blue |
| Magnoflorine iodide | 0* | 0 | 0 | 36* | 70 | 2 | 72 | Yellow | + | Blue |
| Menisperine chloride | 0 | 0 | 2 | 76 | 80 | 0 | 43 | Yellow | + | Blue |
| Morphine | 21 | 3 | 10 | 62 | 83 | 3 | 80 | Yellow | + | Blue |
| Noscapine | 75 | 88 | 92 | 90 | 0 | 54 | 0 | Violet-red | + | Blue |
| Papaverine | 60 | 88 | 83 | 2 | 10 | 37 | 5 | Yellow | + | Blue |
| Phellodendrine iodide | 0 | 0 | 0 | 93 | 47 | 0 | 89 | Yellow | + | Yellow |
| Pilocarpine | 25 | 40 | 55 | 93 | 55 | 2 | 59 | Yellow | + | Blue |
| Physostigmine salicylate | 69 | 68 | 77 | 93 | 90 | 40 | 41 | Yellow | + | Blue |
| Quinine | 52 | 30 | 25 | 3 | 40 | 40 | 8 | Yellow | + | Violet |
| Reserpine | 62 | * | 57 | 95 | 55 | 38 | 0 | Yellow | + | Yellow |
| Scopolamine | 81 | 60 | 76 | 93 | 87 | 30 | 58 | Yellow | + | Yellow |
| Strychnine | 54 | 88 | 41 | 90 | 83 | 40 | 25 | Greenish | + | Violet |
| Tubocurine·HCl | 0 | 1 | 0 | 2* | 83 | 0 | 74 | Violet | + | Blue |
| Yohimbine | 51 | 22 | 35 | 94 | 9 | 10 | 55 | Yellow | + | Yellow |

*Tailing.

(Published with permission of The Chinese Chemical Society).

References p. 48

TABLE 3.8

TLC ANALYSIS OF ALKALOIDS ON ALUMINIUM OXIDE WITHOUT BINDER[4]

Sorbent: aluminium oxide, activity III.
Solvent systems:
- S1 Benzene-ethanol (95:5)
- S2 Benzene-ethanol (9:1)
- S3 Chloroform
- S4 Chloroform-ethanol (98:2)
- S5 Chloroform-ethanol (95:5)
- S6 Chloroform-acetone (1:1)
- S7 Diethyl ether
- S8 Diethyl ether-ethanol (97:3)
- S9 Light petroleum-dioxane (1:1)
- S10 Benzene-ethanol (8:2)
- S11 Diethyl ether-ethanol (95:5)

| Alkaloid | $hR_F$ value | | | | | | | | | | |
|---|---|---|---|---|---|---|---|---|---|---|---|
| | S1 | S2 | S3 | S4 | S5 | S6 | S7 | S8 | S9 | S10 | S11 |
| Strychnine | 27 | 75 | 8 | 35 | 68 | 26 | 4 | 23 | 37 | | 35 |
| Brucine | 19 | 61 | 6 | 28 | 67 | 22 | 3 | 10 | 18 | | 15 |
| Hyoscyamine | 5 | 9 | 0 | 6 | 28 | 9 | 2 | 6 | 11 | | |
| Scopolamine | 14 | 42 | 2 | 18 | 50 | 26 | 5 | 37 | 47 | | |
| Atropine | 7 | 29 | 0 | 5 | 26 | 7 | 3 | 6 | 11 | | 5 |
| (-)-Lobeline | 24 | 71 | 7 | 27 | 66 | 73 | 31 | 80 | | | 82 |
| Lobelanine | 35 | 82 | 28 | 45 | 86 | 82 | 60 | 91 | | | |
| Lobelanidine | 15 | 46 | 0 | 10 | 42 | 15 | 8 | 65 | | | |
| Ethylmorphine | 20 | 48 | 4 | 21 | 54 | 23 | 13 | 36 | 56 | | 52 |
| Codeine | 22 | 46 | 2 | 19 | 53 | 45 | 5 | 25 | 46 | | 25 |
| Quinine | 29 | 43 | 0 | 6 | 56 | 5 | 5 | 33 | | | |
| Quinidine | 30 | 43 | 0 | 6 | 56 | 6 | 8 | 35 | | | |
| Cinchonine | 26 | 41 | 0 | 11 | 49 | 5 | 6 | 32 | | | |
| Cinchonidine | 24 | 39 | 0 | 7 | 46 | 5 | 8 | 31 | | | |
| Ephedrine | 2 | 19 | 0 | 3 | 16 | 0 | 0 | 2 | | | 18 |
| Cocaine | 54 | 74 | 49 | 58 | 91 | 82 | 82 | 87 | | | |
| Ergotoxine | 34 | 49 | 0 | 23 | 72 | 64 | 0 | 37 | 65 | | |
| Ergotamine | 28 | 43 | 0 | 14 | 56 | 35 | 0 | 7 | 32 | | |
| Ergometrine | 11 | 28 | 0 | 3 | 25 | 10 | 0 | 3 | 13 | | |
| Theobromine | | | | | | | | | | 11 | |
| Caffeine | | | | | | | | | | 75 | |
| Protoveratrine | | | | | | | | | | 74 | |
| Emetine | | | | | | | | | | 23 | |
| Morphine | | | | | | | | | | | 8 |
| Papaverine | | | | | | | | | | | 68 |
| Noscapine | | | | | | | | | | | 86 |

(Published with permission of VEB Verlag, Volk und Gesundheit, Berlin).

TABLE 3.9

TLC ANALYSIS OF ALKALOIDS ON CELLULOSE[7]

The plates are impregnated by dipping for a few seconds in the appropriate solution.

TLC systems:

| | | |
|---|---|---|
| S1 | Avicel SF, impregnated with 0.5 *M* potassium chloride | *n*-Butanol-36% hydrochloric acid (98:2) saturated with water at 20°C |
| S2 | Avicel SF, impregnated with 0.2 *M* potassium dihydrogen phosphate | *n*-Butanol saturated with water at 20°C |
| S3 | Avicel SF, impregnated with 0.2 *M* potassium dihydrogen phosphate | *sec.*-Butanol saturated with water at 20°C |
| S4 | Avicel SF, impregnated with 0.5 *M* potassium dihydrogen phosphate | Isopropanol-water (3:1) |
| S5 | Avicel SF | *n*-Butanol-water-acetic acid (10:3:1) |

| Alkaloid | $hR_F$ value | | | | |
|---|---|---|---|---|---|
| | S1 | S2 | S3 | S4 | S5 |
| Papaverine | 96 | 94 | 98 | 91 | |
| Laudanosine | 76 | 62 | 45 | 84 | |
| Armepavine | 80 | 62 | 45 | 87 | |
| Codamine | 92 | 63 | 89 | | |
| Tembetarine | 33 | 39 | 30 | 59 | |
| Magnoflorine | 30 | 22* | 13 | 42 | |
| Noscapine | 75 | 90 | 92 | 93 | |
| Narceine | 83 | 66 | 41 | 73 | |
| Dicentrine | 24* | 27* | 29 | 51 | |
| Ocoteine | 45* | 34** | 35 | 55 | |
| α-*iso*-Sparteine | 91 | 44 | 37 | 84 | |
| Multiflorine | 21 | 10 | 14 | 58 | |
| Lupanine | 50 | 24 | 24 | 85 | |
| Brucine | 41* | 34 | 31 | 50 | |
| Strychnine | 61 | 44 | 78 | 63 | |
| Trilobine | 50** | 17** | 12 | 26 | |
| Isotrilobine | 33* | 31* | 17 | 29 | |
| Repanduline | 41** | 23* | 17* | 33 | |
| Tenuipine | 46** | 38 | 21* | 42 | |
| Sarpagine | 68 | 64 | 46 | 75 | |
| Ajmaline | 95 | 85 | 72* | 89 | |
| Reserpine | 98 | 92 | 81* | 89 | |
| Aspidospermine | 91 | 90 | 91 | 88 | |
| Quebrachamine | 92 | 79 | 92 | | |
| Uleine | 97 | 91 | | | |
| Physostigmine | 86 | 58 | 62 | 75 | |
| Harman | 78* | 59* | 52 | 73 | |
| Morphine | 39 | 13 | 22 | 53 | |
| Codeine | 52 | 22 | 32 | 60 | |
| Heroin | 78 | 46 | 58 | 77 | |
| Thebaine | 84 | 47 | 57 | 74 | |
| Quinine | 92 | 87 | 79 | 90 | |
| Quinidine | 95 | 88 | 83 | 90 | |
| Cinchonine | 95 | 83 | 76 | 90 | |
| Tetrahydroberberine | 98 | 81 | 82** | 75 | 80 |
| Tetrahydropseudoberberine | 75** | 64 | 58** | 61 | 74 |

**References p. 48**

TABLE 3.9 (*continued*)

| Alkaloid | $hR_F$ | | | | |
|---|---|---|---|---|---|
| | S1 | S2 | S3 | S4 | S5 |
| Nandinine | 92* | 66* | 60** | 61 | 54 |
| O-Acetylnandinine | 98 | 94 | 98 | 86 | 88 |
| Allocryptopine | 73** | 51 | 35 | 70 | 75 |
| Protopine | 78** | 52 | 30 | 59 | 72 |
| Fagarine II | 56** | 41 | 34 | 67 | 63 |
| Hunnemanine | 70** | 46 | 26 | 59 | 70 |
| Berberine | 35* | 26** | 21 | 41 | 57 |
| Pseudoberberine | 30 | 22** | 10* | 27 | 45 |
| Palmatine | 36 | 26** | 19 | 39 | 51 |
| Berberrubine | 30 | 22** | 18 | 33 | 45 |
| Chelerythrine | 33* | 37** | 26** | 8 | 54 |
| Sanguinarine | | 41** | 39** | 20 | 45 |

*Oval, but useful spot.
**Long spot, useless for characterization.

TABLE 3.10

TLC ANALYSIS OF ALKALOIDS[3]

Sorbent: silica gel G, activated.
Solvent systems:
S1 Benzene-acetone-diethyl ether-10% ammonia (4:6:1:0.3)
S2 Benzene-acetone-diethyl ether-25% ammonia (4:6:1:0.3)

| Alkaloid | $hR_F$ value | |
|---|---|---|
| | S1 | S2 |
| Atropine | 8 | 20 |
| Morphine | 9 | 9 |
| Hydromorphone | 9 | 13 |
| Sparteine | 11 | 32 |
| Brucine | 10 | 25 |
| Colchicine | 15 | 20 |
| Ergometrine | 14 | 17 |
| Apoatropine | 19 | 43 |
| Dihydrocodeinone | 18 | 25 |
| Codeine | 18 | 24 |
| Strychnine | 20 | 40 |
| Pilocarpine | 25 | 37 |
| Quinine | 23 | 35 |
| Dihydroergotamine | 28 | 29 |
| Ergotamine | 29 | 32 |
| Quinidine | 28 | 43 |
| Scopolamine | 37 | 52 |
| Dihydroergocornine | 47 | 48 |
| Dihydroergocristine | 47 | 48 |
| Dihydroergocryptine | 47 | 48 |
| Thebaine | 46 | 51 |
| Physostigmine | 47 | 60 |
| Meperidine | 48 | 60 |
| Oxycodone | 64 | 69 |
| Papaverine | 64 | 69 |
| Reserpine | 72 | 75 |
| Cocaine | 78 | 80 |
| Noscapine | 78 | 81 |

TABLE 3.11

TLC ANALYSIS OF SOME PHARMACEUTICALS[21]

Sorbent: silica gel GF 254, activated.
Solvent systems:
S1 70% ethanol saturated with ammonia gas
S2 Acetone-cyclohexane-ethyl acetate (1:1:1) saturated with ammonia gas
S3 Chloroform-acetone (1:1) saturated with ammonia gas
S4 Chloroform-methanol-cyclohexane (7:3:1) saturated with ammonia gas

| Compound | $hR_F$ | | | |
|---|---|---|---|---|
| | S1 | S2 | S3 | S4 |
| Acetylsalicylic acid | 80 | 0 | 0 | 28 |
| Atropine | 53 | 20 | 23 | 76 |
| Benzalkonium chloride | 15 | 20 | 11 | 54 |
| Brucine | 57 | 10 | 28 | 85 |
| Cephaeline | 87 | 33 | 25 | 81 |
| Cinchonidine | 88 | 24 | 33 | 89 |
| Cinchonine | 89 | 21 | 20 | 77 |
| Ephedrine | 63 | 30 | 33 | 56 |
| Emetine | 93 | 40 | 49 | 92 |
| Ethylmorphine | 79 | 27 | 24 | 83 |
| Physostigmine | 86 | 42 | 56 | 96 |
| Hyoscyamine | 63 | 17 | 20 | 82 |
| Homatropine | 56 | 21 | 23 | 80 |
| Jervine | 87 | 35 | 40 | 93 |
| Quinidine | 83 | 30 | 29 | 82 |
| Quinine | 81 | 26 | 22 | 80 |
| Codeine | 82 | 27 | 24 | 81 |
| Caffeine | 82 | 41 | 58 | 72 |
| Methylhomatropine | 7 | 0 | 0 | 1 |
| Noscapine | 85 | 77 | 76 | 89 |
| Papaverine | 90 | 60 | 66 | 93 |
| Pilocarpine | 83 | 29 | 40 | 84 |
| Protoveratrine A+B | 91 | 75 | 77 | 92 |
| Scopolamine | 85 | 40 | 49 | 87 |
| Strychnine | 61 | 19 | 40 | 80 |
| Theophylline | 88 | 11 | 8 | 55 |
| Theobromine | 81 | 25 | 27 | 73 |

TABLE 3.12

TLC ANALYSIS OF ALKALOIDS[5]

All solvents were used in combination with 0.5 *M* sodium hydroxide or potassium hydroxide impregnated silica gel, or in combination with silica gel G plates and addition of 5-10% of diethylamine to the solvent systems.

Solvent systems:
- S1 Dichloromethane-methanol (9:1)
- S2 Ethyl acetate-hexane (25:8)
- S3 Benzene-chloroform (7:25)
- S4 Chloroform
- S5 Chloroform-ethanol (9:1)
- S6 Chloroform-acetone (9:1)
- S7 Chloroform-cyclohexane (4:5)

| Alkaloid | $hR_F$ value | | | | | | |
|---|---|---|---|---|---|---|---|
| | S1 | S2 | S3 | S4 | S5 | S6 | S7 |
| Aconitine | | | 75 | 55 | | | |
| Ajmaline | 15 | 30 | | | | 55 | |
| Atropine | | 20 | 20 | | 50 | 35 | |
| Brucine | | | 65 | 10 | | 50 | |
| Caffeine | 0 | | | 40 | | | |
| Cinchonine | 55 | | 20 | | 45 | 40 | |
| Cinchonidine | 5 | | 10 | | | 40 | |
| Cocaine | | 70 | 60 | | 75 | | |
| Codeine | 50 | 35 | 40 | | | | |
| Conessine | 60 | 55 | | | 70 | | |
| Emetine | 58 | 38 | 80 | | | | |
| Ergotamine | | 10 | 20 | 60 | | | |
| Hyoscyamine | | 20 | 20 | | 50 | 35 | |
| Morphine | 20 | 10 | | | 80 | | |
| Narceine | | | | | | 10 | 60 |
| Noscapine | 90 | 70 | | | 72 | | |
| Papaverine | 90 | 60 | | | 68 | | |
| Quinine | 50 | | 15 | | 35 | | |
| Quinidine | 60 | | 5 | | 50 | | |
| Raupine | 60 | 10 | | 30 | | | |
| Rescinnamine | 80 | 60 | 40 | | | | |
| Reserpine | 55 | 70 | 35 | | | | |
| Scopolamine | | 10 | 30 | | | 52 | |
| Sparteine | | 20 | 38 | | | 60 | |
| Strychnine | | | 70 | 30 | | 55 | |
| Thebaine | | | | 30 | | 60 | 45 |
| Yohimbine | 90 | | 45 | | | 70 | 70 |

(Published with permission of Masson, Paris).

References p. 48

Breiter[29] and Breiter and Helger[39] applied a searching scheme for drugs, including some alkaloids. After the sample preparation, a basic fraction containing the alkaloids was separated in ethyl acetate - methanol - ammonia (90:10:1) or chloroform - methanol - ammonia (90:10:1) on silica gel plates.

In addition to the TLC identification schemes mentioned here, similar methods have been described for the analysis of drugs of abuse. In Chapter 12, more specific identification schemes are dealt with.

## 3.2. TLC OF VARIOUS ALKALOIDS

Various groups have used a series of alkaloids to demonstrate the usefulness of a new TLC system or detection reagent. Their results are of interest if alkaloids of totally different structure are to be separated.

Tadjer[20] presented chromatographic data on a number of alkaloids and other organic bases on silica gel plates. In Table 3.5 the results concerning the alkaloids are summarized.

Novakova and Vecerkova[24] described the TLC of a number of alkaloids and local anaesthetics on formamide-impregnated cellulose plates. The results were comparable to those on formamide-impregnated paper, although a higher sensitivity and a shorter development time were obtained with the thin-layer plates. The results are summarized in Table 3.6.

Hsiu et al.[13] used polyamide plates for the separation of a number of alkaloids. The results are given in Table 3.7. Polyamide was also used as stationary phase by Lyakina and Brutko[38] for the separation of a number of alkaloids. Schwarz and Sarsunova[4] used aluminium oxide without a binder for the TLC of various alkaloids and pharmaceutical preparations containing alkaloids (Table 3.8).

Giacopello[7] chromatographed a number of alkaloids using microcrystalline cellulose without a binder (Avicel) on TLC plates (Table 3.9).

Röder et al.[17] used azeotropic mixtures of solvents for TLC. For alkaloids the best results were obtained with the mixtures methanol - benzene (39.1:60.9), methanol - chloroform - methyl acetate (21.6:51.4:27.0) and methanol - chloroform (23.0:34.0).

Zarnack and Pfeifer[3] reported the TLC of a number of alkaloids in two different systems (Table 3.10).

Zádeczky et al.[21] applied TLC to the examination of pharmaceutical mixtures containing, amongst others, alkaloids (Table 3.11).

Paris et al.[5] reported the TLC of a number of alkaloids and plant extracts containing alkaloids. The results are summarized in Table 3.12.

De Zeeuw et al.[41] performed the ion-pair chromatography of basic drugs on silica gel using straight-phase systems. Excellent results were obtained with chloride, bromide or iodide as counter ions, provided that the counter ion concentration was

TABLE 3.13

ION-PAIR ADSORPTION CHROMATOGRAPHY OF BASIC DRUGS[41]

Concentrations of the counter ions or hydroxides in the spreading slurry or in the solvent were 0.1 *M*

Sorbent: silica gel 60 F254 pre-coated plates.

| System | Silica gel prepared or dipped in | Solvent |
|---|---|---|
| S1 | No special treatment | $Br^-$ in methanol |
| S2 | No special treatment | $Cl^-$ in methanol |
| S3 | 0.1 *M* NaOH in methanol | Methanol |
| S4 | No special treatment | Methanol |
| S5 | No special treatment | Methanol-ammonia (100:1.5) |
| S6 | Aqueous phosphate buffer (pH 2), followed by $Br^-$ in methanol | Chloroform-methanol (90:10) |
| S7 | $Br^-$ in methanol | Chloroform-methanol (90:10) |
| S8 | $Cl^-$ in methanol | Chloroform-methanol (90:10) |
| S9 | 0.1 *M* in methanol | Chloroform-methanol (90:10) |
| S10 | No special treatment | Chloroform-methanol (90:10) |

| Drug | $hR_F$ value | | | | | | | | | |
|---|---|---|---|---|---|---|---|---|---|---|
| | S1 | S2 | S3 | S4 | S5 | S6 | S7 | S8 | S9 | S10 |
| Oxycodone | 44 | 39 | 35 | 32 | 76 | 25 | 34 | 34 | 68 | 51 |
| Ethylmorphine | 37 | 30 | 28 | 27 | 59 | 29 | 40 | 30 | 35 | 32 |
| Codeine | 30 | 30 | 24 | 27 | 57 | 21 | 34 | 25 | 32 | 32 |
| Morphine | 25 | 24 | 22 | 24 | 55 | 6 | 16 | 9 | 32 | 31 |
| Quinine | 59 | 55 | 34 | 36 | 74 | 22 | 55 | 41 | 23 | 16 |
| Cocaine | 55 | 47 | 41 | 44 | 89 | 23 | 24 | 30 | 74 | 52 |
| Aminophenazone | 75 | 74 | 75 | 74 | 89 | 37 | 59 | 60 | 65 | 61 |
| Phenazone | 74 | 73 | 73 | 73 | 86 | 51 | 50 | 52 | 59 | 55 |
| Amphetamine | 67 | 65 | 28 | 24 | 61 | 36 | 27 | 29 | 34 | 20 |
| Methylamphetamine | 64 | 59 | 19 | 17 | 51 | 42 | 31 | 31 | 27 | 14 |
| Ephedrine | 67 | 61 | 19 | 18 | 50 | 35 | 23 | 24 | 8 | 7 |
| Amitriptyline | 48 | 41 | 28 | 34 | 77 | 56 | 39 | 38 | 58 | 40 |
| Nortriptyline | 68 | 65 | 11 | 15 | 49 | 56 | 39 | 36 | 37 | 20 |
| Imipramine | 44 | 36 | 24 | 29 | 72 | 61 | 45 | 39 | 55 | 40 |
| Desipramine | 62 | 56 | 11 | 12 | 41 | 61 | 46 | 39 | 30 | 18 |
| Yohimbine | 79 | 77 | 73 | 73 | 93 | 30 | 29 | 31 | 65 | 52 |
| Dextromoramide | 82 | 82 | 82 | 80 | 93 | 69 | 69 | 70 | 94 | 80 |

References p. 48

TABLE 3.14

TLC OF ALKALOIDS ON ION EXCHANGERS[34]

$hR_F$ values of alkaloids on AG1-X4 ($AcO^-$) thin layers. Stationary phase: 3 g of resin + 9 g of cellulose.

| Alkaloid | Eluent | | | |
|---|---|---|---|---|
| | 1 *M* $NH_3$ + 0.05 *M* $CH_3COONa$ | 0.5 *M* $NH_3$ + 0.5 *M* $CH_3COONH_4$ | 0.5 *M* $CH_3COONH_4$ | 0.5 *M* acetate buffer |
| Narceine | 0 | 0 | 0 | 0 |
| Ergocristine | 0 | 0 | 0 | 5 |
| Ergotamine | 0 | 0 | 0 | 6 |
| Papaverine | 1 | 1 | 2 | 46 |
| Ibogaine | 2 | 2 | 3 | 36 |
| Berberine hydrochloride | 3 | 4 | 4 | 3 |
| Reserpine | 0 | 0 | 4 | 19 |
| Boldine | 2 | 2 | 7 | 25 |
| Ergometrine | 2 | 2 | 8 | 21 |
| Noscapine | 10 | 10 | 22 | 72 |
| Hydrastine | 21 | 21 | 36 | 73 |
| Aminophylline | 3 | 3 | 38 | 53 |
| Theophylline | 3 | 3 | 38 | 54 |
| Colchicine | 40 | 40 | 42 | 45 |
| Yohimbine hydrochloride | 7 | 7 | 44 | 58 |
| Quinine | 8 | 10 | 50 | 65 |
| Quinidine sulphate | 13 | 15 | 54 | 67 |
| Brucine | 23 | 26 | 66 | 70 |
| Theobromine | 41 | 52 | 68 | 68 |
| Caffeine | 69 | 69 | 68 | 68 |
| Cinchonine hydrochloride | 0 | 0 | 71 | 74 |
| Cinchonidine | 16 | 20 | 71 | 74 |
| Strychnine | e.s.* | e.s.* | 77 | 79 |
| Ajmaline | 35 | 36 | 78 | 79 |
| Lobeline hydrochloride | 0 | 12 | 82 | 83 |
| Tubocurarine | 74 | 79 | 85 | 85 |
| Cocaine | 25 | 42 | 90 | 91 |
| Atropine | 61 | 88 | 91 | 92 |
| Hyoscyamine | 62 | 89 | 91 | 92 |
| Eucatropine hydrochloride | 62 | 89 | 91 | 92 |
| Emetine hydrochloride | 12 | 21 | 91 | 93 |
| Ethylmorphine | 41 | 47 | 91 | 94 |
| Physostigmine sulphate (1) | 28 | 50 | 93 | 94 |
| (2) | 54 | | | |
| Homatropine | 61 | 90 | 93 | 95 |
| Scopolamine hydrochloride | 68 | 72 | 93 | 95 |
| Arecoline hydrochloride | 79 | 88 | 94 | 95 |
| Hyoscyne | 93 | 96 | 96 | 95 |
| Scopoline | 90 | 96 | 98 | 96 |
| Sparteine sulphate | 94 | 97 | 98 | 97 |
| Tropine | 96 | 97 | 98 | 97 |
| Prostigmine | 96 | 97 | 98 | 97 |

*e.s. = elongates spot.

TABLE 3.15

TLC OF ALKALOIDS ON ION EXCHANGERS[34]

$hR_F$ values of alkaloids on Cellex D and microcrystalline cellulose (mC) thin layers.

| Alkaloid | Eluent | | | | | | | |
|---|---|---|---|---|---|---|---|---|
| | 1 *M* $NH_3$ + 0.5 *M* $CH_3COONa$ | | 0.5 *M* $NH_3$ + 0.5 *M* $CH_3COONH_4$ | | 0.5 *M* $CH_3COONH_4$ | | 0.5 *M* acetate buffer | |
| | Cellex D | mC | Cellex D | mC | Cellex D | mC | Cellex D | mC |
| Narceine | 0 | 6 | 0 | 2 | 0 | 4 | 0 | 8 |
| Reserpine | 0 | 0 | 0 | 0 | 9 | 5 | 41 | e.s.* |
| Ergocristine | e.s. | e.s. | 3 | 1 | 6 | 4 | 41 | 23 |
| Ergotamine | e.s. | e.s. | 12 | e.s. | 17 | 6 | 51 | 31 |
| Boldine | 33 | 65 | 16 | 30 | 32 | 31 | 49 | 35 |
| Ibogaine | 14 | 8 | 18 | 8 | 24 | 12 | 84 | 49 |
| Papaverine | 12 | 32 | 20 | 32 | 37 | 40 | 89 | 82 |
| Berberine hydrochloride | 16 | 0 | 29 | 0 | 36 | 3 | 49 | 5 |
| Ergometrine | 30 | 30 | 35 | 31 | 49 | 24 | 65 | 24 |
| Yohimbine hydrochloride | 35 | 39 | 42 | 45 | 77 | 57 | 86 | 65 |
| Noscapine | 39 | 50 | 45 | 51 | 72 | 56 | 94 | 83 |
| Aminophylline | 47 | 63 | 48 | 63 | 81 | 64 | 85 | 68 |
| Theophylline | 48 | 67 | 49 | 65 | 83 | 66 | 88 | 70 |
| Emetine hydrochloride | 26 | e.s. | 49 | e.s. | 95 | e.s. | 95 | 93 |
| Brucine | 42 | 50 | 55 | 52 | 86 | 67 | 90 | 69 |
| Quinidine sulphate | 37 | 47 | 55 | 49 | 80 | 64 | 89 | 73 |
| Quinine | 38 | 50 | 56 | 54 | 81 | 67 | 89 | 74 |
| Lobeline hydrochloride (1) | 0 | 0 | 56 | 61 | 95 | 86 | 95 | 85 |
| (2) | e.s. | e.s. | | | | | | |
| Hydrastine | 53 | 64 | 64 | 66 | 82 | 70 | 94 | 83 |
| Strychnine | 52 | 63 | 65 | 64 | 87 | 75 | 91 | 79 |
| Cinchonine hydrochloride (1) | 0 | 0 | 70 | 66 | 92 | 79 | 94 | 83 |
| (2) | e.s. | e.s. | | | | | | |
| Cinchonidine | 51 | 65 | 70 | 66 | 92 | 79 | 94 | 83 |
| Ajmaline | 63 | 71 | 73 | 74 | 91 | 79 | 95 | 88 |
| Colchicine | 58 | 87 | 75 | 86 | 84 | 86 | 88 | 88 |
| Ethylmorphine | n.d.** | 82 | 80 | 84 | 95 | 95 | 95 | 95 |
| Theobromine | 83 | 73 | 83 | 73 | 88 | 80 | 89 | 80 |
| Caffeine | 84 | 81 | 88 | 80 | 89 | 81 | 89 | 83 |
| Physostigmine sulphate | 81 | 89 | 90 | 89 | 95 | 95 | 95 | 95 |

*e.s. = elongated spot.
**n.d. = not determined.

References p. 48

TABLE 3.16

TLC ALKALOIDS ON ION EXCHANGERS[35]

$hR_F$ values of alkaloids on alginic acid thin layers. Acetic acid + hydrochloric acid solutions were used as eluent. Acetic acid concentration, 1 mol/l. Abbreviations: e.s. = elongated spot; n.d. = not determined. Stationary phase: 3 g of resin + 9 g of cellulose.

| Alkaloid | pH of eluent | | | | | Amount (µg) |
|---|---|---|---|---|---|---|
| | 2.35 | 2.00 | 1.38 | 1.15 | 0.60 | |
| Spermine | 1 | 1 | e.s. | e.s. | n.d. | 6.0 |
| Spermidine | 1 | 2 | e.s. | 61 | n.d. | 5.0 |
| Quinine | 2 | 6 | e.s. | 69 | 85 | 0.3 |
| Quinidine sulphate | 2 | 6 | e.s. | 69 | 85 | 0.3 |
| Cinchonine hydrochloride | 3 | 8 | 49 | 74 | 91 | 0.7 |
| Cinchonidine | 3 | 8 | 49 | 74 | 91 | 0.7 |
| Tubocurarine | 4 | 10 | 49 | 68 | 83 | 6.0 |
| Ergocristine | 4 | 13 | 28 | 32 | 36 | 0.4 |
| Berberine hydrochloride | 5 | 9 | 13 | 14 | 13 | 0.01 |
| Emetine hydrochloride | 5 | 13 | 54 | 75 | 91 | 4.0 |
| Reserpine | 6 | 13 | 23 | 26 | 31 | 0.5 |
| Ergotamine | 6 | 16 | 28 | 32 | 36 | 0.4 |
| Sparteine sulphate | 8 | 22 | 68 | 86 | 96 | 5.0 |
| Narceine | 12 | 12 | 10 | 9 | 7 | 0.25 |
| Ergometrine | 12 | 18 | 28 | 29 | 25 | 0.4 |
| Boldine | 15 | 23 | 41 | 40 | 39 | 0.4 |
| Yohimbine hydrochloride | 15 | 28 | 62 | 63 | n.d. | 3.0 |
| Strychnine | 17 | 30 | 58 | 65 | 72 | 5.0 |
| Brucine | 18 | 31 | 59 | 67 | 75 | 5.0 |
| Ibogaine | 19 | 33 | 55 | 62 | 64 | 0.4 |
| Papaverine | 21 | 39 | 65 | 75 | 80 | 0.4 |
| Hydrastine | 22 | 41 | 67 | 79 | 88 | 0.7 |
| Noscapine | 23 | 42 | 68 | 80 | 88 | 1.5 |
| Ajmaline | 27 | 46 | 74 | 80 | 81 | 1.0 |
| Morphine | 27 | 49 | 77 | 82 | 88 | 5.0 |
| Lobeline hydrochloride | 28 | 45 | 72 | 77 | 80 | 4.0 |
| Scopoline | 28 | 48 | 78 | 87 | 94 | 5.0 |
| Tropine | 29 | 51 | 80 | 88 | 94 | 7.0 |
| Ethylmorphine | 29 | 51 | 80 | 87 | 92 | 5.0 |
| Cocaine | 30 | 51 | 80 | 87 | 92 | 7.0 |
| Scopolamine hydrochloride | 31 | 52 | 80 | 88 | 91 | 5.0 |
| Physostigmine sulphate | 32 | 56 | 83 | 92 | 96 | 2.5 |
| Arecoline hydrochloride | 33 | 56 | 82 | 91 | 94 | 7.0 |
| Atropine | 33 | 56 | 82 | 91 | 94 | 7.0 |
| Hyoscyamine | 33 | 56 | 82 | 91 | 94 | 7.0 |
| Homoatropine | 33 | 56 | 82 | 91 | 94 | 7.0 |
| Eucatropine hydrochloride | 33 | 56 | 82 | 91 | 94 | 7.0 |
| Hyoscyne | 34 | 56 | 86 | 91 | 94 | 7.0 |
| Prostigmine | 39 | 61 | 89 | 94 | 96 | 7.0 |
| Theophylline | 69 | 70 | 73 | 72 | 76 | 5.0 |
| Aminophylline | 69 | 70 | 73 | 72 | 76 | 5.0 |
| Theobromine | 76 | 77 | 76 | 76 | 76 | 5.0 |
| Caffeine | 81 | 82 | 83 | 81 | 84 | 5.0 |
| Colchicine | 91 | 91 | 93 | 91 | 89 | 3.0 |

TABLE 3.17

TLC OF ALKALOIDS ON ION EXCHANGERS[35]

$hR_F$ values of alkaloids on Rexyn 102 ($H^+$) thin layers. Eluents: (a) 1 *M* acetic acid + hydrochloric acid solutions; (b) 1 *M* acetic acid in water-ethanol mixtures. Stationary phase: 3 g of resin + 9 g of cellulose.

| Alkaloid | pH of eluent | | | Ethanol concentration (%) | |
|---|---|---|---|---|---|
| | 2.35 | 1.00 | 0.30 | 30 | 50 |
| Berberine hydrochloride | 0 | 0 | 1 | 2 | 9 |
| Reserpine | 0 | 2 | 2 | 4 | 20 |
| Ergocristine | 0 | 2 | 2 | 6 | 30 |
| Narceine | 3 | 2 | 2 | 46 | 78 |
| Ergotamine | 1 | 5 | 4 | 8 | 25 |
| Papaverine | 2 | 6 | 4 | 16 | 29 |
| Tubocurarine | 1 | 7 | 6 | 11 | 15 |
| Ibogaine | 1 | 8 | 6 | 12 | 30 |
| Colchicine | 8 | 10 | 14 | 42 | 65 |
| Lobeline hydrochloride | 2 | 11 | 7 | 12 | 32 |
| Boldine | 2 | 11 | 10 | 25 | 41 |
| Ergometrine | 2 | 12 | 10 | 15 | 27 |
| Hydrastine | 2 | 15 | 13 | 16 | 27 |
| Strychnine | 3 | 17 | 14 | 20 | 28 |
| Brucine | 3 | 17 | 14 | 23 | 30 |
| Yohimbine hydrochloride | 3 | 18 | 10 | 18 | 23 |
| Cevadine | 4 | 18 | 12 | 27 | 42 |
| Emetine hydrochloride | 1 | 22 | 19 | 18 | 25 |
| Noscapine | 5 | 22 | 19 | 21 | 30 |
| Ajmaline | 8 | 22 | 28 | 25 | 39 |
| Cocaine | 7 | 26 | 20 | 26 | 44 |
| Hyoscyne | 7 | 27 | 24 | 24 | 39 |
| Protoveratrine A | 6 | 27 | 24 | 24 | 51 |
| Physostigmine sulphate | 12 | 36 | 37 | 37 | 47 |
| Caffeine | 28 | 36 | 42 | 55 | 67 |
| Theophylline | 35 | 40 | 53 | 58 | 68 |
| Aminophylline | 35 | 40 | 54 | 58 | 68 |
| Ethylmorphine | 13 | 42 | 37 | 34 | 46 |
| Theobromine | 36 | 42 | 46 | 60 | 66 |
| Quinine | 2 | 43 | 36 | 13 | 19 |
| Atropine | 14 | 43 | 42 | 36 | 47 |
| Hyoscyamine | 14 | 43 | 42 | 36 | 46 |
| Eucatropine hydrochloride | 14 | 43 | 42 | 36 | 46 |
| Quinidine sulphate | 2 | 45 | 37 | 12 | 21 |
| Prostigmine | 17 | 47 | 47 | 36 | 45 |
| Homatropine | 18 | 53 | 53 | 40 | 50 |
| Cinchonine hydrochloride | 3 | 55 | 46 | 15 | 20 |
| Cinchonidine | 3 | 55 | 47 | 14 | 22 |
| Scopolamine hydrochloride | 21 | 59 | 51 | 42 | 51 |
| Morphine | 23 | 62 | 57 | 43 | 52 |
| Arecoline hydrochloride | 33 | 76 | 77 | 46 | 54 |
| Sparteine sulphate | 18 | 85 | 77 | 30 | 34 |
| Scopoline | 42 | 86 | 82 | 53 | 53 |
| Tropine | 42 | 86 | 83 | 53 | 52 |
| Spermine | 6 | 97 | 97 | 12 | 7 |
| Spermidine | 11 | 97 | 97 | 26 | 18 |

References p. 48

TABLE 3.18

TLC OF ALKALOIDS ON ION EXCHANGERS[35]

$hR_F$ values of alkaloids on Dowex 50-X4 ($H^+$) thin layers. Abbreviations: e.s. = elongated spot; n.d. = not determined. Stationary phase: 3 g of resin + 9 g of cellulose.

| Alkaloid | Eluent | | |
|---|---|---|---|
| | 1 *M* HCl in $H_2O$ | 1 *M* HCl in $H_2O-C_2H_5OH$ (1:1) | 1 *M* HCl in $H_2O-C_2H_5OH$ (1:4) |
| Quinine | 0 | 1 | 1 |
| Quinidine sulphate | 0 | 1 | 1 |
| Cinchonine hydrochloride | 0 | 1 | 1 |
| Cinchonidine | 0 | 1 | 1 |
| Emetine hydrochloride | 0 | 2 | 5 |
| Sparteine sulphate | 4 | 3 | 4 |
| Tubocurarine | 0 | 4 | 4 |
| Berberine hydrochloride | 0 | 4 | 11 |
| Reserpine | 0 | 5 | 33 |
| Ergocristine | 1 | 6 | e.s. |
| Ajmaline | 1 | 6 | 18 |
| Ibogaine | 0 | 6 | 27 |
| Strychnine | 1 | 10 | 23 |
| Physostigmine sulphate | 1 | 10 | 30 |
| Ergotamine | 0 | 11 | e.s. |
| Yohimbine hydrochloride | 2 | 11 | 26 |
| Papaverine | 0 | 11 | 32 |
| Ergometrine | 5 | 12 | 21 |
| Lobeline hydrochloride | 1 | 13 | 34 |
| Brucine | 1 | 14 | 25 |
| Hydrastine | 1 | 16 | e.s. |
| Boldine | 0 | 17 | 31 |
| Ethylmorphine | 4 | 19 | 28 |
| Hyoscyne | 3 | 19 | 30 |
| Noscapine | 2 | 20 | 36 |
| Cocaine | 5 | 21 | 25 |
| Protoveratrine A | 8 | n.d. | n.d. |
| Arecoline hydrochloride | 17 | 22 | 20 |
| Prostigmine | 13 | 25 | 24 |
| Morphine | 13 | 27 | 27 |
| Scopoline | 32 | 29 | 17 |
| Caffeine | 20 | 29 | 30 |
| Eucatropine hydrochloride | 6 | 29 | 39 |
| Tropine | 28 | 30 | 23 |
| Scopolamine hydrochloride | 18 | 30 | 37 |
| Atropine | 7 | 30 | 40 |
| Hyoscyamine | 8 | 31 | 40 |
| Homatropine | 10 | 31 | 40 |
| Aminophylline | 28 | 32 | 31 |
| Theophylline | 29 | 33 | 33 |
| Theobromine | 24 | 38 | 25 |
| Cevadine | 3 | 39 | 65 |
| Colchicine | 1 | 48 | 58 |
| Narceine | 1 | 61 | 95 |

TABLE 3.19

TLC OF ALKALOIDS ON ION EXCHANGERS[35]

$hR_F$ values of alkaloids on CMCNa thin layers. Eluent: 0.5 *M* acetate buffer.

| Alkaloid | $hR_F$ | Alkaloid | $hR_F$ |
|---|---|---|---|
| Boldine | 17 | Papaverine | 67 |
| Berberine hydrochloride | 23 | Hydrastine | 71 |
| Reserpine | 24 | Noscapine | 72 |
| Ergocristine | 29 | Sparteine sulphate | 72 |
| Ergotamine | 29 | Ajmaline | 73 |
| Narceine | 36 | Lobeline hydrochloride | 75 |
| Ergometrine | 38 | Physostigmine sulphate | 87 |
| Quinine | 53 | Ethylmorphine | 89 |
| Quinidine sulphate | 53 | Scopoline | 90 |
| Cinchonine hydrochloride | 54 | Tropine | 91 |
| Cinchonidine | 54 | Theobromine | 94 |
| Brucine | 56 | Caffeine | 95 |
| Yohimbine hydrochloride | 57 | Theophylline | 95 |
| Ibogaine | 58 | Aminophylline | 95 |
| Strychnine | 64 | Colchicine | 95 |

TABLE 3.20

TLC ANALYSIS OF SOME ALKALOIDS WITH NEUTRAL SOLVENT SYSTEMS[33,36,37]

Silica gel 60 (ready made plates). Solvent is introduced into the chromatography tank immediately prior to development.

Solvent systems:
- S1 Dichloromethane-methanol (1:1)
- S2 Methanol
- S3 Chloroform
- S4 Chloroform-methanol (1:1)
- S5 Carbon tetrachloride-methanol (1:1)
- S6 *trans*-dichloro-1,2-ethylene (1:1)
- S7 trichloroethylene
- S8 Trichloroethylene-methanol (1:1)
- S9 1,1,1-Trichloroethylene-methanol (1:1)
- S10 Diisopropyl ether
- S11 Diisopropyl ether-methanol (1:1)
- S12 Ethyl acetate
- S13 Ethyl acetate-methanol (1:1)
- S14 Nitromethane
- S15 Nitromethane-methanol (1:1)

| Alkaloid | $hR_F$ value | | | | | | | | | | | | | | |
|---|---|---|---|---|---|---|---|---|---|---|---|---|---|---|---|
| | S1 | S2 | S3 | S4 | S5 | S6 | S7 | S8 | S9 | S10 | S11 | S12 | S13 | S14 | S15 |
| Morphine | 27 | 21 | 0 | 26 | 21 | 21 | 0 | 22 | 18 | 0 | 11 | 0 | 15 | 0 | 13 |
| Codeine | 31 | 22 | 0 | 33 | 27 | 29 | 0 | 33 | 24 | 0 | 11 | 1 | 16 | 0 | 16 |
| Thebaine | 38 | 24 | 0 | 43 | 36 | 37 | 0 | 49 | 33 | 0 | 15 | 1 | 20 | 0 | 19 |
| Noscapine | 90 | 66 | 0 | 88 | 84 | 89 | 0 | 87 | 85 | 8 | 77 | 62 | 82 | 10 | 84 |
| Brucine | 16 | 5 | 0 | 19 | 11 | 14 | 0 | 17 | 10 | 0 | 3 | 0 | 4 | 0 | 8 |
| Strychnine | 21 | 11 | 0 | 23 | 18 | 19 | 0 | 24 | 15 | 0 | 4 | 0 | 7 | 0 | 10 |
| Ecgonine | 10 | 14 | 0 | 9 | 5 | 5 | 0 | 5 | 5 | 0 | 2 | - | 3 | 0 | 17 |
| Methylecgonine | 18 | 13 | 0 | 16 | 11 | 15 | 0 | 12 | 10 | 1 | 7 | 2 | 10 | 0 | 8 |
| Benzoylecgonine | - | - | - | - | - | - | - | 17 | - | - | - | - | - | - | 42 |
| Cocaine | 46 | 33 | 0 | 45 | 42 | 48 | 0 | 52 | 40 | 3 | 30 | 8 | 34 | 1 | 19 |
| Pseudococaine | 69 | 50 | 0 | 71 | 62 | 70 | 0 | 73 | 63 | 2 | 48 | 9 | 52 | 1 | 52 |
| Tropacocaine | 22 | 14 | 0 | 23 | 21 | 24 | 0 | 28 | 19 | 1 | 11 | 1 | 12 | 0 | 7 |

(Published with permission of F. Vieweg, Wiesbaden).

TABLE 3.21

TLC ANALYSIS OF SOME ALKALOIDS IN ACIDIC SOLVENT SYSTEMS[37]

Silica gel 60 (ready made plates). Solvent is introduced into the chromatography tank immediately prior to development.

Solvent systems:
S1 Dichloromethane-methanol-acetic acid (2:2:1)
S2 Chloroform-methanol-acetic acid (2:2:1)
S3 Carbon tetrachloride-methanol-acetic acid (2:2:1)
S4 *trans*-Dichloro-1,2-ethylene-methanol-acetic acid (2:2:1)
S5 Trichloroethylene-methanol-acetic acid (2:2:1)
S6 1,1,1-Trichloroethane-acetic acid (2:2:1)

| Alkaloid | $hR_F$ value | | | | | |
|---|---|---|---|---|---|---|
| | S1 | S2 | S3 | S4 | S5 | S6 |
| Morphine | 44 | 41 | 34 | 32 | 33 | 32 |
| Codeine | 49 | 48 | 37 | 36 | 41 | 35 |
| Thebaine | 67 | 68 | 55 | 55 | 63 | 53 |
| Noscapine | 87 | 84 | 75 | 79 | 82 | 74 |
| Brucine | 60 | 61 | 43 | 47 | 56 | 43 |
| Strychnine | 61 | 61 | 47 | 48 | 58 | 47 |
| Ecgonine | 24 | - | 14 | 14 | 13 | 14 |
| Methylecgonine | 23 | 20 | 13 | 13 | 12 | 14 |
| Cocaine | 44 | 40 | 33 | 29 | 41 | 34 |
| Pseudococaine | 70 | 67 | 56 | 58 | 65 | 57 |
| Tropacocaine | 55 | 53 | 43 | 43 | 54 | 44 |

(Published with permission of F. Viehweg, Wiesbaden).

at least 0.1 *M*. To avoid the risk of iodine formation, iodide was not used. Perchlorate gave good results but was not used because of the explosion risks. Sulphate and nitrate did not give adequate ion pairs, and organic counter ions such as acetic acid, toluenesulphonic acid, bromothymol blue and bromocresol purple were also found to be unsatisfactory. The results are given in Table 3.13.

Rama Rao and Tandon[40] used silica gel and aluminium oxide impregnated with metal salts for the separation of some alkaloids. Cadmium and zinc nitrate were used as metal salts. The plates were made by mixing one part adsorbent and two parts of a 10% solution of the metal salt.

Lepri et al.[34,35] studied the chromatographic behaviour of a number of alkaloids on thin layers of anion and cation exchangers. The results of these studies are summarized in Tables 3.14-3.19. The authors studied the influence of pH, ionic strength and ethanol concentration in the mobile phase on the retention behaviour of the alkaloids.

Munier and Drapier[33,36,37] investigated the TLC of a number of alkaloids on silica gel plates with binary solvent mixtures, containing one very polar component such as ethanol, methanol or acetone. The effects of exposure of the plates to the vapour of the mobile phase prior to the development of the plates was also studied. Some of the results are summarized in Table 3.20. It was found that in combination with chlorinated solvents methanol gave better results than ethanol.

The use of ethanol-containing mobile phases gave rise to the formation of more elongated spots. The effect of different methanol concentrations in the mobile phases was also studied. Mobile phases containing acetic acid were also found to be useful for the separation of alkaloids (Table 3.21).

REFERENCES


1 D. Waldi, K. Schnackerz and F. Munter, *J. Chromatogr.*, 6 (1961) 61.
2 E. Vidic and J. Schütte, *Arch. Pharm. (Weinheim)*, 295 (1962) 342.
3 J. Zarnack and S. Pfeifer, *Pharmazie*, 19 (1964) 216.
4 V. Schwarz and M. Sarsunova, *Pharmazie*, 19 (1964) 267.
5 R. Paris, R. Rousselet, M. Paris and M.J. Fries, *Ann. Pharm. Fr.*, 23 (1965) 473.
6 E. Marozzi and G. Falzi, *Farmaco, Ed. Prat.*, 20 (1965) 302.
7 D. Giacopello, *J. Chromatogr.*, 19 (1965) 172.
8 S. El Gendi, W. Kisser and G. Machata, *Mikrochim. Acta* (1965) 120.
9 K. Macek, J. Vecerkova and J. Stanislova, *Pharmazie*, 20 (1965) 605.
10 W.W. Fike, *Anal. Chem.*, 38 (1966) 1697.
11 G.J. Dickes, *J. Assoc. Public. Anal.*, 4 (1966) 45.
12 I. Sunshine, W.W. Fike and H. Landesman, *J. Forensic Sci.*, 11 (1966) 428.
13 H.C. Hsiu, J.T. Huang, T.B. Shih, K.L. Yang, K.T. Wang and A.L. Lin, *J. Chin. Chem. Soc.*, 14 (1967) 161.
14 A. Noirfalise and G. Mees, *J. Chromatogr.*, 31 (1967) 594.
15 R.H. Drost and J.F. Reith, *Pharm. Weekbl.*, 102 (1967) 1379.
16 P.E. Haywood and M.S. Moss, *Analyst (London)*, 93 (1968) 737.
17 E. Röder, E. Mutschler and H. Rochelmeyer, *Arch. Pharm.* (*Weinheim*), 301 (1968) 624.

18 R.A. Egli, *Dtsch. Apoth. Ztg.*, 110 (1970) 987.
19 I. Simon and A. Lederer, *J. Chromatogr.*, 63 (1971) 448.
20 G.S. Tadjer, *J. Chromatogr.*, 63 (1971) D44.
21 S. Zádeczky, D. Küttel and M. Takacs, *Acta Pharm. Hung.*, 42 (1972) 7.
22 R.A. Egli, *Z. Anal. Chem.*, 259 (1972) 277.
23 K.F. Ahrend and D. Tiess, *Zbl. Pharm.*, 111 (1972) 933.
24 E. Novakova and J. Vecerkova, *Cesk. Farm.*, 22 (1973) 347.
25 K.F. Ahrend and D. Tiess, *Wiss. Z. Univ. Rostock Math. Naturw. Reihe*, 22 (1973) 951.
26 F. Schmidt, *Dtsch. Apoth. Ztg.*, 114 (1974) 1593.
27 A.C. Moffat and K.W. Smalldon, *J. Chromatogr.*, 90 (1974) 1,9.
28 A.C. Moffat and B. Clare, *J. Pharm. Pharmacol.*, 26 (1974) 66s.
29 J. Breiter, *Kontakte*, 3 (1974) 17.
30 R.J. Armstrong, *N.Z.J. Sci.*, 17 (1974) 15.
31 A.C. Moffat, *J. Chromatogr.*, 110 (1975) 341.
32 E. Vidic and E. Klug, *Z. Rechtsmed.*, 76 (1975) 283.
33 R.L. Munier and A.M. Drapier, *C.R. Acad. Sci., Ser. C*, 283 (1976) 719.
34 L. Lepri, P.G. Desideri and M. Lepori, *J. Chromatogr.*, 116 (1976) 131.
35 L. Lepri, P.G. Desideri and M. Lepori, *J. Chromatogr.*, 123 (1976) 175.
36 R.L. Munier and A.M. Drapier, *Chromatographia*, 10 (1977) 226.
37 R.L. Munier and A.M. Drapier, *Chromatographia*, 10 (1977) 290.
38 M.N. Lyakina and L.I. Brutko, *Khim. Prir. Soedin.*, 4 (1977) 583.
39 J. Breiter and R. Helger, *Med. Labor.*, 30 (1977) 149.
40 N.V. Rama Rao and S.N. Tandon, *J. Chromatogr. Sci.*, 16 (1978) 158.
41 R.A. de Zeeuw, F.J.W. van Mansvelt and J.E. Greving, *J. Chromatogr.*, 148 (1978) 255.

Chapter 4

# ISOLATION OF ALKALOIDS

## 4.1. ISOLATION METHODS

An important goal in alkaloid investigations is the isolation of the alkaloids present in the material to be investigated in their genuine form without any formation of artefacts. Much of the activity in the field of alkaloidal investigations has been directed towards the development of optimum methods for the isolation of alkaloids and alkaloid mixtures with the avoidance of undesirable alterations to them.

As the alkaloids usually occur in plants as salts of organic plant acids and inorganic acids together with often complex mixtures of water-soluble compounds, such as gums, proteins, mineral salts, tannins, lipids (fats and oils) and resins, it is often a great problem to remove all of these non-alkaloidal compounds during the isolation and purification of the alkaloids.

The same kind of problems are encountered in pharmacological and toxicological investigations, where the alkaloids usually are found as salts in complex mixtures of water-soluble compounds of all kind of lipids.

The extraction and isolation of the alkaloids can be carried out in various ways depending on the nature of the alkaloids in question and the material in which they are found. With plant material the primary extraction is usually effected with organic water-immiscible solvents after the liberation of the alkaloidal bases from their salts by treatment with a mineral base. As prolonged contact with strong bases may lead to alterations to many alkaloids, such as hydrolysis of ester alkaloids, and as strong bases also cause the formation of soaps, if fats are present, ammonia is most commonly used. Ammonia is sufficiently basic to liberate most of the common alkaloids without much risk of undesirable reactions. Also, as ammonia is volatile, it can easily be removed afterwards.

In plant materials rich in tannin, the alkaloid tannate salts sometimes have to be decomposed by preliminary treatment by heating with dilute hydrochloric acid. In other instances sodium hydroxide sufficies to cleave the alkaloidal tannate salts.

However, because of problems that may occur when the alkaloidal bases are liberated by treatment with alkali and extracted with organic solvents, such as alterations to the alkaloids under the influence of dichloromethane or chloroform, extractions of the lipids present in the material to be investigated should be

**References p. 57**

performed with other water-immiscible organic solvents.

Extraction of the alkaloids as salts by means of water-alcohol mixtures is often preferred. The risk of alterations to the genuine alkaloids present in the material is less under such conditions. Whereas extractions with most organic solvents, such as benzene, chloroform and diethyl ether, give extracts containing all kinds of lipids and resins, extractions made with water and water-alcohol mixtures will give extracts containing various polar compounds, such as proteins, gums and mineral salts.

In both instances further purification of the alkaloidal extracts is usually necessary before chromatography can be applied. Often a purification is achieved by extracting the alkaloids present in an organic solvent with water and dilute acid, and subsequently alkaloids present in the aqueous solution as salts with an organic solvent after liberating their bases with alkali. Changing the solvent at this point may be advantageous in preventing the extraction of some undesirable constituents still present in the aqueous phase.

Extraction and purification of the alkaloidal bases using a column of silica gel or aluminium oxide may be advantageous in many instances. The bases are liberated by treatment of the alkaloid-containing material with alkali and extracted with an adsorption column. Many non-alkaloidal compounds present in the material are adsorbed on the column. To remove fat and resins, which are extracted together with the alkaloidal bases, the alkaloid-containing extracts can be filtered through a column of phosphoric acid-containing silica gel. The alkaloids will be bonded on the column as salts, whereas neutral lipids can be washed out of the column. Subsequently the alkaloidal bases can be eluted with an organic solvent to which alkali is added.

Because of possible artefact formation when dichloromethane or chloroform is used in the extraction, and because quaternary alkaloids are usually only slightly soluble in non-polar solvents, polar solvents are often preferred when both quaternary and tertiary alkaloids are present in the material.

To obtain the best possible purification of the alkaloids from non-alkaloidal compounds found in an extract obtained with polar solvents, a purification procedure is often carried out by precipitating the alkaloids by means of picric acid, Reinecke's salt or Mayer's reagent. Dissolving the precipitates and running the solution through an anion-exchange column in the chloride form yields the alkaloids as chlorides.

The use of ion-exchange adsorbents for the isolation of alkaloids is often advantageous. Ion-exchange resins may be considered to consist of a three-dimensional network with a large inner surface area, which is electrically charged owing to its dissociated groups. The pores of the network are determined by the degree of cross-linking, and they can act as a sieve for large ions. Therefore, surface adsorption often takes place and not only chemical interactions between the functional groups. When using ion-exchange resins for the isolation of alkaloids from alkaloid-containing solutions, such as extracts of plant materials, it is important to use resins with suitable cross-linking. Resins with low cross-linking are usually preferred for achieving quantitative recoveries of alkaloids fixed on the ion-exchange resin particles. However, many factors govern the adsorption and elution of different alkaloids with different resins. The capacity of a particular resin for a particular group of alkaloids, therefore, has to be investigated carefully in order to obtain optimum results.

The isolation and purification methods to be chosen for an investigation of alkaloids depend to a great extent on the chromatographic technique to be applied. Chromatographic systems may be divided into "open" and "closed" systems. Open chromatographic systems, such as paper and thin-layer chromatography, are of great value when crude extracts or mixtures of unknown composition have to be analysed and little information is available about the components present. Many problems concerning the analysis of alkaloids have been solved by means of these techniques. The use of "closed" systems, such as gas-liquid and high-performance liquid chromatography, is limited to those components in a mixture which can be eluted from the system. Components which are not eluted may gradually change the separation properties of the column and even ruin it completely. The limitations of the application of closed chromatographic systems must, therefore, always be borne in mind.

## 4.2. SOLVENTS AND ARTEFACT FORMATION

During the extraction, isolation and analysis of alkaloids it should be borne in mind that the stability of alkaloids varies widely. Some alkaloids are sensitive to light, others to pH and heat and some even to various organic solvents. Some organic solvents can influence the decomposition rate of sensitive alkaloids, but reactions can also occur between various solvents, or contaminants in solvents, and alkaloids. Because of the extensive use of organic solvents in the extraction, isolation and analysis of alkaloids, some of the possible interactions between organic solvents and alkaloids are discussed below.

The most commonly used organic solvents in alkaloid research belong to various chemical groups:

aliphatic and aromatic hydrocarbons (benzene, toluene, cyclohexane);
alcohols (ethanol, methanol);
ethers (diethyl ether, dioxane, tetrahydrofuran);
esters (ethyl acetate);
ketones (acetone, methyl ethyl ketone);
halogen-containing compounds (chloroform, dichloromethane).

**References p. 57**

In all these groups (except for the hydrocarbons) artefact formation may occur.

Decomposition may take place in all types of solvents. Particularly in chloroform solutions photochemical decomposition may be accelerated. In alcoholic solutions alkaloids are usually more stable[17,23-25]. Reichelt[30] found C-8 epimerization of ergot alkaloids in alcoholic solutions to be faster than in acetone, chloroform or benzene solutions. Alcohols can also react with carbinolamines[12], pseudostrychnine[16] (Fig. 4.1) and akagerine[35] (Fig. 4.2). Reactions with hemiacetals, e.g., Wieland-Gumlich aldehyde[11,18,36], (Fig. 4.3), and derivatives are also possible.

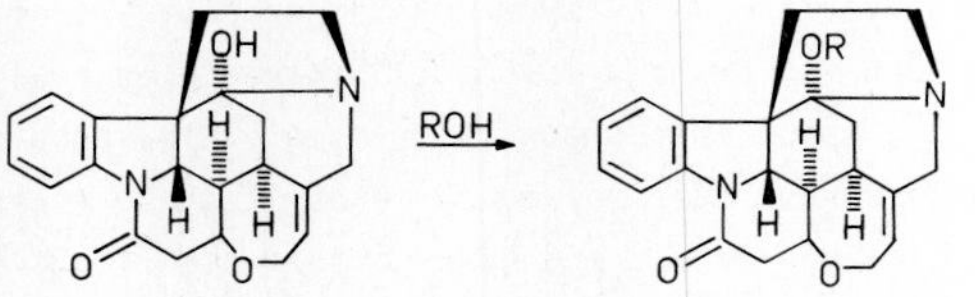

pseudostrychnine ROH → R=$CH_3$ 16-methoxystrychnine

Fig. 4.1.

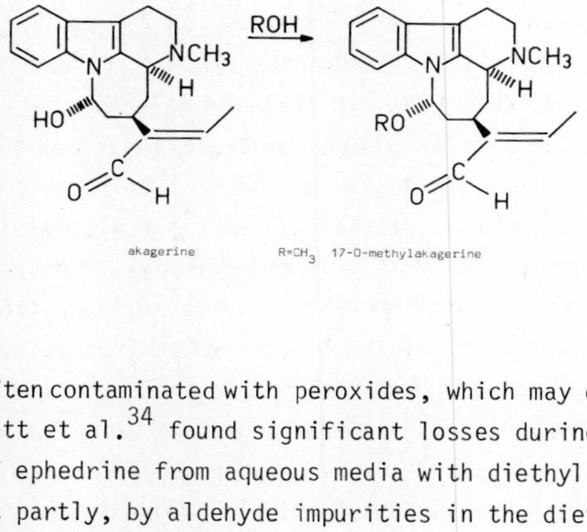

akagerine ROH → R=$CH_3$ 17-O-methylakagerine

Fig. 4.2.

Ethers are often contaminated with peroxides, which may cause oxidation of the alkaloids. Beckett et al.[34] found significant losses during the extraction of small amounts of ephedrine from aqueous media with diethyl ether. The losses were caused, at least partly, by aldehyde impurities in the diethyl ether (formaldehyde, acetaldehyde, propionaldehyde).

Ketones are well known artefact formers and they may react with alkaloids such as 1,2-dehydrobeninine[19] and berberine[1,2,12]. During column chromatography ketones may give condensation products with ammonia, forming compounds with alkaloidal characteristics[15]. Self-condensation of acetone during liquid chromatography is also a problem that may be encountered with this solvent.

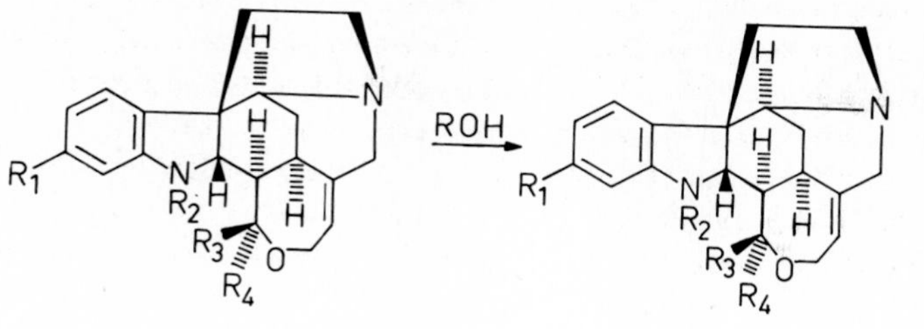

$R_1$=$OCH_3$, $R_2$=Ac, $R_3$=OH, $R_4$=H 11-methoxydiaboline
$R_1$=$OCH_3$, $R_2$=Ac, $R_3$=$OCH_3$, $R_4$=H 17-O-methyl-11-methoxydiaboline
$R_1$=$OCH_3$, $R_2$=Ac $R_3$=H, $R_4$=$OCH_3$, epi-17-O-methyl-11-methoxydiaboline

Fig. 4.3.

Halogen-containing solvents are widely used in alkaloid research. Chloroform in particular is one of the most suitable solvents because of its relatively strong proton donor character[28]. However, the halogen-containing solvents are very active in terms of artefact formation. Even decomposition of alkaloids is accelerated in chloroform solution. Decomposition of reserpine in such solutions has been studied by several groups[4,6,7,8,14,17,20,22]. The major products formed are 3,4-dehydro- and 3,4,5,6-tetradehydroreserpine (lumireserpine). According to Frijns[20], the decomposition is much less in analytical-reagent grade chloroform, provided that the solution is kept in the dark. Phillipson and Bisset[26] reported the formation of the pseudo-compound and the N-oxide of strychnine upon refluxing in chloroform solution.

Dichloromethane reacts readily with tertiary nitrogen atoms, yielding quaternary alkaloids which are insoluble in organic solvents. With strychnine it was found that the tertiary alkaloid was converted completely into the quaternary dichloromethyl compound within 48 h[26]. Besselièvres et al.[21] rightly questioned whether dichloromethane is a solvent or a reagent in connection with the isolation of chloromethyl derivatives of the indole alkaloid tubotaiwine and the steroid alkaloid N-methylparavallarine. Vincze and Gefen[33] found atropine and some other tertiary amines to quaternize readily with dichloromethane.

Chloroform may be contaminated with other halogen-containing compounds (bromochloromethane, dichloromethane)[3,5,9,10,13,26,31]. Williams[9] reported as much as 0.5% of bromochloromethane and 0.1% of dichloromethane; Hansen[31] found 0.04% and 0.03%, respectively. Small amounts of dichloromethane and bromochloromethane may also cause the formation of quaternary chloromethyl compounds of tertiary amines in chloroform solutions. 1,2-Dichloroethane also has alkylating properties[31].

**References p. 57**

Franklin et al.[32] found cyanogen chloride as a contaminant in dichloromethane. With primary and secondary amines the corresponding nitriles are formed. According to the authors, cyanogen chloride was present in most of the samples tested, the amount varying from 0.2 to 11 µg/ml. No evidence for this impurity was found in chloroform, carbon tetrachloride or dichloroethane.

Siek et al.[38] reported the reaction of a secondary amine with ethyl chloroformate, an impurity in chloroform. When using chloroform for the extraction of normeperidine, the ethylcarbamate of the compound was formed owing to reaction with ethyl chloroformate. Ethyl chloroformate, which is formed by the reaction of phosgene, (formed by oxidation of chloroform) and ethanol, which is added as a stabilizer to chloroform, was found in amounts of 0.1-1 ppm.

Not only may impurities in chloroform lead to artefact formation, but chloroform itself can also react with some alkaloids. Protoberberine alkaloids such as berberine and palmatine can react with chloroform yielding berberine-chloroform and palmatine-chloroform[27] (Fig. 4.4).

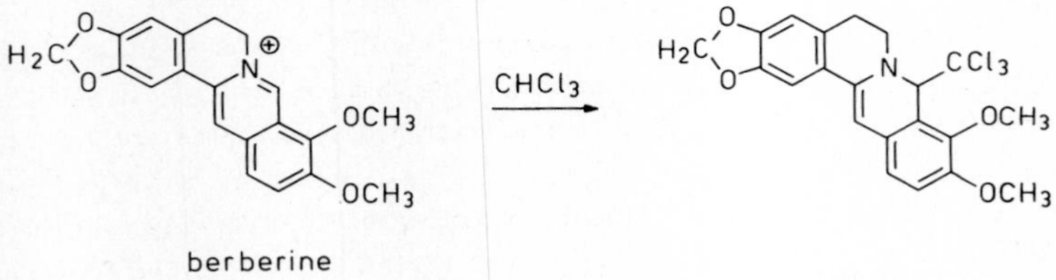

$H_2C$ $OCH_3$ $OCH_3$ berberine $\xrightarrow{CHCl_3}$ $H_2C$ $CCl_3$ $OCH_3$ $OCH_3$

Fig. 4.4.

gentianine

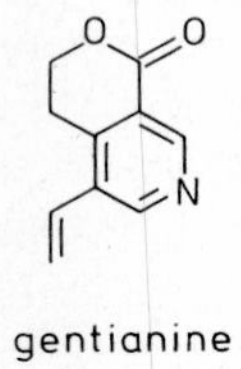

Fig. 4.5.

Basic nitrogen-containing compounds such as ammonia, diethylamine and triethylamine are widely used in the analysis of alkaloids. Ammonia is often employed in connection with the extraction and purification of alkaloids. However, also with ammonia artefacts can be formed in some instances; thus the alkaloid gentianine was formed from the non-nitrogen-containing glycoside sweroside (Fig. 4.5), angustine from vincoside- or isovincoside-lactam[29] and decussine-type alkaloids from akagerine-type alkaloids[37]. In the last two instances ammonia reacts with aldehyde functions, converting a two-nitrogen-containing alkaloid to a three-nitrogen-containing alkaloid.

The conclusions to be drawn from the foregoing are that solvents should be freshly distilled, ethers should be checked for the presence of peroxides, dichloromethane should not be used in alkaloid investigations, and great care should be taken when chloroform is used. In a prolonged column chromatographic separation the use of chloroform may lead to artefact formation. Small amounts of bromochloromethane and dichloromethane and related contaminants in chloroform may lead to quaternization of tertiary alkaloids present. Hydrocarbons are to be preferred because of the small risk of artefact formation.

REFERENCES

1 J. Gadamer, *Arch. Pharm. (Weinheim)*, 243 (1905) 12.
2 J. Gadamer, *Arch. Pharm. (Weinheim)*, 243 (1905) 31.
3 M.E. von Klemperer and F.L. Warren, *Chem. Ind. (London)*, (1955) 1553.
4 A. Leyden, E. Pomerantz and E.F. Bouchard, *J. Pharm. Sci.*, 45 (1956) 771.
5 A.C. Caws and G.E. Foster, *J. Pharm. Pharmacol.*, 9 (1957) 824.
6 E. Kahane and M. Kahane, *Ann. Pharm. Fr.*, 16 (1958) 726.
7 J. Bayer, *Pharmazie*, 13 (1958) 468.
8 S. Ljungberg, *J. Pharm. Belg.*, 14 (1959) 115.
9 H. Williams, *J. Pharm. Pharmacol.*, 11 (1959) 400.
10 D.I. Coomber and B.A. Rose, *J. Pharm. Pharmacol.*, 11 (1959) 703.
11 J.A. Deyrup, H. Schmid and P. Karrer, *Helv. Chim. Acta*, 45 (1962) 2266.
12 D. Beke, in A.R. Katritzky (Editor), *Advances Heterocyclic Chemistry*, I Academic Press, New York, 1963, p. 167.
13 H. Williams, *J. Pharm. Pharmacol.*, 16 (1964) 166T.
14 A. Kaess and C. Mathis, *Ann. Pharm. Fr.*, 23 (1965) 739.
15 D.E. Householder and B.J. Camp, *J. Pharm. Sci.*, 54 (1965) 1676.
16 N.G. Bisset, C.G. Casinovi, C. Galeffi and G.B. Marini-Bettolo, *Ric. Sci.*, 35 (II-B) (1965) 273.
17 E. Ullman and H. Kassalitzky, *Dtsch. Apoth. Ztg.*, 107 (1967) 152.
18 J.R. Hymon, H. Schmid, P. Karrer, A. Boller, H. Els, P. Fahrni and A. Fürst, *Helv. Chim. Acta*, 52 (1969) 1564.
19 V. Agwada, M.B. Patel, M. Hesse and H. Schmid, *Helv. Chim. Acta*, 53 (1970) 1567.
20 J.M.G.J. Frijns, *Pharm. Weekbl.*, 106 (1971) 605.
21 R. Besselièvre, N. Langlois and P. Potier, *Bull. Soc. Chim. Fr.*, (1972) 1477.
22 G.E. Wright and T.Y. Tang, *J. Pharm. Sci.*, 61 (1972) 299.
23 S. Pfeifer, G. Behnsen and L. Kühn, *Pharmazie*, 27 (1972) 639.
24 S. Pfeifer, G. Behnsen and L. Kühn, *Pharmazie*, 27 (1972) 648.
25 S. Pfeifer, G. Behnsen, L. Kühn and R. Kraft, *Pharmazie*, 27 (1972) 734.

26 J.D. Phillipson and N.G. Bisset, *Phytochemistry*, 11 (1972) 2547.
27 G.A. Miana, *Phytochemistry*, 12 (1973) 1822.
28 L.R. Snyder, *J. Chromatogr.*, 92 (1974) 223.
29 J.D. Phillipson, S.R. Hemingway, N.G. Bisset, P.J. Houghton and E.J. Shellard, *Phytochemistry*, 13 (1974) 973.
30 J. Reichelt, *Cesk. Farm.*, 25 (1976) 93.
31 S.H. Hansen, *Arch. Pharm. Chem., Sci. Ed.*, 5 (1977) 194.
32 R.A. Franklin, K. Heatherington, B.J. Morrison, S. Pamela and T.J. Ward, *Analyst (London)*, 103 (1978) 660.
33 A. Vincze and L. Gefen, *Isr. J. Chem.*, 17 (1978) 236.
34 A.H. Beckett, G.R. Jones and D.A. Hollingsbee, *J. Pharm. Pharmacol.*, 30 (1978) 15.
35 W. Rolfsen, L. Bohlin, S.K. Yeboah, M. Geevaratne and R. Verpoorte, *Planta Med.*, 34 (1978) 264.
36 L. Bohlin, W. Rolfsen, J. Strömbon and R. Verpoorte, *Planta Med.*, 35 (1979) 19.
37 W. Rolfsen, A.A. Olaniyi, R. Verpoorte and L. Bohlin, *J. Nat. Prod.*, 44 (1981) 415.
38 T.J. Siek, L.S. Eichmeier, M.E. Caplis and F.E. Esposito, *J. Anal. Toxicol.*, 1 (1977) 211.

# II. SPECIAL PART

CONTENTS

II. SPECIAL PART

II.1. PYRROLIDINE, PYRROLIZIDINE, PYRIDINE, PIPERIDINE AND QUINOLIZIDINE ALKALOIDS

CHAPTER 5.

PYRROLIDINE, PYRROLIZIDINE, PYRIDINE, PIPERIDINE AND QUINOLIZIDINE ALKALOIDS....61
5.1. Solvent systems....61
5.2. Detection....62
5.3. Quantitative analysis....63
5.4. TAS technique and reaction chromatography....64
References....64

CHAPTER 6.

TOBACCO ALKALOIDS....79
6.1. Solvent systems....79
6.2. Detection....80
6.3. Quantitative analysis....80
6.4. TAS technique and reaction chromatography....81
References....81

II.2. TROPANE ALKALOIDS

CHAPTER 7.

TROPINE ALKALOIDS....91
7.1. Solvent systems....91
7.2. Detection....93
7.3. Quantitative analysis....94
7.4. TAS technique and reaction chromatography....94
References....96

CHAPTER 8.

PSEUDOTROPINE ALKALOIDS....113
8.1. Solvent systems....114
8.2. Detection....115
8.3. Quantitative analysis....116
8.4. TAS technique and reaction chromatography....116
References....117

II.3. QUINOLINE ALKALOIDS

CHAPTER 9.

*CINCHONA* ALKALOIDS....131
9.1. Solvent systems....131
9.2. Two-dimensional TLC....134
9.3. Detection....134
9.4. Quantitative analysis....135
9.5. TAS technique and reaction chromatography....136
References....137

Chapter 5

# PYRROLIDINE, PYRROLIZIDINE, PYRIDINE, PIPERIDINE AND QUINOLIZIDINE ALKALOIDS

This heterogeneous group of alkaloids includes some which are readily volatilized and hence very well suited for gas chromatography. Several papers have dealt with the identification of these alkaloids by means of a combination of GC and TLC[22,39] The TLC of nicotine and other tobacco alkaloids is described in a separate part.

## 5.1. SOLVENT SYSTEMS

The separation of *Lobelia* alkaloids on formamide-impregnated cellulose plates was described by Teichert et al.[1] (Table 5.1). Parrak et al.[12] separated these alkaloids and their degradation product acetophenone with solvent S2 (Table 5.1). Kaess and Mathis[23] applied the solvents S3-S6 for the separation of *Lobelia* alkaloids and their reaction products (Table 5.1). Frijns[32] used two-dimensional TLC for the separation of the alkaloids present in *Lobelia* . The first dimension was developed with chloroform - diethylamine (DEA) (9:1) and the second with cyclohexane - chloroform - diethylamine (5:4:1).

Grandolini et al.[16] used chloroform-methanol (7:3) and butanol-hydrochloric acid (95:5) saturated with water in combination with silica gel plates for the separation of cyclopentanopipiridine alkaloids from *Skytanthus*.

For the separation of *Conium* alkaloids, Moll[4] employed as the solvent chloroform - absolute ethanol - 25% ammonia (18:2:2) in combination with activated silica gel plates. Stahl and Schmitt[49] used acetone - diethyl ether - concentrated ammonia (50:50:3) and silica gel plates for the separation of *Conium* alkaloids and the identification of arecoline.

The piperidine alkaloids from *Sedum* species were separated with solvent S1 on 0.5 *M* potassium hydroxide-impregnated plates[7] (Table 5.2), while Bieganowska and Waksmundzki[53] applied solvents S2-S5 and silica gel plates (Table 5.2).

The separation of pyrrolizidine alkaloids from *Crotalaria* species was described by Sharma et al.[21]. Of the 20 TLC systems tested, S1 gave the best results (Table 5.3). Chalmers et al.[22] characterized pyrrolizidine alkaloids by means of GLC, TLC and PC; the results for TLC are summarized in Table 5.3.

Kulakov and Likhosherstov[33] separated the threo and erythro isomers of some of the $\alpha,\beta$-dihydroxy acids present in pyrrolizidine alkaloids. The best results were obtained with silica gel plates impregnated with 4% silver nitrate. Klasek et al.[52] described the identification of necic acids and derivatives, obtained upon hydrolysis of pyrrolizidine alkaloids. Mattocks[28] reported the separation of some pyrrolizidine alkaloids (Table 5.4). For N-oxides and alkaloids, which are not stable as the free base, the organic layer of *n*-butanol-water-acetic acid (4:5:1) was used on silica gel plates.

**References p. 64**

Ramaut[5] analysed Leguminosae alkaloids with solvent S1 (Table 5.5). Several other solvents were also tested. Paris and Paris[6] reported three solvents for the separation of a series of alkaloids, including some Leguminosae alkaloids (Table 5.5).

Gill[11] and Gill and Steinegger[10,13] used cyclohexane - diethylamine in different ratios as the solvent for the separation of quinolizidine alkaloids from *Cytisus* and *Genista* species on silica gel plates (Table 5.5). They also used chloroform - methanol (8:2) as the solvent (Table 5.5). The identification of 22 *Lupine* alkaloids by means of TLC and GC-MS was reported by Cho and Martin[39]. Solvent S8 described by Sharma et al.[21] for the separation of pyrrolizidine alkaloids was found to be suitable for *Lupine* alkaloids also (Table 5.5). Solvent S9 was useful for checking the identities of some alkaloids which have markedly different relative mobilities in this solvent (Table 5.5). Solvent system S10 was based upon previously known paper chromatographic separations (Table 5.5).

Genest[40] used TLC to distinguish poisonous *Arbrus* seeds from *Ormosia* seeds. The latter species contains quinolizidine-type alkaloids. Faugeras and Paris[44] employed a number of solvent systems in connection with the densitometric analysis of Leguminosae alkaloids. Solvent S6[10,11,13] was used for the separation of all type of quinolizidine alkaloids. For alkaloid esters the solvent cyclohexane - dichloromethane - diethylamine (4:4:2) on silica gel plates was employed. For the simple quinolizidine alkaloids toluene - acetone - ethanol - ammonia (60:40:6:2) on silica gel plates was employed, as well as isobutanol - hydrochloric acid - water (7:1:2) on cellulose plates.

Dipiperidine alkaloids could be separated with methanol - acetone - water - ammonia (35:20:40:5) on cellulose plates.

Gafurov et al.[50] used kaolin as the stationary phase for the separation of Leguminosae alkaloids. Wysocka[55] described the separation of epimeric alcohols of some quinolizidine alkaloids (Table 5.5). For the quantitative analysis of *Lupine* alkaloids Karlsson and Peter[58] used the solvent benzene - dichloromethane - diethyl ether - diethylamine (5:5:5:2) on silica gel plates.

## 5.2. DETECTION

Dragendorff's reagent and iodoplatinate are widely used detection reagents for these groups of alkaloids. Dragendorff's reagent had been used in the quantitative analysis of Leguminosae alkaloids[44]. Ramaut[5] reported the colours obtained with Dragendorff's reagent (modification according to Munier) for a number of Leguminosae alkaloids (see Table 5.6).

For the detection of pyrrolizidine alkaloids, Chalmers et al.[22] used iodine vapour. Meissner et al.[42] found 1% cerium(IV) sulphate in 1 *M* sulphuric acid followed by Dragendorff's reagent to be a sensitive detection method for a number of quinolizidine alkaloids.

Karlsson and Peter[58] determined lupine alkaloids quantitatively by measuring the fluorescence after heating the plates for 17 h at 130°C. The fluorescence colour is stable for several weeks. Mattocks[28] described a special detection method for pyrrolizidine alkaloids on TLC plates (no.48). The alkaloids are converted into the N-oxides by means of hydrogen peroxide and subsequently heated. Treatment of the N-oxides with acetic anhydride yields the pyrroles, which with Ehrlich's reagent give blue or mauve spots (Table 5.6). Only alkaloids with an unsaturated ring in the basic moiety give a colour reaction. The method was found to be more sensitive than detection with iodoplatinate reagent, Dragendorff's reagent and iodine. After treatment of the N-oxides with acetic anhydride the alkaloids can be observed as brown spots, with fluorescence in UV light. The reaction with Ehrlich's reagent makes the method more sensitive. Moll[4] described the detection of *Conium* alkaloids by spraying with 0.5% 1-chloro-2,4-dinitrobenzene in ethanol and 0.05% bromothymol blue in ethanol, yielding blue spots on a yellow background. Kaniewska and Borkowski[30] used Labat's and Gaebel's reagents (nos. 20 and 67) for the detection of methylenedioxy-containing alkaloids. Some other alkaloids also gave positive reactions, e.g., arecoline, cytisine, lupinine and sparteine. Menn and McBain[26] described a detection method for cholinesterase inhibitors. Among others the alkaloid lobeline was detected (see Chapter 2, p. 15) (no.19).

## 5.3. QUANTITATIVE ANALYSIS

Massa et al.[36] described the quantitative analysis of alkaloids by means of densitometry after detection with Dragendorff's reagent. Among others sparteine was analysed by measuring in the reflection mode at 400 nm. Faugeras and Paris[44] described the quantitative analysis of a number of Leguminosae alkaloids by means of densitometry. Several solvents were described for specific separations. The alkaloids were determined after spraying with Dragendorff's reagent in both the transmission and reflection modes at a wavelength of 490 nm. The transmission was found to be 2-3 times more sensitive than the reflection mode and more precise. Karlsson and Peter[58] determined alkaloids in *Lupinus* species by means of fluorodensitometry. After development the plates were kept for 17 h at 130°C, after which the alkaloids were observed as blue fluorescent spots, the colour being stable for several weeks. The alkaloids were determined with an excitation wavelength of 360 nm at an emission wavelength of 400 nm.

## 5.4. TAS TECHNIQUE AND REACTION CHROMATOGRAPHY

Jolliffe and Shellard[37] applied the TAS technique to a number of crude drugs, including *Cytisus scoparius* (Broom tops), *Lobelia inflata, Piper nigrum, Punica granatum* and *Areca catechu*. In the Broom tops and the *Piper* only one alkaloid was observed, corresponding to sparteine and piperine, respectively. The results with *Lobelia* were variable and could therefore not be used for identification. The experimental conditions used were a temperature of 275°C for 90 sec, the sample (10-20 mg) being admixed with about 10 mg of calcium hydroxide. Silica gel with a suitable moisture content was used as propellant. Stahl and Schmitt[49] described the TAS technique for the crude drugs from *Arecae semen*, in which arecoline was detected, and *Conii fructus*, in which the alkaloids coniine, N-methylconiine and γ-coneceine were detected. For the latter drug a temperature of 150°C for 120 sec was used for the powdered form, and for the complete fruit a temperature of 220°C. This temperature was also used for the *Areca* seeds in powdered form. As propellant 50 mg of molecular sieve 4Å containing 20% of water was used in all instances.

Reaction chromatography for *Lobelia* alkaloids has been reported by Kaess and Mathis[23,24]. The reactions used for the alkaloids lobeline, lobelanidine and lobelanine were the following: reaction with semicarbazide hydrochloride in the presence of sodium acetate, yielding the carbazones of the keto compounds, reaction with acetyl chloride or acetic anhydride, yielding the acetylated derivatives of the hydroxyl-containing alkaloids, oxidation with chromic acid, yielding the diketo alkaloids of the hydroxyl-containing alkaloids, and reduction with sodium borohydride, yielding the dihydroxy alkaloids from the keto alkaloids.

A summary of the TLC analysis of pyrrolidine, pyrrolizidine, pyridine, piperidine and quinolizidine alkaloids in various materials, with literature references, is given in Tables 5.7 - 5.10.

## REFERENCES


1 K. Teichert, E. Mutschler and H. Rochelmeyer, *Dtsch. Apoth. Ztg.*, 100 (1960) 477.
2 D. Waldi, K. Schnackerz and F. Munter, *J. Chromatogr.*, 6 (1961) 61.
3 E. Vidic and J. Schütte, *Arch. Pharm. (Weinheim)*, 295 (1962) 342.
4 F. Moll, *Arch. Pharm. (Weinheim)*, 296 (1963) 205.
5 J.L. Ramaut, *Bull. Soc. Chim. Belg.*, 72 (1963) 406.
6 R.R. Paris and M. Paris, *Bull. Soc. Chim. Fr.*, (1963) 1597.
7 E. Papp and Z. Szabo, *Herba Hung.*, 2 (1963) 383.
8 C.I. Abou-Chaar, *Leban. Pharm. J.*, 8 (1963) 82; *C.A.*, 61 (1964) 14951f.
9 W. Kamp, W.J.M. Onderberg and W.A. Seters, *Pharm. Weekbl.*, 98 (1963) 993.
10 S. Gill and E. Steinegger, *Sci. Pharm.*, 31 (1963) 135.
11 S. Gill, *Acta Polon. Pharm.*, 21 (1964) 379.
12 V. Parrak, E. Radejova and F. Machovicova, *Chem. Zvesti*, 18 (1964) 369.
13 S. Gill and E. Steinegger, *Pharm. Acta Helv.*, 39 (1964) 508.

14 V. Schwarz and M. Sarsunova, *Pharmazie*, 19 (1964) 267.
15 V.E. Chichiro, *Sb. Nauchn. Tr. Tsentr. Nauchn. Issled. Aptechn. Inst.*, 5 (1964) 167; *C.A.*, 63 (1965) 6008e.
16 G. Grandolini, C. Galeffi, E. Montalvo, C.G. Casinovi and G.B. Marini-Bettolo, in G.B. Marini-Bettolo (Editor), *Thin Layer Chromatography*, Elsevier, Amsterdam, 1964, p. 155.
17 E.B.L. Borio and E.G. Moreira, *Trib. Farm.*, 32 (1964) 64; *C.A.*, 63 (1965) 17798g.
18 Y.N. Forostyan and V.I. Novikov, *Zh. Obshch. Khim.*, 38 (1968) 1222.
19 R. Paris, R. Rousselet, M. Paris and M.J. Fries, *Ann. Pharm. Fr.*, 23 (1965) 473.
20 D. Giacopello, *J. Chromatogr.*, 19 (1965) 172.
21 R.K. Sharma, G.S. Khajuria and C.K. Atal, *J. Chromatogr.*, 19 (1965) 433.
22 A.H. Chalmers, C.J. Culvenor and L.W. Smith, *J. Chromatogr.*, 20 (1965) 270.
23 A. Kaess and C. Mathis, *Ann. Pharm. Fr.*, 24 (1966) 753.
24 A. Kaess and C. Mathis, *Int. Symp. Chromatogr. Electrophor. Lect. Pap. 4th*, (1966) 525.
25 G.J. Dickes, *J. Assoc. Public Anal.*, 4 (1966) 45.
26 J.J. Menn and J.B. McBain, *Nature (London)*, 209 (1966) 1351.
27 E.A. Moreira, *Trib. Farm.*, 34 (1966) 27; *C.A.*, 66 (1967) 8816f.
28 A.R. Mattocks, *J. Chromatogr.*, 27 (1967) 505.
29 A. Noirfalise and G. Mees, *J. Chromatogr.*, 31 (1967) 594.
30 T. Kaniewska and B. Borkowski, *Diss. Pharm. Pharmacol.*, 20 (1968) 111.
31 M. Debackere and L. Laruelle, *J. Chromatogr.*, 35 (1968) 234.
32 J.M.G.J. Frijns, *Pharm. Weekbl.*, 103 (1968) 929.
33 V.N. Kulakov and A.M. Likhosherstov, *Zh. Obshch. Khim.*, 38 (1968) 1715.
34 I.V. Terenteva, A.F. Sholl and V.I. Kogan, *Brevikollin-Alkaloid Osoki Parvskoi*, (1969) 24; *C.A.*, 74 (1971) 45634d.
35 M. Vanhaelen, *J. Pharm. Belg.*, 24 (1969) 87.
36 V. Massa, F. Gal and P. Susplugas, *Int. Symp. Chromatogr. Electrophor. Lect. Pap. 6th*, (1970) 470.
37 G.H. Jolliffe and E.J. Shellard, *J. Chromatogr.*, 48 (1970) 125.
38 V.E. Dauksha, *Khim. Prir. Soedin.*, 6 (1970) 274; *C.A.*, 73 (1970) 117177m.
39 Y.D. Cho and R.O. Martin, *Anal. Biochem.*, 44 (1971) 49.
40 K. Genest, *Forensic Sci. Soc. J.*, 11 (1971) 95.
41 G.P. Londareva and G.B. Tikhomirova, *Khim. Farm. Zh.*, 5 (1971) 43.
42 W. Meissner, H. Pokowinska, H. Kokocinska and W. Stopa, *Pr. Zakresu Towaroznn. Chem., Wyzsza Szk. Ekon. Poznazin, Zesz. Nauk., Ser. 1*, 40 (1971) 41; *C.A.*, 78 (1972) 72406v.
43 K.F. Ahrend and D. Tiess, *Zbl. Pharm.*, 111 (1972) 933.
44 G. Faugeras and M. Paris, *Bull. Soc. Chim. Fr.*, (1973) 109.
45 E. Novakova and J. Vecerkova, *Cesk. Farm.*, 22 (1973) 347.
46 K.F. Ahrend and D. Tiess, *Wiss. Z. Univ. Rostock, Math. Naturw. Reihe*, 22 (1973) 951.
47 F. Schmidt, *Dtsch. Apoth. Ztg.*, 114 (1974) 1593.
48 V.P. Zakharov, Kh.A. Aslanov, A.I. Ishbaev, A.S. Sadykov and B.A. Yankovskii, *Khim. Prir. Soedin.*, 4 (1974) 468; *C.A.*, 82 (1975) 70267j.
49 E. Stahl and W. Schmitt, *Arch. Pharm. (Weinheim)*, 308 (1975) 570.
50 R.G. Gafurov, R.A. Shaimardanov, N.Sh. Kattaev, B.N. Narzieva and E.A. Aripov, *Dokl. Akad. Nauk. Uzb. SSR*, (1975) 40; *C.A.*, 85 (1976) 89510s.
51 E. Spratt, *Toxicol. Annu. 1974*, (1975) 229.
52 A. Klasek, M. Ciesla and S. Dvorackova, *Acta Univ. Palacki, Olomouc., Fac. Med.*, 79 (1976) 47.
53 S. Bieganowska and A. Waksmundzki, *Chromatographia*, 9 (1976) 215.
54 L. Lepri, P.G. Desideri and M. Lepori, *J. Chromatogr.*, 116 (1976) 131.
55 W. Wysocka, *J. Chromatogr.*, 116 (1976) 235.
56 L. Lepri, P.G. Desideri and M. Lepori, *J. Chromatogr.*, 123 (1976) 175.
57 S. Yusupov, A.I. Ishbaev, K.A. Aslanov and A.S. Sadykov, *Khim. Prir. Soedin.*, (1976) 772; *C.A.*, 86 (1977) 111216v.
58 E.M. Karlsson and H.W. Peter, *J. Chromatogr.*, 155 (1978) 218.

TABLE 5.1

*hR<sub>F</sub>* VALUES OF *LOBELIA* ALKALOIDS

TLC systems:

| | | |
|---|---|---|
| S1 | Cellulose impregnated with 20% formamide in acetone | Benzene-heptane-diethylamine (1:60:0.02)[1] |
| S2 | Silica gel G | Chloroform-benzene (1:1) saturated with ammonia, in ammonia atmosphere[12] |
| S3 | Silica gel G, activated | Chloroform-acetone-diethylamine (5:4:1)[23] |
| S4 | Silica gel G, activated | Chloroform-diethylamine (9:1)[23] |
| S5 | Silica gel G, activated | Cyclohexane-chloroform-diethylamine (5:4:0.5)[23] |
| S6 | Silica gel G, activated | Cyclohexane-diethylamine (9:1)[23] |

| Alkaloid | $hR_F$ value | | | | | |
|---|---|---|---|---|---|---|
| | S1 | S2 | S3 | S4 | S5 | S6 |
| Lobeline | 69 | 65 | 72/65 | 78/68 | 27/19 | 21/16 |
| Lobelanine | 92 | 84 | 72 | 80 | 42 | 28 |
| Lobelanidine | 24 | 34 | 66 | 64 | 17 | 19 |
| Acetyllobeline | | | 80 | 85 | 57/52 | 47 |
| Monoacetyllobelanidine | | | 79 | 86 | 50 | 40 |
| Diacetyllobelanidine | | | 79 | 86 | 63 | 58 |
| Acetophenone | | 95 | | | | |

TABLE 5.2

TLC ANALYSIS OF *SEDUM* ALKALOIDS

TLC systems:

| | | |
|---|---|---|
| S1 | Silica gel G, 0.5 *M* potassium hydroxide impregnated | Toluene-methanol-chloroform (9:3:1)[7] |
| S2 | Silica gel G | Chloroform-diethylamine (95:5)[53] |
| S3 | Silica gel G | Chloroform-benzene-diethylamine (45:50:5)[53] |
| S4 | Silica gel G | Chloroform-methanol-benzene (1:3:9)[53] |
| S5 | Silica gel G | Cyclohexane-*n*-propanol-diethylamine (85:10:5)[53] |

| Alkaloid | $hR_F$ value | | | | |
|---|---|---|---|---|---|
| | S1 | S2 | S3 | S4 | S5 |
| Sedamine | 43 | 53 | 33 | 37 | 45 |
| Sedinine | 54 | 45 | 25 | 42 | 48 |
| Sedridine | 6 | | | | |
| Allosedamine | | 53 | 33 | 37 | 45 |

TABLE 5.3

TLC ANALYSIS OF SOME PYRROLIZIDINE ALKALOIDS

TLC systems:

| | | |
|---|---|---|
| S1 | Silica gel G, activated | Chloroform-methanol-ammonia (85:14:1)[21] |
| S2 | Silica gel G, 0.1 *M* sodium hydroxide impregnated | Methanol[22] |

| No. | Base | $hR_F$ values of non-ester pyrrolizidine alkaloids and derivatives | |
|---|---|---|---|
| | | S1 | S2 |
| 1 | 1-Methylenepyrrolizidine | | 5 |
| 2 | Heliotridane (1β-methyl-8α-pyrrolizidine) | | 1 |
| 3 | Anhydroplatynecine | | 5 |
| 4 | 7β-Hydroxy-1-methylene-8α-pyrrolizidine | | 7 |
| 5 | Desoxyretronecine (7β-hydroxy-1-methyl-1,2-dehydro-8α-pyrrolizidine) | | 7 |
| 6 | 7β-Hydroxy-1-methylene-8β-pyrrolizidine | | 14 |
| 7 | Retronecanol (7β-hydroxy-1β-methyl-8α-pyrrolizidine | | 1 |
| 8 | Hydroxyheliotridane (7α-hydroxy-1β-methyl-8α-pyrrolizidine | | 2 |
| 9 | 7α-Hydroxy-1-methyl-1,2-dehydro-8α-pyrrolizidine | | 13 |
| 10 | 1-Methoxymethyl-1,2-dehydro-8α-pyrrolizidine | | 8 |
| 11 | 1-Methoxymethyl-1,2-epoxy-pyrrolizidine | | 23 |
| 12 | Isoretronecanol (1β-hydroxymethyl-8α-pyrrolizidine) | | 2 |
| 13 | Supinidine (1-hydroxymethyl-1,2-dehydro-8α-pyrrolizidine) | | 4 |
| 14 | 1-Hydroxymethyl-1,2-epoxy-8α-pyrrolizidine | | 18 |
| 15 | 7β-Hydroxy-1-methoxymethyl-1,2-dehydro-8α-pyrrolizidine | | 12 |
| 16 | 7β-Acetoxy-1-methoxymethyl-1,2-dehydro-8α-pyrrolizidine | | 30 |
| 17 | Retronecine (7β-hydroxy-1-hydroxy-methyl-1,2-dehydro-8α-pyrrolizidine) | | 7 |
| 18 | Heliotridine (7α-hydroxy-1-hydroxy-methyl-1,2-dehydro-8α-pyrrolizidine) | | 14 |
| 19 | Platynecine (7β-hydroxy-1β-hydroxy-methyl-8α-pyrrolizidine) | | 1 |

| No. | Base | $hR_F$ values of pyrrolizidine esters with monocarboxylic acids | |
|---|---|---|---|
| | | S1 | S2 |
| 20 | 7-Angelylretronecine | | 33 |
| 21 | 7-Angelylheliotridine | | 45 |
| 22 | Heleurine | 49 | 11 |

**References p. 64**

TABLE 5.3 (*continued*)

| No. | Base | $hR_F$ values of pyrrolizidine esters with monocarboxylic acids S1 | S2 |
|---|---|---|---|
| 23 | Supinine | 17 | 10 |
| 24 | Heliotrine | 30 | 30 |
| 25 | Indicine | | 19 |
| 26 | Retronecine trachelanthate | | 19 |
| 27 | Retronecine viridiflorate | | 19 |
| 28 | Rinderine | | 29 |
| 29 | Echinatine | | 30 |
| 30 | Europine | | 29 |
| 31 | Sarracine | | 16 |
| 33 | Echiumine | | 47 |
| 34 | Lasiocarpine | | 54 |
| 35 | Echimidine | | 45 |
| 36 | Heliosupine | | 53 |
| 37 | Latifoline | | 53 |

| No. | Base | $hR_F$ values for macrocyclic diester alkaloids S1 | S2 |
|---|---|---|---|
| 38 | Retusine | | 16 |
| 39 | Fulvine | | 33 |
| 40 | Crispatine | | 29 |
| 41 | Monocrotaline | 44 | 29 |
| 42 | Senecionine | | 40 |
| 43 | Seneciphylline | 56 | 38 |
| 44 | Platyphylline | | 18 |
| 45 | Integerrimine | | 39 |
| 46 | Spectabiline | | 34 |
| 47 | Senkirkine | | 29 |
| 48 | Jacobine | 50 | 38 |
| 49 | Sceleratine | | 34 |
| 50 | Jacozine | | 37 |
| 51 | Jacoline | | 37 |
| 52 | Rosmarinine | | 35 |
| 53 | Jaconine | | 47 |
| 54 | Retrorsine | | 35 |
| 55 | Riddelliine | | 32 |
| 56 | Retusamine | | 30 |
| 57 | Otosenine | | 23 |
| 58 | Grantianine | | 31 |

TABLE 5.4

TLC DETECTION OF PYRROLIZIDINE ALKALOIDS AND OTHER BASES[28]

TLC system:
Silica gel G with chloroform-acetone-ethanol-concentrated ammonia (5:3:1:1).

| Group | Compound | $hR_F$* | Colour of spot (reagent no. 56) |
|---|---|---|---|
| I | Retrorsine | 35 | Blue |
| | Diacetylretrorsine | 70 | Blue |
| | Monocrotaline | 43 | Blue |
| | Senecionine | 63 | Blue |
| | Anacrotine | 33 | Blue |
| | Lasiocarpine | 74 | Blue |
| | Heliotrine | 33 | Blue |
| | Supinine | 28 | Mauve |
| II | Rosmarinine | 33 | Blue |
| III | Retronecine | 9 | Blue |
| | Heliotridine | 5 | Blue |
| IV | Platynecine | 0 | Weak yellow-brown |
| | Retronecanol | - | None |
| V | Strigosine | - | None |
| VI | Brucine | - | None |
| | Strychnine | - | None |
| | Arecoline | 78 | Weak yellow-brown |
| | Quinine | - | None |
| | Nicotine | - | None |
| VII | Benzylamine | - | None |
| | Pyrrolidine | - | None |
| VIII | Indole | 80 | Very weak brown |
| | Pyrrole | - | None |

*$hR_F$ values varied slightly with different batches of solvent and adsorbent. Hence these values should be regarded as relative, not absolute.

TABLE 5.5

$hR_F$ VALUES OF SOME LEGUMINOSAE ALKALOIDS

TLC systems:

| System | Layer | Solvent |
|---|---|---|
| S1 | Silica gel G, activated | Cyclohexanol-cyclohexane-hexane (1:1:1) + 5% diethylamine[5] |
| S2 | Silica gel G | Ethyl acetate-hexane-diethylamine (77.5:17.5:5)[6] |
| S3 | Silica gel G | Benzene-chloroform-diethylamine (20:75:5)[6] |
| S4 | Silica gel G | Hexane-dichloroethylene-diethylamine (20:75:5)[6] |
| S5 | Aluminium oxide G | Hexane-dichloroethylene-diethylamine (20:75:5)[6] |
| S6 | Silica gel G, activated | Cyclohexane-diethylamine (7:3)[10,11,13] |
| S7 | Silica gel G, activated | Chloroform-methanol (8:2)[13] |
| S8 | Silica gel G | Chloroform-methanol-concentrated ammonia (85:14:1)[39] |
| S9 | Aluminium oxide, basic, without binder | Benzene-acetone-methanol (34:3:3)[39] |
| S10 | Cellulose MN300 | Butanol-concentrated hydrochloric acid-water (70:7.5:13.5) (ref. 39) |
| S11 | Silica gel HF 254 | Chloroform-ethanol (3:2)[55] |
| S12 | Silica gel HF 254 | Benzene-methanol (4:1)[55] |
| S13 | Silica gel HF 254 | Ethylacetate-chloroform-benzene (2:2:1)[55] |

| Compound | $hR_F$ value | | | | | | | | | | | | |
|---|---|---|---|---|---|---|---|---|---|---|---|---|---|
| | S1 | S2 | S3 | S4 | S5 | S6 | S7 | S8 | S9 | S10 | S11 | S12 | S13 |
| Adenocarpine | 65 | | | | | | | | | | | | |
| Anagyrine | 61 | 30 | 40 | 37 | 35 | 35 | 60 | 69 | 64 | 18 | | | |
| Angustifoline | | | | | | | | 46 | | 68 | | | |
| Calycotomine | | | | | | 5 | 46 | | | | | | |
| Cytisine | 11 | 0 | 5 | 0 | 0 | 7 | 32 | 28 | 19 | 7 | | | |
| N-Methylcytisine | 46 | 32 | 40 | 36 | 40 | 30 | 63 | 61 | 59 | 10 | | | |
| $\Delta^{11}$-Dehydrolupanine | | | | | | | | 5 | | 32 | | | |
| $\Delta^{5}$-Dehydrolupanine | | | | | | | | 83 | 75 | 53 | | | |
| $\Delta^{5}$-Dehydrolupaninic acid | | | | | | | | 5 | 75 | 19 | | | |
| 13-Hydroxylupanine | | | | | | 19 | 24 | 22 | 44 | 31 | 92 | 93 | 87 |
| *epi*-13-hydroxylupanine | | | | | | | | | | | 76 | 82 | 83 |
| 13-Hydroxy-α-isolupanine | | | | | | | | | | | 83 | 83 | 79 |
| *epi*-13-Hydroxy-α-isolupanine | | | | | | | | | | | 54 | 75 | 66 |
| 4-Hydroxylupanine | | | | | | | | 61 | 3 | 40 | | | |
| Hydroxysparteine | | | | | | 54 | 19 | | | | | | |
| Isophoramine | | | | | | | | 78 | 71 | 26 | | | |
| 8-Ketosparteine | | | | | | | | 88 | 79 | 27 | | | |
| Lamprolobine | | | | | | | | 71 | 76 | 71 | | | |
| Lupanine | 65 | 42 | 50 | 64 | 35 | 61 | 38 | 72 | 74 | 52 | | | |
| Lupaninic acid | | | | | | | | 5 | | 30 | | | |
| α-Isolupanine | | | | | | | | 79 | 71 | 52 | | | |
| Lupinine | 57 | 43 | 55 | 45 | 50 | 63 | 12 | 23 | 56 | 58 | | | |
| Epilupinine | | | | | | 56 | 35 | | | | | | |
| Matrine | | | | | | | | 83 | 73 | 60 | | | |
| Orensine | 65 | | | | | | | | | | | | |
| 17-Oxolupanine | 25 | | | | | | | 82 | 74 | 88 | | | |
| 17-Oxosparteine | | | | | | | | 85 | 76 | 82 | | | |
| Retamine | 83 | 60 | 55 | 70 | 70 | 80 | 30 | | | | | | |
| Rhombifoline | | | | | | | | 85 | 68 | 37 | | | |
| Santiagine | 78 | | | | | | | | | | | | |
| Sarothamnine | | 75 | 50 | 86 | 85 | | | | | | | | |
| Sparteine | 89 | 85 | 70 | 90 | 90 | 93 | 10 | 8 | 67 | 50 | | | |
| L-α-isosparteine | | 38 | 30 | 85 | 48 | 92 | 3 | | | | | | |
| Sphaerocarpine | 30 | | | | | | | | | | | | |
| Thermopsine | | | | | | | | 78 | 68 | 30 | | | |

TABLE 5.6

COLOURS OBTAINED FOR A NUMBER OF LEGUMINOSAE ALKALOIDS AFTER SPRAYING WITH DRAGENDORFF'S REAGENT, MUNIER MODIFICATION (NO. 39b)[5]

| Alkaloid | Colour | Alkaloid | Colour |
|---|---|---|---|
| Cytisine | Orange-red | Retamine | Orange-red |
| Anagyrine | Orange-red | Sphaerocarpine | Orange-red |
| N-methylcytisine | Orange-red | Lupinine | Beige |
| Sparteine | Orange-red | Orensine | Yellow-orange |
| Oxylupanine | Orange-red | Adenocarpine | Yellow-orange |
| Lupanine | Orange-red | Santiaguine | Yellow-orange |

TABLE 5.7

LITERATURE CITED IN CHAPTER 3 WHICH INCLUDES PYRROLIDINE, PYRROLIZIDINE, PYRIDINE, PIPERIDINE AND QUINOLIZIDINE ALKALOIDS

| Alkaloid* | Ref. | Alkaloid* | Ref. |
|---|---|---|---|
| L,spart | 2 | Spart | 29 |
| L | 3 | L,spart | 43 |
| L,lanin,lanid | 14 | L,spart | 45 |
| Spart | 19 | L,spart,arec,con | 46 |
| α-isospart,lupa | 20 | L | 47 |
| L | 25 | L,spart,arec | 54,56 |

*Abbreviations used in Tables 5.7-5.10:

| | | | |
|---|---|---|---|
| adenoc | adenocarpine | lup | lupinine |
| alloseda | allosedamine | lupa | lupanine |
| an | anabasine | monocrot | monocrotaline |
| anac | anacrotine | N | nicotine |
| anag | anagarine | NMe-con | N-methylconiine |
| ang | angustifoline | NMe-conh | N-methylconhydrine |
| ansa | anabasamine | N-Me | N-methylcytisine |
| aph | aphylline | OHisolupa | 13-hydroxy-α-isolupanine |
| aphid | aphyllidine | OHlupa | 13-hydroxylupanine |
| arec | arecoline | OHspart | hydroxysparteine |
| caly | calycotomine | or | orensine |
| Clretc | chlororetromecine | oxylupa | oxylupanine |
| cin | cineverine | pip | piperine |
| con | coniine | plat | platyphylline |
| conc | coniceine | platec | platynecine |
| conh | conhydrine | ret | retamine |
| cyt | cytisine | retc | retronecine |
| diAcretr | diacetylretrorsine | retr | retrorsine |
| epilup | epilupinine | retro | retronecanol |
| hel | heliotrine | ros | rosmarinine |
| heleu | heleurine | sant | santiaguine |
| helid | heliotridine | seda | sedamine |
| isa | isatidine | sedi | sedinine |
| isolupa | α-isolupanine | sedn | sedinon |
| isoor | isoorensine | sedr | sedridine |
| isopel | isopelletierine | sen | seniciphylline |
| jac | jacobine | senio | sencionine |
| L | lobeline | spart | sparteine |
| lanid | lobelanidine | sphaer | sphaerocarpine |
| lanin | lobelanine | strig | strigosine |
| las | lasiocarpine | sup | supinine |

TABLE 5.8

TLC ANALYSIS OF PYRROLIDINE, PYRROLIZIDINE, PYRIDINE, PIPERIDINE AND QUINOLIZIDINE ALKALOIDS IN PLANT MATERIAL

| Alkaloid* | Aim | Adsorbent | Solvent system | Ref. |
|---|---|---|---|---|
| Cyt,NMe-cyt,spart, lup,lupa,oxylupa, anag,or,ret,sphaer, adenoc,sant | Separation of Leguminosae alkaloids (Table 5.5) | $SiO_2$ | Cyclohexanol-cyclohexane-hexane(1:1:1) containing 5% DEA<br>$CHCl_3$-$Me_2CO$-DEA(5:4:1)<br>$CHCl_3$-DEA(9:1)<br>Cyclohexane-$CHCl_3$-DEA (5:4:1)<br>$CHCl_3$-MeOH-DEA(80:20: 0.2) | 5 |
| Sedi,seda,sedr | Separation of *Sedum* alkaloids (Table 5.2) | 0.5 *M* KOH impregnated $SiO_2$ | Toluene-MeOH-$CHCl_3$ (9:3:1) | 7 |
| | Separation of Lupine alkaloids | | No details available | |
| Cyt,NMe-cyt,anag, OHlupa,OHspart,lup, lupa,ret,spart, α-isosparteine | Separation of Leguminosae alkaloids (Table 5.5) | $SiO_2$ | Cyclohexane-DEA(9:1, 8:2, 7:3, 1:1) | 10 |
| Cyt,NMe-cyt,anag, OHlupa,lup,ret,spart | Separation of alkaloids in *Cytisus* and *Genista* species | $SiO_2$ | Cyclohexane-DEA(7:3) | 11 |
| Cyt,NMe-cyt,anag, OHlupa,lupa,lup, epilup,OHspart,ret, spart,α-isospart, caly | Separation of *Cytisus* alkaloids (Table 5.5) | $SiO_2$ | Cyclohexane-DEA(7:3, 9:1, 1:1)<br>$CHCl_3$-MeOH(8:2) | 13 |
| Spart, no details on other alkaloids available | Identification of alkaloids in *Thermopsis* species | $SiO_2$ | $CHCl_3$-$Me_2CO$-MeOH-25% $NH_4OH$(20:20:3:1) | 15 |
| α-,β-,γ- and dehydroskytanthine | Separation of *Skytanthus* alkaloids | $SiO_2$ | $CHCl_3$-MeOH(7:3) | 16 |
| L,lanin,lanid | Separation of alkaloids in *Lobelia* species | $SiO_2$ | Cyclohexane-$CHCl_3$-DEA (50:4:5) | 17 |
| An,lup,aph,aphid | Separation of *Anabis* alkaloids | $SiO_2$<br>$Al_2O_3$ | $CCl_4$-$Me_2CO$-MeOH (3:7:0.5)<br>Benzene-$CHCl_3$-MeOH (8:22:2) | 18 |
| Sup,hel,monocrot, heleu,jac,sen | Separation of *Crotalaria* alkaloids (Table 5.3) | $SiO_2$ | $CHCl_3$-MeOH-$NH_4OH$ (85:14:1) | 21 |
| 58 pyrrolizidine alkaloids | Characterization by TLC, GLC and PC (Table 5.3) | 0.1 *M* NaOH-impregnated $SiO_2$ | MeOH | 22 |
| L,lanin,lanid | Reaction chromatography (Table 5.1) | $SiO_2$ | $CHCl_3$-$Me_2CO$-DEA(5:4:1)<br>$CHCl_3$-DEA(9:1)<br>Cyclohexane-$CHCl_3$-DEA (5:4:0.5)<br>Cyclohexane-DEA(9:1) | 23 |

TABLE 5.8 (*continued*)

TLC ANALYSIS OF PYRROLIDINE, PYRROLIZIDINE, PYRIDINE, PIPERIDINE AND QUINOLIZIDINE ALKALOIDS IN PLANT MATERIAL

| Alkaloids* | Aim | Adsorbent | Solvent systems | Ref. |
|---|---|---|---|---|
| No details available | Alkaloids in *Bacharis* species | $SiO_2$ | $CHCl_3$-cyclohexane(7:3) | 27 |
| L | Identification of *Lobelia* | $SiO_2$ | I. $CHCl_3$-DEA(9:1)<br>II. Cyclohexane-$CHCl_3$-DEA(5:4:1) | 32 |
| L,lanin,lanid | Identification in *Lobelia* | $SiO_2$ | $Me_2CO$-DMFA(38:2) | 35 |
| Spart,pip and non-specified alkaloids | TAS technique for *Conium*, *Lobelia*, *Cytisus*, *Punica* and *Piper* species | $SiO_2$ | $CHCl_3$-$Me_2CO$-DEA(5:4:1)<br>$CHCl_3$-DEA(9:1)<br>Cyclohexane-$CHCl_3$-DEA (5:4:1)<br>Cyclohexane-DEA(9:1)<br>Benzene-EtOAc-DEA (7:2:1) | 37 |
| Sen,plat | Control extraction of alkaloids from *Senecio* species | $SiO_2$ | $Et_2O$-$Me_2CO$-DEA(80:20:5) | 38 |
| 22 lupine alkaloids | Identification of Lupine alkaloids by TLC and GC-MS (Table 5.5) | $SiO_2$<br>$Al_2O_3$<br>cell | $CHCl_3$-MeOH-conc.$NH_4OH$ (95:4:1, 85:14:1)<br>Benzene-$Me_2CO$-MeOH (34:3:3)<br>BuOH-conc.HCl-$H_2O$ (70:7.5:13.5) | 39 |
| Panamine,ormojanine, ormosanine | Identification of *Arbrus* and *Ormosia* species (seeds) | $SiO_2$ | Light petroleum (b.p. 30-60°C)-DEA(7:3)<br>Cyclohexane-DEA(7:3)<br>Benzene-$CHCl_3$-DEA (20:75:15)<br>$Et_2O$-DEA(95:5) | 40 |
| 24 quinolizidine alkaloids | No details available | | | 42 |
| Spart,ret,lupa OHlupa,anag,cyt, NMe-cyt,cin,sant adenoc,isoor | Densitometric analysis of Leguminosae alkaloids | $SiO_2$<br><br><br>cell | Cyclohexane-DEA(7:3)<br>Cyclohexane-$CH_2Cl_2$-DEA (4:4:2)<br>Toluene-$Me_2CO$-EtOH-$NH_4OH$(60:40:6:2)<br>isoBuOH-HCl-$H_2O$(7:1:2)<br>MeOH-$Me_2CO$-$NH_4OH$-$H_2O$ (35:20:5:40) | 44 |
| An,aph,aphid,lup, ansa | Identification in *Anabis* species | | No details available | 48 |
| Con,Nme-con, conc,arec | TAS technique for *Conii fructus* and *Arecae semen* | $SiO_2$ | $Me_2CO$-$Et_2O$-conc.$NH_4OH$ (50:50:3) | 49 |
| No details available | TLC of Leguminosae alkaloids on kaolin | kaolin | $CHCl_3$-MeOH(2:1, 4:1)<br>$CHCl_3$-EtOH(5:1)<br>Benzene-MeOH(5:2) | 50 |
| Pyrrolizidine alkaloids | Identification of necis acid and derivatives | $SiO_2$ | Benzene-MeOH-dioxane-AcOH(45:4:4:4)<br>Benzene-dioxane-AcOH (90:25:4) | 52 |

TABLE 5.8 (*continued*)

TLC ANALYSIS OF PYRROLIDINE, PYRROLIZIDINE, PYRIDINE, PIPERIDINE AND QUINOLIZIDINE ALKALOIDS IN PLANT MATERIAL

| Alkaloids* | Aim | Adsorbent | Solvent systems | Ref. |
|---|---|---|---|---|
| Seda,sedi,allo-seda, isopel,sedr,N,sedn | Identification of alkaloids in *Sedum* species (Table 5.2) | $SiO_2$ | $CHCl_3$-DEA(95:5)<br>$CHCl_3$-benzene-DEA (45:50:5)<br>$CHCl_3$-MeOH-benzene (1:3:9)<br>Cyclohexane-*n*PrOH-DEA (85:10:5) | 53 |
| OHlupa,lupi,lupa, spart,isolupa | Fluorodensitometric quantitative analysis in *Lupinus* species | $SiO_2$ | Benzene-$CH_2Cl_2$-$Et_2O$-DEA(5:5:5:2) | 58 |
| Aph,aphid,an,ansa, lup,pachycarpine | Determination in *Anabis* species (no details available) | $Al_2O_3$ | $Me_2CO$-$H_2O$(100:8)<br>$Et_2O$-$CHCl_3$(10:7) | 57 |

*For abbreviations, see footnote to Table 5.7.

TABLE 5.9

TLC ANALYSIS OF PYRROLIDINE, PYRROLIZIDINE, PYRIDINE, PIPERIDINE AND QUINOLIZIDINE ALKALOIDS IN BIOLOGICAL MATERIALS

| Alkaloid* | Other compounds | Aim | Adsorbent | Solvent system | Ref. |
|---|---|---|---|---|---|
| L,spart | Brucine,strychnine, caffeine,cocaine, quinine,opium alkaloids | Doping control | $SiO_2$ | I. Hexane-$Me_2CO$-DEA (6:3:1)<br>II. $CHCl_3$-MeOH-DEA (95:5:0.05)<br>two-dimensional, I,II | 31 |
| L | Drugs of abuse (chapter 12, ref.385) | Detection drugs of abuse in urine | $SiO_2$ | EtOAc-isoPrOH-$NH_4OH$ (60:40:1)<br>EtOAc-cyclohexane-$NH_4OH$(50:40:0.1) (60:40:1)<br>Dioxane-benzene-$NH_4OH$ (35:60:5)<br>$CHCl_3$-$Me_2CO$(95:5) (unsat. chamber)<br>EtOAc-MeOH-$NH_4OH$ (85:10:5) | 51 |

*For abbreviations, see footnote to Table 5.7.

TABLE 5.10

TLC ANALYSIS OF PYRROLIDINE, PYRROLIZIDINE, PYRIDINE, PIPERIDINE AND QUINOLIZIDINE ALKALOIDS AS PURE COMPOUNDS AND IN PHARMACEUTICAL PREPARATIONS

| Alkaloid* | Other compounds | Aim | Adsorbent | Solvent system | Ref. |
|---|---|---|---|---|---|
| L,lanin,lanid, spart,con,N, an,arec | | Separation (Table 5.1) | FMA-impregnated cell<br>0.05 *M* KOH impregnated $SiO_2$ | Benzene-heptane-DEA(1:6:0.02)<br>$CHCl_3$-96% EtOH(11:1, 9:1, 8:2) | 1 |
| Con,NMe-con, γ-conc,conh, NMe-conh | Synthetic piperidine derivatives, morpholine | Separation | $SiO_2$ | $CHCl_3$-abs.EtOH-25% $NH_4OH$(18:2:2) | 4 |
| Anag,cyt,spart, isospart,lupa, lup,NMe-cyt, ret,sarothamnine | Various steroid and isoquinoline alkaloids | Separation (Table 5.5) | $SiO_2$<br><br><br>$Al_2O_3$ | EtOAc-hexane-DEA(77.5:17.5:5)<br>Benzene-$CHCl_3$-DEA(20:75:5)<br>Hexane-dichloroethylene-DEA(20:75:5)<br>EtOAc-hexane-DEA(77.5:17.5:5) | 6 |
| Spart | Alypin,methadone, pyrilamine,tripilennamine,diphenhydramine,cotarnine, pervitine,meperidine, amylocaine | Identification | $SiO_2$ | I. $CCl_4$-*n*BuOH-MeOH-25% $NH_4OH$(40:30:30:1)<br>II. Light petroleum-$Et_2O$-DEA(20:80:1)<br>two-dimensional: I,II | 9 |
| L,lanin,lanid | Acetophenone | Stability of lobeline (Table 5.1) | $SiO_2$ | $CHCl_3$-benzene(1:1) sat. with $NH_4OH$ in $NH_3$ atmosphere | 12 |
| L,spart,lanin, lanid | | Reaction chromatography | $SiO_2$ | $CHCl_3$-$Me_2CO$-DEA(5:4:1) | 24 |
| L | Physostigmine, ephedrine,homatropine,pilocarpine, reserpine,solanine, solanidine | Detection of cholinesterase inhibitors | 5% Silicone 555 on cellulose | $H_2O$-EtOH-$CHCl_3$(56:42:2) | 26 |

| | | | | | |
|---|---|---|---|---|---|
| Isa,Clretc retr,diAcretr, senio,anac,hel,las, monocrot,sup,ros, retc,helid,plated retro,strig,arec,N | Brucine,strychnine, quinine,benzylamine, pyrrolidine, indole, pyrrole | New detection method (Table 5.4) | $SiO_2$ | $CHCl_3$-$Me_2CO$-EtOH-conc.$NH_4OH$(5:3:1:1)<br>*n*BuOH-AcOH-$H_2O$(4:1:5), organic phase | |
| trachelanthamine, viridiflorine | stereoisomeric dihydroxy acids, fructose | Separation | 4% $AgNO_3$-impregnated $SiO_2$ | $CHCl_3$ sat. with $NH_4OH$<br>EtOH | 28 |
| Brevicolline | | | | Light petroleum-$Et_2O$-MeOH(1:3:5) | 33 |
| | | Determination | $Al_2O_3$ | No details available | 34 |
| Sen,plat, retromecine, platymecine | | Separation | $SiO_2$<br>$Al_2O_3$ | MeOH-$NH_4OH$(95:5)<br>$CHCl_3$-MeOH(8:2) | |
| OHlupa,epi-OHlupa,OHiso-lupa,epiOHiso-lupa | | Separation of epimers (Table 5.5) | $SiO_2$ | $CHCl_3$-EtOH(3:2)<br>Benzene-MeOH(4:1)<br>EtOAc-$CHCl_3$-benzene(2:2:1) | 41<br><br>55 |

*For abbreviations, see footnote to Table 5.7.

# Chapter 6

# TOBACCO ALKALOIDS

Although hundreds of publications have appeared on the analysis of tobacco alkaloids, the number of papers dealing with TLC analysis is limited. The volatile character of these alkaloids makes GLC a more suitable method. Because of the presence of nicotine and its metabolites in the urine of smokers, a number of papers concerning the analysis of drugs in urine have also dealt with nicotine (Tables 6.5-6.7). For the $hR_F$ values of nicotine and some of its metabolites in TLC systems for the analysis of drugs of abuse, see Tables 12.11, 12.12, 12.14 and 12.16[3,21,36]. Ono and Asahina[9] and Goeneckea and Bernhard[28] dealt with the possible interference of nicotine in the analysis of morphine in urine.

## 6.1. SOLVENT SYSTEMS

Teichert et al.[1] used 0.5 *M* potassium hydroxide-impregnated plates in combination with solvent S1 (also in other ratios) (Table 6.1) for the separation of the volatile alkaloids nicotine, anabasine, sparteine, coniine and arecoline. A similar system was used by Fejer-Kossey[5] for the separation of the tobacco alkaloids nicotine, nornicotine, anabasine and nicotyrine, but the ratio of chloroform to ethanol was different (9:1). Winefordner and Moye[7] found silica gel to be less suitable for the separation of tobacco alkaloids, and obtained the best separation with aluminium oxide in combination with solvent S8 (Table 6.1). Papp and Szabo[4] used 0.5 *M* potassium hydroxide-impregnated plates in combination with solvent S2. Hodgson et al.[11] described a two-dimensional separation of nicotine and its metabolites using solvents S3 and S4. Another two-dimensional system for the separation of tobacco alkaloids was described by Fejer-Kossey[18], using solvents S5 and S6. Farkas-Riedel[39] used two consecutive mobile phases in the same direction for the separation of tobacco alkaloids. Similarly to Hodgson et al.[11], he used a basic solvent as the first system and an acidic solvent as the second. Harke et al.[24] analysed nicotine and its metabolites in urine with toluene - isopropanol - ammonia (100:25:1) as the solvent on silica gel plates. The plates were developed three times. Baiulescu and Constantinescu[47] used solvent system S7 in the reaction chromatography of nicotine and related alkaloids

References p. 81

## 6.2. DETECTION

Nicotine and related alkaloids can be detected with the usual alkaloid spray reagents: Dragendorff's reagent and iodoplatinate. With iodoplatinate, nicotine gives a black - blue colour. Holdstock and Stevens[49] studied the recovery of alkaloids from TLC plates after spraying with iodoplatinate. Nicotine remained stable for 7 days on the plate, and its recovery was satisfactory. Constantinescu[25] and Schmidt[40] used iodine for the detection of nicotine. The latter author used also iron(III) chloride followed by iodine solution (no. 60).

A commonly used method for the detection of nicotine and related alkaloids is based on König's reaction. Different procedures have been used. Fejer-Kossey[5] sprayed first with a 2% solution of aniline in ethanol, then the plates were exposed to cyanogen bromide vapour. The same author[18] also reported the use of 1% benzidine in ethanol as a spray reagent followed by exposure to cyanogen bromide vapours; the colours obtained are summarized in Table 6.2. Hodgson et al.[11] used 2% *p*-aminobenzoic acid in methanol + 0.1 *M* phosphate buffer (pH 7.0) (1:1) as a spray reagent and after drying, the plates were exposed to cyanogen bromide vapour (no. 66d). Harke et al.[55] exposed the plates first to cyanogen bromide for 5 min. After 5 min in the open air, the plates were sprayed with 3% 4-chloroaniline in methanol. To obtain a complete reaction this procedure was repeated three times (no. 66a).

Güven and Tekinalp[23] detected nicotine with 10% copper(II) sulphate in 2% ammonia, yielding blue spots. Vinson et al.[50,52] described the use of the TCBI reagent (no. 94) for the detection of drugs of abuse. Nicotine gave a grey-purple colour with this reagent.

For the detection of nornicotin isatin has been used[15,39] (no. 65).

## 6.3. QUANTITATIVE ANALYSIS

Winefordner and Moye[7] analysed nicotine, nornicotine and anabasine in tobacco after separation on aluminium oxide plates. The spots were eluted with 5 ml of ethanol containing 4 drops of diethylamine the alkaloids subsequently being determined phosphorimetrically. Harke et al.[24] determined nicotine and cotinine in animal tissues. After separation on silica gel plates the alkaloids were eluted with water and subsequently treated with aniline and cyanogen bromide (König's reaction). The molar absorptivity was determined at 462 nm. Later Harke et al.[55] determined nicotine and a number of metabolites in urine in a similar manner. However, the colour reaction was carried out on the plate by exposure to cyanogen bromide vapour and subsequently spraying with 3% 4-chloroaniline in methanol. This was repeated three times. The coloured spots were scraped off and the alkaloids eluted with ethanol and determined colorimetrically at 480 nm.

Constantinescu[25] eluted the spots of tobacco alkaloids with alkalinized chloroform from silica gel plates. The alkaloids were determined by means of UV spectroscopy. Lovkova and Minoshedinova[53] eluted the alkaloids from silica gel plates by means of 0.05 *N* sulphuric acid, after which they were measured at 259 nm.

Several authors have described semi-quantitative TLC by comparison of spot areas[5,26,39]. Massa et al.[29] determined a number of alkaloids by means of densitometry, the quenching of UV light at 254 nm being used for the determination of nicotine. Jurzysta[54] used densitometry for the quantitative analysis of nicotine, nornicotine and anabasine in tobacco, after detection with Dragendorff's reagent.

## 6.4. TAS TECHNIQUE AND REACTION CHROMATOGRAPHY

The TAS technique for tobacco has been described by Jolliffe and Shellard[30] and Stahl and Schmitt[48]. The former authors used a temperature of 275°C for 90 sec, the drug was mixed with calcium hydroxide and silica gel with a suitable moisture content was used as propellant. The latter authors used a temperature of 220°C for 120 sec; 50 mg of molecular sieve 4Å with 20% of water was used as propellant. Wilk and Brill[22] exposed the TLC plate, after the alkaloids had been applied, to iodine vapour for 18 h. When the plate was developed a characteristic pattern of spots was observed.

Baiulescu and Constantinescu[47] applied reaction chromatography to a series of tobacco alkaloids. The reactions used were reduction, oxidation, acetylation and saponification.

A summary of the TLC analysis of nicotine and related alkaloids in various materials is given in Table 6.3-6.8.

## REFERENCES


1 K. Teichert, E. Mutschler and H. Rochelmeyer, *Dtsch. Apoth. Ztg.*, 100 (1960) 477.
2 E. Vidic and J. Schütte, *Arch. Pharm. (Weinheim)*, 295 (1962) 342.
3 J. Cochin and J.W. Daly, *Experientia*, 18 (1962) 294.
4 E. Papp and Z. Szabo, *Herba Hung.*, 2 (1963) 383.
5 O. Fejer-Kossey, *Acta Biol. Sci. Hung.*, 15 (1964) 251.
6 I. Schmeltz, R.L. Stedman, W.J. Chamberlain and D. Bubitsch, *J. Sci. Food Agr.*, 15 (1964) 774.
7 J.D. Winefordner and H.A. Moye, *Anal. Chim. Acta*, 32 (1965) 278.
8 R. Paris, R. Rousselet, M. Paris and M.J. Fries, *Ann. Pharm. Fr.*, 23 (1965) 473.
9 M. Ono and H. Asahina, *Eisei Shekenjo Hokoku*, 83 (1965) 16; *C.A.*, 65 (1966) 19151b.
10 E. Marozzi and G. Falzi, *Farmaco, Ed. Prat.*, 20 (1965) 302.
11 E. Hodgson, E. Smith and F.E. Guthric, *J. Chromatogr.*, 20 (1965) 176.

12 L.K. Turner, *J. Forensic Sci. Soc.*, 5 (1965) 94.
13 S. El Gendi, W. Kisser and G. Machata, *Mikrochim. Acta,* (1965) 120.
14 W.W. Fike, *Anal. Chem.*, 38 (1966) 1697.
15 N. Ivanov and A. Boneva, *Bulg. Tyutyum,* 11 (1966) 30; *C.A.*, 65 (1966) 8668f.
16 R.J. Martin and G. Schwartzman, *J. Ass. Offic. Anal. Chem.*, 49 (1966) 766.
17 I. Sunshine, W.W. Fike and H. Landesman, *J. Forensic Sci.*, 11 (1966) 428.
18 O. Fejer-Kossey, *J. Chromatogr.*, 31 (1967) 592.
19 A. Noirfalise and G. Mees, *J. Chromatogr.*, 31 (1967) 594.
20 E.A. Gryaznova and L.A. Podkolzina, *Aktual. Vop. Farm.*, (1968; pub. 1970) 95; *C.A.*, 76 (1972) 56234r.
21 B. Davidow, N.L. Petri and B. Quame, *Amer. J. Clin. Pathol.*, 50 (1968) 714.
22 M. Wilk and U. Brill, *Arch. Pharm. (Weinheim)*, 301 (1968) 282.
23 K.C. Güven and B. Tekinalp, *Eczacilik Bul.*, 10 (1968) 111; *C.A.*, 70 (1969) 6575h.
24 H.P. Harke, B. Frahm and C. Schultz, *Z. Anal. Chem.*, 244 (1968) 119.
25 T. Constantinescu, *Ind. Aliment. (Bucharest)*, 20 (1969) 379; *C.A.*, 72 (1970) 18994y.
26 M.H. Hashmi, S. Parveen and N.A. Chughtai, *Mikrochim. Acta,* (1969) 449.
27 M.Ya. Lovkova and N.S. Minozhedinova, *Prikl. Biokhim. Mikrobiol.*, 5 (1969) 487; *C.A.*, 71 (1969) 105239d.
28 S. Goeneckea and W. Bernhard, *Z. Anal. Chem.*, 246 (1969) 130.
29 V. Massa, F. Gal and P. Susplugas, *Int. Symp. Chromatogr. Electrophor. Lect. Pap. 6th,* (1970) 470.
30 G.H. Jolliffe and E.J. Shellard, *J. Chromatogr.*, 48 (1970) 125.
31 D.J. Berry and J. Grove, *J. Chromatogr.*, 61 (1971) 111.
32 G.S. Tadjer, *J. Chromatogr.*, 63 (1971) D44.
33 A. Viala and M. Estadieu, *J. Chromatogr.*, 72 (1972) 127.
34 K.F. Ahrend and D. Tiess, *Zbl. Pharm.*, 111 (1972) 933.
35 E. Novakova and J. Vecerkova, *Cesk. Farm.*, 22 (1973) 347.
36 S.Y. Yeh, *J. Pharm. Sci.*, 62 (1973) 1827.
37 P.L. Wu and W.R. Sharp, *Ohio J. Sci.*, 73 (1973) 353.
38 K.F. Ahrend and D. Tiess, *Wiss. Z. Univ. Rostock, Math. Naturw. Reihe,* 22 (1973) 951.
39 L. Farkas-Riedel, *Acta Agron. Acad. Sci. Hung.*, 23 (1974) 11.
40 F. Schmidt, *Dtsch. Apoth. Ztg.*, 114 (1974) 1593.
41 P.D. Swaim, V.M. Loyola, H.D. Harlan and M.J. Carlo, *J. Chem. Educ.*, 51 (1974) 331.
42 R.M. Navari, *J. Chem. Educ.*, 51 (1974) 748.
43 A.C. Moffat and K.W. Smalldon, *J. Chromatogr.*, 90 (1974) 1,9.
44 A.C. Moffat and B. Clare, *J. Pharm. Pharmacol.*, 26 (1974) 665.
45 J. Breiter, *Kontakte,* 3 (1974) 17.
46 R.J. Armstrong, *N. Z. J. Sci.*, 17 (1974) 15.
47 G.E. Baiulescu and T. Constantinescu, *Anal. Chem.*, 47 (1975) 2156.
48 E. Stahl and W. Schmitt, *Arch. Pharm. (Weinheim)*, 308 (1975) 570.
49 T.M. Holdstock and H.M. Stevens, *Forensic Sci.*, 6 (1975) 187.
50 J.A. Vinson and J.E. Hooyman, *J. Chromatogr.*, 105 (1975) 415.
51 A.C. Moffat, *J. Chromatogr.*, 110 (1975) 341.
52 J.A. Vinson, J.E. Hooyman and C.E. Ward, *J. Forensic Sci.*, 20 (1975) 552.
53 M.Ya. Lovkova and N.S. Minozhedinova, *Metody Sovrem Biokhem.*, (1975) 107; *C.A.*, 83 (1975) 203397w.
54 A. Jurzysta, *Pamiet. Pulawski,* 62 (1975) 159.
55 H.P. Harke, A. Mauch and B. Frahm, *Z. Anal. Chem.*, 274 (1975) 300.
56 A.N. Masoud, *J. Pharm. Sci.*, 65 (1976) 1585.
57 H. Kroeger, G. Bohn and G. Ruecker, *Dtsch. Apoth. Ztg.*, 117 (1977) 1923.
58 A.N. Masoud, *J. Chromatogr.*, 141 (1977) D9.
59 J. Breiter and R. Helger, *Med. Lab.*, 30 (1977) 149.

TABLE 6.1.

TLC SEPARATION OF NICOTINE AND RELATED ALKALOIDS

TLC systems:

| | | |
|---|---|---|
| S1 | Silica gel G, impregnated with 0.5 *M* KOH, activated | Chloroform-96% ethanol(9:1)[1] |
| S2 | Silica gel G, impregnated with 0.5 *M* KOH, activated | Toluene-methanol-chloroform(9:3:1)[4] |
| S3 | Silica gel G, activated | Chloroform-methanol-ammonia(60:10:1)[11] |
| S4* | Silica gel G, activated | Chloroform-methanol-acetic acid(60:10:1)[11] |
| S5 | Silica gel G, activated | Chloroform-methanol(100:20)[18] |
| S6* | Silica gel G, activated | Chloroform-diethyl ether-tetrahydrofuran(80:15:5)[18] |
| S7 | Silica gel $GF_{254}$, non-activated | Toluene-acetone-methanol-25% ammonia(4:4.5:1:0.5)[47] |
| S8 | Aluminium oxide G, activated | Chloroform-methanol(100:1.5)[7] |

*Solvents S4 and S6 are the second solvents used in two-dimensional chromatography in combination with solvents S3 and S5, respectively.

| Alkaloid | $hR_F$ values | | | | | | | |
|---|---|---|---|---|---|---|---|---|
| | S1 | S2 | S3 | S4 | S5 | S6 | S7 | S8 |
| Nicotine | 44 | 57 | 77 | 8 | 73 | 14 | 68 | 80 |
| Nicotine N-oxide | | | 8 | 5 | 6 | 2 | 5 | |
| Nicotine di-N-oxide | | | | | | | 0 | |
| Nornicotine | | 30 | 34 | 5 | 27 | 5 | 25 | 26 |
| Nicotone | | | | | 74 | 28 | | |
| Nicotyrine | | | 87 | 92 | 81 | 76 | | |
| Cotinine | | | 75 | 76 | | | 54 | |
| Norcotinine | | | 50 | 51 | | | | |
| Anabasine | 16 | 44 | 50 | 6 | 44 | 7 | 51 | 48 |
| Anatabine | | | | | 57 | 7 | | |
| Myosmine | | | | | 74 | 23 | | |
| α,β'-Dipyridyl | | | | | 73 | 40 | | |
| *m*-Nicotine | | | | | 14 | 4 | | |

TABLE 6.2

COLOURS OBTAINED WITH DRAGENDORFF'S REAGENT AND KÖNIG'S REACTION FOR SOME TOBACCO ALKALOIDS[5,18]

| Alkaloid | Colour | | |
|---|---|---|---|
| | Dragendorff[18] | Benzidine+BrCN[18] (no. 66c) | Aniline+BrCN[5] (no. 66b) |
| Nicotine-N-oxide | Red | Raspberry red | |
| *m*-Nicotine | Purple-red | Brownish purple | |
| Nornicotine | Red | Purple | Ochre |
| Anabasine | Red | Carrot | Pink |
| Anatabine | Bright red | Pink | |
| Nicotine | Red | Orange | Citrine |
| Myosmine | Red | Pale yellow | |
| Nicotone | Purple-red | Cyclamen violet | |
| α,β'-Dipyridyl | Red | Pale yellow | |
| Nicotyrine | Red | Cherry | Orange |

TABLE 6.3.

LITERATURE CITED IN CHAPTER 3 WHICH INCLUDES THE ANALYSIS OF NICOTINE AND RELATED ALKALOIDS

| Alkaloid* | Ref. | Alkaloid* | Ref. |
|---|---|---|---|
| N | 2 | N | 43 |
| N | 13 | N | 44 |
| N | 14 | N | 45 |
| N | 17 | N | 46 |
| N | 19 | N | 51 |
| N | 32 | N | 56 |
| N | 34 | N | 58 |
| N | 35 | N | 59 |
| N, norN | 38 | Tobacco | 8 |
| N | 40 | | |

*Abbreviations used in Tables 6.3-6.8:

| | | | |
|---|---|---|---|
| N | nicotine | cot | cotinine |
| norN | nornicotine | norcot | norcotinine |
| An | anabasine | cot N-ox | cotinine n-oxide |
| N N-ox | nicotine N-oxide | | |
| N diN-ox | nicotine di-N-oxide | | |

TABLE 6.4.

TLC ANALYSIS NICOTINE AND RELATED ALKALOIDS IN PLANT MATERIAL (INCLUDING TOBACCO)

| Alkaloid* | Aim | Adsorbent | Solvent system | Ref. |
|---|---|---|---|---|
| N,An,norN | Identification in *Nicotiana tabacum* (Table 6.1) | 0.5 *M* KOH-impregnated $SiO_2$ | Toluene-MeOH-$CHCl_3$(9:3:1) | 4 |
| N,norN,An,nicotyrine | Identification and semi-quantitative analysis in *Nicotiana* species (Table 6.2) | 0.5 *M* KOH-impregnated $SiO_2$ | $CHCl_3$-EtOH(9:1) | 5 |
| N,norN,myosmine, pyridine,α-picoline,2,6-lutidine, β- and γ-picoline, 3-ethylpyridine, 3-vinylpyridine | Identification in tobacco smoke | $SiO_2$ | $CHCl_3$-95% EtOH(9:1)<br>95% EtOH-0.2 *M* acetate buffer (pH 5.6) (1:1) | 6 |
| N,norN,An | Indirect quantitative analysis in tobacco (Table 6.1) | $Al_2O_3$ | $CHCl_3$-MeOH(100:1.5) | 7 |
| N,norN,nicotyrine, nicotinic acid | Determination of norN in N | | No details available | 15 |
| N,*m*-N,norN,An, N N-ox,anatabine, myosmine,nicotone, nicotyrine,α,β'-dipyridyl | Separation of tobacco alkaloids (Table 6.2) | $SiO_2$ | I. $CHCl_3$-MeOH(100:20)<br>II. $CHCl_3$-$Et_2O$-THF(80:15:5)<br>Two-dimensional: I,II | 18 |
| N | Identification | $SiO_2$ | BuOH-AcOH-$H_2O$(3:1:1) | 23 |
| N,norN,An | Identification in tobacco, smoke and other materials, indirect quantitative analysis (UV) | $SiO_2$ | Toluene-$Me_2CO$-MeOH-25% $NH_4OH$(4:4.5:1:0.5) | 25 |
| N,hydroxyN,An, nicotyrine | Identification in tobacco seedlings | $SiO_2$ | $CHCl_3$-EtOH(9:1) | 27 |
| N | TAS technique for tobacco | $SiO_2$ | $CHCl_3$-$Me_2CO$-DEA(5:4:1)<br>$CHCl_3$-DEA(9:1)<br>Cyclohexane-$CHCl_3$-DEA (5:4:1)<br>Cyclohexane-DEA(9:1)<br>Benzene-EtOH-DEA(7:2:1) | 30 |
| N | Determination in tobacco seedlings | $SiO_2$ | I. 80% EtOH-1 *M* HCl(50:2)<br>II. MeOH<br>Two-dimensional: I,II | 37 |
| N,norN,An,N N-ox, nicotyrine,anatabine,α,β'-dipyridyl | Separation of alkaloids in *Nicotiana tabacum* | $SiO_2$ | $CHCl_3$-EtOH-25% $NH_4OH$ (90:10:2),followed by $CHCl_3$-EtOH-AcOH(90:10:2) in the same direction<br>$CHCl_3$-EtOH-FMA(90:10:4), followed by $CHCl_3$-EtOH-AcOH(90:10:2) in the same direction | 39 |
| N | TAS technique for tobacco | $SiO_2$ | $Me_2CO$-$Et_2O$-25% $NH_4OH$ (50:50:3) | 48 |
| N | Separation, indirect quantitative analysis | $SiO_2$ | $CHCl_3$-EtOH(9:1) | 53 |
| N,norN,An | Densitometric analysis in tobacco | 0.5 *M* KOH impregnated $SiO_2$ | $CHCl_3$-MeOH(5:1) | 54 |

*For abbreviations, see footnote to Table 6.3.

TABLE 6.5

NICOTINE AND RELATED ALKALOIDS IN DRUGS OF ABUSE ANALYSIS (SEE REFERENCES CHAPTER 12)

(Some studies of special interest are summarized in Table 6.6)

Numbers of references in Chapter 12:

| | | | | |
|---|---|---|---|---|
| 8 | 141 | 175 | 281 | 369 |
| 22 | 159 | 187 | 289 | 374 |
| 34 | 167 | 202 | 295 | 380 |
| 60 | 168 | 207 | 325 | |
| 68 | 169 | 243 | 328 | |
| 97 | 172 | 244 | 332 | |
| 98 | 174 | 261 | 356 | |

TABLE 6.6.

NICOTINE AND RELATED ALKALOIDS IN DRUGS OF ABUSE ANALYSIS BY TLC

| Alkaloid* | Other compounds | Aim | Adsorbent | Solvent system | Ref. |
|---|---|---|---|---|---|
| N | Opium alkaloids, mescaline, cocaine | Identification in urine (Table 12.11, p.281) | $SiO_2$ | EtOH-pyridine-dioxane-$H_2O$ (50:20:25:5)<br>EtOH-AcOH-$H_2O$(6:3:1)<br>EtOH-dioxane-benzene-$NH_4OH$ (5:40:50:5)<br>MeOH-*n*-BuOH-benzene-$H_2O$ (60:15:10:15) | |
| | | | $Al_2O_3$ | *n*-BuOH-*n*-$Bu_2O$-AcOH(4:5:1)<br>*n*-BuOH-*n*-$Bu_2O$-$NH_4OH$(25:70:5) | 3 |
| N | Morphine, heroin | Interference of N in analysis of narcotic drugs | | No details available | 9 |
| N | Opium alkaloids, amphetamines, stimulants, tranquillizers, antihistamines, analgesics, various alkaloids | Identification in urine (Table 12.12, p.282) | $SiO_2$ | EtOAc-MeOH-25% $NH_4OH$ (85:10:5) | 21 |
| N | Morphine | Interference N in morphine analysis in urine | $SiO_2$ | MeOH-25% $NH_4OH$(25:0.6)<br>Benzene-MeOH-$H_2O$(5:60:5)<br>I. $CHCl_3$-MeOH-25% $NH_4OH$ (60:10:1)<br>II. $CHCl_3$-MeOH-AcOH (60:10:1)<br>V. MeOH<br>EtOH-AcOH-$H_2O$(6:3:1)<br>EtOH-pyridine-dioxane-$H_2O$(50:20:25:5)<br>EtOH-dioxane-benzene-25% $NH_4OH$(5:40:50:5)<br>MeOH-*n*-BuOH-benzene-$H_2O$(60:15:10:15)<br>MeOH-$Me_2CO$-DEA(10:10:0.1)<br>III. *n*-BuOH-*n*-$Bu_2O$-AcOH (4:5:1)<br>IV. *n*-BuOH-*n*-$Bu_2O$-25% $NH_4OH$ (25:70:5)<br>$CHCl_3$-96% EtOH(9:1,8:2)<br>Two-dimensional: I,II or III,IV or V,V | 28 |
| N,cot | Opium alkaloids amphetamines, barbiturates, phenothiazines | Identification in urine | $SiO_2$ | Benzene-dioxane-EtOH-$NH_4OH$ (50:40:5)<br>MeOH- 12 *M* $NH_4OH$(100:1.5), in unsaturated tanks | 31 |
| N,cot | Opium alkaloids, psychotropic drugs, amphetamines, benzodiazepines | Identification of drugs of abuse | $SiO_2$ | I. MeOH-$CHCl_3$-$NH_4OH$<br>II. $Et_2O$-$Me_2CO$-DEA(85:8:7)<br>Two-dimensional: I,II | 33 |

TABLE 6.6. (continued)

| Alkaloid* | Other compounds | Aim | Adsorbent | Solvent system | Ref. |
|---|---|---|---|---|---|
| N | Morphine and its metabolites | Identification morphine metabolites in biological material (Table 12.16, p.288) | $SiO_2$ | EtOAc-MeOH-25% $NH_4OH$<br>*n*-BuOH-AcOH-$H_2O$(35:3:10)<br>*n*-BuOH-*n*-$Bu_2O$-$NH_4OH$ (25:70:2) | 36 |
| N | Opium alkaloids, amphetamines, tranquillizers various alkaloids and basic drugs | Recovery of basic components after detection with iodoplatinate | $SiO_2$<br>5% $NaH_2$ citrate-impregnated cell<br>$Al_2O_3$ | MeOH-25% $NH_4OH$(100:1.5)<br>*n*-BuOH-$H_2O$(43:5:6.5)+ 0.48 g citric acid<br>MeOH-25% $NH_4OH$(100:1.5)<br>MeOAc-25% $NH_4OH$-$H_2O$ (100:1.5:50) upper phase | 49 |

*For abbreviations, see footnote to Table 6.3.

TABLE 6.7

NICOTINE AND RELATED ALKALOIDS IN TOXICOLOGICAL TLC ANALYSIS AND THEIR TLC ANALYSIS IN BIOLOGICAL MATERIALS

| Alkaloid* | Other compounds | Aim | Adsorbent | Solvent system | Ref. |
|---|---|---|---|---|---|
| N | Quinine, reserpine, various drugs | Identification | $SiO_2$ | $CHCl_3$-$Me_2CO$(9:1)<br>MeOH-$n$-BuOH(6:4)<br>MeOH-$NH_4OH$(100:1.5) | 10 |
| N, norN, nicotyrine, An, N N-ox, cot, norcot | | Identification in insects (Table 6.1) | $SiO_2$ | I. $CHCl_3$-MeOH-$NH_4OH$ (60:10:1)<br>II. $CHCl_3$-MeOH-AcOH (60:10:1)<br>Two-dimensional: I,II | 11 |
| N | Opium alkaloids, strychnine, barbiturates, tranquillizers | Identification | $SiO_2$+Celite 545(1:1) | Methyl isobutyl ketone-AcOH-$H_2O$(20:10:5) | 12 |
| N | | Identification in food | $SiO_2$ | $CHCl_3$-96% EtOH(8:2) | 16 |
| N, An | | Identification in biological materials | $SiO_2$ | Benzene-EtOH(95:5) | 20 |
| N, cot | | Indirect quantitative analysis (colorimetric) in animal tissues | $SiO_2$ | Benzene-MeOH-25% $NH_4OH$ (82:18:0.8) | 24 |
| N | | Identification in urine or blood | $SiO_2$ | EtOAc-EtOH-$NH_4OH$(85:10:10) | 42 |
| N, N N-ox, cot, norN, cot N-ox, norcot | | Indirect quantitative analysis (colorimetric) in urine | $SiO_2$ | Toluene-isoprOH-$NH_4OH$ (100:25:1), develop 3x | 55 |
| N | Opium alkaloids, various alkaloids | Identification | $SiO_2$ | MeOH-12 $M$ $NH_4OH$(100:1.5) | 57 |

*For abbreviations, see footnote to Table 6.3.

TABLE 6.8

TLC ANALYSIS OF NICOTINE AND RELATED ALKALOIDS IN COMBINATION WITH OTHER (PURE) COMPOUNDS

| Alkaloid* | Other compounds | Aim | Adsorbent | Solvent system | Ref. |
|---|---|---|---|---|---|
| N,An | Sparteine, coniine, arecoline | Separation (Table 6.1) | 0.5 *M* KOH-impregnated $SiO_2$ | $CHCl_3$-96% EtOH(11:1,9:1 and 8:2) | 1 |
| N | Various alkaloids | Reaction chromatography | $SiO_2$ | Benzene-MeOH-$Me_2CO$-AcOH (70:20:5:5) | 22 |
| N | Various alkaloids | Semi-quantitative circular TLC | $SiO_2$ | $CHCl_3$-MeOH(9:1)<br>$CHCl_3$-EtOH(85:15) | 26 |
| N | Caffeine, quinine, brucine, strychnine | Separation on plates prepared by spraying adsorbent | $SiO_2$ | Benzene-EtOAc-DEA(10:10:3) | 41 |
| N,cot, norN,An, N N-ox, N diN-ox | Opium and tropane alkaloids, lysergic acid | Reaction chromatography (Table 6.1) | $SiO_2$ | Toluene-$Me_2CO$-MeOH-25% $NH_4OH$(4:4.5:1:0.5) | 47 |

*For abbreviations, see footnote to Table 6.3.

# II.2. TROPANE ALKALOIDS

## Chapter 7

## TROPINE ALKALOIDS

The tropane alkaloids can be divided into two groups, tropine and pseudotropine alkaloids. Some alkaloids from the plant family *Solanaceae* belong to the former group, whereas the *Erythroxylon* or Coca alkaloids belong to the latter (see Chapter 8).

### 7.1. SOLVENT SYSTEMS*

Several reviews on the TLC analysis of tropine alkaloids have been published[35,74,102,109]. Stahl and Schorn[35] compared a number of TLC systems for the separation of the alkaloids present in Solanaceae crude drugs as well as apoatropine and tropine. The solvent system S3 (Table 7.1), first described by Büchi and Zimmerman[23], was found to be the best. Polesuk and Ma[74] investigated a series of TLC systems and concluded that acetone - ammonia in combination with silica gel plates was the most versatile system. Using the ratio 97:3 (S15) (Table 7.2) the alkaloids atropine and scopolamine and their degradation products could be separated. To separate tropine and pseudotropine the ratio 8:2 was preferred. An analogous solvent system, S4 (Table 7.1), originally proposed by Oswald and Flück[13] but later modified in several ways, has been extensively used. Puech et al.[95] used solvent system S14 (Table 7.2), a variation of system S12[3] with diethylamine omitted, permitting more sensitive detection.

The degradation products of atropine, scopolamine and homatropine have been analysed in a series of TLC systems[13,16,17,27,33,61,95,104,147,149,153,154]. For the acidic moiety obtained by hydrolysis, acidic solvent systems have been preferred[33,104,149]. Puech and Dupy[104] used chloroform - acetic acid (9:1) and silica gel plates for atropine and its degradation products, and Goeber et al.[149] described a series of systems for the detection of degradation products of scopolamine in ophthalmic solutions.

Atropine and homatropine are difficult to separate. In TLC system S1 the compounds can be separated, whereas in systems S8 and S11 only a partly separation

---

*Because atropine and *l*-hyoscyamine behave similarly in all TLC systems, in the text and tables atropine will be used throughout to describe both alkaloids.

**References p. 96**

is observed (Table 7.1). In solvent system S11, with components in the ratio 45:35:20, the separation of atropine and homatropine is improved. Dijkhuis[59] analysed impurities in eyedrops, including atropine, scopolamine and homatropine (see Table 20.2, p.480). Cyclohexane - chloroform - diethylamine (3:7:1) separated atropine and homatropine on silica gel plates ($hR_F$ 58 and 65, respectively). Eichhorn and Kny[107] separated atropine and homatropine on silica gel plates with methyl ethyl ketone - methanol - 6 *M* ammonia (6:3:1).

The TLC analysis of quaternary compounds, including butylscopolamine and methylscopolamine, by means of ion-pair chromatography has been performed by de Zeeuw et al.[128]. Giebelmann et al.[139] separated quaternary compounds by means of polar acidic solvent systems on silica gel plates. Both methods are dealt with in more detail in the chapter on Quaternary Compounds. Giebelmann[151] used several two-dimensional TLC systems for the identification of some tropane alkaloids and their quaternary analogues (Table 7.2).

Gröningson and Schill[49] studied ion-pair chromatography of alkaloids on cellulose plates impregnated with 0.7 *M* sulphuric acid and either 0.7 *M* sodium chloride, 0.7 *M* sodium bromide, 0.7 *M* sodium perchlorate or 0.7 *M* potassium thiocyanate, using 1-pentanol as the mobile phase. Some tropine alkaloids and their quaternary derivatives were included in the investigations. Jankulov[114] applied the method to the analysis of tropine alkaloids in plant material, but preferred butanol to 1-pentanol. Röder et al.[43] used azeotropic solvent systems for the analysis of alkaloids, including tropine alkaloids. Stahl and Dumont[54] studied the separation of some tropine alkaloids on TLC-plates with a traverse pH-gradient. The optimum pH for obtaining a separation of the alkaloids could be established using this method. In chloroform - methanol the optimum pH ranges were 0.1 - 1 and 7 - 8.8.

Affonso[29] used plaster of Paris as the stationary phase and recommended it particularly for preparative work. Hsiu et al.[131] described the use of polyamide as a stationary phase for TLC alkaloid separations. With two solvent systems (S5 and S7 in Table 3.8, p.32) atropine and homatropine could be separated. Lepri et al.[133,135] studied the separation of alkaloids on thin layers of ion exchangers; on Rexyn 102($H^+$) and with 1 *M* acetic acid + hydrochloric acid (pH = 1 or 0.3), atropine and homatropine could be separated (Table 3.17, p.43).

Most of the more than 100 TLC systems described in the literature for the separation of tropine alkaloids were tested in our laboratories. Eleven of the TLC systems which were found to give best results are listed in Table 7.1, which also gives the $hR_F$ values for some of the tropine alkaloids.

## 7.2. DETECTION

Because of the low molar absorptivity of the tropine alkaloids at 254 nm, UV detection on fluorescent plates is insensitive. The sensitivities of some modifications of Dragendorff's reagent were investigated by Puech et al.[95]. The best results were obtained with the modification according to Munier (no. 39b), followed by spraying with 10% sodium nitrite solution, which improved the contrast of the spots. This procedure gave different colours for atropine and *l*-hyoscyamine. Diethylamine in the solvent system interfered with the detection and led to decreased sensitivity, even after heating the plates. Solvents containing ammonia instead of diethylamine were therefore preferred. In Table 7.3 and 7.4 the colours obtained for a number of tropane alkaloids with Dragendorff's reagent and iodoplatinate reagent, respectively, are summarized. Sita et al.[101] sprayed a sodium thiosulphate solution after the iodoplatinate reagent to improve the sensitivity.

Porges[97] detected tropine alkaloids by exposure of the TLC plates to iodine vapour, followed by spraying with water. Different colours were obtained for atropine and *l*-hyoscyamine (grey-blue and rust brown, respectively). The difference in colour could be used to estimate the percentage of atropine in *l*-hyoscyamine; *l*-hyoscyamine containing 20% of atropine could be distinguished from pure *l*-hyoscyamine.

Schmidt[108] found that atropine and homatropine could be distinguished by spraying first with 2% iron(III) chloride solution followed by 0.1 *N* iodine solution (no. 60). Atropine gave a grey-brown and homatropine a brown-violet colour. Egli[98] observed that atropine and scopolamine gave different colours, violet and blue-green, respectively, if the plates, impregnated with phenothiazine, were exposed to bromine vapour (no. 78). Polesuk and Ma[84] compared some detection methods for tropine alkaloids.

We tested the sensitivity of some detection methods for tropine alkaloids and the results are summarized in Table 7.5.

Sarsunova et al.[129] studied the influence of detection with iodine on the quantitative analysis of the alkaloids detected after elution from the plates. Vinson and co-workers[120,123] reported the detection of different drugs, including atropine, with TCBI reagent (no. 94) (see Table 2.3, p.17) and Rücker and Taha[145] with π-acceptors (see Table 2.2, p.16)(no. 19). Menn and McBain[31] developed a detection method for cholinesterase inhibitors, including atropine (see Chapter 2, p.15).

The non-nitrogenous degradation products of tropine alkaloids (tropic acid and mandelic acid) can be distinguished from the alkaloids by spraying with pH-indicators. Polesuk and Ma[84] used bromophenol blue, methyl red and methyl orange for this purpose.

## 7.3. QUANTITATIVE ANALYSIS

In the indirect quantitative analysis of tropine alkaloids several solvents have been used to elute the alkaloids from the adsorbent. Most commonly used with silica gel is chloroform[65,83,93]. Van Kessel[75] and Dijkhuis[59] also used chloroform, but extracted an alkalinized suspension of the sorbent in dilute ammonia and 4 *M* sodium hydroxide, respectively. Regdon et al.[33] used ethanol, and Baiulescu and Constantinescu[116] used methanol - water (7:3). Mineral acids have also been used for the extraction of the alkaloids: 0.1 *M* hydrochloric acid[53,154] and 86% nitric acid[44]. Karawya et al.[119] eluted the alkaloids with diluted hydrochloric acid, from which the bases were extracted with chloroform after basification.

Ion-pair extraction has also been used to elute the alkaloids from silica gel. After addition of tropeoline 00 in McIlvaine buffer (pH 3.6), the alkaloids were extracted with chloroform[19]. Extraction first with buffered bromothymol blue solution (pH 7.4) and then with ethanol-free chloroform was used by Adamski et al.[36].

Ikram and Bakhsh[9] found that aluminium oxide was better extracted with chloroform than with ethanol. Bican-Fister[83] also used chloroform in connection with aluminium oxide plates. From cellulose plates tropine alkaloids have been eluted as Dragendorff's complex with acetone[37], and ethanol has been used to elute atropine from plaster of Paris[29].

The direct quantitative analysis of tropine alkaloids by measuring spot areas after spraying with Dragendorff's reagent (Munier and Macheboeuf modification) was used by Oswald and Flück[13,16] in connection with TLC system S4. The optimal range for quantitative analysis was 30-40 μg of alkaloid. Büchi and Zimmerman[23] proposed acetone - 10% ammonia (95:5) for the analysis of tropine alkaloids. They were able to analyse amounts of 1-10 μg of alkaloid after spraying with Dragendorff's reagent. Ebel et al.[80] also determined alkaloids, including atropine and homatropine, by measuring spot areas. Garbor[90] used TLC system S1 and planimetric analysis of the alkaloid spots.

Densitometric analysis after spraying with Dragendorff's reagent has been used in several investigations[52,57,60,72] at wavelengths of 530 nm[52,60], 490 nm[57] and 400 nm[72]. Gros-Leban and Debelmas[96] performed densitometry at a wavelength of 400 nm after detection with iodine vapour.

Messerschmidt[50] described a direct fluorimetric method for scopolamine after treatment with concentrated sulphuric acid, which was included in the mobile phase.

## 7.4. TAS TECHNIQUE AND REACTION CHROMATOGRAPHY

Jolliffe and Shellard[73] applied the TAS technique to tropine alkaloids containing crude drugs. An amount of 10 - 20 mg of the crude drug was mixed with

10 mg of calcium hydroxide and heated for 90 sec at 275$^{o}$C, silica gel of suitable moisture content being used as propellant. Stahl and Schmitt[117] established the optimal TAS conditions for a number of alkaloids. The tropine alkaloids proved to be thermolabile. Even at 150$^{o}$C a considerable amount of apoalkaloids was formed and at 250$^{o}$C also tropidine. For crude drugs containing tropine alkaloids (*Atropa, Hyoscyamus* and *Datura* species) the optimal conditions were sample size 10 mg, heating at 220$^{o}$C for 120 sec and 50 mg of molecular sieve 4Å with 20% of water as propellant.

Reaction chromatography has been used for identification purposes. Kaess and Mathis[17,27] described the technique for tropine alkaloids. Several reactions were used. Saponification with 0.1 *M* potassium hydroxide in ethanol at 95$^{o}$C in a capillary tube gave tropine, scopoline and tropic acid. Dehydration by heating the plate with the spotted alkaloids at 120$^{o}$C for 1 h led to the formation of the apo-alkaloids. Oxidation with a 2% solution of *p*-nitroperbenzoic acid in diethyl ether or chloroform in a capillary tube at 100$^{o}$C for 1 h yielded the N-oxides. The reactions mentioned can also be applied to mixtures of tropine alkaloids or tropine alkaloids in plant materials or extracts after removal of chlorophyll.

Wilk and Brill[42] placed silica gel plates with the applied alkaloids for 18 h in iodine vapour. The alkaloids tested, including atropine and scopolamine, gave characteristic patterns after development of the plates. Baiulescu and Constantinescu[116] used reaction chromatography to identify alkaloids. For pure compounds or alkaloid mixtures which can be separated sufficiently, the reaction can be carried out on the plate. Mixtures are first separated on the plate in one direction, then the reactions are carried out and the plates are developed a second time in the direction perpendicular on the first. In complex mixtures the compounds are collected from the plate after TLC separation, then the reactions are carried out, followed by a new TLC analysis. The reactions used are saponification with 10% sodium hydroxide at 110$^{o}$C, oxidation with 30% hydrogen peroxide at 60-70$^{o}$C, acetylation with acetic acid anhydride in pyridine (1:2) at 110$^{o}$C and reduction with 10% acetic acid - 10% hydrochloric acid (1:1) with zinc powder at 110$^{o}$C. The last reaction is slow and incomplete. The alkaloids tested yielded different characteristic patterns of reaction products.

Polesuk and Ma[84] used reaction chromatography to identify tropinon. Tropinon was reduced on the TLC plate with the aid of sodium borohydride, yielding tropine and pseudotropine.

A summary of the TLC analysis of tropine alkaloids in various materials is given in Tables 7.6-7.11.

REFERENCES

1 E. Stahl, G. Schröter, G. Kraft and R. Renz, *Pharmazie*, 11 (1956) 633.
2 K. Teichert, E. Mutschler and H. Rochelmeyer, *Dtsch. Apoth. Ztg.*, 100 (1960) 477.
3 D. Waldi, K. Schnackerz and F. Munster, *J. Chromatogr.*, 6 (1961) 61.
4 J. Baumler and S. Rippstein, *Pharm. Acta Helv.*, 36 (1961) 382.
5 E. Vidic and J. Schütte, *Arch. Pharm. (Weinheim)*, 295 (1962) 342.
6 A. Vegh, R. Budvari, G. Szasz, A. Brantner and P. Gracza, *Acta Pharm. Hung.*, 33 (1963) 67.
7 K. Takahashi, S. Mizumachi and H. Asahina, *Eisei Shikenjo Kenkyu Hokoku*, 81 (1963) 23; *C.A.*, 62 (1965) 8936c.
8 W. Kamp, W.J.M. Onderberg and W.A. van Seters, *Pharm. Wkbl.*, 98 (1963) 993.
9 M. Ikram and M.K. Bakhsh, *Anal. Chem.*, 36 (1964) 111.
10 J. Schnekenburger and I. Hartikainen, *Dtsch. Apoth. Ztg.*, 104 (1964) 1402.
11 I.K. Kaukulov, *Farmatsiya (Sofia)*, 14 (1964) 57; *C.A.*, 62 (1965) 14988g.
12 D. Neumann and H.B. Schröter, *J. Chromatogr.*, 16 (1964) 414.
13 N. Oswald and H. Flück, *Pharm. Acta Helv.*, 39 (1964) 293.
14 J. Zarnack and S. Pfeifer, *Pharmazie*, 19 (1964) 216.
15 V. Schwarz and M. Sarsunova, *Pharmazie*, 19 (1964) 267.
16 N. Oswald and H. Flück, *Sci. Pharm.*, 32 (1964) 136.
17 A. Kaess and C. Mathis, *Ann. Pharm. Fr.*, 23 (1965) 267.
18 R. Paris, R. Rousselet, M. Paris and M.J. Fries, *Ann. Pharm. Fr.*, 23 (1965) 473.
19 R. Zielinska-Sowicka and E. Wojcik, *Diss. Pharm.*, 17 (1965) 555; *C.A.*, 64 (1966) 17355g.
20 K.C. Güven and A. Himcal, *Istanbul Univ. Eczacilik Fak. Mecmuassi*, 1 (1965) 153; *C.A.*, 65 (1966) 8668d.
21 J. Büchi and A. Zimmerman, *Pharm. Acta Helv.*, 40 (1965) 292.
22 J. Büchi and A. Zimmerman, *Pharm. Acta Helv.*, 40 (1965) 361.
23 J. Büchi and A. Zimmerman, *Pharm. Acta Helv.*, 40 (1965) 395.
24 A. Haznagy, K. Szendrei and L. Toth, *Pharmazie*, 20 (1965) 651.
25 D. Vagujvalvi, *Planta Medica*, 13 (1965) 79.
26 W.W. Fike, *Anal. Chem.*, 38 (1966) 1697.
27 A. Kaess and C. Mathis, *Int. Symp. Chromatogr. Electrophor. Lect. Pap. 4th*, (1966) 525.
28 G.J. Dickes, *J. Ass. Public Anal.*, 4 (1966) 45.
29 A. Affonso, *J. Chromatogr.*, 21 (1966) 332.
30 I. Sunshine, W. Fike and H. Landesman, *J. Forensic Sci.*, 11 (1966) 428.
31 J.J. Menn and J.B. McBain, *Nature (London)*, 209 (1966) 1351.
32 F. Wartmann-Hafner, *Pharm. Acta Helv.*, 41 (1966) 406.
33 G. Regdon, B. Selmeczi and G. Kedvessy, *Pharm. Zentralh.*, 105 (1966) 658.
34 I. Juhl and V. Waarst, *Arch. Pharm. Chem.*, 74 (1967) 887.
35 E. Stahl and P.J. Schorn, *Arzneim.-Forsch.*, 17 (1967) 1288.
36 R. Adamski, J. Lutomski and J. Wisniewski, *Dtsch. Apoth. Ztg.*, 107 (1967) 185.
37 A.R. Saint-Firmin and R.R. Paris, *J. Chromatogr.*, 31 (1967) 252.
38 A. Noirfalise and G. Mees, *J. Chromatogr.*, 31 (1967) 594.
39 L.A. Chekryshkina, *Nauch. Tr. Aspir. Ordinatorov, 1-Mosk. Med. Inst.*, (1967) 160; *C.A.*, 70 (1969) 50492c.
40 B. Davidow, N.L. Petri and B. Quame, *Amer. J. Clin. Pathol.*, 50 (1968) 714.
41 P.E. Haywood and M.S. Moss, *Analyst (London)*, 93 (1968) 737.
42 M. Wilk and U. Brill, *Arch. Pharm. (Weinheim)*, 301 (1968) 282.
43 E. Röder, E. Mutschler and H. Rochelmeyer, *Arch. Pharm. (Weinheim)*, 301 (1968) 624.
44 C. Levorato, *Boll. Chim. Farm.*, 107 (1968) 574; *C.A.*, 70 (1969) 22931b.
45 L. Hörhammer, H. Wagner and J. Hölzl, *Dtsch. Apoth. Ztg.*, 108 (1968) 1616.
46 J.A. Delaey and M. Van Ooteghem, *Pharm. Tijdschr. Belg.*, 45 (1968) 241.
47 J.M.G.J. Frijns, *Pharm. Wkbl.*, 103 (1968) 929.

48 A. Puech, M. Jacob, H. Delonca and D. Gaudy, *Trav. Soc. Pharm. Montpellier*, 28 (1968) 211.
49 K. Gröningson and G. Schill, *Acta Pharm. Suecica*, 6 (1969) 447.
50 W. Messerschmidt, *Dtsch. Apoth. Ztg.*, 109 (1969) 199.
51 A. Fiebig, J. Felczak and S. Janicki, *Farm. Pol.*, 25 (1969) 971.
52 Z.P. Kostennikova and V.E. Chichiro, *Farmatsiya (Moscow)*, 18 (1969) 39; *C.A.*, 71 (1969) 128782q.
53 L.A. Chekryshkina and F.M. Shemyakin, *Farm. Zh. (Kiev)*, 24 (1969) 50; *C.A.*, 71 (1969) 53613z.
54 E. Stahl and E. Dumont, *J. Chromatogr. Sci.*, 7 (1969) 517.
55 Gy. Mozsik and E. Toth, *J. Chromatogr.*, 45 (1969) 478.
56 M. Vanhaelen, *J. Pharm. Belg.*, 24 (1969) 87.
57 B.L. Wu Chu, E.S. Mika, M.J. Solomon and F.A. Crane, *J. Pharm. Sci.*, 58 (1969) 1073.
58 M.H. Hashmi, S. Parveen and N.A. Chughai, *Mikrochim. Acta*, (1969) 449.
59 I.C. Dijkhuis, *Pharm. Wkbl.*, 104 (1969) 1317.
60 M.S. Shipalov, V.E. Chichiro and Z.P. Kostennikova, *Prikl. Biokhim. Mikrobiol.*, 5 (1969) 502; *C.A.*, 71 (1969) 116567k.
61 E. Weigert, *Rev. Fac. Farm. Bioquim. Univ. Fed. Santa Maria*, 15 (1969) 61; *C.A.*, 74 (1971) 31873u.
62 G.L. Szendey, *Z. Anal. Chem.*, 244 (1969) 257.
63 G. Szasz and G. Szasz, *Acta Pharm. Hung.*, 40 (1970) 38.
64 B. Srepel, *Acta Pharm. Jugoslav.*, 20 (1970) 99; *C.A.*, 74 (1971) 115936e.
65 B. Pekic, K. Petrovic and M. Gorunovic, *Arh. Farm.*, 19 (1970) 235; *C.A.*, 73 (1970) 18499m.
66 M.L. Bastos, G.E. Kananen, R.M. Young, J.R. Monforte and I. Sunshine, *Clin. Chem.*, 16 (1970) 931.
67 M. Overgaard-Nielsen, *Dan. Tidskr. Farm.*, 44 (1970) 7.
68 R.A. Egli, *Dtsch. Apoth. Ztg.*, 110 (1970) 987.
69 S. Enache and T. Constantinescu, *Farmacia (Bucharest)*, 18 (1970) 149; *C.A.*, 73 (1970) 38579r.
70 S. Gill, *Gdansk Tow. Nauk. Rozpr. Wydz 3*, 7 (1970) 175; *C.A.*, 75 (1971) 64040u.
71 J. Grujic-Vasic, S. Ramic and R. Popovic, *Glas. Hem. Technol. Bosne Hercegovine*, 18 (1970) 41; *C.A.*, 79 (1973) 70260q.
72 V. Massa, F. Gal and P. Susplugas, *Int. Symp. Chromatogr. Electrophor. Lect. Pap. 6th*, (1970) 470.
73 G.H. Jolliffe and E.J. Shellard, *J. Chromatogr.*, 48 (1970) 125.
74 J. Polesuk and T.S. Ma, *Mikrochim. Acta*, 4 (1970) 670.
75 J.F.E. van Kessel, *Pharm. Wkbl.*, 105 (1970) 1293.
76 G. Nagel, *Praktikantenbrief*, 16 (1970) 14.
77 G.F. Lozovaya, *Sud. Med. Ekspertiza*, 13 (1970) 31; *C.A.*, 74 (1971) 74464d.
78 F. Reimers, *Arch. Pharm. Chemi*, 78 (1971) 201.
79 N. Weissman, M.L. Lowe, J.M. Beattie and J.A. Demetriou, *Clin. Chem.*, 17 (1971) 875.
80 S. Ebel, W.D. Mikulla and K.H. Weisel, *Dtsch. Apoth. Ztg.*, 111 (1971) 931.
81 W. Debska and S. Czyszewska, *Farm. Pol.*, 27 (1971) 365; *C.A.*, 75 (1971) 80316u.
82 S.L. Kidman, *J. Ass. Public Anal.*, 9 (1971) 24.
83 T. Bican-Fister, *J. Chromatogr.*, 55 (1971) 417.
84 J. Polesuk and T.S. Ma, *J. Chromatogr.*, 57 (1971) 315.
85 I. Simon and M. Lederer, *J. Chromatogr.*, 63 (1971) 448.
86 A. Tadjer, *J. Chromatogr.*, 63 (1971) D44.
87 C.D. Padha, M.C. Nigam and P.R. Rao, *J. Inst. Chem. Calcutta*, 43 (1971) 5; *C.A.*, 75 (1971) 59560j.
88 G.F. Lozovaya, *Nauch. Tr. Irkutsk, Gos. Med. Inst.*, 113 (1971) 113; *C.A.*, 83 (1975) 54038q.
89 I. Barenze and S.A. Minina, *Rast. Resur.*, 7 (1971) 124; *C.A.*, 74 (1971) 103131k.
90 C. Garbor, *Rev. Asoc. Bioquim. Argent.*, 36 (1971) 149; *C.A.*, 76 (1972) 63173g.
91 S. Zadecky, D. Küttel and M. Takacsi, *Acta Pharm. Hung.*, 42 (1972) 7.
92 M. Gorunovic and P. Lukic, *Acta Pharm. Jugoslav.*, 22 (1972) 69; *C.A.*, 77 (1972) 79590k.

93 O.B. Genius, *Dtsch. Apoth. Ztg.*, 112 (1972) 1261.
94 M. Ono, M. Shimamine and K. Takahashi, *Eisei Shikenjo Hokoku*, 90 (1972) 73; *C.A.*, 79 (1973) 57731z.
95 A. Puech, M. Jacob and D. Gaudy, *J. Chromatogr.*, 68 (1972) 161.
96 C. Gros-Leban and A.M. Debelmas, *Plant. Med. Phytother.*, 6 (1972) 128.
97 M. Porges, *Schweiz. Apoth. Ztg.*, 110 (1972) 703.
98 R.A. Egli, *Z. Anal. Chem.*, 259 (1972) 277.
99 K.F. Ahrend and D. Tiess, *Zbl. Pharm.*, 111 (1972) 933.
100 E. Novakova and J. Vecerkova, *Cesk. Farm.*, 22 (1973) 347.
101 F. Sita, V. Chmelova and K. Chmel, *Cesk. Farm.*, 22 (1973) 234.
102 A. Mucharska, *Chromatogr. Cienkovartswowa Anal. Farm.*, (1973) 157; *C.A.*, 82 (1975) R103196t.
103 H.D. Crone and E.M. Smith, *J. Chromatogr.*, 77 (1973) 234.
104 A. Puech and J. Dupy, *J. Pharm. Belg.*, 28 (1973) 24.
105 T.A. Pletneva, I.S. Simon and Y.V. Shostenko, *Khim. Farm. Zh.*, 7 (1973) 53; *C.A.*, 80 (1974) 30740k.
106 K.F. Ahrend and D. Tiess, *Wiss. Z. Univ. Rostock Math. Naturw. Reihe*, 22 (1973) 951.
107 A. Eichhorn and L. Kny, *Zbl. Pharm.*, 112 (1973) 567.
108 F. Schmidt, *Dtsch. Apoth. Ztg.*, 114 (1974) 1593.
109 N.P. Maksyutina and E.O. Korzhavikh, *Farm. Zh. (Kiev)*, 29 (1974) 20; *C.A.*, 81 (1974) R96480c.
110 O.A. Akopyan, L.V. Vrochinskaya and L.V. Romanchenko, *Farm. Zh. (Kiev)*, 29 (1974) 57; *C.A.*, 82 (1974) 51258t.
111 A.C. Moffat and K.W. Smalldon, *J. Chromatogr.*, 90 (1974) 1, 9.
112 H. Hammerstingl and G. Reich, *J. Chromatogr.*, 101 (1974) 408.
113 R.J. Armstrong, *N. Z. J. Sci.*, 17 (1974) 15.
114 I. Jankulov, *Rastenievud Nauki*, 11 (1974) 59; *C.A.*, 81 (1974) 116705k.
115 E. Spratt, *Toxicol. Annu. 1974*, (1975) 229.
116 G.E. Baiulescu and T. Constantinescu, *Anal. Chem.*, 47 (1975) 2156.
117 E. Stahl and W. Schmitt, *Arch. Pharm. (Weinheim)*, 308 (1975) 570.
118 A.K. Chowdhury and S.A. Chowdhury, *Bangladesh Pharm. J.*, 4 (1975) 11; *C.A.*, 83 (1975) 48262c.
119 M.S. Karawya, S.M. Abdel-Wahab, M.S. Hifnawy and M.G. Ghourab, *J. Ass. Offic. Anal. Chem.*, 58 (1975) 884.
120 J.A. Vinson and J.E. Hooyman, *J. Chromatogr.*, 105 (1975) 415.
121 J.E. Wallace, H.E. Hamilton, H. Schwertner, D.E. King, J.L. McNay and K. Blum, *J. Chromatogr.*, 114 (1975) 433.
122 W.J. Serfontein, D. Botha and L.S. de Villiers, *J. Chromatogr.*, 115 (1975) 507.
123 J.A. Vinson, J.E. Hooyman and C.E. Ward, *J. Forensic Sci.*, 20 (1975) 552.
124 Y. Nunoura and S. Iwagami, *Osaka Furitsu Koshu Eisei Kenkyusho Kenkyu Hokoku Yakuji Shido Hen*, 9 (1975) 45; *C.A.*, 87 (1977) 141333u.
125 I. Hempel and H.D. Woitke, *Pharm. Prax.*, 10 (1975) 223.
126 I.R. da S. Jardim and C.S. Tavares, *Rev. Bras. Farm.*, 56 (1975) 1; *C.A.*, 84 (1976) 35383p.
127 E. Vidic and E. Klug, *Z. Rechtsmed.*, 76 (1975) 283.
128 R.A. de Zeeuw, P.E.W. van der Laan, J.E. Greving and F.J.W. Mansvelt, *Anal. Lett.*, 9 (1976) 831.
129 M. Sarsunova, B. Kakac and M. Ryska, *Cesk. Farm.*, 25 (1976) 156.
130 E.A. Korzkavykh, *Farm. Zh. (Kiev)*, (1976) 43; *C.A.*, 85 (1976) 166692g.
131 H.C. Hsiu, J.T. Huang, T.B. Shih, K.L. Yang, K.T. Wang and A.L. Lin, *J. Chin. Chem. Soc.*, 14 (1976) 161.
132 R. Aigner, H. Spitzy and R.W. Frei, *J. Chromatogr. Sci.*, 14 (1976) 381.
133 L. Lepri, P.G. Desideri and M. Lepori, *J. Chromatogr.*, 116 (1976) 131.
134 J.C. Hudson and W.P. Rice, *J. Chromatogr.*, 117 (1976) 449.
135 L. Lepri, P.G. Desideri and M. Lepori, *J. Chromatogr.*, 123 (1976) 175.
136 A.N. Masoud, *J. Pharm. Sci.*, 65 (1976) 1585.
137 J.D. Phillipson and S.S. Handa, *Phytochemistry*, 15 (1976) 605.

138 I.R. da S. Jardim, M.M. Menezes de Menezes and C.T.G. Soares, *Rev. Braz. Farm.*, 57 (1976) 61; *C.A.*, 86 (1977) 195241h.
139 R. Giebelmann, S.Nagel, Ch. Brunstein and E. Scheibe, *Zbl. Pharm.*, 115 (1976 339.
140 H. Kroeger, G. Bohn and G. Ruecker, *Dtsch. Apoth. Ztg.*, 117 (1977) 1923.
141 K.C. Güven, T. Altinkurt and S. Gulhan, *Eczacilik Bul.*, 19 (1977) 42; *C.A.*, 88 (1978) 66023h.
142 M. Sobiczewska, *Farm. Pol.*, 33 (1977) 365; *C.A.*, 88 (1978) 27838k.
143 I. Ionov and I. Tsankov, *Farmatsiya (Sofia)*, 27 (1977) 25; *C.A.*, 87 (1977) 194891t.
144 J. Vamos, A.I. Lakatos, G. Szasz and A. Brantner, *Gyogyszereszet*, 21 (1977) 206; *C.A.*, 87 (1977) 206564s.
145 G. Rücker and A. Taha, *J. Chromatogr.*, 132 (1977) 165.
146 A.N. Masoud, *J. Chromatogr.*, 141 (1977) D9.
147 A.E. Mair, *J. Clin. Pharm.*, 2 (1977) 101.
148 T.V. Astakhova and S.A. Minina, *Khim. Farm. Zh.*, 11 (1977) 113; *C.A.*, 86 (1977) 161179f.
149 B. Goeber, U. Timm and H. Doehnert, *Zbl. Pharm.*, 116 (1977) 13.
150 U. Timm, B. Goeber and H. Doehnert, *Zbl. Pharm.*, 116 (1977) 151.
151 R. Giebelmann, *Zbl. Pharm.*, 116 (1977) 1011.
152 N.V. Rama Rao and S.N. Tandon, *Chromatographia*, 11 (1978) 227.
153 L.I. Ambrus, J. Vamos, G. Szasz and A. Brantner, *Gyogyszereszet*, 22 (1978) 467; *C.A.*, 90 (1979) 127596n.
154 A. Puech, J. Monleaud Dupy, M. Jacob and M. Jean, *J. Pharm. Belg.*, 33 (1978) 24.
155 *European Pharmacopeia*, published under the direction of the Council of Europe, by Maisonneuve S.A., France, 1971.

TABLE 7.1

TLC SYSTEMS FOR THE ANALYSIS OF SOME TROPINE ALKALOIDS AND $hR_F$ VALUES AS DETERMINED IN OUR LABORATORIES

All on silica gel 60 $F_{254}$ plates (Merck), non-activated, in saturated chromatography chambers. Temperature, 24 ± 2°C; relative humidity, 25 ± 5%.

Solvent systems:
- S1 Chloroform-diethylamine(9:1)[3]
- S2 Acetone-methanol-ammonia(25%)(40:10:2)[53]
- S3 Acetone-water-ammonia(25%)(90:7:3)[23,35]
- S4 Methyl ethyl ketone-methanol-ammonia (7,5%)(30:15:5)[13]
- S5 Carbon tetrachloride-*n*-butanol-methanol-ammonia(10%)(12:9:9:1)[8]
- S6 Chloroform-*n*-butanol-methanol-ammonia(10%)(12:9:9:1)[21]*
- S7 Toluene-acetone-methanol-ammonia(25%)(40:45:10:5)[116]
- S8 Benzene-ethyl acetate-diethylamine(7:2:1)[3]
- S9 Chloroform-methanol-ammonia(25%)(85:14:1)[116]
- S10 Ethyl acetate-ethanol-diethylamine(12:6:2)
- S11 Ethyl acetate-isopropanol-ammonia(25%)(45:35:10)**

*In ref. 21 this solvent is abusevily mentioned as the solvent described in ref. 8.
**In ref. 137 this solvent system was used in the ratio 45:35:15 (with 20% ammonia) for the analysis of N-oxides of tropane alkaloids.

| Alkaloid | $hR_F$ value | | | | | | | | | | | Literature $hR_F$ values | | |
|---|---|---|---|---|---|---|---|---|---|---|---|---|---|---|
| | S1 | S2 | S3 | S4 | S5 | S6 | S7 | S8 | S9 | S10 | S11 | S2[53] | S3[35] | S4[61] |
| Atropine | 35 | 24 | 28 | 28 | 29 | 31 | 32 | 23 | 24 | 46 | 46 | 26 | 30 | 20 |
| *l*-Hyoscyamine | 35 | 24 | 28 | 28 | 29 | 31 | 32 | 23 | 24 | 46 | 46 | 26 | | |
| Scopolamine | 49 | 65 | 64 | 67 | 68 | 78 | 52 | 34 | 58 | 62 | 67 | 74 | 80 | 55 |
| Homatropine | 41 | 25 | 27 | 27 | 27 | 31 | 33 | 26 | 27 | 48 | 44 | | | |
| N-Methylhomatropine | 0 | 0 | 0 | 0 | 0 | 0 | 0 | 0 | 0 | 0 | 0 | | | |
| Apoatropine | 55 | 31 | 38 | 37 | 43 | 46 | 47 | 39 | 46 | 52 | 60 | | 45 | 30 |
| Aposcopolamine | 62 | 63 | 75 | 72 | 82 | 91 | 66 | 58 | 77 | 73 | 78 | | | 70 |
| Belladonnine | | | | | | | | | | | | | | 10 |
| Tropine | | | | | | | | | | | | 5 | | 5 |
| Scopoline | | | | | | | | | | | | | | 40 |
| Development time (min) for a distance of 8 cm | 15 | 12 | 9 | 16 | 34 | 28 | 11 | 14 | 13 | 14 | 17 | | | |

TABLE 7.2

TLC ANALYSIS OF SOME TROPINE ALKALOIDS

TLC systems:

| | | |
|---|---|---|
| S12 | Silica gel G, activated | Chloroform-acetone-diethylamine(5:4:1)[17] |
| S13 | Silica gel G, activated | Chloroform-methanol-diethylamine(9:1:0.5)[17] |
| S14 | Silica gel G | Chloroform-acetone-(ammonia-abs. ethanol, 3:17)(5:4:1)[95] |
| S15 | Silica gel G, activated | Acetone-conc. ammonia(97:3)[74] |
| S16 | Silica gel G, activated | Methyl acetate-isopropanol-conc. ammonia(45:35:15)[74] |
| S17* | Silica gel G | Methanol[151] |
| S18* | Silica gel G | Methanol-1 *M* hydrochloric acid(1:1)[151] |
| S19* | Silica gel G | Methanol-water-acetic acid(16:3:1)[151] |
| S20* | Silica gel G | *n*-Butanol-water-acetic acid(4:5:1)(organic phase)[151] |

| Alkaloid | $hR_F$ value | | | | | | | | |
|---|---|---|---|---|---|---|---|---|---|
| | S12 | S13 | S14 | S15 | S16 | S17 | S18 | S19 | S20 |
| Atropine | 37 | 58 | 20 | 40 | 85 | 7 | 76 | 52 | 30 |
| *l*-Hyoscyamine | | | 20 | | | | | | |
| Scopolamine | 54 | 74 | 70 | 85 | 95 | 48 | 73 | 48 | 26 |
| Homatropine | 36 | 58 | | 40 | 85 | | | | |
| Apoatropine | 52 | 72 | 43 | 60 | 90 | | | | |
| Aposcopolamine | 70 | 90 | 85 | | | | | | |
| Atropine N-oxide | 0 | 19 | 0 | | | | | | |
| Scopolamine N-oxide | 0 | 3 | 0 | | | | | | |
| Homatropine N-oxide | 0 | 12 | | | | | | | |
| Tropine | 20 | 26 | 5 | 10 | 35 | | | | |
| Scopoline | 58 | 79 | 63 | 75 | 85 | | | | |
| Tropine N-oxide | 0 | 24 | | | | | | | |
| Tropic acid | | | | 0 | 10 | | | | |
| Mandelic acid | | | | 0 | 10 | | | | |
| Pseudotropine | | | | 12 | 35 | | | | |
| Tropinon | | | | 70 | 90 | | | | |
| N-Methylatropine | | | | | | 3 | 60 | 30 | 18 |
| N-Butylscopolamine | | | | | | 4 | 66 | 46 | 36 |

*These solvents were used on 0.50 mm thick plates for two-dimensional chromatography. Solvent S17 was used in the first direction and S18, S19 or S20 in the second direction.

TABLE 7.3

COLOURS OBTAINED FOR SOME TROPINE ALKALOIDS BY SPRAYING WITH DRAGENDORFF'S REAGENT (MUNIER MODIFICATION) FOLLOWED BY SPRAYING WITH 10% SODIUM NITRITE SOLUTION

Solvent system:
chloroform-acetone-(ammonia-abs. ethanol 3:17) (5:4:1) on silica gel G plates[95].

| Alkaloid | Colour | | |
|---|---|---|---|
| | Dragendorff's reagent | $NaNO_2$ | After drying |
| Atropine | Orange | Brown | Grey |
| *l*-Hyoscyamine | Orange | Brown | Brown-red |
| Scopolamine | Orange | Brown | Orange |
| Tropine | Violet | Grey | Brown vanishing |
| Scopine | Violet | Grey | Brown vanishing |
| Apoatropine | Orange | Brown | Disappeared |
| Aposcopolamine | Orange | Brown | Disappeared |
| Atropine N-oxide | Orange | Brown | Disappeared |
| Scopolamine N-oxide | Orange | Brown | Disappeared |

TABLE 7.4

COLOURS OBTAINED FOR SOME TROPINE ALKALOIDS BY SPRAYING WITH IODOPLATINATE REAGENT[17]

| Alkaloid | Colour | Alkaloid | Colour |
|---|---|---|---|
| Atropine | Blue-violet | Tropine | Grey-blue |
| *l*-Hyoscyamine | Blue-violet | Scopine | Grey-blue |
| Scopolamine | Blue-violet | Atropine N-oxide | Pink-violet |
| Homatropine | Blue-violet | Scopolamine N-oxide | Blue-violet |
| Apoatropine | Blue-violet | Homatropine N-oxide | Blue-violet |
| Aposcopolamine | Blue-violet | Tropine N-oxide | Grey-blue |

TABLE 7.5

SENSITIVITY OF SOME DETECTION REAGENTS FOR TROPINE ALKALOIDS

As determined in our laboratories with atropine, l-hyoscyamine, scopolamine, homatropine and methylhomatropine.

| Reagent | No. | Sensitivity (μg) | Colour | | Ref. |
|---|---|---|---|---|---|
| | | | Background | Alkaloid | |
| Quenching 254 nm UV | | 0.1-1 | | | |
| Dragendorff's modifications: | | | | | |
| Munier-Machebouef | 39c | 0.1-1 | Yellow | Orange-red | |
| Vágujfalvi | 39f | 0.1-1 | White-light yellow | Orange | 25 |
| Munier | 39b | 0.01-0.1 | Light yellow | Orange-red | |
| Bregoff-Delwiche | 39a | 0.1-1 | Light yellow | Orange | |
| Munier+$NaNO_2$ | 39h | 0.01 | White | Grey-reddish brown | 95 |
| Bouchardat(acidified) | 52c | | Light yellow | Yellowbrown-dark brown | 8 |
| Iodine in methanol | 51d | 0.01-0.1 | White | Orange brown | 26,111 |
| | | 0.1-1 | Yellow | Brown | 78 |
| Iron(III) chloride,$KI_3$ | 60 | 0.01 | Yellow-light green | Grey-brown | 108 |
| Iodine vapour | | 0.01 | Yellow | Brown | 97 |
| Methyl red after evaporation $I_2$ | | 1-10 | Pink | Yellow | 68 |
| Next with iodoplatinate | | 0.1-1 | Violet | Dark violet | 68 |
| Iodine vapour,pyrrole vapour | 54 | 1 | Yellow | Grey-brown-green | 98 |
| Phenothiazine,Iodine vapour | 78 | 0.1-0.01 | Light brown | Dark brown | 98 |
| Phenothiazine,bromine vapour | | 1-10 | Violet | Yellow | |
| Then ammonia vapour | | 10 | Violet | Yellow | 98 |
| Iodoplatinate | 56d | 0.1-1 | Violet | Dark violet-brown | |
| Bromophenol blue | 9 | 10 | Blue | Dark blue | 74 |
| Bromecresol green | 8 | 0.1-1 | Blue-green | Blue | 51 |
| Methyl orange | 71 | 0.1 | Orange-yellow | Orange-red | 74 |
| Cobalt(II) thiocyanate | 26b | 1-10 | Purple | Blue | |
| Citric acid in acetic anhydride | 22 | 10 | Light brown | Violet | 74 |
| Tannin | 93 | 10 | Light brown | Light brown | 74 |
| 0.5% alkaline $KMnO_4$ | 83 | 0.1 | Purple | Yellow | |

References p. 96

TABLE 7.6

LITERATURE CITED IN CHAPTER 3 WHICH INCLUDES THE ANALYSIS OF TROPINE ALKALOIDS

| Alkaloid* | Ref. | Alkaloid* | Ref. |
|---|---|---|---|
| A,S,H,apoA | 3 | A,S | 98 |
| A,S,Sbut | 5 | A,S | 99 |
| A,S | 14 | A,S,H,AMe,Sbut | 100 |
| A,S | 15 | A,S,H,Sbut | 106 |
| A,S | 18 | A,H,S | 108 |
| A,S | 26 | A,S | 111 |
| A,H | 28 | A,S | 113 |
| A,S | 30 | S | 127 |
| A,S | 38 | A,S,H | 131 |
| A | 41 | A,S,H | 133 |
| S | 43 | A,S,H | 135 |
| A,S | 68 | A,S | 136 |
| A,S,H | 85 | A,S | 146 |
| A,S | 86 | A | 152 |
| A,S,H,HMe | 91 | | |

*Abbreviations used in Tables 7.6-7.11:

| | | | |
|---|---|---|---|
| A | atropine | scop | scopine |
| S | scopolamine | scopl | scopoline |
| H | homatropine | trop ac | tropic acid |
| AMe | N-methylatropine | tropon | tropinon |
| SMe | N-methylscopolamine | tropd | tropidine |
| Sbut | N-butylscopolamine | pstrop | pseudotropine |
| HMe | N-methylhomatropine | man | mandelic acid |
| B | belladonnine | atrop ac | atropic acid |
| trop | tropine | | |

TABLE 7.7

TLC ANALYSIS OF TROPINE ALKALOIDS IN DRUGS OF ABUSE (SEE CHAPTER 12)

| Alkaloid* | Ref. in this Chapter | Ref. No. in Chapter 12 | Alkaloid* | Ref. in this Chapter | Ref. No. in Chapter 12 |
|---|---|---|---|---|---|
| A | 4 | 8 | A | | 207 |
| A,S | 7 | 24 | A | 121 | 322 |
| A | | 60 | A | 122 | 324 |
| A,S | 40 | 68,102 | A | 123 | 328 |
| A,S | 66 | 141 | A,S | 115 | 332 |
| A,MeA | 78 | 156 | A,S | 136,146 | 356,379 |
| A | 79 | 159 | A,S | 140 | 369 |
| | | | A | 134 | 35(Ch10) |

*For abbreviations, see footnote to Table 7.6.

TABLE 7.8

TLC ANALYSIS OF TROPINE ALKALOIDS IN PLANT MATERIAL

| Alkaloid* | Aim | Adsorbent | Solvent system | Ref. |
|---|---|---|---|---|
| A,S,B,apoA,trop | Separation | 0.5 *M* KOH-impregnated $SiO_2$<br>20% FMA-impregnated cell | EtOAc-EtOH-DMFA-DEA (12:6:1:1)<br>70% EtOH-25% $NH_4OH$(99:1)<br>Heptane-DEA(500:1) followed by Benzene-heptane-$CHCl_3$-DEA (6:5:1:0.02) | 2 |
| A | Indirect quantitative analysis(titrimetric) in *Datura* species | $Al_2O_3$ | MeOH | 9 |
| A,S | Control of quantitative analysis | $SiO_2$ | EtOAc-EtOH-DMFA-DEA (12:6:1:1) | 10 |
| A,S | Separation | $SiO_2$ | Benzene-MeOH(1:1) | 11 |
| trop,tropon,pstrop | Separation | $SiO_2$ | EtOH-25% $NH_4OH$ | 12 |
| A,S,apoA,apoS,trop, B,scop,scopl | Separation,and direct quantitative analysis (planimetric)(Table 7.1) | $SiO_2$ | MeEtCO-MeOH-7.5% $NH_4OH$ (30:15:5)<br>MeEtCO-MeOH-10% $NH_4OH$ (30:10:5) | 13,16 |
| A,S | Separation on $Al_2O_3$ without binder | $Al_2O_3$ | Benzene-MeOH(9:1)<br>$Et_2O$-EtOH(97:3) | 15 |
| A,S,H,apoA,apoS, trop,scop,tropN-ox, A N-ox | Identification with reaction chromatography (Table 7.2) | $SiO_2$ | $CHCl_3$-$Me_2CO$-DEA(5:4:1)<br>$CHCl_3$-MeOH-DEA(9:1:0.5) | 17,27 |
| A,S | Indirect quantitative analysis(colorimetric) in *Scopolia* species | $SiO_2$ | $Me_2CO$-$CHCl_3$-MeOH-25% $NH_4OH$ (20:20:3:1) | 19,65 |
| A,S | Control quantitative analysis (Table 7.1) | $SiO_2$ | $Me_2CO$-$Et_2O$-10% $NH_4OH$ (95:100:5) | 21,22 |
| A,S | Direct quantitative analysis(planimetric) in Belladonna (Table 7.1) | $SiO_2$ | $Me_2CO$-10% $NH_4OH$(95:5) | 23 |
| A,S | Adulterants in Altheae | pH 7.6 impregnated $SiO_2$ | $CHCl_3$-MeOH(1:1) | 24 |
| A,S,apoA,B | Identification in *Atropa*, *Datura* and *Hyoscyamus* | $SiO_2$ | MeEtCO-MeOH-7.5% $NH_4OH$ (30:15:5) | 32 |
| A,S,apoA,trop | Review and comparison of TLC separations of alkaloids in plant material (Table 7.1) | $SiO_2$ | $Me_2CO$-$H_2O$-25% $NH_4OH$ (90:7:3) | 35 |
| A,S | Indirect quantitative analysis(colorimetric) in *Hyoscyamus* species | Cellulose | isoBuOH-$H_2O$-conc. HCl (7:2:1) | 37 |
| A,S | Indirect quantitative analysis(colorimetric) in pharmaceuticals containing Solanaceous extracts | $SiO_2$ | EtOH-$H_2O$-triethanolamine (1:1:0.4) | 44 |

TABLE 7.8 (continued)

| Alkaloid* | Aim | Adsorbent | Solvent system | Ref. |
|---|---|---|---|---|
| Flavonoids | Characterization in Solanaceae drugs | $SiO_2$ | EtOAc-$H_2O$-HCOOH(100:3:2) | 45 |
| A,S | Differentiation between Belladonna and Hyoscyamus | $SiO_2$ | $CHCl_3$-$Me_2CO$-DEA(5:4:1) | 46 |
| A,S | Identification of Solanaceous drugs | $SiO_2$ | MeEtCO-MeOH-7% $NH_4OH$ (6:3:1) | 47 |
| A,S,A N-ox,S N-ox, apoA,apoS | Identification in Belladonna | $SiO_2$ | $CHCl_3$-$Me_2CO$-DEA(5:4:1) | 48 |
| A,S | Identification in Solanaceae drugs | $SiO_2$ | MeEtCO-$CHCl_3$(6:2)<br>MeEtCO-$CHCl_3$-MeOH-17% $NH_4OH$ (6:2:1.5:0.5) | 56 |
| A,S | Densitometric analysis | $SiO_2$ | MeOH-10% $NH_4OH$(200:1) | 57 |
| A,S | Densitometric analysis in Belladonna | $SiO_2$ | MeOH-$CHCl_3$-25% $NH_4OH$ (20:10:1) | 60 |
| A,S | Indirect quantitative analysis(spectrophotometric) in *Scopolia* species | $SiO_2$ | MeEtCO-MeOH-7.5% $NH_4OH$ (15:30:5) | 64 |
| Not specified | TAS technique for Solanaceous drugs | $SiO_2$ | $CHCl_3$-$Me_2CO$-DEA(5:4:1)<br>$CHCl_3$-DEA(9:1)<br>Cyclohexane-$CHCl_3$-DEA (5:4:1)<br>Cyclohexane-DEA(9:1)<br>Benzene-EtOAc-DEA(7:2:1) | 73 |
| A,S,apoA,apoS, trop,trop ac, tropon,scop, scopl,pstrop, man | Comparison of different TLC systems and detection methods (Table 7.2) | $SiO_2$ | $Me_2CO$-25% $NH_4OH$(97:3)<br>MeOAc-isoprOH-25% $NH_4OH$ (45:35:15) | 74 |
| A,S,apoA | Indirect quantitative analysis in Belladonna preparations | $SiO_2$ | $Me_2CO$-3 *M* $NH_4OH$(9:1) | 75 |
| trop,pstrop, tropon | Reaction chromatography | $SiO_2$ | $Me_2CO$-25% $NH_4OH$(8:2) | 84 |
| A,S | Indirect quantitative analysis in Belladonna | $SiO_2$ | Benzene-MeOH(6:4) | 87 |
| A,S,cuscohygrine | Identification in *Scopolia* species | $Al_2O_3$ | No details available | 89 |
| A | Direct quantitative analysis(planimetric) in Belladonna preparations | $SiO_2$ | $CHCl_3$-DEA(9:1) | 90 |
| A,S,cuscohygrine | Identification in *Scopolia* species | $SiO_2$ | $Me_2CO$-$CHCl_3$-MeOH-20% $NH_4OH$ (20:20:3:1) | 92 |
| A,S,apoA,apoS, A N-ox,S N-ox, trop,scop | Identification in Belladonna (Table 7.2) | $SiO_2$ | $CHCl_3$-$Me_2CO$-($NH_4OH$-abs. EtOH, 3:17)(5:4:1) | 95 |
| A,S | Densitometric analysis in *Datura* species | $SiO_2$ | MeOH,MeOH-$NH_4OH$(200:1 or 200:5) | |

TABLE 7.8 (continued)

| Alkaloid* | Aim | Adsorbent | Solvent system | Ref. |
|---|---|---|---|---|
| A,S | Identification in Belladonna preparations | $SiO_2$ | $MeEtCO-MeOH-H_2O$-25% $NH_4OH$ (60:30:7:3) | 97 |
| A,S | Indirect quantitative analysis (spectrophotometric) in Belladonna | $SiO_2$ | $Me_2CO$-10% $NH_4OH$ (95:5) | 105 |
| Not specified | Ion-pair chromatography | Cellulose impregnated with 0.7 *M* $H_2SO_4$+0.7 *M* NaCl | *n*-BuOH | 114 |
| A,apoA,tropd | TAS technique for solanaceous drugs | $SiO_2$ | $Me_2CO-H_2O$-25% $NH_4OH$ (90:7:3) | 117 |
| A,S | Indirect quantitative analysis (colorimetric) in Solanaceous drugs | $SiO_2$ | 70% EtOH-$NH_4OH$ (95:5) | 119 |
| A,S | Identification in Belladonna preparations | $SiO_2$ | $CHCl_3$-MeOH (40:7) in $NH_3$ atmosphere | 130 |
| A,A N-ox | Identification of N-oxides in Belladonna | $SiO_2$ | $Me_2CO-H_2O$-25% $NH_4OH$ (90:7:3)<br>EtOAc-isoprOH-20% $NH_4OH$ (45:35:15) | 137 |
| A,S | Identification | | No details available | 141 |
| A,S | Optimal extraction conditions in *Scopolia* species | | No details available | 148 |
| A,S | Identification Solanaceous drugs | $SiO_2$ | $Me_2CO-H_2O$-25% $NH_4OH$ (90:7:3) | 155 |

*For abbreviations, see footnote to Table 7.6.

TABLE 7.9

TLC ANALYSIS OF TROPINE ALKALOIDS IN PHARMACEUTICAL PREPARATIONS

| Alkaloid* | Other compounds | Aim | Adsorbent | Solvent system | Ref. |
|---|---|---|---|---|---|
| A,S | | | $SiO_2$ | *n*-BuOH sat. with AcOH | 1 |
| A | Papaverine, aminopyrine | Identification | $SiO_2$ | Benzene-MeOH(1:2) | 6 |
| A,H,HMe, trop, tropMe | | Identification | $Al_2O_3$ | EtOH-pyridine-$H_2O$(1:6:4) | 20 |
| A,apoA, trop, trop ac | | Stability control of A in injections, indirect quantitative analysis | $SiO_2$ | EtOH-*n*-BuOH-10% $NH_4OH$ (1:1:1)<br>Cyclohexane-$CHCl_3$-AcOH (6:2:2) | 33 |
| A,S | Ergotamine, phenobarbital | Identification in tablets | $SiO_2$ | EtOH-$CHCl_3$-10% $NH_4OH$ (70:19:11) | 34 |
| A,S,trop, scop, trop ac | | Indirect quantitative analysis in injections | 1 *M* KOH-impregnated $SiO_2$ | 80% EtOH-25% $NH_4OH$ | 36 |
| A | Ethylmorphine, morphine,papaverine | Identification | | No details available | 39 |
| S | | Fluorodensitometric analysis | $SiO_2$ | 98% EtOH-conc. $H_2SO_4$(93:7) | 50 |
| A,H,S, apoA, apoS,trop | Cocaine,ecgonine | Detection of degradation products in eyedrops | 0.5 *M* NaOH-impregnated $SiO_2$ | $CHCl_3$-MeOH-25% $NH_4OH$ (8:2:0.2) | 51 |
| A | Morphine, papaverine | Densitometric and planimetric quantitative analysis | $SiO_2$ | $CHCl_3$-MeOH-25% $NH_4OH$ (10:20:1) | 52 |
| A | | Purity control (Table 7.1) | $SiO_2$ | $Me_2CO$-25% $NH_4OH$(95:5)<br>96% EtOH-$CHCl_3$-25% $NH_4OH$ (80:20:4)<br>$CHCl_3$-MeOH-25% $NH_4OH$ (30:60:2)<br>$Me_2CO$-MeOH-25% $NH_4OH$ (40:10:2) | 53 |
| A,HMe | Isopropamide, gastrixone | Separation in parasympatholytica | 1 *M* NaOH-impregnated $SiO_2$<br>$Al_2O_3$ | $CHCl_3$-EtOH(1:1,7:3)<br>$CHCl_3$-EtOH(1:1,7:3) | 55 |
| A,H,S, AMe,HMe | Pilocarpine,cocaine,physostigmine,ethylmorphine | Detection of impurities in eyedrops | $SiO_2$ | Benzene-$Me_2CO$-$Et_2O$-5% $NH_4OH$ (40:60:10:3.4)<br>Cyclohexane-$CHCl_3$-DEA (3:7:1) | 59 |
| A,AMe | Xanthines, strychnine,codeine,papaverine,ephedrine, procaine,phenobarbital, chlorpromazine, tripelennamine, melipramin | Separation | $SiO_2$ | $Me_2CO$-cyclohexane-EtOAc (1:1:1) in $NH_3$ atmosphere | 63 |

TABLE 7.9 (continued)

| Alkaloid* | Other compound | Aim | Adsorbent | Solvent system | Ref. |
|---|---|---|---|---|---|
| A,S | Cocaine, morphine, meperidine, strychnine | Separation on microslides | $SiO_2$ | $CHCl_3$-DEA(9:1)<br>$CHCl_3$-$Me_2CO$-DEA(5:4:1) | 67 |
| A | Codeine, papaverine, morphine, phenobarbital, aminopyrine | Identification | $SiO_2$ or $Al_2O_3$ | Benzene-MeOH-AcOH(7:1:2)<br>Benzene-$Me_2CO$-EtOH-25% $NH_4OH$ (5:4:0.5:0.5) | 69 |
| A,S | | Separation | $SiO_2$+$Al_2O_3$+ Kieselguhr (1:1:1) | $CHCl_3$-EtOH(9:1) | 70 |
| A,S | Local anaesthetics | Separation | pH=6.6 impregnated cellulose | *n*-BuOH sat. with $H_2O$ | 71 |
| A,S | Aminopyrine, papaverine | Identification in suppositoria | $SiO_2$ | $Me_2CO$-7.5% $NH_4OH$(9:1) | 76 |
| A,AMe | Morphine, papaverine, noscapine | Identification | $SiO_2$ | $Me_2CO$-5 *M* $NH_4OH$(8:2) | 78 |
| A,H | Physostigmine, pilocarpine | Semi-quantitative analysis in eyedrops (planimetric) | $SiO_2$ | $CHCl_3$-$Me_2CO$-DEA(5:4:1)<br>$CHCl_3$-DEA(9:1) | 80 |
| A,S | Morphine | Separation | $SiO_2$ | $CHCl_3$-MeOH(3:2) | 81 |
| A | Cyclizine, dimenhydranate, meclozine, mepyramine, promethazine | Identification | $SiO_2$ | MeOH | 82 |
| A,S | Morphine, ergotamine, phenobarbital | Indirect quantitative analysis (colorimetric) | $Al_2O_3$<br>$SiO_2$ | $CHCl_3$-$NH_4OH$(100:0.15)<br>$CHCl_3$-benzene-EtOH(4:1:2) | 83 |
| A | Ergot alkaloids | Separation | $SiO_2$ | EtOH-25% $NH_4OH$ | 88 |
| A | Ephedrine | Indirect quantitative analysis in tablets | $SiO_2$ | $CHCl_3$-MeOH-AcOH(25:65:5) | 93 |
| A,S | Noscapine, papaverine | Identification | $SiO_2$ | $CHCl_3$-$Me_2CO$(3:1)<br>$CHCl_3$-MeOH(4:1)<br>MeOH-28% $NH_4OH$(100:1.5)<br>$CHCl_3$-DEA(9:1) | 94 |
| A,S | | Densitometric analysis | $SiO_2$ | MeOH<br>MeOH-$NH_4OH$(200:1,200:5) | 96 |
| A, trop, trop ac, trop atrop ac | | Identification | $SiO_2$ | $CHCl_3$-AcOH(9:1) | 104 |
| A,S,H, apoA,B | | Purity control and identification | $SiO_2$ | MeEtCO-MeOH-6 *M* $NH_4OH$ (6:3:1) | 107 |
| A, | Morphine, pseudomorphine | Quality control of injection | $SiO_2$ | $Me_2CO$-EtOH-benzene-25% $NH_4OH$ (32.5:35:35:2.5) | 112 |

TABLE 7.9 (continued)

| Alkaloid* | Other compound | Aim | Adsorbent | Solvent system | Ref. |
|---|---|---|---|---|---|
| A | Cotarnine, strychnine,brucine | Identification | No details available | | 118 |
| A | | Direct quantitative analysis | No details available | | 124 |
| AMe | Theophylline, ethophylline, papaverine, phenobarbital | Identification in suppositories | $SiO_2$ | $MeOH-Me_2CO$-conc. HCl (90:10:4) | 125 |
| A | Physostigmine | Indirect quantitative analysis | $SiO_2$ | $CHCl_3-Me_2CO$-DEA | 126 |
| A,S | Amidopyrine, novocaine,thiamine | Identification | $SiO_2$ | $CHCl_3$-MeOH(40:7) in $NH_3$ atmosphere | 130 |
| A,S,H, apoA | Ergot alkaloids, barbiturates, xanthine derivatives | Separation on $AgNO_3$-impregnated plates | $SiO_2$,$AgNO_3$-impregnated | Mobile phase not specified | 132 |
| A | Butethamate, papaverine | Identification | No further details available | | 138 |
| AMe,S | Morphine | Indirect quantitative analysis (colorimetric) | $SiO_2$ | MeOH<br>$EtOH-H_2O$(95:5) | 142 |
| A,S,H,HMe | | Purity control | $SiO_2$ | $Me_2CO-H_2O$-25% $NH_4OH$ (90:4:6,90:7:3) | 144 |
| A,apoA, trop,trop ac,atrop ac | | Detection of degradation products | $SiO_2$ | $Me_2CO$-7.5% $NH_4OH$(9:1) | 147 |
| S,apoS, scop,trop ac | | Detection of degradation products in ophthalmic solutions | $SiO_2$ | MeEtCO-MeOH-6 *M* $NH_4OH$ (6:3:1)<br>$CHCl_3$-DEA(9:1)<br>70% EtOH-25% $NH_4OH$ (97.5:2.5,99:1)<br>Benzene-dioxane-AcOH (75.6:21:3.4) | 149<br>150 |
| A,S,AMe, Sbut | | Identification (Table 7.2) | $SiO_2$ | Two-dimensional:<br>I. MeOH<br>II. MeOH-1 *M* HCl(1:1) or $MeOH-H_2O$-AcOH(16:3:1)<br>*n*-BuOH-AcOH-$H_2O$(4:5:1) | 151 |
| A,S,apoA, scop,trop, H,HMe,AMe | | Purity control | Cellulose | *n*-BuOH-AcOH-$H_2O$(40:4:20), upper phase | 153 |
| A,trop ac, atrop ac | | Indirect quantitative analysis in aqueous solutions (UV) | $SiO_2$ | $CHCl_3-Me_2CO$-ammoniacal EtOH (pH 12.4)(4:4:1) | 154 |

*For abbreviations, see footnote to Table 7.6.

TABLE 7.10

TLC ANALYSIS OF TROPINE ALKALOIDS AS PURE COMPOUNDS AND IN COMBINATION WITH OTHER COMPOUNDS

| Alkaloid* | Other compounds | Aim | Adsorbent | Solvent system | Ref. |
|---|---|---|---|---|---|
| A,S | Veratrine, strychnine, tetracaine | Separation (Table 7.1) | $SiO_2$ | $CHCl_3$-DEA(9:1)<br>$CCl_4$-*n*-BuOH-MeOH-10% $NH_4OH$ (12:9:9:1) | 8 |
| A | Brucine,aconitine,codeine | Separation on plaster of Paris | $CaSO_4$ | $CHCl_3$-AmOH-toluene-conc. HCl (50:1.5:1.5:0.25) | 29 |
| H | Solanine,solanidine,physostigmine,ephedrine,lobeline, pilocarpine,reserpine | Detection of cholinesterase inhibitors | Cellulose impregnated with 5% silicone 555 | $H_2O$-EtOH-$CHCl_3$(56:42:2) | 31 |
| A,S | Various alkaloids | Reaction chromatography | $SiO_2$ | Benzene-MeOH-$Me_2CO$-AcOH (70:20:5:5) | 42 |
| A,S,AMe, SMe,Sbut | Strychnine,brucine,papaverine | Ion-pair chromatography | Cellulose impregnated with 0.7 *M* $H_2SO_4$+0.7 *M* NaCl (or NaBr, KSCN,$NaClO_4$) | AmOH,AmOH-$CHCl_3$(1:1) | 49 |
| A,S,B | Brucine | Separation on plates with traverse pH gradient | $SiO_2$ impregnated with pH gradient | $CHCl_3$-MeOH(8:2) | 54 |
| A | Various alkaloids | Semi-quantitative circular TLC | $SiO_2$ | $CHCl_3$-MeOH(9:1)<br>$CHCl_3$-MeOH(85:15) | 58 |
| A,S,apoA, apoS,trop, scop,B | | Separation (Table 7.1) | $SiO_2$ | MeEtCO-MeOH-7.5% $NH_4OH$ (6:3:1) | 61 |
| Sbut,SMe | Quaternary N-compounds | Separation with ion-pair chromatography | $SiO_2$<br>NaBr-impregnated $SiO_2$ | 0.5 *M* NaBr in MeOH<br>0.5 *M* NaI in $CHCl_3$-MeOH(2:8)<br>$CHCl_3$-MeOH(7:3) | 88 |
| A | Tubocurarine, ephedrine,physostigmine | Separation of quaternary N-drugs | $SiO_2$ | 1 *M* HCl<br>EtOH-1 *M* HCl(1:1) | 103 |
| A,S | Opium alkaloids,tobacco alkaloids,LSD | Reaction chromatography (Table 7.1) | $SiO_2$ | Toluene-$Me_2CO$-MeOH-25% $NH_4OH$ (4:4.5:1:0.5) | 116 |
| A,H,S | Procaine,cocaine | Effect of detection with iodine | $Al_2O_3$ | Benzene-EtOH(9:1) | 129 |
| AMe,Sbut | Tubocurarine, various quaternary N-drugs | Separation | $SiO_2$ | $Me_2CO$-1 *M* HCl(1:1)<br>EtOH-1 *M* HCl(1:1)<br>$Me_2CO$-AcOH-25% HCl(10:85:5)<br>MeOH-AcOH-25% HCl(10:85:5)<br>Pyridine-AcOH-$H_2O$-MeOH (5:10:10:75) | 139 |
| A,H,S, AMe,HMe, trop | Various alkaloids | Detection with π acceptors | $SiO_2$ | $Me_2CO$-toluene-MeOH-$NH_4OH$ (45:40:10:5) | 145 |

*For abbreviations, see footnote to Table 7.6.

**References p. 96**

TABLE 7.11

TLC ANALYSIS OF TROPINE ALKALOIDS IN BIOLOGICAL MATERIAL (SEE ALSO TABLE 7.7)

| Alkaloid* | Other compounds | Aim | Adsorbent | Solvent system | Ref. |
|---|---|---|---|---|---|
| A,S,trop, various tropine derivatives | Physostigmine, papaverine, ethylpapaverine, xanthines,choline,acetylcholine | Study of enzymatic hydrolysis | $SiO_2$ | $EtOAc-H_2O-HCOOH$(10:3:2) | 62 |
| A | Neonal,phenobarbital,ergotamine,ergotoxine | Detection after intoxication | $SiO_2$ | $CHCl_3$-EtOH-25% $NH_4OH$ (20:5:1) | 77 |
| A,apoA, trop, trop ac | | Detection of metabolites | No details available | | 110 |
| A | Strychnine, brucine,cocaine | Identification | $SiO_2$ | Cyclohexane-$Me_2CO$-DMA (6:3:0.2) | 143 |

*Abbreviations, see footnote to Table 7.6.

# Chapter 8

## PSEUDOTROPINE ALKALOIDS

Cocaine is an ester of benzoic acid and methylecgonine and in plant material it occurs in combination with chemically related alkaloids. These have an ecgonine, tropine or hygrine skeleton and may occur as esters of cinnamic, α- or β-truxillic or benzoic acid.

Most investigations on the analysis of cocaine and related alkaloids have been carried out in connection with the abuse of cocaine as a narcotic. Cocaine is excreted from the body mainly as benzoylecgonine and the detection and determination of this compound in urine has therefore been the subject of several studies. The polar and amphoteric character of it necessitates special methods for its extraction from biological material. Such methods have been described by Meola and co-workers[40,49], who used charcoal to isolate drugs of abuse from biological material, and Bastos et al.[43], who used ethanol for the extraction of urine after saturation with potassium carbonate. The extract obtained was treated with concentrated sulphuric acid and *n*-butanol to give the butyl derivatives of ecgonine, benzoylecgonine and benzoylnorecgonine. The butyl derivatives were identified by means of TLC (see Table 8.1). Mulé[27] used solvent extraction with chloroform containing 25% of ethanol to extract benzoylecgonine from urine. He found that acid hydrolysis of urine, which usually is performed for the analysis of morphine-like alkaloids, destroys cocaine and its metabolites, so that they cannot be detected. Other solvent extraction methods were described by Valanju et al.[34], who used chloroform - isopropanol - 1,2-dichloroethane (8:1:3). The same solvent, in the ratio 4.5:0.9:4.5, was used by Kaistha and Tadrus[70] after saturation of the urine with sodium hydrogen carbonate. They also described a single- and a two-step isolation procedure, using ion-exchange paper. Müller et al.[72] used $XAD_2$ for the isolation of cocaine and benzoylecgonine, Wallace et al.[53] used a method similar to that of Mulé[27], and Koontz et al.[35] modified the salting-out procedure proposed by Bastos et al.[43] by using potassium hydrogen phosphate to saturate the urine before extraction.

Several reviews on the analysis of biological material for cocaine and metabolites have been published[42,58,71]. Wallace et al.[68] evaluated enzyme immunoassays and GLC and TLC analysis[53] for their ability to detect cocaine and its major metabolite benzoylecgonine in human urine. A similar study was made by Mulé et al.[66].

**References p. 117**

## 8.1. SOLVENT SYSTEMS

Because of the different aims of the TLC analysis of cocaine and related alkaloids, the solvents used are of different character. For the analysis of the alkaloids from plant material, Wartmann-Hafner[11] tested several solvent systems in combination with silica gel plates. Ethyl acetate - ethanol - dimethylformamide diethylamine (75:20:5:2) was found to be useful for cocaine, whereas methanol - diethylamine (95:5) also separated the more polar compounds (Table 8.2), but not of cocaine and cinnamylcocaine. Stahl and Schmitt[48] used toluene - acetone (85:15) on basic aluminium oxide plates to separate cocaine and cinnamylcocaine and the decomposition product methylecgonidine. Ragazzi et al.[4] used magnesium oxide for the TLC analysis of cocaine.

Munier and Drapier[59,64,65] studied a number of alkali-free solvent systems on silica gel plates. Different chlorine-containing solvents were used in combination with methanol. It was found that the selectivity of the system was determined by the chlorine-containing solvent. Increasing amounts of methanol increased the mobility of all alkaloids in the same way. Exposure of the TLC plate to the vapour of the mobile phase prior to its development with different neutral solvent systems was also investigated. The best separation of cocaine-type alkaloids was obtained with methanol - trichloroethylene - acetic acid (2:2:1) (see Table 3.21, p.47).

Fiebig et al.[17] and Dijkhuis[19] analysed alkaloid-containing eyedrops. The $hR_F$ values reported by the latter are summarized in Table 20.2, p.480.

Illicit samples of cocaine are often adulterated with synthetic local anaesthetics, and their separation has therefore been an analytical problem[6,25,26,36] (see Table 8.3). The method developed by Brown et al.[36] was discussed by Woodford[56].

To separate cocaine and its metabolites isolated from biological material solvent systems more polar than those mostly used in the analysis of drugs of abuse are needed, because of the amphoteric character of the metabolites. Mulé[27] used methanol - ammonia (99:1) on silica gel. However cocaine and benzoylecgonine were not separated. Their separation was possible in ethyl acetate - methanol - ammonia (85:10:10). Misra et al.[33] used a series of solvents for the separation of cocaine and its metabolites (Table 8.4), and Valanju et al.[34] described two solvents, which were run in succession for the same purpose (Table 8.5). Bastos et al.[43] overcame the problem of the separation of amphoteric compounds by derivatization, using butylation of the acidic groups. The TLC of the butylated compounds was performed with several solvents (Table 8.1). If morphine was present in a sample, a combination of two solvent systems was found to be necessary; S1 followed by S7 gave satisfactory results. Wallace et al.[53] described three solvent systems for the analysis of cocaine and benzoylecgonine on silica gel plates: *n*-butanol - sulphuric acid (95:5) saturated with water needed a long development time and chloroform - methanol - concentrated ammonia (100:20:1 or 60:60:1) was preferred (Table 8.6).

Jain et al.[55] used a number of solvent systems for silica gel plates to analyse drugs of abuse in urine and discussed which systems should be used in screening for a specific compound and how its identity should be proved by TLC. For cocaine and benzoylecgonine, ethyl acetate - methanol - diethylamine (90:10:1.6) was found to be suitable, with dichloromethane - methyl ethyl ketone - concentrated ammonia (90:10:0.8) as a good alternative. To prove the presence of cocaine the latter solvent system or acetone - methanol (6:4 and 7:3) gave satisfactory results, and for the presence of benzoylecgonine, methanol - chloroform - concentrated ammonia (74:25:0.8) or acetone - methanol (6:4) was suitable. Kaistha and Tadrus[70] used chloroform - methanol - water - concentrated ammonia (70:30:0.5:1) on silica gel plates for the detection of benzoylecgonine and Müller et al.[72] described four solvent systems for the same purpose. In ethyl acetate - methanol - concentrated ammonia (10:5:1) on silica gel plates the $R_F$ value was less dependent on the concentration of the benzoylecgonine to be analysed than in the other solvents. A number of drugs which did not interfere in the analysis with the solvents mentioned were given.

## 8.2. DETECTION

In the screening of drugs of abuse, cocaine is usually detected with iodoplatinate or Dragendorff's reagent. For the detection of cocaine and derivatives in drug of abuse screening, see also Chapter 12.

Wallace et al.[53] compared Dragendorff's reagent (Munier and Macheboeuf modification) with iodoplatinate reagent in the detection of cocaine and benzoylecgonine. Although iodoplatinate reagent is reasonably selective because of the differences in colour obtained for various alkaloids, it is not as sensitive as Dragendorff's reagent (see Table 8.7). Detection with Dragendorff's reagent could be made even more sensitive by subsequently spraying with sulphuric acid (20%). A doubling of the sensitivity for cocaine and a 5- to 10-fold enhancement of the sensitivity for benzoylecgonine was achieved. A brief exposure (60-90 sec) to iodine vapour led to a further enhancement of the sensitivity, but longer exposure to iodine decreased the sensitivity. The sensitivity of the iodoplatinate reagent could also be enhanced by subsequently spraying with sulphuric acid. Addition of sulphuric acid to Dragendorff's reagent enhanced the sensitivity, but decreased the stability of the reagent.

Brown et al.[36] sprayed the plates with acidified iodoplatinate and also with *p*-dimethylaminobenzaldehyde. The latter reagent gave colours with some synthetic local anaesthetics (Table 8.3). The colours obtained with iodoplatinate reagent are listed in Table 8.8.

**References p. 117**

Holdstock and Stevens[51] described the recovery of alkaloids, including cocaine, from plates after spraying with iodoplatinate reagent (see Chapter 2, p.14). Vinson and Hocyman[52] used TCBI reagent (no. 94) for cocaine (see Table 2.3, p.17). Detection with the aid of $\pi$-acceptors was described by Rücker and Taha[69] (see Table 2.2, p.16). Kaniewska and Borkowski[16] used chromotropic acid (no. 20) in the detection of methylenedioxy-containing compounds. The reagent also gave various colours with a number of other alkaloids, including ecgonine and tropacocaine. Quenching of UV light on fluorescent plates at 254 nm was found to be 40 times less sensitive than Dragendorff's or iodoplatinate reagent in the detection of cocaine and related alkaloids[11].

According to Mulé[27], cocaine could be detected with 1% iodine in methanol, whereas ecgonine and benzoylecgonine did not react. Kaistha and Tadrus[70] sprayed with iodine - potassium iodide solution after spraying with Dragendorff's reagent (Munier modification) for the detection of benzoylecgonine. The spots disappeared in less than 1 min, but reappeared on spraying with iodine - potassium iodide solution.

Grant[62] used cobalt thiocyanate (no. 26a) for the detection of various compounds, including cocaine. Sarsunova et al.[57] studied the effect of a detection with iodine on the quantitative TLC determination of cocaine.

## 8.3. QUANTITATIVE ANALYSIS

Hashmi et al.[18] carried out the semi-quantitative analysis of some alkaloids, including cocaine, by means of circular TLC. Ragazzi et al.[4] used magnesium oxide as the sorbent in the TLC analysis of alkaloids. For quantitative analysis the spots of the alkaloids can be collected and dissolved in a dilute acid in which both the alkaloids and the sorbent are soluble. Massa et al.[22] determined a number of alkaloids densitometrically. Cocaine was analysed by measuring the quenching of UV light at 254 nm on fluorescent plates.

## 8.4. TAS TECHNIQUE AND REACTION CHROMATOGRAPHY

The TAS technique for the identification of Coca leaves was first described by Jolliffe and Shellard[23]. A 10-20 mg amount of the crude drug was admixed with 10 mg of calcium hydroxide and steam distilled at 275$^{o}$C for 90 sec. Silica gel with a suitable moisture content was used as a propellant. Hiermann and Still[30] used heating at 200$^{o}$C for 90 sec, with ammonium carbonate as propellant, for the identification of cocaine in Coca leaves. Stahl and Schmitt[48] used a temperature of 220$^{o}$C for 120 sec, and 50 mg of molecular sieve 4Å containing 20% water as propellant. Both cocaine and cinnamylcocaine and the decomposition product methylecgonidine were found on the TLC plate. An increase in temperature led to increased decomposition of the alkaloids.

Reaction chromatography of cocaine was described by Kaess and Mathis[8]. Upon heating at 100°C cocaine was completely transformed into benzoylecgonine and ecgonine. With potassium hydroxide in water at 100°C cocaine yielded benzoylecgonine. Treatment with ethanolic potassium hydroxide resulted in complete conversion into ecgonine.

Eskes[75] separated *d*- and *l*-cocaine by conversion into the diastereoisomer ecgonine-2-octanol. Cocaine was first hydrolysed to ecgonine, which was then esterificated with *d*-, *l*- or *dl*-2-octanol. The two diastereoisomers were separated on silica gel plates with methanol as the mobile phase. Bastos et al.[43] prepared the butyl derivatives of ecgonine, benzoylecgonine and benzoylnorecgonine prior to TLC separation (Table 8.1).

Schmidt[37] exposed TLC-plates with the applicated compounds for 18 hrs. to iodine vapours. After development of the plates characteristic patterns were observed for the investigated compounds, e.g. cocaine.

A summary of the TLC analysis of pseudotropine alkaloids in various materials is given in Tables 8.9-8.14.

## REFERENCES


1 D.Waldi, K. Schnackerz and F. Munter, *J. Chromatogr.*, 6 (1961) 61.
2 E. Vidic and J. Schütte, *Arch. Pharm. (Weinheim)*, 295 (1962) 342.
3 V. Schwarz and M. Sarsunova, *Pharmazie*, 19 (1964) 267.
4 E. Ragazzi, G. Veronese and C. Giacobazzi, in G.B. Marini-Bettòlo (Editor), *Thin-Layer Chromatography*, Elsevier, Amsterdam, 1964, p. 154.
5 R. Paris, R. Rousselet, M. Paris and M.J. Fries, *Ann. Pharm. Fr.*, 23 (1965) 473.
6 S. El-Gendi, W. Kisser and G. Machata, *Mikrochim. Acta*, (1965) 120.
7 W.W. Fike, *Anal. Chem.*, 38 (1966) 1697.
8 A. Kaess and C. Mathis, *Int. Symp. Chromatogr. Electrophor. Lect. Pap. 4th*, (1966) 525.
9 G.J. Dickes, *J. Ass. Public Anal.*, 4 (1966) 45.
10 I. Sunshine, W.W. Fike and H. Landesman, *J. Forensic Sci.*, 11 (1966) 428.
11 F. Wartmann-Hafner, *Pharm. Acta Helv.*, 41 (1966) 406.
12 W.W. Fike, *Anal. Chem.*, 39 (1967) 1019.
13 H.C. Hsiu, J.T. Huang, T.B. Shih, K.L. Yang, K.T. Wang and A.L. Lin, *J. Chin. Chem. Soc.*, 14 (1967) 161.
14 A. Noirfalise and G. Mees, *J. Chromatogr.*, 31 (1967) 594.
15 P.E. Haywood and M.S. Moss, *Analyst (London)*, 93 (1968) 737.
16 T. Kaniewska and B. Borkowski, *Diss. Pharm. Pharmacol.*, 20 (1968) 111.
17 A. Fiebig, J. Felczak and S. Janicki, *Farm. Pol.*, 25 (1969) 971.
18 M.H. Hashmi, S. Parveen and N.A. Chughtai, *Mikrochim. Acta*, (1969) 449.
19 I.C. Dijkhuis, *Pharm. Wkbl.*, 104 (1969) 1317.
20 M. Sarsunova, B. Kakac, K. Macek and G. Hudakova, *Z. Anal. Chem.*, 245 (1969) 154.
21 M. Overgaard-Nielsen, *Dan. Tidsskr. Farm.*, 44 (1970) 7.
22 V. Massa, F. Gal and P. Susplugas, *Int. Symp. Chromatogr. Electrophor. Lect. Pap. 6th*, (1970) 470.

23 G.H. Jolliffe and E.J. Shellard, *J. Chromatogr.*, 48 (1970) 125.
24 M. Vanhaelen, *J. Pharm. Belg.*, 25 (1970) 175.
25 A. Cavallaro, G. Elli and G. Bandi, *Boll. Lab. Chim. Prov.*, 22 (1971) 813; *C.A.*, 77 (1972) 66230w.
26 M.L. Pavielo, B.D. De Simoni and E.N. Negro, *Jornados Argent. Toxicol. Anal. Actas 1st, 1971*, (1972) 3, 351 and 353; *C.A.*, 79 (1973) 88001r.
27 S.J. Mulé, *J. Chromatogr.*, 55 (1971) 255.
28 I. Simon and A. Lederer, *J. Chromatogr.*, 63 (1971) 448.
29 G.S. Tadjer, *J. Chromatogr.*, 63 (1971) D44.
30 A. Hiermann and F. Still, *Oester. Apoth. Ztg.*, 26 (1972) 337.
31 M. Sarsunova, *Cesk. Farm.*, 22 (1973) 259.
32 E. Novakova and J. Vecerkova, *Cesk. Farm.*, 22 (1973) 347.
33 A.L. Misra, R.B. Pontani and S.J. Mulé, *J. Chromatogr.*, 81 (1973) 167.
34 N.N. Valanju, M.M. Baden, S.N. Valanju, D. Mulligan and S.K. Verma, *J. Chromatogr.*, 81 (1973) 170.
35 S. Koontz, D. Besemer, N. Mackey and R. Phillips, *J. Chromatogr.*, 85 (1973) 75.
36 J.K. Brown, R.H. Schingler, M.G. Chaubal and M.H. Malone, *J. Chromatogr.*, 87 (1973) 211.
37 F. Schmidt, *Krankenhaus-Apoth.*, 23 (1973) 10.
38 J. Paul and F. Conine, *Microchim. J.*, 18 (1973) 142.
39 K.F. Ahrend and D. Tiess, *Wiss. Z. Univ. Rostock, Math. Naturw. Reihe*, 22 (1973) 951.
40 J.M.Meola and M.Vanko, *Clin. Chem.*, 20 (1974) 184.
41 F. Schmidt, *Dtsch. Apoth. Ztg.*, 114 (1974) 1593.
42 M.L. Bastos and D.B. Hoffman, *J. Chromatogr. Sci.*, 12 (1974) 269.
43 M.L. Bastos, D. Jukofsky and S.J. Mulé, *J. Chromatogr.*, 89 (1974) 335.
44 A.C. Moffat and K.W. Smalldon, *J. Chromatogr.*, 90 (1974) 1.
45 A.C. Moffat and K.W. Smalldon, *J. Chromatogr.*, 90 (1974) 9.
46 A.C. Moffat and B. Clare, *J. Pharm. Pharmacol.*, 26 (1974) 665.
47 R.J. Armstrong, *N. Z. J. Sci.*, 17 (1974) 15.
48 E. Stahl and W. Schmitt, *Arch. Pharm. (Weinheim)*, 308 (1975) 570.
49 J.M. Meola and H.H. Brown, *Clin. Chem.*, 21 (1975) 945.
50 D. Smarzynska, A. Cegielska and J. Wolynski, *Farm. Pol.*, 31 (1975) 219.
51 T.M. Holdstock and H.M. Stevens, *Forensic Sci.*, 6 (1975) 187.
52 J.A. Vinson and J.E. Hooyman, *J. Chromatogr.*, 105 (1975) 415.
53 J.E. Wallace, H.E. Hamilton, H. Schwertner, D.E. King, J.L. McNay and K. Blum, *J. Chromatogr.*, 114 (1975) 433.
54 W.J. Serfontein, D. Botha and L.S. de Villiers, *J. Chromatogr.*, 115 (1975) 507.
55 N.C. Jain, W.J. Leung, R.D. Budd and T.C. Sneath, *J. Chromatogr.*, 115 (1975) 519.
56 W.J. Woodford, *J. Chromatogr.*, 115 (1975) 678.
57 M. Sarsunova, B. Kakac and M. Ryska, *Cesk. Farm.*, 25 (1976) 156.
58 P. Jatlow, in S.J. Mulé (Editor), *Cocaine: Chemical, Biological, Clinical, Social and Treatment Aspects*, CRC, West Palm Beach, 1976, p. 61.
59 R.L. Munier and A.M. Drapier, *C. R. Acad. Sci., Ser. C*, 283 (1976) 719.
60 M.S. Dahiya and G.C. Jain, *Indian J. Criminol.*, 4 (1976) 131; *C.A.*, 89 (1978) 37467w.
61 L. Lepri, P.G. Desideri and M. Lepori, *J. Chromatogr.*, 116 (1976) 131.
62 F.W. Grant, *J. Chromatogr.*, 116 (1976) 230.
63 L. Lepri, P.G. Desideri and M. Lepori, *J. Chromatogr.*, 123 (1976) 175.
64 R.L. Munier and A.M. Drapier, *Chromatographia*, 10 (1977) 226.
65 R.L. Munier and A.M. Drapier, *Chromatographia*, 10 (1977) 290.
66 S.J. Mulé, D. Jukofsky, M. Kogan, A. De Pace and K. Verebey, *Clin. Chem.*, 23 (1977) 796.
67 I. Ionov and I. Tsankov, *Farmatsiya (Sofia)*, 27 (1977) 25.
68 J.E. Wallace, H.E. Hamilton, J.G. Christenson, E.L. Shimek, P. Land and S.C. Harris, *J. Anal. Toxicol.*, 1 (1977) 20.

69 G. Rücker and A. Taha, *J. Chromatogr.*, 132 (1977) 165.
70 K.K. Kaistha and R. Tadrus, *J. Chromatogr.*, 135 (1977) 385.
71 K.K. Kaistha, *J. Chromatogr.*, 141 (1977) 145.
72 M.A. Müller, S.M. Adams, D.L. Lewand and R.I.H. Wang, *J. Chromatogr.*, 144 (1977) 101.
73 E. Marozzi, E. Cozza, A. Pariali, V. Gambaro, F. Lodi and E. Saligari, *Farmaco, Ed. Prat.*, 33 (1978) 195.
74 R.A. de Zeeuw, F.J.W. Mansvelt and J.E. Greving, *J. Chromatogr.*, 148 (1978) 255.
75 D. Eskes, *J. Chromatogr.*, 152 (1978) 589.

TABLE 8.1

TLC ANALYSIS OF BUTYLATED COCAINE DERIVATIVES[43]

0.25 mm Polygram silica gel sheets. Solvent systems: S1 = ethyl acetate-methanol-water (7:2:1); S2 = ethyl acetate-methanol-ammonia (15:4:1); S3 = chloroform-acetone-diethylamine (5:4:1); S4 = chloroform-acetone-ammonia (5:4, saturated); S5 = methanol-ammonia (100:1.5); S6 = benzene-ethyl acetate-methanol-ammonia (80:20:1.2:0.1); S7 = chloroform-acetone-ammonia (5:94:1).

| Solvent system No. | $hR_F$ value | | | |
|---|---|---|---|---|
| | Morphine | Butylated ecgonine | Butylated benzoyl-ecgonine | Butylated benzoyl-norecgonine |
| 1 | 7 | 8 | 45 | 49 |
| 2 | 31 | 82 | 92 | 78 |
| 3 | 7 | 84 | 91 | 79 |
| 4 | 12 | 87 | 85 | 87 |
| 5 | 34 | 52 | 64 | 58 |
| 6 | 0 | 14 | 32 | 7 |
| 7 | 9 | 78 | 75 | 64 |
| 1 + 2 | 36 | 87 | - | - |
| 1 + 3 | 20 | 92 | - | - |
| 1 + 4 | 19 | 88 | - | - |
| 1 + 5 | 44 | 55 | - | - |
| 1 + 7 | 23 | 83 | - | - |

TABLE 8.2

TLC ANALYSIS OF COCAINE AND RELATED ALKALOIDS IN PLANT MATERIAL, ACCORDING TO WARTMANN-HAFNER[11]

Silica gel G. Solvent system: methanol - diethylamine (95:5).

| Alkaloid | $hR_F$ value |
|---|---|
| Cocaine | 76 |
| Cinnamylcocaine | 76 |
| Ecgonine | 62 |
| Benzoylecgonine | 50 |
| Tropacocaine | 68 |

TABLE 8.3

TLC ANALYSIS OF COCAINE, HEROIN AND SELECTED LOCAL ANAESTHETICS[36]

Activated pre-coated silica gel plates. Solvent system: ethyl acetate - *n*-propanol - conc. ammonia (40:30:3)

| Drug | Minimum amount for detection (μg) | Colour with | | $hR_F$ value |
|---|---|---|---|---|
| | | PDAB* | AIPA** | |
| Heroin*** | ? | - | Dark brown | 45 |
| Tetracaine | 2 | - | Grey-violet | 55 |
| Procaine | 1 | Yellow | Blue-violet | 70 |
| Cocaine | 10 | - | Dark purple | 79 |
| Lidocaine | 1 | - | Light blue | 87 |
| Butacaine | 1 | Yellow | Dark blue | 89 |
| Benzocaine | 1 | Yellow | - | 89 |
| Holocaine | 5 | - | Purple-violet | 93 |

*Colour immediate upon spraying with *p*-dimethylaminobenzaldehyde (no. 36g).
**Colours ca. 5 min after spraying with acidic iodoplatinate reagent (no. 56e).
***A known sample of "Mexican Brown" heroin of unknown strength.

TABLE 8.4

TLC ANALYSIS OF COCAINE AND SOME OF ITS METABOLITES AND CONGENERS[33]

Gelman ITLC, silica gel plates. Solvent systems: S1 = Chloroform-acetone-conc. ammonia (5:94:1); S2 = benzene-ethyl acetate-methanol-conc. ammonia (80:20:1.2:0.1); S3 = chloroform-acetone-diethylamine (5:4:1); S4 = ethyl acetate-methanol-conc. ammonia (17:2:1); S5 = ethyl acetate-methanol-conc. ammonia (15:4:1); S6 = *n*-butanol-acetic acid-water (35:3:10).

| Compound | $hR_F$ value | | | | | |
|---|---|---|---|---|---|---|
| | S1 | S2 | S3 | S4 | S5 | S6 |
| Cocaine | 98 | 98 | 98 | 98 | 98 | 83 |
| Benzoylecgonine | 0 | 0 | 29 | 50 | 73 | 88 |
| Ecgonine methylester | 98 | 64 | 98 | 98 | 98 | 66 |
| Ecgonine | 0 | 0 | 33 | 11 | 28 | 48 |
| Benzoyl norecgonine | 0 | 0 | 12 | 30 | 46 | 95 |
| Norecgonine | 0 | 0 | 12 | 4 | 15 | 75 |

TABLE 8.5

TLC DATA FOR BASIC DRUGS EXTRACTED FROM URINE[34]

Silica gel G plates. Solvent systems: S1 = ethyl acetate-methanol (170:20) and 20 ml of 50% ammonium hydroxide solution in a beaker; S2 = chloroform-methanol (100:100) and 20 ml of 50% ammonium hydroxide in a beaker.

| Drug | $hR_F$ value | |
|---|---|---|
| | S1 | S2 |
| D-Amphetamine | 31 | Solvent front |
| D-Methamphetamine | 23 | Solvent front |
| Benzoylecgonine | Origin | 50 |
| Caffeine | 49 | Solvent front |
| Cocaine | 73 | -* |
| Codeine | 18 | Solvent front |
| Dihydrocodeine | 9 | Solvent front |
| Ecgonine | Origin | 20 |
| Imipramine | 49 | Solvent front |
| Nicotine | 51 | Solvent front |
| Meperidine | 35 | Solvent front |
| Methadone | 68 | -* |
| Morphine | 8 | Solvent front |
| Quinine | 18 | Solvent front |
| Procaine | 52 | Solvent front |
| Propoxyphene | 74 | -* |
| Pentazocine | 58 | Solvent front |
| Thorazine | 17 | Solvent front |
| | 40 | Solvent front |
| | 50 | Solvent front |

*Above solvent front.

TABLE 8.6

TLC ANALYSIS OF COCAINE AND BENZOYLECGONINE IN URINE, ACCORDING TO WALLACE et al.[53]

Silica gel G. Solvent system: chloroform - methanol - concentrated ammonia (100:20:1).

| Alkaloid | $hR_F$ value | Alkaloid | $hR_F$ value |
|---|---|---|---|
| Benzoylecgonine | 20 | Propoxyphene | 93 |
| Cocaine | 87 | Chlorpromazine | 89 |
| Morphine | 43 | Atropine | 33 |
| Methadone | 77 | Amitryptiline | 90 |
| Meperidine | 87 | Antistine | 66 |

TABLE 8.7

EVALUATION OF VARIOUS REAGENTS FOR THE DETECTION OF COCAINE AND BENZOYLECGONINE IN URINE[53]

5 ml urine specimens were extracted and chromatographed by the method described in[53].

| Detection reagent(s) | Cocaine (μg/ml) | | | Benzoylecgonine (μg/ml) | | |
|---|---|---|---|---|---|---|
| | 0.5 | 1.0 | 2.0 | 0.5 | 1.0 | 2.0 |
| Dragendorff's* | 100** | 100 | 100 | 12 | 55 | 67 |
| Dragendorff's + $H_2SO_4$ + $I_2$ | 100 | 100 | 100 | 83 | 100 | 100 |
| Iodoplatinate, iodoplatinate + Dragendorff's, or Dragendorff's + iodoplatinate | 13 | 75 | 92 | 0 | 0 | 62 |
| Iodoplatinate + $H_2SO_4$ | 50 | 50 | 75 | 0 | 0 | 75 |

*Munier and Macheboeuf modification.
**The values given are the percentages of specimens of that concentration determined to be positive. The number of specimens examined per combination of reagents and concentration was ten to fifteen, except for iodoplatinate followed by sulphuric acid, for which only four determinations were made.

TABLE 8.8

COLOURS OF COCAINE AND RELATED ALKALOIDS OBTAINED WITH IODOPLATINATE REAGENT[27]

| Alkaloid | Colour |
|---|---|
| Cocaine | Red-blue |
| Ecgonine | Blue |
| Benzoylecgonine | Red-blue |

TABLE 8.9

PSEUDOTROPINE ALKALOIDS IN DRUGS OF ABUSE ANALYSIS DEALT WITH IN CHAPTER 12

Reference numbers correspond to the references in the Chapter 12.

| Number of references in Chapter 12 | | | |
|---|---|---|---|
| 16 | 150 | 216 | 281 |
| 22 | 168 | 220 | 289 |
| 57 | 169 | 226 | 319 |
| 78 | 172 | 235 | 325 |
| 97 | 173 | 240 | 327 |
| 102 | 187 | 243 | 328 |
| 121 | 195 | 247 | 337 |
| 125 | 202 | 250 | 339 |
| 128 | 203 | 261 | 356 |
| 129 | 207 | 272 | 369 |
| 130 | 208 | 273 | |
| 141 | 213 | 275 | |

TABLE 8.10

LITERATURE CITED IN CHAPTER 3 WHICH INCLUDES THE ANALYSIS OF PSEUDOTROPINE ALKALOIDS

| Alkaloid* | Ref. |
|---|---|
| Coc | 1,2,3,5,6,7,9,10,12,13,14,15,28,29,32,39,41,44,45,46,47,60,61,63,73,74 |
| Coc,Me-ecg,pseudococ,tropacoc,ecg | 59,64,65 |

*Abbreviations used in Tables 8.10-8.14:

| | | | |
|---|---|---|---|
| benzecg | benzoylecgonine | ecg | ecgonine |
| benznorecg | benzoylnorecgonine | Meecg | methylecgonine |
| cincoc | cinnamylcocaine | pseudococ | pseudococaine |
| coc | cocaine | tropacoc | tropacocaine |

TABLE 8.11

TLC ANALYSIS OF PSEUDOTROPINE ALKALOIDS IN PLANT MATERIAL

| Alkaloid* | Aim | Adsorbent | Solvent system | Ref. |
|---|---|---|---|---|
| Coc,cincoc,ecg, benzecg,tropacoc | Separation of alkaloids in plant material and extracts (Table 8.2) | $SiO_2$ | EtOAc-EtOH-DMFA-DEA (75:20:5:2)<br>MeOH-DEA(95:5)<br>*tert.*-BuOH-*n*-BuOH-7% $H_2SO_4$-$H_2O$(50:50:6:24)<br>Benzene-EtOAc-DEA(7:2:1)<br>$CHCl_3$-$Me_2CO$-DEA(5:4:1) | 11 |
| Not specified | TAS technique for Coca leaves | $SiO_2$ | $CHCl_3$-$Me_2CO$-DEA(5:4:1)<br>$CHCl_3$-DEA(9:1)<br>Cyclohexane-$CHCl_3$-DEA (5:4:1)<br>Cyclohexane-DEA(9:1)<br>Benzene-EtOAc-DEA(7:2:1) | 23 |
| Coc | Identification in plant material | $SiO_2$ | MeEtCO-$CHCl_3$-MeOH-17% $NH_4OH$ (59:30:15:5) | 24 |
| Coc | TAS technique for Coca leaves | $SiO_2$ | MeOH-25% $NH_4OH$(100:1.5) | |
| Coc,cincoc,Me-ecgonidine | TAS technique for Coca leaves | $Al_2O_3$,basic | Toluene-$Me_2CO$(85:15) | 48 |

*For abbreviations, see footnote to Table 8.10.

TABLE 8.12

TLC ANALYSIS OF PSEUDOTROPINE ALKALOIDS IN PHARMACEUTICAL PREPARATIONS AND IN COMBINATION WITH OTHER COMPOUNDS

| Alkaloid* | Other compounds | Aim | Adsorbent | Solvent system | Ref. |
|---|---|---|---|---|---|
| Coc | | Separation on MgO | 2.5% $CaCl_2$-impregnated MgO | *n*-Hexane-EtOAc(9:1) | 4 |
| Coc | | Reaction chromatography | $SiO_2$ | $CHCl_3$-$Me_2CO$-DEA(5:4:1) | 8 |
| Coc,ecg | Atropine,apoatropine,scopolamine,aposcopolamine,homatropine,tropine | Detection of degradation products in eyedrops | 0.5 *M* NaOH-impregnated $SiO_2$ | $CHCl_3$-MeOH-25% $NH_4OH$ (8:2:0.2) | 17 |
| Coc | Various alkaloids | Semi-quantitative circular TLC | $SiO_2$ | $CHCl_3$-MeOH(9:1)<br>$CHCl_3$-EtOH(8.5:1.5) | 18 |
| Coc | Atropine,homatropine,scopolamine,Me-atropine Me-homatropine pilocarpine,ethylmorphine,physostigmine | Detection of impurities in eyedrops | $SiO_2$ | Benzene-$Me_2CO$-$Et_2O$-5% $NH_4OH$ (40:60:10:3.4)<br>Cyclohexane-$CHCl_3$-DEA(3:7:1) | 19 |
| Coc | Atropine | Reproducibility, influence of mobile phase | $SiO_2$ | Benzene-EtOH(9:1) | 20 |
| Coc | Atropine,homatropine,scopolamine,morphine,meperidine,strychnine | TLC on microslides | $SiO_2$ | $CHCl_3$-DEA(9:1)<br>$CHCl_3$-$Me_2CO$-DEA(5:4:1) | 21 |
| Coc | Atropine | Detection in ointments | $SiO_2$ | $CHCl_3$-$Me_2CO$-DEA(5:4:1) | 31 |
| Coc | Phenothiazines | Identification | $SiO_2$ | MeOH-25% $NH_4OH$(100:1.5)<br>$Me_2CO$-25% $NH_4OH$(99:1)<br>$CHCl_3$-dioxane-EtOAc-25% $NH_4OH$(15:60:10:5)<br>$CHCl_3$-MeOH(1:1) | |
| | | | 0.1 *M* NaOH-impregnated $SiO_2$ | $Me_2CO$ | 50 |
| Coc | Atropine,homatropine,scopolamine,procaine | Effect of detection with iodine | $Al_2O_3$ | Benzene-ethanol(9:1) | 57 |

*For abbreviations, see footnote to Table 8.10.

TABLE 8.13

TLC ANALYSIS OF PSEUDOTROPINE ALKALOIDS IN BIOLOGICAL MATERIAL

| Alkaloid* | Aim | Adsorbent | Solvent system | Ref. |
|---|---|---|---|---|
| Coc,ecg,benzecg | Detection in urine | $SiO_2$ | MeOH-$NH_4OH$(99:1)<br>EtOAc-MeOH-$NH_4OH$(85:10:10) | 27 |
| Coc,ecg,Me ecg,benzecg, norecg,benznorecg | Detection cocaine metabolites (Table 8.4) | $SiO_2$ | $CHCl_3$-$Me_2CO$-conc. $NH_4OH$(5:94:1)<br>Benzene-EtOAc-MeOH-conc. $NH_4OH$(80:20:1.2:0.1)<br>$CHCl_3$-$Me_2CO$-DEA(5:4:1)<br>EtOAc-MeOH-conc. $NH_4OH$(15:4:1),(17:2:1)<br>$n$-BuOH-AcOH-$H_2O$(35:3:10) | 33 |
| Coc,ecg,benzecg | Detection of cocaine metabolites in urine of drug abusers (Table 8.5) | $SiO_2$ | EtOAc-MeOH(170:20)+20 ml conc. $NH_4OH$ in a beaker placed in chromatography tank | 34 |
| Benzecg | Preparative TLC as sample clean-up for GLC | $SiO_2$ | MeOH-$NH_4OH$(100:1.5) | 35 |
| Coc,benzecg | Extraction of drugs of abuse from urine by means of charcoal | $SiO_2$ | $CHCl_3$-$Me_2CO$(9:1) followed by<br>$CHCl_3$-$Me_2CO$-$NH_4OH$(90:10:2)<br>$CHCl_3$-$n$-BuOH-$NH_4OH$(90:10:2) followed by<br>EtOAc-MeOH-$NH_4OH$(85:10:5) | 40,49 |
| Coc,ecg,benzecg, benznorecg | Detection in urine, TLC of butyl derivatives (Table 8.1) | $SiO_2$ | 1. EtOAc-MeOH-$H_2O$(7:2:1)<br>2. EtOAc-MeOH-$NH_4OH$(15:4:1)<br>3. $CHCl_3$-$Me_2CO$-DEA(5:4:1)<br>4. $CHCl_3$-$Me_2CO$(5:4) sat. with $NH_4OH$<br>5. MeOH-$NH_4OH$(100:1.5)<br>6. Benzene-EtOAc-MeOH-$NH_4OH$(80:20:1.2:0.1)<br>7. $CHCl_3$-$Me_2CO$-$NH_4OH$(5:94:1)<br>also subsequently development in 1+2,3,4,5 or 7<br>Two-dimensional: 1+7 | |
| Coc,benzecg | Identification in urine (Table 8.6) | $SiO_2$ | $n$-BuOH-$H_2SO_4$(95:5) sat. with $H_2O$<br>$CHCl_3$-MeOH-conc. $NH_4OH$(100:2:1,60:60:1) | 53,68 |

TABLE 8.13 (continued)

| Alkaloid* | Aim | Adsorbent | Solvent system | Ref. |
|---|---|---|---|---|
| Coc,codeine,caffeine, atropine | Identification in biological material | $SiO_2$ | $CHCl_3$-$Me_2CO$(9:1)<br>MeOH-$NH_4OH$(100:1.5)<br>Two-dimensional: I. EtOAc-MeOH-$NH_4OH$(85:10:5)<br>II. *n*-BuOH-AcOH-$H_2O$(7:2:1) | 54 |
| Coc,benzecg | Detection of basic drugs in urine (see also Chapter 12, ref. 325) | $SiO_2$ | EtOAc-MeOH-DEA(90:10:1.6)<br>$CH_2Cl_2$-MeEtCO-conc. $NH_4OH$(74:25:0.8)<br>MeOH-$CHCl_3$-conc. $NH_4OH$(90:10:0.7)<br>$Me_2CO$-MeOH(6:4) | 55 |
| Coc,ecg | Comparison of methods of urine analysis | $SiO_2$ | $CHCl_3$-MeOH-$NH_4OH$(100:20:1) | 66 |
| Coc,atropine,strychnine, brucine | Detection in biological material | $SiO_2$ | Cyclohexane-$Me_2CO$-DMA(6:3:0.2) | 67 |
| Coc,benzecg | Detection in urine (see also Chapter 12, refs. 319 and 376) | $SiO_2$ | $CHCl_3$-MeOH-$H_2O$-conc. $NH_4OH$(70:30:0.5:1) | 70 |
| Benzecg | Detection in urine | $SiO_2$ | $CHCl_3$-conc. $NH_4OH$(50:0.1)<br>Benzene-hexane-DEA(25:10:1)<br>EtOAc-MeOH-conc. $NH_4OH$(10:5:1)<br>$Me_2CO$-DEA(30:1) | 72 |

*For abbreviations, see footnote to Table 8.10.

TABLE 8.14

TLC ANALYSIS OF COCAINE IN DRUG SEIZURES

| Alkaloid* | Other compounds | Aim | Adsorbent | Solvent system | Ref. |
|---|---|---|---|---|---|
| Coc | Quinine, local anaesthetics, analgesics, amphetamines, caffeine | Identification | $SiO_2$ | $CHCl_3$-DEA(9:1)<br>Cyclohexane-DEA(9:1)<br>Cyclohexane-benzene-DEA (75:15:10)<br>Cyclohexane-benzene-$Me_2CO$-DEA(50:25:17:8)<br>MeOH-DEA(95:5) | 25 |
| Coc | Procaine, lidocaine, amphetamine | Identification in the presence of adulterants | $SiO_2$ | *n*-BuOH-MeOH-HCl(40:60:2.8) | 26 |
| Coc | Opium, heroin, LSD, peyotl, hashish, amphetamine | TAS technique | $SiO_2$ | MeOH-25% $NH_4OH$(100:1.5) | 30 |
| Coc | Heroin, local anaesthetics | Screening procedure for street drugs (Table 8.3) | $SiO_2$ | EtOAc-*n*-prOH-28% $NH_4OH$ (40:30:3) | 36,56 |
| Coc | Codeine, heroin, morphine, quinine, 6-O-Ac-morphine | Separation on microslides | 1 *M* NaOH-impregnated $SiO_2$ | $Me_2CO$-benzene(1:1) | 38 |
| *d*-coc, *l*-coc | | Separation of *d*- and *l*-cocaine as diastereoisomers | $SiO_2$ | MeOH | 75 |

*For abbreviations, see footnote to Table 8.10.

# II.3. QUINOLINE ALKALOIDS

## Chapter 9

## *CINCHONA* ALKALOIDS

Of the *Cinchona* alkaloids the two stereoisomeric alkaloids quinine and quinidine are in use as drugs. Quinine is also used in food and beverages for its bitter taste and it is often found as an adulterant of drugs of abuse (Tables 9.9 and 9.10). Hence these two alkaloids have been studied in more detail than the other *Cinchona* alkaloids. The other major *Cinchona* alkaloids are cinchonine and cinchonidine, which lack the aromatic methoxy group present in quinine and quinidine. Of these four alkaloids, referred to as the parent alkaloids, a number of derivatives are known: the epi-alkaloids, which are stereoisomers of the parent alkaloids, and the dihydro-derivatives of both parent alkaloids and epi-alkaloids. The dihydro derivatives are commonly found as impurities in the parent and epi-alkaloids.

### 9.1. SOLVENT SYSTEMS

The literature demonstrates that some solvent systems are used more often than others for the separation of the parent alkaloids. The solvents (all on silica gel plates) that have been used most often are S1[2], S3[2], S9[1], S14[13] and chloroform - methanol - diethylamine in different ratios[6,10,11,12,21,28]. Solvent system S3 was adopted in the USP XVIII[121] for the quality control of *Cinchona* alkaloids, whereas the British Pharmacopoeia 1973[73] adopted solvent system S14. With chloroform - methanol - diethylamine in combination with base-impregnated silica gel plates, a vinyl - dihydro alkaloid separation was also obtained[12,21]. The other solvents mentioned above did not separate the vinyl and the dihydro alkaloids, and for this purpose various other solvents have been used: acetone - ammonia (58:2)[19], acetone - water - 25% ammonia (80:20:1)[104], methyl ethyl ketone - ammonia (58:2)[19], methyl ethyl ketone - methanol - water (6:2:1) on 0.1 *M* sodium hydroxide-impregnated silica gel plates[44,45,105,115], acetone - methanol - diethylamine (50:50:1)[43,74], methanol - ammonia (100:1)[79] and chloroform - acetone - methanol - ammonia (60:20:20:1)[79,112], all on silica gel plates. Böhme and Bitsch[44] postulated that polar solvents would be able to separate vinyl and dihydro alkaloids, whereas non-polar solvents would not. Stöver[42] used the reaction of the vinyl alkaloids with mercury acetate to separate the vinyl from the dihydro al-

References p. 137

kaloids. The vinyl alkaloids gave polar mercury derivatives, which did not move in the solvent system used (S14), whereas the dihydro alkaloids were not affected.

To separate the epi-alkaloids from the parent alkaloids, TLC system S3 and methanol - ammonia (100:1) were used by Smith et al.[79], whereas Storck and co-workers[43,74] used solvent system S1.

According to Andary[58], system S6 gave a complete separation of the parent alkaloids. This solvent system was later used by Massa et al.[101] in the quantitative analysis of the parent alkaloids. For the separation of quinine and quinidine Pound and Sears[104] used benzene - diethylamine (1:1) in combination with silica gel plates.

Verpoorte et al.[125] made a comparative study of the TLC systems described for the separation of *Cinchona* alkaloids. Eighteen solvents were found to be suitable for the analysis of the 24 alkaloids investigated. The results of this study are summarized in Tables 9.1-9.3. None of the TLC systems tested was able to separate all 24 alkaloids in one run, but with TLC systems S6, S12, S13, S14 and S15 an optimal separation of all 24 alkaloids was obtained. Some general observations can be made.

(1) Silica gel plates in combination with basic solvents or base-impregnated silica gel plates together with neutral solvents give the best results without tailing, except for the epi-alkaloids, which in all solvents tested showed some tailing.

(2) The use of ammoniacal solvents or of base-impregnated plates usually leads to the separation of the vinyl alkaloids from the dihydro alkaloids. An increase in the pH of the mobile or stationary phase leads to an improvement in the vinyl-dihydro alkaloid separation. In addition to solvents S2, S4, S5, S10, S11, S16 and S18 in Table 9.1 the solvents listed in Table 9.2 can be used successfully for the vinyl-dihydro alkaloid separation. On comparing ammonia-containing solvents with those containing diethylamine, it is observed that in the former both Cd and epiCd have $R_F$ values equal or higher than those of C and epiC, whereas the opposite is true in the latter. The same observation is made for Q,Cd and Qd,C. Although the differences in the $R_F$ values are smaller, a similar behaviour is observed for the epi-alkaloids.

(3) Chromatography with ammonia-containing systems leads to deterioration of the separation of the four stereoisomers of each group, such as Q,Qd, EpiQ and epiQd.

(4) If diethylamine in solvent S3 is replaced with ammonia (S4), an improved separation of the vinyl and dihydro alkaloids is obtained, but the separation of the parent alkaloids and the epi-alkaloids deteriorates in comparison with that obtained with the original diethylamine-containing solvent. Solvents containing diethylamine are in general suitable for the separation of the various compounds within each group of stereoisomers.

Considering the specific separation problems of *Cinchona* alkaloids, the following conclusions were drawn (see also Table 9.3):

(1) An almost complete separation of the parent alkaloids can be achieved with solvents S6, S7, S8 and S13. Solvents S1, S3, S9, S12, S14, S15 and S17 give a nearly complete separation of three of the alkaloids and a partial separation of the fourth. None of these systems give a complete "baseline" separation of Qd and Cd, which is achieved only in solvent S2. The solvents methanol and chloroform - methanol (9:1) on base-impregnated plates are able to separate the pair cinchonidine-cinchonine from the pair quinine-quinidine.

(2) The vinyl alkaloids can be separated from the dihydro alkaloids in solvents S2, S4, S5, S10, S11, S16 and S18 and in the solvents given in Table 9.2.

(3) The epi-compounds usually have $R_F$ values higher than those of the parent alkaloids. The best separation from the parent alkaloids is obtained with solvents S1, S3, S6, S7, S8, S12, S13, S14, S15 and S18 and the best separation of the epi-alkaloids is obtained with solvents S6, S12, S13 and S17. In solvent S17 the epi-compounds have $R_F$ values between those of Q and Qd and it is therefore less suitable for the separation of the epi-alkaloids from the parent alkaloids.

(4) Separation of the parent alkaloid from its dihydro, epi-vinyl and epi-dihydro derivative is achieved with solvents S16 and S18 (not for HCd-epiCd) and partly with solvents S2 and S6. Solvents S16 and S18 give poor resolutions of Q and Qd and of C and Cd, which make them unsuitable for the separation of the stereoisomers of each group.

(5) Separation of the four possible stereoisomers of each group (Q, Qd, epiQ and epiQd, for example) can be obtained with solvents S1, S6, S12, S13, S14 and S15.

(6) Separation of the alkaloids with the same stereochemistry as quinine (Q, HQ, Cd and HCd) can best be accomplished with solvents S2, S5 and S15 and partly with solvents S7, S8, S13 and S16. The quinidine group (Qd, HQd, C and HC) can be separated best with solvents S10 and S16 and partly with solvents S2, S5, S6, S7, S8, S13 and S18. The group of epiquinine (epiQ, epiHQ, epiCd, epiHCd) can be only partly resolved with solvents S9, S12, S14 and S18. The epiquinidine group (epiQd, epiHQd, epiC, epiHC) can be partly resolved with solvents S10 and S17.

(7) The phenolic alkaloids cupreine and cupreidine were not separated in any of the solvents, nor were their dihydro derivatives separated from each other. The same was the case with quinidinone and cinchoninone. The separation of the cupreine series of alkaloids, quinidinone and cinchoninone from the other alkaloids does not give any problem, as can be seen from Table 9.4. The keto compounds can be separated from the other alkaloids in solvents S4, S8, S11, S12,

**References p. 137**

S14, S15 and S16. For the cupreine series low $R_F$ values were found in all of the systems tested, except for solvents S11 and S16. In the latter solvent the alkaloids of the cupreine series coincide with several of the other alkaloids, which makes this solvent less useful for the separation of all alkaloids.

## 9.2. TWO-DIMENSIONAL TLC

Two-dimensional TLC has been used to obtain a complete separation of the four parent alkaloids. Van Severen[6] used cyclohexane - cylohexanol - hexane (1:1:1) + 5% diethylamine and chloroform - methanol - diethylamine (80:2:0.2) in combination with silica gel plates. Kamp et al.[8a] modified the method by first running in chloroform - methanol - diethylamine (80:20:1) followed by Van Severen's first solvent. This was done because of difficulties in drying the plates after using the latter mobile phase for the first run. Kamp et al. used also two other combinations for two-dimensional TLC on silica gel plates: chloroform - *n*-butanol (1:1) saturated with 10% ammonia followed by kerosine - acetone - diethylamine (23:9:9) or Van Severen's first solvent.

Böhme and Bitsch[44] used methyl ethyl ketone - methanol - water (6:2:1) followed by benzene - isopropanol - diethylamine (4:2:1) on 0.1 *M* sodium hydroxide-impregnated silica gel plates. Instead of the latter solvent, Vermes[69] used benzene - diethyl ether - diethylamine (20:12:5) as the second solvent. Wijesekera et al.[116] described a combination of chloroform - methanol - 17% ammonia (24:6:0.05) and diethyl ether - diethylamine (17:1) on 0.1 *M* sodium hydroxide-impregnated silica gel plates for the separation of the alkaloids present in *Cinchona* bark. Pound and Sears[104] proposed the combination of the solvents acetone - water - 25% ammonia (80:20:1) and benzene - diethylamine (1:1) to separate the parent alkaloids and the dihydro bases of quinine and quinidine. Chmel and Chmelova-Hlavata[105] used methyl ethyl ketone - methanol - water (6:2:1) on 0.1 *M* potassium hydroxide-impregnated silica gel plates as the first solvent and chloroform - acetone - diethylamine (5:4:1) as the second solvent.

## 9.3. DETECTION

The commonest detection method for the *Cinchona* alkaloids makes use of the fluorescence of these compounds in acidic conditions. After development the plate is sprayed with dilute sulphuric acid or 25% formic acid in water, immersed in a mixture of diethyl ether - concentrated sulphuric acid (95:5)[57] or exposed to formic acid vapour. The aryl methoxyl-containing alkaloids show a strong blue fluorescence, stronger in 366-nm than in 254-nm light. The unsubstituted aryl-group alkaloids have a weaker dark blue fluorescence. As can be seen from Table

9.4, the modifications of Dragendorff's reagent according to Munier and Munier and Machebouef are the most sensitive of the other detection methods. The iodoplatinate reagent gave different colours for some of the alkaloids (Table 9.5) and it is also sensitive. Holdstock and Stevens[107] studied the recovery of alkaloids after detection with iodoplatinate reagent (see Chapter 2). Among the alkaloids tested was quinine.

Detection with the TCBI reagent (no. 94) of quinine, among other alkaloids, was described by Vinson and Hocyman[108] (see Table 2.3, p.17). Rücker and Taha[120] described detection of quinine with $\pi$-acceptors (see Table 2.2, p.16).

## 9.4. QUANTITATIVE ANALYSIS

Several indirect quantitative methods have been described based on the elution of the spots followed by fluorimetric[23,79] or spectrophotometric determination[43,45,47,72,74,82,122] of the alkaloids. The alkaloids were extracted with different solvents: chloroform[72,82], absolute ethanol[43], methanol - 25% ammonia (9:1)[44,45], 0.1 *M* hydrochloric acid[47,122], 0.1 *N* sulphuric acid[79,112] and ethanol - acetone (1:1)[37,40]. To dissolve both the sorbent and the alkaloids, dilute hydrochloric acid and 4 *N* sulphuric acid have been used when magnesium oxide was used as the sorbent[17,27] and 0.1 *M* hydrochloric acid when calcium carbonate and calcium sulphate were used[119]. For the indirect fluorimetric determination an excitation wavelength of 350 nm was used[23,37,79,112] and the emission was measured in dilute acid at 450 nm[23] and 455 nm[79,112] for quinidine and quinine, respectively. For the spectrophotometric analysis the wavelengths 332 nm[44,45], 324 nm[43,74] and 366 nm[47] were used for quinine and 316 nm for cinchonine[43,74].

To obtain a complete separation of quinine from the other *Cinchona* alkaloids for quantitative analysis, Böhme and Bitsch[44,45] improved the separation by first developing the plates with a polar solvent (in which quinine had a high $R_F$ value) followed by a reversed development (the plate is turned through 180°) with a less polar solvent in which quinine had a low $R_F$ value. In this way quinine could be separated from quinidine, cinchonidine, dihydroquinidine and dihydroquinine.

Massa and co-workers[50,101] studied the optimum conditions for direct quantitative analysis. The four main alkaloids were separated and, after spraying with ethanol - concentrated sulphuric acid (9:1), the absorption maximum for quinine and quinidine was found by photodensitometry to be 330 nm and for cinchonine and cinchonidine 288 nm. The detection limit was about 100 ng. After spraying with 1% sulphuric acid in ethanol, quinine and quinidine could be determined fluorimetrically by using an excitation wavelength of 365 nm and measuring at 450 nm; the detection limit was 1 ng. An excitation wavelength of 313 nm permitted determinations of cinchonine and cinchonidine at a wavelength of 410 nm with a detection limit of 5 ng (the fluorescence maximum of cinchonidine is actually at 420 nm). Quinine and quinidine are also excited with light of 313 nm, but have their fluorescence maxima at 450 nm. Direct fluorimetric analysis of TLC plates after acidification has also been performed with excitation wavelengths of 361 nm[78] and 345 nm[79] for quinine and 335 nm[111], 365 nm[83] and 366 nm[117] for quinidine and emission wavelengths of 438 nm[78] and 430 nm[79] for quinine and 455 nm[83], 430 nm[111] and 456 nm[117] for quinidine. Röder et al.[57] described a direct fluorimetric analysis of the four parent alkaloids on TLC plates after immersion in diethyl ether - concentrated sulphuric acid (95:5). Quinine and quinidine had a fluorescence maximum at 460 nm (excitation at 365 nm). Under these conditions cinchonine and cinchonidine did not interfere. Cinchonine and cinchonidine could be determined without interference from quinine and quinidine by excitation at 313 nm and measurement of the emission at 390 nm. In this way the four main alkaloids could be determined quantitatively without a complete separation. Ebel and Herold[86] used this method for the determination of quinine.

Wesley-Hadzija and Mattocks[121] determined quinidine in biological fluids by means of densitometry. The absorbance was measured at 278 nm. Hey[67] performed the semi-quantitative analysis of quinine in foods by comparing the fluorescence intensities of the spots.

Okumura et al.[109] described a modified FID method for the quantitative analysis of a number of alkaloids, e.g., quinine, after separation on glass rods covered with sintered silica gel or aluminium oxide.

## 9.5. TAS TECHNIQUE AND REACTION CHROMATOGRAPHY

Jolliffe and Shellard[54] found that the results with thermofractography of the alkaloids from *Cinchona* bark were inconsistent. The investigations of Stahl and Schmitt[89,102] showed that the pure *Cinchona* alkaloids could be volatilized at temperatures over 180°C without decomposition. However, with *Cinchona* bark only decomposition products were obtained. They were identified as simple quinoline derivatives and could be used to characterize *Cinchona* bark by the TAS technique. Chmel and Chmelova-Hlavata[115] succeeded in isolating the alkaloids present in *Cinchona* bark by thermofractography. They used a temperature of 300°C, mixed 20-40 mg of bark with 20 mg of lithium hydroxide and used 100 mg of $Ni(NH_3)Cl_2$ as propellant. Reaction chromatography of *Cinchona* alkaloids was described by Wilk and Brill[33]. Before development of the TLC plates they were exposed to iodine vapour for 18 h. After development characteristic patterns of spots were observed for the alkaloids. Schmidt[81] used this method for the identification of drugs, including quinine.

Kaess and Mathis[25] used several reagents in connection with *Cinchona* alkaloids. Chromic acid led to the formation of the less polar ketones, but in small yields, and potassium ethanolate (0.1 *N*) to the formation of a number of unidentified compounds from quinine. With acetyl chloride or acetic anhydride *Cinchona* alkaloids give the less polar acetyl compounds. According to the authors the reactions can be helpful in the identification of the alkaloids.

A summary of the TLC analysis of *Cinchona* alkaloids in various materials is given in Tables 9.6-9.11.

REFERENCES


1 K.H. Müller and H. Honerlagen, *Mitt. Dtsch. Pharm. Ges.*, 30 (1960) 202.
2 D. Waldi, K. Schnackerz and F. Munter, *J. Chromatogr.*, 6 (1961) 61.
3 J. Baümler and S. Rippstein, *Pharm. Acta Helv.*, 36 (1961) 382.
4 E. Vidic and J. Schütte, *Arch. Pharm. (Weinheim)*, 295 (1962) 342.
5 H. Feltkamp, *Dtsch. Apoth. Ztg.*, 102 (1962) 1269.
6 R. van Severen, *J. Pharm. Belg.*, 17 (1962) 40.
7 R.R. Paris and M. Paris, *Bull. Soc. Chim. Fr.*, (1963) 1597.
8 R. Neu, *J. Chromatogr.*, 11 (1963) 364.
8a W. Kamp, W.J.M. Onderberg and W.A. Seters, *Pharm. Wkbl.*, 98 (1963) 993.
9 M. von Schantz, *Thin-Layer Chromatogr. Proc. Symp. Rome*, (1963) 122; *C.A.*, 62 (1965) 7084g.
10 A. Suszko-Purzycka and W. Trzebny, *Chem. Anal. (Warsaw)*, 9 (1964) 1103; *C.A.*, 63 (1965) 431c.
11 P. Braeckman, R. van Severen and L. de Jaeger-van Moseke, *Dtsch. Apoth. Ztg.*, 104 (1964) 1211.
12 A. Suszko-Purzycka and W. Trzebny, *J. Chromatogr.*, 16 (1964) 239.
13 N. Oswald and H. Flück, *Pharm. Acta Helv.*, 39 (1964) 293.
14 J. Zarnack and S. Pfeifer, *Pharmazie*, 19 (1964) 216.
15 V. Schwarz and M. Sarsunova, *Pharmazie*, 19 (1964) 267.
16 P. Vacha, P. Cuba, V. Preininger, L. Hruban and F. Santavy, *Planta Med.*, 12 (1964) 406.
17 E. Ragazzi, G. Veronese and C. Giacobazzi, in G.B. Marini-Bettōlo (Editor), *Thin-Layer Chromatography*, Elsevier, Amsterdam, 1964, p. 149.
18 R.R. Paris, R. Rousselet, M. Paris and J. Fries, *Ann. Pharm. Fr.*, 23 (1965) 473.
19 M. Petkovic, *Arh. Farm.*, 15 (1965) 437; *C.A.*, 66 (1967) 3180m.
20 E. Marozzi and G. Falzi, *Farmaco, Ed. Prat.*, 20 (1965) 302.
21 S. Suszko-Purzycka and W. Trzebny, *J. Chromatogr.*, 17 (1965) 114.
22 D. Giacopello, *J. Chromatogr.*, 19 (1965) 172.
23 E. Ragazzi and G. Veronese, *Mikrochim. Acta*, (1965) 966.
24 W.W. Fike, *Anal. Chem.*, 38 (1966) 1697.
25 A. Kaess and C. Mathis, *Int. Symp. Chromatogr. Electrophor. Lect. Pap. 4th*, (1966) 525.
26 G.J. Dickes, *J. Ass. Public Anal.*, 4 (1966) 45.
27 F. Wartmann-Hafner, *Pharm. Acta Helv.*, 41 (1966) 406.
28 A. Suszko-Purzycka and W. Trzebny, *Poznan. Tow. Przyj. Nauk, Pr. Kom. Farm.*, 4 (1966) 43; *C.A.*, 65 (1966) 8668e.
29 M. Petkovic, *Arh. Farm.*, 17 (1967) 193; *C.A.*, 69 (1968) 54299x.
30 E. Stahl, *Dünnschichtchromatographie*, Springer, Berlin, 2nd ed., p. 815.
31 H.C. Hsiu, J.T. Huang, T.B. Shih, K.L. Yang, K.T. Wang and A.L. Lin, *J. Chin. Chem. Soc.*, 14 (1967) 161.
32 A. Noirfalise and G. Mees, *J. Chromatogr.*, 31 (1967) 594.
33 M. Wilk and U. Brill, *Arch. Pharm. (Weinheim)*, 301 (1968) 282.

34 E. Röder, E. Mutschler and H. Rochelmeyer, *Arch. Pharm. (Weinheim)*, 301 (1968) 624.
35 R. Adamski and J. Bitner, *Farm. Pol.*, 24 (1968) 17; *C.A.*, 69 (1968) 46085j.
36 M. Debackere and L. Laruelle, *J. Chromatogr.*, 35 (1968) 234.
37 E. Härtel and A. Korhonen, *J. Chromatogr.*, 37 (1968) 70.
38 J.M.G.J. Frijns, *Pharm. Wkbl.*, 103 (1968) 929.
39 G. Cesaire, F. Fauran, C. Pellissier, J. Goudote and J. Mondain, *Bull. Mem. Fac. Mixte Med. Pharm. Dakar*, 17 (1969) 245; *C.A.*, 79 (1973) 97027f.
40 G. Härtel and A. Harjanne, *Clin. Chim. Acta*, 23 (1969) 289.
41 M.H. Hashmi, S. Parveen and N.A. Chughtai, *Mikrochim. Acta*, (1969) 449.
42 D.J. Stöver, *Pharm. Wkbl.*, 104 (1969) 738.
43 J. Storck, J.P. Papin and D. Plas, *Ann. Pharm. Fr.*, 28 (1970) 25.
44 H. Böhme and R. Bitsch, *Arch. Pharm. (Weinheim)*, 303 (1970) 418.
45 H. Böhme and R. Bitsch, *Arch. Pharm. (Weinheim)*, 303 (1970) 456.
46 V. Vukecovic, *Bull. Sci. Cons. Acad. Sci. Arts RSF Yougoslav. Sect. A*, 15 (1970) 238; *C.A.*, 73 (1970) 123554y.
47 L. Hörhammer, H. Wagner and J. Hölzl, *Dtsch. Apoth. Ztg.*, 110 (1970) 227.
48 R.A. Egli, *Dtsch. Apoth. Ztg.*, 110 (1970) 987.
49 S. Gill, *Gdansk. Tow. Nauk. Rozpr. Wydz*, 37 (1970) 175; *C.A.*, 75 (1971) 64040u.
50 V. Massa, *Int. Symp. Chromatogr. Electrophor. Lect. Pap. 6th, 1970*, (1971) 470.
51 Hung-Cheh Chiang and Chu-Chi-Liu, *J. Chin. Chem. Soc.*, 17 (1970) 101; *C.A.*, 73 (1970) 69909c.
52 Hung-Cheh Chiang and Tzong-Min Chiang, *J. Chromatogr.*, 47 (1970) 128.
53 J.G. Montalvo, E. Klein, D. Eyer and B. Harper, *J. Chromatogr.*, 47 (1970) 542.
54 G.H. Jolliffe and E.J. Shellard, *J. Chromatogr.*, 48 (1970) 125.
55 G. Kananen, I. Sunshine and J. Monforte, *J. Chromatogr.*, 52 (1970) 291.
56 M. Vanhaelen, *J. Pharm. Belg.*, 25 (1969) 87.
57 K. Röder, E. Eich and E. Mutschler, *Pharm. Ztg.*, 115 (1970) 1430.
58 C. Andary, *Trav. Soc. Pharm. Montpellier*, 30 (1970) 307.
59 *United States Pharmacopeia XVIII*, United States Pharmacopeial Convention, Bethesda, 1970, pp. 581 and 582.
60 F. Pellerin and D. Mancheron, *Int. Symp. Chromatogr. Electrophor. Lect. Pap. 6th, 1970*, (1971) 536.
61 G.S. Tadjer and A. Lustig, *J. Chromatogr.*, 56 (1971) D44-D47.
62 S. Zadeczky, D. Küttel and M. Takacsi, *Acta Pharm. Hung.*, 42 (1972) 7.
63 R.L. Neman, *J. Chem. Educ.*, 49 (1972) 834.
64 K.H. Müller and H. Honerlagen, *Mitt. Dtsch. Pharm. Ges.*, 30 (1972) 202.
65 R.A. Egli, *Z. Anal. Chem.*, 259 (1972) 277.
66 M. Sarsunova, B. Kakac and L. Krasnec, *Z. Anal. Chem.*, 260 (1972) 291.
67 H. Hey, *Z. Lebensm.-Unters.-Forsch.*, 148 (1972) 1.
68 K.F. Ahrend and D. Tiess, *Zbl. Pharm.*, 111 (1972) 933.
69 M.V. Vermes, *Acta Pharm. Hung.*, 43 (1973) 25; *C.A.*, 78 (1973) 133371d.
70 M. Vermes-Vincze and Z. Vincze, *Acta Pharm. Hung.*, 43 (1973) 49; *C.A.*, 79 (1973) 5481y.
71 M. Petkovic, *Arh. Farm.*, 23 (1973) 1; *C.A.*, 80 (1974) 63802k.
72 G. Bärwald and J. Prucha, *Brauwissenschaften*, 26 (1973) 299.
73 *British Pharmacopoeia 1973*, Her Majesty's Stationery Office, London, 1973, pp. 407 and 409.
74 J. Storck and J.P. Papin, *Bull. Soc. Chim. Fr.*, (1973) 105.
75 F. Pellerin, D. Dumitrescu-Mancheron and C. Chabrelie, *Bull. Soc. Chim. Fr.*, (1973) 123.
76 E. Novakova and J. Vecerkova, *Cesk. Farm.*, 22 (1973) 347.
77 K.C. Güven and N. Güven, *Eczacilik Bull.*, 15 (1973) 77; *C.A.*, 80 (1974) 52412g.
78 P.J. Beljaars and P.J. Koken, *J. Ass. Offic. Anal. Chem.*, 56 (1973) 1284.
79 E. Smith, S. Barkan, B. Ross, M. Maienthal and J. Levine, *J. Pharm. Sci.*, 62 (1973) 1151.
80 A.M. Guyot-Hermann and H. Robert, *J. Pharm. Belg.*, 28 (1973) 557.
81 F. Schmidt, *Krankenhaus-Apoth.*, 23 (1973) 10.
82 G. Bärwald and J. Prucha, *Monatsschr. Brau.*, 26 (1973) 190.
83 C. Mulder and D.B. Faber, *Pharm. Wkbl.*, 108 (1973) 289.

84 H. Thielemann, *Sci. Pharm.*, 41 (1973) 47.
85 K.F. Ahrend and D. Tiess, *Wiss. Z. Univ. Rostock, Math.-Naturwiss. Reihe*, 22 (1973) 951.
86 S. Ebel and G. Herold, *Z. Anal. Chem.*, 266 (1973) 281.
87 A. Eichhorn and L. Kny, *Zbl. Pharm.*, 112 (1973) 567.
88 M. Petkovic, *Acta Pharm. Jugoslav.*, 24 (1974) 23; *C.A.*, 81 (1974) 25833j.
89 E. Stahl and W. Schmitt, *Arch. Pharm. (Weinheim)*, 307 (1974) 925.
90 H. Sybirska and H. Gajkzinska, *Bromatol. Chem. Toksykol.*, 7 (1974) 189; *C.A.*, 81 (1974) 164121p.
91 F. Schmidt, *Dtsch. Apoth. Ztg.*, 114 (1974) 1593.
92 D.D. Datta and C. Ghosh, *East. Pharm.*, 17 (1974) 113.
93 D.W. Chasar and G.B. Toth, *J. Chem. Educ.*, 51 (1974) 22.
94 P.D. Swaim, V.M. Loyola, H.D. Harlan and M.J. Carlo, *J. Chem. Educ.*, 51 (1974) 331.
95 A.C. Moffat, K.W. Smalldon and C. Brown, *J. Chromatogr.*, 90 (1974) 1.
96 A.C. Moffat and K.W. Smalldon, *J. Chromatogr.*, 90 (1974) 9.
97 A.C. Moffat and B. Clare, *J. Pharm. Pharmacol.*, 26 (1974) 665.
98 F. Conine and J. Paul, *Mikrochim. Acta*, (1974) 443.
99 R.J. Armstrong, *N. Z. J. Sci.*, 17 (1974) 15.
100 M. Sarsunova and J. Hrivnak, *Pharmazie*, 29 (1974) 608.
101 V. Massa, P. Susplugas and R. Taillade, *Trav. Soc. Pharm. Montpellier*, 32 (1974) 141.
102 E. Stahl and W. Schmitt, *Arch. Pharm. (Weinheim)*, 308 (1975) 570.
103 D. Zivanov-Stakic, D. Radulovic and V. Brzulja, *Arh. Farm.*, 25 (1975) 29; *C.A.*, 83 (1975) 152434w.
104 N.J. Pound and R.W. Sears, *Can. J. Pharm. Sci.*, 10 (1975) 122.
105 K. Chmel and V. Chmelova-Hlavata, *Cesk. Farm.*, 24 (1975) 433.
106 T. Inoue, M. Tatsuzawa, T. Ishii and Y. Inoue, *Eisei Shikenjo Hokoku*, 93 (1975) 31; *C.A.*, 85 (1976) 10490d.
107 T.M. Holdstock and H.M. Stevens, *Forensic Sci.*, 6 (1975) 187.
108 J.A. Vinson and J.E. Hooyman, *J. Chromatogr.*, 105 (1975) 415.
109 T. Okumura, T. Kadono and A. Iso'o, *J. Chromatogr.*, 108 (1975) 329.
110 A.C. Moffat, *J. Chromatogr.*, 110 (1975) 341.
111 J.M. Steyn and H.K.L. Hundt, *J. Chromatogr.*, 111 (1975) 463.
112 *United States Pharmacopeia XIX*, United States Pharmacopeial Convention, Rockville, 1975, pp. 434 and 436.
113 E. Vidic and E. Klug, *Z. Rechtsmed.*, 76 (1975) 283.
114 L. Lepri, P.G. Desideri and M. Lepori, *J. Chromatogr.*, 116 (1976) 131.
115 K. Chmel and V. Chmelova-Hlavata, *J. Chromatogr.*, 118 (1976) 276.
116 R.O.B. Wijesekera, L.S. Rajapakse and D.W. Chelvarajan, *J. Chromatogr.*, 121 (1976) 388.
117 J. Christiansen, *J. Chromatogr.*, 123 (1976) 57.
118 L. Lepri, P.G. Desideri and M. Lepori, *J. Chromatogr.*, 123 (1976) 175.
119 B.E. Aarø and K.E. Rasmussen, *Medd. Norsk Farm. Selsk.*, 38 (1976) 13.
120 G. Rücker and A. Taha, *J. Chromatogr.*, 132 (1977) 165.
121 B. Wesley-Hadzija and A.M. Mattocks, *J. Chromatogr.*, 144 (1977) 223.
122 S.M. Karawya and A.M. Diab, *J. Pharm. Sci.*, 66 (1977) 1317.
123 R.A. de Zeeuw, F.J.W. Mansvelt and J.E. Greving, *J. Chromatogr.*, 148 (1978) 255.
124 R.E. Kates, D.W. McKennon and T.J. Comstock, *J. Pharm. Sci.*, 67 (1978) 269.
125 R. Verpoorte, Th. Mulder-Krieger, J.J. Troost and A. Baerheim Svendsen, *J. Chromatogr.*, 184 (1980) 79.

TABLE 9.1

TLC SEPARATION OF *CINCHONA* ALKALOIDS[125]

The $hR_F$ values found in the literature are also given. The $hR_F$ values were calculated from at least six chromatograms run under the following conditions: plates, silica gel Si 60 $F_{254}$ pre-coated aluminium sheets, 20 x 20 cm (Merck); temperature, 24±2°C, relative humidity, 25±5%; normal chromatography chamber, saturated for 30 min before use.

Solvent systems:

- S1 Chloroform-diethylamine(9:1)[2,7,18,43,74,105]
- S2 Chloroform-methanol-25% ammonia(85:14:1)[48,65]
- S3 Chloroform-acetone-diethylamine(5:4:1)[2,59,61,79,100,105]
- S4 Chloroform-acetone(25% ammonia-absolute ethanol, 3:17)(5:4:1)
- S5 Chloroform-acetone-methanol-25% ammonia(60:20:20:1)[79]
- S6 Chloroform-ethyl acetate-isopropanol-diethylamine(20:70:4:6)[58,101]
- S7 Chloroform-dichloromethane-diethylamine(20:15:5)[35]
- S8 Dichloromethane-diethyl ether-diethylamine(20:15:5)[35]
- S9 Kerosine-acetone-diethylamine(23:9:9)[1,5,38,44,84]
- S10 Acetone-25% ammonia(58:2)[29]
- S11 Ethyl acetate-isopropanol-25% ammonia(45:35:5)
- S12 Toluene*-ethyl acetate-diethylamine(7:2:1)[2,47]
- S13 Toluene*-ethyl acetate-diethylamine(10:10:3)[94]
- S14 Toluene*-diethyl ether-diethylamine(20:12:5)[13,27,42,45,69,73,92]
- S15 Toluene*-diethyl ether-dichloromethane-diethylamine(20:20:20:8)[57]
- S16 Carbon tetrachloride-*n*-butanol-methanol-10% ammonia(12:9:9:1)
- S17 Cyclohexanol-cyclohexane-*n*-hexane(1:1:1) + 5% diethylamine[6,8a,11]
- S18 Methanol-25% ammonia(100:1)[79]

*The original solvent described in the literature contains the more toxic benzene. Direct comparison of benzene- and toluene-containing solvents did not show any major difference.

References p. 137

TABLE 9.1 (continued)

| Alkaloid | $hR_F$ value | | | | | | | | | | | | | | | | | | | | | | | | | | | |
|---|---|---|---|---|---|---|---|---|---|---|---|---|---|---|---|---|---|---|---|---|---|---|---|---|---|---|---|---|
| | S1 | Ref. 43 | S2 | Ref. 65 | S3 | Ref. 79 | S4 | S5 | Ref. 79 | S6 | S7 | S8 | S9 | Ref. 38 | S10 | S11 | S12 | Ref. 2 | S13 | S14 | Ref. 27 | S15 | Ref. 57 | S16 | S17 | Ref. 6 | S18 | Ref. 79 |
| Quinine(Q) | 17 | 16 | 44 | 64 | 17 | 24 | 21 | 37 | 42 | 11 | 22 | 23 | 32 | 35 | 32 | 49 | 12 | 17 | 18 | 18 | 25 | 20 | 19 | 67 | 41 | 56 | 45 | 50 |
| Dihydro-Q | 14 | 16 | 36 | | 15 | 23 | 17 | 31 | 32 | 10 | 19 | 21 | 32 | | 24 | 43 | 11 | | 17 | 17 | | 17 | | 60 | 41 | | 38 | 39 |
| Quinidine(Qd) | 28 | 36 | 44 | 66 | 26 | 45 | 26 | 41 | 46 | 21 | 34 | 35 | 41 | 45 | 37 | 55 | 20 | 25 | 28 | 26 | 44 | 29 | 37 | 71 | 60 | 74 | 46 | 52 |
| Dihydro-Qd | 24 | 29 | 35 | | 24 | 39 | 18 | 31 | 34 | 18 | 31 | 32 | 41 | | 27 | 49 | 18 | | 26 | 25 | 35 | 27 | | 62 | 58 | | 37 | 39 |
| Cinchonidine(Cd) | 25 | 29 | 38 | | 24 | 40 | 23 | 35 | 40 | 17 | 31 | 31 | 39 | 45 | 33 | 52 | 19 | | 27 | 25 | 41 | 27 | 32 | 67 | 57 | 71 | 43 | 48 |
| Dihydro-Cd | 21 | 29 | 30 | | 22 | 37 | 16 | 26 | 27 | 15 | 28 | 29 | 40 | | 25 | 46 | 18 | | 25 | 24 | 35 | 25 | | 57 | 57 | | 35 | 35 |
| Cinchonine(C) | 32 | 47 | 37 | | 32 | 54 | 23 | 34 | 38 | 24 | 40 | 40 | 44 | 50 | 34 | 53 | 24 | 27 | 33 | 31 | 54 | 33 | 54 | 65 | 67 | 80 | 39 | 43 |
| Dihydro-C | 26 | 41 | 28 | | 28 | 48 | 15 | 24 | 23 | 20 | 35 | 38 | 44 | | 24 | 45 | 22 | | 30 | 29 | 49 | 31 | | 52 | 65 | | 29 | 30 |
| Epi-Q | 52 | 68 | 48 | | 42 | 64 | 32 | 37 | 41 | 26 | 56 | 46 | 39 | | 41 | 48 | 29 | | 35 | 30 | | 38 | | 53 | 37 | | 30 | 31 |
| Dihydroepi-Q | 51 | | 37 | | 41 | | 24 | 26 | | 40 | 56 | 47 | 42 | | 31 | 40 | 30 | | 36 | 31 | | 39 | | 39 | 38 | | 19 | |
| Epi-Qd | 55 | 73 | 49 | | 44 | 68 | 33 | 39 | 43 | 31 | 59 | 50 | 42 | | 39 | 48 | 33 | | 41 | 35 | | 42 | | 52 | 44 | | 29 | 32 |
| Dihydroepi-Qd | 53 | | 38 | | 43 | | 23 | 24 | | 29 | 57 | 49 | 43 | | 30 | 39 | 32 | | 40 | 35 | | 41 | | 39 | 41 | | 19 | |
| Epi-Cd | 54 | | 46 | | 44 | | 34 | 37 | | 31 | 57 | 49 | 43 | | 43 | 49 | 34 | | 41 | 36 | | 42 | | 53 | 46 | | 30 | |
| Dihydroepi-Cd | 52 | | 34 | | 44 | | 25 | 25 | | 30 | 56 | 50 | 45 | | 33 | 40 | 34 | | 41 | 37 | | 43 | | 39 | 45 | | 20 | |
| Epi-C | 55 | | 47 | | 45 | | 33 | 38 | | 33 | 58 | 51 | 45 | | 41 | 48 | 36 | | 43 | 39 | | 44 | | 51 | 49 | | 30 | |
| Dihydroepi-C | 53 | | 35 | | 43 | | 25 | 25 | | 30 | 57 | 51 | 46 | | 31 | 40 | 36 | | 42 | 39 | | 44 | | 39 | 47 | | 19 | |
| Cupreine(Cu) | 1 | | 19 | | 1 | | 8 | 22 | | 1 | 1 | 3 | 8 | | 13 | 29 | 2 | 0 | 3 | 2 | | 3 | | 51 | 9 | | 43 | |
| Dihydro-Cu | 1 | | 15 | | 1 | | 5 | 15 | | 1 | 1 | 2 | 9 | | 8 | 22 | 2 | | 3 | 3 | | 3 | | 41 | 9 | | 34 | |
| Cupreidine(Cud) | 1 | | 20 | | 1 | | 7 | 21 | | 1 | 2 | 3 | 8 | | 11 | 29 | 2 | | 3 | 2 | | 3 | | 50 | 9 | | 42 | |
| Dihydro-Cud | 1 | | 14 | | 1 | | 4 | 12 | | 1 | 1 | 3 | 8 | | 7 | 21 | 2 | | 3 | 2 | | 3 | | 36 | 8 | | 29 | |
| Quinidinone | 60 | | 71 | | 53 | 76 | 53 | 65 | 74 | 41 | 63 | 57 | 49 | | 59 | 59 | 44 | | 51 | 47 | | 51 | | 83 | 63 | | 54 | |
| Cinchoninone | 60 | | 69 | | 52 | | 53 | 64 | | 41 | 61 | 56 | 49 | | 58 | 59 | 44 | | 50 | 47 | | 51 | | 81 | 63 | | 54 | 64 |
| HQd ar-N-oxide | 12 | | 28 | | 12 | | 9 | 22 | | 6 | 16 | 14 | 16 | | 10 | 25 | 6 | | 10 | 6 | | 10 | | 53 | 41 | | 29 | |
| HCd ar-N-oxide | 9 | | 23 | | 11 | | 6 | 16 | | 5 | 13 | 11 | 15 | | 8 | 22 | 5 | | 8 | 5 | | 7 | | 41 | 28 | | 25 | |
| Development time (min/8 cm) | 15 | | 13 | | 15 | | 13 | 13 | | 13 | 12 | 11 | 15 | | 10 | 17 | 14 | | 12 | 11 | | 12 | | 28 | 45 | | 13 | |

TABLE 9.2

SOLVENT SYSTEMS SUITABLE FOR THE SEPARATION OF VINYL AND DIHYDRO *CINCHONA* ALKALOIDS (SEE ALSO TABLE 9.3)

| Solvent system | Plate |
|---|---|
| Chloroform-methanol-17% ammonia(24:6:0.5) | Silica gel, not activated |
| Acetone-benzene-diethyl ether-25% ammonia (6:4:1:0.3) | |
| Chloroform saturated with ammonia | |
| Methanol | 0.1 *M* sodium hydroxide-impregnated silica gel |
| Chloroform-methanol(9:1) | |

TABLE 9.3

SOLVENT SYSTEMS USED FOR THE TLC SEPARATION OF *CINCHONA* ALKALOIDS

The numbers refer to solvents S1-S18 in Table 9.1 which can be used to obtain a separation within a group of alkaloids or between two groups of alkaloids or between two alkaloids. Numbers in parentheses indicate that complete baseline separation was not obtained.

| Alkaloid* | Parent alkaloids | | | | Dihydro alkaloids | | | | Epi-alkaloids | | | | Epi-dihydro alkaloids | | | |
|---|---|---|---|---|---|---|---|---|---|---|---|---|---|---|---|---|
| | Q | Qd | Cd | C | HQ | HQd | HCd | HC | epiQ | epiQd | epiCd | epiC | epiHQ | epiHQd | epiHCd | epiHC |
| Parent alkaloids | 1,3, 6-9, 12-15,17 | | | | 2,4, 5,10, 11, 16,18 | | | | 1,3, 6-8, 12 (2x), 13-15,18 | | | | 1,3 5-8, 10,12 (2x), 13-16,18 | | | |
| Q | | 1,3 (4), 6-9, (11), 12-15,17 | 1-3, 5-10, (11), 12-15,17 | 1-3, 5-10, (11), 12-15, 17, (18) | 2,4 5, (6), 10, 11, 16,18 | 1-10, 12-18 | 1-10, (11), 12-18 | 1-10, (11), 12-18 | 1, (2), 3,6-9,12-16,18 | 1, (2), 3,6-9, (5), 12-18 | 1,3, 6-10, 12-18 | 1,3, 6-10, 12-18 | 1-3, 5-10, 12-16,18 | 1-3, 5-10, 12-16, (17), 18 | 1-3, 5-10, 12-18 | 1-3, 5-10, 12-18 |
| Qd | | | 2, (4), 5,6, (7), (8), 10, (13), (16) | 1-3, (4), 5-8, (9), 10, 12-17, (18) | 1-18 | (1) 2,4, 5, (6), 10, 11, 16,18 | 1,2, (3), 4-8, 10, 11, 13, (15), 16,18 | 2,4, 5, (8), (9), 10-14, 16-18 | 1, (2), 3,5-8,11-18 | 1, (2), 3,6-8,11-18 | 1,3, (4), 5-8, 11-18 | 1,3, 5-9, 11-18 | 1-3, (4), 5,6, 7,8, 10-18 | 1-3, (4), 5-8, 10-18 | 1-3, (4), 5-18 | 1-3, (4), 5-18 |

TABLE 9.3 (continued)

| Alkaloid* | Parent alkaloids | | | | Dihydro alkaloids | | | | Epi-alkaloids | | | | Epi-dihydro alkaloids | | | |
|---|---|---|---|---|---|---|---|---|---|---|---|---|---|---|---|---|
| | Q | Qd | Cd | C | HQ | HQd | HCd | HC | epiQ | epiQd | epiCd | epiC | epiHQ | epiHQd | epiHCd | epiHC |
| Cd | | | | 1,3 6-9 12- 15,17 | 1,3- 15, (16), 17,18 | 4,5 10, (16), 18 | 2,4 5,10, 11, 16,18 | (1) 2-14, (15), 16-18 | 1-4, 6-8, 10, (11), 12-18 | 1-8 10, (11), 12-18 | 1-4 (5), 6-8, (9), 10, (11), 12-18 | 1-10, (11), 12-18 | 1,3 5-8, 10-18 | 1,3 5-8, (9), 10-18 | 1,3 5-18 | 1,3 5-18 |
| C | | | | | 1,3, (5), 6-15, (16), 17, (18) | 1,3, (5), 6,7, 8,10, (11), 12- 15, (16), 17, (18) | 1-3, 5-18 | 1,2, 5,6, (8) 10, 11, (12), (13), (15), 16,18 | 1-3, (5), 6,8- 12, (13), 15-18 | 1-3, (5), 6-8, 10-18 | 1-3, 6-8, 10-18 | 1-3, 5,6- 8,10- 18 | 1,3, 5-8, (9), 10- 13, 15-18 | 1,3, 5-8, 10-18 | 1,3, 5-8, 10-18 | 1,2, 5-8, 10-18 |
| Dihydro alkaloids | | | | | 1,3, 6-9, 12- 15,17 | | | | 1,3, 5-8, 10, 12 (2x), 13-15 | | | | 1,3, 6,7, 8,12 (2x), 13- 16,18 | | | |
| HQ | | | | | | 1,3, 6-9, (11), 12- 15,17 | 1-3, 5-9, 12- 15, (16), 17 | 1-3, 5-9, (10), 12-18 | 1-10, 12- 16, (18) | 1-10, 12- 17, (18) | 1-10, 12- 17, (18) | 1-10, 12- 17, (18) | 1,3, (4), 5-9, (11), 12- 16,18 | 1,3, (4), 5-9, (11), 12- 16,18 | 1,3, (4), 5-9, (10), (11), 12-18 | 1,3, (4), 5-9, (11), 12-18 |

| | | | | | | | | | | | | |
|---|---|---|---|---|---|---|---|---|---|---|---|---|
| HQd | 2,5, (8), (10), (11), (15), (16) | (1) 2, (3), 5, (6), (7), 8, (9), 10, (11), 12-14, (15), 16-18 | | 1,3-8,10, 12-17, (18) | 1,3-8,10, 12-17, (13) | 1,3-8, (9), 10, 12-17, (18) | 1,3-10, 12-17, (18) | | 1,3, 5-8, 11-18 | 1,3, 5-8, (9), 11-18 | 1,3, (4), 5-9, 11-18 | 1,3, (4), 5-9, 11-18 |
| HCd | | 1,3, 6-9, 12-15, (16), 17,18 | | 1-8, 10, 12-15, (16), 17 | 1-8, (9), 10, 12-15, (16), 17 | 1-8, (9), 10, 12-15, (16), 17 | 1-10, 12-15, (16), 17 | | 1-3, (4), 6-8, (10), 11-18 | 1-3, (4), 6-8, (9), (10), 11-18 | 1-4, 6-18 | 1-4, 6-9, (10), 11-18 |
| HC | | | | 1-10, 12, 13, 15, (16), 17 | 1-8, 10, 12-15,17 | 1-8, 10, 12-15, 17, (18) | 1-8, 10, 12-15,17 | | 1-4, 6-8, 10-13, 15-18 | 1-4, 6-8, 10-18 | 1-4, 6-8, 10-18 | 1-4, 6-8, 10-18 |
| Epi-alkaloids | | | (9), 12 (2x), 13-15,17 | | | | | 2, (3), 4,5, 10, 11, 16,18 | | | | |
| epiQ | | | | | (1), 6,8, (9), 12-15,17 | 6, (8), 9, (10), 12-15,17 | 1, (3), 6, (7), 8,9, 12-15,17 | | 2,4 5,8, (9), 10, 11, 16,18 | (1), 2,4, 5,8-18 | (1), 2,4-6,8-18 | (1), 2,4-6,8-18 |

TABLE 9.3 (continued)

| Alkaloid* | Parent alkaloids | | | | Dihydro alkaloids | | | | Epi-alkaloids | | | | Epi-dihydro alkaloids | | | |
|---|---|---|---|---|---|---|---|---|---|---|---|---|---|---|---|---|
| | Q | Qd | Cd | C | HQ | HQd | HCd | HC | epiQ | epiQd | epiCd | epiC | epiHQ | epiHQd | epiHCd | epiHC |
| epiQd | | | | | | | | | | | (10) | (9), (10), 12, (13), 14, (15) | 1-6, 8,10-16,18 | 2,4, 5,10, 11, 16, (17), 18 | 2,4, 5, (9), 10, 11, 16,18 | 2,4, 5,9-12, (13), 14, (15), 16, (17), 18 |
| epiCd | | | | | | | | | | | | (1), 2, (9), (12), (13), (14), (15) | (1), 2-6, 8,10-14, (15), 16-18 | 2,4, 5,10, 11, 16, (17), 18 | 2,4, 5, (9), 10, 11, 16,18 | 2,4,5, 9-11, (12), (13), 14, (15), 16,18 |
| epiC | | | | | | | | | | | | | 1-6, (7), 8, (9), 10-18 | (1), 2,4-6, (9), 10-14, (15), 16-18 | (1), 2,4, 5, (6), 10, 11, (12), (13), (15), 16,18 | 2,4,5, (6), 10,11, 16,18 |
| Epi-dihydro alkaloids | | | | | | | | | | | | | (9), 12 (2x), 13, 14, 15,17 | | | |

References p. 137

| | | | |
|---|---|---|---|
| epiHQ | (1), (3), 8, (9), (6), 12-14, (15), 17 | (1), (3), 6,8, (9), (10), 12-15, 17 | (1), (3), 6,8,9, 12-15, 17 |
| epiHQd | | (10), (12), (14), (17), | (9), 12, (13), 14,(15), 17 |
| epiHCd | | | (12), (13), |
| epiHC | | | (14), (15) |

*For abbreviations, see Table 9.1.

TABLE 9.4

TLC DETECTION OF *CINCHONA* ALKALOIDS[125]

| Reagent | No. | Sensitivity* (μg) | Background colour | Colour with parent alkaloids | Ref. |
|---|---|---|---|---|---|
| Quenching, 254 nm | | 0.1 | | | |
| Fluorescence, 366 nm (formic acid or sulphuric acid spray) | | 0.01 Qd,Q;0.1 C,Cd | | Q,Qd light blue;C,Cd dark blue | |
| Dragendorff's modification: | | | | | |
| Munier-Macheboeuf | 39c | 0.1 | Yellow | Orange-red | |
| Munier | 39b | 0.01 | Light yellow | Orange-red | |
| Munier, $NaNO_2$ | 39h | 0.01 | Light yellow-white | Brown | |
| Văgujfalvi | 39f | 0.1-1 | Light yellow | Orange | |
| Bregoff-Delwiche | 39a | 0.1 | Light yellow | Orange | |
| Iodine vapour | | 0.1 | Yellow-white | Brown | |
| Iodine in KI | 52c | 1 | White | Brown | |
| Iodine in methanol | 51d | 0.1 | Light yellow | Brown | 28 |
| Iodine vapour, pyrrole vapour | 54 | 1 | Yellow | Brown | 65 |
| Iron(III) chloride, iodine in KI | 60 | 0.1 | Light green-yellow | Brown | 91 |
| Iodoplatinate | 56d | 0.01-0.1 | Violet | Q,Qd violet;C,Cd blue | |
| Iodoplatinate, acidified | 56e | 0.1 | Dark violet | Q,Qd violet;C,Cd blue | 74 |
| Iron(III) hexacyanoferrate(III) | 80a | 10 | Light green-blue | Dark green-blue | 41 |
| Iron(III) chloride-perchloric acid | 46c | 1 | Yellow-white | Violet | |
| Methyl orange | 71 | 10 | Light orange | Orange | |
| Tetraphenylborate, quercetin | 98 | 10 | | In UV:Q,Qd blue;C,Cd yellow | 8 |
| Phenothiazine, iodine vapour | | 0.1 | Violet | Brown | 65 |
| Phenothiazine, bromine vapour (ammonia vapour) | 78 | 1 | Violet | Q,light brown;Qd green;C yellow; Cd red-brown | 65 |

*As tested in our laboratories for the parent alkaloids Q, Qd, C and Cd. For abbreviations see Table 9.1.

TABLE 9.5

DETECTION OF *CINCHONA* ALKALOIDS: COLOURS OBTAINED AFTER SPRAYING WITH DILUTE SULPHURIC ACID AND IODOPLATINATE REAGENT

| Alkaloid | Fluorescence colour (366 nm) | Iodoplatinate reagent[43] |
|---|---|---|
| Quinine | Light blue | Violet-brown |
| Quinidine | Light blue | Violet-brown |
| Dihydroquinine | Light blue | Violet-brown |
| Dihydroquinidine | Light blue | Violet-brown |
| Cinchonine | Dark blue | Blue-violet-brown |
| Cinchonidine | Dark blue | Blue |
| Dihydrocinchonine | Dark blue | Blue-violet |
| Dihydrocinchonidine | Dark blue | Blue-violet |
| Epiquinine | Light blue | Violet-brown |
| Epiquinidine | Light blue | Violet-brown |
| Dihydroepiquinine | Light blue | Violet-brown |
| Dihydroepiquinidine | Light blue | Violet-brown |
| Epicinchonine | Dark blue | Blue-violet-brown |
| Dihydroepicinchonine | Dark blue | Blue-violet |
| Epicinchonidine | Dark blue | Blue-violet-brown |
| Dihydroepicinchonidine | Dark blue | Blue-violet |
| Quinidinone | Yellow-green | Yellow-violet |
| Cinchoninone | Yellow-green | Yellow-violet |
| Cupreine | Orange-red | Light blue-violet |
| Dihydrocupreine | Orange-red | Light blue-violet |
| Cupreidine | Orange-red | Blue-violet-brown |
| Dihydrocupreidine | Orange-red | Blue-violet-brown |

TABLE 9.6

LITERATURE CITED IN CHAPTER 3 WHICH INCLUDES THE ANALAYSIS OF *CINCHONA* ALKALOIDS

| Alkaloid* | Ref. | Alkaloid* | Ref. |
|---|---|---|---|
| Q,Qd,C,cupreine | 2 | Q | 61 |
| Q,Qd | 4 | Q,Qd,C,Cd | 62 |
| Q,Qd | 14 | Q,Qd | 65 |
| Q,Qd,C,Cd | 15 | Q,Qd | 68 |
| Q,Qd,C,Cd | 18 | Q,Qd | 76 |
| Q | 20 | Qd | 85 |
| Q,Qd,C | 22 | Q | 91 |
| Q,C | 24 | Q,C | 95,96,97 |
| Q | 26 | Q | 99 |
| Q,C | 31 | C | 110 |
| Q,Qd | 32 | Q,Qd | 113 |
| Q,Qd | 34 | Q,Qd,C,Cd | 114,118 |
| Q | 48 | Q | 123 |

*For abbreviations, see Table 9.1.

TABLE 9.7

TLC ANALYSIS OF *CINCHONA* ALKALOIDS IN PLANT MATERIAL AND AS PURE SUBSTANCES

| Alkaloid* | Aim | Adsorbent | Solvent system | Ref. |
|---|---|---|---|---|
| Q,Qd,C,Cd | Analysis of *Cinchona* bark | $SiO_2$ | Kerosine-$Me_2CO$-DEA(23:9:9) | 1 |
| Q,Qd | Separation on glass rods | $SiO_2$ | Kerosine-$Me_2CO$-DEA(23:9:9) | 5 |
| Q,Qd,C,Cd | Identification (Table 9.1) | $SiO_2$ | I. Cyclohexane-cyclohexanol-hexane(1:1:1)+5% DEA<br>II. $CHCl_3$-MeOH-DEA(80:20:0.2)<br>Two-dimensional: I,II | 6,11 |
| Q,Qd,C,Cd | Separation | $SiO_2$ | $CHCl_3$-DEA(9:1) | 7 |
| Q,Qd,C,Cd,HQ,EthylQ | Separation | $SiO_2$ | I. $CHCl_3$-*n*-BuOH(1:1) sat. with 10% $NH_4OH$<br>II. Kerosine-$Me_2CO$-DEA(23:9:9)<br>III. $CHCl_3$-MeOH-DEA(80:20:1)<br>IV. Cyclohexane-cyclohexanol-hexane(1:1:1)+5% DEA<br>Two-dimensional: I,II or III,IV or I,IV | 8a |
| Q,Qd,C,Cd,HQ,HQd,HC,HCd | Separation of vinyl and dihydro alkaloids, quality control | 0.1 *M* NaOH-impregnated $SiO_2$ | $CHCl_3$-MeOH-DEA(80:20:1,50:50:1) | 10,12<br>21,28 |
| Q,Qd,C,Cd | Direct quantitative analysis (spot areas) | $SiO_2$ | Benzene-isoprOH-DEA(4:2:1)<br>Benzene-$Et_2O$-DEA(20:12:5), developed twice | 13 |
| Q,Qd,C,Cd,EpiQ,EpiQd, quininone | Analysis of *Cinchona* bark | $Al_2O_3$<br>$SiO_2$ | *n*-Hexane-$CCl_4$-DEA(5:4:1)<br>Benzene-MeOH(8:2,6:4) | 16 |
| Q,Qd,C,Cd | Separation on MgO | 2.5% $CaCl_2$-impregnated MgO | EtOAc-$Me_2CO$(4:1)<br>EtOAc-$Me_2CO$-$CHCl_3$(2:2:1) | 17 |
| Q | Reaction chromatography | $SiO_2$ | $CHCl_3$-$Me_2CO$-DEA(5:4:1) | 25 |
| Q,Qd,C,Cd,HQ,HC,HCd | Identification (Table 9.1) | $SiO_2$ | Benzene-$Et_2O$-DEA(20:12:5), developed twice | 27 |
| Q,HQ | Separation | $SiO_2$ | $Me_2CO$-$NH_4OH$(58:2)<br>MeEtCO-$NH_4OH$(58:2) | 29 |
| Q,Qd,C,Cd,HQ,HQd,HC, Hcd | Separation | $SiO_2$ | $CH_2Cl_2$-$Et_2O$-DEA(20:15:5)<br>$CHCl_3$-$CH_2Cl_2$-DEA(20:15:5)<br>$CH_2Cl_2$-$CHCl_3$-$Et_2O$-DEA(35:20:15:5) | 35 |

| | | | | |
|---|---|---|---|---|
| Q,Qd,C,Cd | Identification in *Cinchona* bark (Table 9.1) | $SiO_2$ | Kerosine-$Me_2CO$-DEA(23:9:9) | 38 |
| Q,Qd,HQ,HQd | Vinyl-dihydro separation after reaction with $Hg^{2+}$ | $SiO_2$ | Benzene-$Et_2O$-DEA(20:12:5) | 42 |
| Q,Qd,C,Cd,HQ,HQd,HC, HCd,epiQ,epiQd,quininone, quinicine | Separation and indirect quantitative analysis(UV),vinyl-dihydro separation (II) (Table 9.1) | $SiO_2$ | I. $CHCl_3$-DEA(9:1)<br>II. $Me_2CO$-MeOH-DEA(50:50:1) | 43,74 |
| Q,Qd,C,Cd,HQ,HQd | Separation and purification | $SiO_2$<br>0.1 *M* NaOH-impregnated $SiO_2$ | Kerosine-$Me_2CO$-DEA(23:9:9)<br>I. MeEtCO-MeOH-$H_2O$(6:2:1)<br>II. Benzene-isoprOH-DEA(4:2:1)<br>Two-dimensional: I,II | 44 |
| Q,Qd,C,Cd,HQ,HQd | Indirect quantitative analysis in *Cinchona* bark | 0.1 *M* NaOH-impregnated $SiO_2$ | MeEtCO-MeOH-$H_2O$(6:2:1), turn the plate through 180° and develop with benzene-isoprOH-DEA(4:2:1) or benzene-$Et_2O$-DEA(20:12:5) or $CHCl_3$-$Me_2CO$-DEA (5:4:1) | 45 |
| Q,Qd,C,Cd | Indirect quantitative analysis in *Cinchona* bark | $SiO_2$ | Benzene-EtOAc-DEA(7:2:1) | 47 |
| Q,C,Cd | Separation | $SiO_2$-$Al_2O_3$-Kieselguhr (1:1:1) | $CHCl_3$-EtOH(9:1) | 49 |
| Q,C,Cd | Identification in *Cinchona* bark | $SiO_2$ | Cyclohexane-$Me_2CO$-$CHCl_3$-$isopr_2O$-DEA(20:10:5:5:5) | 56 |
| Q,Qd,C,Cd | Fluorodensitometric analysis (Table 9.1) | $SiO_2$ | Benzene-$Et_2O$-$CH_2Cl_2$-DEA(20:20:20:8) | 57,86 |
| Q,Qd,C,Cd | Separation | $SiO_2$ | $CHCl_3$-EtOAc-isoprOH-DEA(20:70:4:6) | 58 |
| Q,Qd,C,Cd | Influence of $I_2$ detection on recovery of alkaloids | $Al_2O_3$ | Benzene-EtOH(95:5) | 66 |
| Q,Qd,C,Cd | Separation | $SiO_2$ | I. MeEtCO-MeOH-$H_2O$<br>II. Benzene-$Et_2O$-DEA(20:12:5)<br>Two-dimensional: I,II | 69 |
| Q,Qd,C,Cd | Quantitative analysis | | No details available | 70 |

TABLE 9.7 (continued)

| Alkaloid* | Aim | Adsorbent | Solvent system | Ref. |
|---|---|---|---|---|
| Q,Qd,C,Cd,HQ,HCu,optochin,eucopin | Relationship between adsorption and structure | $SiO_2$ | MeOH,abs. EtOH,MeEtCO-conc. $NH_4OH$(29:1),MeOAc-conc. $NH_4OH$(10:1),EtOAc-conc. $NH_4OH$(10:1),EtOAc-$Me_2CO$-conc. $NH_4OH$(30:5:3),$Me_2CO$-DEA(29:1,14:1,9:1), MeEtCO-DEA(9:1),MeOAc-DEA(9:1),EtOAc-$Me_2CO$-DEA (9:10:1),*n*-BuOAc-$Me_2CO$-DEA(9:10:1),*n*-AmOAc-$Me_2CO$-DEA(9:10:1),$CHCl_3$-$Me_2CO$-DEA(9:10:1),$CCl_4$-$Me_2CO$-DEA(9:10:1),dichloroethane-$Me_2CO$-DEA(9:10:1), trichloroethylene-$Me_2CO$-DEA(9:10:1) | 71 |
| Q,Qd | Identification | $SiO_2$ | $CHCl_3$-$Me_2CO$-DEA(50:45:5)<br>BuOH-$H_2O$-AcOH(3:1:1) | 77 |
| Q,Qd,C,Cd,HQ,HQd,HC, HCd,epiQ,epiQd,quininone,quinicine | Direct and indirect quantitative analysis (Table 9.1) | $SiO_2$ | $CHCl_3$-$Me_2CO$-MeOH-$NH_4OH$(60:20:20:1)<br>MeOH-$NH_4OH$(100:1)<br>$CHCl_3$-$Me_2CO$-DEA(5:4:1) | 79 |
| Q,Qd,HQd | Separation | $SiO_2$ | Benzene-$Me_2CO$-DEA(8:2:1)<br>$CHCl_3$-$Me_2CO$-DEA(50:45:5) | 84 |
| Q,Qd,C,Cd,HQ,HQd | Separation, purity control | $SiO_2$ | $CHCl_3$-$Me_2CO$-DEA(50:35:15), developed twice | 87 |
| Q,Qd,C,Cd | Separation of diastereoisomers | $SiO_2$ | MeOH, abs. EtOH,MeEtCO-conc. $NH_4OH$(29:1),EtOAc-conc. $NH_4OH$(10:1)<br>$Me_2CO$-DEA(29:1),MeEtCO-DEA(9:1),MeEtCO-DEA(9:1), *n*-AmOAc-$Me_2CO$-DEA(9:10:1),$CHCl_3$-$Me_2CO$-DEA(9:10:1) | 88 |
| Q,Qd,C,Cd | TAS technique for *Cinchona* alkaloids | $SiO_2$ | Benzene-$Et_2O$-DEA(55:35:10), developed twice | 89 |
| Q,Qd,C,Cd | Control homeopathic *Cinchona* mother tincture | $SiO_2$ | Benzene-$Et_2O$-DEA(20:12:5) | 92 |
| Q,Qd,C,Cd | TLC and GLC *Cinchona* alkaloids | $SiO_2$ | $CHCl_3$-$Me_2CO$-DEA(5:4:1) | 100 |
| Q,Qd,C,Cd | Densitometric analysis (UV and fluorescence) | $SiO_2$ | EtOAc-$CHCl_3$-isoprOH-DEA(70:20:4:6) | 101 |
| Q | TAS technique for *Cinchona* bark | $SiO_2$ | Benzene-$Me_2CO$(9:1) | 102 |
| Q,Qd,C,Cd,HQ,HQd | Separation | $SiO_2$ | $Me_2CO$-$H_2O$-25% $NH_4OH$(80:20:1)<br>Benzene-DEA(1:1) | 104 |

| | | | | |
|---|---|---|---|---|
| Q,Qd,C,Cd,HQ,HQd | Separation | 0.1 *M* KOH-impregnated $SiO_2$ | I. MeEtCO-MeOH-$H_2O$(6:2:1)<br>II. $CHCl_3$-$Me_2CO$-DEA(5:4:1) or (50:35:15)<br>III. $CHCl_3$-DEA(9:1)<br>Two-dimensional: I,II | 105 |
| Q,Qd,C,Cd | TAS technique for *Cinchona* bark | 0.05 *M* KOH-impregnated $SiO_2$ | MeEtCO-MeOH-$H_2O$(6:2:1) | 115 |
| Q,Qd,C,Cd | Separation | 0.1 *M* NaOH-impregnated $SiO_2$ | I. $CHCl_3$-MeOH-17% $NH_4OH$(24:6:0.05)<br>II. $Et_2O$-DEA(17:1)<br>Two-dimensional: I,II | 116 |
| Q,Qd | Indirect quantitative analysis in biological samples, pharmaceutical preparations and *Cinchona* bark (colorimetric) | $SiO_2$ | Benzene-EtOAc-DEA(2:2:1) | 122 |

*For abbreviations, see Table 9.1.

TABLE 9.8

TLC ANALYSIS OF *CINCHONA* ALKALOIDS IN PHARMACEUTICAL PREPARATIONS AND IN COMBINATION WITH OTHER PURE COMPOUNDS

| Alkaloid* | Other compounds | Aim | Adsorbent | Solvent system | Ref. |
|---|---|---|---|---|---|
| Q | Various alkaloids | Toxicological analysis | $SiO_2$ | $Me_2CO$-MeOH(1:1)<br>$Me_2CO$-MeOH-triethanolamine(1:1:0.03) | 3 |
| Q,Qd,C,Cd, HQ | Hydrastinine,etacridine | Separation | $SiO_2$ | I. $CHCl_3$-MeOH-DEA(80:20:1)<br>II. Kerosine-$Me_2CO$-DEA(23:9:9)<br>III. $CHCl_3$-*n*-BuOH(1:1) sat. with 10% $NH_4OH$<br>Two-dimensional: I,II or III,II | 8a |
| Q | Morphine,papaverine | Circular TLC | $SiO_2$ | MeOH-$Me_2CO$-25% $NH_4OH$(100:45:5) | 9 |
| Q | | Quality control | $SiO_2$ | $Me_2CO$-$H_2O$(8:2) | 19 |
| Q | Hydrastine,reserpine | Indirect quantitative fluorimetric analysis | 2.5% $CaCl_2$-impregnated MgO | EtOAc-$Me_2CO$(4:1) | 23 |
| Q,C | Aminoquinoline derivatives | Identification of antimalaria drugs | 0.02 *M* NaOAc-impregnated $SiO_2$ | MeOH-conc. $NH_4OH$(98:2)<br>Dichloroethylene-MeOH-conc. $NH_4OH$(90:10:1) | 39 |
| Q,C | Various alkaloids | Semi-quantitative circular TLC | $SiO_2$ | $CHCl_3$-EtOH(9:1,85:15) | 41 |
| Q | Analgesics | Identification | | No details available | 46 |
| Q | Antipyretics | Identification | $SiO_2$-polyamide | No details available | 51 |
| Q | Antipyretics | Identification | Polyamide, Kieselguhr, or a mixture of these(1:5) | $CHCl_3$-cyclohexane-dioxane-AcOH(40:60:10:1)<br>$CHCl_3$-cyclohexane-AcOH(40:60:1) | 52 |
| Q | | Comparison of adsorbents from different manufacturers and different methods of development | $SiO_2$ | 96% EtOH<br>EtOAc-MeOH-conc. $NH_4OH$(85:10:5)<br>Cyclohexane-benzene-DEA(75:15:10) | 55 |

| | | | | | |
|---|---|---|---|---|---|
| Q | Analgesics and anti-pyretics | Identification and indirect quantitative analysis(UV) | $SiO_2$ | Benzene-$Et_2O$-MeOH-AcOH(120:120:1:18)<br>$CHCl_3$-96% EtOH(99:1)<br>Cyclohexane-$CHCl_3$-pyridine(20:60:5) | 60,75 |
| Q | Acetylsalicylic acid, caffeine,pholcodine, khelline,promethazine | Identification | $SiO_2$ | I. BuOH-$H_2O$-AcOH(4:5:1) organic phase<br>II. 0.15% aq. tartaric acid<br>Two-dimensional: I,II | 80 |
| Q | Brucine,caffeine | Separation on microslides | $SiO_2$ | MeOH | 93 |
| Q | Caffeine, brucine, nicotine,strychnine | Separation | $SiO_2$ | Benzene-EtOAc-DEA(10:10:3) | 94 |
| Q,Qd | Papaverine,phenobarbital,procainamide | Identification on microslides | | No details available | 103 |
| Q | Aminopyrine,caffeine, phenacetin | Direct quantitative analysis | | No details available | 106 |
| Q | Codeine | Indirect quantitative analysis | $CaCO_3$+$CaSO_4$ (1:1) or $CaSO_4$ | Cyclohexane-DEA(9:1) | |
| | | | $SiO_2$ | $CHCl_3$-MeOH-DEA(6:3:1) | 119 |

*For abbreviations, see Table 9.1.

TABLE 9.9

LITERATURE CITED IN CHAPTER 12 WHICH INCLUDES THE ANALYSIS OF QUININE (ANALYSIS OF DRUGS OF ABUSE IN BIOLOGICAL MATERIAL)

Numbers given are references from Chapter 12, in parentheses numbers of references in this chapter.

| | | | | | | |
|---|---|---|---|---|---|---|
| 22 | 141 | 174 | 207 | 243 | 296 (98) | 339 |
| 68 | 147 (53) | 175 | 210 | 244 | 307 | 348a |
| 78 | 159 | 183 | 213 | 255 | 318 | 374 |
| 87 | 167 | 187 | 220 | 261 | 319 | 380 |
| 102 | 168 | 195 | 226 | 273 | 329 | |
| 112 (36) | 169 | 200 (63) | 235 | 275 | 332 | |
| 128 | 172 | 202 | 241 | 289 | 337 | |
| 129 | 173 | 204 | 242 | 295 | 338 | |

TABLE 9.10

LITERATURE CITED IN CHAPTER 12 WHICH INCLUDES THE ANALYSIS OF QUININE (ANALYSIS OF DRUGS OF ABUSE IN DRUG SEIZURES)

Numbers given are references from Chapter 12.

| | |
|---|---|
| 57 | 328 |
| 60 | 356 |
| 150 | 369 |
| 190 | 379 |
| 255 | 384 |
| 312 | 390 |

References p. 137

TABLE 9.11

TLC ANALYSIS OF *CINCHONA* ALKALOIDS IN BIOLOGICAL MATERIAL AND FOOD

| Alkaloid* | Other compounds | Aim | Adsorbent | Solvent system | Ref. |
|---|---|---|---|---|---|
| Q | Strychnine, brucine, caffeine, cocaine, opium alkaloids, lobeline, sparteine | Detection of doping in biological specimens from horses | $SiO_2$ | I. $CHCl_3$-MeOH-DEA(95:5:0.05)<br>II. Hexane-$Me_2CO$-DEA(6:3:1)<br>Two-dimensional: I,II | 36 |
| Qd,HQd, metabolites | | Indirect quantitative analysis in plasma (fluorimetric) | $SiO_2$ | MeOH-$Me_2CO$(4:1) | 37,40 |
| Q | | Densitometric analysis in food | $SiO_2$ | Benzene-MeOH-DEA(45:6:2) | 67 |
| Q | | Indirect quantitative analysis(UV) in tonic and beer | $SiO_2$ | $CHCl_3$-$Me_2CO$-DEA(4:4:1) | 72,82 |
| Q,Qd,HQd, metabolites | | Fluorodensitometric analysis in plasma | $SiO_2$ | $CHCl_3$-MeOH-AcOH(75:15:10) | 83 |
| Q | Glutethimide, hydroxyzine, aminopyrine, caffeine | Identification in cadaveric material | $SiO_2$<br>$Al_2O_3$ | MeOH-$NH_4OH$(100:1)<br>$CHCl_3$-$Et_2O$(1:1) | 90 |
| Qd | | Fluorodensitometric analysis in human serum | $SiO_2$ | Not mentioned | 111 |
| Qd | | Fluorodensitometric analysis in serum | $SiO_2$ | Benzene-dioxane-EtOH-25% $NH_4OH$(50:40:5:3) | 117 |
| Qd,HQd, metabolites | | Densitometric analysis in biological fluids | $SiO_2$ | MeOH-$Me_2CO$(5:1) | 121 |
| Q,Qd | | Indirect quantitative analysis (colorimetric) in blood | $SiO_2$ | Benzene-EtOAc-DEA(2:2:1) | 122 |
| Qd,HQd | | Comparison of HPLC and TLC-fluorimetric analysis in plasma | | Not specified | 124 |

*For abbreviations, see Table 9.1.

II.4. PHENYLETHYLAMINE AND ISOQUINOLINE ALKALOIDS

Chapter 10

# CACTUS ALKALOIDS

10.1. Solvent systems ......................................................... 159
10.2. Detection ......................................................... 160
10.3. Quantitative analysis ......................................................... 161
10.4. TAS technique and reaction chromatography ......................................................... 161
References ......................................................... 152

Most of the work carried out on the separation of cactus alkaloids concerns the identification of mescaline and peyote, the cactus in which mescaline occurs.

In a number of investigations on the identification of drugs of abuse in urine or narcotic seizures, mescaline is also included (Tables 10.6 and 10.7). The cactus alkaloids are derived biosynthetically from the amino acid tyrosine. Via tyramine a number of alkaloids are formed, with primary, secondary and tertiary nitrogens, and one or more aromatic methoxy or hydroxy groups. Because of the primary amine group present in a number of cactus alkaloids, including mescaline, the TLC of these alkaloids is often described in combination with other primary amines.

## 10.1. SOLVENT SYSTEMS

McLaughlin and Paul[5] used a number of solvents for the identification of the alkaloids in the peyote cactus. TLC system S1 was found to be the most effective (Table 10.1), and S2 and S5 were used for the further identification of phenolic alkaloids. Methanol - carbon tetrachloride - glacial acetic acid (28:12:1) was recommended for the separation of the quaternary alkaloids. Two-dimensional TLC was also used. Lundström and Agurell[9] carried out extensive studies on the TLC of cactus alkaloids (Table 10.1).

Derivatization of primary amines and phenolic amines to give coloured or fluorescent compounds, such as dansyl derivatives, has often been used in TLC. Seiler and Weichmann[3,6,7] described the preparation and separation of a number of dansyl derivatives of amines, including cactus alkaloids. Ho et al.[17] identified some drugs of abuse as dansyl derivatives. The dansylated compounds, including mescaline, were separated on 3 x 3 cm polyamide plates. Two-dimensional TLC of some dansylated cactus alkaloids was carried out by Neal and McLaughlin[21], because it was difficult to separate the alkaloids as such (Table 10.2).

Beckett and Choulis[2] observed multiple spot formation with amines, including β-phenethylamines, when neutral or weakly acidic solvent systems were used for the separation of the amine salts, or when salts were present in the sample. Similar results were reported by Wesley-Hadzija[26]. The phenomenon was explained by the formation of ion pairs, which have $hR_F$ values different from those of the free amines.

## 10.2. DETECTION

The cactus alkaloids can be detected with the common alkaloid spray reagents, iodoplatinate and Dragendorff's reagent. To obtain more sensitive detection a number of different derivatives of the primary and secondary amines have been prepared, including the dansyl derivatives[5,21] (no. 28). Dansyl chloride spray reagent was found to be more sensitive than other alkaloid spray reagents: the alkaloids were observed as yellow-orange fluorescent spots in UV light. However, N-acetylmescaline did not react and lophophorine and pellotine required about 30 min for the fluorescence to develop. N-Acetylmescaline did not react with any of the alkaloid spray reagents, but could be detected by spraying with a 20% antimony(V) chloride chloroform solution, giving a pale gold colour[5].

For the phenolic alkaloids McLaughlin and Paul[5] used tetrazotized benzidine (no. 32). This reagent could be used to distinguish *p*-hydroxy phenols (yellow) from 8-hydroxytetrahydroisoquinoline phenols (red).

Genest and Hughes[11] detected some hallucinogenic phenethylamines, including mescaline, by consecutive spraying with 10% sodium acetate in water and 1% 2,6-dibromo-*p*-benzoquinone-4-chlorimide in ethanol, then placing the plates in iodine vapour, giving yellow spots. This method of detection was found to be more sensitive than the methods using the common alkaloid spray reagents, including dansyl chloride and ninhydrin.

Heacock and Forrest[25] tested a number of electron-acceptor reagents for the detection of some hallucinogenics. Only fluoranil (no. 41) gave positive reactions for mescaline (purple). TCBI reagent (no. 94)[33] has also been used for the detection of mescaline (see Chapter 2, p.17).

Lundström and Agurell[9] used *o*-dianisidine in dilute hydrochloric acid (no. 31) for the detection of cactus alkaloids and Ranieri and McLaughlin[32] used fluorescamine (no. 42). The latter reagent could distinguish secondary from primary amines and did not interfere with consecutive spraying with dansyl chloride, iodoplatinate reagent or tetrazotized benzidine. The conjugates formed with fluorescamine and primary amines did not change colour on spraying the TLC plates with dansyl chloride, but secondary amines became yellow fluorescent. In this way it is also possible to distinguish between primary and secondary amines (Table 10.3). Phenols and imidazoles do not react with fluorescamine.

Stahl[13] used anisaldehyde-sulphuric acid in combination with molybdophosphoric acid and heating to detect some peyote alkaloids (no. 5). In the detection of drugs of abuse ninhydrin reagent has often been used for the detection of amines. This reagent is generally used in a series of consecutive spray reagents[15,22,30,31,36-38] (see Chapter 12).

## 10.3. QUANTITATIVE ANALYSIS

Seiler and Weichmann[6] determined amines quantitatively by direct and indirect analysis as dansyl derivatives. Direct quantitative analysis was performed by separation of the amines as dansyl derivatives. After development of the plates they were sprayed with triethanolamine - isopropanol (2:8) and the fluorescent spots of the amines were determined by fluorodensitometry. For indirect quantitative analysis methanol - concentrated ammonia (95:5), benzene - triethylamine (95:5) and benzene - acetic acid (99:1) have been used for the elution of the dansyl derivatives from the plate. Because the dansyl derivatives are light sensitive, the best results are obtained if the plates are kept in the dark.

Neal and McLaughlin[21] eluted the dansyl derivatives of cactus alkaloids from the plate with ethyl acetate and identified them by means of mass spectrometry.

Lum and Lebish[30] described a method for the elution of the alkaloids from TLC plates for further identification by means of colour reactions, IR spectroscopy or UV spectroscopy. The eluent used was 0.1 *M* hydrochloric acid for UV spectroscopy and chloroform - methanol - concentrated ammonia (80:20:1) for IR spectroscopy.

## 10.4. TAS TECHNIQUE AND REACTION CHROMATOGRAPHY

The TAS technique for peyote was first described by Stahl[13]. A 25-mg amount of plant material was heated at 250°C for 90 sec, without a propellant. Hiermann and Still[23] used the same conditions, but used ammonium carbonate as a propellant.

Reaction chromatography of cactus alkaloids has been applied by a number of workers. The most usual reaction is the dansylation (with 1-dimethylaminonaphthalene-5-sulphonic acid) of primary or secondary amine groups or phenolic groups[3,6,7,17,21]. The dansyl derivatives have strong fluorescence and permit quantitative analysis with a sensitivity of $10^{-8}$-$10^{-12}$ *M*[6]. Seiler and Weichmann[6] prepared the dansyl derivatives prior to the TLC separation by treating the amine with dansyl chloride in aqueous acetone solution (10-30% water) at a pH of about 8. Alcohols should not be present, because of the formation of dansyl esters which

**References p. 162**

may interfere. McLaughlin and Paul[5] used the dansyl reagent also for the detection of cactus alkaloids after development of the TLC plate.

Jart and Bigler[8] separated primary and secondary amines as 4-(phenylazo)benzenesulphonamides, whereas Hudson and Rice[35] identified amines in illicit preparations by means of their NDB chloride derivatives (reaction with a 1% solution of 4-chloro-7-nitrobenzo-2,1,3-oxadiazole in 0.1 *M* sodium hydrogen carbonate solution). The derivatives were prepared prior to the TLC analysis.

## REFERENCES


1 J. Cochin and J.W. Daly, *Experientia,* 18 (1962) 294.
2 A.H. Beckett and N.H. Choulis, *J. Pharm. Pharmacol.,* 15 (1963) 236T.
3 N. Seiler and M.Weichmann, *Experientia,* 21 (1965) 293.
4 J.A. Steele, *J. Chromatogr.,* 19 (1965) 300.
5 J.L. McLaughlin and A.G. Paul, *Lloydia,* 29 (1966) 315.
6 N. Seiler and M. Weichmann, *Z. Anal. Chem.,* 220 (1966) 109.
7 N. Seiler and M. Weichmann, *J. Chromatogr.,* 28 (1967) 351.
8 A. Jart and A.J. Bigler, *J. Chromatogr.,* 29 (1967) 255.
9 J. Lundström and S. Agurell, *J. Chromatogr.,* 30 (1967) 271.
10 J.J. Thomas and L. Dryon, *J. Pharm. Belg.,* 22 (1967) 163.
11 K. Genest and D.W. Hughes, *Analyst (London),* 93 (1968) 485.
12 A. Noirfalise, *J. Pharm. Belg.,* 23 (1968) 387.
13 E. Stahl, *Analyst (London),* 94 (1969) 723.
14 S.J. Mulé, *J. Chromatogr.,* 39 (1969) 302.
15 S.J. Mulé, M.L. Bastos, D. Jukofsky and E. Saffer, *J. Chromatogr.,* 63 (1971) 289.
16 M. Steinigen, *Pharm. Ztg.,* 116 (1971) 2072.
17 I.K. Ho, H.H. Loh and E.L. Way, *Proc. West. Pharmacol. Soc.,* 14 (1971) 183.
18 E. Röder and K.H. Surborg, *Z. Anal. Chem.,* 256 (1971) 362.
19 M. Steinigen, *Deut. Apoth.-Ztg.,* 112 (1972) 51.
20 J.K. Brown, L. Shapazian and G.D. Griffin, *J. Chromatogr.,* 64 (1972) 129.
21 J.M. Neal and J.J. McLaughlin, *J. Chromatogr.,* 73 (1972) 277.
22 K.K. Kaistha and J.H. Jaffe, *J. Pharm. Sci.,* 61 (1972) 679.
23 A. Hiermann and F. Still, *Oester. Apoth.-Ztg.,* 26 (1972) 337.
24 R.A. van Welsum, *J. Chromatogr.,* 78 (1973) 237.
25 R.A. Heacock and J.E. Forrest, *J. Chromatogr.,* 78 (1973) 241.
26 B. Wesley-Hadzija, *J. Chromatogr.,* 79 (1973) 243.
27 M.L. Bastos, D. Jukofsky and S.J. Mulé, *J. Chromatogr.,* 81 (1973) 93.
28 P.M. Kullberg and C.W. Gorodetzky, *Clin. Chem.,* 20 (1974) 177.
29 R.J. Bussey and R.C. Backer, *Clin. Chem.,* 20 (1974) 302.
30 P.W. Lum and P. Lebish, *J. Forensic Sci. Soc.,* 14 (1974) 63.
31 K.K. Kaistha, R. Tadrus and R. Janda, *J. Chromatogr.,* 107 (1975) 359.
32 R.L. Ranieri and J.L. McLaughlin, *J. Chromatogr.,* 111 (1975) 234.
33 J.A. Vinson, J.E. Hooyman and C.E. Ward, *J. Forensic Sci.,* 20 (1975) 552.
34 E. Spratt, *Toxicol. Annu. 1974,* (1975) 229.
35 J.C. Hudson and W.P. Rice, *J. Chromatogr.,* 117 (1976) 449.
36 A.N. Masoud, *J. Pharm. Sci.,* 65 (1976) 1585.
37 K.K. Kaistha, *J. Chromatogr.,* 141 (1977) 145.
38 A.N. Masoud, *J. Chromatogr.,* 141 (1977) D9.

TABLE 10.1

TLC SEPARATION OF CACTUS ALKALOIDS

TLC systems:

| | | |
|---|---|---|
| S1 | Silica gel H, impregnated with pH 9.2 buffer(48 g boric acid and 15 g NaOH in 1000 ml) | Methyl ethyl ketone-dimethylformamide-conc. ammonia (13:1.9:0.1)[5] |
| S2 | Silica gel G, activated | Chloroform-acetone-diethylamine(5:4:1)[5] |
| S3 | Silica gel G, activated | Chloroform-diethylamine(9:1)[5] |
| S4 | Silica gel G, activated | Cyclohexane-chloroform-diethylamine(5:4:1)[5] |
| S5 | Silica gel G, 0.1 *M* NaOH impregnated | Benzene-ethyl acetate-diethylamine(7:2:1)[5] |
| S6 | Silica gel G, activated | Chloroform-ethanol-diethylamine(85:5:10)[9] |
| S7 | Silica gel G, activated | Chloroform-ethanol-diethylamine(85:10:5)[9] |
| S8 | Silica gel G, activated | Chloroform-ethanol-conc. ammonia(85:15:0.4)[9] |
| S9 | Silica gel G, activated | Chloroform-*n*-butanol-conc. ammonia(50:50:2.5)[9] |
| S10 | Silica gel G, activated | Pyridine-conc. ammonia(9:1)[9] |

| Alkaloid | $hR_F$ value | | | | | | | | | |
|---|---|---|---|---|---|---|---|---|---|---|
| | S1 | S2 | S3 | S4 | S5 | S6 | S7 | S8 | S9 | S10 |
| Phenolic alkaloids: | | | | | | | | | | |
| Anhalamine | 8 | 4 | 3 | 1 | 2 | 11 | 20 | | | 40 |
| Anhalidine | 68 | 36 | 26 | 15 | 23 | 55 | 65 | | | 72 |
| Anhalonidine | 23 | 21 | 16 | 7 | 13 | 39 | 51 | | | 51 |
| Hordenine | 56 | 42 | 30 | 14 | 23 | 51 | 56 | | | 60 |
| N-Methyltyramine | 10 | 19 | 13 | 6 | 10 | 31 | 31 | | | 32 |
| Pellotine | 63 | 51 | 40 | 25 | 35 | 63 | 70 | | | 69 |
| Pilocereine | 90 | 84 | 89 | 80 | 81 | | | | | |
| Tyramine | 54 | 43 | 16 | 8 | 11 | 34 | 33 | | | 42 |
| Non-phenolic alkaloids: | | | | | | | | | | |
| Anhalinine | 22 | | | | | | | 30 | 41 | 48 |
| Anhalonine | 49 | | | | | | | 45 | 58 | 56 |
| Lophophorine | 82 | | | | | | | 68 | 80 | 72 |
| Mescaline | 55 | | | | | | | 24 | 31 | 36 |
| N-Acetylmescaline | 72 | | | | | | | 82 | 95 | 68 |
| N-Methylmescaline | 10 | | | | | | | 22 | 20 | 25 |
| O-Methylanhalonidine | 34 | | | | | | | 33 | 45 | 50 |

TABLE 10.2

TWO-DIMENSIONAL TLC OF SOME DANSYLATED β-PHENETHYLAMINES[21]

Silica gel G. First dimension, chloroform-butyl acetate (10:1); second dimension, benzene-triethylamine (10:1).

| Dansylated conjugate | $hR_F$ value | |
|---|---|---|
| | First dimension | Second dimension |
| β-Phenethylamine | 53 | 38 |
| N-Methyl-β-phenethylamine | 77 | 55 |
| 4-Methoxy-β-phenethylamine | 41 | 34 |
| N-Methyl-4-methoxy-β-phenethylamine | 67 | 52 |
| 3,4-Dimethoxy-β-phenethylamine | 25 | 26 |
| N-Methyl-3,4-dimethoxy-β-phenethylamine | 46 | 45 |
| Mescaline | 14 | 45 |
| N-Methylmescaline | 29 | 44 |

**References p. 162**

TABLE 10.3

COLOURS OF ALKALOIDS WITH SEQUENCE OF SPRAY REAGENTS[32]

Silica gel plates developed with diethyl ether-methanol-ammonia(17:2:1). The developed plates were sprayed first with fluorescamine, second with dansylchloride and last with iodoplatinate.

| Alkaloid | Fluorescamine (under UV light) | Dansyl Cl (under UV light) | Iodoplatinate (visible) |
|---|---|---|---|
| Primary amines (β-phenethylamines): | | | |
| β-Hydroxymescaline | Aquamarine | Aquamarine | Yellow-brown |
| Mescaline | Aquamarine | Aquamarine | Yellow-brown |
| Norepinephrine* | Yellow | Yellow | Yellow-brown |
| Normetanephrine* | Aquamarine | Aquamarine | Yellow-brown |
| Octopamine* | Aquamarine | Aquamarine | Yellow-brown |
| β-Phenethylamine | Aquamarine | Aquamarine | Yellow-brown |
| Tyramine* | Aquamarine | Aquamarine | Yellow-brown |
| Secondary amines (β-phenethylamines, tetrahydroisoquinolines, and proline): | | | |
| Anhalonine | Dark purple | Yellow | Yellow-brown |
| Ephedrine | Dark purple | Faint yellow | Yellow-brown |
| Metanephrine* | Dark purple | Yellow | Yellow-brown |
| 7-Methoxy-1,2,3,4-tetrahydroisoquinoline | Dark purple | Yellow | Faint purple |
| 8-Methoxy-1,2,3,4-tetrahydroisoquinoline | Dark purple | Yellow | Faint purple |
| N-Methylmescaline | Dark purple | Yellow | Yellow-brown |
| N-Methyl-β-phenethylamine | Dark purple | Yellow | Yellow-brown |
| N-Methyltyramine* | Dark purple | Yellow | Yellow-brown |
| Phenylephrine* | Dark purple | Yellow | Yellow-brown |
| Proline | Dark purple | Yellow | Yellow-brown |
| Salsolidine | Dark purple | Faint yellow | Yellow-brown |
| Salsoline* | Dark purple | Yellow | Yellow-brown |
| Synephrine* | Dark purple | Yellow | Yellow-brown |
| Tertiary amines (β-phenethylamines and tetrahydroisoquinolines): | | | |
| Carnegine | - | - | Purple |
| Corypalline* | - | Yellow | Purple |
| N,N-Dimethyl-β-phenethylamine | - | - | Purple |
| Hordenine* | - | Yellow | Purple |
| Lophophorine | - | - | Purple |
| N-Methyl-5-hydroxy-1,2,3,4-tetrahydroisoquinoline* | - | Yellow | Blue |
| N-Methyl-6-hydroxy-1,2,3,4-tetrahydroisoquinoline* | - | Yellow | Purple |
| N-Methyl-7-hydroxy-1,2,3,4-tetrahydroisoquinoline* | - | Yellow | Purple |
| N-Methyl-8-hydroxy-1,2,3,4-tetrahydroisoquinoline* | - | Yellow | Blue |
| Quaternary amine: | | | |
| Candicine* | - | Yellow | Purple |
| Amide: | | | |
| N-Acetylmescaline | - | - | - |
| Imidazoles: | | | |
| Dolichotheline | | Yellow | Yellow |
| Histamine | Aquamarine | Aquamarine | Yellow |
| Histidine | Aquamarine | Aquamarine | Yellow |

*Phenolic compound.

TABLE 10.4

TLC ANALYSIS CACTUS ALKALOIDS IN PLANT MATERIAL

| Alkaloid* | Aim | Adsorbent | Solvent system | Ref. |
|---|---|---|---|---|
| anham,anhid,anhod, hord,pel,pil,tyr, Metyr,anhin,anhon, lop,N-AcMSC,N-MeMSC, O-Meanhod,bet,cand, chol | Identification in *Lophophora* (Table 10.1) | $SiO_2$ | I. $CHCl_3$-$Me_2CO$-DEA (5:4:1)<br>II. $CHCl_3$-DEA(9:1)<br>III. Cyclohexane-$CHCl_3$-DEA (5:4:1) | |
| | | 0.1 *M* NaOH-impregnated $SiO_2$ | IV. Benzene-EtOAc-DEA (7:2:1)<br>Two-dimensional: V. benzene-MeOH-5% $NH_4OH$(10:15:2), followed by I. | |
| | | pH 9.2-impregnated $SiO_2$ | VI. MeEtCO-DMFA-conc. $NH_4OH$(13:1.9:0.1) | |
| | | $Al_2O_3$ | MeOH-$CCl_4$-AcOH(28:12:1) | 5 |
| anham,anhod,anhin, anhad,anhon,OMe-anhod,tyr,Metyr, hord,pel,MSC, N-MeMSC,N-AcMSC, lop | Identification in plant material (Table 10.1) | $SiO_2$ | $CHCl_3$-EtOH-DEA(85:5:10), (85:10:5)<br>$CHCl_3$-EtOH-conc. $NH_4OH$ (85:10:0.4)<br>$CHCl_3$-*n*-BuOH-conc. $NH_4OH$ (50:50:2.5)<br>Pyridine-conc. $NH_4OH$(9:1) | 9 |
| gram,ser,MSC, trypt,indole | TAS technique for *Lophophora* | $SiO_2$ | $Me_2CO$-conc. $NH_4OH$(99:1) | 13 |
| β-phea,N-Mephea, 4OMe-phea,N-Me-4-OMephea,3,4OMe-phea,N-Me-3,4OMe-phea,MSC,N-MeMSC | Separation as dansyl derivatives | $SiO_2$ | Two dimensional:<br>I. $CHCl_3$-BuOAc(10:1)<br>II. Benzene-TrEa(10:1) | 21 |
| MSC | TAS technique for peyote | $SiO_2$ | MeOH-$NH_4OH$ (25%) | 23 |
| MSC,anhon,lop, pey | Identification of peyote cactus | $SiO_2$ | $CHCl_3$-*n*-BuOH-conc. $NH_4OH$ (50:50:2.5) | 30 |

*Abbreviations used in Tables 10.4-10.7:

| | | | |
|---|---|---|---|
| MSC | mescaline | N-AcMSC | N-acetylmescaline |
| anhan | anhalamine | N-MeMSC | N-methylmescaline |
| anhad | anhalidine | O-Meanhod | O-methylanhalonidine |
| anhod | anhalonidine | bet | betaine |
| hord | hordenine | cand | candicine |
| Metyr | N-methyl-tyramine | chol | choline |
| pel | pellotine | gram | gramine |
| pil | pilocereine | ser | serotonine |
| tyr | tyramine | trypt | tryptamine |
| anhin | anhalinine | β-phea | β-phenethylamine |
| anhon | anhalonine | pey | peyopherine |
| lop | lophophorine | | |

References p. 162

TABLE 10.5

TLC OF CACTUS ALKALOIDS AS PART OF THE SEPARATION OF AMINES

| Alkaloid* | Other compounds | Aim | Adsorbent | Solvent system | Ref. |
|---|---|---|---|---|---|
| tyr,hord,ser, MSC,dopamine, 3,4-dimethoxy- and 3-methoxy-4-hydroxy-phenethylamine | Various amines | Separation as dansyl derivatives | $SiO_2$ | Two dimensional: I. EtOAc-cyclohexane (75:50) II. Benzene-MeOH-cyclohexane(85:5:10) or benzene-TrEA(10:2) | 3,6 |
| MSC,hord,tyr, dopamine, 3,4-dimethoxy- and 3-methoxy-4-hydroxy-phenylethyl-amine | 71 amines | Separation as dansyl derivatives | $SiO_2$ | Two dimensional: I. $CHCl_3$-BuOAc(10:2) II. a)$(isopr)_2O$ (2x) b)$(isopr)_2O$-TrEA(10:2) (2x) | 7 |
| MSC | 35 amines | Separation as 4-(phenylazo)benzene sulphonamide derivatives | $Al_2O_3$ | EtOAc-light petroleum (b.p. 62-82°C)(25:100), sat. with $H_2O$ | 8 |

*For abbreviations, see footnote to Table 10.4.

TABLE 10.6

TLC ANALYSIS OF MESCALINE IN COMBINATION WITH OTHER DRUGS OF ABUSE IN URINE

| Alkaloid* | Other compounds | Aim | Adsorbent | Solvent system | Ref. |
|---|---|---|---|---|---|
| MSC | Opium alkaloids, cocaine,nicotine | Identification in urine (see Chapter 12) | $SiO_2$ | EtOH-pyridine-dioxane-$H_2O$ (50:20:25:5)<br>EtOH-AcOH-$H_2O$(6:3:1)<br>EtOH-dioxane-benzene-$NH_4OH$ (5:40:50:5)<br>MeOH-*n*-BuOH-benzene-$H_2O$ (60:15:10:15) | |
| | | | $Al_2O_3$ | *n*-BuOH-*n*-$Bu_2O$-AcOH(4:5:1)<br>*n*-BuOH-*n*-$Bu_2O$-$NH_4OH$(25:70:5) | 1 |
| MSC | LSD,ergot alkaloids | Identification | $SiO_2$ | Benzene-light petroleum-$Me_2CO$-$NH_4OH$(35:35:35:1) | 10 |
| MSC | Amphetamines, phenmetrazine, methylphenidate | Identification | $SiO_2$ | $Me_2CO$-$NH_4OH$(99:1)<br>$CHCl_3$-MeOH(1:1) | 12 |
| MSC | Opium alkaloids, quinine,LSD, marihuana,cocaine,benzodiazapines,psilocybin,chlorpromazine | Comparison of direct solvent extraction and ion-paper extraction from urine | $SiO_2$ | EtOH-dioxane-benzene-$NH_4OH$ (5:40:50:5)<br>EtOAc-MeOH-$NH_4OH$(85:10:5)<br>MeOH-*n*-BuOH-benzene-$H_2O$ (60:15:10:15)<br>EtOH-pyridine-dioxane-$H_2O$ (50:20:25:5)<br>*tert.*-AmOH-*n*-$Bu_2O$-$H_2O$ (80:7:13) | 14 |
| MSC | Amphetamines, opium alkaloids, barbiturates, phenothiazines, antihistamines, caffeine,quinine, nicotine,cocaine | Identification | $SiO_2$ | $CHCl_3$-MeOH-$NH_4OH$(90:10:1)<br>EtOAc-MeOH-$H_2O$-$NH_4OH$ (85:10:3:1) | 15, 27 |
| MSC | Opium alkaloids, amphetamines, catecholamines, LSD,marihuana | TLC as dansyl derivatives | polyamide | $H_2O$-HCOOH(100:1.5)<br>Benzene-AcOH(9:1)<br>$H_2O$-AcOH(50:1)<br>EtOH-$H_2O$-*n*-BuOH-HCOOH (93:150:4:3) | 17 |
| MSC | Opium alkaloids, amphetamines, benzodiazepines, LSD,quinine, cocaine, various others | Identification (see Chapter 12) | $SiO_2$ | EtOAc-cyclohexane-dioxane-MeOH-$H_2O$-$NH_4OH$(50:50:10:10:0.5:1.5,50:50:10:10:1.5:0.5)<br>EtOAc-cyclohexane-MeOH-$H_2O$-$NH_4OH$(70:15:8:0.5:2) | 22 |
| MSC | Opium alkaloids, amphetamine, caffeine,cocaine | Identification | $SiO_2$ | EtOAc-MeOH-$NH_4OH$(180:17:7) | 28 |
| MSC | Primary amines | Identification | $SiO_2$ | EtOAc-cyclohexane-MeOH-$H_2O$-$NH_4OH$(70:15:8:0.5:2),followed by EtOAc-MeOH-$H_2O$ (80:15:5)<br>MeOH-$NH_4OH$(100:1.5)<br>EtOAc-MeOH-$NH_4OH$(80:10:5) | 29 |

TABLE 10.6 (continued)

| Alkaloid* | Other compounds | Aim | Adsorbent | Solvent system | Ref. |
|---|---|---|---|---|---|
| MSC | Opium alkaloids, sedatives, hypnotics,cocaine, quinine,LSD, amphetamines, various others | Identification | $SiO_2$ | EtOAc-cyclohexane-MeOH-$NH_4OH$ (70:15:10:5) EtOAc-cylohexane-$NH_4OH$ (50:40:0.1) | 31 |
| MSC | Amphetamines, barbiturates, opium alkaloids, antihistamines, phenothiazines, sedatives,tropane alkaloids, quinine,caffeine, lobeline,nicotine | Identification | $SiO_2$ | EtOAc-isoprOH-$NH_4OH$ (60:40:1) Dioxane-benzene-$NH_4OH$ (35:60:5) | 34 |

*For abbreviation, see footnote to Table 10.4.

TABLE 10.7

TLC ANALYSIS OF MESCALINE IN DRUG OF ABUSE SEIZURES

| Alkaloid* | Other compounds | Aim | Adsorbent | Solvent system | Ref. |
|---|---|---|---|---|---|
| MSC | Opium alkaloids, cocaine,quinine | Identification | | See Table 12.15 | 4 |
| MSC | Opium alkaloids, amphetamines, tryptamines,LSD, psilocin,psilocybin | Identification | $SiO_2$ | MeOH-$NH_4OH$(25%)(100:1.5) | 16, 19 |
| MSC | Opium alkaloids, LSD | Separation with azeotropic solvents | $SiO_2$ | Benzene-MeEtCO-*n*-butylamine (75:15:10)<br>IsoprOH-1,2-dichloroethane-diisopropylamine(5:4:1)<br>1-Chlorobutane-ethanol-*n*-butylamine(8:1:1) | 18 |
| MSC | Strychnine, amphetamines, tryptamines,LSD, psilocybin, phencyclidine | Identification | $SiO_2$ | EtOAc-*n*-prOH-28% $NH_4OH$ (40:30:3) | 20 |
| MSC | Opium alkaloids, amphetamines, barbiturates, tryptamines,cocaine,caffeine, psilocin,psilocybin,LSD | Identification | $SiO_2$ | $CHCl_3$-$Et_2O$-MeOH-25% $NH_4OH$ (75:25:5:1) | 24 |
| MSC | Atropine,phenylethylamines, ephedrine,various basic drugs | Separation as NDB-Cl derivatives | $SiO_2$ | EtOAc-cyclohexane(2:3,3:2)<br>$Et_2O$-benzene(1:1) | 35 |
| MSC | Amphetamines, tryptamines, 4-methyl-2,5-dimethoxy-α-methylphenethylamine | Identification | $SiO_2$<br><br>$Al_2O_3$ | EtMeCO-DMFA-$NH_4OH$ (13:1.9:0.1)<br>$CHCl_3$-MeOH-AcOH(75:20:5)<br>$CHCl_3$-MeOH(1:1) | 11 |

*For abbreviation, see footnote to Table 10.4.

Chapter 11

# ISOQUINOLINE ALKALOIDS

11.1. Protoberberine and protopine alkaloids
11.1.1. Solvent systems ... 172
11.1.2. Detection ... 172
11.1.3. Quantitative analysis ... 173
11.2. Benzophenanthridine alkaloids ... 174
11.2.1. Solvent systems ... 174
11.2.2. Detection ... 174
11.2.3. Quantitative analysis ... 175
11.3. Hydrastis (phthalide) alkaloids and benzylisoquinoline alkaloids ... 175
11.3.1. Solvent systems ... 175
11.3.2. Detection ... 175
11.3.3. Quantitative analysis ... 176
11.3.4. TAS technique and reaction chromatography ... 176
11.4. Bisbenzylisoquinoline alkaloids ... 177
11.4.1. Tertiary alkaloids ... 177
11.4.1.1. Solvent systems ... 177
11.4.1.2. Detection ... 177
11.4.2. Quaternary alkaloids: tubocurarine and related alkaloids ... 178
11.4.2.1. Solvent systems ... 178
11.4.2.2. Detection ... 178
11.4.2.3. Quantitative analysis ... 179
11.5. Aporphine alkaloids ... 179
11.5.1. Solvent systems ... 180
11.5.2. Detection ... 180
11.5.3. Quantitative analysis ... 180
11.5.4. Reaction chromatography ... 181
11.6. Rhoeadine alkaloids ... 181
11.6.1. Solvent systems ... 181
11.6.2. Detection ... 181
11.7. Ipecacuanha alkaloids ... 182
11.7.1. Solvent systems ... 182
11.7.2. Detection ... 182
11.7.3. Quantitative analysis ... 182
11.7.4. TAS technique and reaction chromatography ... 183
11.8. Various isoquinoline alkaloids ... 184
References ... 184
... 184

The isoquinoline alkaloids represent a very heterogeneous group of alkaloids and in dealing with this group a subdivision into basic structures is used, based on the systematic arrangement found in Kametani's survey on the chemistry of the isoquinoline alkaloids[50]. However, plants may contain alkaloids belonging to different chemical groups. Therefore, in some instances a botanical subdivision has been applied, i.e., the ipecacuanha and the opium alkaloids are dealt with separately.

The chromatography of methylenedioxyphenyl compounds has been reviewed by Fishbein and Falk[51], who also dealt with a number of isoquinoline alkaloids.

Giacopello[20] described the TLC of 48 alkaloids on cellulose; 30 isoquinoline alkaloids were part of this study, which is dealt with in Chapter 3.

Other studies of the TLC of alkaloids in general, including some isoquinoline alkaloids, are summarized in Table 11.32.

## 11.1. PROTOBERBERINE AND PROTOPINE ALKALOIDS

### *11.1.1. Solvent systems*

In the groups of protoberberine and protopine alkaloids the quaternary alkaloids have received most attention. In particular, the separation of the alkaloids berberine, palmatine, jatrorrhizine, columbamine and magnoflorine (an aporphine-type alkaloid) has been the subject of several investigations. Because of the polar character of the quaternary alkaloids, partition chromatography on cellulose was used on several occasions[20,29,56] (Table 11.1). Polar acidic solvents such as *n*-butanol-acetic acid-water mixtures in combination with silica gel plates have been used by several workers (Tables 11.2-11.4). Several investigations of the separation of quaternary alkaloids have also included one or more of the protoberberine-type alkaloids[8,73]. In the Chapter 21, on quaternary ammonium compounds, the particular problems with the separation of such compounds are dealt with in more detail.

To achieve a complete separation of the alkaloids berberine, palmatine, jatrorrhizine, columbamine and magnoflorine a combination of solvents is needed. To separate berberine and palmatine *n*-butanol-acetic acid-water (4:1:1) on silica gel plates[30], *n*-butanol-saturated with 2 *M* hydrochloric acid on cellulose plates[56] (Table 11.1), chloroform-methanol-ammonia (25%) (3:3:1), methyl ethyl ketone-methanol-formic acid (8:1:1) and cyclohexane-diethylamine (9:1), all on silica gel plates, are suitable. Borkowski and Kaniewska[67] described the TLC of *Thalictrum* alkaloids, i.e. a number of tertiary alkaloids (see Table 11.16) and six quaternary alkaloids (see Table 11.2). For the separation of berberine, palmatine and columbamine TLC system S6 was found to be suitable. MacLean and Jewers[73] used basic or acidic aluminium oxide as the stationary phase for the TLC of some quaternary alkaloid chlorides and other quaternary compounds. The counter ion of the quaternary compounds was found to influence the retention behaviour of the compounds. The use of basic aluminium oxide in combination with chloroform-methanol (85:15) gave the best results for the protoberberine alkaloids.

Cooper and Dugal[85] used continuous flow TLC in a BN-chamber for the separation of a number of isoquinoline alkaloids (Table 11.3). If the amounts of the alkaloids did not exceed 5 µg a good separation of all compounds was obtained.

For the separation of the alkaloids palmatine, jatrorrhizine and columbamine good results were obtained with the solvent 96% ethanol-17% ammonia (25:5) on silica gel plates[62]. Wa et al.[42] used the solvent methanol-water-ammonia (8:1:1) for the separation of palmatine, jatrorrhizine, columbamine and magnoflorine (Table 11.2). Two-dimensional TLC of these alkaloids with solvent S1 and *n*-propanol-water-ammonia (8:1:1) on silica gel plates was also described.

Golkiewicz and Wawrzynowicz[58] discussed the TLC behaviour of some tertiary protoberberine and protopine alkaloids in terms of the proton donor and acceptor properties of the solvents used.
A number of publications have dealt with some protoberberine and protopine alkaloids in connection with the identification of Papaveraceae alkaloids[14,27,35,48,53]. The results are summarized in Table 11.26.

### *11.1.2. Detection*

The protoberberine and protopine alkaloids can be detected by means of general alkaloid spray reagents, such as Dragendorff's reagent in its different modifications and iodoplatinate reagent. The colours obtained with the latter reagent for some alkaloids are summarized in Table 11.6.

The quaternary berberine-type of alkaloids have a yellow colour and are characterized by a strong yellow fluorescence. According to Cooper and Dugal[85] detection of these alkaloids by means of their fluorescence is more sensitive than with Dragendorff's or iodoplatinate reagent. For the phenolic alkaloids the fluorescence colour depends on the pH, which means that the solvent system used can influence the colours observed.

In Table 11.5 the colours observed for some alkaloids[85] after development in a basic or acidic solvent are summarized. Borkowski and Kaniewska[67] used diazotized *p*-nitraniline (no. 76) for the detection of phenolic groups in a number of isoquinoline alkaloids and Labats reagent (no. 20) for the detection of dioxymethylene groups.

The chromotropic acid reagent (no. 20) as described by Beroza[6] for the detection of compounds containing the dioxymethylene group can also be applied to the alkaloids of the protopine and protoberberine group containing this functional group.

Golkiewicz and Wawrzynowicz[58] used bis-diazotized benzidine (Wachtmeister's reagent) (no. 32) and concentrated sulphuric acid for the selective detection of *Fumaria* alkaloids belonging to the protoberberine and protopine group. Santavy[38] described the use of cerium(IV) sulphate reagent (no. 14c) for the detection of a series of isoquinoline alkaloids, including protoberberine and protopine alkaloids. The colours observed are summarized in Table 11.6. Genest and Hughes[49] used cerium(IV) ammonium sulphate (no. 12), potassium permanganate in sodium carbonate solution (no. 83) and 2,4,7-trinitrofluorenone (no. 100a) for the detection of *Hydrastis* alkaloids, including berberine.

### *11.1.3. Quantitative analysis*

For the indirect quantitative analysis of berberine several solvents have been described for eluting the alkaloid from silica gel, viz., 0.05 *M* sulphuric acid in ethanol[65], 0.05 *M* sulphuric acid[84] and methanol-0.1 *M* hydrochloric acid (8:2)[103].

Kaniewska and Borkowski[45] added the reagent used for the colorimetric analysis of berberine [1 ml of chromotropic acid (0.1 g of the sodium salt in 1 ml of water) solution + 5 ml of concentrated sulphuric acid] to the adsorbent collected from the plate to avoid an extraction step. Kiryanov et al.[108] determined tetrahydropalmatine quantitatively by means of UV spectrometry (280 nm) after elution with ethanol from the adsorbent. Indirect spectrophotometric analysis of berberine and hydrastine was described by Datta et al.[70], whereby the alkaloids were extracted with 0.1 *M* hydrochloric acid from aluminium oxide, prior to the quantitative determination.

Hattori et al.[109] determined berberine colorimetrically after a two-dimensional TLC separation and elution of the alkaloid with 1% hydrochloric acid in methanol.

The direct quantitative analysis of berberine by means of fluorimetric densitometry (excitation 352 nm, emission 530 nm) was described by Messerschmidt[52]. A similar method was described by Hashimoto et al.[102] (excitation 353 nm, emission 510 nm). Wu et al.[92] used densitometry for the determination of berberine.

## 11.2. BENZOPHENANTHRIDINE ALKALOIDS

### *11.2.1. Solvent systems*

In the group of benzophenanthridine alkaloids most attention has been paid to *Chelidonium*-type alkaloids and sanguinarine and related alkaloids.

According to Balderstone and Dyke[107], aluminium oxide is not suitable for the analysis of sanguinarine and related alkaloids because of the risk of the quaternary alkaloids being transformed into the pseudo-alkaloids. Although sanguinarine is a quaternary alkaloid, the solvents described in the literature are of low polarity and neutral. The quaternary benzophenanthridine alkaloids are apparently much less polar than the quaternary alkaloids of the protoberberine type, as can be seen from Table 11.8.

### *11.2.2. Detection*

The benzophenanthridine alkaloids can be detected with the general alkaloid reagents such as Dragendorff's reagent and iodoplatinate reagent in their different modifications. The colours obtained with the latter reagent for some alkaloids are summarized in Table 11.9. Further, these alkaloids can be detected by their fluorescence. Santavy[38] described the use of cerium(IV) sulphate reagent (no. 14c) for the detection of some of these alkaloids. Chromotropic acid (no. 20) may also be applied as a reagent for dioxymethylene-containing alkaloids[6,45].

### *11.2.3. Quantitative analysis*

The quantitative analysis of benzophenanthridine alkaloids has been described by Scholz et al.[101] and Balderstone and Dyke[107]. The former workers separated the alkaloids from *Radix chelidonii* by means of solvent S7 (Table 11.8) and eluted chelidonine by means of methanol-water (95:5) from the adsorbent, and chelerythrine and sanguinarine by means of methanol-water (95:5) containing 1% of tartaric acid, prior to UV spectrometric determination. Balderstone and Dyke[107] determined sanguinarine quantitatively in *Argemone* oil. The alkaloid was reduced to the dihydro compound which allowed extraction by means of organic solvents. Also, decomposition of sanguinarine into the dihydro and oxy compound was avoided by this procedure. After development of the chromatogram with solvent S1 (Table 11.8) the dihydro compound was converted into sanguinarine by irradiation with long-wavelength UV light. The alkaloid was than extracted with ethanol containing a drop of concentrated hydrochloric acid and the alkaloid was determined spectrophotometrically at 330 nm.

## 11.3. *HYDRASTIS* (PHTHALIDE) ALKALOIDS AND BENZYLISOQUINOLINE ALKALOIDS

To the group of phthalide and benzylisoquinoline alkaloids also belong a number of opium alkaloids, which will be dealt with separately. Most of the remaining publications on these classes of alkaloids deal with the *Hydrastis* alkaloids.

### *11.3.1. Solvent systems*

For the separation of the major components present in *Hydrastis* solvent S1 (Table 11.11) was reported by Vanhaelen[54]. Zwaving and de Jong-Havenga[74] also found this solvent to be useful, but recommended solvent S2 (Table 11.11) as suitable for separating hydrastine from the other alkaloids present and the degradation products meconine, opianic acid and hydrastinine, allowing quantitative

**References p. 184**

analysis. The solvent cyclohexane-chloroform-diethylamine (5:4:1) on silica gel plates[47] did not give a satisfactory separation of hydrastine and hydrastinine but, because of the fluorescence of the latter alkaloid, the two spots could be distinguished. Pfeifer[27,53] dealt with the identification of a number of Papaveraceae alkaloids, including some benzylisoquinoline alkaloids (see Table 11.26).

### *11.3.2. Detection*

For the detection of the phthalide and bisbenzylisoquinoline alkaloids the general alkaloid detection methods can be used. The colours obtained with iodoplatinate reagent are summarized in Table 11.12[38,74]. Santavy[38] used cerium(IV) sulphate and Genest and Hughes[49] used cerium(IV) ammonium sulphate for the detection of these alkaloids (nos. 14c and 12, respectively).

For the detection of hydrastine on magnesium oxide plates, Ragazzi and Veronese[21] used cobalt rhodanide, which reveals the alkaloids as blue spots against a pink background. Alkaloids containing the dioxymethylene groups can be detected with chromotropic acid (no. 20)[6,45].

Vanhaelen[54] oxidized hydrastine to hydrastinine by subsequently spraying with sulphuric acid and 1% potassium permanganate solution, followed by exposure of the plate to hydrogen sulphide to decolorize excess of potassium permanganate. The hydrastine spot was observed as a blue fluorescent spot.

Genest and Hughes[49] used 0.5% potassium permanganate solution in 2% aqueous sodium carbonate and 2,4,7-trinitrofluorenone (0.5% in benzene) (no. 100a) for the detection of *Hydrastis* alkaloids. Grant[97] described the detection of a series of alkaloids, including laudanosine, with cobalt thiocyanate (no. 26a).

### *11.3.3. Quantitative analysis*

Kaniewska and Borkowski[45] determined hydrastine indirectly by means of colorimetry (see quantitative analysis of protoberberine alkaloids, p.174). Ragazzi and co-workers[15,21] used magnesium oxide as adsorbent to separate hydrastine from the other *Hydrastis* alkaloids. For the indirect quantitative analysis the alkaloid and adsorbent were dissolved in nitric acid and heated at 50$^{o}$C for 30 min. The hydrastinine formed in this way was determined by means of fluorimetry.

Datta et al.[70] determined hydrastine and berberine indirectly in *Hydrastis* extracts by means of spectrophotometry after elution of the spots with 0.1 *M* hydrochloric acid, The direct quantitative analysis of hydrastine has been described by Stanislas et al.[43]. Zwaving and de Jong-Havenga[74] described the direct quantitative analysis of hydrastine in *Hydrastis* and compared it with other quantitative methods such as indirect quantitative TLC. For the extraction of the alkaloid from the adsorbent they used chloroform, then the alkaloid was determined spectrophotometrically at 298 nm. In the densitometric method the alkaloids are also measured at 298 nm. In both methods care has to be taken to prevent oxidation of hydrastine into hydrastinine, particularly under the influence of UV light. The results of the two methods in general showed good agreement.

#### *11.3.4. TAS technique and reaction chromatography*

Jolliffe and Shellard used the TAS technique for *Hydrastis* plant material[61]. Whether the alkaloids are distilled unchanged or not is not known. Several alkaloid-positive spots were observed on TLC. Genest and Hughes[49] described the reaction chromatography of hydrastine. They used the oxidative hydrolysis, which yields hydrastinine and opianic acid. The reaction can be performed by exposure to UV light or daylight or treatment with potassium permanganate, chloramine-T, hydrogen peroxide or nitric acid. The last reagent was found to be the most suitable.

### 11.4. BISBENZYLISOQUINOLINE ALKALOIDS

Among the bisbenzylisoquinoline alkaloids there are two major groups for which most of the chromatographic data have been reported, namely the tertiary alkaloids from *Thalictrum* species, known particularly through the tumour-inhibiting alkaloid tetrandrine, and the quaternary curare alkaloids, particularly known through the muscle-relaxant alkaloid tubocurarine. These two groups will be dealt with separately, owing to the different separation problems involved.

#### *11.4.1. Tertiary alkaloids*

##### *11.4.1.1. Solvent systems*

The tertiary bisbenzylisoquinoline alkaloids usually have a strongly basic character and must hence be chromatographed in basic solvents if silica gel is used as the stationary phase or the plates are to be impregnated with base[4] (Table 11.14).

Bhatnagar and Bhattacharji[19] described four solvent systems for the separation of a number of bisbenzylisoquinoline alkaloids (Table 11.15). None of the solvents was capable of separating all of the alkaloids, but a combination of two or more of them could effect a good resolution. According to Borkowski and Kaniewska[67], *Thalictrum* alkaloids could be separated by means of two-dimensional TLC, using the combination of the solvents S1 and S2 or S1 and S3 (Table 11.16). In this way the 16 alkaloids in Table 11.16 could be separated.

**References p. 184**

Verpoorte et al.[111] found the solvents ethyl acetate-isopropanol-25% ammonia (45:35:5) in combination with chloroform-cyclohexane-diethylamine (5:4:1) to be very useful in the identification of isomeric alkaloids of the tetrandrine type. The system toluene-ethyl acetate-diethylamine (7:2:1) also proved to be very useful in the analysis of the tetrandrine-type of alkaloids. All three solvents systems were used in combination with silica gel plates.

The analysis of thalicarpine has been dealt with by several workers. The alkaloid has been detected in drugs[106] and in biological materials[110]. Yankulov and Eustatieva[94] described the quantitative analysis of thalicarpine in *Thalictrum* species by comparison of spot sizes.

*11.4.1.2. Detection*

For the detection of bisbenzylisoquinoline alkaloids the general alkaloid reagents can be used. Reagents for phenolic groups have also been used. Borkowski and Kaniewska[67] used diazotized *p*-nitraniline (no. 76).

Bhatnager and Bhattacharji[19] used iodine in potassium iodide solution for the detection, because the colours obtained with iodoplatinate and Dragendorff's reagent faded rapidly. Verpoorte et al.[111] found iron(III) chloride in perchloric acid (no. 62c) to be valuable for the identification of berbaman-type alkaloids. By observing the colours immediately after spraying and the colour changes during heating with hot air, isomeric alkaloids could be distinguished (Table 11.17).

*11.4.2. Quaternary alkaloids: tubocurarine and related alkaloids*

Tubocurarine is a monoquaternary alkaloid and polar solvent systems are therefore needed for its TLC analysis. Most of the chromatography described has been in connection with the analysis of curarimetics, usually synthetic bisquaternary nitrogen compounds. For the problems involved in the TLC of quaternary alkaloids, see also Chapter 21.

*11.4.2.1. Solvent systems*

Wollmann et al.[34] tested a number of polar solvents for the separation of curarimetics. They found that basic solvent systems or acetic acid containing solvents were unable to remove the ammonium bases from the start on silica gel plates. Only highly polar solvents such as methanol, acetone, ethanol, isopropanol and dioxane in combination with hydrochloric acid (0.1-2 *M*) were suitable. In particular the solvents acetone-1 *M* hydrochloric acid and dioxane-1 *M* hydrochloric acid (1:1) gave good results.

Fiori and Marigo[41] used aluminium oxide for the TLC of curarimetics, and found that quaternary compounds did not move on silica gel. Acidic aluminium oxide was found to be particularly suitable. Stevens and Fox[66] used TLC to see if tubocurarine isolated from tissues contained thymine and tyramine, both of which may interfere in the quantitative analysis of the alkaloid. Williamson[79] used TLC to separate tubocurarine from its tertiary and bisquaternary analogue (Table 11.19). The method developed by Crone and Smith[81] can only be applied on microslides. Using pre-coated plastic sheets erratic results were obtained owing to the poor wetting of the plate by the polar solvent system.

As part of a study of the TLC analysis of quaternary alkaloids and N-oxides, Verpoorte and Baerheim Svendsen[99] found combinations of methanol and ammonium salt solutions with or without the addition of ammonia to be a very useful solvent in combination with silica gel (see Chapter 21, p.484). Giebelmann et al.[104] found acetone-1 *M* hydrochloric acid (1:1) and ethanol-1 *M* hydrochloric acid (1:1) to be suitable solvents for the TLC analysis of quaternary compounds on silica gel (see Chapter 21, p.488). De Zeeuw et al.[93] described the chromatography of quaternary nitrogen-containing compounds by means of ion-pair TLC on silica gel plates (see Chapter 21, p.489).

*11.4.2.2. Detection*

The general alkaloid spray reagents were used in most of the investigations. Wollmann et al.[34] applied several other spray reagents for the detection of curarimetics; tubocurarine could be detected with the iodopalladate reagent (no. 55) and Millons reagent (no. 72), the colours observed being brown and red-brown, respectively. With iodoplatinate a red colour was observed for tubocurarine, Giebelmann et al.[104] used, in addition to the general alkaloid reagents, iodine in chloroform (no. 51b). The combination of this reagent with Dragendorff's reagent or iodoplatinate reagent was found to be most sensitive in the detection of quaternary compounds. As it has a phenolic hydroxyl group, tubocurarine can also be detected by means of specific reagents for this group. Tyfczynska[37] used diazotized *p*-nitraniline[76].

*11.4.2.3. Quantitative analysis*

TLC has not been used for either direct or indirect quantitative analysis. Stevens and Fox[66] determined tubocurarine quantitatively by means of UV spectroscopy after extraction from tissues, TLC being used only for identification and to monitor tyramine or thymine when present in the extracts, because these substances might interfere in the quantitative analysis.

Fiori and Marigo[41] used TLC for the identification of curarimetics, but used paper chromatography for their indirect UV spectroscopic determination.

## 11.5. APORPHINE ALKALOIDS

### *11.5.1. Solvent systems*

Papers dealing with the TLC of aporphine-type alkaloids mainly concern the alkaloid magnoflorine, usually in combination with protoberberine-type alkaloids or the alkaloid boldine. Various aporphine alkaloids were dealt with by Santavy[38]; the solvents used are summarized in Table 11.26. Pfeifer[27,53] and Maturova et al.[35] studied a number of aporphine alkaloids in connection with the identification of Papaveraceae alkaloids (see Table 11.26). Calderwood and Fish[56] analysed some *Fagara* alkaloids, including some aporphines (see Table 11.1).

Cooper and Dugal[85] used contineous flow TLC in a BN-chamber for the separation of a number of isoquinoline alkaloids, including magnoflorine (see Table 11.3). The two solvent systems mentioned allowed the separation of the alkaloids and in both systems magnoflorine was strongly retained. The amount of alkaloid should not exceed 5 μg, otherwise overlapping of the alkaloids may occur. Genest and Hughes[44], in connection with the identification of the alkaloids in *Peumus boldus*, described several solvents; either the stationary phase of the mobile phase contained a base (see Table 11.21). Vanhaelen[54] used two successive developments with benzene-chloroform-methanol-acetone-formamide (8:7:4:3:0.05) for the identification of boldine in *Boldus* preparations.

### *11.5.2. Detection*

The general alkaloid reagents can be used for the detection of the aporphine alkaloids and proaporphine alkaloids. Vanhaelen[54] used a solvent containing formamide which interfered in the alkaloid detection methods. To avoid this interference, the plates were first sprayed with nitrous acid to decompose the formamide. Santavy[38] used cerium(IV) sulphate (no. 14c) in addition to the general reagents. Some of the colours obtained are summarized in Table 11.22. Magnoflorine can be detected by its blue fluorescence (see Tables 11.1 and 11.5)[56,85].

Boldine has been detected by means of Gibb's reagent (no. 34) (see Table 11.21). Di Renzo[72] used a 0.25% 2,6-dichloro- or -dibromoquinone chlorimide solution in methanol (no. 35) for the detection of boldine; exposure to UV light also led to the formation of a fluorescent product that can be used for densitometric analysis[72]. Fleischmann et al.[90] sprayed with 0.5% aqueous magnesium acetate solution, followed by drying at 110°C for 20 min and exposure to UV light (365 nm) for 1 h for the quantitative determination of boldine.

*11.5.3. Quantitative analysis*

Direct quantitative analyses of boldine in plant material and pharmaceutical preparations have been described by Di Renzo[72] and Fleischmann et al.[90]. Di Renzo sprayed with 0.25% 2,6-dichloro- or -dibromoquinone chlorimide in methanol or exposed the plates to UV light (254 nm) prior to densitometric determination. Fleischmann et al. sprayed with magnesium acetate, dried the plates at 110$^{o}$C and exposed the plates to UV light (365 nm) for at least 1 h, after which the boldine was determined by densitometry at 430 or 465 nm.

*11.5.4. Reaction chromatography*

Genest and Hughes[44] described the reaction chromatography of boldine and *Boldo* alkaloids. Acetylation was performed by means of a mixture of acetic anhydride (5 ml) and perchloric acid (70-72%, 0.24 ml). This reagent can be used prior to application of the alkaloids on the plates or by application of the reagent on the alkaloidal spot on the plate. The latter method may lead to distorted spots, however.

11.6. RHOEADINE ALKALOIDS

Although the rhoeadine-type alkaloids are not strictly isoquinoline alkaloids, they are usually dealt with in connection with the isoquinoline alkaloids, because they are biogenetically derived from them.

*11.6.1. Solvent systems*

The rhoeadine alkaloids are found in Papaveraceae plant species. Most of the chromatographic data are from the publications of Pfeifer and co-workers[14,27,48,53]. The systems used are summarized in Table 11.26.

Maturova et al.[35] described several systems for the identification of alkaloids in Papaveraceae. For the identification of papaverrubines they used the same system as Pfeifer and Banerjee[14]. Vanhaelen[62] described the identification of rhoeadines in pharmaceutical preparations by means of silica gel plates developed twice with benzene-methanol (9:1). Cjevic et al.[76] used talc as the stationary phase for the TLC of papaverrubine alkaloids, and ethyl acetate-methanol (9:1) as the solvent system.

References p. 184

### *11.6.2. Detection*

For the rhoeadine type of alkaloids the general alkaloid reagents can be used for detection. More specific detection can be obtained by exposing the plates to hydrochloric acid vapour, which gives a red colour[14,35]. The colours obtained for some rhoeadine-type alkaloids with iodoplatinate reagent are listed in Table 11.24.

A summary of the TLC-analysis of rhoeadine-type alkaloids is given in Table 11.25. In Table 11.26 the separation of various Papaveraceae alkaloids is summarized.

## 11.7. IPECACUANHA ALKALOIDS

Ipecacuanha preparations are widely used in pharmaceutical formulations and hence the identification of the main alkaloids, emetine and cephaeline, has been subject of several publications.

### *11.7.1. Solvent systems*

The most often encountered solvent system for the analysis of ipecacuanha alkaloids is S1 (Table 11.27) originally described by Stahl[38]. Solvent system S2 (Table 11.27) described by Waldi et al.[2] has also been used by several workers (refs 5, 46). According to Ghosh et al.[46], this solvent should give better results than system S1, with less tailing and better separation. Habib and Harkiss (refs 55, 63, 91) modified this system slightly by using toluene-benzene-ethyl acetate-diethylamine (35:35:20:10), and obtained the best results when the plates were not activated but only allowed to dry in air. Kamp and Onderberg[33] applied two-dimensional TLC for the separation of a mixture of ipecacuanha and opium alkaloids. System S3 (Table 11.27) gave a good separation of cephaeline and emetine but caused some tailing of the alkaloids. Frei and co-workers[87,98] separated emetine and cephaeline from codeine, morphine, narcotine and ephedrine after derivatization with dansyl chloride.

### *11.7.2. Detection*

The ipecacuanha alkaloids can be detected with the traditional alkaloid spray reagents such as Dragendorff's and iodoplatinate reagents in their different modifications. Emetine and cephaeline have no fluorescence of their own, but they are readily oxidized to O-methylpsychotrine and psychotrine, respectively, which fluoresce[18]. This oxidation also occurs on the plate after development, which explains why the spots of emetine and cephaeline are sometimes described as having fluorescence in 366-nm UV light. The oxidation of emetine and cephaeline to fluorescent products has been used by several workers for the detection of these alkaloids. The oxidizing agent used is a 0.5% solution of iodine in chloroform[33] (no. 51b) or in carbon tetrachloride (no. 51a)[38,46,54,55,63] or a saturated solution of chloramine in 10% acetic acid (no. 15)[18].

According to Kamp and Onderberg[33], Bouchardat's reagent is less suitable for the detection of emetine and cephaeline.

### *11.7.3. Quantitative analysis*

For the indirect quantitative analysis of ipecacuanha alkaloids several solvents have been described for eluting the alkaloids from the plates. Machovicova and Parrak[12] and Ghosh et al.[46] used ethanol for the elution of emetine. Ludwicki[57] used 0.1 *M* hydrochloric acid for the elution of cephaeline, Schuyt et al.[105] 0.2 *M* hydrochloric acid for emetine and Habib[91] ammoniated methanol for cephaeline and emetine. After the elution, emetine and cephaeline were determined spectrophotometrically at 283 nm[46,91,105]. Ludwicki[57] determined cephaeline colorimetrically after treatment with Folin-Ciocalteau reagent. According to Schuyt et al.[105], the spectrofluorimetric method is more sensitive than the spectrophotometric method. They used an excitation wavelength of 284 nm and measured the emission at 318 nm for emetine. This method permitted the determination of emetine in the presence of decomposition products. Direct quantitative analysis has been described by Habib and Harkiss[55,63,91], who measured the spot areas of emetine and cephaeline after spraying with iodine solution in carbon tetrachloride and heating at 60$^{o}$C for 20 min. Emetine showed up as a yellow spot in daylight with a yellow fluorescence in 366-nm UV light, and cephaeline as a brown spot with blue fluorescence. According to Habib[91], the indirect method gives better results if compared with the spot area method. Massa and co-workers[60,64] used direct spectrofluorimetry for the determination of cephaeline and emetine. The plates were treated with iodine, whereby the alkaloids were oxidized to yellow fluorescent compounds. The fluorescence was measured at 592 nm after excitation at 358 nm.

Frei and co-workers[87,98] prepared the dansyl derivatives of emetine and cephaeline prior to TLC separation. The spots were determined by direct spectrofluorimetry (excitation wavelenght 360 nm, emission measured at 500-510 nm). This method permitted the analysis of the alkaloids in the presence of 10-100-fold excess of other drugs in pharmaceutical preparations. The method was applied to pharmaceuticals containing ephedrine, morphine, codeine, noscapine and the ipecacuanha alkaloids emetine and cephaeline. Of these compounds, noscapine and codeine do not react with dansyl chloride.

**References p. 184**

### *11.7.4. TAS technique and reaction chromatography*

Jolliffe and Shellard[61] described the TAS technique for the identification of alkaloid-containing crude drugs, including ipecacuanha preparations. After TLC separation and spraying with Dragendorff's reagent characteristic patterns were obtained. However, it was not discussed if the spots observed were due to the main alkaloids or decomposition products. According to Stahl and Schmitt[88], cephaeline and emetine decompose when the TAS technique is applied. However, the thermolysis products could also be used for the characterization of ipecacuanha preparations. Reaction chromatography of the ipecacuanha alkaloids has been described by Kaess and Mathis[24,26,28]. Both emetine and cephaeline were oxidized by mercury acetate to give yellow oxidation products after heating for 1 h at 160$^{o}$C. The oxidation product of cephaeline turns red-violet in the presence of ammonia or diethylamine. Acetylation with acetic anhydride or acetyl chloride led to the formation of monoacetylcephaeline, whereas emetine remained unchanged.

Frei and co-workers[87,98] described the derivatization of emetine and cephaeline with dansyl chloride, leading to the monodansylemetine and didansylcephaeline, respectively, both having strong fluorescence. The dansylated compounds were separated by means of TLC. Because of the increased selectivity and sensitivity the compounds could be analysed quantitatively in the presence of 10-100-fold excesses of other drugs.

## 11.8. VARIOUS ISOQUINOLINE ALKALOIDS

The TLC of pavine and isopavine alkaloids has been described by several workers in connection with more extensive studies of the TLC of Papaveraceae alkaloids (refs 4, 38, 48).

Simple isoquinoline alkaloids such as cotarnine are dealt with in Table 11.26 (TLC) and in Table 11.12 (detection)[38]. The anhalonium alkaloids are dealt with in connection with other cactus alkaloids, which usually belong to the phenylethylamine alkaloids[31,40]. The TLC analysis of Amaryllidaceae alkaloids has been studied by Döpke[4]. The results are summarized in Table 11.31.

## REFERENCES

1 K. Teichert, E. Mutschler and H. Rochelmeyer, *Deut. Apoth.-Ztg.*, 100 (1960) 477.
2 D. Waldi, K. Schnackerz and F. Munster, *J. Chromatogr.*, 6 (1961) 61.
3 E. Vidic and J. Schütte, *Arch. Pharm. (Weinheim)*, 295 (1962) 342.
4 W. Döpke, *Arch. Pharm. (Weinheim)*, 295 (1962) 605.
5 M. Bacchini and A. Bonati, *Fitoterapia*, 33 (1962) 40.
6 M. Beroza, *Agr. Food Chem.*, 11 (1963) 51.
7 R.R. Paris and M. Paris, *Bull. Soc. Chim. Fr.*, (1963) 1597.

8 D. Waldi, *Naturwissenschaften*, 50 (1963) 614.
9 W. Kamp, W.J.M. Onderberg and W.A. Seters, *Pharm. Weekbl.*, 98 (1963) 993.
10 H. Gertig, *Acta Pol. Pharm.*, 21 (1964) 59; *C.A.*, 62 (1965) 12069e.
11 H. Gertig, *Acta Pol. Pharm.*, 21 (1964) 127; *C.A.*, 62 (1965) 13507d.
12 F. Machovicova and V. Parrak, *Cesk. Farm.*, 13 (1964) 200.
13 V. Schwarz and M. Sarsunova, *Pharmazie*, 19 (1964) 267.
14 S. Pfeifer and S.K. Banerjee, *Pharmazie*, 19 (1964) 286.
15 E. Ragazzi, G. Veronese and C. Giacobazzi, in G.B. Marini-Bettolo (Editor), *Thin-layer Chromatography*, Elsevier, Amsterdam, 1964, p. 149.
16 R. Paris, R. Rousselet, M. Paris and M.J. Fries, *Ann. Pharm. Fr.*, 23 (1965) 473.
17 M.R. Gasco and G. Gatti, *Boll. Chim. Farm.*, 104 (1965) 639.
18 H. Schilcher, *Deut. Apoth.-Ztg.*, 105 (1965) 1067.
19 A.K. Bhatnager and S. Bhattacharji, *Indian J. Chem.*, 3 (1965) 43.
20 D. Giacopello, *J. Chromatogr.*, 19 (1965) 172.
21 E. Ragazzi and G. Veronese, *Mikrochim. Acta*, (1965) 966.
22 G. Kurono, K. Ogura and K. Sasaki, *Yakugaku Zasshi*, 85 (1965) 262; *C.A.*, 62 (1965) 15991d.
23 E. Hultin, *Acta Chem. Scand.*, 20 (1966) 1588.
24 A. Kaess and C. Mathis, *Ann. Pharm. Fr.*, 24 (1966) 753.
25 W. Golkiewicz, L. Jusiak and E. Soczewinski, *Diss. Pharm. Pharmacol.*, 18 (1966) 485.
26 A. Kaess and C. Mathis, *Int. Symp. Chromatogr. Electrophor. Lect. Pap. 4th*, (1965) 525.
27 S. Pfeifer, *J. Chromatogr.*, 24 (1966) 364.
28 A. Mathis and A. Kaess, *J. Pharm. Belg.*, 21 (1966) 561.
29 J.M. Calderwood and F. Fish, *J. Pharm. Pharmacol.*, 18 (1966) 119S.
30 G.H. Constantine, Jr., M.R. Vitek, K. Sheth, P. Catalfomo and L.A. Scinchetti, *J. Pharm. Sci.*, 55 (1966) 982.
31 J.L. McLaughlin and A.G. Paul, *Lloydia*, 29 (1966) 315.
32 F. Wartmann-Hafner, *Pharm. Acta Helv.*, 41 (1966) 406.
33 W. Kamp and W.J.M. Onderberg, *Pharm. Weekbl.*, 101 (1966) 1077.
34 Ch. Wollmann, S. Nagel and E. Scheibe, *Pharmazie*, 21 (1966) 665.
35 M. Maturova, D. Pavlaskova and F. Santavy, *Planta Med.*, 14 (1966) 22.
36 I. Iwasa, S. Naruto and Y. Utsui, *Yakugaku Zasshi*, 86 (1966) 396; *C.A.*, 65 (1966) 5302c.
37 J. Tyfczynska, *Diss. Pharm. Pharmacol.*, 19 (1967) 585.
38 F. Santavy, in E. Stahl (Editor), *Dünnschichtschromatographie, ein Laboratoriumshandbuch*, Springer Verlag, Berlin, 1967, p. 424.
39 H.C. Hsiu, J.T. Huang, T.B. Shih, K.L. Yang, K.T. Wang and A.L. Lin, *J. Clin. Chem. Soc.*, 14 (1967) 161.
40 J. Lundström and S. Agurell, *J. Chromatogr.*, 30 (1967) 271.
41 A. Fiori and M. Marigo, *J. Chromatogr.*, 31 (1967) 171.
42 M.T. Wa, J.L. Beal and R.W. Doskotch, *Lloydia*, 30 (1967) 245.
43 E. Stanislas, R. Rouffiac, J. Gleye and P. Theron, *Bull. Soc. Pharm. Marseille*, 17 (1968) 97; *C.A.*, 72 (1970) 15783t.
44 K. Genest and D.W. Hughes, *Can. J. Pharm. Sci.*, 3 (1968) 84.
45 T. Kaniewska and B. Borkovski, *Diss. Pharm. Pharmacol.*, 20 (1968) 111.
46 D. Ghosh, D.D. Datta and P.C. Bose, *J. Chromatogr.*, 32 (1968) 774.
47 J.M.G.J. Frijns, *Pharm. Weekbl.*, 103 (1968) 929.
48 S. Pfeifer and H. Döhnert, *Pharmazie*, 23 (1968) 585.
49 K. Genest and D.W. Hughes, *Can. J. Pharm. Sci.*, 4 (1969) 41.
50 T. Kametani, *The Chemistry of Isoquinoline Alkaloids*, Elsevier, Amsterdam, 1969.
51 L. Fishbein and H.L. Falk, *Chromatogr. Rev.*, 11 (1969) 1.
52 W. Messerschmidt, *J. Chromatogr.*, 39 (1969) 90.
53 S. Pfeifer, *J. Chromatogr.*, 41 (1969) 127.
54 M. Vanhaelen, *J. Pharm. Belg.*, 24 (1969) 87.
55 M.S. Habib and K.J. Harkiss, *J. Pharm. Pharmacol.*, 21S (1969) 57S.
56 J.M. Calderwood and F. Fish, *J. Pharm. Pharmacol.*, 21S (1969) 126S.
57 H. Ludwicki, *Acta Pol. Pharm.*, 27 (1970) 469.

58 W. Golkiewicz and T. Wawrzynowicz, *Chromatographia*, 3 (1970) 356.
59 M.R. Verma, J. Rai and A. Ram, *Int. Symp. Chromatogr. Electrophor. Lect. Pap. 6th*, (1970) 396.
60 V. Massa, F. Gal and P. Susplugas, *Int. Symp. Chromatogr. Electrophor. Lect. Pap. 6th*, (1970) 470.
61 G.H. Jolliffe and E.J. Shellard, *J. Chromatogr.*, 48 (1970) 125.
62 M. Vanhaelen, *J. Pharm. Belg.*, 25 (1970) 175.
63 M.S. Habib and K.J. Harkiss, *Planta Med.*, 18 (1970) 270.
64 V. Massa, F. Gal, P. Susplugas and G. Maestre, *Trav. Soc. Pharm. Montpellier*, 30 (1970) 301.
65 L. Csupor, *Deut. Apoth.-Ztg.*, 111 (1971) 481.
66 H.M. Stevens and R.H. Fox, *Forensic Sci. Soc. J.*, 11 (1971) 177.
67 B. Borkowski and T. Kaniewska, *J. Chromatogr.*, 59 (1971) 222.
68 I. Simon and M. Lederer, *J. Chromatogr.*, 63 (1971) 448.
69 A. Tadjer, *J. Chromatogr.*, 63 (1971) D44.
70 D.D. Datta, P.C. Bose and D. Ghosh, *Planta Med.*, 19 (1971) 258.
71 S. Zadecky, D. Küttel and M. Takacsi, *Acta Pharm. Hung.*, 42 (1972) 7.
72 N. Di Renzo, *Boll. Chim. Farm.*, 111 (1972) 450.
73 W.F.H. MacLean and K. Jewers, *J. Chromatogr.*, 74 (1972) 297.
74 J.H. Zwaving and E.H.J. de Jong-Havenga, *Pharm. Weekbl.*, 107 (1972) 137.
75 M. Pitea, P. Petcu, T. Goina and N. Preda, *Planta Med.*, 21 (1972) 177.
76 A. Cjevic, O. Gasic, M. Pergal and C. Canic, *Zb. Prir. Nauhe, Matica Srp.*, 43 (1972) 185; *C.A.*, 79 (1973) 42717a.
77 R.A. Egli, *Z. Anal. Chem.*, 259 (1972) 277.
78 E. Novakova and J. Vecerkova, *Cesk. Farm.*, 22 (1973) 347.
79 D.E. Williamson, *Chromatographia*, 6 (1973) 281.
80 W. Debska and R. Walkowiak, *Farm. Pol.*, 29 (1973) 695; *C.A.*, 80 (1974) 63784f.
81 H.D. Crone and E.M. Smith, *J. Chromatogr.*, 77 (1973) 234.
82 T. Mura and T. Tominga, *Shoyakugaku Zasshi*, 27 (1973) 135; *C.A.*, 81 (1974) 29487x.
83 K.F. Ahrend and D. Tiess, *Wiss. Z. Univ. Rostock Math. Naturw. Reihe*, 22 (1973) 951.
84 A.I. Tsesko and E.Ya. Ladygina, *Farmatsiya (Moscow)*, 23 (1974) 27; *C.A.*, 82 (1975) 28038k.
85 S.F. Cooper and R. Dugal, *J. Chromatogr.*, 101 (1974) 395.
86 T. Mura and T. Tominaga, *Shoyakugaku Zasshi*, 27 (1974) 63; *C.A.*, 81 (1974) 29486w.
87 F. Nachtmann, H. Spitzey and R.W. Frei, *Anal. Chim. Acta*, 76 (1975) 57.
88 E. Stahl and W. Schmitt, *Arch. Pharm. (Weinheim)*, 308 (1975) 570.
89 A.K. Chowdhury and S.A. Chowdhury, *Bangladesh Pharm. J.*, 4 (1975) 11; *C.A.*, 83 (1975) 48262c.
90 L. Fleischmann, G. Maffi and P.G. Mezzanzanica, *Boll. Chim. Farm.*, 114 (1975) 534.
91 M.S. Habib, *Planta Med.*, 27 (1975) 294.
92 T.M. Wu, S.F. Lee, Y.P. Chen and H.Y. Hsu, *Taiwan Yao Hsueh Tsa Chih*, 27 (1975) 17; *C.A.*, 86 (1977) 27410w.
93 R.A. de Zeeuw, P.E.W. van der Laan, J.E. Greving and F.J.W. Mansvelt, *Anal. Lett.*, 9 (1976) 831.
94 I. Yankulov and L. Eustatieva, *Dokl. Bolg. Akad. Nauk*, 29 (1976) 1345; *C.A.*, 86 (1977) 52298g.
95 S. Kuroda and Y. Kochi, *Iyakuhin Kenkyu*, 7 (1976) 154; *C.A.*, 88 (1978) 141747c.
96 L. Lepri, P.G. Desideri and M. Lepori, *J. Chromatogr.*, 116 (1976) 131.
97 F.W. Grant, *J. Chromatogr.*, 116 (1976) 230.
98 R.W. Frei, W. Santi and M. Thomas, *J. Chromatogr.*, 116 (1976) 365.
99 R. Verpoorte and A. Baerheim Svendsen, *J. Chromatogr.*, 124 (1976) 152.
100 L. Lepri, P.G. Desideri and M. Lepori, *J. Chromatogr.*, 123 (1976) 175.
101 C. Scholz, R. Hänsel and C. Hille, *Pharm. Ztg.*, 121 (1976) 1571.
102 Y. Hashimoto, K. Ando and M. Mizuno, *Shoyakugaku Zasshi*, 30 (1976) 127; *C.A.*, 87 (1977) 29066h.

103 N.H. Paik, M.K. Park and B.R. Lim, *Soul Taehakkyo Yakhak Nonmunjip*, 1 (1976) 121; *C.A.*, 87 (1977) 122850u.
104 R. Giebelmann, S. Nagel, C. Brunstein and E. Scheibe, *Zentralbl. Pharm.*, 115 (1976) 339.
105 C. Schuyt, G.M.J. Beijersbergen van-Henegouwen and K.W. Gerritsma, *Analyst (London)*, 102 (1977) 298.
106 Kh. Duchevska, S. Filipov, V. Khristov, R. Tarandzhiiska and D. Chobanov, *Farmatsiya (Sofia)*, 27 (1977) 15; *C.A.*, 89 (1978) 48938b.
107 P. Balderstone and S.F. Dyke, *J. Chromatogr.*, 132 (1977) 359.
108 A.A. Kiryanov, B.A. Krivut and M.E. Perelson, *Khim.-Farm. Zh.*, 11 (1977) 124; *C.A.*, 86 (1977) 127351w.
109 T. Hattori, M. Inoue and M. Hayakawa, *Yakugaku Zasshi*, 97 (1977) 1263; *C.A.*, 88 (1978) 126386r.
110 M. Smellic, M. Corder and J.P. Rosazza, *J. Chromatogr.*, 155 (1978) 439.
111 R. Verpoorte, A.H.M. van Rijzen, J. Siwon and A. Baerheim Svendsen, *Planta Med.*, 34 (1978) 274.
112 G. Verzar-Petri and P.T. Minh-Hoang, *Sci. Pharm.*, 46 (1978) 169.
113 V. Cavrini, A. Ferranti and G. Fabbri, *Farmaco, Ed. Prat.*, 34 (1979) 15.

TABLE 11.1

TLC SEPARATION OF *FAGARA* ALKALOIDS[56]

TLC systems:
- S1 Cellulose 0.1 *M* hydrochloric acid
- S2 Cellulose *n*-Butanol saturated with 2 *M* hydrochloric acid
- S3 Cellulose *n*-Butanol-pyridine-water (6:4:3)

| Alkaloid | $hR_F$ values* | | | Fluorescence after $NH_3$ | Colour with iodoplatinate reagent |
|---|---|---|---|---|---|
| | S1 | S2 | S3 | | |
| Candicine | 92 | 45 | 47 | - | Purple |
| Coryneine | 88 | 28 | 64 | - | Pale blue |
| Tembetarine | 85 | 55 | 57 | - | Green |
| Magnoflorine | 24 | 30 | 24 | Blue | Purple |
| N-Methylisocorydine | 77 | 53 | 49 | Blue | Purple |
| N-Methylcorydine | 74 | 57 | 65 | Blue | Purple |
| Laurifoline | 11 | 19 | 42 | Blue | Purple |
| Xanthoplanine | 34 | 42 | 63 | Blue | Purple |
| Palmatine | 15 | 28 | 45 | Yellow | Brown |
| Berberine | 13 | 36 | 49 | Lime green | Brown |
| Chelerythrine | 7 | 21 | 92 | Yellow | Brown |
| Nitidine | 0 | 0 | 40 | Green | Brown |

*Mean values of 10-20 runs.

TABLE 11.2

TLC SEPARATION OF SOME QUATERNARY PROTOBERBERINE ALKALOIDS

TLC systems:
- S1 Silica gel G — Methanol-water-ammonia (8:1:1)[42]
- S2 Silica gel G, activated — *n*-Butanol-water-acetic acid (4:5:1), upper phase[30]
- S3 Silica gel G, activated — *n*-Propanol-water-ammonia (2:1:1)[30]
- S4 Silica gel G, activated — *n*-Butanol-water-ammonia (8:1:1)[30]
- S5 Silica gel G, activated — Chloroform-methanol-ammonia (75:30:5)[67]
- S6 Silica gel G, activated — Chloroform-methanol-pyridine-monoethylamine (50% in water) (100:40:10:5)[67]

| Alkaloid | $hR_F$ values | | | | | |
|---|---|---|---|---|---|---|
| | S1 | S2 | S3 | S4 | S5 | S6 |
| Berberine | | 36 | 68 | 32 | 75 | 48 |
| Palmatine | 7 | 34 | 68 | 26 | 75 | 40 |
| Columbamine | 32 | | | | 75 | 15 |
| Magnoflorine | 24 | 24 | 64 | 12 | 43 | |
| Jatrorrhizine | 42 | 42 | 66 | 33 | 65 | |
| Thalifendine | | | | | 88 | |

TABLE 11.3

$hR_{St.}$ VALUES OF SOME ISOQUINOLINE ALKALOIDS OBTAINED WITH CONTINOUS FLOW TLC IN A BN-CHAMBER[85]

TLC systems:
S1 Silica gel HF254, activated — Methyl acetate-methanol-conc. ammonia (67:25:8)
S2 Silica gel HF254, activated — Ethanol-chloroform-acetic acid (67:30:3)

| Alkaloid | $hR_{St.}$ value relative to berberrubine chloride in solvent system S1 (basic) | $hR_{St.}$ value relative to thalifendine chloride in solvent system S2 (acidic) |
|---|---|---|
| Demethyleneberberine iodide (non-alkaloidal isoquinoline base) | 0 | 0 |
| Magnoflorine iodide | 17 | 5 |
| Columbamine chloride | 44 | 57 |
| Tetradehydrocheilanthifoline chloride | 46 | 39 |
| Jatrorrhizine chloride | 54 | 79 |
| Palmatine chloride | 61 | 45 |
| Coptisine chloride | 75 | 32 |
| Berberine sulphate | 80 | 68 |
| Thalifendine chloride | 90 | 100 |
| Berberrubine chloride | 100 | 84 |

TABLE 11.4

TLC SEPARATION OF *CORYDALIS* ALKALOIDS[36]

TLC systems:
S1 Silica gel G, activated — Benzene-chloroform-diethylamine (75:20:5)
S2 Silica gel G, activated — Cyclohexane-ethyl acetate-diethylamine (80:15:5)
S3 Silica gel G, activated — *n*-Butanol-water-acetic acid (4:5:1), upper phase
S4 Silica gel G, activated — Diisopropyl ether-ethanol-diethylamine (70:25:5)
S5 Silica gel G, activated — Ethyl acetate-chloroform-ethanol-diethylamine (70:15:10:5)

| Alkaloid | $hR_F$ values | | | | | Fluorescence colour under UV light |
|---|---|---|---|---|---|---|
| | S1 | S2 | S3 | S4 | S5 | |
| *l*-Tetrahydrocolumbamine | 19 | 5 | 47 | Tailing | 53 | Pale blue |
| Corydalmine | 14 | 7 | 49 | Tailing | 52 | Blue |
| *d*-Corydaline | 64 | 42 | 44 | 83 | 77 | Yellowish green |
| *l*-Tetrahydrocoptisine | 63 | 49 | 53 | 84 | 77 | Yellow |
| *dl*-Tetrahydropalmatine | 52 | 26 | 49 | 76 | 76 | Yellow |
| Protopine | 48 | 27 | 52 | 67 | 71 | Pale yellow |
| α-Allocryptopine | 32 | 14 | 41 | 45 | 56 | Pale yellow |
| *dl*-Canadine | 51 | 46 | 53 | 86 | 76 | Yellow |
| *dl*-Thalictricavine | 61 | 60 | 52 | 90 | 79 | Yellowish green |
| Coptisine nitrate | 15 | 28 | 32 | 29 | 29 | Bright yellow |
| Dehydrocorydaline nitrate | 1 | 1 | 34 | 2 | 1 | Yellowish green |
| Palmatine chloride | 0 | 0 | 35 | 1 | 0 | Dark yellow |
| Columbamine nitrate | 2 | 1 | 32 | 2 | 2 | Yellow |
| Berberine chloride | 3 | 4 | 30 | 7 | 11 | Yellow |

TABLE 11.5

FLUORESCENCE COLOURS OF SOME ISOQUINOLINE ALKALOIDS (365 nm)[85]

| Alkaloids | Fluorescence after development with solvent system S1 (basic)* | Fluorescence after development with solvent system S2 (acidic)* |
|---|---|---|
| Demethyleneberberine iodide (non-alkaloidal isoquinoline base) | Yellow | Yellow |
| Magnoflorine iodide | Blue | Blue |
| Columbamine chloride | Yellow | Light yellow |
| Tetradehydrocheilanthifoline chloride | Golden yellow | Golden yellow |
| Jatrorrhizine chloride | Yellow | Light yellow |
| Palmatine chloride | Green | Olive green |
| Coptisine chloride | Bright yellow | Bright yellow |
| Berberine sulphate | Green | Olive green |
| Thalifendine chloride | Greenish blue | Olive green |
| Berberrubine chloride | Olive green | Greenish yellow |

*Solvents as in Table 11.3.

TABLE 11.6

COLOURS OF SOME PROTOBERBERINE AND PROTOPINE ALKALOIDS WITH THE SPRAY REAGENTS IODOPLATINATE AND CERIUM(IV) SULPHATE AND FLUORESCENCE COLOURS IN LONGWAVE UV LIGHT[38]

| Alkaloid | Colour* | | |
|---|---|---|---|
| | Iodoplatinate (no. 56a) | Cerium(IV) sulphate (no. 14c) | Fluorescence (365 nm) |
| Coptisine | rd-br | Faint yl | Gold-yl |
| Berberine | Faint br | yl-br | yl |
| Stylopine | Beige | | |
| Tetrahydroberberine | yl-br - yl | | Faint yl[74] |
| Tetrahydropalmatine | yl-br - yl | | |
| Tetrahydrocorysamine | yl-br - yl | | |
| Thalictricavine | yl | | |
| Corydaline | yl | | |
| Hunnefolline | Colourless-orange | | |
| Mecambridine | or-br | | |
| Protopine | vio | yl-br | or |
| Hunnemannine | yl | | |
| Cryptopine | vio | rd-vio | |
| Allocryptopine | vio - rd-br | yl-br | Faint or |
| Muramine | vio - rd-br | yl-br | Faint or |

*rd = red; br = brown; yl = yellow; vio = violet; or = orange.

TABLE 11.7

TLC ANALYSIS OF PROTOBERBERINE AND PROTOPINE-TYPE ALKALOIDS

| Alkaloid* | Aim | Adsorbent | Solvent system | Ref. |
|---|---|---|---|---|
| Protopine,allocryptopine | Separation of Papaveraceae alakloids (Table 11.26) | $SiO_2$ | $CHCl_3$-EtOAc-MeOH(2:2:1) | 4 |
| α-, β- and γ-fagarine | Separation of various alkaloids | $SiO_2$ | EtOAc-hexane-DEA(77.5:17.5:5)<br>Benzene-$CHCl_3$-DEA(20:75:5)<br>Hexane-dichloroethylene-DEA(20:75:5) | |
| | | $Al_2O_3$ | EtOAc-hexane-DEA(77.5:17.5:5) | 7 |
| berb | Separation of quaternary compounds also neostigmine,padisal and choline | $Al_2O_3$ | Cyclohexane-$CHCl_3$-AcOH(45:45:10)<br>Cyclohexane-$CHCl_3$-EtOH-AcOH(40:30:20:10) | 8 |
| Protopine,allocryptopine,berb,copt | Separation of alkaloids in *Escholtzia* species, also benzophenanthridine type of alkaloids | $SiO_2$ | *n*-BuOH sat. with $H_2O$<br>*n*-BuOH-$H_2O$-AcOH(10:3:1)<br>$CCl_4$-$CHCl_3$-benzene(4:2:1)<br>$CHCl_3$-benzene-MeOH-FMA(40:50:10:0.5) | 10, 11 |
| berb,palm | Direct quantitative analysis | $SiO_2$ | *n*-BuOH-$H_2O$-AcOH(7:2:1) | 22 |
| Protopine,cryptopine, allocryptopine,muramine,berb,copt,palm, sinactine | Identification Papaveraceae alkaloids (Table 11.26) | $SiO_2$ | Benzene-$Me_2CO$-MeOH(7:2:1)<br>Benzene-EtOAc-DEA(5:4:1)<br>Cyclohexane-DEA(8:2)<br>Benzene-$Me_2CO$-$Et_2O$-25% $NH_4OH$(4:6:1:0.3) | |
| | | $Al_2O_3$ | Heptane-$CHCl_3$-$Et_2O$(4:5:1)<br>Heptane-$CHCl_3$(1:9) | 27, 53 |
| Allocryptopine,N-methylcanadine | Separation of *Fagara* alkaloids | Cellulose | I. *n*-BuOH-$H_2O$-AcOH(10:3:1)<br>II. *tert*.-amylOH-isoamylOH-$H_2O$-HCOOH (1:1:5:1)<br>Two-dimensional: I,II | |
| | | $Al_2O_3$ | $CHCl_3$-EtOH(98:2)(I), (96:4)(II)<br>Two-dimensional: I,II | 29 |
| berb,palm,jatro,magno | Identification in *Aquilegia* species (Table 11.2) | $SiO_2$ | *n*-BuOH-$H_2O$-AcOH(4:1:1)<br>*n*-BuOH-$H_2O$-AcOH(4:5:1), upper phase<br>*n*-prOH-$H_2O$-$NH_4OH$(2:1:1) | 30 |
| Allocryptopine,alpinone, copt,cryptopine,muramine,protopine | Identification of Papaveraceae alkaloids (Table 11.26) | $SiO_2$ | Benzene-EtOAc-DEA(5:4:1)<br>Cyclohexane-DEA(8:2) | 35 |

TABLE 11.7 (*continued*)

| | | | | |
|---|---|---|---|---|
| 14 *Corydalis* alkaloids | Identification of *Corydalis* alkaloids (Table 11.4) | $SiO_2$ | Benzene-$CHCl_3$-DEA(75:20:5)(I)<br>Cyclohexane-EtOAc-DEA(80:16:5)(II)<br>*n*-BuOH-$H_2O$-AcOH(4:5:1), upper phase<br>$(isopr)_2O$-EtOH-DEA(70:25:5)<br>EtOAc-$CHCl_3$-EtOH-DEA(70:15:10:5)<br>Two-dimensional: I,II | 36 |
| berb,copt,styl,THberb, THpalm,THcorysamine, thalictricavine,corydaline,hunnefolline, mecambridine | Separation of protoberberine alkaloids (Tables 11.6 and 11.26) | $SiO_2$ | Cyclohexane-DEA(9:1)<br>Xylene-MeEtCO-MeOH-DEA(20:20:3:1) | 38 |
| Protopine,hunnemannine, cryptopine,allocryptopine,muramine | Separation of protopine alkaloids (Table 11.26) | $SiO_2$ | Cyclohexane-DEA(9:1)<br>Cyclohexane-$CHCl_3$-DEA(7:2:1)<br>Xylene-MeEtCO-MeOH-DEA(20:20:3:1) | 38 |
| palm,col,jatro,magno | Separation of alkaloids in *Stephania* species (Table 11.2) | $SiO_2$ | I. MeOH-$H_2O$-$NH_4OH$(8:1:1)<br>II. *n*-prOH-$H_2O$-$NH_4OH$(7:2:1)<br>Two-dimensional: I,II | 42 |
| berb,hyd,noscapine, tacetine,licorine | Indirect colorimetric quantitative analysis | $SiO_2$ | $CHCl_3$-MeOH-$NH_4OH$(75:20:5) | 45 |
| berb,canadine,hyd, hydine | Detection in plant material | $SiO_2$ | Cyclohexane-$CHCl_3$-DEA(5:4:1) | 47 |
| Cryptopine,alborine, protopine,muramine, oreophiline | Separation of Papaveraceae alkaloids (Table 11.26) | $SiO_2$<br><br>$Al_2O_3$ | Benzene-$Me_2CO$-MeOH(7:2:1)<br>Benzene-$Me_2CO$-$Et_2O$-25% $NH_4OH$(4:6:1:0.3)<br>Heptane-$CHCl_3$-$Et_2O$(4:5:1)<br>Heptane-$CHCl_3$(1:9) | 48 |
| berb,canadine,hyd, hydine | Detection in *Hydrastis* species | $SiO_2$<br>$Al_2O_3$<br>0.1 *M* NaOH-impregnated $SiO_2$ | Cyclohexane-DEA(9:1)<br>Cyclohexane-$CHCl_3$-AcOH(45:45:10)<br><br>Benzene-MeOH(8:2) | 49 |
| berb | Fluorodensitometric analysis | $SiO_2$ | *n*-BuOH-AcOH-$H_2O$(7:1:2) | 52 |
| berb,palm,(Table 11.1) | Separation of quaternary alkaloids in *Fagara* species (Table 11.1) | Cellulose | 0.1 *M* HCl<br>*n*-BuOH sat. with 2 *M* HCl<br>*n*-BuOH-pyridine-$H_2O$(6:4:3) | 56 |

| | | | | |
|---|---|---|---|---|
| styl,sinactine, aurotensine,protopine, cryptocavine | Identification in *Fagara* species | $SiO_2$ | $Me_2CO$, cyclohexane-$Me_2CO$(3:2)<br>Cyclohexane-DEA(45:5)<br>Cyclohexane-BuOAc-DEA(35:10:5)<br>Cyclohexane-*n*-prOH(3:2), (2:3)<br>$CCl_4$-MeOH(3:2), (35:15)<br>Cyclohexane-*n*-prOH-DEA(40:5:5)<br>$CHCl_3$-MeOH(45:5) | 58 |
| palm,jatro,col | Detection in plant material, extracts and tinctures | $SiO_2$ | EtOH (94%)-17% $NH_4OH$(25:5) | 62 |
| berb | Indirect spectrophotometric analysis in *Radix berberidis* | $SiO_2$ | *n*-BuOH-$H_2O$-AcOH(4:1:1) | 65 |
| berb,palm,col,jatro, magno,thalifendine, tetrahydrothalifendine | Identification of *Thalictrum* alkaloids (Tables 11.2 and 11.16) | $SiO_2$ | $CHCl_3$-MeOH-$NH_4OH$(75:30:5)<br>Benzene-$CHCl_3$-isoprOH-$NH_4OH$(90:90:16:20)<br>Benzene-EtOAc-MeOH(75:75:100)<br>Benzene-$CHCl_3$-MeOH-EtOAc(2:7:3:1) | |
| | | $Al_2O_3$ | Benzene-$CHCl_3$(1:9)<br>$Et_2O$ | 67 |
| berb,hyd | Indirect quantitative analysis in *Hydrastis* species | $Al_2O_3$ | Cyclohexane-$CHCl_3$-AcOH(9:9:2) | 70 |
| berb,palm,jatro,copt | Separation of quaternary compounds (see also Chapter 21 on Quaternary Alkaloids) | $Al_2O_3$(basic) | $Me_2CO$-$H_2O$(85:15)<br>$CHCl_3$-MeOH-$NH_4OH$(6:3:1) | |
| | | $Al_2O_3$ (acidic) | $CHCl_3$-MeOH(85:15) | 73 |
| berb,canadine | Separation of *Hydrastis* alkaloids (Table 11.11) | $SiO_2$ | Benzene-EtOH-EtOAc-17% $NH_4OH$(4:4:2:1)<br>Benzene-MeOH(8:2) | 74 |
| berb,palm,oxyacanthine, berbamine | Identification in *Berberis* species | $SiO_2$ | Benzene-MeOH(8:2) in $NH_3$ atm.<br>$CHCl_3$-MeOH-AcOH(7:2.5:0.25) | 75 |
| berb | Identification in *Coptis* species | | No details available | 82 |
| berb | Indirect quantitative analysis in *Berberis* species | $SiO_2$ | $CHCl_3$-EtOH-25% $NH_4OH$(3:3:1) | 84 |

References p. 184

TABLE 11.7 (*continued*)

| Alkaloid* | Aim | Adsorbent | Solvent system | Ref. |
|---|---|---|---|---|
| berb,palm,col,jatro, copt,berberrubine,thalifendine,demethyleneberb,tetradehydrocheilanthiofoline,magno | Separation with continuous solvent flow TLC in BN-chamber (Tables 11.3 and 11.5) | $SiO_2$ | MeOAc-MeOH-conc. $NH_4OH$(67:25:8)<br>EtOH-$CHCl_3$-AcOH(67:30:3) | 85 |
| berb | Indirect quantitative analysis (spectrophotometric) in *Phellodendron* bark | $SiO_2$ | *n*-BuOH-$H_2O$-AcOH(7:2:1), followed by cyclohexane-DEA(9:1) | 86 |
| berb | Densitometric quantitative analysis in *Coptis* species | | No details available | 92 |
| berb,copt | Densitometric quantitative analysis in *Coptis* and *Phellodendron* species | $SiO_2$ | Cyclohexane-DEA(9:1) | 95 |
| berb,cryptopine | Quantitative analysis of *Chelidonium* alkaloids | $SiO_2$ | Toluene-MeOH(9:1) | 101 |
| berb | Fluorodensitometric quantitative analysis in plant material | $SiO_2$ | *n*-BuOH-$H_2O$-AcOH(7:2:1), followed by cyclohexane-DEA(9:1) | 102 |
| berb | Indirect quantitative analysis (spectrophotometric), in combination with furazolidine | $SiO_2$ | MeOH-DEA(24:1) | 103 |
| berb,protopine | Quantitative analysis of sanguinarine | $SiO_2$ | Benzene-MeOH(6:1), (3:2)<br>Benzene-$Me_2CO$-MeOH(7:2:1) | 107 |
| THpalm | Indirect quantitative analysis (spectrophotometric) in *Stephania* species | $SiO_2$ | Benzene-$CHCl_3$-$Me_2CO$(5:5:1) | 108 |
| berb | Indirect quantitative analysis (colorimetric) in *Coptis* rhizome | $SiO_2$ | I. *n*-BuOH-$H_2O$-AcOH(7:2:1)<br>II. Cyclohexane-DEA(9:1)<br>Two-dimensional: I,II | 109 |
| Corydaline,thalyctricavine,THpalm,bulbocapnine | Separation of *Corydalis* alkaloids | $SiO_2$ | Benzene-MeOH(8:2) | 112 |

*Abbreviations used in Tables:

| | | | | | |
|---|---|---|---|---|---|
| berb | berberine | hyd | hydrastine | styl | stylopine |
| ce | cephaeline | hydine | hydrastidine | THberb | tetrahydroberberine |
| col | columbamine | jatro | jatrorrhizine | THpalm | tetrahydropalmatine |
| copt | coptisine | magno | magnoflorine | tub | tubocurarine |
| em | emetine | palm | palmatine | | |

TABLE 11.8

TLC SEPARATION OF SOME BENZOPHENANTHRIDINE ALKALOIDS

TLC systems:

| | | |
|---|---|---|
| S1 | Silica gel G | Benzene-methanol (6:1)[107] |
| S2 | Silica gel G | Benzene-methanol (3:2)[107] |
| S3 | Silica gel G | Benzene-acetone-methanol (7:2:1)[107] |
| S4 | Silica gel G, activated | Chloroform-ethyl acetate-methanol (2:2:1)[38] |
| S5 | Silica gel G, activated | Cyclohexane-diethylamine (9:1)[38] |
| S6 | Silica gel G, activated | Cyclohexane-chloroform-diethylamine (7:2:1)[38] |
| S7 | Silica gel F254, ready made plates | Toluene-methanol (9:1)[101] |
| S8 | Silica gel G, activated | Chloroform-benzene-methanol-formamide (40:50:10:0.5)[10,11] |

| Alkaloid | $hR_F$ values | | | | | | | |
|---|---|---|---|---|---|---|---|---|
| | S1 | S2 | S3 | S4 | S5 | S6 | S7 | S8 |
| Sanguinarine | 78 | | | 70 | 78 | | 93 | 71 |
| Chelerythrine | 76 | 72 | 67 | 59 | 74 | | 84 | 60 |
| Chelidonine | | | | | 40 | 63 | 78 | |
| Norchelidonine | | | | | 16 | 42 | | |
| Homochelidonine | | | | | 36 | 64 | | |
| Chelilutine | | | | 64 | | | | 64 |
| Chelirubine | | | | 77 | | | | 77 |
| Berberine | 0 | 0 | 0 | | | | 3 | |
| Protopine | 35 | 27 | 25 | | | | | 21 |
| Allocryptopine | | | | | | | | 8 |

TABLE 11.9

COLOURS OF SOME BENZOPHENANTHRIDINE ALKALOIDS WITH IODOPLATINATE REAGENT AND CERIUM(IV) SULPHATE REAGENT AND THE FLUORESCENT COLOUR IN LONG-WAVELENGTH UV LIGHT

| Alkaloid | Colour* | | |
|---|---|---|---|
| | Iodoplatinate (no. 56a)[35,38,56] | Cerium(IV) sulphate (no. 14c)[38] | Fluorescence (366 nm) |
| Chelidonine | vio-yl - br | or - pink | or |
| Norchelidonine | vio-yl - br | br - pink | Grey |
| Homochelidonine | vio-yl - br | br-rd - yl | Grey-yl |
| Chelerythrine | Bronze | rd-br | Gold-yl |
| Sanguinarine | br | White | Grey-rose[38,10,56] |
| Nitidine[56] | br | | gr |
| Chelirubine[10] | | | rd-vio |
| Chelilutine[10] | | | Pink |
| Oxysanguinarine[35] | Dark rd - br | | |

*vio = violet; yl = yellow; br = brown; rd = red; or = orange.

TABLE 11.10

TLC ANALYSIS OF BENZOPHENANTHRIDINE ALKALOIDS

| ALKALOID | AIM | ADSORBENT | SOLVENT SYSTEM | REF. |
|---|---|---|---|---|
| *Chelidonium* alkaloids (not specified) | Separation of alkaloids in *Chelidonium* species | $SiO_2$ | $CHCl_3$-96% EtOH(35:1) | 1 |
| Chelerythrine,chelidonine,sanguinarine | Separation of various alkaloids | $SiO_2$ | EtOAc-hexane-DEA(77.5:17.5)<br>Benzene-$CHCl_3$-DEA(20:75:5)<br>Hexane-dichloroethylene-DEA(20:75:5) | |
| | | $Al_2O_3$ | EtOAc-hexane-DEA(77.5:17.5:5) | 7 |
| Chelerythrine,chelilutine,sanguinarine, chelirubine | Identification in *Escholtzia* species, also protoberberine and protopine-type alkaloids (Table 11.8) | $SiO_2$ | *n*-BuOH-$H_2O$-AcOH(10:3:1)<br>*n*-BuOH sat. with $H_2O$<br>$CCl_4$-$CHCl_3$-benzene(4:2:1)<br>$CHCl_3$-benzene-MeOH-FMA(40:50:10:0.5) | 10, 11 |
| Chelerythrine,sanguinarine,oxysanguinarine | Identification of Papaveraceae alkaloids (Table 11.26) | $SiO_2$ | Benzene-$Me_2CO$-MeOH(7:2:1)<br>Benzene-EtOAc-DEA(5:4:1)<br>Cyclohexane-DEA(8:2)<br>Benzene-$Me_2CO$-$Et_2O$-25% $NH_4OH$(4:6:1:0.3) | |
| | | $Al_2O_3$ | Heptane-$CHCl_3$-EtO(4:5:1)<br>Heptane-$CHCl_3$(1:9) | 27, 53 |
| Angoline,angolinine | Detection in *Fagara* species | $Al_2O_3$ | $CHCl_3$-EtOH(98:2)(I), (96:4)(II)<br>Two-dimensional: I,II | 29 |
| Oxysanguinarine, sanguinarine | Identification of Papaveraceae alkaloids (Table 11.26) | $SiO_2$ | Benzene-EtOAc-DEA(5:4:1)<br>Cyclohexane-DEA(8:2) | 35 |
| Cnelidonine,norchelidonine,homochelidonine,chelerythrine,chelilutine,chelirubine, sanguinarine | Separation of Papaveraceae alkaloids (Tables 11.8 and 11.26) | $SiO_2$ | $CHCl_3$-EtOAc-MeOH(2:2:1)<br>Cyclohexane-DEA(9:1)<br>Cyclohexane-$CHCl_3$-DEA(7:2:1)<br>Xylene-MeEtCO-MeOH-DEA(20:20:3:1) | 38 |
| Sanguinarine | Separation of Papaveraceae alkaloids (Table 11.26) | $SiO_2$ | Benzene-$Me_2CO$-MeOH(7:2:1)<br>Benzene-$Me_2CO$-$Et_2O$-25% $NH_4OH$(4:6:1:0.3) | |
| | | $Al_2O_3$ | Heptane-$CHCl_3$-$Et_2O$(4:5:1)<br>Heptane-$CHCl_3$(1:9) | 48 |

| | | | | |
|---|---|---|---|---|
| Chelerythrine,nitidine | Separation of quaternary alkaloids in *Fagara* species (Table 11.1) | Cellulose | 0.1 *M* HCl<br>*n*-BuOH sat. with 2 *M* HCl<br>*n*-BuOH-pyridine-$H_2O$(6:4:3) | 56 |
| Sanguinarine,dihydrosanguinarine | Detection in *Argemone* oil | $SiO_2$ | Cyclohexane-benzene(1:1)<br>Cyclohexane-$CHCl_3$(1:1) | 59 |
| Chelidonine,chelerythrine,sanguinarine | Identification in *Chelidonium* species | $SiO_2$ | $CHCl_3$-$Me_2CO$-$Et_2O$-toluene(10:5:5:5) | 80 |
| Chelidonine,chelerythrine,sanguinarine | Indirect quantitative analysis (spectrophotometric) in *Chelidonium* (Table 11.8) | $SiO_2$ | Toluene-MeOH(9:1) | 101 |
| Chelerythrine,sanguinarine,dihydrosanguinarine,norsanguinarine | Indirect quantitative analysis (spectrophotometric) of sanguinarine in *Argemone* oil (Table 11.8) | $SiO_2$ | Benzene-MeOH(6:1), (3:2)<br>Benzene-$Me_2CO$-MeOH(7:2:1)<br>Two-dimensional: I. benzene<br>II. benzene-MeOH(6:1) | 107 |

TABLE 11.11

TLC SEPARATION OF *HYDRASTIS* (PHTHALIDE) ALKALOIDS (SEE ALSO TABLE 11.26)

TLC systems:

| | | |
|---|---|---|
| S1 | Silica gel G | Benzene-ethanol-ethyl acetate-17% ammonia (4:4:2:1)[54,74] |
| S2 | Silica gel G | Benzene-methanol (8:2)[74] |
| S3 | Aluminium oxide G | Cyclohexane-chloroform-acetic acid (45:45:10)[49] |
| S4 | Silica gel G, impregnated with 0.1 *M* NaOH | Benzene-methanol (8:2)[49] |

| Alkaloid | $hR_F$ values | | | |
|---|---|---|---|---|
| | S1 | S2 | S3 | S4 |
| Hydrastine | 60 | 38 | 62 | 40 |
| Hydrastinine | 34 | 7 | 5 | 2 |
| Canadine | 66 | 54 | 85 | 53 |
| Berberine | 19 | 4 | 33 | 12 |
| Meconine | 56 | 45 | | |
| Opianic acid | 12 | 18 | | |

TABLE 11.12

DETECTION OF *HYDRASTIS* (PHTHALIDE) ISOQUINOLINE ALKALOIDS AND BENZYLISOQUINOLINE ALKALOIDS

| Alkaloid | Colour* | | | |
|---|---|---|---|---|
| | 2,4,7-Trinitro-fluorenone (no. 100a)[49] | Iodoplatinate (no. 56a)[38] | Cerium(IV) sulphate (no. 14c)[38] | Fluorescence (365 nm)[38,74] |
| Hydrastine | or-br | br-vio | Light-yl | Faint bl |
| Hydrastinine | gr-br | r-vio | Light-yl | Bright bl |
| Canadine | dgr-br | yl-br - yl | Pink** | Faint yl |
| Meconine | | Faint yl | | vio |
| Opianic acid | | - | | Faint bl-gr |
| Cotarnine | | br-vio | White | yl-gr |
| Armepavine | | vio-blvio | yl-br | Light bl |
| Laudanosine | | vio - rd-vio | yl - pink | - |
| Latericine | | bl-vio | Pink | bl |
| Bicuculine | | vio | Light yl | Light bl |
| Adlumidine | | vio | Light yl | Faint light bl |
| Capnoidine | | vio | Light yl | Faint light bl |
| Corlumine | | vio | Light yl | Light bl |
| Corlumidine | | rvio - yl | Light or | or |
| *d*-Adlumine | | rvio - br | Light yl | Light bl |
| Berberine | or-yl | | red** | |

*bl = blue; br = brown; dgr = darkgreen; gr = green; or = orange; rvio = red-violet; vio = violet; yl = yellow.
**ref. 49 uses cerium(IV) ammonium sulphate in phosphoric acid (no. 12).

References p. 184

TABLE 11.13

TLC ANALYSIS OF *HYDRASTIS* (PHTHALIDE) ALKALOIDS AND BENZYLISOQUINOLINE ALKALOIDS

| Alkaloid* | Aim | Adsorbent | Solvent system | Ref. |
|---|---|---|---|---|
| Hydine | Separation in mixtures with quinine, quinidine, dihydroquinine, euchinine, aristochine, atebrine and rivanolacetate | $SiO_2$ | I. $CHCl_3$-*n*-BuOH(1:1) sat. with 10% $NH_4OH$<br>II. Kerosine-$Me_2CO$-DEA(23:9:9)<br>III. Cyclohexanol-cyclohexane-hexane(1:1:1)+5% DEA<br>Two-dimensional: I,II or I,III | 9 |
| hyd | Indirect quantitative analysis in plant material | 2.5% $CaCl_2$-impregnated MgO | Benzene-$Me_2CO$(4:1), (9:1)<br>*n*-Hexane-$Me_2CO$(4:1), (9:1) | 15, 21 |
| Various alkaloids (Table 11.26) | Separation of Papaveraceae alkaloids (Table 11.26) | $SiO_2$<br>$Al_2O_3$ | Benzene-$Me_2CO$-MeOH(7:2:1)<br>Heptane-$CHCl_3$-$Et_2O$(4:5:1) | 27, 53 |
| Tembetarine | Detection in *Fagara* species | Cellulose | I. *n*-BuOH-$H_2O$-AcOH(10:3:1)<br>II. *tert.*-amylOH-isoamylOH-$H_2O$-HCOOH(1:1:5:1)<br>Two-dimensional: I,II | 29 |
| Various alkaloids (Table 11.26) | Separation of Papaveraceae alkaloids (Tables 11.12 and 11.26) | $SiO_2$ | Xylene-MeEtCO-MeOH-DEA(20:20:3:1)<br>Cyclohexane-DEA(9:1) | 38 |
| hyd | Direct quantitative analysis | | No details available | 43 |
| Hydine, berb, noscapine, tacetine, licorine | Indirect quantitative analysis (colorimetric) | $SiO_2$ | $CHCl_3$-MeOH-$NH_4OH$(75:20:5) | 45 |
| hyd, hydine, berb, canadine | Identification in plant material | $SiO_2$ | Cyclohexane-$CHCl_3$-DEA(5:4:1) | 47 |
| hyd, hydine, berb, canadine | Identification in *Hydrastis* species (Tables 11.11 and 11.12) | $SiO_2$<br>$Al_2O_3$<br>0.1 *M* NaOH-impregnated $SiO_2$ | Cyclohexane-DEA(9:1)<br>Cyclohexane-$CHCl_3$-AcOH(45:45:10)<br>Benzene-MeOH(8:2) | 49 |
| Tembetarine | Separation of *Fagara* alkaloids (Table 11.1) | Cellulose | 0.1 *M* HCl<br>*n*-BuOH sat. with 2 *M* HCl<br>*n*-BuOH-pyridine-$H_2O$(6:4:3) | 56 |
| hyd | TAS technique for *Hydrastis* | $SiO_2$ | Solvents as in ref. 2 | 61 |

TABLE 11.13 (*continued*)

| Alkaloid* | Aim | Adsorbent | Solvent system | Ref. |
|---|---|---|---|---|
| Thalifendlerine, laudanosine | Identification of *Thalictrum* alkaloids (Table 11.16) | $SiO_2$ | Benzene-$CHCl_3$-isoprOH-$NH_4OH$(90:90:16:20)<br>Benzene-EtOAc-MeOH(75:75:100)<br>Benzene-$CHCl_3$-MeOH-EtOAc(2:7:3:1) | |
| | | $Al_2O_3$ | Benzene-$CHCl_3$(1:9)<br>$Et_2O$ | 67 |
| hyd,berb | Indirect quantitative analysis (spectrophotometric) in *Hydrastis* extracts | $Al_2O_3$ | Cyclohexane-$CHCl_3$-AcOH(9:9:2) | |
| hyd,hydine,berb,canadine | Direct and indirect quantitative analysis of hyd in *Hydrastis* (Table 11.11 and 11.12) | $SiO_2$ | Benzene-EtOH-EtOAc-17% $NH_4OH$(4:4:2:1)<br>Benzene-MeOH(8:2) | 70 |
| | | $Al_2O_3$ | Cyclohexane-$CHCl_3$-AcOH(45:45:10) | 74 |

*For abbreviations, see footnote to Table 11.7.

TABLE 11.14

TLC SEPARATION OF SOME BISBENZYLISOQUINOLINE ALKALOIDS

TLC system:
0.1 *M* NaOH-impregnated silica gel G    Ethyl acetate-chloroform-methanol (2:2:1)[4]

| Alkaloid | $hR_F$ values |
|---|---|
| Isotetrandrine | 49 |
| Pycnamine | 47 |
| Phaeantine | 55 |
| Oxyacanthine | 35 |
| Berbamine | 42 |

TABLE 11.15

TLC SEPARATION OF SOME BISBENZYLISOQUINOLINE ALKALOIDS[19]

TLC systems:

| | | |
|---|---|---|
| S1 | Silica gel G, activated | Chloroform-methanol (5:1) |
| S2 | Silica gel G, activated | Chloroform-diethylamine (9:1) |
| S3 | Silica gel G, activated | Benzene-ethyl acetate-diethylamine (7:2:1) |
| S4 | Silica gel G, activated | Ethyl acetate-ethanol-dimethylformamide-diethylamine (6:12:1:1) |

| Alkaloid | $hR_F$ values | | | |
|---|---|---|---|---|
| | S1 | S2 | S3 | S4 |
| N-Methylcoclaurine | 66 | 82 | 64 | 70 |
| Laudanosine | 88 | 93 | 88 | 81 |
| Papaverine | 91 | 93 | 84 | 83 |
| Dauricine | 66 | 82 | 64 | 70 |
| Magnoline | 38 | 26 | 21 | 68 |
| Berbamine | 61 | 71 | 70 | 74 |
| Isotetrandrine | 58 | 71 | 69 | 72 |
| Phaeanthine | 85 | 86 | 89 | 72 |
| Cepharanthine | 82 | 86 | 91 | 81 |
| Isochondrodendrine | 52 | 58 | 38 | 60 |
| Cycleanine | 78 | 87 | 89 | 77 |
| Bebeerine | 65 | 66 | 47 | 70 |
| Hayatin | 64 | 68 | 44 | 78 |
| Hayatinin | 74 | 88 | 66 | 73 |
| Isotrilobine | 94 | 86 | 86 | 77 |
| Hayatidin | 76 | 87 | 69 | 76 |

TABLE 11.16

TLC SEPARATION OF TERTIARY *THALICTRUM* ALKALOIDS

TLC systems:

| | | |
|---|---|---|
| S1 | Silica gel G, activated | Benzene-chloroform-isopropanol-ammonia (90:90:16:20) |
| S2 | Silica gel G, activated | Benzene-ethyl acetate-methanol (75:75:100) |
| S3 | Silica gel G, activated | Benzene-chloroform-methanol-ethyl acetate (2:7:3:1) |
| S4 | Aluminium oxide, activated | Benzene-chloroform (1:9) |
| S5 | Aluminium oxide, activated | Diethyl ether |

| Alkaloid | $hR_F$ values | | | | |
|---|---|---|---|---|---|
| | S1 | S2 | S3 | S4 | S5 |
| Thalifendlerine | 43 | 69 | 77 | 10 | 37 |
| Laudanosine* | 97 | 49 | 72 | 78 | 71 |
| Isocorydine | 95 | 72 | 85 | 89 | 89 |
| Glaucine | 98 | 69 | 89 | 87 | 85 |
| Obamegine | 35 | 23 | 48 | 10 | 8 |
| Berbamine | 48 | 29 | 56 | 10 | 22 |
| Hernandesine | 79 | 45 | 73 | 65 | 0 |

TABLE 11.16 (*continued*)

| Alkaloid | $hR_F$ values | | | | |
|---|---|---|---|---|---|
| | S1 | S2 | S3 | S4 | S5 |
| Thalidasine | 80 | 78 | 89 | 55 | 80 |
| Thalidesine | 50 | 0 | 63 | 0 | 0 |
| Thalmetine | 60 | 79 | 85 | 0 | 0 |
| O-Methylthalmetine | 70 | 83 | 68 | 69 | 15 |
| Thalmelatine | 55 | 30 | 59 | 9 | 9 |
| Thalicarpine | 70 | 30 | 67 | 70 | 70 |
| Thalmineline | 45 | 71 | 81 | 21 | 10 |
| Tetrahydrothalifendine | 50 | 43 | 24 | 0 | 0 |
| Veronamine | 7 | 54 | 51 | 0 | 0 |

*Has not been discovered in *Thalictrum* sp. up to the present.

TABLE 11.17

COLOUR REACTIONS OF SOME BISBENZYLISOQUINOLINE ALKALOIDS WITH IRON(III) CHLORIDE-PERCHLORIC ACID SPRAY REAGENT (NO. 46c)[111]

| Alkaloid | Immediately | Hot air from hairdrier after | |
|---|---|---|---|
| | | 5 min | 10 min |
| Phaeantine | - | Purple-violet | Yellow-brown |
| Isotetrandrine | - | Purple-violet | Yellow-brown |
| Tetrandrine | - | Purple-violet | Yellow-brown |
| Pycnamine | Yellow | Yellow-brown yellow in the centre | Yellow |
| Berbamine | Yellow | Yellow-brown yellow in the centre | Yellow |
| Penduline | Yellow | Yellow-brown yellow in the centre | Yellow |
| 2-N-Norberbamine | Yellow | Yellow-brown yellow in the centre | Yellow |
| Limacine | Light red-brown with yellow-white centre | Green-blue with yellow centre | Red-brown |
| Isofangchinoline | Light red-brown with yellow white centre | Green-blue with yellow centre | Red-brown |
| Fangchinoline | Light red-brown with yellow-white centre | Green-blue with yellow centre | Red-brown |
| 2-N-Norobamegine | Grey with yellow-white centre | Grey with yellow-white centre | Grey with yellow-white centre |
| Obamegine | Grey with yellow-white centre | Grey with yellow white centre | Grey with yellow-white centre |
| Repanduline | Yellow | Purple | Grey-brown |

TABLE 11.18

TLC ANALYSIS OF BISBENZYLISOQUINOLINE ALKALOIDS (EXCEPT TUBOCURARINE AND RELATED ALKALOIDS)

| Alkaloid | Aim | Adsorbent | Solvent system | Ref. |
|---|---|---|---|---|
| Isotetrandrine, phaeantine, pycnamine, oxyacanthane, berbamine | Separation (Table 11.15) | 0.1 *M* NaOH-impregnated $SiO_2$ | EtOAc-$CHCl_3$-MeOH(2:2:1) | 4 |
| Cephantarine | Separation of various alkaloids | $SiO_2$ | EtOAc-hexane-DEA(77.5:17.5:5)<br>Benzene-$CHCl_3$-DEA(20:75:5)<br>Hexane-dichloroethylene-DEA(20:75:5) | |
| | | $Al_2O_3$ | EtOAc-hexane-DEA(77.5:17.5:5) | 7 |
| Various bisbenzylisoquinoline alkaloids (Table 11.16) | Separation | $SiO_2$ | $CHCl_3$-MeOH(5:1)<br>$CHCl_3$-DEA(9:1)<br>Benzene-EtOAc-DEA(7:2:1)<br>EtOAc-EtOH-DMFA-DEA(6:12:1:1) | 19 |
| Isotetrandrine, chondrocurine, curine, isochondrodendrine and their methylated and demethylated derivatives | Separation in plant material | $SiO_2$ | $CHCl_3$-BuOH-conc. $NH_4OH$(10:50:4)<br>$CHCl_3$-$Me_2CO$-DEA(5:4:1)<br>Benzene-MeOH(8:2)<br>EtOH-$H_2O$-HCOOH-FMA(5:1:1:3) | 23 |
| Various bisbenzylisoquinoline alkaloids (Table 11.14) | Separation of *Thalictrum* alkaloids | $SiO_2$ | Benzene-$CHCl_3$-isoprOH-$NH_4OH$(90:90:16:20)<br>Benzene-EtOAc-MeOH(75:75:100)<br>Benzene-$CHCl_3$-MeOH-EtOAc(2:7:3:1) | |
| | | $Al_2O_3$ | Benzene-$CHCl_3$(1:9)<br>$Et_2O$ | 67 |
| Oxyacanthine, berbamine, berberine, palmatine | Separation in *Berberis* species | $SiO_2$ | Benzene-MeOH(8:2) in $NH_3$ atm.<br>$CHCl_3$-MeOH-AcOH(70:25:2.5) | 75 |
| Thalicarpine | Direct quantitative analysis (planimetric) in *Thalictrum* species | $SiO_2$ | $CHCl_3$-benzene-MeOH(4:4:2) | 94 |
| Thalicarpine | Determination in drugs | | No details available | 106 |
| Thalicarpine, dehydrohernandaline, hernandalinol | Detection in urine | $SiO_2$ | Benzene-MeOH-conc. $NH_4OH$(80:30:0.1)<br>$Me_2CO$-abs. EtOH(50:1) | 110 |
| Various berman-type alkaloids | Identification in *Pycnarrhena* species (Table 11.17) | $SiO_2$ | EtOAc-isoprOH-25% $NH_4OH$(45:35:5)<br>$CHCl_3$-cyclohexane-DEA(5:4:1) | 111 |

TABLE 11.19

SEPARATION OF TUBOCURARINE AND ITS TERTIARY AND BISQUATERNARY ANALOGUES[79]

TLC systems:
S1 Silica gel GF254 Methyl ethyl ketone-water-formic acid (8:1:1)
S2 Silica gel GF254 Chloroform-methanol-ammonia (50:50:1)

| Compounds | $hR_F$ values | |
|---|---|---|
| | S1 | S2 |
| *d*-Chondrocurarine | 17 | |
| *d*-Tubocurarine | 30 | 9 |
| *d*-Tubocurine | | 50 |

TABLE 11.20

TLC ANALYSIS OF TUBOCURARINE AND RELATED ALKALOIDS

| Alkaloid* | Aim | Adsorbent | Solvent system | Ref. |
|---|---|---|---|---|
| tub | Separation of synthetic curarimetics | Carboxymethyl-cellulose | 0.075 *M* aq. NaCl | 17 |
| Chondrocurine, curine, isochondrodendrine and methylated and demethylated derivatives | Separation in plant material | $SiO_2$ | $CHCl_3$-BuOH-conc. $NH_4OH$(10:50:4)<br>$CHCl_3$-$Me_2CO$-DEA(5:4:1)<br>Benzene-MeOH(8:2)<br>EtOH-$H_2O$-HCOOH-FMA(5:1:1:3) | 23 |
| tub | Identification of curarimetics | $SiO_2$ | $Me_2CO$-2 *M* HCl(1:1)<br>$Me_2CO$-1 *M* HCl(1:1)<br>MeOH-1 *M* HCl(1:1)<br>96% EtOH-1 *M* HCl(1:1)<br>Dioxane-$Me_2CO$-1 *M* HCl (0.5:0.5:1)<br>Dioxane-1 *M* HCl(1:1)<br>isoprOH-1 *M* HCl(1:1) | 34 |
| tub | Decomposition in injections | $Al_2O_3$ | MeOH | 37 |
| tub | Identification of curarimetics in biological materials | Acidic $Al_2O_3$<br>Basic $Al_2O_3$<br>$Al_2O_3$ | $CHCl_3$-MeOH(8:2)<br>$CHCl_3$-MeOH(8:2)<br>MeOH-$H_2O$-AcOH(92:5:3) | 41 |
| tub | Determination in biological materials, separation from thymine and tyramine | $Al_2O_3$ | MeOAc-dil. $NH_4OH$(2:1) | 66 |
| tub, chondrocurarine, tubocurine | Separation of bis- and monoquaternary alkaloids (Table 11.19) | $SiO_2$ | MeEtCO-$H_2O$-HCOOH(8:1:1)<br>$CHCl_3$-MeOH-$NH_4OH$(50:50:1) | 79 |
| tub | TLC of quaternary nitrogen compounds on microslides | $SiO_2$ | 1 *M* HCl<br>EtOH-1 *M* HCl(1:1) | 81 |
| diMe-tub | Separation of quaternary nitrogen compounds | $SiO_2$<br><br>0.5 *M* NaBr-impregnated $SiO_2$ | 0.5 *M* NaBr in MeOH, unsat. tank<br>0.5 *M* NaI in $CHCl_3$-MeOH(2:8), unsat. tank<br>$CHCl_3$-MeOH(7:3) | 93 |
| tub | Separation of quaternary alkaloids and N-oxides | $SiO_2$ | MeOH-0.2 *M* $NH_4NO_3$(3:2)<br>MeOH-2 *M* $NH_4OH$-1 *M* $NH_4NO_3$(7:2:1) | 99 |

TABLE 11.20 (*continued*)

| Alkaloid* | Aim | Adsorbent | Solvent system | Ref. |
|---|---|---|---|---|
| tub | TLC of quaternary nitrogen-containing drugs | $SiO_2$ | $Me_2CO$-1 *M* HCl(1:1)<br>96% EtOH-1 *M* HCl(1:1)<br>$Me_2CO$-AcOH-25% HCl(10:85:5)<br>MeOH-$H_2O$-pyridine-AcOH(75:10:5:10) | 104 |
| tub | Detection in plasma | $SiO_2$<br>$Al_2O_3$<br>Cellulose<br>Polyamide | MeOH-$Me_2CO$-conc. HCl(90:10:4)<br>*n*-BuOH-$H_2O$-HCOOH(35:10:5)<br>*n*-BuOH-$H_2O$-HCOOH(5:1:1)<br>MeEtCO-$H_2O$-isoprOH(4:2:4) | 113 |

*tub = tubocurarine.

TABLE 11.21

TLC ANALYSIS AND DETECTION OF SOME ALKALOIDS FROM *PEUMUS BOLDUS*[44]

TLC systems:

| | | |
|---|---|---|
| S1 | Silica gel G, 0.1 *M* sodium hydroxide impregnated, activated | Chloroform-methanol(85:15) |
| S2 | Silica gel G, 0.1 *M* sodium hydroxide impregnated, activated | Ethyl acetate-acetone-methanol(5:2:3) |
| S3 | Silica gel G, 0.1 *M* sodium hydroxide impregnated, activated | Ethyl acetate-methyl ethyl ketone-methanol(3:1:1) |
| S4 | Silica gel G, activated | Ethyl acetate-acetone-methanol-diethylamine (45:30:20:5) |

| Alkaloid | $hR_F$ values | | | | Colour | |
|---|---|---|---|---|---|---|
| | S1 | S2 | S3 | S4 | Gibb's reagent (no. 34) | Iodoplatinate |
| Boldine | 47 | 58 | 42 | 53 | Brown-purple | Dark purple-brown |
| Norisocorydine | 57 | 35 | 25 | 62 | Blue | Green |
| N-Methyllaurotetanine | 61 | 55 | 36 | 61 | Grey-green | Dark brown |
| Isocorydine | 73 | 56 | 33 | 78 | Blue | Green |

TABLE 11.22

COLOURS OF SOME APORPHINE AND PROAPORPHINE ALKALOIDS WITH IODOPLATINATE AND CERIUM(IV) SULPHATE SPRAY REAGENTS AND FLUORESCENCE COLOURS IN LONG-WAVELENGTH UV LIGHT

| Alkaloid | Colour* | | |
|---|---|---|---|
| | Fluorescence (365 nm) | Iodoplatinate (no. 56) | Cerium(IV) sulphate (no. 14c) |
| Isothebaine | or | br-red - gr | vio-red |
| Glaucine | | yl-br | |
| Corydine | | bl-gr | |
| Isocorydine | - | vio-gr | br-red |
| Corytuberine | | colless-grey | |
| Roemerine | | yl with br edge | |
| Domesticine | | colless-red-br | |
| Bulbocapnine | | colless-gr | |
| Mecambrine | red | vio-br-dark-vio | bl-vio - red-br |
| Mecambroline | | colless | |

*bl = blue; br = brown; gr = green; or = orange; vio = violet; yl = yellow; colless = colourless.

TABLE 11.23

TLC ANALYSIS APORPHINE ALKALOIDS (EXCLUDING OPIUM ALKALOIDS)

| Alkaloid | Aims | Adsorbent | Solvent systems | Ref. |
|---|---|---|---|---|
| Amuroline,amuronine | Separation of Papaveraceae alkaloids (Table 11.26) | $SiO_2$ | $CHCl_3$-EtOAc-MeOH(2:2:1) | 4 |
| Various aporphines and proaporphines | Identification of Papaveraceae alkaloids (Table 11.26) | $SiO_2$ | Benzene-$Me_2CO$-MeOH(7:2:1)<br>Benzene-EtOAc-DEA(5:4:1)<br>Cyclohexane-DEA(8:2)<br>Benzene-$Me_2CO$-$Et_2O$-25% $NH_4OH$(4:6:1:0.3) | |
| | | $Al_2O_3$ | Heptane-$CHCl_3$-$Et_2O$(4:5:1)<br>Heptane-$CHCl_3$(1:9) | 27,53 |
| Magnoflorine | In combination with protoberberines in *Aquilegia* (see also protoberberines) | $SiO_2$ | *n*-BuOH-$H_2O$-AcOH(4:1:1)<br>*n*-BuOH-$H_2O$-AcOH(4:5:1), upper phase<br>*n*-prOH-$H_2O$-$NH_4OH$(2:1:1) | 30 |
| Isocorydine,mecambridine,oridine | Identification of Papaveraceae alkaloids (Table 11.26) | $SiO_2$ | Benzene-EtOAc-DEA(5:4:1)<br>Cyclohexane-DEA(8:2) | 35 |
| Various aporphines | Separation (Table 11.26) | $SiO_2$ | Cyclohexane-DEA(9:1)<br>Cyclohexane-$CHCl_3$-DEA(7:2:1)<br>Xylene-MeEtCO-MeOH-DEA(20:20:3:1) | 38 |
| Magnoflorine | In combination with protoberberines in *Stephania* (see also protoberberines) | $SiO_2$ | I. MeOH-$H_2O$-$NH_4OH$(8:1:1)<br>II. prOH-$H_2O$-$NH_4OH$(7:1:2)<br>Two-dimensional: I,II | 42 |
| Boldine,norisocorydine,isocorydine, N-methyllaurotetanine | Identification in plant materials, reaction chromatography (Table 11.21) | 0.1 *M* NaOH-impregnated $SiO_2$ | $CHCl_3$-MeOH(85:15)<br>EtOAc-$Me_2CO$-MeOH(50:20:30)<br>EtOAc-MeEtCO-MeOH(60:20:20) | |
| | | $SiO_2$ | EtOAc-$Me_2CO$-MeOH-DEA(45:30:20:5) | 44 |
| Boldine | Identification in plant materials | $SiO_2$ | Benzene-$CHCl_3$-MeOH-$Me_2CO$-FMA(8:7:4:3:0.05) (2x) | 54 |
| Various *Fagara* alkaloids | Separation (Table 11.1) | Cellulose | 0.1 *M* HCl<br>*n*-BuOH- sat. with 2 *M* HCl<br>*n*-BuOH-pyridine-$H_2O$(6:4:3) | 56 |
| Magnoflorine | Separation of *Thalictrum* alkaloids (Table 11.2) | $SiO_2$ | $CHCl_3$-MeOH-$NH_4OH$(75:30:5) | 67 |

| | | | | |
|---|---|---|---|---|
| Boldine | Densitometric quantitative analysis | | Xylene-MeEtCO-MeOH-DEA(20:20:3:1) | 72 |
| Magnoflorine | Separation of isoquinoline alkaloids with continuous flow TLC in BN-chamber (Table 11.3) | $SiO_2$ | MeOAc-MeOH-conc. $NH_4OH$(67:25:8)<br>EtOH-$CHCl_3$-AcOH(67:30:3) | 85 |
| Boldine | Densitometric quantitative analysis | $SiO_2$ | $CHCl_3$-DEA(75:25) | 90 |

TABLE 11.24

COLOURS OF SOME RHOEADINE-TYPE ALKALOIDS WITH IODOPLATINATE REAGENT[35]

| Alkaloid | Colour |
|---|---|
| Alpinine | Beige-pink |
| Alpinigenine | Beige-pink |
| Glaucamine | Brown-beige |
| Isorhoeadine | White-beige |
| Rhoeadine | Beige |
| Rhoeagenine | Beige |

References p. 184

TABLE 11.25

TLC ANALYSIS OF RHOEADINE-TYPE ALKALOIDS

| Alkaloid | Aim | Adsorbent | Solvent system | Ref. |
|---|---|---|---|---|
| Glaucamine,rhoeadine | Separation of Papaveraceae alkaloids (Table 11.26) | $SiO_2$ | $CHCl_3$-EtOAc-MeOH(2:2:1) | 4 |
| Papaverrubines A - F | Detection in plant material | $SiO_2$ | Benzene-$Me_2CO$-MeOH(7:2:1) | 14 |
| Alkaloids in *Papaver rhoeas* | Separation | $SiO_2$ | I. $CHCl_3$-EtOAc-MeOH(50:40:5)<br>II. Cyclohexane-PrOH-DEA(85:5:10)<br>Two-dimensional: I,II | 25 |
| Various rhoeadines | Separation of Papaveraceae alkaloids (Table 11.26) | $SiO_2$ | Benzene-$Me_2CO$-MeOH(7:2:1)<br>Benzene-EtOAc-DEA(5:4:1)<br>Cyclohexane-DEA(8:2)<br>Benzene-$Me_2CO$-$Et_2O$-25% $NH_4OH$(4:6:1:0.3) | |
| | | $Al_2O_3$ | Heptane-$CHCl_3$-$Et_2O$(4:5:1)<br>Heptane-$CHCl_3$(1:9) | 27, 53 |
| Alpinine,alpinigenine, glaucamine,rhoeadine isorhoeagenine,papaverrubine | Separation of Papaveraceae alkaloids (Table 11.26) | $SiO_2$ | Benzene-EtOAc-DEA(5:4:1)<br>Cyclohexane-DEA(8:2)<br>Benzene-$Me_2CO$-MeOH(7:2:1) | 35 |
| Various rhoeadines | Separation of Papaveraceae alkaloids (Table 11.26) | $SiO_2$ | Benzene-$Me_2CO$-MeOH(7:2:1)<br>$CHCl_3$-MeOH(1:1) | 38 |
| Various rhoeadines | Separation of Papaveraceae alkaloids (Table 11.26) | $SiO_2$ | Benzene-$Me_2CO$-MeOH(7:2:1)<br>Benzene-$Me_2CO$-$Et_2O$-25% $NH_4OH$(4:6:1:0.3)<br>$CHCl_3$-MeOH(1:1) | |
| | | $Al_2O_3$ | Heptane-$CHCl_3$-$Et_2O$(4:5:1)<br>Heptane-$CHCl_3$(1:9) | 48 |
| Rhoeadine,rhoeagine, isorhoeadine | Identification in plant material and extracts | $SiO_2$ | Benzene-MeOH(9:1) (2x) | 62 |
| Papaverrubine A, D and E | Separation on talc | Talc | EtOAc-MeOH(9:1) | 76 |

References p. 184

TABLE 11.26

TLC SEPARATION OF PAPAVERACEAE ALKALOIDS

TLC systems:
S1 Silica gel G, activated — Chloroform-benzene (4:5), sat. with formamide + 10% methanol[38]
S2 Silica gel G, activated — Chloroform-ethyl acetate-methanol (2:2:1)[4,38]
S3 Silica gel G, activated — Cyclohexane-diethylamine (9:1)[38]
S4 Silica gel G, activated — Cyclohexane-chloroform-diethylamine (7:2:1)[38]
S5 Silica gel G, activated — Xylene-methyl ethyl ketone-methanol-diethylamine (20:20:3:1)[38]
S6 Silica gel G, activated — Benzene-acetone-methanol (7:2:1) (S6a[27,53]; S6b[14,48])
S7 Silica gel G, activated — Benzene-ethyl acetate-diethylamine (5:4:1)[35,53]
S8 Silica gel G, activated — Cyclohexane-diethylamine (8:2)[35,53]
S9 Silica gel G, activated — Benzene-acetone-diethyl ether-25% ammonia (4:6:1:0.3)[48,53]
S10 Silica gel G, activated — Chloroform-methanol (1:1)[48]
S11 Aluminium oxide G, activated — Heptane-chloroform-diethyl ether (4:5:1)[27,48,53]
S12 Aluminium oxide G, activated — Heptane-chloroform (1:9)[48,53]

| *Alkaloid* | *$hR_F$ values* | | | | | | | | | | | | | |
|---|---|---|---|---|---|---|---|---|---|---|---|---|---|---|
| | S1 | S2 | S3a | S3b | S4 | S5 | S6a | S6b | S7 | S8 | S9 | S10 | S11 | S12 |
| Tetrahydroisoquinoline alkaloids: | | | | | | | | | | | | | | |
| Hydrastine | | | | 77 | | 46 | | | | | | | | |
| Cotarnine | | | | 65 | | 47 | | | | | | | | |
| Hydrocotarnine | | | | | | | 39 | | | | | | 35 | |
| Benzylisoquinoline alkaloids: | | | | | | | | | | | | | | |
| Armepavine | | | 7 | 8 | 26 | 54 | 15 | | | | | | 0 | |
| Papaverine | | 83 | 9 | 18 | 48 | 71 | | | | | | | | |
| Laudanosine | | | | 37 | | 59 | 23 | | | | | | 21 | |
| Latericine | | | | 0 | | 9 | 5 | | | | | | 0 | |
| Papaveraldine | | | | | | | 66 | | | | | | 36 | |
| Laudanine/laudanidine | | | | | | | 18 | | | | | | 2 | |
| Reticuline | | | | | | | 20 | | | | | | 0 | |
| Codamine | | | | | | | 28 | | | | | | 6 | |
| Palaudine | | | | | | | 50 | | | | | | 0 | |
| Phthalide isoquinoline alkaloids: | | | | | | | | | | | | | | |
| Bicuculine | | | | 29 | | 82 | | | | | | | | |
| Adlumidine | | | | 23 | | 90 | | | | | | | | |
| Capnoidine | | | | 22 | | 90 | | | | | | | | |
| Corlumine | | | | 16 | | 73 | | | | | | | | |

TABLE 11.26 (*continued*)

| | | | | | | | | | | | | | |
|---|---|---|---|---|---|---|---|---|---|---|---|---|---|
| Corlumidine | | | 2 | | 73 | | | | | | | | |
| *d*-Adlumine | | | 23 | | 86 | | | | | | | | |
| β-Hydrastine | | | 28 | | 75 | | | | | | | | |
| α-Narcotine | 91 | 23 | 39 | 55 | 84 | 70 | | | | | | 34 | |
| Narcotoline | | | 5 | | 80 | 60 | | | | | | 0 | |
| β-Narcotine | | | 40 | | 85 | | | | | | | | |
| Gnoscopine | | | | | | 69 | | | | | | 26 | |
| Narceine | | | 0 | | 4 | 0 | | | | | | 0 | |
| Nornarceine | | | 0 | | 0 | | | | | | | | |
| Aporphine and proaporphine alkaloids: | | | | | | | | | | | | | |
| Isothebaine | 70 | | 30 | | 62 | 46 | | | | | | 26 | |
| Glaucine | | 27 | | 62 | | 43 | | | | | | 37 | |
| Corydine | | 19 | | 53 | | | | | | | | | |
| Isocorydine | | 25 | 33 | 56 | 51 | 47 | | 65 | 35 | | | 34 | |
| Corytuberine | | 0 | | 0 | | 0 | | | | | | 0 | |
| Roemerine | | 55 | | 71 | | 55 | | | | | | 47 | |
| Amurine | 66 | | | | | | 31 | | | 41 | | | 62 |
| Domesticine | | 14 | | 43 | | | | | | | | | |
| Bulbocapnine | | 15 | | 48 | | | | | | | | | |
| Amuroline | 43 | | | | | 19 | | | | | | 8 | |
| Amuronine | 52 | | | | | 35 | | | | | | 21 | |
| Mecambrine | | 23 | 34 | 50 | 59 | 44 | | | | | | 24 | |
| Mecambroline | | 8 | | 24 | | 43 | | | | | | 9 | |
| Nuceferine | | | | | | 48 | | | 57 | | | 49 | |
| Oridine | | | | | | 2 | | | | 10 | | 0 | |
| N-Methyloridine | | | | | | 7 | | | | 32 | | 0 | 10 |
| Nuciferoline | | | | | | 39 | | 36 | 9 | | | 7 | |
| Magnoflorine | | | | | | 0 | | | | | | 0 | |
| Isoboldine | | | | | | 26 | | | | | | 0 | |
| Pronuciferine | | | | | | 37 | | | | | | 17 | |
| Glaziovine | | | | | | 22 | | | | 39 | | 10 | |
| Orientalinone | | | | | | 19 | | | | | | 5 | |
| Bracteoline | | | | | | 16 | | | | | | 0 | |
| Homolinearisine | | | | | | 24 | | | | 25 | | 3 | |
| Protoberberine alkaloids: | | | | | | | | | | | | | |
| Coptisine | | | 75 | | 53 | 0 | | | 63 | | | 0 | |
| Berberine | | | 61 | | 22 | 0 | | | | | | 0 | |
| Stylopine | | 56 | | | | | | | | | | | |
| Tetrahydroberberine | | 51 | | | | | | | | | | | |
| Tetrahydropalmatine | | 36 | | | | | | | | | | | |
| Tetrahydrocorysamine | | 68 | | | | | | | | | | | |
| Isocorypalmine (tetrahydrocolumbamine) | | | | | | 68 | | | | | | 40 | |

References p. 184

| | | | | | | | | | | | | | | |
|---|---|---|---|---|---|---|---|---|---|---|---|---|---|---|
| Thalictricavine | | | 68 | | | | | | | | | | | |
| Corydaline | | | 50 | | | | | | | | | | | |
| Hunnefolline | | | 4 | | 22 | | | | | | | | | |
| Mecambridine (oreophiline) | | | 12 | | 35 | | 48 | 40 | 64 | 25 | 63 | | 7 | 47 |
| Palmatine | | | | | | | 0 | | | | | | 0 | |
| Sinactine | | | | | | | 70 | | | | | | 50 | |
| Alborine | | | | | | | 7 | 0 | | | 0 | | 0 | 0 |
| Scoulerine | | | | | | | 54 | | | | | | 5 | |
| Protopine alkaloids: | | | | | | | | | | | | | | |
| Protopine | 18 | 31 | 43 | 54 | 64 | 71 | 23 | 18 | 85 | 49 | 72 | | 7 | 64 |
| Hunnemannine | | | 0 | | 9 | | | | | | | | | |
| Cryptopine | | | | 38 | | 58 | 11 | 15 | 78 | 35 | 69 | | 14 | 65 |
| Allocryptopine | 7 | 14 | 27 | 35 | 56 | 53 | 11 | | 70 | 32 | | | 0 | |
| Muramine | | | | 34 | | 52 | 6 | 5 | 73 | 32 | 56 | | 5 | 46 |
| Oxomuramine | | | | | | | 3 | | | | | | 0 | |
| Alpinone | | | | | | | | | 21 | 0 | | | | |
| Morphinane alkaloids: | | | | | | | | | | | | | | |
| Morphine | | 15 | | | | | 6 | | | | | | 0 | |
| Codeine | | | | | | | 11 | | | | | | 6 | |
| Neopine | | | | | | | 6 | | | | | | 6 | |
| Thebaine | | 43 | | | | | 26 | | 63 | 45 | | | 19 | |
| Amurine | | | | | | | 37 | | | | | | 8 | |
| Nudaurine | 37 | | | | | | 6 | 9 | | | 25 | | 0 | 35 |
| Salutaridine | | | | | | | 27 | | | | | | 8 | |
| Pseudomorphine | | | | | | | 0 | | | | | | 0 | |
| Benzophenanthridine alkaloids: | | | | | | | | | | | | | | |
| Chelidonine | | | 30 | 40 | 63 | 87 | | | | | | | | |
| Norchelidonine | | | 11 | 16 | 42 | 86 | | | | | | | | |
| Homochelidonine | | | 28 | 36 | 64 | 84 | | | | | | | | |
| Chelerythrine | | 59 | | 74 | | 93 | 66 | | | | | | 52 | |
| Sanguinarine | | 70 | | 78 | | 95 | 79 | 71 | 87 | 69 | 88 | | 51 | 86 |
| Chelilutine | | 64 | | | | | | | | | | | | |
| Chelirubine | | 77 | | | | | | | | | | | | |
| Oxysanguinarine | | | | | | | 0 | | 0 | 0 | | | 0 | |
| Pavine and isopavine alkaloids: | | | | | | | | | | | | | | |
| Argemonine | | 55 | 16 | | 26 | | | | | | | | | |
| Norargemonine | | 47 | 4 | | 16 | | | | | | | | | |
| Bisnorargemonine | | | 0 | | 2 | | | | | | | | | |

| | | | | | | | | | |
|---|---|---|---|---|---|---|---|---|---|
| Amurensine | 26 | 17 | 17 | 46 | 15 | 33 | | 3 | 20 |
| Amurensinine | | 23 | 20 | 69 | 42 | 52 | | 21 | 68 |
| Rhoeadine alkaloids: | | | | | | | | | |
| Rhoeadine | 82 | 57 | 70* | 89 | 57 | | 71 | 37 | |
| Rhoeagenine | | 62 | 60* | 87 | 30 | | 66 | 14 | |
| Isorhoeadine | | 87 | 80* | 89 | 61 | | 88 | 52 | |
| Glaudine | | 72 | 73* | | | | 79 | 46 | |
| Glaucamine | 88 | 67 | | 74 | 22 | | | 17 | |
| Oreodine | | 48 | 64* | | | | | 31 | |
| Oreogenine | | 57 | | | | | | 16 | |
| Papaverrubine A | | 71 | 65* | | | | 82 | 36 | |
| Papaverrubine B | | 64 | 59* | | | | | 31 | |
| Papaverrubine C | | 60 | 54* | | | | | 0 | |
| Papaverrubine D | | 56 | 48* | | | 65 | 54 | 11 | 57 |
| Papaverrubine E | | 46 | 34* | | | | 59 | 15 | |
| Papaverrubine F | | 37 | 26* | | | | | 20 | |
| Alpinine | | 82 | 69 | 81 | 54 | 80 | 75 | 53 | 84 |
| Alpigenine | | 68 | 58 | 70 | 35 | 74 | 55 | 15 | 65 |
| Papaverrubine G | | 65 | 55 | | | 74 | | 39 | 81 |
| Epipapaverrubine G | | | 59 | | | 75 | | | 81 |
| Epialpinine | | 86 | 72 | | | 81 | | 53 | 84 |
| Isorhoeagenine | | 74 | 60* | | | | 68 | 17 | |
| Glaugenine | | | | | | | 44 | | |
| N-Methyl-14-O-desmethylepiporphyroxin | | 63 | | | | | | 0 | |

$hR_F$ values according to ref. 38.

TABLE 11.27

TLC SEPARATION OF IPECACUANHA ALKALOIDS

TLC systems:

- S1 Silica gel G, activated — Chloroform-methanol (85:15), developed twice[38]
- S2 Silica gel G, activated — Benzene-ethyl acetate-diethylamine (7:2:1)[2]
- S3 Silica gel G, activated — Light petroleum-diethyl ether-ethanol-diethylamine (4:16:2:1)[33]
- S4 Silica gel G, activated — Carbon tetrachloride-*n*-butanol-methanol-10% ammonia (40:30:30:2)[33]
- S5 Silica gel G, activated — Xylene-methyl ethyl ketone-methanol-diethylamine (20:20:3:1)[33]
- S6 Silica gel G, activated — Cyclohexane-chloroform-diethylamine (5:4:1)[2]

| Alkaloid | $hR_F$ values | | | | | |
|---|---|---|---|---|---|---|
| | S1 | S2 | S3 | S4 | S5 | S6 |
| Cephaeline | 13 | 23 | 23 | 76 | 47 | 20 |
| Emetine | 28 | 45 | 76 | 87 | 64 | 50 |
| Psychotrine | 16 | | | | | |
| O-Methylpsychotrine | 47 | | | | | |
| Emetamine | 67 | | | | | |
| Protoemetine | 63 | | | | | |
| 2-Dehydroisoemetine | 18 | | | | | |
| 2-Dehydroemetine | 21 | | | | | |

TABLE 11.28

COLOURS OF IPECACUANHA ALKALOIDS WITH SOME SPRAY REAGENTS AND IN 366-nm UV LIGHT

| Alkaloid | Colour* | | | |
|---|---|---|---|---|
| | Iodoplatinate[47] | UV light (366 nm)[2,47,55,63] | Iodine in $CHCl_3$, 20 min at 60°C[38,46,54,55,63,91] | |
| | | | Day light | UV light (366 nm) |
| Cephaeline | White-violet | bl | br | Light bl |
| Emetine | Red-violet | yel-bl | yel | yel-bl |
| Psychotrine | | | br | yel with bl edge |
| O-Methylpsychotrine | | | yel | yel |
| Emetamine | | | Beige | yel |
| Protoemetine | | | Faint beige | yel |
| 2-Dehydroisoemetine | | | Beige | Turquoise |
| 2-Dehydroemetine | | | Beige | Turquoise |

*bl = blue; br = brown; yel = yellow.

TABLE 11.29

LITERATURE CITED IN CHAPTER 3 WHICH INCLUDES THE ANALYSIS OF IPECACUANHA ALKALOIDS

| Alkaloid* | Ref. | Alkaloid* | Ref. |
|---|---|---|---|
| em,ce | 2 | em,ce | 71 |
| em | 3 | em | 77 |
| em | 13 | em | 78 |
| em | 16 | em | 83 |

*em = emetine; ce = cephaeline.

TABLE 11.30

TLC ANALYSIS OF IPECACUANHA ALKALOIDS

| Alkaloid* | Aim | Adsorbent | Solvent system | Ref. |
|---|---|---|---|---|
| em,ce | Identification in ipecacuanha preparations | $SiO_2$ | Benzene-EtOAc-DEA(7:2:1) | 5 |
| em,ce,psychotrine,OMe-psychotrine,rubremetine | Stability in injections | $SiO_2$ | IsoamylOH-benzene(1:1) sat. with 25% $NH_4OH$ | 12 |
| em,ce,psychotrine,OMe-psychotrine | Identification in ipecacuanha preparations | alkali-impregnated $SiO_2$ | $CHCl_3$-2-5% MeOH | 18 |
| em,ce,rubremetine | Reaction chromatography | $SiO_2$ | $CHCl_3$-$Me_2CO$-DEA(5:4:1)<br>MeEtCO-EtOH-20% $NH_4OH$(5:4:1) | 24 |
| em,ce | Identification in sirups | $SiO_2$ | $CHCl_3$-$Me_2CO$-DEA(5:4:1) | 28 |
| em,ce,psychotrine,OMe-psychotrine,emetamine | Identification in ipecacuanha preparations | $SiO_2$ | $CHCl_3$-MeOH(85:15) | 32 |
| em,ce | Identification in mixtures of ipecacuanha and opium (Table 11.27) | $SiO_2$ | I. Light petroleum-$Et_2O$-EtOH-DEA(4:16:2:1)<br>II. $CCl_4$-*n*-BuOH-MeOH-10% $NH_4OH$(40:30:30:2)<br>III. Xylene-MeEtCO-MeOH-DEA(20:20:3:1)<br>Two-dimensional: II,I or II,III | 33 |
| em,ce | Indirect quantitative analysis in ipecacuanha | $SiO_2$ | Benzene-EtOAc-DEA(7:2:1) | 46 |
| em,ce | Identification in ipecacuanha preparations | $SiO_2$ | Cyclohexane-$CHCl_3$-DEA(5:4:1) | 47 |
| em,ce | Identification in ipecacuanha preparations | $SiO_2$ | $CHCl_3$-MeOH(85:15) (2x) | 54 |
| em,ce | Direct and indirect quantitative analysis (spot area and UV) in ipecacuanha | $SiO_2$ | Toluene-benzene-EtOAc-DEA(35:35:20:10) | 55,63,91 |
| ce | Indirect quantitative analysis in ipecacuanha | $SiO_2$ | $CHCl_3$-MeOH(85:15) | 57 |

TABLE 11.30 (*continued*)

| Alkaloid* | Aim | Adsorbent | Solvent system | Ref. |
|---|---|---|---|---|
| em,ce | Fluorodensitometric analysis in ipecacuanha | $SiO_2$ | BuOH-$H_2O$-AcOH(60:25:15)<br>$CHCl_3$-MeOH(85:15)<br>Cyclohexane-$CHCl_3$-DEA(5:4:1)<br>Cyclohexane-$CHCl_3$(3+7) + 1 drop DEA | 64 |
| em,ce,morphine, codeine,ephedrine | Derivatization with dansyl chloride prior to TLC | $SiO_2$ | Benzene-MeOH(3:1) | 87 |
| em,ce,morphine, ephedrine | Derivatization with dansyl chloride prior to fluorodensitometric analysis | $SiO_2$ | Toluene-MeOH-$Me_2CO$(9:1:1) | 98 |
| em | Indirect quantitative analysis in the presence of decomposition products | $SiO_2$ | $CCl_4$-*n*-BuOH-MeOH-$NH_4OH$(40:30:30:2) | 105 |

*em = emetine; ce = cephaeline.

TABLE 11.31

TLC ANALYSIS OF AMARYLLIDACEAE ALKALOIDS[4]

TLC system:
Silica gel G, activated    Ethyl acetate-chloroform-methanol (2:2:1)

| Alkaloid | $hR_F$ | Alkaloid | $hR_F$ |
|---|---|---|---|
| Lycorine | 35 | Lycorenine | 28 |
| Galanthine | 48 | Hippeastrine | 52 |
| Caranine | 38 | Masonine | 44 |
| Narcissidine | 18 | Clivonine | 73 |
| Nartazine | 53 | Nerinine | 68 |
| Nerispine | 71 | Narcissamine | 29 |
| Parkamine | 57 | Bulphanamine | 49 |
| Pluviine | 47 | Crinamidine | 50 |
| Criwelline | 74 | Undulatine | 62 |
| Tazettine | 72 | Crinine | 51 |
| Haemanthamine | 39 | Powellamine | 81 |
| Crinamine | 37 | Flexinine | 45 |
| Haemanthidine | 42 | Crinine | 66 |
| Hippawine | 43 | Flexine | 56 |
| Galanthamine | 24 | Flexamine | 63 |
| Narwedine | 26 | Neflexine | 65 |
| Crinidine | 33 | Annapawine | 13 |
| Elwesine | 36 | Brunsdonnine | 32 |
| Penareine | 46 | Krelagine | 27 |
| Homolycorine | 40 | Krepowine | 59 |
| Powelline | 34 | | |
| Oxo-Powelline | 41 | | |

TABLE 11.32

LITERATURE CITED IN CHAPTER 3 WHICH INCLUDES THE ANALYSIS OF ISOQUINOLINE ALKALOIDS

For ipecacuanha alkaloids see Table 11.29; for opium alkaloids see Chapter 12.

| Alkaloid | Ref. | Alkaloid | Ref. |
|---|---|---|---|
| Boldine, bulbocapnine, cotarnine, hydrastinine | 2 | Cotarnine, hydrastine | 68 |
| 30 isoquinoline alkaloids among a series of other alkaloids | 20 | Berberine | 69 |
| Cyclanoline, phellodendrine, menispermine, jatrorrhizine, magnoflorine, tubocurine | 39 | Berberine, tubocurarine, hydrastinine | 83 |
| | | Berberine, boldine, hydrastine, tubocurarine | 96,100 |

TABLE 11.33

TLC ANALYSIS OF VARIOUS ISOQUINOLINE ALKALOIDS

| Alkaloid | Aim | Adsorbent | Solvent system | Ref. |
|---|---|---|---|---|
| Amaryllidaceae alkaloids, | Separation (Table 11.31) | $SiO_2$ | EtOAc-$CHCl_3$-MeOH(2:2:1) | 4 |
| Amurensine, argemonine, norargemonine | Separation of Papaveraceae alkaloids (Table 11.26) | $SiO_2$ | EtOAc-$CHCl_3$-MeOH(2:2:1) | 4 |
| Hydrocotarnine | Identification of Papaveraceae alkaloids (Table 11.26) | $SiO_2$<br>$Al_2O_3$ | Benzene-$Me_2CO$-MeOH(7:2:1)<br>Heptane-$CHCl_3$-$Et_2O$(4:5:1) | 27 |
| Anhalonium alkaloids | See Chapter 10 | | | 31,40 |
| Amurensine, amurensinine, cotarnine | Identification of Papaveraceae alkaloids (Table 11.26) | $SiO_2$ | Benzene-EtOAc-DEA(5:4:1)<br>Cyclohexane-DEA(8:2) | 35 |
| Argemonine, norargemonine, bisnorargemonine, amurensine | Separation of Papaveraceae alkaloids (Table 11.26) | $SiO_2$ | $CHCl_3$-EtOAc-MeOH(2:2:1)<br>Cyclohexane-DEA(9:1)<br>Cyclohexane-$CHCl_3$-DEA(7:2:1) | 38 |
| Amurensine, amurensinine | Separation of Papaveraceae alkaloids (Table 11.26) | $SiO_2$ | Benzene-$Me_2CO$-MeOH(7:2:1)<br>Benzene-$Me_2CO$-$Et_2O$-25% $NH_4OH$(4:6:1:0.3)<br>$CHCl_3$-MeOH(1:1) | |
| | | $Al_2O_3$ | Heptane-$CHCl_3$-$Et_2O$(4:5:1)<br>Heptane-$CHCl_3$(1:9) | 48 |
| Cotarnine | In tablets with hyoscyamine, strychnine and brucine | | No details available | 89 |

Chapter 12

# OPIUM ALKALOIDS

12.1 Opium alkaloids....221
12.1.1 Solvent system....222
12.1.2 Two-dimensional TLC....225
12.1.3 Detection....226
12.1.4 Quantitative analysis....227
12.1.4.1 Indirect analysis....227
12.1.4.2 Direct analysis....231
12.1.5 TAS technique and reaction chromatography....232
12.2 Drugs of abuse....233
12.2.1 Solvent systems....233
12.2.2 Two-dimensional TLC....235
12.2.3 Detection....235
12.2.4 Quantitative analysis....238
12.2.5 TAS technique and reaction chromatography....239
12.2.6 Isolation from biological material....239
12.2.6.1 Solvent extraction....240
12.2.6.2 Isolation by means of ion-exchange resins....240
12.2.6.3 Isolation by adsorption....240
12.2.7 Extraction procedures for the isolation of drugs of abuse from urine....241
12.3 References....245

## 12.1 OPIUM ALKALOIDS

Thin-layer chromatography of opium alkaloids has been performed for several purposes:

(1) Analysis of the alkaloids present in plant material and opium. Such analyses concern mainly the six major alkaloids, but sometimes also a number of minor alkaloids and other plant constituents.

(2) Analysis of opium alkaloids present in pharmaceutical preparations. Here the alkaloids are usually present as pure compounds, but often in mixtures with various other compounds.

(3) Analysis of opium alkaloids in biological materials, most often identification of drugs of abuse in addicts.

(4) Analysis, often identification, of opium alkaloids in drug seizures ("street drugs"), often in mixtures with other drugs: cocaine, amphetamine, LSD and adulterants such as caffeine and strychnine.

(5) Quality control of pure alkaloids; detection and identification of impurities and/or decomposition products by TLC.

(6) Studies of the metabolism of opium alkaloids in animals and man.

**References p. 245**

Reviews on the TLC of opium alkaloids have been published by Stahl et al.[118], Fischbein and Falk[127] and Massa et al.[154]. In this chapter the alkaloids present in *Papaver somniferum* and *Papaver bracteatum* are dealt with, whereas the analysis of alkaloids present in other *Papaver* species are described separately in the chapter on isoquinoline alkaloids (p.171).

*12.1.1 Solvent systems*

The first paper on the TLC separation of opium alkaloids was published by Borke and Kirsch in 1953[1]. On a layer of silica gel, magnesium oxide and calcium sulphate (10:10:4) impregnated with phosphate buffer (pH 6.6) they separated the alkaloids with dioxane.

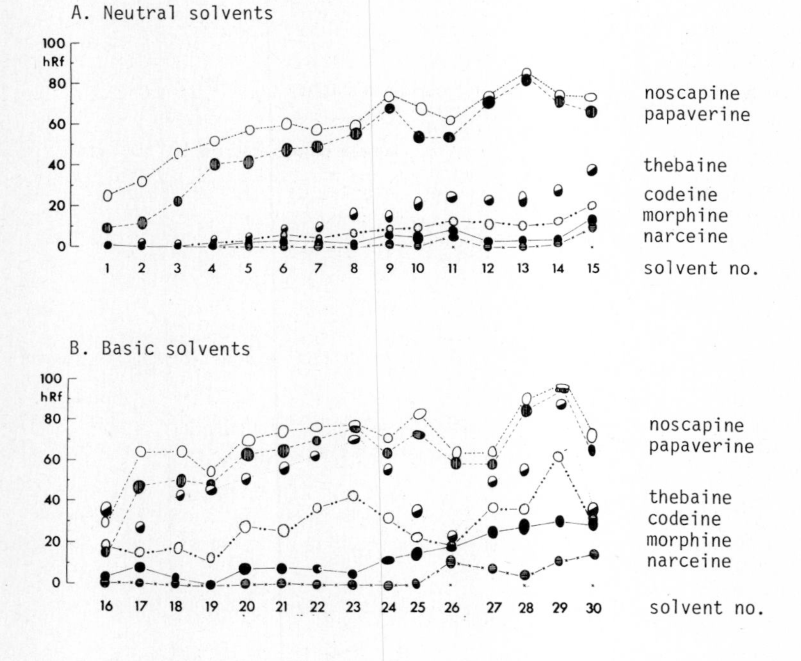


Fig. 12.1. Comparison of solvents (see Table 12.1) for the separation of the major opium alkaloids[118]. Standard conditions: Silica gel $HF_{254}$, 0.25 mm, activated and stored in a dry atmosphere. Saturated normal chromatography chambers, temperature 21°C. Distance of development: 15 cm.

Later an enormous number of different solvent systems have been tried in order to obtain optimal separations of the opium alkaloids. Stahl et al.[118] compared

58 solvent systems for the analysis of the alkaloids present in opium and concluded that aluminium oxide was less suitable than silica gel as an adsorbent. Impregnation of it with alkali was found to be advantageous only for some special separations. Impregnated cellulose did not give better analytical results. The results obtained with some of the solvent systems tested (Table 12.1) are shown in Fig. 12.1.

A modification of a solvent system originally proposed by Bayer[37], but using ammonia instead of diethylamine because of its greater volatility, was chosen as the best solvent system. By means of this solvent, toluene-acetone-95% ethanol-25% ammonia (40:40:6:2), the main opium alkaloids could be separated[115,118]. A number of variations of the solvent system have been tried, using xylene or benzene instead of toluene, methyl ethyl ketone instead of acetone, methanol instead of ethanol and diethylamine instead of ammonia[64,86,91,107,109,114,117,257,259,260,262,263,267,291,300,330,331,375]. However, none of these changes had any clear effect on the separation.

Replacement of acetone with dioxane[87,259,277] does not influence the selectivity of the solvent system much, because both solvents belong to the same class in Snyder's classification system.

Eichhorn and Kny[267] tested some solvent systems for opium alkaloid separations and found that TLC systems S3 and S7[17] (Table 12.2) gave better separations of the major and the minor alkaloids than the solvent system proposed by Bayer[37]. The former solvent system was the best for the separation of both major and minor alkaloids. It can also be used for the quality control of pure opium alkaloids.

The neutral solvent systems, which were used particularly in the early 1960s, gave satisfactory results, according to Stahl et al.[118]. Brochmann-Hanssen and co-workers[26,42] tried a number of neutral solvent systems and their results are given in Table 12.2. Benzene-methanol (4:1), proposed by Neubauer and Mothes[9], was less suitable for quantitative analysis. However, combinations of benzene and an alcohol give complete separations of the five main opium alkaloids. Benzene-95% ethanol (4:1) can be used for the separation of papaverine and noscapine for quantitative purposes, whereas chloroform-methanol (9:1) gives good results for the separation of codeine, morphine and thebaine.

Poethke and Kinze[29] studied some acidic solvent systems in combination with aluminium oxide layers for opium alkaloids. Fairbairn and El-Masry[95] used ethyl acetate to separate noscapine and papaverine on silica gel; Vignoli et al.[70] used water-saturated diethyl ether-acetone-diethylamine (85:8:7) for the same purpose, and methanol-chloroform-25% ammonia (85:15:0.7) to move narceine from the starting point.

Puech et al.[265] applied chloroform-acetone-ethanolic ammonia (pH 12.4) (10:8:2) for the TLC of opiate-containing syrups, as a diethylamine-free variation of the solvent system proposed by Waldi et al.[7] (chloroform-acetone-diethylamine, 5:4:1).

Munier and Drapier[340,366,367] studied the TLC behaviour of a number of alkaloids in chlorinated solvent systems, including some opium alkaloids (see chapter 3, p.48).

For the determination of laudanine in opium, Genest and Belec[90] used 0.1 *M* sodium hydroxide-impregnated plates and chloroform-methanol (9:1 or 87:15) as the solvent system, and Danos[108] used benzene-methanol (9:1) followed by chloroform-ethanol-acetone-ethyl acetate (6:2:1:1) to separate 18 alkaloids from *Papaver somniferum*.

Paris and Sarsunova[98] applied a number of solvent systems for some difficult separations, such as codeine-ethylmorphine-dihydrohydroxycodeinone, noscapine-thebaine-papaverine and codeine-thebaine. The results obtained are given in Table 12.2.

In an investigation of the TLC separation of papaverine and its decomposition products, Gundermann and Pohloudek-Fabini[399] obtained good results with silica gel plates and chloroform-acetone-25% ammonia (75:25:1).

In addition to silica gel and aluminium oxide, other adsorbents have also been used. Affonso[81] described the use of calcium sulphate for the separation of four alkaloids, including codeine. Ebel et al.[170] preferred calcium carbonate for the separation of a number of opium alkaloids and related compounds, and used chloroform-ethyl acetate (1:1) as the mobile phase. However, the results obtained were not reproducible because of differences in the batches of calcium carbonate used. With magnesium oxide good results were observed, whereas magnesium carbonate was found to be less suitable. Magnesium oxide was also used as adsorbent by Ragazzi et al.[48], who described a series of solvent systems for the separation of opium alkaloids. Aarø and Rasmussen[358] preferred calcium sulphate or calcium sulphate plus calcium carbonate (1:1) for the determination of codeine in pharmaceutical preparations.

The advantage of the type of adsorbents mentioned is their solubility in acids, which faciliates the subsequent quantitative analysis of the alkaloids.

Polyamide was applied for the separation of opium alkaloids by Hsiu et al.[92] and Huang et al.[93]. In many analyses cellulose[56,236] and cellulose impregnated with formamide[5,233] have also been used. Gasco and Gatti[73] tried carboxymethyl-cellulose, whereas Rao and Tandon[393] used silica gel and aluminium oxide impregnated with metal salts for alkaloid separations in general; some opium alkaloids were included in their studies. The best results were obtained with cadmium nitrate and zinc nitrate in an amount of 10%. Talc and talc impregnated with formamide have also been used as stationary phase[99,268].

Stahl and Dumont[126] applied gradient TLC to opium alkaloids. With a pH gradient perpendicular to the direction of flow, characteristic curves were obtained for basic and acidic compounds. The results of gradient TLC can be used to determine the optimal conditions for a specific separation problem.

Röder and co-workers[104,138,166] reported the use of isobaric (azeotropic) solvent systems for the separation of alkaloids, including a number of opium alkaloids.

Ion-pair chromatography of opium alkaloids was proposed by Gröningson and Schill[116,137] in an investigation on alkaloids in general. Both straight-phase and reversed-phase ion-pair chromatography were applied for the analysis. De Zeeuw et al.[394] also applied ion-pair TLC to a series of basic drugs, including some opium alkaloids.

Some further TLC systems for the analysis of opium alkaloids and related compounds in pharmaceuticals are listed in Table 12.3.

### *12.1.2 Two-dimensional TLC*

Two-dimensional TLC has been used by many workers in opium alkaloid analysis. Poethke and Kinze[29,33] used benzene-methanol (9:1) + 1 drop of formic acid (80%) and benzene-acetic acid (9:1) on aluminium oxide for the separation of the main alkaloids present in opium. Kamp et al.[28] used chloroform-ethanol (9:1) saturated with 10% ammonia, or light petroleum-diethyl ether-ethanol-diethylamine (4:16:2:1) for development in one direction, and carbon tetrachloride-*n*-butanol-methanol-10% ammonia (40:30:30:2) for the other direction. Opium alkaloids and some derivatives were chromatographed. Kamp and Onderberg[86] applied two-dimensional TLC in connection with the analysis of opium and ipecacuanha alkaloids in pharmaceutical preparations. Light petroleum-diethyl ether-ethanol-diethylamine (4:16:2:1) was used for the first direction and xylene-methyl ethyl ketone-methanol-diethylamine (20:20:3:1) for the other direction, or carbon tetrachloride-*n*-butanol-methanol-10% ammonia (40:30:30:2) for both directions. Chloroform-methanol-diethyl ether-25% ammonia (60:25:25:1) and *n*-butanol-acetic acid (100:14) saturated with water was preferred by Nonclerq and Nijs[40] for the identification of codeine and ethylmorphine in syrups. With methanol-chloroform-25% ammonia (85:15:0.7) followed by diethyl ether (saturated with water)-acetone-diethylamine(85:8:7) Vignoli et al.[70] obtained a complete separation of the opium alkaloids and some derivatives. Lewandowska and Soklowska[107] applied the same combination for the analysis of opium alkaloids in pharmaceutical preparations.

Sobiczewska and Borkowski[143] applied a two-dimensional TLC of the alkaloids present in opium prior to a quantitative analysis. Their solvent systems were chloroform-methanol (3:1) and chloroform-methanol-25% ammonia (150:50:4) and they used basic silica gel plates. Röder et al.[155] also used two-dimensional TLC in connection with the direct determination of the main opium alkaloids. Benzene-ethanol (9:1) and subsequently chloroform-ethanol-ethyl acetate-acetone (60:20:10:10) and silica gel plates were used.

**References p. 245**

### *12.1.3 Detection*

For the detection of opium alkaloids in TLC, Dragendorff's reagent in its various modifications, iodoplatinate reagent and iodine are mostly used.

Vagujfalvi[66] modified Dragendorff's reagent to obtain more sensitive detection in TLC (see Chapter 2 on Detection). Fike[69] and Puech et al.[265] used Dragendorff's reagent, but subsequently sprayed with a sodium nitrite solution to increase the sensitivity of the reaction. Diethylamine containing solvent systems should, however, be avoided. Poethke and Kinze[29] used Thies and Reuter's modification of Dragendorff's reagent. It was found to be more sensitive after subsequent spraying with 1% sodium hydroxide solution. The advantage of the modification according to Thies and Reuter was stated to be the white background obtained. It was observed that morphine and thebaine gave less intense colours than the other opium alkaloids.

Iodoplatinate reagent has the advantage that different colours are usually obtained with different alkaloids (Table 12.4), whereas Dragendorff's reagent gives only slight differences in colour.

Iodine has often been used to detect alkaloids in TLC. Sarsunova and Kakac[191] found that iodine causes irreversible changes of the alkaloids, and iodine is therefore not suitable as a non-destructive detection reagent. Egli[221] and Rao and Murty[293] modified the iodine detection technique by subsequently exposing the chromatogram to pyrrole vapour (no. 54), thereby giving different colours for codeine, morphine and thebaine. Spraying with iodine in chloroform (no. 51b) followed by heating of the TLC plate at 60°C for 15 min was recommended by Kamp and Onderberg[86]. A light green fluorescence was observed for morphine and green-yellow for noscapine, whereas codeine, papaverine and thebaine did not give any fluorescent spots.

Enache and Constantinescu[144] combined the reaction with iodine with spraying with phenothiazine solution for the detection of codeine, papaverine and morphine in pharmaceutical preparations.

A number of other reagents have also been used for the TLC detection of opium alkaloids (Table 12.5). They can be divided into two groups: (1) reagents specific for particular functional groups, including phenolic hydroxy groups or dioxymethylene groups, and (2) reagents that give different colours with various types of opium alkaloids. Examples of reagents specific for phenolic groups (present in morphine, laudanine, laudaninine and reticuline) are Gibb's reagent (nos. 34, 35a, 35b)[42,90,349], by means of which the phenolic alkaloids can be observed as violet spots on a green background, and Fast Blue B salt (no. 40b)[90], which detects the phenolic compounds as blue spots on a white background. Chrastil[323] sprayed the TLC plate with a solution of 5% ammonium cerium(IV) nitrate in acetone followed by a solution of 5% hydroxylamine in 80% acetone, and dried it with hot air (no.11). Teichert et al.[5] and Kamp and Onderberg[86] used Pauly's reagent (nos. 33a, 33b) to detect laudanidine, narcotoline and reticuline, whereas Kupferberg et al.[38] used a mixture of potassium hexacyanoferrate(II) and potassium hexacyanoferrate(III) (no. 81) for alkaloids with a phenanthrene skeleton and phenolic functions. The alkaloids were observed as fluorescent spots. The pH of the TLC plate should be 8-9 in order to obtain optimal results. Often a solution of sodium nitrite has been applied as spray reagent. Morphine yields nitrosomorphine, which is orange-brown[63,72,154]. Chromotropic acid reagent (no. 20) has been used for the detection of alkaloids containing dioxymethylene groups, including noscapine, narcotoline and narceine[127].

In connection with an investigation by Paris and Faugeras[72] to find a suitable method for the densitometric determination of morphine, the best results were achieved with sodium nitrite reagent. Other reagents have also been tried: Marquis reagent (no. 45), Fröhde's reagent (no. 90), *p*-dimethylaminobenzaldehyde in sulphuric acid (Wasicky's reagent) (purple-violet spots with morphine), vanillin-sulphuric acid (no. 104) (yellow spots) and *p*-nitraniline (weak yellow spots).

For the detection of opium alkaloids in general, *p*-dimethylaminobenzaldehyde in sulphuric acid has also been used[12,113,270,309], as well as a series of classic alkaloid colour reagents[39,63,70,72,91,218,270]. The colours observed for a series of opium alkaloids are listed in Table 12.6.

Massa et al.[154] used 0.001 *M* potassium permanganate containing a few drops of sulphuric acid to detect narceine, papaverine and noscapine. An intense blue fluorescence was observed for the three alkaloids mentioned. A weak fluorescence could be seen for the other opium alkaloids. Enache et al.[91] sprayed with phosphotungstic acid solution and Thielemann[400] with phosphomolybdic acid solution to detect some opium alkaloids and derivatives. Rao and Tandon used a solution of citric acid in acetic anhydride (no. 22)[373] for the detection of the five major opium alkaloids. They were observed as coloured spots in daylight (5 μg) and as fluorescent spots in UV light after heating at 80°C for 10 min (0.5 μg) (Table 12.4).

Spraying with vanillin-sulphuric acid reagent (no. 104) followed by heating of the TLC plate at 100°C for 1 min was used by Hammerstingl and Reich[291] for morphine and pseudomorphine. The colours obtained were reddish violet and green, respectively. Stahl et al.[118] compared several detection methods for opium alkaloids and found that only papaverine and narceine could be detected by their own fluorescence in UV light at 366 nm, and morphine could be observed only after conversion into pseudomorphine. In addition to detection with Dragendorff's reagent (Vagujfalvi modification, followed by spraying with 0.05 or 0.1 *M* sulphuric acid) and iodoplatinate reagent (which was found to be slightly less sensitive), two other detection methods were found to be useful: (1) spraying with anisaldehyde in sulphuric acid (no. 4), followed by heating at 120°C for 15 min (for morphine, codeine, thebaine, narceine and noscapine in amounts of 0.4-1 μg, but negative for papaverine (2) spraying with a saturated methanolic solution of mercury(II) acetate, drying at 80°C and subsequently spraying with anisaldehyde-sulphuric acid + 10% methanolic phosphomolybdic acid (9:1) (no. 68) and heating at 120°C for 15 min.

Güven and co-workers[196,236,344] detected opium alkaloids in TLC by spraying with (1) Deniges reagent (no. 30) (brown spots), (2) a 5% solution of 3,5-dinitrobenzoyl chloride in ethyl acetate (no. 37) (yellow spots) and (3) 3% hydrogen peroxide solution followed by heating at 100°C for 10 min and spraying again with 5% potassium hexacyanoferrate(II) solution (no. 70) (brown spots). The sensitivity of the reagent mentioned was in the range 10-100 μg of alkaloid. Rao[368] used a spray reagent consisting of 4% mercury(II) nitrate in 3% nitric acid for the detection of opium alkaloids. After heating at 110°C for 15 min reddish brown-yellow spots were obtained The reaction was claimed to be more sensitive than the reaction with Dragendorff's reagent.

Klug[222] developed a specific detection technique for morphine and heroin. By spraying with a 5% solution of sodium iodate in 1% ammonia solution (no. 87) the alkaloids could be seen as light green fluorescent spots in UV light at 366 nm. On drying, the TLC plate turned blue; on spraying with water the light green fluorescence returned.

A number of other spray reagents, used for alkaloid detection in TLC in general, such as π-acceptors[375] and TCBI reagent (no. 94)[315], have been used in TLC opium analysis, and are dealt with in Chapter 2 (p.17). Cobalt thiocyanate (no. 26a) has been used in some instances[330,342,343,352]. The alkaloids are observed as blue spots on a light blue background. Instead of cobalt, copper(II) can also be used without loss in sensitivity[343]; for codeine the sensitivity increases. Petkovic[335,346] analysed a series of analgesics, including codeine, by spraying first with an ethanolic solution of methyl red or methyl orange, and then with 2% mercury(II) chloride solution. Yellow-white spots on a yellow background were obtained.

Okumura et al.[320] developed a flame-ionization detector for thin-layer chromatography. This method was tested on some opium alkaloids.

*12.1.4 Quantitative analysis*

*12.1.4.1 Indirect analysis*

Methods have been developed for indirect determinations of alkaloids present in plant material, opium and pharmaceutical preparations and concern mainly the determination of morphine, but sometimes also codeine, thebaine, papaverine and noscapine have been included in the determinations.

Quantitative determinations of morphine were performed by Heusser[53,62,111], the morphine spots being scraped off the TLC plate and added to a solution of the reagent used for the spectrophotometric determination of morphine as nitrosomorphine. By centrifugation the silica gel was removed and the determination carried out at 470 nm. Gaevskii and Loshkarev[214] eluted the morphine with 0.1 *M* hydrochloric acid prior to the determination as nitrosomorphine, whereas Sobiczewska and Borkowski[143] used Folin-Ciocalteu's reagent for the determination.

A TLC method specific for the determination of noscapine and papaverine was developed by Fairbairn and El-Masry[95]. They applied a solvent system that separated noscapine and papaverine, whereas the other alkaloids remained in the starting spot. The spots of noscapine and papaverine were eluted with chloroform-ethanol-concentrated ammonia (90:10:1.5), the solution was evaporated to dryness and the residue dissolved in 0.1 *M* hydrochloric acid. The determination was performed by UV spectrometry; noscapine was determined at 313 and 295 nm and papaverine at 251 and 230 nm. No blank for the adsorbent was needed in this method.

Determinations of the principal opium alkaloids, morphine, codeine, thebaine, noscapine and papaverine, present in opium, were performed by Mary and Brochmann-Hanssen[26]. Papaverine and noscapine were separated in benzene-96% ethanol (4:1) and morphine, codeine and thebaine in chloroform-methanol (9:1). The alkaloid spots were eluted with methanol, and the compounds determined by UV spectrometry (morphine at 286 nm, codeine at 215 nm, noscapine at 312 nm, papaverine at 279 nm and thebaine at 285 nm). The same method was applied by Jensen[157] for the determination of the alkaloids in tetrapon (a mixture of morphine, codeine, noscapine and papaverine). Farkas and Bayer[109] determined morphine, codeine, thebaine, noscapine and papaverine by UV spectrometry after TLC separation. Dzhumashev and Aimukhamedova[215] used methanol to elute morphine from the adsorbent prior to UV spectrometric determination. Poethke and Kinze[33] described the determination of morphine, codeine, thebaine, noscapine and papaverine in opium. Noscapine and papaverine were separated by two-dimensional TLC and the alkaloids eluted from the adsorbent with chloroform by Soxhlet extraction. Thebaine and papaverine were determined spectrophotometrically with cerium(IV) sulphate, the other alkaloids with picric acid. Neubauer and Mothes[9] performed quantitative analyses of opium alkaloids by means of UV spectrometry at 285 or 287 nm after elution of the alkaloid-containing spots from the TLC plate. For routine analysis, however, they preferred a direct method. For the determination of the alkaloids in opium Teichert et al.[10] used silica gel washed several times with sulphuric acid in ethanol to remove impurities.

Von Schantz[47] introduced circular TLC for the indirect determination of morphine and papaverine. The spots were eluted with 0.1 *M* hydrochloric acid and the alkaloids subsequently determined spectrophotometrically.

**References p. 245**

Ragazzi et al.[48] preferred magnesium oxide as the adsorbent, because it allowed a quantitative analysis by dissolving the adsorbent (and the alkaloid) in dilute acid. A similar method was used by Aarø and Rasmussen[358], calcium carbonate-calcium sulphate (1:1) being used in determinations of codeine in pharmaceutical preparations. After dissolving the spots in 1 *M* hydrochloric acid the alkaloids were determined by UV spectrometry at 285 nm.

Lewandowska and Soklowska[107] determined codeine, thebaine, noscapine and papaverine spectrophotometrically (at 575 nm) with tropeoline 00 after TLC separation and elution of the spots with chloroform (three times). Narceine could also be determined at 570 nm after similar elution, using sulphuric acid - chromotropic acid as the reagent. Morphine was eluted from the adsorbent with chloroform-isopropanol (3:1) (four times) and determined spectrophotometrically with sulphanilic acid at 515 nm. The alkaloids were separated by two-dimensional TLC from artificial mixtures of the alkaloids. Blagojevic and Skrlj[139] used ethanol to elute morphine, codeine, noscapine and papaverine from the TLC plate for the determination of alkaloids in pharmaceutical preparations. Sobiczewska[371] used ethanol-water (95:5) to elute morphine from the adsorbent after TLC separation of a mixture containing morphine, atropine methylnitrate and scopolamine. Morphine was determined spectrophotometrically after addition of bromothymol blue at 428 nm.

By derivatization of morphine to its strongly fluorescent dansyl derivative prior to TLC analysis with dansyl chloride (5-dimethyl-1-aminonaphthalenesulphonyl chloride), Nachtmann et al.[301] and Frei et al.[353] were able to increase the sensitivity of the determination 7-fold compared with determination without derivatization. Codeine and noscapine did not react. The dansyl derivative of morphine has also been used for HPLC analysis. The derivative can be determined directly on the TLC plate by fluorodensitometry[353] or after elution[301,353] with benzene-triethylamine (19:1). An area of adsorbent equal to that of the morphine spot was used as a blank in the fluorimetric analysis after extraction (excitation at 365 nm, emission at 510 nm).

Wullen and Thielemans[113] eluted codeine and ethylmorphine from the TLC adsorbent with chloroform-methanol-17% ammonia (75:25:0.5) in connection with their determination in pharmaceutical preparations, whereas Chekryshkina[179,180] performed a spectrophotometric determination of papaverine with brilliant green or bromothymol blue after elution from the adsorbent with methanol. Habashy and Farid[264] also eluted papaverine and its oxidation products with methanol prior to their UV spectrophotometric determination. Belesova and Zahradnicek[232] eluted papaverine and codeine with 0.1 *M* hydrochloric acid and determined them by UV spectrophotometry at 308 and 235 nm, respectively.

*12.1.4.2 Direct analysis*

The semi-quantitative analysis of opium alkaloids by measuring spot areas has been described by Hashmi et al.[133] and Blagojevic et al.[189].

Several workers have described the densitometric determination of opium alkaloids in pharmaceutical preparations, plant material or opium after spraying with Dragendorff's reagent[43,122,146,154,188,219]. Massa and co-workers[146,154] measured at 400 nm, whereas Kostennikova and Chichiro[122] measured at 530 nm. Poethke and Kinze[43] used Thies-Reuter's modification of Dragendorff's reagent for the analysis of the main opium alkaloids because it gave a colourless background. They also found that immersion of the plate in the reagent gave better results than spraying. Paris and Faugeras[72] found Dragendorff's reagent less suitable for the determination of morphine in *Papaver* capsules, because too many spots were obtained that interfered with the detection. Of the different methods tested they found detection with sodium nitrite to be the most suitable. Later Paris et al.[188] compared the determination of morphine in plant extracts by means of densitometry using Dragendorff's reagent with a spectrophotometric and a gas chromatographic method. Massa et al.[154] compared different detection methods in the densitometric analysis of the main opium alkaloids. The methods used were quenching of fluorescence at 254 nm; fluorescence (excitation 358 nm, emission 529 nm) after oxidation with acidified 0.001 *M* potassium permanganate solution; spraying with Dragendorff's reagent and measuring at 400 nm; and, with morphine, spraying with sodium nitrite solution followed by measuring at 400 nm. Papaverine and narceine could be determined best with the quenching or fluorescence method, morphine with nitrosation and codeine with Dragendorff's reagent.

Hashiba et al.[365] determined noscapine by means of fluorodensitometry after heating the developed plate for 5 h at 120$^{o}$C. A fluorodensitometric method for codeine was developed by Haefelfinger[354]. After the separation the plates were sprayed with a mixture of 65% nitric acid and 96% sulphuric acid (4:1) and subsequently heated for 10 min at 110$^{o}$C. The fluorescence was determined at 578 nm after excitation at 300 nm.

Frei et al.[353] used the dansyl derivatives for the fluorodensitometric analysis of morphine, ephedrine, emetine and cephaeline in pharmaceutical preparations. Dansyl derivatization also gave good results in connection with elution of the alkaloids both after TLC and in HPLC. The direct method was about five times faster than the indirect TLC method.

Neubauer and Mothes[9] determined the main alkaloids in opium both directly and indirectly. For the direct method the plates were sprayed with Dragendorff's reagent (Munier modification) and subsequently photographed. The quantitative analysis was performed by measuring the blackness of the spots on the negatives. Genest and Belec[90] determined laudanine in opium. Two spray reagents were compared: Gibb's

References p. 245

reagent and Fast Blue B salt. The spots produced by Gibb's reagent were found to be more stable for densitometric purposes. The spots were measured at 545 nm, and the spots obtained with Fast Blue B salt at 465 nm. Quantitative densitometric analyses of opium alkaloids by means of reflection have been performed for codeine[259] and for codeine and noscapine[246] in pharmaceutical preparations, and also for the main alkaloids in opium[155]. The alkaloids in opium were separated by two-dimensional TLC and the analysis was carried out at 285 nm. Borkowski and Bluzniewska[388] determined meprobamate and codeine in post mortem material by means of densitometry.

*12.1.5 TAS technique and reaction chromatography*

The TAS technique has been used for the identification of opium by Jolliffe and Shellard[148]. In combination with the TLC systems of Waldi et al.[7], the TAS technique gives characteristic patterns that can be used for the identification of opium. Whether the opium alkaloids were changed or not was not discussed. The conditions used were temperature 275°C, distillation time 90 sec, sample (10-20 mg) admixed with 10 mg of calcium hydroxide and silica gel of suitable moisture content (indicator pink) as the propellant.

Bican-Fister and Grdinic[185] used the TAS technique for the application of compounds including codeine present in analgesic tablets on TLC plates. Stahl and Schmitt[303] identified the major alkaloids in opium by means of TAS. The condition used were 5 mg of sample, temperature 220°C, distillation time 2 min and 50 mg of molecular sieve charged with water as the propellant . The alkaloids identified were morphine, codeine, noscapine, papaverine and thebaine.

Reaction chromatography of opium alkaloids and related compounds was described by Kaess and Mathis[50,76,84]. Acetylation was achieved with morphine, codeine, ethylmorphine and pholcodine. Oxidation with *p*-nitroperbenzoic acid, yielding the N-oxides or epoxides of the alkaloids, was particularly suitable for distinguishing between papaverine and noscapine. Morphine, codeine, ethylmorphine, thebaine and pholcodine also reacted with this reagent. Reduction with sodium borohydride was possible only for narceine. Reaction chromatography has also been applied to opium alkaloids present in syrups[84].

Wilk and Brill[103] exposed alkaloids applied to a TLC plate to iodine vapours for 18 h. After TLC development each alkaloid gave a characteristic pattern of spots that could be used for identification. Schmidt[254] used the same method for some opium alkaloids.

Frei et al.[353] and Nachtmann et al.[301] derivatized morphine with dansyl chloride prior to TLC or HPLC in order to increase the sensitivity in the quantitative analysis (see quantitative analysis). Baiulescu and Constantinescu[300] used reaction chromatography after TLC separation in one direction, followed by TLC of the reaction products in the second direction. With complex mixtures the compounds were eluted from the plate. The reactions used were saponification with 10% sodium hydroxide solution at 110$^{o}$C, oxidation with 30% hydrogen peroxide at 60-70$^{o}$C, acetylation with acetic anhydride in pyridine (1:2) at 110$^{o}$C and reduction with 10% acetic acid - 10% hydrochloric acid (1:1) and zinc powder at 110$^{o}$C. By means of these reactions and TLC, identification of the major opium alkaloids heroin and 6-O-acetylmorphine was possible.

## 12.2 DRUGS OF ABUSE

The analysis of drugs of abuse fall into two categories, one dealing with drug seizures and the other with the detection of drugs of abuse in biological material. In the former instance the separation of the drugs of interest (alkaloids) from other compounds often present, such as amphetamines, barbiturates and other hypnotics, sedatives, tranquillizers, local anaesthetics, analgesics, etc., is the main problem. In the latter, the extraction of small amounts of the drug from the biological material and the clean-up procedure prior to TLC are the main problems. In this chapter the morphine type of alkaloids and related drugs will be dealt with. Cocaine, LSD, mescaline and psilocybe alkaloids are dealt with in separate chapters, and here they will be mentioned only when they have been analysed in connection with the narcotic analgesics. The problems in the analysis of drugs of abuse are well illustrated in a paper by Sunshine[245], in which the results of analyses of some test samples performed by 163 laboratories were presented. Only 72 laboratories analysed all samples correctly.

The merits of the different methods available in the analysis of drugs of abuse (radioimmunoassay, UV spectroscopy, fluorimetry, GLC and TLC) have been dealt with by various authors[225,227,271,276,307,327,380]. A number of general reviews on the analysis of drugs of abuse have also been published[124,127,171,178,192,194,201,212,225,227,280,282,378].

### *12.2.1 Solvent systems*

Drugs of abuse represent a heterogeneous group of compounds, and the separation problems are fairly complicated. TLC systems must be found for the separation of complex mixtures of compounds, containing combinations of acidic, neutral and basic compounds (including barbiturates, amphetamines and alkaloids) and often having different polarities.

Cochin and Daly[16] applied several solvents for the separation of basic compounds (analgesics) extracted from urine (Table 12.11). Other workers have used the same

**References p. 245**

solvents[31,57,75,88,128,151,173,201,204,316,336,363]. Schweda[88] tested TLC system S3 (Table 12.11) with different ready-made TLC plates and sheets in normal chromatography tanks and in sandwich chambers.

Davidow and co-workers[68,102] and Dole et al.[78] used ethyl acetate-methanol-ammonia (85:10:5) for the separation of a number of drugs on silica gel (see Table 12.12). This solvent system was later widely used in the analysis of drugs of abuse. Less polar versions have been described by Kaistha and co-workers[213,319] (Table 12.13). Jain et al.[325] replaced the ammonia in this solvent with diethylamine and also presented a number of solvents to be used in the confirmation of the identity of drugs of abuse. Spratt[332] used a modification of the Davidow and Dole solvent, replacing methanol with isopropanol. He also listed a series of compounds that can lead to a false-positive identification, through similar $R_F$ values or colour reactions. Methanol-ammonia (in various ratios) is also widely used in the analysis of drugs of abuse[18,135,168,173,241,261,283,324,328,369,374,38…]. This polar solvent system has the advantage that compounds with very different polarities can be separated in one TLC run. The use of azeotropic solvents in the analysis of drugs of abuse was described by Röder and co-workers[121,184].

Masoud[356] used five solvents for the identification of drugs of abuse by means of TLC; several of the solvents had been applied earlier[203,240]. The results are given in Table 12.14.

Steele[57] used eight solvents for the identification of opiates in drug seizures. The results are summarized in Table 12.15.

The separation of morphine and heroin from their metabolites was performed by Yeh[253] and Yeh et al.[381]; the results are summarized in Table 12.16.

A number of acidic solvents have been used with good results, such as *n*-butanol-acetic acid-water in different ratios[31,169,201,207,253,381,382,398]. In Table 12.17 the results obtained with two acidic solvents are presented.

Loh and co-workers[182,239] described solvents for the separation of the dansyl derivatives of morphine, naloxone, LSD and amphetamine on polyamide mini-plates. The strong fluorescence of the dansyl derivatives permitted a sensitive detection.

Broich et al.[242] used dual-bed TLC for the detection of drugs of abuse. The lower half of the plate was covered with a weak anion-exchange material and the upper part with silica gel. The advantage of these plates is that urine samples can be analysed without prior clean-up procedures. The urine is spotted on the anion-exchange part of the plate, which was then run in a normal solvent system. Contaminants remained mainly on the ion-exchange part of the plate. To obtain better separations of some pairs of compounds that are difficult to separate, a two-stage development proved valuable[202,213,226,244,273,286,319]. Several workers deal with compounds that may interfere in the identification of drugs of abuse[54,135,212,213,318,325,332,337] and presented alternative solvents that may be used to prevent such interferences. The risk of false-positive or false-negative identifications was illustrated by Sunshine[245].

Confirmation of an identification made by TLC should always be made with other analytical techniques (GLC, UV or IR spectroscopy, microchemical reactions).

### *12.2.2 Two-dimensional TLC*

Vignoli et al.[70,71] used two-dimensional TLC for the separation of opium alkaloids on silica gel plates. The solvents used were methanol-chloroform-25% ammonia (85:15:0.7) and water-saturated diethyl ether-acetone-diethylamine (85:8:7). The same method has also been applied to the analysis of illicit samples of morphine and heroin[132,345] and for the detection of drugs of abuse in urine[210]. Serfontein et al.[324] used ethyl acetate-methanol-ammonia (85:10:5) for the first run and *n*-butanol-acetic acid-water (7:2:1) for the second run in the identification of drugs in biological material. Debackere and Laruelle[112] used two-dimensional TLC for the detection of drugs used in doping in biological material. Two-dimensional TLC on polyamide mini-plates was applied by Ho et al.[182] for the separation of dansyl derivatives of some drugs of abuse.

### *12.2.3 Detection*

For the detection of drugs of abuse the same reagents can be used as for opium alkaloids (listed in Table 12.4-12.6). In the screening and identification of drugs of abuse in biological material and in drug seizures it is often necessary to use a number of consecutive spray reagents, because compounds of different classes are to be detected. In some instances specific reagents are used to confirm the presence of a compound.

The two methods most widely used are those described by Dole et al.[78] and Davidow et al.[102].

Dole et al.[78] used the following consecutive spray reagents for the detection of the compounds present in an alkaloidal extract:

(1) 0.5% sulphuric acid in water, followed by examination of the plate in long-wave UV light to detect quinine;

(2) acidified iodoplatinate reagent, which reveals a variety of narcotic drugs and tranquillizers;

(3) ammoniacal silver nitrate solution (no. 1) and heating on a hot-plate for 2 min, or Marquis reagent (no. 45) to detect morphine, dihydromorphinone and other phenolic phenanthrene alkaloids; Marquis reagent is less sensitive than silver nitrate reagent and further testing is not possible afterwards;

(4) 0.02 *M* potassium permanganate in water, to confirm the presence of phenolic phenanthrene alkaloids.

**References p. 245**

The colours observed for some drugs are summarized in Table 12.18[128].

Davidow et al.[102] developed a method for detection of acidic, neutral and basic compounds, using the following consecutive spray reagents:

(1) 0.1% ninhydrin in acetone; after spraying the TLC plate was placed under UV light for 2 min to detect amphetamines;

(2) 0.01% diphenylcarbazone in acetone-water (1:1) followed by spraying with 0.25% mercury(II) sulphate in 10% sulphuric acid to detect barbiturates, heating for 2 min at 75°C to detect phenothiazines and observation under UV light to detect quinine;

(3) iodoplatinate reagent after cooling of the plate;

(4) Dragendorff's reagent (Munier and Macheboeuf modification) after drying the plate.

The colours observed are summarized in Table 12.12.

Heaton and Blumberg[129] used the procedure described by Dole et al. but sprayed first with bromocresol green solution to detect amphetamines and larger amounts of alkaloids. Kaistha and Jaffe[213] used also a modification of the Dole et al. method. They started with ninhydrin solution when amphetamines were present.

Debackere and Laruelle[112] used Dragendorff's reagent (Munier and Macheboeuf modification) followed by spraying with a saturated solution of silver nitrate in 10% sulphuric acid and Dragendorff's reagent again to detect alkaloids. Very sharp spots were obtained in this way. Using this method caffeine was also detected. Gorodetzky[220] determined the sensitivity of a spraying method using consecutively iodoplatinate, ammoniacal silver nitrate and potassium permanganate reagents. For most of the compounds tested (16 opiates, cocaine and quinine) increased sensitivity was observed.

Bastos et al.[141,174] used two series of consecutive spray reagents. One series started with spraying with ninhydrin in acetone, followed by exposure to 350-nm UV light for 10 min. Then the plate was placed for 10 min in a chamber saturated with dithiocarbamate, sprayed with copper(II) chloride solution and heated at 90°C for 15 min. Subsequently the plate was sprayed with iodoplatinate reagent, followed by spraying with concentrated hydrochloric acid and finally with *p*-dimethylaminobenzaldehyde (1% in 1 *M* hydrochloric acid). The other series of spray-reagents started with ninhydrin in acetone, heating for 5 min at 90°C, followed by spraying with iodoplatinate reagent, concentrated hydrochloric acid and finally with diazotized *p*-nitroaniline (no. 76). The colours observed for a number of compounds were given. For the detection of primary amines, Bussey and Backer[274] used as the first spray reagent ninhydrin-phenylacetaldehyde in ethanol (no. 75a) (yellow spots). After heating at 75°C for 10 min the primary amines were observed as green fluorescent spots in longwave UV light. Consecutive spraying with *p*-dimethylaminobenzaldehyde led to bright yellow colours for the aromatic primary amines. Finally the plate was sprayed with chromotropic acid to distinguish between amphetamine and methylenedioxyamphetamine.

Masoud[356] also used ninhydrin-phenylacetaldehyde solution as the first spray reagent in an identification scheme for drugs of abuse, followed by iodoplatinate reagent. Iodoplatinate reagent could also be used after spraying with *p*-dimethylaminobenzaldehyde reagent. Kroeger et al.[369] started with Fast Blue B salt spray reagent and continued with 1 *M* sodium hydroxide solution and iodoplatinate.

After a two-stage development of the plate a series of consecutive spray reagents were used by Kaistha and co-workers[319,376,378]. The upper, middle and lower parts of the plate were sprayed with different spray reagents. (The spray reagents used in this procedure are basicaly the same as described by Dole et al.[78] and Davidow et al.[102]). The colours obtained were listed in an extensive table. Codeine was detected more specifically: exposure to UV light and to heat, spraying with ninhydrin, then spraying with mercury(II) sulphate and further heating; an orange spot was observed.

Iodoplatinate spray reagent has been used extensively. It has the advantage that different colours are obtained for different compounds, but the colours obtained are influenced by the solvent used for the development of the plates and the kind of plates used. The colours obtained for different alkaloids as described in the literature therefore vary. The detection of morphine can be improved if the plate, after spraying, is examined with high-intensity light shining through the plate[338].

Yoshimura et al.[75] developed a special detection method for morphine: after spraying with iodoplatinate reagent the plate was exposed to ammonia vapour until the colour of the morphine spot had faded, then set aside for 5 min. In this way morphine was converted into pseudomorphine, which could be detected as a fluorescent spot under UV light.

Spratt[332] listed the compounds encountered in drugs of abuse screening and the colours obtained with common spray reagents. The recovery of alkaloids, including a number of drugs of abuse, after spraying with iodoplatinate reagent was described by Holdstock and Stevens[312]. According to Berry and Grove[173], acidified iodoplatinate spray reagent gives less distinct colour differences than the neutral reagent. With the neutral reagent careful drying of the plate is necessary in order to remove the ammonia used in the TLC solvent system.

A number of more specific spray reagents have also been used in the TLC analysis of drugs of abuse. Vinson and co-workers[315,328] used TCBI reagent (no. 94) for the identification of street drugs. Various colours were observed for all classes of compounds. The minimum detectable amount was 0.05-5 μg (see Chapter 3, p.17). Thielemann[387] tested the sensitivity of the following reagents: 0.25% Rhodamine B in ethanol (no. 85) followed by examination under UV light; cobalt thiocyanate in water (no. 26b); iodine vapour; and Dragendorff's reagent (Munier and Macheboeuf modification) in combination with different types of ready-made and home-made plates. Dragendorff's reagent was found to be the most sensitive.

Harrison and Cook[125,130] used Marquis reagent (no. 45) for the identification of morphine, dextromethorphane and cyclizine. Uhlmann[97] described the reaction of a number of drugs of abuse with Pauly's reagent (no. 33a) and Fast Blue B salt (with or without base). Kupferberg et al.[38] sprayed with potassium hexacyanoferrate(III) and potassium hexacyanoferrate(II) in water (no. 81) to obtain blue fluorescent spots for phenolic hydroxyl-containing phenanthrene alkaloids. For optimal results the pH should be between 8 and 9. Fisher et al.[202] detected morphi by spraying with ammoniacal silver nitrate (no. 1) and methadone with 0.25% potassium hexacyanoferrate(III) in water (white spot for methadone). Spraying with 2% aqueous iron(III) chloride gave a grey spot for methadone and blue spots for morphine and codeine. The authors also presented technical data for photographing the chromatograms. After drying of the plate in hot air, morphine was observed as a blue fluorescent spot in longwave UV light[210], owing to the formation of pseudomorphine.

Stahl and Brombeer[390] used Mandelin reagent (no. 91) followed by heating at 105-110°C for the detection of opium alkaloids and derivatives. Gagliardi et al.[38] used 2% $HIO_3$ in sulphuric acid (80%) to detect heroin, morphine and strychnine.

*12.2.4 Quantitative analysis*

Choulis[250] determined drugs of abuse present in drug seizures. After TLC separation the spots were eluted with appropriate solvents and quantitative analysis was performed by means of UV spectroscopy. For determination of morphine, extraction of the alkaloid from silica gel plates with methanol[290,345] or saturated sodium borate solution[318] has been performed, combined with a fluorimetric determination after conversion to pseudomorphine (excitation at 254 nm, emission at 410 nm).

Genest and Farmilo[41] determined LSD and lysergic acid in narcotic seizures by UV spectroscopy after TLC separation and elution with methanol.

Ishii et al.[77] determined plasma levels of morphine after esterification of morphine with radioactive $^{35}S$-labelled pipsyl chloride. Klug and Toffel[401] extracted morphine from silica gel plates with chloroform-isopropanol (8:2) and determined morphine quantitatively by means of EMIT. Sherma et al.[285] determined morphine and amphetamine quantitatively in urine samples by means of densitometry. Three methods were used:

(1) measuring (transmission) of quenching at 254 nm;

(2) fluorescence of morphine (excitation at 360 nm, emission at 470 nm, reflectance mode);

(3) transmission at 675 nm after exposure to hydrogen chloride vapour and spraying with aqueous 2% iron(III) chloride-2% potassium hexacyanoferrate(III) (1:1). The second method was the most sensitive.

Gold et al.[211] determined heroin, morphine and some derivatives by means of direct reflectance spectrophotometry at 279 nm. Ahmed et al.[151] analysed phenanthrene alkaloids on TLC plates by means of spot area measurement, and obtained satisfactory results. The results were compared with those of a densitometric and a spectroscopic analysis after elution from the plate.

*12.2.5 TAS technique and reaction chromatography*

For the use of the TAS technique and reaction chromatography the section on this subject for the opium alkaloids (p.232). should also be consulted.

The TAS technique was described by Hiermann and Still[216,256] for the analysis of drugs of abuse. For alkaloids and related compounds heating at 250$^{o}$C for 90 sec with ammonium carbonate as the propellant was used. Tablets were also analysed. Stahl and Brombeer[390] applied the TAS technique to the analysis of psilocybin. Heating at 240$^{o}$C for 120 sec with 20% water in molecular sieve as the propellant converted psilocybin into psilocin, which was then analysed by TLC.

Ho and co-workers[182,239] analysed some drugs of abuse containing phenolic hydroxyl groups as their dansyl derivatives. The highly fluorescent derivatives allowed more sensitive detection. This technique was later applied to some drugs with phenolic hydroxyl groups in pharmaceutical preparations[301,353]. Ishii et al.[77] determined morphine, normorphine and nalorphine after esterification with radioactive $^{35}$S-labelled pipsyl chloride and TLC purification.

*12.2.6 Isolation of drugs of abuse from biological material*

Drugs of abuse can be isolated from urine by extraction with organic solvents, by means of ion-exchange resins or by adsorption on a suitable sorbent.

One of the problems in the isolation of drugs of abuse is that they can have basic, neutral or acidic characters. With some of the methods all of these types of compounds are extracted in a single step, whereas with others a more selective isolation is obtained.

In some instances acidic[160,168,171,207,210,396] or enzymatic hydrolysis[210,396] of urine prior to extraction of the drugs is preferred. In the following some of the most widely used methods for the isolation of drugs of abuse are discussed. The isolation of such compounds from biological materials other than urine, including post mortem materials, has been described by several workers[204,283,317,363,372,388]. Isolation from bile has also been reported[206,269].

References p. 245

*12.2.6.1 Solvent extraction*

Solvent extraction is the oldest method. By extraction at different pH values a selective isolation of different groups of drugs can be achieved. Such extractions are, however, time consuming.

Bastos and co-workers[141,286] extracted urine with ethanol after saturating with potassium carbonate. This method is particularly advantageous in the extraction of cocaine metabolites, which are not extracted with less polar organic solvents or the $XAD_2$ extraction procedure. Speaker[284] extracted with ethyl acetate after saturation of the urine with ammonium sulphate at pH 6.6. Yoshimura used continous extraction to increase the sensitivity of the detection of morphine in urine[74].

Stoner and Parker[275] developed an ion-pair extraction method for the extraction of different classes of drugs at a single pH. Thunell[374] also used ion-pair extraction for the screening of urine for the presence of drugs of abuse. Other papers have also dealt with drug screening procedures for urine using solvent extraction[102,187,213]. The extraction of drugs from lyophilized urine was reported by Broich et al.[175].

Predmore et al.[396] studied the recovery of morphine from biological samples using solvent extraction. Breiter and co-workers[295,348a,384] developed a solvent extraction method in which the body fluid (urine, blood) was adsorbed on a column of dry solid support material. Subsequently the column was extracted with organic solvents, immiscible with water, at different pHs. The method was compared with $XAD_2$ extraction and a conventional solvent extraction method[348a].

*12.2.6.2 Isolation by means of ion-exchange resins*

Dole et al.[78] developed a method for the extraction of drugs of abuse using ion-exchange paper. The method has been tested and compared with other methods[128,171,212,234,235]. The ion-exchange paper is easy to handle, can be stored easily and, if necessary, can be sent by post. Elution from the ion-exchange paper can be performed in one or more steps. Several applications of this method in the screening for drugs in urine have been reported[129,213,319,376].

Broich et al.[242] described the use of a TLC plate that is half covered with anion-exchange material and half with silica gel. The urine sample is applied on the anion-exchange part of the plate and the plate is subsequently developed. Further sample clean-up is not needed.

*12.2.6.3 Isolation by adsorption*

The adsorbents most widely used are $XAD_2$ and charcoal. $XAD_2$ was first used by Fujimoto and Wong[153]. Further studies on the use of $XAD_2$ were described by, e.g., Mulé and co-workers[174,243], Weissman et al.[159], Hetland et al.[195], Kullberg and Gorodetzky[272], Miller et al.[230] and Pranitis and Stolman[329]. By using different solvents the drugs of abuse can be recovered from the $XAD_2$ in different groups[243, 329]. The use of charcoal for the concentration of drugs of abuse was described by Meola and Vanko[273]. The drugs could be eluted selectively with different solvents. Cordova and Banford[327] found the isolation procedure with charcoal to be superior to other methods.

### *12.2.7 Extraction procedures for the isolation of drugs of abuse from urine*

*Procedure 1. Solvent extraction (Davidow et al.[102])*

For analgesics, antihistamines, hypnotics, sedatives, opium alkaloids and derivatives, stimulants, tranquillizers and other compounds, including some alkaloids.

*Extracting solvent.* Chloroform-isopropanol (96:4).

*Buffer solution.* pH 9.5; saturated ammonium chloride solution, 100 ml, concentrated ammonia solution to give pH 9.5.

*Procedure.* Ten millilitres of a urine sample and 1 ml of buffer are added to 50 ml of the extraction solvent in a 125-ml glass-stoppered bottle. A control urine sample containing 10 μg each of morphine, codeine, and quinine per 10 ml of urine, and 50 μg each of amphetamine, pentobarbital, phenobarbital, glutethimide and chlorpromazine per 10 ml of urine, are prepared and analysed in the same way as the urine sample for every plate used. The bottles are agitated in a shaking machine (3-in. stroke, 180 cycles/min) for 5 min. The organic layer is separated and partially dried by being filtered through filter-paper into a 100-ml beaker. One drop of 0.1 *N* hydrochloric acid is added to the solvent after the ammonia has been boiled off. The solvent is then evaporated to dryness on a steam-bath. The sides and the bottom of the beaker are washed with a fine stream of methanol. The methanol is then allowed to evaporate to approximately 25 μl while the beakers are tilted at an angle of $45^0$. The entire extract is transferred in 5-μl aliquots to the thin-layer plate with a disposable capillary pipette.

*Procedure 2. Ion-exchange paper isolation method (Dole and co-workers[78,128])*

*Citrate buffer.* pH 2.2; 980 ml of 0.1 *M* citric acid mixed with 20 ml of 0.2 *M* disodium hydrogen orthophosphate solution.

*Borate buffer.* pH 9.3; 950 ml of saturated borax solution mixed with 50 ml of 0.3 *M* sodium hydroxide solution.

*Carbonate buffer.* pH 11.0; 12.5 ml of 0.2 *M* sodium hydrogen carbonate solution mixed with 500 ml of 0.2 *M* sodium carbonate solution.

*Procedure.* Volumes of 50-ml of undiluted urine (pH 5-6) were transferred into 4-oz jars containing Reeve-Angel SA-2 ion-exchange paper (5 x 5 cm). The urines were decanted and the ion-exchange papers washed twice with distilled water.

*Barbiturates*. To each jar containing the SA-2 paper 20 ml of citrate buffer (pH 2.2) and 10 ml of chloroform were added. The samples were shaken for 10 min either intermittently by hand or on an International Shaker. The lower organic phases were separated from the aqueous phases in a separating funnel. The papers were re-extracted with 10 ml of chloroform by shaking the samples by hand for 1 min. The organic phases were combined and evaporated to dryness under nitrogen or air on a water-bath. The residues were dissolved in 25-50 µl of either methanol or chloroform.

*Narcotic analgesics and psychoactive drugs*. To each jar containing the papers 20 ml of borate buffer (pH 9.3) and 20 ml of chloroform-isopropanol (3:1) were added. The samples were shaken for 10 min intermittently by hand or on an International Shaker. The phases were separated in a separating funnel and the aqueous phases discarded. The organic phases were evaporated to dryness and the residues dissolved in 50 µl of methanol.

*d-Amphetamine and analogues*. To each jar containing the papers 20 ml of carbonate buffer (pH 11.0) and 20 ml of chloroform were added. The samples were shaken intermittently by hand for 10 min or on an International Shaker, the phases were separated and 50-100 µl of glacial acetic acid were added to each organic phase. All organic phases were evaporated to dryness and the residues dissolved in 25-50 µl of either methanol or chloroform.

*Procedure 3. Ion-pair extraction (Stoner and Parker[275])*

For hypnotics, sedatives, tranquillizers, stimulants, alkaloids, opium alkaloids and related compounds.

*Extraction solvent*. Chloroform-isopropanol (3:1).

*Bromocresol purple buffer*. Dissolve 1.7 g of anhydrous disodium hydrogen phosphate, 12.0 g of potassium dihydrogen phosphate and 0.4 g of the sodium salt of bromocresol purple in 500 ml of water, and dilute the mixture to 1 l with water. The pH should be 6.0.

*Extraction*. Add 15 ml of urine and 2.0 ml of bromocresol purple buffer to a Silber-Porter centrifuge tube. The mixture should now have a pH of 6.0 ± 0.3, as evidenced by a reddish colour. If necessary adjust the pH with either dilute sodium hydroxide or hydrochloric acid. Almost fill the tube with extraction solvent (about 25 ml), stopper and mix vigorously on a mechanical shaker for 5 min. Centrifuge for 10 min at 400 *g*, aspirate the aqueous layer and discard it. Place the tube in a water-bath at 70°C and insert a vacuum line into the mouth of each tube. When all the solvent has evaporated, dissolve the residue by adding 100 µl of methanol down the sides of the tube while it is being rotated.

*Procedure 4. Solvent extraction of basic drugs in combination with salting out of urine (Bastos et al.*[141]*)*

*Buffer.* Dissolve 53.5 g of ammonium chloride in 950 ml of water. Add 18 ml of concentrated ammonia solution and adjust the pH to 8.5 with ammonia solution or hydrochloric acid. This pH must be checked daily.

*Procedure.* To 1.0 ml of ethanol in a screw-capped test-tube add 10 ml of urine and mix. Add solid potassium carbonate until the solution is saturated (about 12 g). After capping the tube, mix the contents immediately by inversion. Agitate the tube with a Vortex R mixer for 20 sec. As the process is exothermic, the tube must be securely capped. Place the tube in an upright position. After 5 min, a creamy dark-brown suspension should float to the surface. Centrifuge the tube for 3 min. Usually 0.8-0.9 ml of ethanol separates from the aqueous phase; if less separates, more potassium carbonate should be added and the tube capped, mixed and recentrifuged.

Some urines form a film of protein between the water and ethanol phases. If this film is present, the tip of the Pasteur pipette should be used to push the film against the wall of the test-tube. Then aspirate the highly coloured ethanol concentrate into the pipette with a rubber bulb, rejecting any aqueous phase that comes into the pipette by squeezing the bulb. Transfer the ethanol concentrate into a 10-ml test-tube. If this volume is less than 0.8 ml, add several drops of ethanol to the original tube to wash the protein layer and add this wash solution to the ethanol concentrate in the test-tube. Add 5 ml of diethyl ether and 1.0 ml buffer to the ethanol concentrate and agitate the mixture thoroughly with a Vortex mixer. Two layers form on standing. The (lower) ethanol-buffer mixture contains the water-soluble fraction, which includes any morphine glucuronides that may be present. The (upper) ethanol-ether layer contains any basic organic drugs, and is transferred into a second test-tube. Save both for subsequent testing.

*Procedure 5. Isolation by adsorption on charcoal (Meola and Vanko*[273]*)*

For barbiturates, amphetamines and narcotic alkaloids.

*Adsorption of drugs.*

*(A) Preparation of charcoal.* To about 500 mg of charcoal add 200 ml of the carbonate buffer and mix thoroughly with a magnetic stirrer. Let the charcoal settle for about 10 min. Aspirate as much buffer as possible.

*(B) Adsorption.*

(1) Into appropriately labelled 16 x 150 mm screw-capped culture tubes, add 10 ml of urine, 5 ml of carbonate buffer and 0.5 ml of the pre-treated charcoal slurry.

(2) Mix thoroughly by repeated gentle inversion, then allow to stand for a few minutes.

**References p. 245**

(3) Centrifuge at about 2000 rpm for 5 min.

(4) Aspirate and discard as much of the supernatant as possible.

*Elution of drugs.*

*(A) Barbiturates, glutethimide, cocaine.*

(1) To the extraction tube, add 2.0 ml of diethyl ether (a glassy syringe is helpful).

(2) Shake vigorously by hand to ensure that the charcoal is dispersed and thoroughly extracted.

(3) When the charcoal has settled, decant the diethyl ether into a 10 x 75 mm tube.

(4) Add a further 0.5 ml of diethyl ether to the tube containing the charcoal and shake again.

(5) Pour the diethyl ether into the same 10 x 75 mm tube.

(6) Evaporate to dryness at room temperature with the aid of compressed air or nitrogen.

*(B) Amphetamines, narcotic alkaloids, and other drugs.*

(1) To the same extraction tube that contains the charcoal, add 2.0 ml of the chloroform-isopropanol reagent.

(2) Shake vigorously and filter through Whatman No. 1 filter paper (5.5-cm diameter) into a 5-ml beaker. The filter-paper should be pre-treated with the solvent before adding the extract.

(3) Add a further 0.5 ml of chloroform-isopropanol to the charcoal, mix again and add it to the same beaker with the filter-paper.

(4) Into another 10 x 75 mm tube, add a small drop of dilute hydrochloric acid (0.1 *M*) and pour the contents of the beaker into it.

(5) Use 2-4 drops of chloroform-isopropanol to rinse the beaker, and pour it into the same 10 x 75 mm tube.

(6) Evaporate to dryness at room temperature under air or nitrogen.

*Procedure 6. Isolation by adsorption on $XAD_2$ (Weismann et al.[159])*

For hypnotics, sedatives, amphetamines, narcotics and analgesics.

*Resin.* Amberlite $XAD_2$, 20-50 mesh (Mallinckrodt). Wash the resin by stirring it with five bed-volumes of methanol for 2 h, allow to settle, then decant the methanol. Repeat this process twice. Rinse the resin three times with ten bed-volumes of water and store in the refrigerator under methanol-water (30:70). The size of the resin column is 1 x 16 cm.

*Procedure.* Centrifuge about 5 ml of urine at 500 *g* for 5 min to remove suspended matter, which may clog the resin column. Place a small plug of glass-wool in the bottom of the glass column. Pour the resin suspension, prepared as directed, into the column to a height of 6 cm, and wash it with 20 ml of de-ionized water. Decant the centrifuged urine into the column and allow the urine to drain into the resin bed. Allow 60 ml of de-ionized water to pass through the column, then aspirate as much residual water as possible from the column with a water aspirator applied to the outlet of the chromatographic tube. Add 10 ml of methanol to the column and collect the eluate in a conical glass tube.

Evaporate the methanol in a water-bath at 45°C under a stream of nitrogen (compressed air may oxidize amphetamines or other drugs, and is usually contaminated with water and oil from the compressor). Rinse the walls of the conical tube with about 1 ml of methanol before the eluate is completely evaporated; this prevents the formation of an insoluble deposit on the tube wall. Dissolve the dry residue in about 30 µl of methanol (2-3 drops from a dropping bottle).

## 12.3. REFERENCES


1 M.L. Borke and E.R. Kirsch, *J. Amer. Pharm. Ass.*, 42 (1953) 627.
2 E. Nürnberg, *Arch. Pharm. (Weinheim)*, 292 (1959) 610.
3 A. Mariani and O. Mariani-Marelli, *Rend. Ist. Super. Sanita*, 22 (1959) 759; *C.A.*, 54 (1961) 11374g.
4 H. Gänshirt and A. Malzacher, *Arch. Pharm.(Weinheim)*, 293 (1960) 925.
5 K. Teichert, E. Mutschler and H. Rochelmeyer, *Deut. Apoth.-Ztg.*, 100 (1960) 477.
6 G. Machata, *Mikrochim. Acta*, (1960) 79.
7 D. Waldi, K. Schnackerz and F. Munter, *J. Chromatogr.*, 6 (1961) 61.
8 J. Baumler and S. Rippstein, *Pharm. Acta Helv.*, 36 (1961) 382.
9 D. Neubauer and K. Mothes, *Planta Med.*, 9 (1961) 466.
10 K. Teichert, E. Mutschler and H. Rochelmeyer, *Z. Anal. Chem.*, 181 (1961) 325.
11 T. Bican-Fister, *Acta Pharm. Jugoslav.*, 12 (1962) 73; *C.A.*, 58 (1963) 13721d.
12 R. Wasicky, *Anal. Chem.*, 34 (1962) 1346.
13 E. Vidic and J. Schütte, *Arch. Pharm. (Weinheim)*, 295 (1962) 342.
14 W. Döpke, *Arch. Pharm. (Weinheim)*, 295 (1962) 605.
15 H. Feltkamp, *Deut. Apoth.-Ztg.*, 102 (1962) 1269.
16 J. Cochin and J.W. Daly, *Experientia*, 18 (1962) 294.
17 J.A.C. van Pinxteren and M.E. Verloop, *Pharm. Weekbl.*, 97 (1962) 1.
18 W. Poethke and W. Kinze, *Pharm. Zentralhalle*, 101 (1962) 685.
19 T. Kartnig and H. Weixlbaumer, *Sci. Pharm.*, 30 (1962) 87.
20 A. Vegh, A. Brantner, G. Szasz, Zs. Budvari and K. Gracza, *Acta Pharm. Hung.*, 33 (1963) 67.
21 G. Szasz, L. Khin and R. Budvari, *Acta Pharm. Hung.*, 33 (1963) 245.
22 I. Sunshine, *Amer. J. Clin. Pathol.*, 40 (1963) 576.
23 V.E. Chichiro, *Aptechn. Delo*, 12 (1963) 36; *C.A.*, 61 (1964) 10534e.
24 K. Takahashi, S. Mizumachi and H. Asahina, *Eisei Shikenjo Kenkyu Hokoku*, 81 (1963) 23; *C.A.*, 62 (1965) 8936f.
25 M. Ikram, G.A. Miana and M. Islam, *J. Chromatogr.*, 11 (1963) 260.
26 N.Y. Mary and E. Brochmann-Hanssen, *Lloydia*, 26 (1963) 223.
27 J.A.C. van Pinxteren and M.E. Verloop, *Pharm. Acta Helv.*, 38 (1963) 437.
28 W. Kamp, W.J.M. Onderberg and W.A. van Seters, *Pharm. Weekbl.*, 98 (1963) 993.
29 W. Poethke and W. Kinze, *Pharm. Zentralhalle*, 102 (1963) 692.
30 B. Danos, *Acta Pharm. Hung.*, 34 (1964) 221.
31 S.J. Mulé, *Anal. Chem.*, 36 (1964) 1907.
32 R.K. Moiseev, *Aptechn. Delo*, 13 (1964) 29; *C.A.*, 62 (1965) 6341b.
33 W. Poethke and W. Kinze, *Arch. Pharm. (Weinheim)*, 297 (1964) 593.
34 H. Eberhardt and O. Norden, *Arzneim.-Forsch.*, 14 (1964) 1354.

35 U. Fumagalli, V. Ambrogi and G. Balestra, *Boll. Chim. Farm.*, 103 (1964) 911; *C.A.*, 62 (1965) 8936g.
36 A. Penna-Herreros, *J. Chromatogr.*, 14 (1964) 536.
37 I. Bayer, *J. Chromatogr.*, 16 (1964) 237.
38 H.J. Kupferberg, A. Burkhalter and E. Leong Way, *J. Chromatogr.*, 16 (1964) 558.
39 L. Dryon, *J. Pharm. Belg.*, 19 (1964) 19.
40 M. Nonclerq and C. Nijs, *J. Pharm. Belg.*, 19 (1964) 421.
41 K. Genest and C.G. Farmilo, *J. Pharm. Pharmacol.*, 16 (1964) 250.
42 E. Brochmann-Hanssen and T. Furuya, *J. Pharm. Sci.*, 53 (1964) 1549.
43 W. Poethke and W. Kinze, *Pharm. Zentralhalle*, 103 (1964) 577.
44 J. Zarnack and S. Pfeifer, *Pharmazie*, 19 (1964) 216.
45 V. Schwarz and M. Sarsunova, *Pharmazie*, 19 (1964) 267.
46 M.M. Lopes, M. Leal, R.H. Rosario and A.J.C. Ralha, *Rev. Port. Farm.*, 14 (1964) 256; *C.A.*, 64 (1966) 4868c.
47 M. von Schantz, in G.B. Marini-Bettōlo (Editor), *Thin-layer Chromatography*, Elsevier, Amsterdam, 1964, p. 122.
48 E. Ragazzi, G. Veronese and C. Giacobazzi, in G.B. Marini-Bettōlo (Editor), *Thin-layer Chromatography*, Elsevier, Amsterdam, 1964, p. 149.
49 R. Paris, R. Rousselet, M. Paris and M.J. Fries, *Ann. Pharm. Fr.*, 23 (1965) 473.
50 A. Kaess and C. Mathis, *Ann. Pharm. Fr.*, 23 (1965) 739.
51 J.H. Thiel, *Bol. Soc. Quim. Peru*, 31 (1965) 31; *C.A.*, 64 (1966) 11029h.
52 L. Vignoli, J. Guillot, F. Gouezo and J. Catalin, *Bull. Trav. Soc. Pharm. Lyon*, 9 (1965) 291; *C.A.*, 65 (1966) 17345d.
53 D. Heusser and E. Jackwerth, *Deut. Apoth.-Ztg.*, 105 (1965) 107.
54 M. Ono and H. Asahina, *Eisei Shikenjo Hokoku*, 83 (1965) 16; *C.A.*, 65 (1966) 19151b.
55 J.L. Emmerson and R.C. Andersen, *J. Chromatogr.*, 17 (1965) 495.
56 D. Giacopello, *J. Chromatogr.*, 19 (1965) 172.
57 J.A. Steele, *J. Chromatogr.*, 19 (1965) 300.
58 A. Noirfalise, *J. Chromatogr.*, 20 (1965) 61.
59 L.K. Turner, *J. Forensic Sci. Soc.*, 5 (1965) 94.
60 S. El Gendi, W. Kisser and G. Machata, *Mikrochim. Acta*, (1965) 120.
61 W. Kinze, *Pharmaz. Praxis*, (1965) 241.
62 D. Heusser, *Pharm. Tijdschr. Belg.*, 42 (1965) 263.
63 F. Kavka, J. Trojanek and Z. Cekan, *Pharmazie*, 20 (1965) 434.
64 Vl. Preininger, P. Vrublovsky and Vl. Stastny, *Pharmazie*, 20 (1965) 439.
65 K. Macek, J. Vecerkova and J. Stanislavova, *Pharmazie*, 20 (1965) 605.
66 D. Vagujfalvi, *Planta Med.*, 13 (1965) 79.
67 F. Huidobro and H. Miranda, *Acta Physiol. Lat. Amer.*, 16 (1966) 335; *C.A.*, 67 (1967) 20182x.
68 B. Davidow, N. Li Petri, B. Quame, B. Searle, E. Fastlich and J. Savitzky, *Amer. J. Clin. Pathol.*, 46 (1966) 58.
69 W.F. Fike, *Anal. Chem.*, 38 (1966) 1697.
70 L. Vignoli, J. Guillot, F. Gouezo and J. Catalin, *Ann. Pharm. Fr.*, 24 (1966) 461.
71 L. Vignoli, L. Guillot, F. Gouezo and J. Catalin, *Ann. Pharm. Fr.*, 24 (1966) 529.
72 R.R. Paris and G. Faugeras, *Ann. Pharm. Fr.*, 24 (1966) 613.
73 M.R. Gasco and G. Gatti, *Atti Accad. Sci. Torino, Cl. Sci. Fis. Mat. Nat.*, 100 (1966) 683; *C.A.*, 68 (1968) 98661t.
74 H. Yoshimura, K. Oguri and H. Tsukamoto, *Chem. Pharm. Bull.*, 14 (1966) 62.
75 H. Yoshimura, K. Oguri and H. Tsukamoto, *Chem. Pharm. Bull.*, 14 (1966) 1286.
76 A. Kaess and C. Mathis, *Int. Symp. Chromatogr. Electrophor. Lect. Pap. 4th*, (1966) 525.
77 H. Ishii, Y. Aizawa and T. Naka, *Jikeikai Med. J.*, 13 (1966) 153; *C.A.*, 69 (1968) 104746p.
78 V.P. Dole, W.K. Kim and I. Eglitis, *J. Amer. Med. Ass.*, 198 (1966) 349.
79 G.R. Nakamura, *J. Ass. Offic. Anal. Chem.*, 49 (1966) 1086.

80 G.J. Dickes, *J. Ass. Public Anal.*, 4 (1966) 45.
81 A. Affonso, *J. Chromatogr.*, 21 (1966) 332.
82 I. Sunshine, W.W. Fike and H. Landesman, *J. Forensic Sci.*, 11 (1966) 428.
83 J.I. Thornton and D.J. Dillon, *J. Forensic Sci. Soc.*, 6 (1966) 42.
84 A. Mathis and A. Kaess, *J. Pharm. Belg.*, 21 (1966) 561.
85 F. Wartmann-Hafner, *Pharm. Acta Helv.*, 41 (1966) 406.
86 W. Kamp and W.J. Onderberg, *Pharm. Weekbl.*, 101 (1966) 1077.
87 V.P. Dole, W.K. Kim and I. Englitis, *Psychopharmacol. Bull.*, 3 (1966) 45.
88 P. Schweda, *Anal. Chem.*, 39 (1967) 1019.
89 F. Reimers, *Arch. Pharm. Chemi*, 74 (1967) 531.
90 K. Genest and G. Belec, *Can. J. Pharm. Sci.*, 2 (1967) 44.
91 S. Enache, T. Constantinescu and V. Ignat, *Farmacia (Bucharest)*, 15 (1967) 723; *C.A.*, 68 (1968) 53285g.
92 H.C. Hsiu, J.T. Huang, T.B. Shih, K.L. Yang, K.T. Wang and A.L. Lin, *J. Chin. Chem. Soc.*, 14 (1967) 161.
93 J.T. Huang, H.C. Hsiu and K.T. Wang, *J. Chromatogr.*, 29 (1967) 391.
94 A. Noirfalise and G. Mees, *J. Chromatogr.*, 31 (1967) 594.
95 J.W. Fairbairn and S. El-Masry, *J. Pharm. Pharmacol.*, 19S (1967) 93S.
96 V.M. Pechennikov, *Nauchn. Tr. Aspir. Ordinatorov, 1-i Most. Med. Inst.*, (1967) 145; *C.A.*, 70 (1969) 99647m.
97 H.J. Uhlmann, *Pharm. Ztg.*, 112 (1967) 936.
98 R.R. Paris and M. Sarsunova, *Pharmazie*, 22 (1967) 483.
99 E. Grigorescu and A. Verbuta, *Rev. Med.*, 13 (1967) 349; *C.A.*, 68 (1968) 107914g.
100 S.B. Poey and B.G.K. Hwa, *Suara Pharm.*, 10 (1967) 77; *C.A.*, 68 (1968) 6203u.
101 S.B. Poey and A. Abimanju, *Suara Pharm.*, 10 (1967) 91; *C.A.*, 69 (1968) 21936z.
102 B. Davidow, N.L. Petri and B. Quame, *Amer. J. Clin. Pathol.*, 50 (1968) 714.
103 M. Wilk and U. Brill, *Arch. Pharm. (Weinheim)*, 301 (1968) 282.
104 E. Röder, E. Mutschler and H. Rochelmeyer, *Arch. Pharm. (Weinheim)*, 301 (1968) 624.
105 F. Reimers, *Arch. Pharm. Chemi*, 75 (1968) 1064.
106 V. Vukcevic-Kovacevic, *Arh. Farm.*, 18 (1968) 3; *C.A.*, 69 (1968) 99424a.
107 I. Lewandowska and I. Soklowska, *Diss. Pharm. Pharmacol.*, 20 (1968) 81.
108 B. Danos, *Herba Hung.*, 7 (1968) 27.
109 S. Farkas and I. Bayer, *Herba Hung.*, 7 (1968) 37.
110 H. Halpaap, *J. Chromatogr.*, 33 (1968) 144.
111 D. Heusser, *J. Chromatogr.*, 33 (1968) 400.
112 M. Debackere and L. Laruelle, *J. Chromatogr.*, 35 (1968) 234.
113 H. Wullen and H. Thielemans, *J. Pharm. Belg.*, 23 (1968) 307.
114 J.M.G.J. Frijns, *Pharm. Weekbl.*, 103 (1968) 929.
115 E. Stahl and H. Jork, *Z. Anal. Chem.*, 234 (1968) 12.
116 K. Gröningson and G. Schill, *Acta Pharm. Suecica*, 6 (1969) 447.
117 T. Kaniewska and H. Rafalowska, *Acta Pol. Pharm.*, 26 (1969) 529.
118 E. Stahl, H. Jork, E. Dumont, H. Bohrmann and H. Vollmann, *Arzneim.-Forsch.*, 19 (1969) 194.
119 M. Ono, B.F. Engelke and C. Fulton, *Bull. Narcotics*, 21 (1969) 31.
120 A.K. Kristensen, *Dan. Tidsskr. Farm.*, 43 (1969) 220; *C.A.*, 72 (1970) 70638z.
121 E. Röder, E. Mutschler and H. Rochelmeyer, *Deut. Apoth.-Ztg.*, 109 (1969) 1219.
122 Z.P. Kostennikova and V.E. Chichiro, *Farmatsiya (Moscow)*, 18 (1969) 39; *C.A.*, 71 (1969) 128782q.
123 B. Botev, *Farmatsiya (Sofia)*, 19 (1969) 5; *C.A.*, 71 (1969) 53617d.
124 E.G.C. Clarke, *Isolation and Identification of Drugs in Pharmaceuticals, Body Fluids and Post-Mortem Material*, Pharm. Press, London, Vol. 1, 1969.
125 A.J. Harrison and A. Cook, *J. Ass. Public. Anal.*, 7 (1969) 47.
126 E. Stahl and E. Dumont, *J. Chromatogr. Sci.*, 7 (1969) 517.
127 L. Fishbein, H.L. Falk, *Chromatogr. Rev.*, 11 (1969) 1.
128 S.J. Mulé, *J. Chromatogr.*, 39 (1969) 302.
129 A.M. Heaton and A.G. Blumberg, *J. Chromatogr.*, 41 (1969) 367.
130 A.J. Harrison and A. Cook, *J. Forensic Sci. Soc.*, 9 (1969) 165.
131 M. Vanhaelen, *J. Pharm. Belg.*, 24 (1969) 87.

132 A. Viala, F. Gouezo and J. Catalin, *Med. Lég. Dommage Corpor.*, 2 (1969) 21.
133 H.H. Hashmi, S. Parveen and N.A. Chughtai, *Mikrochim. Acta*, (1969) 449.
134 F. Schmidt, *Pharm. Ztg.*, 114 (1969) 1523.
135 S. Goeneckea and W. Bernhard, *Z. Anal. Chem.*, 246 (1969) 130.
136 G. Szasz and G. Szasz, *Acta Pharm. Hung.*, 40 (1970) 38.
137 K. Gröningson, *Acta Pharm. Suecica,* 7 (1970) 635.
138 E. Röder, *Arch. Pharm. (Weinheim)*, 303 (1970) 176.
139 Z. Blagojevic and M. Skrlj, *Arh. Farm.*, 20 (1970) 109; *C.A.*, 73 (1970) 112999w.
140 V. Vukecevic-Kovacevic, *Bull. Sci. Cons. Acad. Sci. Arts RSF Yougosl., Sect.A,* 15 (1970) 238; *C.A.*, 73 (1970) 123554y.
141 M.L. Bastos, G.E. Kananen, R.M. Young, J.R. Monforte and I. Sunshine, *Clin. Chem.*, 16 (1970) 931.
142 M. Overgaard Nielsen, *Dan. Tidsskr. Farm.*, 44 (1970) 7.
143 M. Sobiczewska and B. Borkowski, *Farm. Pol.*, 26 (1970) 539.
144 S. Enache and T. Constantinescu, *Farmacia (Bucharest)*, 18 (1970) 149; *C.A.*, 73 (1970) 38579r.
145 O.B. Stepanenko and F.M. Shemyakin, *Farmatsiya (Moscow)*, 19 (1970) 37; *C.A.*, 72 (1970) 136459a.
146 V. Massa, F. Gal and P. Susplugas, *Int. Symp. Chromatogr. Electrophor. Lect. Pap. 6th, 1970*, (1971) 470.
147 J.G. Montalvo Jr., E. Klein, D. Eyer and B. Harper, *J. Chromatogr.*, 47 (1970) 542.
148 G.H. Jolliffe and E.J. Shellard, *J. Chromatogr.*, 48 (1970) 125.
149 P. Bose, *J. Inst. Chem. Calcutta,* 42 (1970) 113; *C.A.*, 74 (1971) 2767g.
150 A.S. Curry and D.A. Patterson, *J. Pharm. Pharmacol.*, 22 (1970) 198.
151 Z.F. Ahmed, Z. El-Darawy, G.M. Wassel and S.H. El-Sayed, *U.A.R. J. Pharm. Sci.,* 11 (1970) 163.
152 S.P. Pfeifer, G. Behnsen and L. Kühn, *Pharmazie,* 25 (1970) 529.
153 J.M. Fujimoto and J.H. Wong, *Toxicol. Appl. Pharmacol.*, 16 (1970) 186.
154 V. Massa, F. Gal, P. Susplugas and G. Maestre, *Trav. Soc. Pharm. Montpellier,* 30 (1970) 273.
155 K. Röder, E. Eich and E. Mutschler, *Arch. Pharm. (Weinheim)*, 304 (1971) 297.
156 F. Reimers, *Arch. Pharm. Chemi*, 78 (1971) 201.
157 K. Jensen, *Arch. Pharm. Chemi*, 78 (1971) 249.
158 H. Fischer, H.G. Eulenhoefer and R. Kraft, *Arzneim.-Forsch.*, 21 (1971) 169.
159 N. Weissman, M.L. Lowe, J.M. Beattie and J.A. Demetriou, *Clin. Chem.,* 17 (1971) 875.
160 J.T. Payte, J.E. Wallace and K. Blum, *Curr. Ther. Res. Clin. Exp.*, 13 (1971) 412.
161 W. Debska and S. Czyszewska, *Farm. Pol.*, 27 (1971) 365; *C.A.*, 75 (1971) 80316n.
162 V.E. Chichiro, Z.P. Kostennikova and S.D. Mekhtikhanov, *Farmatsiya (Moscow)*, 20 (1971) 37; *C.A.*, 76 (1972) 49988m.
163 B. Botev, *Farmatsiya (Sofia),* 21 (1971) 8; *C.A.*, 75 (1971) 9899f.
164 B. Botev, *Farmatsiya (Sofia),* 21 (1971) 27; *C.A.*, 76 (1972) 131565t.
165 V.E. Chichiro and A.V. Suranov, *Farm. Zh. (Kiev)*, 26 (1971) 38; *C.A.*, 76 (1972) 90092c.
166 E. Röder, *Int. Symp. Chromatogr. Electrophor. Lect. Pap. 6th, 1970* (1971) p. 194.
167 R.J. Coumbis, C.C. Fulton, J.P. Calise and C. Rodriguez, *J. Chromatogr.*, 54 (1971) 245.
168 S.J. Mulé, *J. Chromatogr.*, 55 (1971) 255.
169 R.C. Baselt and L.J. Casarett, *J. Chromatogr.*, 57 (1971) 139.
170 S. Ebel, E. Bahr and E. Plate, *J. Chromatogr.*, 59 (1971) 212.
171 K.K. Kaistha and J.H. Jaffe, *J. Chromatogr.*, 60 (1971) 83.
172 J.D. Broich, D.B. Hoffman, S. Andryauskas, L. Galante and C.J. Umberger, *J. Chromatogr.*, 60 (1971) 95.
173 D.J. Berry and J. Grove, *J. Chromatogr.*, 61 (1971) 111.
174 S.J. Mulé, M.L. Bastos, D. Jukofsky and E. Saffer, *J. Chromatogr.*, 63 (1971) 289.

175 J.R. Broich, D.B. Hoffman, S.J. Goldner, S. Andryauskas and C.J. Umberger, *J. Chromatogr.*, 63 (1971) 309.
176 I. Simon and M. Lederer, *J. Chromatogr.*, 63 (1971) 448.
177 G.S. Tadjer, *J. Chromatogr.*, 63 (1971) D44.
178 J.F. Taylor, in D.H. Clouet (Editor), *Narcotic Drugs: Biochemical Pharmacology*, Plenum Press, New York, 1971, p. 17.
179 L.A. Chekryshkina, *Nauchn. Tr. Permsk. Farm. Inst.*, (1971) 71; *C.A.*, 79 (1973) 149335b.
180 L.A. Chekryshkina and G.I. Oleshko, *Nauchn. Tr. Permsk. Farm. Inst.*, (1971) 73; *C.A.*, 79 (1973) 149334a.
181 M. Steinigen, *Pharm. Ztg.*, 116 (1971) 2072.
182 I.K. Ho, H.H. Loh and E.L. Way, *Proc. West. Pharmacol. Soc.*, 14 (1971) 183.
183 G.A. Jansen and I. Bickers, *S. Med. J.*, 64 (1971) 1072.
184 E. Röder and K.H. Surborg, *Z. Anal. Chem.*, 256 (1971) 362.
185 T. Bican-Fister and S. Grdinic, *Zbl. Pharm.*, 110 (1971) 1247.
186 S. Zadeczky, D. Küttel and M. Takacsi, *Acta Pharm. Hung.*, 42 (1972) 7.
187 M.M. Baden, N.N. Valanju, S.K. Verma and S.N. Valanju, *Amer. J. Clin. Pathol.*, 57 (1972) 43.
188 M. Paris, J.P. Gramond and R.R. Paris, *Ann. Pharm. Fr.*, 32 (1972) 97.
189 Z. Blagojevic, M. Skrij and V. Bulajic, *Arh. Farm.*, 22 (1972) 97; *C.A.*, 79 (1973) 70261r.
190 L. May and L.C.T. Kno, *Bull. Narcotics*, 24 (1972) 35.
191 M. Sarsunova and B. Kakac, *Cesk. Farm.*, 21 (1972) 102.
192 S.J. Mulé and H. Brill (Editors), *Chemical and Biological Aspects of Drug Dependence*, Chemical Rubber Co., Cleveland, Ohio, 1972.
193 L.R. Sellier and V. Torre, *Cienc. Ind. Farm.*, 4 (1972) 146; *C.A.*, 82 (1975) 160280u.
194 D. Sohn and J. Simon, *Clin. Chem.*, 18 (1972) 405.
195 L.B. Hetland, D.A. Knowlton and D. Couri, *Clin. Chim. Acta*, 36 (1972) 473.
196 K.C. Güven and N. Güven, *Eczacilik Bul.*, 14 (1972) 75.
197 M. Ono, M. Shimamine and K. Takahashi, *Eisei Shikenjo Hokoku*, (1972) 73; *C.A.*, 79 (1973) 57731z.
198 E. Pawelczyk, Z. Plotkowiakowa and T. Malesza, *Farm. Pol.*, 28 (1972) 263; *C.A.*, 77 (1972) 39350d.
199 G. Chams, N. Kheradmandan, I. Yaraghtchi and V. Chahmaneche, *Int. Crim. Police Rev.*, 27 (1972) 162; *C.A.*, 78 (1973) 67745a.
200 R.L. Neman, *J. Chem. Educ.*, 49 (1972) 834.
201 S.J. Mulé, *J. Chromatogr. Sci.*, 10 (1972) 275.
202 W.T. Fisher, A.D. Baitsholts and G.S. Gran, *J. Chromatogr. Sci.*, 10 (1972) 303.
203 J.K. Brown, L. Shapazian and G.D. Griffin, *J. Chromatogr.*, 64 (1972) 129.
204 G.N. Christopoulos and E.R. Kirch, *J. Chromatogr.*, 65 (1972) 507.
205 I.K. Ho, H.H. Loh and E. Leong Way, *J. Chromatogr.*, 65 (1972) 577.
206 D.B. Hoffman, C.J. Umberger, S. Goldner, S. Andryauskas, D. Mulligam and J.R. Broich, *J. Chromatogr.*, 66 (1972) 63.
207 J.E. Wallace, J.D. Biggs, J.H. Merritt, H.E. Hamilton and K. Blum, *J. Chromatogr.*, 71 (1972) 135.
208 L.T. Kenison, E.L. Loveridge, J.A. Grounlund and A.A. Elmowafi, *J. Chromatogr.*, 71 (1972) 165.
209 A.L. Misra, R.B. Pontani and S.J. Mulé, *J. Chromatogr.*, 71 (1972) 554.
210 A. Viala and M. Estadieu, *J. Chromatogr.*, 72 (1972) 127.
211 E.W. Gold, J.B. Murray and G. Smith, *J. Pharm. Pharmacol.*, 24S (1972) 143p.
212 K.K. Kaistha, *J. Pharm. Sci.*, 61 (1972) 655.
213 K.K. Kaistha and J.H. Jaffe, *J. Pharm. Sci.*, 61 (1972) 679.
214 A.V. Gaevskii and P.M. Loshkarev, *Khim.-Farm. Zh.*, 6 (1972) 54; *C.A.*, 77 (1972) 79578r.
215 A. Dzhumashev and G.B. Aimukhamedova, *Lek. Veshchestva Rast. Syr'ya Kirg.*, (1972) 102; *C.A.*, 79 (1973) 35194z.
216 A. Hiermann and F. Still, *Oester. Apoth. Ztg.*, 26 (1972) 337.

217 K. Thassler and W. Kross, *Pharm. Ztg.*, 117 (1972) 253.
218 F. Eiden and G. Kammash, *Pharm. Ztg.*, 117 (1972) 1994.
219 M. Dsoneydi, St. Michailova, I. Ivanova and O. Budewski, *Pharmazie*, 27 (1972) 657.
220 C.W. Gorodetzky, *Toxicol. Appl. Pharmacol.*, 23 (1972) 511.
221 R.A. Egli, *Z. Anal. Chem.*, 259 (1972) 277.
222 E. Klug, *Z. Anal. Chem.*, 260 (1972) 31.
223 K.F. Ahrend and D. Tiess, *Zbl. Pharm.*, 111 (1972) 933.
224 M. Petkovic, *Acta Pharm. Jugosl.*, 23 (1973) 23; *C.A.*, 79 (1973) 9936y.
225 G.D. Lathrop, H.L. Kaplan and J.E. Wallace, *Agard Conf. Proc. (Agard-CP-108)*, A14 (1973) 1; *C.A.*, 78 (1973) 155215h.
226 M.K. Brandt, *Amer. J. Med. Technol.*, 39 (1973) 217.
227 J. Simon, M.A. Hanna, G.V. Ghali, R.A. Tolba and V. Melkonian, *Anal. Chem.*, 45 (1973) 1498.
228 Z. Blagojevic, L. Glisovic and D. Zivanov-Stakic, *Arh. Farm.*, 23 (1973) 9; *C.A.*, 80 (1974) 63888t.
229 M. Petkovic, *Arh. Farm.*, 23 (1973) 309; *C.A.*, 83 (1975) 84904x.
230 W.L. Miller, M.P. Kullberg, M.E. Banning, L.D. Brown and B.P. Doctor, *Biochem. Med. J.*, 7 (1973) 145.
231 A. Viala, J. Catalin and F. Gouezo, *Bull. Soc. Chim. Fr.*, (1973) 97.
232 M. Belesova and M. Zahradnicek, *Cesk. Farm.*, 22 (1973) 75.
233 E. Novakova and J. Vecerkova, *Cesk. Farm.*, 22 (1973) 347.
234 C.W. Gorodetzky, *Clin. Chem.*, 19 (1973) 753.
235 R.E. Juselius and F. Barnhart, *Clin. Toxicol.*, 6 (1973) 53.
236 K.C. Güven and B. Aran, *Eczacilik Bul.*, 15 (1973) 28.
237 B. Botev, *Farmatsiya (Sofia)*, 23 (1973) 19; *C.A.*, 80 (1974) 78271k.
238 B. Botev, *Farmatsiya (Sofia)*, 23 (1973) 23; *C.A.*, 79 (1973) 97031c.
239 H.H. Loh, I.K. Ho, T.M. Cho and W. Lipscomb, *J. Chromatogr.*, 76 (1973) 505.
240 R.A. van Welsum, *J. Chromatogr.*, 78 (1973) 237.
241 D.J. Berry and J. Grove, *J. Chromatogr.*, 80 (1973) 205.
242 J.R. Broich, S. Goldner, G. Gourdet, S. Andryauskas, C.J. Umberger and D.B. Hoffman, *J. Chromatogr.*, 80 (1973) 275.
243 M.L. Bastos, D. Jukofsky and S.J. Mulé, *J. Chromatogr.*, 81 (1973) 93.
244 N.N. Valanju, M.M. Baden, S.N. Valanju, D. Mulligan and S.K. Verma, *J. Chromatogr.*, 81 (1973) 170.
245 I. Sunshine, *J. Chromatogr.*, 82 (1973) 125.
246 W. Schlemmer and E. Kammerl, *J. Chromatogr.*, 82 (1973) 143.
247 J.K. Brown, R.H. Schlingler, M.G. Chaubal and M.H. Malone, *J. Chromatogr.*, 87 (1973) 211.
248 R.V. Smith, M.R. Cook and A.W. Stocklinski, *J. Chromatogr.*, 87 (1973) 294.
249 P.C. Maiti, A. Chatterjee and S. Mookherjee, *J. Indian Acad. Forensic Sci.*, 12 (1973) 19; *C.A.*, 80 (1974) 141560w.
250 N.H. Choulis, *J. Pharm. Sci.*, 62 (1973) 112.
251 I.J. Holcomb, R.B. Luers and S.A. Fusari, *J. Pharm. Sci.*, 62 (1973) 1505.
252 S.E. Hays, L.T. Grady and A.V. Kruegel, *J. Pharm. Sci.*, 62 (1973) 1509.
253 S.Y. Yeh, *J. Pharm. Sci.*, 62 (1973) 1827.
254 F. Schmidt, *Krankenhaus-Apotheke*, 23 (1973) 10.
255 J. Paul and F. Conine, *Microchem. J.*, 18 (1973) 142.
256 F. Still and A. Hiermann, *Oesterr. Apoth. Ztg.*, 27 (1973) 270.
257 W. Best and H. Olberg, *Pharm. Ztg.*, 118 (1973) 282.
258 F. Eiden and G. Khammash, *Pharm. Ztg.*, 118 (1973) 638.
259 E. Kammerl and E. Mutschler, *Pharm. Ztg.*, 118 (1973) 1905.
260 A. Gyeresi and G. Racz, *Pharmazie*, 28 (1973) 271.
261 T. VuDuc and A. Vernay, *Praeventivmedizin*, 18 (1973) 309.
262 A. Gyeresi and G. Racz, *Rev. Med. (Tirgu-Mures, Rom.)*, 19 (1973) 49; *C.A.*, 79 (1973) 45884p.
263 A. Gyeresi and G. Racz, *Rev. Med. (Tirgu-Mures, Rom.)*, 19 (1973) 384; *C.A.*, 80 (1974) 124801h.
264 G.M. Habashi and N.A. Farid, *Talanta*, 20 (1973) 699.

265 A. Puech, M. Jacob and D. Gaudy, *Trav. Soc. Pharm. Montpellier,* 33 (1973) 515.
266 K.F. Ahrend and D. Tiess, *Wiss. Z. Univ. Rostock, Math. Naturw. Reihe,* 22 (1973) 951.
267 A. Eichhorn and L. Kny, *Zentralbl. Pharm.*, 112 (1973) 567.
268 D. Radulovic, Z. Blagojevic and D. Zivanov-Stakic, *Acta Pharm. Jugosl.*, 4 (1974) 173; *C.A.*, 82 (1975) 35064u.
269 S.J. Goldner, C.J. Umberger, D.B. Hoffman, G. Gourdet, S. Andryauskas and J.R. Broich, *Biochem. Med.*, 10 (1974) 79.
270 H. Tobolska, R. Kanarkrowski and W. Dutkiewicz, *Bromatol. Chem. Toksykol.*, 7 (1974) 415; *C.A.*, 82 (1975) 129322g.
271 H.E. Sine, N.P. Kubasik and T.A. Rejent, *Clin. Biochem.*, 7 (1974) 102.
272 P.M. Kullberg and C.W. Gorodetzky, *Clin. Chem.*, 20 (1974) 177.
273 J.M. Meola and M. Vanko, *Clin. Chem.*, 20 (1974) 184.
274 R.J. Bussey and R.C. Backer, *Clin. Chem.*, 20 (1974) 302.
275 R.E. Stoner and C. Parker, *Clin. Chem.*, 20 (1974) 309.
276 C.W. Gorodetzky, C.R. Angel, D.J. Beach, D.H. Catlin and S.Y. Yeh, *Clin. Pharmacol. Ther.*, 15 (1974) 461.
277 G.C. Jain and M.S. Dahiya, *Curr. Sci.*, 43 (1974) 444.
278 B. Unterhalt, *Deut. Apoth.-Ztg.*, 114 (1974) 1017.
279 F. Schmidt, *Deut. Apoth.-Ztg.*, 114 (1974) 1593.
280 E.G.C. Clarke, *Forensic Toxicol. Proc. Symp.*, 1972, (1974) 28; *C.A.*, 83 (1975) 53023e.
281 A. Wislocki, P. Martel, R. Ito, W.S. Dunn and C.D. McGuire, *Health Lab. Sci.*, 11 (1974) 13.
282 S.J. Mulé, *J. Chromatogr. Sci.*, 12 (1974) 245.
283 A.E. Robinson and A.T. Holder, *J. Chromatogr. Sci.*, 12 (1974) 281.
284 J.H. Speaker, *J. Chromatogr. Sci.*, 12 (1974) 297.
285 J. Sherma, M.F. Dobbins and J.C. Touchstone, *J. Chromatogr. Sci.*, 12 (1974) 300.
286 M.L. Bastos, D. Jukofsky and S.J. Mulé, *J. Chromatogr.*, 89 (1974) 335.
287 A.C. Moffat and K.W. Smalldon, *J. Chromatogr.*, 90 (1974) 1.
288 A.C. Moffat and K.W. Smalldon, *J. Chromatogr.*, 90 (1974) 9.
289 K.G. Blass, R.J. Thibert and T.F. Draisey, *J. Chromatogr.*, 95 (1974) 75.
290 D.J. Doedens and R.B. Forney, *J. Chromatogr.*, 100 (1974) 225.
291 H. Hammerstingl and G. Reich, *J. Chromatogr.*, 101 (1974) 408.
292 G.J. Digregorio and C. O'Brien, *J. Chromatogr.*, 101 (1974) 424.
293 N.V. Rama Rao and H.R.K. Murty, *J. Indian Acad. Forensic Sci.*, 13 (1974) 18.
294 A.C. Moffat and B. Clare, *J. Pharm. Pharmacol.*, 26 (1974) 665.
295 J. Breiter, *Kontakte,* 3 (1974) 17.
296 F. Conine and J. Paul, *Mikrochim. Acta*, (1974) 443.
297 R.J. Armstrong, *N.Z. J. Sci.*, 17 (1974) 15.
298 A.F. Rubstov and E.M. Salomatin, *Sud.-Med. Ekspert.,* 17 (1974) 45; *C.A.*, 81 (1974) 99154k.
299 M. Przyborowska, *Acta Pol. Pharm.*, 32 (1975) 173; *C.A.*, 84 (1976) 44472n.
300 G.E. Baiulescu and T. Constantinescu, *Anal. Chem.*, 47 (1975) 2156.
301 F. Nachtmann, H. Spitzy and R.W. Frei, *Anal. Chim. Acta*, 76 (1975) 57.
302 E. Curea and M. Martinovici, *Ann. Pharm. Fr.*, 33 (1975) 505.
303 E. Stahl and W. Schmitt, *Arch. Pharm. (Weinheim)*, 308 (1975) 570.
304 D. Zivanov-Stakic, D. Radulovic and V. Brzulja, *Arh. Farm.,* 25 (1975) 29; *C.A.*, 83 (1975) 152434w.
305 D. Radulovic, Z. Blagojevic and D. Zivanov-Stakic, *Arh. Farm.*, 24 (1974) 215; *C.A.*, 83 (1975) 103334e.
306 M. Lastovkova, *Cesk. Farm.*, 24 (1975) 212.
307 R.J. Kokoski and M. Jain, *Clin. Chem.*, 21 (1975) 417.
308 E. Curea and M. Martinovici-Fagarasan, *Clujul Med.*, 48 (1975) 253; *C.A.*, 84 (1976) 184977n.
309 H. Tobolska, R. Kanarkowski and W. Songin, *Farm. Pol.*, 31 (1975) 205; *C.A.*, 83 (1975) 108071b.

310 K. Szymkowska, Z. Legowska and M. Piekarewicz, *Farm. Pol.*, 31 (1975) 211; *C.A.*, 83 (1975) 72833h.
311 F.E. Kagan, F.A. Mitchenko, L.O. Kirichenko and T.A. Koget, *Farm. Zh. (Kiev)*, 30 (1975) 75; *C.A.*, 83 (1975) 120940s.
312 T.M. Holdstock and H.M. Stevens, *Forensic Sci.*, 6 (1975) 187.
313 A. Brantner, J. Vamos, E. Jeney and G. Szasz, *Gyogyszereszet*, 19 (1975) 10; *C.A.*, 83 (1975) 33099f.
314 G.J. Down and S.A. Gwyn, *J. Chromatogr.*, 103 (1975) 208.
315 J.A. Vinson and J.E. Hooyman, *J. Chromatogr.*, 105 (1975) 415.
316 P. Liras, *J. Chromatogr.*, 106 (1975) 238.
317 G.N. Christopoulos, N.Wu Chen and A.J. Thomas, *J. Chromatogr.*, 106 (1975) 446.
318 P.A.F. Pranitis and A. Stolman, *J. Chromatogr.*, 106 (1975) 485.
319 K.K. Kaistha, R. Tadrus and R. Janda, *J. Chromatogr.*, 107 (1975) 359.
320 T. Okumura, T. Kadono and A. Iso'o, *J. Chromatogr.*, 108 (1975) 329.
321 A.C. Moffat, *J. Chromatogr.*, 110 (1975) 341.
322 J.E. Wallace, H.E. Hamilton, H. Schwertner, D.E. King, J.L. McNay and K. Blum, *J. Chromatogr.*, 114 (1975) 423.
323 J. Chrastil, *J. Chromatogr.*, 115 (1975) 273.
324 W.J. Serfontein, D. Botha and L.S. de Villiers, *J. Chromatogr.*, 115 (1975) 507.
325 N.C. Jain, W.J. Leung, R.D. Budd and T.C. Sneath, *J. Chromatogr.*, 115 (1975) 519.
326 W.J. Woodford, *J. Chromatogr.*, 115 (1975) 678.
327 V.F. Cordova and T.A. Banford, *J. Forensic Sci.*, 20 (1975) 58.
328 J.A. Vinson, J.E. Hooyman and C.E. Ward, *J. Forensic Sci.*, 20 (1975) 552.
329 P.A.F. Pranitis and A. Stolman, *J. Forensic Sci.*, 20 (1975) 726.
330 H. Thielemann and F. Groh, *Pharmazie*, 30 (1975) 255.
331 A. Gyeresi and G. Racz, *Rev. Med. (Tirgu-Mures, Rom.)*, 21 (1975) 37; *C.A.*, 84 (1976) 49866k.
332 E. Spratt, *Toxicol. Annu. 1974*, (1975) 229.
333 V.E. Chichiro, Z.P. Kostennikova, L.S. Semenova and V.V. Drozhzhina, *Tr. Tsentr. Aptechn. Nauchno-Issledovatel'skogo Inst.*, N11 (1975) 214; *C.A.*, 84 (1976) 79755r.
334 E. Vidic and E. Klug, *Z. Rechtsmed.*, 76 (1975) 283.
335 M. Petkovic, *Arh. Farm.*, 25 (1976) 435; *C.A.*, 87 (1977) 90786s.
336 F. Goc-Pietras and E. Lotysz, *Bromatol. Chem. Toksykol.*, 9 (1976) 369; *C.A.*, 85 (1976) 187323c.
337 D.R. Wilkinson, P. Jensen and D. Winsley, *Clin. Chem.*, 22 (1976) 393.
338 J.A. McIntyre and A.E. Armandi, *Clin. Chem.*, 22 (1976) 396.
339 K.K. Kaistha and R. Tadrus, *Clin. Chem.*, 22 (1976) 1936.
340 R.L. Munier and A.M. Drapier, *C.R. Acad. Sci., Ser. C*, 283 (1976) 719.
341 M. Caldini, M.A. Bianchi and T. Valenza, *Cron. Chim.*, 48 (1976) 3; *C.A.*, 86 (1977) 50420h.
342 N.V.R. Rao, H.R.K. Murty and T.R. Baggi, *Curr. Sci.*, 45 (1976) 332.
343 N.V.R. Rao and H.R.K. Murty, *Curr. Sci.*, 45 (1976) 410; *C.A.*, 85 (1976) 57610h.
344 K.C. Güven and N. Güven, *Eczacilik Bul.*, 18 (1976) 14.
345 A. Viala and M. Estadieu, *Eur. J. Toxicol. Environ. Hyg.*, 9 (1976) 75.
346 M. Petkovic, *Farm. Glas.*, 32 (1976) 363; *C.A.*, 86 (1977) 96075c.
347 V.V. Drozhzhina, V.E. Chichiro and T.I. Bulenkov, *Farmatsiya (Moscow)*, 25 (1976) 44; *C.A.*, 86 (1977) 96046u.
348 Yu.V. Shostenko, V.A. Danel'yants and L.Yu. Chernysh, *Farmatsiya (Moscow)*, 25 (1976) 74; *C.A.*, 85 (1976) 112776u.
348a J. Breiter, R. Helger and H. Lang, *Forensic Sci.*, 7 (1976) 131.
349 T.R. Baggi, R.N.V. Rao and H.R.K. Murty, *Forensic Sci.*, 8 (1976) 265.
350 L. Lepri, P.G. Desideri and M. Lepori, *J. Chromatogr.*, 116 (1976) 131.
351 P.W. Erhardt, R.V. Smith, T.T. Sayther and J.E. Keiser, *J. Chromatogr.*, 116 (1976) 218.
352 F.W. Grant, *J. Chromatogr.*, 116 (1976) 230.
353 R.W. Frei, W. Santi and M. Thomas, *J. Chromatogr.*, 116 (1976) 365.

354 P. Haefelfinger, *J. Chromatogr.*, 124 (1976) 351.
355 L. Lepri, P.G. Desideri and M. Lepori, *J. Chromatogr.*, 123 (1976) 175.
356 A.N. Masoud, *J. Pharm. Sci.*, 65 (1976) 1585.
357 F. Chrobok, *Krim. Forensische Wiss.*, 24 (1976) 157; *C.A.*, 88 (1978) 1083h.
358 B.E. Aarø and K.E. Rasmussen, *Medd. Nor. Farm. Selsk.*, 38 (1976) 13.
359 T. VuDuc, A. Vernay and C. Nicole, *Pharm. Acta Helv.*, 51 (1976) 126.
360 J.H. Rengerink and I.C. Dijkhuis, *Pharm. Weekbl.*, 111 (1976) 701.
361 P.C. Barett, *Proc. Anal. Div. Chem. Soc.*, 12 (1975) 271; *C.A.*, 85 (1976) 83270s.
362 I.R. da Silva Jardim, M.M. Menezes de Menezes and C.T.G. Soares, *Rev. Bras. Farm.*, 57 (1976) 61; *C.A.*, 86 (1977) 195241k.
363 V.V. Zimnukhov, N.M. Kisvyantseva, V.F. Nikitenko and A.A. Shandyba, *Sud.-Med. Ekspert.*, 19 (1976) 34; *C.A.*, 86 (1977) 134354c.
364 S.N. Tewari and D.N. Sharma, *Z. Anal. Chem.*, 281 (1976) 381.
365 S. Hashiba, M. Tatsusawa and A. Ejima, *Bunseki Kagaku (Jap. Anal.)*, 26 (1977) 804; *C.A.*, 88 (1978) 126387s.
366 R.L. Munier and A.M. Drapier, *Chromatographia*, 10 (1977) 226.
367 R.L. Munier and A.M. Drapier, *Chromatographia*, 10 (1977) 290.
368 N.V.R. Rao, *Curr. Sci.*, 46 (1977) 637; *C.A.*, 88 (1978) 32641t.
369 H. Kroeger, G. Bohn and G. Ruecker, *Deut. Apoth.-Ztg.*, 117 (1977) 1923.
370 F. Machovicova, L. Mesarosova and W. Stalmach, *Farm. Obz.*, 46 (1977) 351; *C.A.*, 90 (1979) 61292b.
371 M. Sobiczewska, *Farm. Pol.*, 33 (1977) 365; *C.A.*, 88 (1978) 27838h.
372 M.A. Peat and A. Sengupta, *Forensic Sci.*, 9 (1977) 21.
373 N.V.R. Rao and S.N. Tandon, *Forensic Sci.*, 9 (1977) 103.
374 S. Thunell, *J. Chromatogr.*, 130 (1977) 209.
375 G. Rücker and A. Taha, *J. Chromatogr.*, 132 (1977) 165.
376 K.K. Kaistha and R. Tadrus, *J. Chromatogr.*, 135 (1977) 385.
377 F.F. Wu and R.H. Dobberstein, *J. Chromatogr.*, 140 (1977) 65.
378 K.K. Kaistha, *J. Chromatogr.*, 141 (1977) 145.
379 A.N. Masoud, *J. Chromatogr.*, 141 (1977) D9.
380 M. Oellerich, W.R. Kuelpmann, R. Haeckel, F. Behrends, I. Isberner and K. Petry, *J. Clin. Chem. Clin. Biochem.*, 15 (1977) 275.
381 S.Y. Yeh, R.L. McQuinn and C.W. Gorodetzky, *J. Pharm. Sci.*, 66 (1977) 201.
382 S.Y. Yeh, C.W. Gorodetzky and H.A. Krebs, *J. Pharm. Sci.*, 66 (1977) 1288.
383 B.I. Chumburidze, A.E. Mshvidobadze, O.V. Sardzhveladze, R.V. Makharadze and L.Sh. Kunchuliya, *Kromatogr. Metody Farm.*, (1977) 17; *C.A.*, 90 (1979) 67290w.
384 J. Breiter and R. Helger, *Med. Lab.*, 30 (1977) 149.
385 S. Enache and R. Vasilliev, *Rev. Chim. (Bucharest)*, 28 (1977) 1103; *C.A.*, 89 (1978) 12231b.
386 L. Gagliardi, A. Amato, G. Ricciardi and S. Chiavarelli, *Riv. Tossicoli Sper. Clin.*, 7 (1977) 191; *C.A.*, 88 (1978) 32648a.
387 H. Thielemann, *Sci. Pharm.*, 45 (1977) 240.
388 T. Borkowski and A. Bluzniewska, *Z. Zagadnien Krym.*, 12 (1977) 57; *C.A.*, 90 (1979) 34561g.
389 F. Chrobok and W. Gubala, *Z. Zagadnien Krym.*, 12 (1977) 66; *C.A.*, 90 (1979) 34562h.
390 E. Stahl and J. Brombeer, *Deut. Apoth.-Ztg.*, 118 (1978) 1527.
391 E. Marozzi, E. Cozza, A. Pariali, V. Gamgaro, F. Lodi and E. Saligari, *Farmaco, Ed. Prat.*, 33 (1978) 195.
392 A.B. Narbutt-Mering and W. Weglowska, *Farm. Pol.*, 34 (1978) 525; *C.A.*, 90 (1979) 61235q.
393 N.V.R. Rao and S.N. Tandon, *J. Chromatogr. Sci.*, 15 (1978) 158.
394 R.A. de Zeeuw, F.J.W. van Mansvelt and J.E. Greving, *J. Chromatogr.*, 148 (1978) 255.
395 J.J. Manura, J.M. Chao and R. Saferstein, *J. Forensic Sci.*, 23 (1978) 44.
396 D.B. Predmore, G.D. Christian and T.A. Loomis, *J. Forensic Sci.*, 23 (1978) 481.
397 Y. Maruma, T. Inoue, T. Niwase and T. Niwaguchi, *Kagaku Keisatsu Kenkyusho Hokoku*, 31 (1978) 280; *C.A.*, 90 (1979) 162779w.

398 J.B. Lopez, J.E. Buttery and G.F. de Witt, *Mod. Med. Asia*, 14 (1978) 7; *C.A.* 89 (1978) 140103y.
399 P. Gundermann and R. Pohloudek-Fabini, *Pharmazie*, 33 (1978) 205.
400 H. Thielemann, *Sci. Pharm.*, 46 (1978) 322.
401 E. Klug and P. Toffel, *Z. Anal. Chem.*, 294 (1979) 46.

TABLE 12.1

TLC SYSTEMS FOR THE SEPARATION OF OPIUM ALKALOIDS[118]

(A) Neutral solvents in combination with silica gel $HF_{254}$ (activated) plates:

| Solvent system No. | Components |
|---|---|
| 1 | Benzene - chloroform - acetone (70:15:15) |
| 2 | Benzene - tetrahydrofuran (8:2) |
| 3 | Benzene - dioxane (8:2) |
| 4 | Benzene - propanol (8:2) |
| 5 | Chloroform - acetone (75:25) |
| 6 | Dioxane |
| 7 | Benzene - 90% ethanol (8:2) |
| 8 | Chloroform - *n*-hexane - methanol (52:40:8) |
| 9 | Chloroform - ethyl acetate - methanol (4:4:2) |
| 10 | Benzene - acetone - methanol (7:2:1) |
| 11 | Benzene - methanol (8:2) |
| 12 | Chloroform - *n*-hexane - methanol (65:25:10) |
| 13 | Chloroform - ethanol (9:1) |
| 14 | Chloroform - methanol (9:1) |
| 15 | Benzene - methanol (7:3) |

(B) Basic solvents in combination with silica gel $HF_{254}$ (activated) plates:

| Solvent system No. | Components |
|---|---|
| 16 | Cyclohexane - diethylamine (8:2) |
| 17 | Xylene - methyl ethyl ketone - methanol - diethylamine (40:60:6:2) |
| 18 | Benzene - ethyl acetate - diethylamine (5:4:1) |
| 19 | Cyclohexane - chloroform - diethylamine (5:4:1) |
| 20 | Dioxane - chloroform - ethyl acetate - 25% ammonia (60:25:10:5) |
| 21 | Benzene - dioxane - 90% ethanol - 25% ammonia (50:40:5:5) |
| 22 | Chloroform - acetone - diethylamine (5:4:1) |
| 23 | Chloroform - diethylamine (9:1) |
| 24 | Dioxane - light petroleum (b.p. 30-60$^{o}$C) - benzene - chloroform - ethanol - ethyl acetate - 25% ammonia (50:15:10:10:5:5:5) |
| 25 | Ethyl acetate - 94% ethanol - dimethylformamide - diethylamine (75:20:5:2) |
| 26 | Methanol - acetone - triethanolamine (50:50:1.5) |
| 27 | Acetone - chloroform - triethylamine - methanol (40:30:20:10) |
| 28 | Carbon tetrachloride - *n*-butanol - methanol - 6 *M* ammonia (40:50:30:2) |
| 29 | Dichloromethane - methanol - 10% ammonia (85:15:2) |
| 30 | Methanol - acetone - triethylamine (50:50:1.5) |

TABLE 12.2

TLC ANALYSIS OF OPIUM ALKALOIDS

TLC systems:

| System | Stationary phase | Mobile phase |
|---|---|---|
| S1 | Silica gel G | Chloroform-methanol (9:1)[42] |
| S2 | Silica gel G | Benzene-ethanol (4:1)[42] |
| S3 | Silica gel G | Dichloromethane-methanol-6 *M* ammonia (85:15:2)[267] |
| S4 | Silica gel G | Ethyl acetate-ethanol(96%)-dimethylformamide-diethylamine (75:20:5:2)[85] |
| S5 | Silica gel G, activated | Benzene-methanol (4:1)[9] |
| S6 | Silica gel G | Light petroleum-diethyl ether-ethanol-diethylamine (4:16:2:1)[86] |
| S7 | Silica gel G | Carbon tetrachloride-*n*-butanol-methanol-10% ammonia (40:30:30:2)[86] |
| S8 | Silica gel G, activated | Xylene-methyl ethyl ketone-methanol-diethylamine (20:20:3:1)[37] |
| S9 | Silica gel F254, ready-made plate, activated | Dioxane-ethanol-diethylamine (98:1:1)[98] |
| S10 | Silica gel F254, ready-made plate, activated | Chloroform-ethanol-diethylamine (89:10.5:0.5)[98] |
| S11 | Silica gel G, activated | Chloroform-ethyl acetate - ethanol (5:4:1)[98] |
| S12 | Silica gel F254, ready-made plate, activated | Chloroform-*n*-butanol-diethylamine (40:5:5)[98] |
| S13 | Aluminium oxide G, activated | Dioxane-ethanol (98:2)[98] |
| S14 | Silica gel G, impregnated with $Na_2CO_3$ solution ($d$=1.33), activated | Dioxane-ethanol (98:2)[98] |
| S15 | Silica gel G, impregnated with $Na_2CO_3$ solution ($d$=1.33), activated | Ethyl acetate-chloroform-ethanol (6:5:1)[98] |
| S16 | Silica gel F254, ready-made plate, activated | Diisopropyl ether-ethanol-diethylamine (97:2:1)[98] |
| S17 | Silica gel F254, ready-made plate, activated | Chloroform-acetone-20% ammonia (8:1:1)[98] |
| S18 | Silica gel G, activated | Methanol-chloroform-23% ammonia (85:15:0.7)[98] |
| S19 | Silica gel G | Methanol-25% ammonia (100:1.5)[154] |
| S20 | Silica gel G, activated | Chloroform-acetone-diethylamine (5:4:1)[50] |
| S21 | Silica gel G, activated | Chloroform-cyclohexane-diethylamine (7:2:1)[50] |

| Alkaloid | $hR_F$ values | | | | | | | | | | | | | | | | | | | | |
|---|---|---|---|---|---|---|---|---|---|---|---|---|---|---|---|---|---|---|---|---|---|
| | S1 | S2 | S3 | S4 | S5 | S6 | S7 | S8 | S9 | S10 | S11 | S12 | S13 | S14 | S15 | S16 | S17 | S18 | S19 | S20 | S21 |
| Codeine | 35 | 15 | 59 | 31 | 21 | 29 | 49 | 26 | 71 | 55 | 5 | 58 | 75 | 71 | 22 | 7 | 13 | 48 | 35 | 36 | 38 |
| Morphine | 12 | 8 | 36 | 18 | 11 | 8 | 37 | 12 | 51 | 8 | 2 | 15 | 20 | 27 | 10 | 2 | 0 | 45 | 34 | 8 | 3 |
| Narceine | 9 | 3 | 5 | 2 | | | | | 0 | 0 | 0 | 0 | 0 | 0 | 0 | 0 | 0 | 34 | 33 | 0 | 0 |
| Noscapine | 97 | 87 | 92 | 73 | 68 | 76 | 91 | 74 | 95 | 65 | 85 | 65 | 95 | 97 | 95 | 60 | 82 | 81 | | 73 | 71 |
| Papaverine | 97 | 77 | 86 | 65 | 63 | 59 | 91 | 59 | 90 | 60 | 75 | 64 | 90 | 95 | 90 | 25 | 72 | 79 | 66 | 67 | 73 |
| Thebaine | 65 | 38 | 77 | 41 | 40 | 59 | 72 | 45 | 75 | 60 | 20 | 60 | 83 | 67 | 66 | 25 | 20 | 50 | 62 | 64 | 69 |
| 10-Hydroxycodeine | 17 | 15 | | | | | | | | | | | | | | | | | | | |
| Reticuline | 28 | 18 | | | | 13 | | | | | | | | | | | | | | | |
| Neopine | 38 | 12 | | | | | | | | | | | | | | | | | | | |
| Protopine | 46 | 42 | | | 38 | | | | | | | | | | | | | | | | |
| Laudanine | 47 | 28 | | | 26 | | | | | | | | | | | | | | | | |
| Laudanidine | 47 | 28 | | | | 22 | | | | | | | | | | | | | | | |
| Cryptopine | 48 | 40 | | | 34 | | | | | | | | | | | | | | | | |
| Laudanosine | 74 | 36 | | | 42 | | | | | | | | | | | | | | | | |
| Narcotoline | 88 | 71 | 77 | | 58 | 15 | | | | | | | | | | | | | | | |
| Porphyroxine | 93 | 65 | | | | | | | | | | | | | | | | | | | |
| Apomorphine | | | | | | | | | 84 | 55 | 30 | 57 | 90 | 75 | 60 | | | | | | |
| Papaverinol | | | 6 | | | | | | | | | | | | | | | | | | |
| Papaveraldine | | | 59 | | | | | | | | | | | | | | | | | | |
| Methylcodeine | | | 76 | | | | | | | | | | | | | | | | | | |
| Diethylmorphine | | | 77 | | | | | | | | | | | | | | | | | | |
| Ethylmorphine | | | 63 | | | | | | 76 | 54 | 8 | 58 | 80 | 76 | 29 | | 29 | | | 43 | 38 |
| Dihydrocodeine | | | 51 | | | | | | | | | | | | | | | | | | |
| Dihydrohydroxycodeinone | | | | | | | | | 90 | 62 | 10 | 60 | 85 | 80 | 82 | | 47 | | | | |
| Pholcodine | | | | | | | | | | | | | | | | | | | | 35 | 26 |
| Cotarnine | | | | | | | | | | | | | | | | | | | | 60 | 26 |
| Monoacetylmorphine | | | | | | | | | | | | | | | | | | | | 46 | 46 |
| Heroin | | | | | | | | | | | | | | | | | | | | 63 | 65 |
| Acetylpholcodine | | | | | | | | | | | | | | | | | | | | 60 | 68 |
| Acetylethylmorphine | | | | | | | | | | | | | | | | | | | | 64 | 69 |
| Acetylcodeine | | | | | | | | | | | | | | | | | | | | 63 | 69 |

TABLE 12.3

FURTHER TLC ANALYSIS OF OPIUM ALKALOIDS AND RELATED DRUGS

TLC systems:

| | | |
|---|---|---|
| S22 | Silica gel G, activated | Acetone-methanol-17% ammonia (50:50:1)[39] |
| S23 | Silica gel G, activated | Chloroform-ethanol-17% ammonia (9:1:1)[39] |
| S24 | Silica gel G, activated | Chloroform-ethanol-17% ammonia (50:50:1)[39] |
| S25 | Silica gel G, activated | Chloroform-96% ethanol (9:1)[5] |
| S26 | Silica gel G, impregnated with 0.5 *M* KOH | Chloroform-ethanol (8:2)[5] |
| S27 | Silica gel G, activated | Ethyl acetate-ethanol-dimethylformamide-diethylamine (75:20:5:2)[5] |
| S28 | Cellulose, impregnated with formamide (20% in acetone) | Benzene-heptane-chloroform-diethylamine (6:5:1:0.02)[5] |
| S29 | Silica gel G, activated | Methanol-acetone-triethanolamine (1:1:0.03)[8] |

| Alkaloid | $hR_F$ values | | | | | | | |
|---|---|---|---|---|---|---|---|---|
| | S22 | S23 | S24 | S25 | S26 | S27 | S28 | S29 |
| Codeine | 42 | 83 | 28 | 12 | 33 | 41 | | 43 |
| Morphine | 31 | 40 | 0 | 2 | 2 | 27 | 0 | 40 |
| N-Allyl-nor-morphine | 52 | 47 | 18 | | | | | |
| Dihydromorphinone | 17 | 60 | 7 | 5 | 13 | 27 | 6 | 28 |
| 3-Hydroxy-N-methylmorphinan | 27 | 87 | 30 | | | | | 28 |
| Dextromethorphane | 28 | 98 | 50 | | | | | 23 |
| Dihydrocodeinone | 28 | 89 | 30 | 10 | 28 | 34 | 63 | 29 |
| Oxycodone | 66 | 96 | 94 | 47 | 70 | 79 | 75 | |
| Ethylmorphine | 44 | 86 | 30 | 14 | 37 | 44 | 57 | 37 |
| Pholcodine | 34 | 76 | 12 | | | | | |
| Heroin | 58 | 96 | 64 | | | | | |
| Dihydrocodeine | 34 | 96 | 24 | 6 | 22 | 34 | | |
| Acetyldihydrocodeine | 41 | 98 | 65 | | | | | |
| Codeine-N-oxide | 20 | 34 | 0 | | | | | |
| Acetyldihydrocodeinone | 32/54 | 98 | 42 | 24 | 59 | | 90 | 31 |
| Meperidine | 78 | 95 | 67 | | | | | 56 |
| Alphaprodine | 74 | 94 | 64 | | | | | |
| Carbetidine | 88 | 91 | 60 | | | | | |
| Ketobemidone | 64 | 80 | 22 | | | | | 56 |
| Methadone | 53 | | 88 | | | | | 48 |
| Normethadone | 61 | | 70 | | | | | |
| Dextromoramide | 87 | | in solvent front | | | | | 87 |
| Noscapine | 84 | | | | | 92 | 94 | 82 |
| Narceine | 19 | | | 78 | 81 | | | 23 |
| Papaverine | | | | 74 | 78 | 86 | 89 | 82 |
| Thebaine | | | | | | | 85 | 41 |

TABLE 12.4

COLOURS OF SOME OPIUM ALKALOIDS AND THEIR DERIVATIVES WITH IODOPLATINATE REAGENT[50], IN UV LIGHT (365 nm)[50] AND AFTER SPRAYING WITH CITRIC ACID IN ACETIC ANHYDRIDE (no. 22) AND HEATING AT 80°C FOR 10 min[373]

| Alkaloid | Colour Iodoplatinate | Fluorescence (365 nm) | 4% citric acid in acetic anhydride | |
|---|---|---|---|---|
| | | | Colour | Fluorescence (365 nm) |
| Morphine | Blue-grey | Blue | Yellow | Yellow-green |
| Codeine | Blue-violet | - | Pink | Bright green |
| Thebaine | Brown | - | Pink | Yellow-pink |
| Papaverine | Yellow | Yellow | Yellow | Pink-red |
| Noscapine | Yellow | Blue | Yellow | Blueish |
| Narceine | Blue | Blue | | |
| Cotarnine | Blue-violet | Yellow | | |
| Apomorphine | Green | Blue | | |
| Monoacetylmorphine | Blue-grey | Blue | | |
| Heroin | Blue-grey | Blue | | |
| Acetylcodeine | Blue-violet | - | | |
| Pholcodine | Blue-violet | - | | |
| Acetylpholcodine | Blue-violet | - | | |
| Acetylcodeine | Blue-violet | - | | |
| Ethylmorphine | Blue-violet | - | | |
| Acetylethylmorphine | Blue-violet | - | | |

TABLE 12.5

DETECTION METHODS FOR OPIUM ALKALOIDS

| Reagent | No. | Sensitivity | Background colour | Alkaloid colour | Ref. |
|---|---|---|---|---|---|
| Dragendorff's modifications: | | | | | |
| Vagujfalvi | 39f | 0.01-0.05 | White | Orange | 66 |
| Thies-Reuter | 39e | | White | Orange | 18,29,43 |
| Munier | 39b | | Yellow | Orange-red | |
| Munier, followed by $NaNO_3$ | 39h | | White | Brown | 265 |
| Munier + Machebouef | 39c | | Yellow | Orange-red | 364 |
| Iodoplatinate | | 0.7-1.0 | | | 118 |
| $KMnO_4$ (acidic) | | 0.5-2.5 | | Fluorescence; blue | 154 |
| $KMnO_4$ (basic) | | | | | 218 |
| Iodine vapour | | | Yellow | Brown | 191 |
| Iodine in $CHCl_3$ | 51b | | Light yellow | Brown, fluorescent spots after heating | 86 |
| Iodine vapour, followed by pyrrole vapour | 54 | | Yellow | Thebaine red; codeine yellow | 221,293 |
| Iodine, followed by phenothiazine | | | Light brown | Brown | 144,221 |
| Bouchardat | | | Dark yellow | Brown | 86,309 |
| Marquis | 45 | 1 | | Various (Table 12.6) | 39,63,70, 72,218,270, 310 |
| Fröhde | 90 | | | Various for some alkaloids (Table 12.6) | 70,72,390 |
| Mandelin | 91 | | | Various (Table 12.6) | 70,270, |
| Mandelin-Marquis | | 50 | | | 63 |
| Folin-Ciocalteu | 43 | 1 | | | 63 |
| Rahman | | 5 | | | 63 |
| Husemann | | 50 | | | 63 |
| Phosphotungstic acid | | | | | 91,196 |
| Phosphomolybdate | 79 | | | | 144,400 |
| *p*-Dimethylaminobenzaldehyde in $H_2SO_4$ (Wasicky) | | | | Morphine purple-violet | 12,72,113, 270,309 |
| Vanillin-$H_2SO_4$ | 104 | 1 | | Morphine red-violet; pseudomorphine green | 72,291 |
| *p*-Nitraniline | | | | Morphine yellow | 72 |
| Methyl orange/$HgCl_2$ | | | Yellow | Yellow-white | 346 |
| Methyl red/$HgCl_2$ | | | | | 335 |
| Tropeoline 00 | | | | | 311 |
| Na-Co nitrite | | | | | 342 |
| $Co(CNS)_2$ | 26a | | Light blue | Blue | 330,343 |
| $FeCl_3$-$K_3Fe(CN)_6$ | 80a | | | | 196 |
| $H_2O_2$-$K_4Fe(CN)_6$ | 49 | 10 | | Brown | 196 |
| Denigès | 30 | 25 | | Brown | 236 |
| 3,5-Dinitrobenzoyl chloride | 37 | 100 | | Yellow | 344 |
| $HgNO_3$+$HNO_3$ | 70 | 2 | | Red-brown-yellow | 368 |
| $NaNO_2$ | 89 | 1 | | Orange-brown | 63,72,154 |
| Fast Blue B salt | 40b | 0.5 | Grey | Violet-orange (phenolic OH) | 90,97 |
| Gibb's reagent | 34,35 | 0.5 | White | Blue (phenolic OH) | 90,101,349 |
| $(NH_4)Ce(NO_3)_6$-hydroxylamine | 11 | 1-5 | | Brown (phenolic OH) | 323 |
| Na iodate | 87 | | | Fluorescence (light green), morphine, heroin | 222 |

TABLE 12.5 (*continued*)

| Reagent | No. | Sensitivity | Background colour | Alkaloid colour | Ref. |
|---|---|---|---|---|---|
| Pauly's reagent | 33a | | | Orange-red (phenolic OH) | 5,86,97 |
| TCBI reagent | 94 | 5 | | Various | 315,328 |
| $\pi$-Acceptors | | 0.5-10 | | Various (see Table 2.2, p.16 | 375 |
| 4-Aminopyrine | | | | | 348 |
| Anisaldehyde in $H_2SO_4$ | 4 | 0.4-1 | | | 118 |
| $Hg(OAc)_2$-MeOH, followed by anisaldehyde/$H_2SO_4$ + phosphomolybdic acid in MeOH (10%) | 68 | 0.1-0.4 | | Blue-grey | 118 |
| $K_4Fe(CN)_6$+$K_3Fe(CN)_6$ | 81 | 0.1-10 | | Fluorescent spots (phenanthrene + phenolic OH) | 38 |
| Citric acid in acetic anhydride | 22 | 0.5 | White | Fluorescent spots | 373 |

TABLE 12.6

COLOUR REACTIONS OF SOME OPIUM ALKALOIDS[70]

Colours: bl = blue; gr = green; vio = violet; yel = yellow.

| Spray reagent | No. | Codeine | Morphine | Narceine | Noscapine | Papaverine | Thebaine | Cryptopine | Heroin | Acetyl-codeine | O-acetyl-morphine |
|---|---|---|---|---|---|---|---|---|---|---|---|
| Fröhde's reagent (sulphomolybdic acid in $H_2SO_4$) | 90 | gr | vio | grey-gr | - | - | yel | - | - | - | - |
| Mandelin's reagent (sulphovanadic acid in $H_2SO_4$) | 91 | grey-bl | grey-vio | pink-vio | pink-vio | grey | ochre | bl | - | vio | - |
| Lafon's reagent (ammonium selenite in $H_2SO_4$) | 2 | bl-gr | bl→gr | yel | - | - | yel | vio | bl→gr→grey | grey-gr | light gr |
| Marquis reagent (formaldehyde in $H_2SO_4$) | 45 | bl-vio | vio | yel-br | - | bl(weak) | yel | bl(100°C) | grey-bl | vio | bl(weak) |
| D'Aloy and Valdiguié reagent (formaldehyde in $H_2SO_4$ + $FeCl_3$) | 46 | bl | vio | yel | - | bl(weak) | yel | yel | grey-gr | - | grey-bl |
| Kieffer's reagent [$K_3Fe(CN)_6$+$FeCl_3$] | 80a | - | bl | - | - | - | - | - | - | - | bl |
| Vincent and Schwal reagent (hydroxylamine-$FeCl_3$) | 50 | - | - | - | - | - | - | - | bl-vio | bl-vio | bl-vio |
| Sulphuric acid at 100°C | | grey | grey | orange | pink | pink(weak) | red-br | grey | vio | grey-bl | vio |

TABLE 12.7

LITERATURE CITED IN CHAPTER 3 WHICH INCLUDES THE ANALYSIS OF OPIUM ALKALOIDS

| Alkaloids* | Ref. | Alkaloids* | Ref. |
|---|---|---|---|
| C,M,Na,No,P,cot,DiHMone,DiHC,DiHCone | 7 | P,EtM | 221 |
| C,M,No,P,DiHMone,DiHCone | 13 | C,M,DiHMone,DiHC,oxyC,EtM,H,meth,nalo | 223 |
| C,M,No,P,T,DiHMone,DiHCone,oxyC | 44 | C,M,Na,No,P,T,DiHMone,DiHCone,DiHC,EtM,H,oxyC | 233 |
| C,M,No,P,EtM | 45 | C,M,Na,No,P,T,EtM,H,oxyC,meth,nalo | 266 |
| C,M,Na,No,P,T | 49 | C,M,P,DiHMone,DiHCone,oxyC,EtM,EtP,AcdiHCone,nalo | 279 |
| C,M,Na,No,P,T,L,H | 56 | C,M,P,H,meth | 287,288 |
| C,M,No,P,DiHMone,EtM | 60 | C,M,P,H,meth | 294 |
| C,M,Na,No,P,T,DiHMone,DiHCone,DiHC,oxyC,apoM,H | 65 | C,M,P,DiHM,oxyC,EtM,meth | 295,384 |
| C,M,P,DiHMone,DiHCone,meth | 69 | C,M,T,H, meth,mep | 297 |
| C,M,No,H | 80 | C,M,DiHCone,oxyC,EtM | 299 |
| C,M,P,DiHMone,DiHCone,H,meth | 82 | C | 321 |
| C,M,No,P,DiHC,EtM,H | 92 | DiHC,EtM,H,AcdiHCone | 334 |
| C,M,Na,No,P,T,AcDiHCone,apoM,H,meth | 94 | C,M,No,T | 340,366,367 |
| C,No,P | 104 | C,M,No,P,DiHMone,DiHCone,DiHC,oxyC,EtM,apoM,H,AcdiHCone,meth | 369 |
| M,Na,No,P,cot,apoM,H | 176 | C,M,P,H | 379 |
| C,M,P,DiHCone,EtM,H,meth | 177 | M,Na,No,P,EtM | 350,355 |
| C,No,P,EtM | 186 | C,M,No,P,T | 393 |
| | | C,M,EtM,oxyC | 394 |

*Abbreviations:

| | | | |
|---|---|---|---|
| A | atropine | M | morphine |
| Acc | acetylcodeine | MeA | methylatropine |
| AcdiHC | acetyldihydrocodeine | MediHMone | methyldihydromorphinone |
| AcdiHCone | acetyldihydrocodeinone (thebacone) | mep | meperidine |
| acetoM | acetomorphine | mesc | mescaline |
| AcSalac | acetylsalicylic acid | meth | methadone |
| amp | aminopyrine | Na | narceine |
| apoC | apocodeine | Naline | narcotoline |
| apoM | apomorphine | nalo | nalorphine |
| atp | antipyrine | nalox | naloxone |
| B | brucine | nic | nicotine |

barb barbital
benzylM benzylmorphine
C codeine
caf caffeine
C N-ox codeine N-oxide
coc cocaine
Cone codeinone
cot cotarnine
crypt cryptopine
dex dextromoramide
diHC dihydrocodeine
diHCone dihydrocodeinone
diHM dihydromorphine
diHMone dihydromorphinone
Dmethphan dextromethorphan
eph ephedrine
EtM ethylmorphine
H heroin
14OHC 14-hydroxycodeine
14OHCone 14-hydroxycodeinone
14OHM 14-hydroxymorphine
14OHMone 14-hydroxymorphinone
keto ketobemidone
L laudanine
lev levorphanol
leval levallorphan
Losine laudanosine
No noscapine
norC norcodeine
normep normeperidine
normeth normethadone
3OAcM 3-O-acetylmorphine
6OAcM 6-O-acetylmorphine
oxyC oxycodone
oxyM oxymorphone
P papaverine
par paracetamol
Ph pholcodine
phb phenobarbital
phen phenacetine
physo physostigmine
prop propoxyphene
prot protopine
pseudoM pseudomorphine
Q quinine
Qd quinidine
RSP reserpine
S strychnine
Salac salicylic acid
Salam salicylamide
scopo scopolamine
T thebaine
thbr theobromine
thph theophylline

TABLE 12.8

TLC ANALYSIS OF OPIUM ALKALOIDS IN PLANT MATERIAL AND OPIUM

| Alkaloid* | Aim | Adsorbent | Solvent system | Ref. |
|---|---|---|---|---|
| C,M,Na,No,P,meconic acid | In opium and opium preparations | $SiO_2$+MgO+$CaSO_4$ (10:10:4), impregnated with phosphate buffer (pH=6.6) | Dioxane | 1 |
| C,M,No,P,T | Separation alkaloids in opium samples | $Al_2O_3$ impregnated with acetate buffer (pH=5) | BuOH-EtOH-$H_2O$(9:1:1) | 3 |
| C,M,No,P,T,Naline,crypt, prot,L,Losine | Direct and indirect quantitative analysis in opium (Table 12.2) | $SiO_2$ | Benzene-MeOH(8:2) | 9 |
| M | Indirect quantitative analysis in opium | Alkaline $SiO_2$ Cellulose, FMA impregnated | $CHCl_3$-EtOH(9:1, 3:1)<br>Benzene-heptane-$CHCl_3$-DEA(6:5:3:0.03)<br>Benzene-$CHCl_3$-DEA (2:3:0.02) | 10 |
| Various Papaveraceae alkaloids, see Table 26 Isoquinoline alkaloids | Separation | $SiO_2$ | $CHCl_3$-EtOAc-MeOH(2:2:1) | 14 |
| C,M,Na,No,P,T,pseudoM, Naline | Identification in opium | $SiO_2$ | $CCl_4$-BuOH-MeOH-6 *M* $NH_4OH$(40:30:30:3.3, 40:30:30:2) | 17,27 |
| C,M,No,P,T | Separation | $Al_2O_3$ | Benzene-MeOH(8:2) | 18 |
| M | Study of quantitative analysis of M in opium | $SiO_2$ | $CH_2Cl_2$-MeOH-10% $NH_4OH$(85:15:2) | 19 |
| C,M,No,P | Separation in opium preparations | $SiO_2$ | $CHCl_3$-$Me_2CO$-25% $NH_4OH$(12:24:1)<br>$CHCl_3$-$Me_2CO$-MeOH-25% $NH_4OH$(20:20:3:1) | 23 |
| C,M,No,P | Separation | $Al_2O_3$ | $Me_2CO$,$CHCl_3$,benzene-$CHCl_3$-$Me_2CO$(70:15:15) | 25 |
| C,M,No,P,T | Indirect quantitative analysis (spectrophotometric | $SiO_2$ | Benzene-MeOH(9:1, 4:1), benzene-PrOH(4:1), benzene-*n*-BuOH(3:1), EtOH-dioxane-benzene-$NH_4OH$ (1:8:10:1), THF-benzene(1:4), benzene-EtOH(4:1), $CHCl_3$-MeOH(9:1), $CHCl_3$-isoprOH(4:1), $CHCl_3$-$Me_2CO$(3:1) | 26 |

| | | | | |
|---|---|---|---|---|
| C,M,No,P,T, papaverinol | Separation, quality control of C and P | $Al_2O_3$ | I. Benzene-MeOH(9:1)+1 drop 80% HCOOH<br>II. Benzene-AcOH(9:1)<br>Two-dimensional: I,II or II,I | 29 |
| C,M,Na,No,P,T,neopine, L,Losine,prot,crypt,Naline | Identification in *Papaver somniferum* | $SiO_2$ | Benzene-MeOH(9:1), subsequently $CHCl_3$-EtOH-$Me_2$CO-MeOAc(6:2:1:1) | 30 |
| C,M,P,T | Separation alkaloids from oliferous poppy | $SiO_2$ | Benzene-EtOH(9:1) | 32 |
| C,M,No,P,T | Indirect quantitative analysis in opium (spectrophotometric) | $Al_2O_3$ | I. Benzene-MeOH(9:1)+2 drops 80% HCOOH<br>II. Benzene-AcOH(9:1)<br>Two-dimensional: I,II | 33,61 |
| C,M,No,P,T | Identification in opium | $SiO_2$ | Benzene-MeOH-DEA(80:19:1) | 35 |
| C,M,No,P,T | Separation (Table 12.2) | $SiO_2$ | Xylene-MeEtCO-MeOH-DEA(20:20:3:1) | 37 |
| 16 alkaloids (Table 12.2) | Separation | $SiO_2$ | $CHCl_3$-MeOH(9:1)<br>Benzene-EtOH(4:1) | 42 |
| C,M,No,P,T | Densitometric analysis of main opium alkaloids | $Al_2O_3$ | Benzene-MeOH(9:1)+2 drops 80% HCOOH | 43 |
| opium alkaloids, no details available | | $SiO_2$ | $CHCl_3$-EtOH(9:1) | 46 |
| C,M,Na,No,P,T,H,apoM,ph, EtM | Reaction chromatography (Tables 12.2 and 12.4) | $SiO_2$ | $CHCl_3$-$Me_2$CO-DEA(5:4:1)<br>$CHCl_3$-cyclohexane-DEA(7:2:1) | 50,76 |
| C,M,T | Identification in opium extracts | $SiO_2$ | $CHCl_3$-EtOH(8:2) | 51 |
| M | Indirect quantitative analysis in opium | $SiO_2$ | $CHCl_3$-$Me_2$CO-TriEA(3:4:1:2) | 53,111 |
| C,M,No,P,T | Indirect quantitative analysis in vegetable drugs | $SiO_2$ | $CHCl_3$-$Me_2$CO-TriEA(1:2:3:4) | 62 |
| M | Methods for detection in opium | $Al_2O_3$ | Benzene - 5-10% EtOH | 63 |
| C,M,No,P,T,rhoeadine | Identification in *Papaver somniferum* seeds | $SiO_2$ | Xylene-MeEtCO-MeOH-DEA(45.5:45.5:7:2)<br>*n*-Hexane-$CHCl_3$-MeOH(25:65:10, 40:52:8) | 64 |
| C,M,Na,No,P,T,H,crypt, OAcM,Acc | Separation of opium alkaloids and some acetyl derivatives (Table 12.6) | $SiO_2$<br><br>$Al_2O_3$ | $Et_2O$(sat. with $H_2O$)-$Me_2$CO-DEA(85:8:7)<br>MeOH-$CHCl_3$-23% $NH_4OH$(85:15:0.7)<br>Benzene-$CHCl_3$-$Me_2$CO(70:15:15) sat. with 3.5% $NH_4OH$ | 70 |
| M | Densitometric analysis in poppies | NaOH-impregnated $SiO_2$ | Toluene-MeOH-$CHCl_3$(10:4:2) | 72 |

| | | | | |
|---|---|---|---|---|
| C,M,No,P,T,L,apoM | Separation | Carboxymethylcellulose | 0.2 or 0.05 *M* NaCl | 73 |
| C,M,No,P,T | Identification in poppies | $SiO_2$ | Benzene-dioxane-EtOH-$NH_4OH$(50:40:5:5) | 83 |
| C,M,Na,No,P,T | Identification in opium and its preparations (Table 12.2) | $SiO_2$ | EtOAc-96% EtOH-DMFA-DEA(75:20:5:2) | 85 |
| C,M,No,P,T,L,Naline, reticuline,papaverrubine D | Densitometric analysis of laudanine in opium | 0.1 *M* NaOH-impregnated $SiO_2$ | $CHCl_3$-MeOH(87:15) | 90 |
| No,P | Indirect quantitative analysis in opium and pharmaceutical preparations | $SiO_2$ | EtOAc | 95 |
| C,M,Na,No,P,T,EtM,apoM, oxyC | Separation in 18 different TLC systems | | See Table 12.2 | 98 |
| C,M,No,P | Reaction chromatography | $SiO_2$ | Benzene-MeOH-$Me_2CO$-AcOH(70:20:5:5) | 103 |
| C,M,Na,No,P,T,L,Losine, crypt,prot,neopine, reticuline,codamine,cotarnoline,naline,papaverrubine D,isocorypalmine, papaveraldine | Separation of *Papaver somniferum* alkaloids | $SiO_2$ | Benzene-MeOH(9:1), subsequently $CHCl_3$-EtOH-$Me_2CO$-EtOAc(6:2:1:1) | 108 |
| C,M,No,P,T | Indirect quantitative (UV) analysis of poppy alkaloids | $SiO_2$ | Xylene-MeEtCO-MeOH-DEA(20:20:3:1) | 109 |
| C,M,Na,No,P,T | Identification in opium and its preparations | $SiO_2$ | Xylene-MeEtCO-MeOH-DEA(20:20:3:1) | 114 |
| C,M,No,P,T | Identification in opium and quality control of M | $SiO_2$ | Toluene-$Me_2CO$-95% EtOH-25% $NH_4OH$(45.5:45.5:6.5:2.5) $CH_2Cl_2$-MeOH-10% $NH_4OH$(84:14:2) | 115 |
| C,M,Na,No,P,T | Identification in opium | $SiO_2$ | Benzene-MeOH(8:2), twice | 131 |
| C,M,Na,No,P,T | Separation with isobaric mobile phases | $SiO_2$ | $(Isopr)_2NH$-dichloroethylene-isoprOH(1:4:5) | 138 |

| | | | | |
|---|---|---|---|---|
| M,Mep | Identification in tinctures, on micro-slides | $SiO_2$ | $CHCl_3$-$Me_2CO$-DEA(5:4:1)<br>$CHCl_3$-DEA(9:1)<br>Benzene-EtOAc-DEA(5:4:1)<br>Benzene-EtOAc(8:2) | 142 |
| M | Indirect quantitative analysis (spectrophotometric) in opium and tinctures | Basic $SiO_2$ | $CHCl_3$-MeOH(3:1), subsequently $CHCl_3$-MeOH-25% $NH_4OH$ (150:50:4) | 143 |
| C,M,No,P,T | Separation opium alkaloids | 0.1 *M* $Na_2CO_3$-impregnated $SiO_2$ | $CHCl_3$-benzene-$Me_2CO$-MeOH(9:7:2:2) | 145 |
| C,M,Na,No,P | Densitometric analysis (Table 12.2) | $SiO_2$ | $CHCl_3$-MeOH(9:1), benzene-EtOH(4:1), $Me_2CO$, *n*-BuOH-AcOH-$H_2O$(10:1.5:3), MeOH-$NH_4OH$(100:1.5), $CHCl_3$-$Me_2CO$-DEA(5:4:1) | 146,154 |
| Not specified | TAS technique | $SiO_2$ | $CHCl_3$-$Me_2CO$-DEA(5:4:1)<br>$CHCl_3$-DEA(9:1)<br>Cyclohexane-$CHCl_3$-DEA(5:4:1)<br>Cyclohexane-DEA(9:1)<br>Benzene-EtOAc-DEA(7:2:1) | 148 |
| C,M,Na,No,P,T | Identification of opium in tea | | No details available | 149 |
| C,M,No,P,T | Densitometric analysis in opium | $SiO_2$ | Xylene-$CH_2Cl_2$-$Me_2CO$-EtOH-10% $NH_4OH$(40:20:20:6:2.5)<br>Two dimensional: I. benzene-EtOH(9:1);<br>II. $CHCl_3$-EtOH-EtOAc-$Me_2CO$(6:2:1:1) | 155 |
| C,M,No,P,T | Densitometric analysis in tinctures | $SiO_2$ | $CHCl_3$-$Me_2CO$-25% $NH_4OH$(12:24:1) | 162 |
| Not specified | Characterization of opium and hashish by TLC, using isobaric solvents | $SiO_2$ | Benzene-MeEtCO-$BuNH_2$(75:15:10) | 166 |
| M | Densitometric analysis in plant extracts, compared with UV and GLC analysis | $SiO_2$ | Toluene-$Me_2CO$-96% EtOH-10% $NH_4OH$(30:40:12:4) | 188 |
| C,M,No,P,T | New detection method | $SiO_2$ | *n*-BuOH-$H_2O$-AcOH(3:1:1) | 196 |
| M | Indirect quantitative analysis in poppy heads (spectrophotometric) | $SiO_2$ | $CHCl_3$-isoprOH-25% $NH_4OH$(30:10:1) | 214 |
| M | Indirect quantitative analysis (UV) in poppy heads | 0.1 *M* $NaHCO_3$-impregnated $SiO_2$ | MeOH-benzene(85:15) | 215 |

| | | | | |
|---|---|---|---|---|
| C,M,No,P,T | Identification of opium-yielding papavers | | No details available | 249 |
| C,M,Na,No,P,T | Separation | $SiO_2$ | $Me_2CO$-*p*-xylene-MeOH-25% $NH_4OH$(49.5:41.5:5:4) | 260 |
| C,M,Na,No,P,T | Separation | $SiO_2$ | $Me_2CO$-*p*-xylene-MeOH-25% $NH_4OH$(50:40:6:5) | 262 |
| C,M,P | Separation | $SiO_2$ | Benzene-dioxane-EtOH-$NH_4OH$(10:8:1:1)<br>Benzene-dioxane-EtOH-40% KOH(56:40:2:2)<br>EtOAc-MeOH-conc. $NH_4OH$(17:2:1) | 277 |
| C,M,No,P,T | Identification of opium and its alkaloids | 2% $Na_2CO_3$-impregnated $SiO_2$ | $CHCl_3$-EtOH(8:2) | 293 |
| C,M,No,P,T | TAS technique | $SiO_2$ | Toluene-$Me_2CO$-EtOH-conc. $NH_4OH$(45:45:7:3) | 303 |
| C,M,Na,No,P,T | Separation | $SiO_2$ | Xylene-MeEtCO-MeOH-25% $NH_4OH$(40:40:6:4) | 330 |
| C,M,No,P,T | Detection of M in adulterated opium | $SiO_2$ | $CHCl_3$-MeOH(1:2) | 364 |
| C,M,No,P,T | Detection method | $SiO_2$ | Benzene-MeOH(8:2) | 368 |
| C,M,Na,No,P,T,Naline, L,Losine | TLC control in determination of M in opium | $SiO_2$ | Xylene-$CH_2Cl_2$-$Me_2CO$-EtOH-10% $NH_4OH$(40:20:20:6:2.5) | 370 |
| T | Quantitative HPLC in *Papaver bracteatum* | $SiO_2$ | Toluene-$Me_2CO$-EtOH-conc. $NH_4OH$(20:20:3:1)<br>Benzene-$Me_2CO$-MeOH(7:2:1) | 377 |
| | | $Al_2O_3$ | Benzene-EtOH(9:1) | |

*For abbreviations, see footnote to Table 12.7.

References p. 245

TABLE 12.9

TLC ANALYSIS IN THE QUALITY CONTROL OF OPIUM ALKALOIDS

| Alkaloid* | Aim | Adsorbent | Solvent system | Ref. |
|---|---|---|---|---|
| C,P | Detection of M in C and papaverinol in P | $Al_2O_3$ | Benzene-MeOH(9:1) + 1 drop 80% HCOOH<br>Benzene-MeOH(8:2)<br>Benzene-AcOH(9:1) | 29 |
| C | | $SiO_2$ | $Me_2CO$-5 *M* $NH_4OH$(9:1) | 105 |
| M | | $SiO_2$ | $CH_2Cl_2$-MeOH-10% $NH_4OH$(84:14:2) | 115 |
| Nalo | Decomposition in injections | $SiO_2$ | EtOH-dioxane-benzene-25% $NH_4OH$(4:8:7:1) | 120 |
| EtM | Detection of diEtM in EtM | $Al_2O_3$ | $CHCl_3$-*n*-heptane(9:1)<br>Benzene-EtOH(95:5) | 152 |
| M | Detection of C in M | $SiO_2$ | $CHCl_3$-$Me_2CO$-25% $NH_4OH$(12:24:1) | 165 |
| M | C and pseudoM in M | | No details available | 229 |
| M | Assay of M in injections (M N-oxide, pseudoM, atropine, methylparaben) | $SiO_2$ | $CHCl_3$-MeOH-DEA(80:15:5) | 251 |
| M,H | Purity profiles (impurities: C,AcC,3OAcM,6OAcM, pseudoM) | $SiO_2$ | EtOH-AcOH-$H_2O$(6:3:1)<br>MeOH-25% $NH_4OH$(200:3)<br>EtOH-pyridine-dioxane-$H_2O$(10:4:5:1)<br>Benzene-dioxane-EtOH-25% $NH_4OH$(10:8:1:1)<br>EtOAc-DMFA(3:1)<br>MeOH<br>IsoprOH-$H_2O$-AcOH(8:1:1) | 252 |
| | | cellulose | MeOH | |
| P | Detection of oxidation products (papaverinol, papaveraldine) and indirect quantitative analysis (UV) | $SiO_2$ | $CHCl_3$ sat. with $NH_3$ | 264 |
| C,M,Na,No,P,T, DiHC,apoM,EtM, Naline | Purity control (including diEtM, papaverinol, papaveraldine) (Table 12.2) | $SiO_2$ | Xylene-MeEtCO-MeOH-DEA(40:40:6:2)<br>$CCl_4$-*n*-BuOH-MeOH-6 *M* $NH_4OH$(40:30:30:2)<br>$CH_2Cl_2$-MeOH-6 *M* $NH_4OH$(85:15:2) | 267 |
| P | Detection of oxidation products (papaverinol, papaveraldine), semi-quantitative, also in injections | $SiO_2$ | Cyclohexane-$CHCl_3$-MeOH-DEA(70:20:5:10)<br>Xylene-MeEtCO-MeOH-DEA(20:20:3:1) | 306 |

References p. 245

| | | | | |
|---|---|---|---|---|
| M | Detection of foreign alkaloids | | No details available | 333 |
| C,M,P | DiMeM in C, C in M | $SiO_2$ | Toluene-EtOAc-DEA(7:2:1) | 361 |
| P | Decomposition of P (papaverinol,papaveraldine) | $SiO_2$ | Benzene-dioxane(8:2)<br>$CHCl_3$-cyclohexane-DEA(4:5:1)<br>$CHCl_3$-*n*-hexane-MeOH(52:40:8)<br>EtOAc-benzene-25% $NH_4OH$(60:35:5)<br>$CHCl_3$-$Me_2CO$-25% $NH_4OH$(75:25:1)<br>Benzene-$CHCl_3$-$Me_2CO$-25% $NH_4OH$(70:15:15:1) | 399 |

*For abbreviations, see footnote to Table 12.7.

TABLE 12.10

TLC ANALYSIS OF OPIUM ALKALOIDS IN PHARMACEUTICAL PREPARATIONS AND AS PURE COMPOUNDS

| Alkaloid* | Other compounds | Aim | Adsorbent | Solvent system | Ref. |
|---|---|---|---|---|---|
| C | Ilvin,decentan,ascorbic acid,atp,salac,N-Me-ephedrine | Separation | $SiO_2$ | MeOH-acetate buffer (pH 4.62) (3:7) | 2 |
| C,P | thph,caf | Identification | $SiO_2$ | Benzene-EtOH-AcOH(80:12:5) | 4 |
| C,M,No,P,T,diHMone, diHCone,diHC,oxyC, EtM | | Separation (Table 12.3) | $SiO_2$ | $CHCl_3$-96% EtOH(8:2)<br>EtOAc-EtOH-DMFA-DEA(75:20:5:2) | 5 |
| | | | 0.5 *M* KOH-impregnated $SiO_2$ | $CHCl_3$-96% EtOH(8:2) | |
| | | | FMA-impregnated cellulose | Benzene-heptane-$CHCl_3$-DEA(6:5:1:0.02) | |
| C,EtM | caf,amp,atp,phen,phb, Na benzoate,belladona extract | Identification | $SiO_2$ | $Me_2CO$-cyclohexane(5:4)<br>$Me_2CO$-hexane(5:4)<br>EtOH-$NH_4OH$(4:1)<br>$CHCl_3$-EtOH(4:1)<br>$Me_2CO$-MeOH-DEA(1:1:0.03, 2:1:0.03)<br>Benzene-dioxane-AcOH(5:4:1)<br>$Me_2CO$-$CHCl_3$-DEA(2:1:0.03)<br>$CHCl_3$-benzene-AcOH(5:4:1)<br>EtOAc-benzene(11:5)<br>EtOAc-MeOH-AcOH(8:1:1) | 6 |
| C | amp,phen,caf,phb,Acsalac | Separation | $SiO_2$ | $Et_2O$, EtOAc, $CHCl_3$-EtOH(100:1) | 11 |
| M | | TLC on microslides | $SiO_2$ | *n*-BuOH-AcOH(9:1), sat. with $H_2O$ | 12 |
| C,M,P | | TLC on glass rods | 0.5 *M* NaOH-impregnated $SiO_2$ | $CHCl_3$-EtOH(9:1) | 15 |
| P | amp,atropine | Identification | $SiO_2$ | Benzene-MeOH(1:2) | 20 |

References p. 245

| | | | | | |
|---|---|---|---|---|---|
| C,P,EtM | caf,thph,Me-homatropine | Separation | $SiO_2$ | MeOH-$NH_4OH$(99:1)<br>$CHCl_3$-EtOH(9:1)<br>$CHCl_3$-$NH_4OH$(95:5)<br>MeOH-$CHCl_3$-$NH_4OH$(90:5:5)<br>MeOH-light petroleum-$Et_2O$(1:1:1, 25:100:100)<br>MeOH-AcOH(9:1) | 21 |
| C,M,Na,No,P,T,H,<br>diHMone,diHCone,<br>AcdiHCone,oxyC,EtM | | Separation | $SiO_2$ | I. $CHCl_3$-EtOH(9:1) sat. with 10% $NH_4OH$<br>II. $CCl_4$-*n*-BuOH-MeOH-10%$NH_4OH$(40:30:30:2)<br>III. Light petroleum-$Et_2O$-EtOH-DEA(4:16:2:1)<br>Two-dimensional: I,II or III,II | 28 |
| M,norM,Nalo | | Separation | $SiO_2$ | $CHCl_3$-isoprOH(1:3) | 36 |
| M,H,norM,diHM,6OAcM,<br>nalo,meth,mep | | Specific detection of phenolic phenanthrene derivatives | NaOH-impregnated $SiO_2$<br>(pH 8.5) | MeOH-*n*-BuOH-benzene-$H_2O$(6:1.5:1:1.5) | 38 |
| C,M,Na,No,H,diHMone,<br>diHCone,diHC,oxyC,<br>EtM,AcdiHC, C N-ox,<br>nalo,AcdiHCone,<br>meth,normeth,mep,<br>dex,keto,dromoran,<br>Dmethphan | Alphaprodine,<br>carbetidine | Identification<br>(Table 12.3) | $SiO_2$ | $Me_2CO$-MeOH-17% $NH_4OH$(50:50:1)<br>$CHCl_3$-EtOH-17% $NH_4OH$(9:1:1)<br>$CHCl_3$-$Me_2CO$-17% $NH_4OH$(50:50:1)<br>MeOH-$CHCl_3$-17% $NH_4OH$(85:15:1)<br>Benzene-$Me_2CO$-light petroleum-17% $NH_4OH$<br>(35:35:35:1)<br>Benzene-EtOH-17% $NH_4OH$(85:15:1) | 39 |
| C,EtM | | Identification in syrups | $SiO_2$ | I. $CHCl_3$-MeOH-$Et_2O$-25% $NH_4OH$(60:25:25:1)<br>II. *n*-BuOH-AcOH(100:14) sat. with $H_2O$<br>Two-dimensional: I,II | 40 |
| C,P | amp,atp,phen,par,AcSalac,<br>Salam,caf,quinine,methampyrone,phenylbutazone | Identification | $SiO_2$ | BuOAc-$CHCl_3$-85% HCOOH(6:4:2)<br>BuOAc-$Me_2CO$-*n*-BuOH-10% $NH_4OH$(5:4:3:1) | 44 |
| M,P | Quinine | Indirect quantitative analysis, circular TLC | $SiO_2$ | MeOH-$Me_2CO$-25% $NH_4OH$(100:45:5)<br>$CHCl_3$-EtOH-25% $NH_4OH$(3:1:0.2) | 47 |
| C,M,Na,No,P,T,EtM | | Separation and indirect quantitative analysis | $CaCl_2$-impregnated MgO | Hexane-$Me_2CO$(3:1)<br>EtOAc-pyridine(10:1)<br>$CHCl_3$-EtOAc(3:1, 1:1, 1:3) | 48 |
| | | | $MgSO_4$-impregnated MgO | EtOAc, $CHCl_3$-EtOAc(3:1, 1:1) | |
| C,M,Na,No,P,T,H,<br>apoM,EtM,Ph | | Reaction chromatography | $SiO_2$ | $CHCl_3$-$Me_2CO$-DEA(5:4:1)<br>$CHCl_3$-cyclohexane-DEA(7:2:1) | 50 |
| C | Atropine,aconitine,<br>brucine | Preparative TLC on $CaSO_4$ | $CaSO_4$ | $CHCl_3$-AmOH-toluene-conc.HCl(50:1.5:1.5:0.25) | 81 |

TABLE 12.10 (*continued*)

| Alkaloid* | Other compounds | Aim | Adsorbent | Solvent system | Ref. |
|---|---|---|---|---|---|
| C,No,P,T | Emetine,cephaeline, atropine | Identification in syrups, reaction chromatography | $SiO_2$ | $CHCl_3$-$Me_2CO$-DEA(5:4:1) | 84 |
| C,M,No,P,T,Naline, reticuline, laudanidine | Emetine,cephaeline | Identification (Table 12.2) | $SiO_2$ | Light petroleum-$Et_2O$-EtOH-DEA(4:16:2:1)<br>$CCl_4$-*n*-BuOH-MeOH-10% $NH_4OH$(40:30:30:2)<br>Xylene-MeEtCO-MeOH-DEA(20:20:3:1) | 86 |
| C | Analgesics,antipyretics | Separation of C from interfering substances | $SiO_2$ | $Me_2CO$-5 *M* $NH_4OH$(9:1) | 89 |
| C,M,No,P,EtM | | Identification | $SiO_2$ | Benzene-$Me_2CO$-EtOH-25% $NH_4OH$(5:4:0.5:0.5) | 91 |
| C,M,No,P,H,diHC,EtM | | Separation on polyamide | Polyamide | $H_2O$-abs.EtOH-DMA(88:12:0.1)<br>Cyclohexane-EtOAc-PrOH-DMA(30:2.5:0.9:0.1) | 93 |
| No,P | | Indirect quantitative analysis (UV) | $SiO_2$ | EtOAc | 95 |
| C | amp,phb,caf,methampyrone | Identification | | No details available | 96 |
| C,P,EtM | Procaine | Identification | Talc | $H_2O$-$Me_2CO$-EtOH-$NH_4OH$(30:1:10:5)<br>$H_2O$-EtOH-$CHCl_3$-$NH_4OH$(30:10:5:1) | 99 |
| C,M,Na,No,P,T,H, diHMone,diHCone, oxyC,EtM | | Separation | 0.5 *M* KOH-impregnated $SiO_2$ | $CHCl_3$-MeOH-$Me_2CO$(7:5:1) | 100 |
| C,M,Na,No,P,T | | Separation in opialum | $SiO_2$ | Benzene-MeOH-isoprOH(9:1:1) | 101 |
| C | amp,atp,phen,caf, AcSalac | Identification in analgesics | $SiO_2$ | $Me_2CO$, 0.5 *M* NaOH, $CHCl_3$, $H_2O$<br>BuOH-AcOH(4:1:5)<br>BuOH-$H_2O$(1:1) | 106 |
| C,M,Na,No,P,T | | Identification | $SiO_2$ | I. Benzene-EtOAc-DEA(35:10:8)<br>II. Xylene-MeEtCO-MeOH-DEA(20:20:3:1)<br>III. $Et_2O$ ($H_2O$ sat.)-$Me_2CO$-DEA(85:8:7)<br>IV. MeOH-$CHCl_3$-23% $NH_4OH$(85:15:0.7) | 107 |

| | | | | | |
|---|---|---|---|---|---|
| C,No,P | | Identification | $SiO_2$ | $CHCl_3$-MeOH(9:1) in $NH_3$ atm.<br>Light petroleum-$CHCl_3$(2:8) in $NH_3$ atm. | 110 |
| | | | 5% paraffin-impregnated $SiO_2$ | MeOH-$H_2O$(6:4) | |
| | | | 20% ethylene glycol-impregnated $SiO_2$ | Light petroleum-$CHCl_3$(9:1) | |
| C,EtM | eph | Indirect quantitative analysis, identification | $SiO_2$ | $CHCl_3$-$Et_2O$-MeOH-17% $NH_4OH$(15:15:4:12.5, 15:2:2:0.2) in $NH_3$ atm. | 113 |
| P | Atropine,scopolamine, strychnine,brucine | Ion pair chromatography | $Cl^-$ in 0.7 *M* $H_2SO_4$-impregnated cellulose | $CHCl_3$ | 116 |
| Leverphanol,nalo, phenazocine | | Identification | $SiO_2$ | BuOH-$H_2O$-AcOH(100:48:20)<br>Benzene-$Me_2CO$-$Et_2O$-10% $NH_4OH$(40:60:10:3)<br>$CHCl_3$-benzene-isoprOH-conc. $NH_4OH$(45:45:8:10) | 117 |
| M,P | Atropine,procaine | Densitometric analysis | $SiO_2$ | MeOH-$CHCl_3$-25% $NH_4OH$(20:10:1)<br>$Me_2CO$-$CHCl_3$-25% $NH_4OH$(24:12:1) | 122 |
| M,Naline,H,pseudoM, EtM,diHMone,nalo, lemoran | Atropine,scopolamine | Separation | $SiO_2$ | $CCl_4$-isoBuOH-MeOH-20% $NH_4OH$(40:30:30:4, 40:15:15:4, 20:25:5:2)<br>$CCl_4$-*n*-BuOH-MeOH-20% $NH_4OH$(20:30:30:3)<br>$Cl_2CHMe$-MeOH-20% $NH_4OH$(60:20:3, 10:90:4)<br>$Cl_2CHMe$-isoBuOH-MeOH-20% $NH_4OH$(30:15:15:3)<br>$Cl_2CHMe$-*n*-BuOH-MeOH-20% $NH_4OH$(60:15:15:1)<br>$Cl_2CHMe$-$Me_2CO$-MeOH-20% $NH_4OH$(50:30:10:3) | 123 |
| C,M,Na,P,T | | pH-gradient TLC | $SiO_2$ impregnated with pH gradient | $CHCl_3$-MeOH(75:25)<br>$CHCl_3$-MeOH(8:2), twice | 126 |
| M,No,P | Various alkaloids | Semi-quantitative circular TLC of various alkaloids | $SiO_2$ | $CHCl_3$-MeOH(9:1)<br>$CHCl_3$-EtOH(85:15) | 133 |
| C | phen,AcSalac | Identification in suppositories | $SiO_2$ | Benzene-$Et_2O$-AcOH(40:30:9) | 134 |
| C,P | Strychnine,atropine, methylatropine,xanthines, eph,procaine,phb,chlorpromazine,tripelennamine, melipramin | Separation | $SiO_2$ | $Me_2CO$-cyclohexane-EtOAc(1:1:1) in $NH_3$ atm. | 136 |

References p. 245

TABLE 12.10 (*continued*)

| Alkaloid* | Other compounds | Aim | Adsorbent | Solvent system | Ref. |
|---|---|---|---|---|---|
| P | Strychnine | Ion-pair chromatography | 10% acetylcellulose impregnated with 0.5 g/g *n*-BuOH | 0.1 *M* NaCl | 137 |
| | | | 10% acetylcellulose impregnated with 0.5 g/g AmOH | 0.1 *M* NaCl, NaBr or $NaClO_4$ in 0.5 *M* $H_2SO_4$ | |
| C,M,No,P | | Indirect quantitative analysis (UV) | $SiO_2$ | Cyclohexane-$CHCl_3$-DEA(5:4:1) | 139 |
| C | amp,phen,phb,AcSalac, caf,quinine,barb | Identification | $SiO_2$ | $Me_2CO$<br>NaOH | 140 |
| C,M,P | amp,phb,atropine | Identification | $SiO_2$ | Benzene-MeOH-AcOH(7:1:2)<br>Benzene-$Me_2CO$-EtOH-25% $NH_4OH$(5:4:0.5:0.5) | 144 |
| C,M,No,P | | Indirect quantitative analysis (UV) | $SiO_2$ | Benzene-EtOH(4:1)<br>$CHCl_3$-MeOH(9:1) | 157 |
| M | Atropine,scopolamine | Identification | $SiO_2$ | $CHCl_3$-MeOH(3:2) | 161 |
| C,M,diHMone,diH diHCone,diHC,AcC, AcdiHCone | | Separation | $SiO_2$ | $CCl_4$-isoBuOH-MeOH-20% $NH_4OH$(40:30:30:4)<br>$CCl_4$-BuOH-MeOH-20% $NH_4OH$(40:15:15:4, 20:25:5:2)<br>$Cl_2CHMe$-isoBuOH-MeOH-20% $NH_4OH$(30:15:15:3)<br>$Cl_2CHMe$-MeOH-20% $NH_4OH$(60:20:3, 10:90:4)<br>$Cl_2CHMe$-BuOH-MeOH-20% $NH_4OH$(60:15:15:4)<br>$Cl_2CHMe$-$Me_2CO$-MeOH-20% $NH_4OH$(50:30:10:3) | 163 |
| C,P,diHC,oxyC | amp,phb,caf,eph,phen, scopolamine,pentylenetetrazole | Identification | $SiO_2$ | $CCl_4$-isoBuOH-MeOH-20% $NH_4OH$(40:30:30:4, 40:15:15:4, 20:25:5:2)<br>$CCl_4$-*n*-BuOH-MeOH-20% $NH_4OH$(20:30:30:3)<br>$Cl_2CHMe$-isoBuOH-MeOH-20% $NH_4OH$(30:15:15:3)<br>$Cl_2CHMe$-MeOH-20% $NH_4OH$(60:20:3, 10:90:4)<br>$Cl_2CHMe$-*n*-BuOH-MeOH-20% $NH_4OH$(60:15:15:4)<br>$Cl_2CHMe$-$Me_2CO$-MeOH-20% $NH_4OH$(50:30:10:3)<br>$Cl_2CHMe$-MeOH(9:1) | 164 |
| C,M,No,P,diHMone, diHcone,diHC,oxyC, AcdiHCone,meth, mep,EuP | | Separation on $CaCO_3$, MgO and $MgCO_3$ | 2.5% $CaCl_2$-impregnated $CaCO_3$ | $CHCl_3$-EtOAc(1:1) | 170 |

| | | | | | |
|---|---|---|---|---|---|
| P | Salsoline | Indirect quantitative analysis (spectrophotometric) | $SiO_2$ | $Me_2CO$-conc. $NH_4OH$(95:5)<br>EtOH-$CHCl_3$(3:2)<br>$CHCl_3$-BuOH(9:1)<br>$CHCl_3$-MeOH-conc. $NH_4OH$(15:30:1)<br>$Me_2CO$-MeOH-conc. $NH_4OH$(20:5:1) | 179 |
| P | Diphenhydramine | Indirect quantitative analysis (spectrophotometric) | $SiO_2$ | $CHCl_3$-BuOH(9:1)<br>$Me_2CO$-MeOH(1:2)<br>EtOH-$CHCl_3$(2:3) | 180 |
| C | amp,phen,phb,caf | TAS technique | | No TLC details | 185 |
| C,M,No,P | | Semi-quantitative analysis (spot-size) | | No details available | 189 |
| P | phb,eph,xanthine derivatives,benzocaine, prednisone | Separation | $SiO_2$ | $CH_2Cl_2$-MeOH-AcOH(90:10:3) | 193 |
| No,P | Atropine,scopolamine | Identification | $SiO_2$ | $CHCl_3$-$Me_2CO$(3:1), $CHCl_3$-MeOH(4:1), $CHCl_3$-DEA(9:1), MeOH-28% $NH_4OH$(100:1.5) | 197 |
| C,P | amp,phb,xanthines | TLC on microslides | $SiO_2$ | MeOH<br>Benzene-$Me_2CO$ in $NH_3$atm. | 198 |
| M,meth | Quinine | Separation | $SiO_2$ | EtOAc-MeOH-conc. $NH_4OH$(425:50:25) | 200 |
| C,P,mep | amp,phen,phb,AcSalac, caf | Identification | $SiO_2$ | $Et_2O$, EtOAc in $NH_3$atm.<br>EtOAc-$CHCl_3$-HCOOH (98%)(6:4:2)<br>EtOAc-EtOH-25% $NH_4OH$(100:5:5)<br>70% EtOH-25% $NH_4OH$(97.5:2.5) | 217 |
| C,Na,No,P,EtM, normeth,benzylM | Various | Identification in cough preparations (see also ref. 258) | $SiO_2$ | $(Isopr)_2O$-$Me_2CO$(1:1) in $NH_3$ atm.<br>IsoprOH in $NH_3$ atm.<br>Benzene-dioxane-AcOH(5:4:1)<br>Cyclohexane-EtOAc-DEA(65:30:5) | 218 |
| C,P | amp,atp,phen,phb,caf, quinine,monobromocamphor, *Belladonna* extract | Densitometric analysis | $SiO_2$ | $CHCl_3$-$Me_2CO$(2:1)<br>Cyclohexane-$Me_2CO$(4:5) | 219 |
| C,M,No,H,diHMone, diHCone,oxyC,meth, mep,dex | | Specific detection of M | $SiO_2$ | $CHCl_3$-isoprOH-DEA(18:20:10)<br>MeOH-25% $NH_4OH$(100:1) | 222 |
| C,M,T,diHC,oxyC, EtM | | Adsorption mechanism on $SiO_2$ | $SiO_2$ | 31 different mobile phase, no details available | 224 |
| P | Xanthines | Separation on microslides | | No details available | 228 |

References p. 245

TABLE 12.10 (*continued*)

| Alkaloid* | Other compounds | Aim | Adsorbent | Solvent system | Ref. |
|---|---|---|---|---|---|
| C,M,Na,No,P,T, crypt,H,AcC, 3OAcM,6OAcM | | Separation | $SiO_2$ | $Et_2O$ (sat. with $H_2O$)-$Me_2CO$-DEA(85:8:7) | 231 |
| C,P | | Indirect quantitative analysis (UV) | $SiO_2$ | EtOAc-benzene-$NH_4OH$(60:35:5)<br>$CHCl_3$-MeOH-$Et_2O$-$NH_4OH$(15:2:2:0.2) | 232 |
| C,M,Na,No,diHMone, diHCone,diHC,EtM | | New detection method | Cellulose | *n*-BuOH-AcOH-$H_2O$(10:1:3) | 236 |
| M,meth,mep,lev,dex | Promedol,isopromedol | TLC characteristics | | No details available | 237 |
| EtM,benzylM,ph,Naline | | Chromatographic behaviour | | No details available | 238 |
| C,No | phb,caf,diphenhydramine, phenylephedrine,thiamine, etophylline | Densitometric analysis | $SiO_2$ | Cyclohexane-$CHCl_3$-DEA(90:18:12) | 246 |
| P,diHC | caf,quinine,cocaine, pilocarpine,various others | Reaction chromatography | $SiO_2$ | Benzene-MeOH-$Me_2CO$-AcOH(70:20:5:5) | 254 |
| C | Butethamate | Separation | $SiO_2$ | Toluene-$Me_2CO$-EtOH-25% $NH_4OH$(45:45:7:3) | 257 |
| C,Na,diHC,EtM, benzylM | Various | Identification in cough preparations | $SiO_2$ | IsoprOH in $NH_3$ atm.<br>MeOH<br>Ligh petroleum(b.p. 40-60°C)-$Et_2O$-abs.EtOH-DEA(16:64:8:4)<br>Pyridine-$Me_2CO$(15:85) | 258 |
| C,meth,mep | Reserpine | Densitometric analysis, content uniformity test | $SiO_2$ | $CHCl_3$-$Me_2CO$-25% $NH_4OH$(32:65:7)<br>Benzene-dioxane-EtOH-25% $NH_4OH$(50:40:5:5) | 259 |
| C,M,Na,No,P,T | | Identification | $SiO_2$ | $Me_2CO$-xylene-MeOH-$NH_4OH$(50:40:6:5) | 263 |
| C,M,P,EtM | | Identification in syrups | $SiO_2$ | $CHCl_3$-$Me_2CO$-EtOH containing $NH_4OH$(pH 12.4) (10:8:2) | 265 |
| C | amp,phen,phb,caf,barb | Separation on microslides | FMA-impregnated talc | Benzene, toluene | 268 |
| OxyC | Parkopan | Identification | $SiO_2$ | MeOH-$NH_4OH$(100:15)<br>EtOH-AcOH-$H_2O$(6:3:1) | 270 |

| | | | | | |
|---|---|---|---|---|---|
| M,pseudoM | Atropine | Decomposition of M in injections | $SiO_2$ | Benzene-$Me_2CO$-70% EtOH-25% $NH_4OH$ (35:32.5:35:2.5) | 291 |
| C,M,H,6OAcM | Quinine,cocaine,caf, AcSalac | Separation | $SiO_2$ | $CHCl_3$-xylene-isoprOH-EtOH(50:25:12.5:12.5) | 296 |
| C,M,No,P,H,6OAcM, EtM | | Reaction chromatography | $SiO_2$ | Toluene-$Me_2CO$-MeOH-25% $NH_4OH$(40:45:10:5) | 300 |
| C,M,No | eph,emetine, cephaeline | Indirect quantitative analysis as dansyl derivatives | $SiO_2$ | Benzene-MeOH(3:1)<br>Toluene-MeOH-$Me_2CO$(9:1:1) | 301,353 |
| C | caf,phb,promethazine | Identification | $SiO_2$ | Hexane-95% EtOH-DEA(25:10:5) | 302,308 |
| P | phb,quinine,quinidine | Identification on microslides | | No details available | 304 |
| C | amp,phen,phb,caf,ergotamine,mecloxamine | Identification on microslides | | No details available | 305 |
| M | Parkopan | Identification | $SiO_2$ | $CHCl_3$-MeOH(1:1) | 309 |
| P | phb,caf,salsoline, platyphylline,theobromine | Identification | $Al_2O_3$ | Benzene-MeOH(9:1) | 311 |
| C,P,EtM | | Separation with single-component mobile phases | $SiO_2$ | MeOH, EtOH, PrOH, BuOH, AmOH, $Me_2CO$, MeEtCO, MePrCO, AmOAc, $Et_2O$, $isopr_2O$, $Bu_2O$, $CH_2Cl_2$, $CHCl_3$, $CCl_4$, benzene, cyclohexane | 313 |
| C,meth | Caf | TLC-MS for identification | $SiO_2$ | EtOH-$H_2O$-AcOH(50:20:30) | 314 |
| C,T | Various alkaloids | FID for TLC | Sintered $SiO_2$<br>Sintered $Al_2O_3$ | $CHCl_3$-DEA(30:1)<br>Benzene-$CHCl_3$-DEA(36:8:1) | 320 |
| C,M,No,P | | Identification in syrups | $SiO_2$ | $Me_2CO$-xylene-MeOH-$NH_4OH$(50:40:6:5) | 331 |
| C | amp,phen,phb,caf,barb | New detection methods | | No details available | 335,346 |
| C,M,Na,No,P,T,H, EtM | | Detection method | $SiO_2$ | $CHCl_3$-benzene-MeOH-AcOH(4:3:2:1)<br>MeOH-$H_2O$-AcOH(6:2:2)<br>$CHCl_3$-DEA(8:2)<br>Benzene-$CHCl_3$-EtOAc-DEA(6:2:1:1)<br>Benzene-heptane-$CHCl_3$-DEA(4:2:3:1)<br>EtOAc-EtOH-DMFA-DEA(12:6:1:1) | 344 |
| C | Strychnine | Identification in pills | $SiO_2$ | $CHCl_3$-$Me_2CO$-conc. $NH_4OH$(12:24:1) | 347 |
| C | Chlorpheniramine | Densitometric analysis in antitussives and plasma | $SiO_2$ | MeOH, followed by EtOAc-EtOH-conc. $NH_4OH$(40:4:3) | 354 |

References p. 245

TABLE 12.10 (*continued*)

| Alkaloid* | Other compounds | Aim | Adsorbent | Solvent system | Ref. |
|---|---|---|---|---|---|
| P | phb,caf | Indirect quantitative analysis | | No details available | 357 |
| C | | Indirect quantitative analysis in syrups | $CaSO_4$ or $CaSO_4$-$CaCO_3$ (1:1) | Cyclohexane-DEA(9:1) | 358 |
| P | Atropine,butethamate | | | No details available | 362 |
| No | | Fluorodensitometric analysis | $SiO_2$ | $Me_2CO$-benzene-$CHCl_3$(1:3:3) | 365 |
| M | Methylatropine, scopolamine | Indirect quantitative analysis (spectrophotometric) | $SiO_2$ | MeOH<br>EtOH-$H_2O$(95:5) | 371 |
| C,M,P | Various alkaloids | Detection with π-acceptors | $SiO_2$ | $Me_2CO$-toluene-MeOH-$NH_4OH$(45:40:10:5) | 375 |
| C,M,diHCone | | Separation | | No details available | 383 |
| C | eph,nipagin,romergan, ascorbic acid | Identification in syrups | $SiO_2$ | I. Benzene-$Me_2CO$-MeOH(6:1.5:2.5)<br>II. Benzene-$Me_2CO$-EtOH-25% $NH_4OH$(5:4:0.5:0.5)<br>Two dimensional: I,II | 385 |
| C,M,diHC,EtM | | Identification | $SiO_2$ | $CHCl_3$-$Me_2CO$-DEA(5:4:1) | 400 |

*For abbreviations, see footnote to Table 12.7.

TABLE 12.11

TLC ANALYSIS OF VARIOUS COMPOUNDS EXTRACTABLE FROM URINE AT pH 9.0 THAT REACT WITH IODOPLATINATE REAGENT[16]

TLC systems:

S1 Silica gel G, activated — Ethanol-pyridine-dioxane-water (50:20:25:5)
S2 Silica gel G, activated — Ethanol-acetic acid-water (6:3:1)
S3 Silica gel G, activated — Ethanol-dioxane-benzene-ammonia (5:40:50:5)
S4 Silica gel G, activated — Methanol-*n*-butanol-benzene-ammonia (60:15:10:15)
S5 Aluminium oxide G, activated — *n*-Butanol-di-*n*-butyl ether-acetic acid (4:5:1)
S6 Aluminium oxide G, activated — *n*-Butanol-di-*n*-butyl ether-ammonia (25:70:5)

| Compound | $hR_F$ values | | | | | |
|---|---|---|---|---|---|---|
| | S1 | S2 | S3 | S4 | S5 | S6 |
| Mescaline | 18 | 81 | 30 | 30 | 36 | 53 |
| Morphine | 40 | 55 | 17 | 34 | 59 | 18 |
| Normorphine | 14 | 76 | 5 | 16 | 33 | 11 |
| Meperidine | 51 | 64 | 90 | 44 | 77 | 98 |
| Cocaine | 74 | 45 | 98 | 39 | 61 | 96 |
| Heroin | 47 | 54 | 80 | 43 | 72 | 83 |
| Nicotine | 58 | 27 | 90 | 44 | 72 | 90 |
| Codeine | 40 | 52 | 46 | 32 | 72 | 65 |
| Levorphanol | 34 | 75 | 62 | 29 | 76 | 91 |
| Methadone | 53 | 78 | 96 | 20 | 78 | 98 |
| Chlorpromazine | 48 | 79 | 93 | 35 | 80 | 98 |
| Tripelennamine | 50 | 46 | 93 | 28 | 76 | 98 |
| Phenazocine | 90 | 93 | 93 | 59 | 86 | 98 |
| Nalorphine | 82 | 71 | 34 | 75 | 72 | 20 |
| Propoxyphene | 80 | 82 | 97 | 53 | 84 | 100 |
| Dihydromorphinone | 19 | 28 | 22 | 18 | 65 | 25 |

TABLE 12.12

TLC DATA FOR VARIOUS DRUGS EXTRACTABLE FROM URINE[102]

TLC system:
Silica gel G Ethyl acetate-methanol-conc. ammonia (85:10:5)

| Compound | $hR_F$* | Colour** | | | | |
|---|---|---|---|---|---|---|
| | | Ninhydrin | Diphenyl-carbazone-$HgSO_4$ | Oven | Ultraviolet light (254 nm) | Iodopla-tinate-Dragendorff |
| Analgesics: | | | | | | |
| Acetaminophen | NR | | | | | |
| Cocaine | 96 | | | | | OR |
| Phenacetin | 98 | | | | | Br |
| Procaine | 97 | | | | | RV |
| Propoxyphene | 90 | | | | | O |
| Salicylic acid | NR | | | | | |
| Antihistamines: | | | | | | |
| Bromodiphenhydramine | 95 | | | | | RBr |
| Carbinoxamine | 75 | | | | B | BV |
| Chlorpheniramine | 88 | | | | | RV |
| Chlorcyclizine | 99 | | | | | BV |
| Diphenhydramine | 90 | | | | | V |
| Dimenhydrinate | 80 | | | | | OR |
| Dimethindene | 82 | | | | | BV |
| Diphenylpyraline | 95 | | | | | OR |
| Meclizine | 95 | | | | | Pu |
| Methapyrilene | 94 | | | | | Pu |
| Pheniramine | 92 | | | | | RV |
| Phenyltoloxamine | 99 | | | | | RV |
| Promethazine | 95 | | | O | | BV |
| Pyrathiazine | 97 | | | P | | BV |
| Pyrilamine | 90 | | | | | Br |
| Pyrrobutamine | 99 | | | | | LR |
| Rotoxamine | 92 | | | | LB | Br |
| Thenyldiamine | 94 | | | | | Pu |
| Tripelennamine | 96 | | | | | RV |
| Tripolidine | 90 | | | | LB | Pu |
| Hypnotics and sedatives: | | | | | | |
| Amobarbital | 75 | | P | | | |
| Butabarbital | 73 | | P | | | |
| Carbromal | NR | | | | | |
| Diallylbarbituric acid | 73 | | P | | | |
| Diphenylhydantoin | 75 | | P | | | |
| Ethinamate | NR | | | | | |
| Glutethimide | 99 | | P | | | |
| Heptabarbital | 72 | | V | | | |
| Hexobarbital | 67 | | P | | | |
| Methyprylon | 90 | | | | | P |
| Pentabarbital | 75 | | Pu | | | |
| Phenobarbital | 46 | | P | | | |
| Secobarbital | 75 | | P | | | |
| Talbutal | 82 | | V | | | |
| Thiamylal | 75 | | P | | | |

TABLE 12.12 (*continued*)

| Compound | $hR_F$* | Colour** | | | | |
|---|---|---|---|---|---|---|
| | | Ninhydrin | Diphenyl-carbazone-$HgSO_4$ | Oven | Ultraviolet light (254 nm) | Iodoplatinate-Dragendorff |
| Opium alkaloids: | | | | | | |
| Apomorphine | NR | | | | | |
| Codeine | 54 | | | | | RV |
| Dihydrocodeine | 50 | | | | | Pu |
| Dihydrocodeinone | 50 | | | | | Pu |
| Ethylmorphine | 63 | | | | | V |
| Morphine | 32 | | | | | DB |
| Noscapine | 99 | | | | | OPu |
| Papaverine | 99 | | | | | RV |
| Synthetic narcotics: | | | | | | |
| Heroin | 80 | | | | | RV |
| Meperidine | 90 | | | | | RV |
| Methadone | 99 | | | | | OR |
| Stimulants: | | | | | | |
| Amphetamine | 78 | P | P | P | | Br |
| Caffeine | 80 | | | | | B |
| Ephedrine | NR | | | | | |
| Methamphetamine | NR | | | | | |
| Methylphenidate | 92 | | | | | V |
| Phenylephrine | NR | | | | | |
| Phenylpropanolamine | 60 | V | V | | | |
| Strychnine | 63 | | | | | RV |
| Tranquillizers: | | | | | | |
| Acetophenazine | 70 | | | P | Br | BV |
| Amitriptyline | 98 | | | | | Br |
| Chlordiazepoxide | 88 | | | | B | RV |
| Chlormezanone | NR | | | | | |
| Chlorpromazine | 96 | | | P | | BV |
| Chlorprothixene | 99 | | | | Y | Pu |
| Diazepam | 98 | | | | | V |
| Ectylurea | 95 | | | | | LPu |
| Imipramine | 95 | | | | B | Pu |
| Mepazine | 95 | | | P | | Br |
| Meprobamate | 75 | | PF | | | Br |
| Nortryptyline | 75 | | | | | RV |
| Oxanamide | NR | | | | | |
| Perphenazine | 84 | | | P | | BV |
| Prochlorperazine | 82 | | | P | | |
| Promazine | 96 | | | O | | BV |
| Thiopropazate | 85 | | | P | | Pu |
| Thioridazine | 97 | | | P | | OBr |
| Trifluoperazine | 99 | | | P | | Pu |
| Trimeprazine | 95 | | | P | | V |
| Miscellaneous: | | | | | | |
| Atropine | 43 | | | | | Pu |
| Chloroquine | 60 | | | | | Br |
| Dicyclomine | 97 | | | | | RBr |
| Scopolamine | 84 | | | | | Pu |
| Nicotine | 90 | | | | | B |
| Quinacrine | 97 | | | | YG | Br |
| Quinidine | 65 | | | | LB | RV |
| Quinine | 65 | | | | LB | RV |
| Reserpine | NR | | | | | |

*NR = no reaction.
**B, blue; Br, brown; D, dark; G, green; L, light; p, pink; Pu, purple; R, red; O, orange, V, violet; Y, yellow; F, fades.

**References p. 245**

TABLE 12.13

TLC ANALYSIS OF SOME DRUGS[213]

Silica gel, Gelman pre-coated glass microfibre sheets.
Solvent systems:
S1 Ethyl acetate-cyclohexane-dioxane-methanol-water-ammonia (50:50:10:10:1.5:0.5)
S2 Ethyl acetate-cyclohexane-dioxane-methanol-water-ammonia (50:50:10:10:0.5:1.5)
S3 Ethyl acetate-cyclohexane-ammonia-methanol-water (70:15:2:8:0.5)
S4 Ethyl acetate-cyclohexane-methanol-ammonia (70:15:10:5)

| Drug | $hR_F$ values* | | | |
|---|---|---|---|---|
| | S1 | S2 | S3 | S4 |
| Acetylmethadol | 81,93** | 95 | 98 | 96 |
| Amphetamine | 31 | 53 | 52 | 83 |
| Chlordiazepoxide | 55 | 64 | 72 | 74 |
| Chlorpheniramine | 44 | 58 | 70 | 91 |
| Chlorpromazine | 71 | 87 | 88 | 92 |
| Cocaine | 88 | 93 | 95 | 99 |
| Codeine | 24 | 34 | 42 | 76 |
| Cyclazocine | 60 | 70 | 72 | 97 |
| Diazepam | 89 | 92 | 92 | 93 |
| Diphenhydramine (Benadryl) | 71 | 83 | 86 | 91 |
| Ephedrine | 17 | 26 | 34 | 68 |
| Hydromorphone | 15 | 15 | 24 | 25 |
| Imipramine | 62 | 77 | 85 | 96 |
| Iproniazid | 25,50** | 60,74** | 62,89** | 36,67** |
| Isoniazid | 22 | 28 | 35 | 38 |
| Lysergic acid diethylamide | 55 | 64 | 66 | - |
| Meperidine (Demerol) | 62 | 66 | 72 | 95 |
| Mescaline | 15 | 15 | 22 | - |
| Methadone | 83 | 91 | 94 | 98 |
| Methamphetamine | 22 | 46 | 48 | 81 |
| Methapyrilene (Histadyl) | 70 | 78 | 86 | 97 |
| Methaqualone | 88 | 91 | 95 | - |
| Methylphenidate (Ritalin) | 80 | 80 | 85 | 96 |
| Morphine | 13 | 14 | 24 | 45 |
| Naloxone | 78 | 74 | 69 | 76 |
| Pentazocine (Talwin) | 90 | 85 | 90 | 98 |
| Phenmetrazine (Preludin) | 43 | 48 | 55 | 91 |
| Pipradrol | 91 | 96 | 98 | 99 |
| Promazine | 60 | 73 | 77 | 86 |
| Propoxyphene (Darvon) | 88 | 95 | 98 | 93 |
| Quinine | 34 | 38 | 50 | 71 |
| Tetracycline | 0 | 0 | 0 | 0 |
| Thioridazine (Mellaril) | 73 | 80 | 82 | 97 |
| Trifluoperazine (Eskazine, Stelazine) | 77 | 79 | 83 | 85 |

*Each developing solvent was allowed to travel a distance of 10 cm.
**Showed two spots.

TABLE 12.14

IDENTIFICATION OF DRUGS OF ABUSE[356]

Silica gel $GF_{254}$, activated.
Solvent systems:
S1 Chloroform-diethyl ether-methanol-conc. ammonia (75:25:5:1)
S2 Ethyl acetate-*n*-propanol-conc. ammonia (40:30:3)
S3 Methanol-conc. ammonia (100:1.5)
S4 Ethanol-acetic acid-water (6:3:1)
S5 Benzene

| Compound | $hR_F$ values* | | | | | Visibility** | | | Spray reagents | | | | | | |
|---|---|---|---|---|---|---|---|---|---|---|---|---|---|---|---|
| | S1 | S2 | S3 | S4 | S5 | Visible | Longwave UV | Shortwave UV | Ninhydrin-phenyl-acetaldehyde (no. 75a) | Iodoplatinate | Ninhydrin-phenylacet-aldehyde-iodoplatinate | Ehrlich's | Ehrlich's-iodoplatinate | Mercuric chloride-diphenylcarbazone | Diazotized Benzidine |
| Alkaloids: | | | | | | | | | | | | | | | |
| Atropine | 0 | 11 | 18 | 47 | 0 | | | | | + | + | | | | |
| Caffeine | 45 | 46 | 72 | 68 | 0 | | | + | | | | | + | | |
| Cocaine hydrochloride | 81 | 81 | 65 | 47 | 0 | | | + | | + | + | | | | |
| Codeine phosphate | 27 | 25 | 38 | 47 | 0 | | | + | | + | + | | | | |
| Ephedrine sulphate | 2 | 14 | 32 | 63 | 0 | | | + | | | | | + | | |
| Heroin | 48 | 40 | 46 | 52 | 0 | | | + | | + | + | | | | |
| Lysergide | 42 | 56 | 73 | 69 | 0 | | + | + | | | | + | | | |
| Mescaline hydrochloride | 22 | 17 | 26 | 69 | 0 | | | + | + | + | + | | + | | |
| Morphine sulphate | 5 | 15 | 36 | 53 | 0 | | | + | | + | + | | | | |
| Nicotine salicylate | 68 | 47 | 64 | 34 | 0 | | | + | | + | + | | | | |
| Opium | 6 | 16 | 36 | 53 | 0 | | | + | | + | + | | | | |
| Papaverine | 86 | 72 | 72 | 65 | 0 | | + | + | | + | + | | | | |
| Physostigmine salicylate | 64 | 52 | 62 | 55 | 0 | | | + | | + | + | | | | |
| Psilocin | 22 | 33 | 36 | 60 | 0 | + | | + | + | + | + | + | + | | |
| Psilocybin | 0 | 41 | 5 | 31 | 0 | | | + | | | | | + | | |
| Quinine sulphate | 19 | 52 | 53 | 63 | 0 | | + | + | | + | + | | | | |
| Scopolamine hydrobromide | 21 | 27 | 67 | 52 | 0 | | | | | + | + | | | | |
| Strychnine | 28 | 77 | 16 | 51 | 0 | | | + | | + | + | | | | |
| Yohimbine hydrochloride | 43 | 79 | 83 | 71 | 0 | | + | + | | + | + | | | | |
| Non-alkaloids that give positive alkaloidal reactions: | | | | | | | | | | | | | | | |
| Lidocaine hydrochloride | 80 | 82 | 84 | 60 | 0 | | | + | | + | + | | | | |
| Meperidine | 65 | 65 | 53 | 60 | 0 | | | + | | + | + | | | | |
| Methadone hydrochloride | 64 | 77 | 38 | 72 | 0 | | | + | | + | + | | | | |
| Methapyrilene hydrochloride | 28 | 46 | 62 | 48 | 0 | | | + | | + | +*** | + | | | |
| Methaqualone | 85 | 80 | 81 | 91 | 0 | | | + | | | | | + | | |
| Methylphenidate hydrochloride | 68 | 73 | 65 | 74 | 0 | | | + | | + | + | | | | |
| Pentazocine hydrochloride | 50 | 83 | 60 | 82 | 0 | | | + | | + | + | | | | |
| Phencyclidine hydrochloride | 83 | 82 | 48 | 70 | 0 | | | + | | + | + | | | | |
| Procaine hydrochloride | 31 | 60 | 71 | 54 | 0 | | | + | | | | | + | | |
| Propoxyphene napsylate | 77 | 81 | 80 | 73 | 0 | | | ± | | + | + | | | | |
| Barbiturates: | | | | | | | | | | | | | | | |
| Amobarbital | 65 | 82 | 86 | 95 | 0 | | | + | | | | | | + | |
| Phenobarbital sodium | 49 | 77 | 85 | 95 | 0 | | | + | | | | | | + | |
| Secobarbital | 70 | 84 | 84 | 94 | 0 | | | + | | | | | | + | |

TABLE 12.14 (*continued*)

| Compound | $hR_F$ values* | | | | | Visibility** | | | Spray reagents | | | | | | |
|---|---|---|---|---|---|---|---|---|---|---|---|---|---|---|---|
| | S1 | S2 | S3 | S4 | S5 | Visible | Longwave UV | Shortwave UV | Ninhydrin-phenyl-acetaldehyde (no. 75a) | Iodoplatinate | Ninhydrin-phenylacet-aldehyde-iodoplatinate | Ehrlich's | Ehrlich's-iodoplatinate | Mercuric chloride-diphenylcarbazone | Diazotized Benzidine |
| Amphetamines: | | | | | | | | | | | | | | | |
| Dextroamphetamine sulphate | 15 | 30 | 41 | 69 | 0 | | | + | + | | | | | | |
| Methamphetamine hydrochloride | 26 | 34 | 26 | 70 | 0 | | | + | | + | + | | | + | |
| Miscellaneous: | | | | | | | | | | | | | | | |
| Aspirin | 0 | 25 | 71 | 92 | 10 | | ± | + | | | | | | + | |
| Benzocaine | 79 | 78 | 84 | 92 | 14 | | | + | | | | + | + | | |
| Cannabidiol | 84 | 86 | 82 | 98 | 50 | | | + | | | | | | | + |
| Cannabinol | 82 | 86 | 81 | 98 | 45 | | | + | | | | | | | + |
| Glutethimide | 77 | 86 | 85 | 95 | 0 | | | ± | | | | | | + | |
| Meprobamate | 30 | 76 | 72 | 86 | 0 | | | | | | | + | | | |
| Phenytoin sodium | 48 | 83 | 86 | 95 | 0 | | | ± | | | | | | + | |
| $\Delta^9$-Tetrahydrocannabinol | 83 | 85 | 83 | 98 | 42 | | | | | | | | | | + |
| Thiopental | 82 | 82 | 90 | 94 | 10 | | | + | + | + | | + | + | + | |

*The $R_F$ values are averages of three determinations.

**+ = easily detected spot, and ± = very faint spot. The empty spaces indicate negative reactions with the exception of the last three columns. These reagents were not tested on the alkaloids, the compounds that give positive alkaloidal reactions and the cannabinoids.

***Thiopental produced a faint white spot 15-20 min after the application of iodoplatinate.

TABLE 12.15

TLC ANALYSIS NARCOTIC SEIZURES[57]

Silica gel G, activated.
Solvent systems:
S1 Ethanol-dioxane-benzene-ammonia (5:40:50:5).
S2 Chloroform-dioxane-ethyl acetate-ammonia (25:60:10:5)
S3 Ethanol-chloroform-dioxane-light petroleum (b.p. 30-60°C)-benzene-ammonia-ethyl acetate (5:10:50:15:10:5:5)
S4 Ethyl acetate-benzene-ammonia (60:35:5)
S5 Ethyl acetate-di-*n*-butyl ether-ammonia (60:35:5)
S6 Ethyl acetate-benzene-acetonitrile-ammonia (50:30:15:5)
S7 Acetonitrile-chloroform-ethyl acetate-ammonia (40:30:25:5)
S8 Acetonitrile-benzene-ethyl acetate ammonia (40:30:25:5)

| Compound | $hR_F$ values | | | | | | | |
|---|---|---|---|---|---|---|---|---|
| | S1 | S2 | S3 | S4 | S5 | S6 | S7 | S8 |
| Morphine | 18 | 20 | 23 | 2 | 2 | 5 | 8 | 7 |
| Heroin | 74 | 67 | 73 | 19 | 11 | | | |
| Numorphan | 56 | 56 | 58 | 15 | 13 | | | |
| Nisentil | 83 | 79 | 87 | 25 | 22 | | | |
| Levo Dromoran | 57 | 54 | 44 | 20 | 17 | | | |
| Alvodine | 97 | 98 | 97 | 93 | 78 | | | |
| Phenazocine | 96 | 96 | 95 | 80 | 70 | | | |
| Anileridine | 95 | 95 | 93 | 77 | 67 | | | |
| Codeine | 28 | 42 | 40 | 5 | 5 | 11 | 24 | 21 |
| Dihydromorphine | 9 | 16 | 17 | 1 | 2 | | | |
| Meperidine | 63 | 70 | 70 | 27 | 28 | | | |
| Methadon | 79 | 81 | 82 | 58 | 60 | | | |
| Narcotine | 80 | 83 | 85 | 72 | 57 | 80 | 89 | 85 |
| Thebaine | 65 | 73 | 74 | 23 | 15 | 34 | 62 | 54 |
| Papaverine | 73 | 77 | 77 | 53 | 35 | 61 | 81 | 72 |
| Benzylmorphine | 53 | 51 | 52 | 9 | 7 | | | |
| Dihydromorphinone | 18 | 19 | 21 | 2 | 1 | | | |
| Methyldihydromorphinone | 28 | 30 | 34 | 3 | 2 | | | |
| Monoacetylmorphine | 50 | 55 | 56 | 11 | 10 | | | |
| Ethylmorphine | 40 | 46 | 47 | 5 | 5 | | | |
| Cocaine | 86 | 84 | 84 | 68 | 62 | | | |
| Dihydrocodeinone | 36 | 35 | 42 | 5 | 5 | | | |
| Mescaline* | 21 | 23 | 27 | 4 | 12 | | | |
| Procaine* | 67 | 69 | 68 | 41 | 46 | | | |
| Tetracaine* | 68 | 70 | 70 | 33 | 39 | | | |
| Quinine* | 41 | 42 | 46 | 4 | 7 | | | |

*These compounds are not classified as opiates but sometimes appear as diluents in narcotic seizures.

TABLE 12.16

TLC ANALYSIS MORPHINE AND ITS METABOLITES[253]

TLC systems:

| | | |
|---|---|---|
| S1 | Silica gel, instant TLC sheet (Gelman, ITLCC-SG type) | Ethyl acetate-methanol-ammonia (17:2:1) |
| S2 | Silica gel, instant TLC sheet (Gelman, ITLCC-SG type) | *n*-Butanol-water-acetic acid (35:10:3) |
| S3 | Silica gel, instant TLC sheet (Gelman, ITLCC-SG type) | *n*-Butanol-di-*n*-butyl ether-ammonia (25:70:2) |
| S4 | Silica gel pre-coated plates | Ethyl acetate-methanol-ammonia (17:2:1) |

| Compound | $hR_F$ values | | | |
|---|---|---|---|---|
| | S1 | S2 | S3 | S4 |
| Codeine | 98 | 98 | 70 | 60 |
| Morphine | 97 | 98 | 62 | 34 |
| Morphine N-oxide | 40 | 90 | 23 | 3 |
| Norcodeine | 97 | 98 | 64 | 34 |
| Normorphine | 85 | 97 | 50 | 15 |
| Morphine 3-ethereal sulphate | 47 | 67 | 19 | 3 |
| Morphine glucuronide | 0 | 40 | 0 | 0 |
| Pseudomorphine | 10-15 | 30 | 0 | 1 |
| Morphine-N-methyl iodide | 0 | 75 | 3 | 0 |
| Nalorphine | 98 | 90 | 81 | 50 |
| Nicotine and its metabolites | 0, 8, 27, 50 70, 90 | 90, 96 | 0-23, 65-80 | 4, 8, 34 64, 83 |

TABLE 12.17

TLC DATA FOR VARIOUS NARCOTIC ANALGESICS ARRANGED ACCORDING TO CHEMICAL TYPE[31]

Silica gel G, activated.
Solvent systems:
- S1 Ethanol-pyridine-dioxane-water (50:20:25:5)
- S2 Ethanol-acetic acid-water (6:3:1)
- S3 Ethanol-dioxane-benzene-ammonia (5:40:50:5)
- S4 Methanol-*n*-butanol-benzene-water (60:15:10:15)
- S5 *tert.*-Amyl alcohol-di-*n*-butyl ether-water (80:7:13)
- S6 *n*-Butanol-acetic acid-water (4:1:2)
- S7 *n*-Butanol-conc. hydrochloric acid (9:1) saturated with water

| Compound (free base) | $hR_F$ values | | | | | | | | |
|---|---|---|---|---|---|---|---|---|---|
| | S1 | S2 | S3 | S4 | | S5 | | S6 | S7 |
| Iminoethanophenanthrofurans: | | | | | | | | | |
| Morphine | 29 | 27 | 11 | 21 | 86* | 7 | 85* | 54 | 34 |
| Normorphine | 8 | 48 | 4 | 7 | 48 | STR** | 25 | 66 | 62 |
| Codeine | 30 | 29 | 39 | 25 | 86 | 8 | 91 | 53 | 30 |
| Norcodeine | 12 | 50 | 13 | 9 | 59 | 6 | 56 | 63 | 49 |
| Heroin | 37 | 35 | 76 | 35 | 90 | 15 | 95 | 61 | 32 |
| Nalorphine | 71 | 55 | 35 | 67 | 88 | 25 | 96 | 59 | 41 |
| Methyldihydromorphinone | 16 | 24 | 25 | 15 | 76 | STR | 92 | 45 | 26 |
| Dihydromorphinone | 11 | 21 | 17 | 13 | 65 | STR | 85 | 41 | 25 |
| Ethylmorphine | 33 | 25 | 46 | 27 | 84 | 8 | 96 | 53 | 33 |
| Dihydrohydroxymorphinone | 46 | 29 | 34 | 24 | 63 | 10 | 81 | 45 | 28 |
| Dihydromorphine | 15 | 21 | 10 | 10 | 67 | STR | 73 | 43 | 29 |
| Dihydrocodeinone | 17 | 25 | 41 | 19 | 76 | STR | 94 | 42 | 23 |
| Dihydrohydroxycodeinone | 46 | 24 | 87 | 29 | | 16 | | 32 | 34 |
| 6-O-Acetylmorphine | 38 | 40 | 64 | 29 | | 19 | | 37 | 37 |
| Iminoethanophenanthrenes: | | | | | | | | | |
| *l*-3-Hydroxy-N-methylmorphinan | 11 | 47 | 80 | 10 | | 7 | | 51 | 60 |
| *l*-3-Hydroxymorphinan | 5 | 68 | 19 | 10 | | 8 | | 72 | 80 |
| *l*-3-Methoxy-N-methylmorphinan | 13 | 43 | 91 | 8 | | 7 | | 55 | 59 |
| *l*-3-Methoxymorphinan | 7 | 65 | 38 | STR | | STR | | 66 | 81 |
| *l*-3-Hydroxy-N-allylmorphinan | 65 | 70 | 98 | 41 | | 44 | | 64 | 73 |
| Diarylalkoneamines: | | | | | | | | | |
| *dl*-Methadone | 34 | 59 | 99 | 17 | | 17 | | 55 | 62 |
| *l*-Acetylmethadol | 64 | 60 | 99 | 40 | | 38 | | 52 | 62 |
| *d*-Propoxyphene | 73 | 68 | 97 | 54 | | 56 | | 53 | 61 |
| Arylpiperidines: | | | | | | | | | |
| Meperidine | 42 | 41 | 97 | 36 | | 20 | | 46 | 44 |
| Normeperidine | 12 | 65 | 51 | 10 | | 11 | | 58 | 63 |
| Ketobemidone | 31 | 39 | 47 | 24 | | 12 | | 42 | 40 |
| *dl*-Alphaprodine | 39 | 40 | 93 | 34 | | 20 | | 42 | 40 |
| Piminodine | 88 | 73 | 99 | 85 | | 76 | | 69 | 58 |
| Benzomorphans: | | | | | | | | | |
| *dl*-2'-Hydroxy-5,9-dimethyl-2-phenethyl-6,7-benzomorphan | 88 | 87 | 97 | 82 | | 70 | | 76 | 77 |
| *l*-2'-Hydroxy-2,5,9-trimethyl-6,7-benzomorphan | 12 | 36 | 56 | 8 | | 5 | | 43 | 51 |
| 2'-Hydroxy-5,9-dimethyl-2-(3,3-dimethylallyl)-6,7-benzomorphan | 73 | 81 | 96 | 25 | | 34 | | 65 | 77 |
| 2'-Hydroxy-5,9-dimethyl-2-cyclopropylmethyl-6,7-benzomorphan | 45 | 71 | 92 | 15 | | 16 | | 55 | 67 |

*$hR_F$ values on 0.1 *M* phosphate buffer (pH 8.0)-impregnated cellulose plates.
**STR = compound streaks.

**References p. 245**

TABLE 12.18

COLOUR REACTIONS OBSERVED FOR SOME DRUGS OF ABUSE AFTER CONSECUTIVE SPRAYING WITH DIFFERENT SPRAY REAGENTS[128]

| Drug | Colour* | | | |
|---|---|---|---|---|
| | Iodoplatinate | Ammoniacal silver nitrate (no.1) | 0.02 *M* aq. potassium permanganate | Dragendorff's reagent (no.39c) |
| Morphine | Dark purple | Black | Black-yellow | Rust-orange |
| Codeine | Purple | Purple | Purple-yellow | Rust-orange |
| Dihydrohydroxycodeinone | Purple | Red-brown | Yellow | Brown-orange |
| Dihydrocodeinone | Dark purple | Light black | Yellow | Rust-orange |
| Oxymorphone | Purple | Black | Black-yellow | Brown-orange |
| Methadone | Red-brown | Red-brown | Yellow | Rust-orange |
| Meperidine | Purple | Yellow | Yellow | Brown-orange |
| Propoxyphene | Red-purple | Red-purple | Light-yellow | Brown-orange |
| Nalorphine | Dark blue | Black | Black-yellow | Brown-orange |
| Naloxone | Light blue | Blue-black | Yellow | Rust-orange |
| Quinine | Purple | Yellow | Yellow | Rust-orange |
| Marihuana | Tan | Tan | Brown | - |
| Mescaline | Dark purple | Purple | Dark purple | - |
| LSD-25 | Blue-purple | Blue | Rust-brown | - |
| Glutethimide | Brown | Dark brown | Dark brown | - |
| Psilocybin | Blue-purple | Purple | Purple | - |
| Diazepam | Reddish brown | Brown | Dark brown | - |
| Chlordiazepoxide | Light purple | Purple | Purple | Orange |
| Chlorpromazine | Dark purple | Purple | Yellow | Orange |
| Cocaine | Purple | Purple | Purple | - |

*With iodoplatinate most compounds were detected at a minimum level of 0.1-1 μg, except for glutethimide and psilocybin (20 μg) and marihuana (100 μg).

References p. 245

TABLE 12.19

TLC ANALYSIS DRUGS OF ABUSE IN BIOLOGICAL MATERIAL

| Alkaloid* | Other compounds | Aim | Adsorbent | Solvent system | Ref. |
|---|---|---|---|---|---|
| C,M,H,NorM,meth, diHMone,nalo, phen, prop,lev,coc,nic | Chlorpromazine, tripelennamine | Identification of analgesics in urine (Table 12.11) | $SiO_2$ | EtOH-dioxane-pyridine-$H_2O$(50:25:20:5)<br>EtOH-AcOH-$H_2O$(6:3:1)<br>EtOH-dioxane-benzene-$NH_4OH$(5:40:50:5)<br>MeOH-*n*-BuOH-benzene-$H_2O$(60:15:10:15) | 16 |
| | | | $Al_2O_3$ | *n*-BuOH-*n*-$Bu_2O$-AcOH(4:5:1)<br>*n*-BuOH-*n*-$Bu_2O$-$NH_4OH$(25:70:5) | |
| C,M,P,H,diHMone, mep,meth,Q,S,nic | Propiomazine | Identification | $SiO_2$ | MeOH-$NH_4OH$(100:1.5)<br>$CHCl_3$-DEA(9:1)<br>Cyclohexane-DEA(9:1) | 22 |
| C,M,H,norM,norC, diHmone,diHcone, MediHMone,oxyM,oxyC, 6OAcM,EtM,nalo,mep, normep,meth,keto, lev,various benzo-morphans | | Detection in biological material (Table 12.17) | $SiO_2$ | EtOH-dioxane-pyridine-$H_2O$(50:25:20:5)<br>EtOH-AcOH-$H_2O$(6:3:1)<br>EtOH-dioxane-benzene-$NH_4OH$(5:40:50:5)<br>*n*-BuOH-AcOH-$H_2O$(4:1:2)<br>*n*-BuOH-conc. HCl(9:1) sat. with $H_2O$ | 31,201 |
| | | | 0.1 *M* phosphate buffer (pH 8)-impregnated cellulose or $SiO_2$ | MeOH-*n*BuOH-benzene-$H_2O$(60:15:10:15)<br>*tert.*-AmOH-*n*-$Bu_2O$-$H_2O$(80:7:13) | |
| C,M,diHMone,diHCone, AcdiHCone,oxyC, 3OHNMeM,mep,meth, normeth,keto,dex,nic | | Identification | $SiO_2$ | EtOAc-DMFA(3:1) | 34 |
| M,H,AcC,nic | | Interference of AcC in detection of H | | No details available | 54 |
| M,nalo | | Detection in urine | | No details available | 67 |
| C,M,No,P,H,apoM, diHCone,diHC,EtM, mep,meth,coc,caf, nic,A,Q,Qd,RSP,S | Amphetamines,antihistamines,hypnotics,sedatives,tranquillizers, various others | TLC-screening in urine (Table 12.12) | $SiO_2$ | EtOAc-MeOH-$NH_4OH$(17:2:1) | 68,102 |
| C,M,oxyC,nalo | | Detection of M in urine | $SiO_2$ | EtOH-dioxane-benzene-conc. $NH_4OH$(5:40:50:5)<br>Dioxane-$NH_4OH$(60:5) | 74,75 |

| | | | | | |
|---|---|---|---|---|---|
| M,H,norM,nalo | | Indirect quantitative analysis after esterification with labelled compound | $SiO_2$ | $CHCl_3$-MeOH(9:1) | 77 |
| M,mep,meth,diHMone, coc,Q | Amphetamines,barbiturates, tranquillizers | Detection in urine | $SiO_2$ | EtOAc-MeOH-conc. $NH_4OH$(85:10:5) | 78,87 |
| C,M,No,P,diHMone, diHCone,diHC,EtM, oxyC,AcdiHCone,meth, normeth,mep,dex,lev, levometh,racemorphan,keto,coc,nic | Amphetamines | Detection in urine | $SiO_2$ | isoprOH-25% $NH_4OH$(9:1) | 97 |
| C,M,P,H,coc,caf,Q, S,B,lobeline, sparteine | | Identification of doping in horse urine | $SiO_2$ | I. Hexane-$Me_2CO$-DEA(6:3:1)<br>II. $CHCl_3$-MeOH-DEA(95:5:0.05)<br>Two dimensional: I,II | 112 |
| C,M,norC,diHMone, meth,Q | Amphetamines | Identification in urine | $SiO_2$ | $CHCl_3$-MeOH(4:1) | 119 |
| C,M,oxyM,diHCone, oxyC,mep,meth,nalo, nalox,prop,coc, eph,Q,LSD,mesc, psilocybin | Amphetamines, barbiturates,marihuana, various psychoactive drugs | Comparison of direct solvent extraction and ion-exchange paper extraction from urine (Table 12.18) | $SiO_2$ | EtOH-dioxane-benzene-$NH_4OH$(5:40:50:5)<br>EtOAc-MeOH-$NH_4OH$(85:10:5)<br>MeOH-*n*-BuOH-benzene-$H_2O$(60:15:10:15)<br>EtOH-dioxane-pyridine-$H_2O$(50:25:20:5)<br>*tert*.-AmOH-*n*-$Bu_2O$-$H_2O$(80:7:13) | 128 |
| C,M,meth,coc,Q | Amphetamines,barbiturates, psychotropics | Detection in urine | $SiO_2$ | EtOAc-MeOH-28% $NH_4OH$(85:10:5) | 129 |
| M,nic | | Interference of nic in analysis of M in urine | $SiO_2$ | MeOH-25% $NH_4OH$(100:2.4) | 135 |
| C,M,P,H,diHMone, mep,meth,dex,lev, nic,eph,coc,caf,A, scopo,S,Q,Qd,cinchonine,physo,yohimbine, thbr,thph | Amphetamines,benzodiazepines,phenothiazines, imipramines,tranquillizers | Detection of basic drugs and their metabolites in urine | $SiO_2$ | $CHCl_3$-MeOH(9:1)<br>$Isopr_2O$-EtOH(8:2)<br>MeOH-conc. $NH_4OH$(100:1.5) | 141 |
| C,M,meth,Q | Amphetamine,phenobarbital | Masking of C by Q metabolites | $SiO_2$ | EtOAc-MeOH-conc. $NH_4OH$(85:10:5) | 147 |

| | | | | | |
|---|---|---|---|---|---|
| C,M,T | | Direct quantitative analysis (spot areas) | $SiO_2$ | $CHCl_3$-MeOH(4:1), benzene-MeOH(9:1, 4:1), benzene-*n*-BuOH(3:1), EtOH-dioxane-benzene-$NH_4OH$(1:8:10:1), benzene-dioxane(4:1), benzene-EtOH(4:1), $CHCl_3$-isoprOH(4:1), $CHCl_3$-$Me_2CO$(3:1), $CHCl_3$-$Me_2CO$-MeOH(5:4:1) | 151 |
| | | | $Al_2O_3$ | $CHCl_3$-MeOH(1:1), $CHCl_3$-$Me_2CO$(1:1) | |
| C,M,mep,meth,prop, Qd,S,A,nic | Amphetamines, hypnotics, sedatives | Screening in urine | $SiO_2$ | EtOAc-MeOH-$NH_4OH$(85:10:5) | 159 |
| M,meth,prop,Q,nic | Procaine,chlorpromazine, trimethobenzamide | Identification drugs after TLC separation | $SiO_2$ | EtOAc-MeOH-$NH_4OH$(85:10:5)<br>EtOH-AcOH-$H_2O$(24:12:4) | 167 |
| C,M,mep,meth,Q, nic,coc,ecgonine, benzoylecgonine | Tranquillizers,phenothiazines | Routine identification | $SiO_2$ | EtOAc-MeOH-$NH_4OH$(85:10:5)<br>EtOAc-MeOH-$H_2O$-$NH_4OH$(85:10:3:1)<br>MeOH-$NH_4OH$(99:1)<br>$CHCl_3$-MeOH-$NH_4OH$(90:10:1) | 168 |
| C,M,meth,Q,nic, coc | | Detection in urine for meth treatment programmes | $SiO_2$ | BuOH-$H_2O$-AcOH(4:5:1), top layer, develop plates twice | 169 |
| C,M,diHMone,mep, meth,nalo,Q,Qd,nic, coc | Amphetamines,barbiturates, benzodiazepines,local anaesthethics | Screening in urine | $SiO_2$ | EtOAc-MeOH-conc. $NH_4OH$(85:10:1) | 172 |
| C,M,mep,meth, normeth,cotinine, coc,nic,Q | Amphetamines,barbiturates, phenothiazines | Screening in urine | $SiO_2$ | Benzene-dioxane-EtOH-$NH_4OH$(50:40:5:5)<br>MeOH-12 *M* $NH_4OH$(100:1.5), unsat. chamber | 173 |
| M,mep,meth,methorphan,Q,coc,nic,caf, eph,mesc | Amphetamines,barbiturates, phenothiazines,antihistamines,minor tranquillizers | Identification in urine | $SiO_2$ | $CHCl_3$-MeOH-$NH_4OH$(90:10:1)<br>EtOAc-MeOH-$H_2O$-$NH_4OH$(85:10:3:1) | 174,243 |
| C,M,mep,meth,prop, Q,nic | Amphetamines,barbiturates, tranquillizers | Extraction from lyophilized urine | $SiO_2$ | EtOAc-MeOH-$NH_4OH$(85:10:3)<br>Hexane-EtOH(92:8, 85:15) | 175 |
| M,LSD,mesc | Amphetamines,$\Delta^9$-THC, catecholamines | Mini-TLC of dansylated compounds | Polyamide | I. $H_2O$-HCOOH(100:1.5)<br>II. Benzene-AcOH(9:1)<br>III. $H_2O$-AcOH(50:1)<br>IV. EtOH-*n*-BuOH-$H_2O$-HCOOH(93:4:150:3)<br>Two-dimensional: I,III or I,(III+I) | 182 |

References p. 245

| | | | | | |
|---|---|---|---|---|---|
| C,M,mep,meth,prop, Q,coc | | Detection in urine | $SiO_2$ | EtOAc-MeOH-$NH_4OH$(85:10:2.5) | 183 |
| C,M,diHC,mep,meth, prop,Q,coc,nic,caf | Amphetamines,hypnotics, various others | Detection in urine | $SiO_2$ | EtOAc-MeOH(85:10) in atm. sat. with 5 ml of 0.5 *M* $NH_4OH$<br>$CHCl_3$-MeOH(8:2) | 187 |
| C,M,diHMone,mep, meth,prop,Q,coc,S | Amphetamines,barbiturates, benzodiazepines,various others | Detection of drugs at therapeutic dosages in urine | $SiO_2$ | EtOAc-MeOH-conc. $NH_4OH$(85:10:5) | 195 |
| C,M,EtM | | Detection in biological material | $SiO_2$ | IsoBuOH-AcOH-$Me_2CO$-$H_2O$(no further details) | 199 |
| C,M,EtM,mep,meth, Q,nic,coc,caf | | Use of pre-coated sheets in drug screening | $SiO_2$ | EtOAc-MeOH-$NH_4OH$(85:10:5, 7:2:1), followed by EtOAc-MeOH(95:5) | 202 |
| C,M,H,apoM, pentazocine,Q | | Recovery of M from various post mortem tissues | $SiO_2$ | EtOAc-MeOH-$NH_4OH$(85:10:5)<br>EtOH-dioxane-pyridine-$H_2O$(50:25:20:5)<br>EtOH-benzene-dioxane-$NH_4OH$(5:50:40:5)<br>$CHCl_3$-DEA(9:1)<br>MeOH-BuOH-benzene-$H_2O$(60:15:10:15)<br>EtOH-AcOH-$H_2O$(6:3:1) | 204 |
| C,M,mep,meth | Amphetamine | Mini-TLC | $SiO_2$ | EtOAc-MeOH-$NH_4OH$(85:10:5) | 205 |
| C,M,mep,meth, Dmethphan,Q,nic, coc,caf,A | Various | Determination of M in urine | $SiO_2$ | *n*-BuOH-$H_2O$-AcOH(4:2:1)<br>*n*-BuOH-conc. HCl(9:1)<br>MeOH-$H_2O$-AcOH-benzene(80:15:2:5) | 207 |
| C,M,meth,coc | Amphetamines,barbiturates | Screening in urine | $SiO_2$ | $CHCl_3$-MeOH-$NH_4OH$(85:10:1)<br>EtOAc-MeOH-$NH_4OH$(85:10:1.5) | 208 |
| C,M,H,6OAcM,AcC, meth,nalo,cotinine, Q,nic,coc | Amphetamines,benzodiazepines,psychotropic drugs | Detection in urine | $SiO_2$ | I. MeOH-$CHCl_3$-$NH_4OH$(85:15:0.7)<br>II. $Et_2O$-$Me_2CO$-DEA(85:8:7)<br>Two-dimensional: I,II | 210 |
| C,M,diHMone,meth, naloxone,coc,eph, mesc,LSD,Q | Amphetamines,barbiturates, benzodiazepines,CNS stimulants,psychotropics | Screening of urine (Table 12.13) | $SiO_2$ | EtOAc-cyclohexane-dioxane-MeOH-$H_2O$-$NH_4OH$ (50:50:10:10:0.5:1.5)<br>EtOAc-cyclohexane-MeOH-$H_2O$-$NH_4OH$ (70:15:8:0.5:2)<br>EtOAc-cyclohexane-MeOH-$NH_4OH$(70:15:10:5)<br>EtOAc-cyclohexane-$NH_4OH$(50:40:0.1) | 213 |
| C,M,diHMone,oxyM, oxyC,mep,meth,normep,naloxone,lev, leval,prop,nalo, pentazocine,cyclazocine,Q,coc | | Sensitivity of detection | $SiO_2$ | EtOAc-MeOH-$NH_4OH$(85:10:5) | 220 |

| | | | | | |
|---|---|---|---|---|---|
| C,M,mep,meth,Q, coc | | Screening in urine | $SiO_2$ | EtOAc-MeOH-$NH_4OH$(75:24:1) followed by EtOAc-MeOH(9:1) | 226 |
| M | | Quantitative extraction from urine with $XAD_2$ | | | 230 |
| M | | Comparison of two common screening methods[78,128] | $SiO_2$ | EtOAc-MeOH-$NH_4OH$(85:10:5) | 234 |
| M,naloxone,LSD | Amphetamine | Mini-TLC of dansylated compounds | Polyamide | $H_2O$-EtOAc-EtOH(50:25:25)<br>Toluene-AcOH-DMFA(200:10:4)<br>Heptane-*n*-BuOH-AcOH(30:40:2.5)<br>AcOH-DMFA-EtOH-$H_2O$(25:20:20:50)<br>Benzene-AcOH(9:1)<br>Benzene-HCOOH(200:3) | 239 |
| C,M,Q | | Emergency toxicological screening | $SiO_2$ | MeOH-$NH_4OH$(100:1.5) | 241 |
| C,M,diHC,mep,meth, Q,nic,caf,coc, ecgonine,benzoylecgonine | Amphetamines,procaine, various others | Detection of cocaine metabolites in urine | $SiO_2$ | EtOAc-MeOH(17:2) in $NH_3$ atm., followed by $CHCl_3$-MeOH(1:1) in $NH_3$ atm. | 244 |
| M,mep,meth,prop,Q | Amphetamines,barbiturates, tranquillizers | Dual-bed TLC, to avoid sample clean-up | Cellex PAB weak ion exchanger and $SiO_2$, each half of the plate | EtOAc-MeOH-$NH_4OH$(85:10:1, 85:10:10) | 242 |
| C,M,norC,nalo,nic, M metabolites | | Detection in biological material (Table 12.16) | $SiO_2$ | EtOAc-MeOH-conc. $NH_4OH$(85:10:5)<br>*n*-BuOH-AcOH-$H_2O$(35:3:10)<br>*n*-BuOH-*n*-$Bu_2O$-$NH_4OH$(25:70:2) | 253 |
| C,M,P,mep,meth,Q, nic,coc,eph | Amphetamines,barbiturates, various others | Detection in urine | $SiO_2$ | MeOH-conc. $NH_4OH$(100:1.5)<br>$CHCl_3$-MeOH(9:1)<br>EtOAc-MeOH-$NH_4OH$(85:10:5) | 261 |
| M | | Extraction from bile | $SiO_2$ | EtOAc-MeOH-$NH_4OH$(85:10:1) | 269 |

References p. 245

| | | | | | |
|---|---|---|---|---|---|
| C,M,mep,meth,nalo, nalox,caf,coc,mesc | Amphetamines,barbiturates, various others | Use of $XAD_2$ in extraction of drugs from urine | $SiO_2$ | EtOAc-MeOH-$NH_4OH$(180:17:7) | 272 |
| C,M,mep,meth,Q, coc,eph | Barbiturates,phenothiazines,various others | Concentration from urine with charcoal | $SiO_2$ | $CHCl_3$-$Me_2CO$(90:10), followed by the same solvent to which 2 ml of $NH_4OH$ is added<br>$CHCl_3$-*n*-BuOH-$NH_4OH$(190:10:3), followed by EtOAc-MeOH-$NH_4OH$(85:10:5) | 273 |
| mesc | Primary amines (including amphetamines) | Detection in urine | $SiO_2$ | EtOAc-cyclohexane-MeOH-conc. $NH_4OH$ (70:15:8:0.5:2)<br>EtOAc-MeOH-$H_2O$(80:15:5, 80:10:5)<br>MeOH-$NH_4OH$(100:1.5) | 274 |
| C,M,mep,meth,Q,coc, eph | Amphetamines,barbiturates, benzodiazepines,various others | Ion-pair extraction method | $SiO_2$ | EtOAc-MeOH-conc. $NH_4OH$(85:10:5) | 275 |
| M,H | | Comparison of different analytical methods(RIA,FRAT, fluorimetry,TLC) | $SiO_2$ | EtOAc-MeOH-$NH_4OH$(85:10:5) | 276 |
| C,M,meth,prop,caf, nic,coc | Amphetamines,barbiturates | Detection in urine | $SiO_2$ | EtOAc-MeOH-$NH_4OH$(85:10:5)<br>$CHCl_3$-$Me_2CO$-$NH_4OH$(90:9:1), lower phase | 281 |
| M,diHMone,mep,meth, leval | Amphetamines,barbiturates | Detection in autopsy specimens | $SiO_2$ | MeOH-$NH_4OH$(100:1.5)<br>EtOAc-MeOH-$NH_4OH$(85:10:5) | 283 |
| M | Amphetamines | (Fluoro)densitometric analysis | $SiO_2$ | $CHCl_3$-$Me_2CO$-MeOH-*tert.*-butylamine(3:4:1:2) | 285 |
| M,coc,ecgonine, benzoylecgonine | | Routine analysis of coc. metabolites in urine | $SiO_2$ | I. EtOAc-MeOH-$H_2O$(7:2:1)<br>II. EtOAc-MeOH-$NH_4OH$(15:4:1)<br>III. $CHCl_3$-$Me_2CO$-DEA(5:4:1)<br>IV. $CHCl_3$-$Me_2CO$(5:4) sat. with $NH_4OH$<br>V. MeOH-$NH_4OH$(100:1.5)<br>VI. Benzene-EtOAc-MeOH-$NH_4OH$(80:20:1.2:0.1)<br>VII. $CHCl_3$-$Me_2CO$-$NH_4OH$(5:95:1)<br>Subsequently developed: I+II,III,IV,V or VII<br>Two-dimensional: I,VII | 286 |
| C,M,mep,meth,Q,coc, caf,nic,S | Amphetamines,barbiturates | Screening in urine | $SiO_2$ | EtOAc-MeOH-$NH_4OH$(85:10:5) | 289 |
| M | | Confirmation of M by fluorimetry after TLC | $SiO_2$ | EtOAc-MeOH-conc. $NH_4OH$(85:10:5) | 290 |

| | | | | | |
|---|---|---|---|---|---|
| M,meth,nalox, naltrexone | Cyclazocine | Detection of narcotic antagonists | $SiO_2$ | EtOAc-MeOH-$NH_4OH$(85:10:5) | 292 |
| C,M,P,H,diHMone, oxyC,EtM,mep,meth, pentazocine,Q,coc, nic,LSD | Amphetamines,hypnotics, phenothiazines,benzodiazepines,tranquillizers | Identification in urine | $SiO_2$ | $CHCl_3$-EtOH-25% $NH_4OH$(80:15:5)<br>$CHCl_3$-MeOH-25% $NH_4OH$(90:10:1)<br>EtOAc-MeOH-25% $NH_4OH$(85:10:5) | 295 |
| M | | Detection in blood and urine | $SiO_2$ | EtOAc-MeOH-$NH_4OH$(85:10:5) | 298 |
| M,Q | | Comparison of RIA and TLC in detection of M in urine | $SiO_2$ | EtOAc-MeOH-$NH_4OH$(85:10:5) | 307 |
| M | Barbiturates | Separation from putrified post mortem material | $SiO_2$ | EtOAc-MeOH-$NH_4OH$(85:10:5) | 317 |
| M,Q,Q metabolites | | Specialized TLC systems for some common drug identification problems,indirect quantitative analysis | $SiO_2$ | $Et_2O$-light petroleum (b.p. 30-60$^{o}$C)-EtOH-conc. $NH_4OH$(70:22:3:5)<br>EtOAc-MeOH-conc. $NH_4OH$(85:10:5) | 318 |
| C,M,P,H,diHMone, diHCone,meth,nalox, naltrexone,Dmethphan, Q,coc,caf,eph,mesc, LSD | Sedatives,hypnotics, CNS stimulants,various others | Simultanous detection of drugs of abuse in urine screening | $SiO_2$ | EtOAc-cyclohexane-MeOH-$H_2O$-$NH_4OH$ (70:15:8:0.5:2)<br>EtOAc-cyclohexane-MeOH-$NH_4OH$(70:15:10:5)<br>EtOAc-cyclohexane-$NH_4OH$(50:40:0.1) | 319,339 |
| M,mep,meth,prop,A, coc,benzoylecgonine | Sedatives | Detection of coc and benzoylecgonine in urine | $SiO_2$ | $CHCl_3$-MeOH-conc. $NH_4OH$(100:20:1, 60:60:1) | 322 |
| C,coc,caf,A | Analgesics,antihistamines,sedatives, hypnotics,tranquillizers | Routine identification | $SiO_2$ | $CHCl_3$-$Me_2CO$(9:1)<br>MeOH-$NH_4OH$(100:1.5)<br>Two-dimensional: I. EtOAc-MeOH-$NH_4OH$ (85:10:5) II. *n*-BuOH-AcOH-$H_2O$(7:2:1) | 324 |

References p. 245

| | | | | | |
|---|---|---|---|---|---|
| C,M,mep,meth,coc, benzoylecgonine,nic | Benzodiazepines,various others | Screening and identification in urine | $SiO_2$ | EtOAc-MeOH-DEA(90:10:1.6)<br>$CH_2Cl_2$-MeEtCO-conc. $NH_4OH$(74:25:0.8)<br>MeOH-$CHCl_3$-conc. $NH_4OH$(74:25:0.8)<br>EtOAc-$CH_2Cl_2$-conc. $NH_4OH$(90:10:0.7)<br>EtOAc-benzene-$CHCl_3$(2:2:1)<br>$CH_2Cl_2$-$n$-BuOH-conc. $NH_4OH$(85:15:0.2)<br>$Me_2CO$-MeOH-DEA(90:10:0.7)<br>$Me_2CO$-MeOH(6:4)<br>$Me_2CO$-$CHCl_3$(7:3) | 325 |
| C,M,meth,coc | Amphetamines,barbiturates | Identification in urine | $SiO_2$ | Dioxane-$NH_4OH$(95:5) | 327 |
| C,M,mep,meth,Q,coc, caf,eph | Amphetamines,hypnotics, sedatives | Differential elution of drugs from $XAD_2$ | $SiO_2$ | EtOAc-MeOH-$NH_4OH$(85:10:5)<br>$CHCl_3$-MeOH-$NH_4OH$(90:10:1) | 329 |
| C,M,diHMone,diHCone, oxyC,EtM,mep,dex, prop,pentazocine,Q, coc,caf,nic,A,scopo, mesc,lobeline | Amphetamines,hypnotics, sedatives,antihist-amines,phenothiazines | Interfering compounds in urine screening | $SiO_2$ | EtOAc-isoprOH-$NH_4OH$(60:40:1)<br>EtOAc-cyclohexane-$NH_4OH$(50:40:0.1, 60:40:1)<br>Dioxane-benzene-$NH_4OH$(35:60:5)<br>$CHCl_3$-$Me_2CO$(95:5), unsat. chamber<br>EtOAc-MeOH-$NH_4OH$(85:10:5) | 332 |
| C,M,meth and metabolites | | Identification in urine | $SiO_2$ | EtOAc-MeOH-$NH_4OH$(85:10:5) | 341 |
| M | | Detection in urine | $SiO_2$ | Two dimensional: I. MeOH-$CHCl_3$-$NH_4OH$ (85:15:0.7) II. $Et_2O$ (sat. with $H_2O$)-MeEtCO-DEA(85:8:7) | 345 |
| C,M,No,P,T | | Detection in urine | $SiO_2$ | Toluene-$Me_2CO$-EtOH-$NH_4OH$(45:45:7:3)<br>EtOAc-MeOH-$NH_4OH$(85:10:10) in sandwich chambers | 359 |
| M | Amphetamines | Routine control of urine | $SiO_2$ | EtOAc-MeOH-25% $NH_4OH$(85:10:5) | 360 |
| C | | In cadaveric material | $SiO_2$ | EtOH-dioxane-benzene-25% $NH_4OH$(1:8:10:1) | 363 |
| C,M,diHMone,diHCone, oxyC,EtM,mep,meth, dex,keto,nalo,ph, Q,caf,nic | CNS stimulants, psychotropic drugs | Identification | $SiO_2$ | MeOH-12 $M$ $NH_4OH$(49:1)<br>$CHCl_3$ sat. with $NH_4OH$ | 374 |
| Benzoylecgonine | Amphetamines,barbiturates, various others | Extraction method for benzoylecgonine | $SiO_2$ | EtOAc-cyclohexane-MeOH-conc. $NH_4OH$ (70:15:10:5)<br>EtOAc-cyclohexane-conc. $NH_4OH$(50:40:0.1) | 376 |

| | | | | | |
|---|---|---|---|---|---|
| C,M,mep,meth, normeth,keto,lev, levallorphan,Q,S, nic | Amphetamines,hypnotics, sedatives,various others | Comparison of EMIT and TLC in drug screening | $SiO_2$ | EtOAc-MeOH-conc. $NH_4OH$(85:10:5) | 380 |
| C,M,H,AcC,norC, 6OAcM,norM,C-6-glucuronide,M-3-glucuronide,norM-glucuronide,6OAcM-3-glucuronide,M-3-ethereal sulphate | | Identification of H-metabolites | $SiO_2$ | *n*-BuOH-$H_2O$-AcOH(35:10:3)<br>EtOAc-MeOH-$NH_4OH$(85:10:5) | 381 |
| M | | Indirect quantitative analysis (UV) | $SiO_2$ | MeOH-25% $NH_4OH$ | 389 |
| M | | Detection of M | $SiO_2$ | *n*-BuOH-$H_2O$-AcOH(4:2:1)<br>*n*-BuOH-conc. HCl(9:1)<br>MeOH-$H_2O$-AcOH-benzene(80:15:2:5) | 398 |
| M | | Combined EMIT-TLC determination | $SiO_2$ | $CHCl_3$-isoprOH-DEA(18:2:1) | 401 |

*For abbreviations, see footnote to Table 12.7.

TABLE 12.20

TLC ANALYSIS OF METABOLITES OF OPIUM ALKALOIDS AND OPIUM ALKALOIDS IN BIOLOGICAL MATERIAL

| Alkaloid* | Aim | Adsorbent | Solvent system | Ref. |
|---|---|---|---|---|
| C | Densitometric analysis of C and chlorphenamine | $SiO_2$ | Benzene-dioxane-EtOH-25% $NH_4OH$(50:40:5:5) | 158 |
| T,C,M,norC,norM,Cone,glucuronides | Detection of T metabolites in rat brains or urine | $SiO_2$ | Benzene-EtOAc-MeOH-conc. $NH_4OH$(80:20:6.5:0.1)<br>*n*-Hexane-EtOAc-conc. $NH_4OH$(60:40:0.1)<br>*n*-BuOH-*n*-$Bu_2O$-conc. $NH_4OH$(25:70:2)<br>*n*-BuOH-AcOH-$H_2O$(35:3:10) | 209 |
| ApoC,apoM,norapoC,norapoM | Quantitative analysis metabolites apoM | $SiO_2$ | $Me_2CO$-MeOH(1:1) | 248 |
| M,14OHM,14OHMone,oxyM,diHMone,C, Cone,14OHCone,14OHC,oxyC | TLC and GLC of C and M metabolites produced by bacteria | $SiO_2$ | EtOAc-MeOH-conc. $NH_4OH$(86:10:4)<br>$CHCl_3$-MeOH(9:1)<br>EtOH-pyridine-dioxane-$H_2O$(50:20:25:5)<br>EtOH-dioxane-benzene-conc. $NH_4OH$(10:40:50:10)<br>MeOH-BuOH-benzene-$H_2O$(60:15:10:15)<br>Plates sat. with solvent vapour prior to development | 316 |
| ApoM,apoC,isoapoC,apoMdiMe ether, 10,11-diOH-6-*n*propylnoraporphine, 10-OH-aporphine,10-$OCH_3$-aporphine, 10-OH-6-*n*-propylaporphine,apoM, orthoquinone | TLC of apoM metabolites | $SiO_2$ | Benzene-MeOH(4:1)<br>$Me_2CO$-MeOH(1:1)<br>Benzene-EtOAc-DEA(6:3:1)<br>$CHCl_3$-$Me_2CO$(8:2) | 351 |
| C,M,mep,meth,norC,diHC,prop,norprop, methadol,nic | C and diHC in post mortem materials | $SiO_2$<br>0.1 *M* NaOH-impregnated $SiO_2$ | EtOAc-MeOH-conc. $NH_4OH$(85:10:5)<br>$CHCl_3$-MeOH(4:1) | 372 |
| C,M,H,AcC,norC,6OAcM,norM,C-6-glucuronide,M-3-glucuronide,norM-glucuronide,6OAcM-3-glucuronide, M-3-ethereal sulphate | Separation of H metabolites | $SiO_2$ | *n*-BuOH-AcOH-$H_2O$(35:3:10)<br>EtOAc-MeOH-$NH_4OH$(17:2:1) | 381 |

| | | | | |
|---|---|---|---|---|
| M,M-3- and M-6-glucuronide,M-3,6-di-glucuronide,norM,norM-6-glucuronide, M-3-ethereal sulphate | Separation of metabolites | $SiO_2$ | *n*-BuOH-AcOH-$H_2O$(35:3:10)<br>EtOAc-MeOH-$NH_4OH$(17:2:1) | 382 |
| C | Densitometric analysis in post mortem material | $SiO_2$ | MeOH-25% $NH_4OH$(98:2) | 388 |

*For abbreviations, see footnote to Table 12.7.

References p. 245

TABLE 12.21

TLC ANALYSIS OF DRUGS OF ABUSE IN DRUG SEIZURES AND SEPARATION OF MIXTURES OF PURE COMPOUNDS

| Alkaloid* | Other compounds | Aim | Adsorbent | Solvent system | Ref. |
|---|---|---|---|---|---|
| C,M,Na,No,P,T, diHMone,diHCone,EtM, AcdiHCone,meth,mep, keto,Dex,Dmethphan | | Separation (Table 12.3) | $SiO_2$ | MeOH-$Me_2$CO-triethanolamine(1:1:0.03) | 8 |
| C,M,No,P,T,A,scopo, emetine | | Identification | $SiO_2$ | $CHCl_3$-$Me_2$CO-MeOH(5:4:1)<br>$CHCl_3$-$Me_2$CO-DMFA(no further details available) | 24 |
| C,M,No,P,T,H,LSD, lysac,ergot alkaloids | Barbiturates, amphetamines | Identification of LSD and Lysac in drug seizures | 0.1 *M* NaOH-impregnated $SiO_2$ | $CHCl_3$-MeOH(9:1)<br>$CHCl_3$-MeOH-conc. $NH_4$OH(4:4:2) | 41 |
| C,M,Na,No,P,T,H, OAcM | | Analysis of illicit samples of M and H | $SiO_2$ | $CHCl_3$-EtOH(9:1) sat. with 14% $NH_4$OH<br>$Et_2$O-$Me_2$CO-DEA(85:8:7) sat. with $H_2O$ | 52 |
| | | | $Al_2O_3$ | Benzene-$CHCl_3$-$Me_2$CO(70:15:15) sat. with 3.5 % $NH_4$OH | |
| C,M,norM,mep,prop, Methphan,anileridine, ethoheptazine,coc | | Separation | 0.5 *M* LiOH-impregnated $SiO_2$ | Light petroleum (b.p.30-60$^o$C), $CCl_4$,$CH_2Cl_2$, isopr$_2$O,benzene,ethylene dichloride, $CHCl_3$, $Et_2$O,EtOAc,*n*-BuOH,isoprOH,$Me_2$CO,MeOH; all solvents in $NH_3$ atm. | 55 |
| C,M,No,P,T,H, diHMone,diHCone, diHM,EtM,oxyM,OAcM, MediHMone,benzylM, mep,meth,lev, alphaprodine,coc, mesc,Q | Various narcotic analgesics,procaine, tetracaine | Identification of opiates (Table 12.15) | $SiO_2$ | EtOH-dioxane-benzene-$NH_4$OH(5:40:50:5)<br>$CHCl_3$-dioxane-EtOAc-$NH_4$OH(25:60:10:5)<br>EtOH-$CHCl_3$-dioxane-light petroleum (b.p.30-60$^o$C)-benzene-EtOAc-$NH_4$OH(5:10:50:15:10:5:5)<br>EtOAc-benzene-$NH_4$OH(60:35:5)<br>EtOAc-*n*-$Bu_2$O-$NH_4$OH(60:35:5)<br>EtOAc-benzene-actonitrile-$NH_4$OH(50:30:15:5)<br>Acetonitrile-$CHCl_3$-EtOAc-$NH_4$OH(40:30:25:5)<br>Acetonitrile-benzene-EtOAc-$NH_4$OH(40:30:25:5) | 57 |
| C,M,H,diHCone,EtM, mep,meth,dex, acetoM | CNS stimulants and depressors | Identification | $SiO_2$ | $Me_2$CO-conc. $NH_4$OH(99:1)<br>MeOH-conc. $NH_4$OH(99:1)<br>$CHCl_3$-MeOH(1:1) | 58 |
| C,M,S,nic | Barbiturates, tranquillizers | Testing fast-running TLC plates | $SiO_2$-Celite 545 (1:1) | MeEtCO-AcOH-$H_2$O(20:10:5) | 59 |

References p. 245

| | | | | | |
|---|---|---|---|---|---|
| C,M,No,P,diHMone, EtM,mep,meth,S,B, Q,A,nic,eph,coc, caf,thph | Antihistamines,local anaesthetics,barbiturates,tranquillizers | | $SiO_2$ | $CHCl_3$-$Me_2CO$(10:1)<br>MeOH<br>$CHCl_3$-EtOH(10:1, 10:2) | 60 |
| C,M,Na,No,P,T,H, crypt,AcC,6OAcM, 3OAcM | | Application ref.70 method on illicit samples of M and H | $SiO_2$<br><br><br>$Al_2O_3$ | I. $Et_2O$ (sat. with $H_2O$)-$Me_2CO$-DEA(85:8:7)<br>II. MeOH-$CHCl_3$-23% $NH_4OH$(85:15:0.7)<br>Two-dimensional: II,I<br>Benzene-$CHCl_3$-$Me_2CO$(70:15:15) sat. with 3.5% $NH_4OH$ | 71,132 |
| C,M,H,AcC,6OAcM | | Separation of AcC and H | $SiO_2$ | $n$-BuOH-$n$-$Bu_2O$-$NH_4OH$(25:70:5), upper layer, bottom layer for equilibration of atm. in tank, 24 h prior to use | 79 |
| C,M,No,diHMone, EtM,mep,coc,thph | Barbiturates, tranquillizers, various other drugs | Comparison of different types of ready-made plates | $SiO_2$ | $CHCl_3$-$n$-BuOH-conc. $NH_4OH$(70:40:5)<br>Benzene-dioxane-EtOH-conc. $NH_4OH$(50:40:5:5)<br>$CHCl_3$-cylohexane-DEA(5:4:1) | 88 |
| M,diHMone, diHCone,oxyC,mep, meth,AcdiHCone,lev, dex,normeth,keto, fentanyl | Diphenoxylate, amphetamines | Use of azeotropic solvents | 0.1 $M$ $Na_2CO_3$-impregnated $SiO_2$ | $CHCl_3$-MeOH(87.4:12.6)<br>Benzene-MeOH(62.0:38.0)<br>$Me_2CO$-$CCl_4$(87.4:12.6)<br>Benzene-EtOH(92.7:7.3)<br>$CH_2Cl_2$-MeOH(68.3:31.7) | 121 |
| M,H,Dmethphan, dipipanone,coc | Cyclizine | Identification | $SiO_2$ | $Me_2CO$-$CHCl_3$-conc. $NH_4OH$(94:5:1) | 125,130 |
| M,H,diHM,diHC,AcC, 6OAcM,mep,meth,dex, prop,Dmethphan,caf, Q,coc | Various | Analysis of illicit samples of H | 5% $NaH_2$-citrate-impregnated cellulose | 4.8 g citric acid in 130 ml of $H_2O$ + 870 ml of $n$-BuOH | 150 |
| M,No,P,diHMone, diHCone,oxyC,mep, meth,AcdiHCone,A, MeA | | Separation and identification | $SiO_2$ | MeOH-5 $M$ $NH_4OH$(99:1)<br>$CHCl_3$-MeOH-5 $M$ $NH_4OH$(29:20:1)<br>MeOH-$H_2O$-AcOH(17:5:3)<br>$Me_2CO$-5 $M$ $NH_4OH$(8:2)<br>$Me_2CO$-MeOH-5 $M$ HCl(5:4:1) | 156 |
| C,M,H,diHMone, diHCone,EtM,oxyC, AcdiHCone,mep,meth, keto,lev,dex,mesc, psilocin,psilocybin, LSD | Amphetamines,DOM, dimethyltryptamine | Analysis of street drugs | $SiO_2$ | MeOH-25% $NH_4OH$(100:1.5)<br>Benzene-dioxane-EtOH-25% $NH_4OH$(50:40:5:5)<br>EtOH-AcOH-$H_2O$(6:3:1) | 181 |
| M,H,LSD,mesc | THC | Separation with azeotropic solvents | $SiO_2$ | Benzene-MeEtCO-$n$-butylamine(75:15:10)<br>IsoprOH-1,2-dichloroethane-diisopropylamine (5:4:1)<br>1-Chlorobutane-EtOH-$n$-butylamine(8:1:1) | 184 |

TABLE 12.21 (*continued*)

| Alkaloid* | Other compound | Aim | Adsorbent | Solvent system | Ref. |
|---|---|---|---|---|---|
| H,meth,Q | Sucrose,maltose, lactose | Identification | $SiO_2$ | $CHCl_3$-EtOH(9:1) | 190 |
| Mesc,LSD,S, psilocybin | Amphetamines, tryptamines | Screening of street drugs | $SiO_2$ | EtOAc-prOH-conc. $NH_4OH$(40:30:3) | 203,326 |
| M,H,3OAcM,6OAcM | | Densitometric analysis of H | $Al_2O_3$ | Benzene-MeOH-EtOAc-conc. $NH_4OH$(400:50:50:1) | 211 |
| C,M,T,H,mep, meth,LSD,coc,mesc | Amphetamines, marihuana | Detection with TAS technique | $SiO_2$ | $Me_2CO$-MeOH-triethanolamine(50:50:1.5)<br>MeOH-conc. $NH_4OH$(100:1.5) | 216 |
| C,M,Na,No,P,T,H, 6OAcM,S,caf,coc, eph,LSD,psilocin, psilocybin | Amphetamines, barbiturates, tryptamines | Identification | $SiO_2$ | $CHCl_3$-$Et_2O$-MeOH-25% $NH_4OH$(75:25:5:1) | 240 |
| H,coc | Local anaesthetics | Screening | $SiO_2$ | EtOAc-prOH-conc. $NH_4OH$(40:30:3) | 247,326 |
| C,M,meth,coc,eph, LSD | Amphetamines, barbiturates, marihuana | Separation and quantitation of drugs of abuse | $SiO_2$ | $CHCl_3$-$Me_2CO$(9:1)<br>EtOH-MeOH-conc. $NH_4OH$(85:15:5)<br>MeOH-conc. $NH_4OH$(100:2)<br>$CHCl_3$-dioxane-EtOAc-conc. $NH_4OH$(25:60:10:5) | 250 |
| C,M,H,6OAcM,coc, Q | | Separation on microslides | 1.0 *M* NaOH-impregnated $SiO_2$ | $Me_2CO$-benzene(1:1) | 255 |
| M,diHCone,EtM,mep, meth,normeth,dex, scopo | Various other drugs | Detection in tablets using TAS technique | $SiO_2$ | MeOH-conc. $NH_4OH$(100:1.5) | 256 |
| C,M,H,diHMone, diHCone,oxyC, AcdiHCone,mep,meth | Amphetamines | Identification | $SiO_2$ | Light petroleum (b.p. $40^oC$)-$Et_2O$-EtOH-DEA (20:80:10:5) | 278 |
| C,M,meth,prop,nic, coc,S,Q,eph, ergotamine | Amphetamines, tranquillizers, various basic drugs | Recovery of basic compounds after detection with iodoplatinate | $SiO_2$<br>$Al_2O_3$<br><br>cellulose | MeOH-25% $NH_4OH$(100:1.5)<br>MeOH-25% $NH_4OH$(100:1.5)<br>MeOAc-25% $NH_4OH$-$H_2O$(100:2.5:50), upper layer<br>*n*-BuOH-$H_2O$(87:13) + 0.48 g of citric acid | 312 |

| | | | | | |
|---|---|---|---|---|---|
| C,M,H,apoM,mep,meth, DMethphan,prop,mesc, LSD,A,S,eph,Q,Qd, nic,coc,ibogaine | Amphetamines, barbiturates, tranquillizers,local anaesthetics,various others | Identification with TCBI spray reagent | $SiO_2$ | EtOAc-MeOH-$NH_4OH$(100:18:1.5)<br>MeOH-$NH_4OH$(100:1.5) | 328 |
| H | Phenobarbital | Identification | $SiO_2$ | $CHCl_3$-isoprOH-$NH_4OH$(45:45:10)<br>EtOH-pyridine-dioxane-$H_2O$(10:4:5:1)<br>$CHCl_3$-MeOH(9:1) | 336 |
| C,M,P,H,mep,meth, prop,A,scopo,coc, caf,eph,lysergide, nic,physo,mesc, psilocybin,psilocin,Q,S,yohimbine | Amphetamines, barbiturates, local anaesthetics, various others | Systematic identification (Table 12.14) | $SiO_2$ | Benzene<br>$CHCl_3$-$Et_2O$-MeOH-conc. $NH_4OH$(75:25:5:1)<br>EtOAc-prOH-conc. $NH_4OH$(40:30:3)<br>MeOH-conc. $NH_4OH$(100:1.5)<br>EtOH-AcOH-$H_2O$(6:3:1) | 356,379 |
| C,M,No,P,H,EtM, diHMone,diHCone, diHC,oxyC,apoM, AcdiHCone,mep,meth, normeth,dex,lev, keto,Ph,A,S,Q, scopo,nic,coc,LSD | | Identification by TLC and GLC | $SiO_2$ | MeOH-12 *M* $NH_4OH$ | 369 |
| C,M,No,P,T | | Detection method | $SiO_2$ | Benzene-MeOH(8:2) | 373 |
| C,M,H,caf,S | | Quantitative analysis of S | $SiO_2$ | MeOH-28% $NH_4OH$(100:1.5)<br>$CHCl_3$-$Et_2O$-MeOH-28% $NH_4OH$(75:25:5:1) | 386 |
| H | | Detection limit of H | $SiO_2$<br>cellulose | Benzene-MeOH(66:34)<br>PrOH-$NH_4OH$(95:5) | 387 |
| C,M,Na,No,P,T,H,AcC, 6OAcM,3OAcM,Q,S,caf, LSD,psilocin, psilocybin | | Rapid detection | $SiO_2$ | Toluene-$Me_2CO$-EtOH-conc. $NH_4OH$(45:45:7:3)<br>$CHCl_3$-MeOH(9:1)<br>*n*-BuOH-AcOH-$H_2O$(65:13:22) | 390 |
| C,M,H,EtM,mep,meth, coc,S | Amphetamines, barbiturates, procaine,amp, pentazocine | Identification | $SiO_2$ | MeOH-$NH_4OH$(100:1.5)<br>Toluene-$Me_2CO$-EtOAc-$NH_4OH$(45:45:7:3) | 391 |
| M,H,mep,S,eph,coc | Amphetamines | Identification | | No details available | 392 |

References p. 245

TABLE 12.21 (*continued*)

| Alkaloid* | Other compound | Aim | Adsorbent | Solvent system | Ref. |
|---|---|---|---|---|---|
| C,M,H,oxyM,oxyC, diHMone,diHCone, acetyl,propionyl and butyl derivatives of C and EtM, 3-O- and 6-O-acetyl,propionyl and butyryl derivatives of M (mono- and di-),lev, racemorphan, racemethorphan | | Forensic identification of H | $SiO_2$ | Benzene-dioxane-EtOH-$NH_4OH$(50:40:5:5)<br>EtOAc-MeOH-$NH_4OH$(85:10:5)<br>MeOH-$NH_4OH$(100:1.5)<br>EtOH-AcOH-$H_2O$(6:3:1) | 395 |
| H | | Direct quantitative analysis of H | | No details available | 397 |

*For abbreviations, see footnote to Table 12.7.

II.5. INDOLE ALKALOIDS

# Chapter 13

# TERPENOID INDOLE ALKALOIDS AND SIMPLE INDOLE ALKALOIDS

13.1. *Strychnos* alkaloids .... 307
13.1.1. Solvent systems .... 308
13.1.2. Detection .... 309
13.1.3. Quantitative analysis .... 309
13.1.4. TAS technique and reaction chromatography .... 310
13.2. *Rauwolfia* alkaloids .... 310
13.2.1. Solvent systems .... 311
13.2.2. Detection .... 312
13.2.3. Quantitative analysis .... 313
13.2.4. TAS technique and reaction chromatography .... 315
13.3. Heteroyohimbine and oxindole alkaloids (*Mitragyna* and *Uncaria* alkaloids .... 316
13.3.1. Solvent systems .... 317
13.3.2. Detection .... 317
13.3.3. Quantitative analysis .... 317
13.4. *Vinca* and *Catharanthus* alkaloids .... 318
13.4.1. Solvent systems .... 319
13.4.2. Detection .... 320
13.4.3. Quantitative analysis .... 320
13.5. Physostigmine .... 321
13.5.1. Solvent systems .... 321
13.5.2. Detection .... 322
13.5.3. Quantitative analysis .... 322
13.5.4. TAS technique and reaction chromatography .... 323
13.6. Various indole alkaloids .... 323
References .... 324

In this survey indole alkaloids are divided into two major groups: (1) terpenoid indole alkaloids and (2) ergot indole alkaloids. The terpenoid indole alkaloids dealt with in this chapter are classified into subgroups according to their botanical origin.

In Table 13.1 those publications on the analysis of terpenoid indole alkaloids and physostigmine are summarized which already have been dealt with in Chapter 3. In Table 13.2 publications on the analysis of drugs of abuse, including terpenoid indole alkaloids and physostigmine, are listed.

## 13.1. *STRYCHNOS* ALKALOIDS

About 300 *Strychnos* alkaloids are known. Of these, strychnine and brucine have been analysed most extensively by TLC, as present in plant material or plant extracts (Table 13.8), in pharmaceutical preparations (Table 13.9), in toxicol-

References p. 324

ogical analysis (Table 13.10) or as adulterants in drugs of abuse (Table 13.2). Because strychnine and brucine are easy to separate, this pair of alkaloids has often been used to illustrate new separation methods.

*13.1.1. Solvent systems*

Some solvent systems that have been used in the analysis of *Strychnos* alkaloids are summarized in Table 13.3. Because of the strongly basic properties of strychnine and related alkaloids, the use of basic solvent systems in combination with silica gel is preferred in order to obtain optimal results. The closely related solvent systems S2, S3 and S4 and ethyl acetate - isopropanol - concentrated ammonia (45:35:10, 80:15:5: or 100:2:1) in combination with silica gel plates are useful for the separation of the different types of *Strychnos* alkaloids[158,200,201]. The TLC behaviour of 30 *Strychnos* alkaloids on silica gel and aluminium oxide plates has been studied by Phillipson and Bisset[124]. Seven solvents with and seven similar solvents without diethylamine were used. The authors made an attempt to explain the adsorption of the alkaloids in terms of their structural features. Bisset and Phillipson[155,200] described a combination of four TLC systems for the identification of alkaloids in *Strychnos* species. Ethyl acetate - isopropanol - 5.5% ammonia (45:35:20) in combination with silica gel plates was used for the separation of polar alkaloids, for example the N-oxides from their parent alkaloids. Ethyl acetate - isopropanol - concentrated ammonia (80:15:5) in combination with silica gel was used for the separation of the "normal" series (including strychnine and brucine) from the less polar "pseudo" series (including pseudostrychnine and pseudobrucine) and N-methyl-*sec.*-pseudo series of alkaloids (including icajine, novacine and vomicine). *n*-Butanol - 0.1 *M* hydrochloric acid + 7.4% aqueous potassium hexacyanoferrate(III) (100:15:34) in combination with silica gel plates was used to separate the pseudo and N-methyl-*sec.*-pseudo series of alkaloids, although the spots of the pseudo alkaloids tended to be elongated. In ethyl acetate - isopropanol - concentrated ammonia (100:2:1) in combination with silica gel plates the spots of alkaloids of the two series partially overlap.

Verpoorte and Baerheim Svendsen[196] separated a series of quaternary alkaloids, N-oxides and tertiary alkaloids (Table 13.4). Grandolini et al.[39] described a two-dimensional TLC separation of a number of *Strychnos* alkaloids. Quirin et al.[41] used base-impregnated silica gel plates and two different developing solvents in the same direction to separate the alkaloids present in the leaves of *Strychnos nux vomica*. Rama Rao and Tandon[213,215] separated a series of alkaloids, including strychnine and brucine, on silica gel and aluminium oxide plates impregnated with metal salts (see Chapter 3, p.48). Munier and Drapier[192,202,203] claimed that alkaloids, including strychnine and brucine, can be separated on silica gel or cellulose plates with the use of a mobile phase containing chlorinated organic solvents (see Chapter 3, p.48).

### *13.1.2. Detection*

Many different detection methods have been used for *Strychnos* alkaloids. Of the non-specific methods, quenching of UV light of 254 nm on fluorescent plates is the most sensitive (0.01-0.1 μg). With iodine the detection limit is 0.01-0.1 μg. Of the general alkaloid reagents, Dragendorff's reagent (Munier or Munier and Machebouef modification) is the most sensitive (0.01-0.1 μg). Potassium iodoplatinate reagent has a sensitivity of 0.1-1 μg and gives different colours for some of the *Strychnos* alkaloids. This may be useful in the identification (see Table 13.5). The recovery of the alkaloids after spraying with iodoplatinate reagent was studied by Holdstock and Stevens[175] (see Chapter 2, p.14).

In Table 13.6 a survey is presented of the detection methods described for indole alkaloids. The reagents were tested in our laboratories on a series of indole alkaloids. Iron(III) chloride in perchloric acid[155] was found to be useful as selective spray reagent for *Strychnos* alkaloids.

The colours obtained for a number of *Strychnos* alkaloids are summarized in Table 13.7. Cerium(IV) sulphate in sulphuric acid also proved to be useful as a selective spray reagent. In particular, dimeric *Strychnos* alkaloids give specific colours with this reagent, but also a number of monomers can be detected. For the detection of dimeric alkaloids cinnamic aldehyde-hydrochloric acid is also suitable[1].

Ammonium vanadate in nitric acid has also been used as selective spray reagent[166]. Mathis and Dequénois described colour reactions of some *Strychnos* alkaloids for paper chromatography[19].

Vinson and co-workers[179,183] applied TCBI reagent (no. 94) for the identification of a series of alkaloids, including some indole alkaloids (see Chapter 2, p.17) and Grant[195] used cobalt thiocyanate reagent (no.26a). Kaniewska and Borkowski[89] used chromotropic acid (no.20) for the detection of some alkaloids. Okumura et al.[180] developed a flame-ionization detection method for TLC. The method was tested on some alkaloids, including strychnine and brucine.

### *13.1.3. Quantitative analysis*

For the indirect quantitative analysis of strychnine and brucine dilute hydrochloric acid[177], 2% aqueous acetic acid[36], chloroform[51] and ethanol[62,69,140] have been used as extraction solvents. Ragazzi et al.[38] used magnesium oxide as the sorbent, which enabled them to dissolve the sorbent and the alkaloids to be determined in aqueous mineral acids. Direct quantitative analysis of strychnine and brucine by measuring the spot size was used by Chowdhury and Chowdhury[172] and by Metwally[188] by visually comparing the spot areas. Massa et al.[121] determined a number of alkaloids, including strychnine and brucine, densitometrically using quenching of 254-nm UV light. Okumura et al.[180] used a modified flame-ionization detector for the quantitative analysis of alkaloids, including strychnine and brucine, on sintered silica gel or aluminium oxide rods.

### *13.1.4. TAS technique and reaction chromatography*

The TAS technique has been applied to *Strychnos nux vomica* seeds by Jolliffe and Shellard[123]. Strychnine and brucine were detected but the pattern of the other compounds was not consistent. Stahl and Schmitt[171] also analysed strychnine and brucine by the TAS technique, but they did not specify the experimental conditions.

Kaess and Mathis applied reaction chromatography to some alkaloids present in *Strychnos nux vomica*[56,122]. Strychnine could be reduced with sodium borohydride, the other alkaloids not. The reduced strychnine was detected with sodium nitrite. α-Colubrine, β-colubrine and brucine could be oxidized with nitric acid, whereas strychnine and vomicine remained unchanged. The oxidation products of α-colubrine and β-colubrine were separated by TLC whereas the mother alkaloids were difficult to separate. The reactions mentioned were also applied to the identification of the alkaloids isolated from plant material and extracts. Wilk and Brill[86] exposed the alkaloids, after application on silica gel plates, to iodine vapour for 18 h. After removal of the excess of iodine the plates were developed in benzene-methanol-acetone-acetic acid (70:20:5:5). Strychnine and brucine gave characteristic patterns of oxidation products.

## 13.2. *RAUWOLFIA* ALKALOIDS

Most of the *Rauwolfia* alkaloids belong to the β-carboline type of alkaloids. Both mixtures of *Rauwolfia* alkaloids and pure alkaloids (including reserpine and rescinnamine) are used as drugs. The TLC analyses described for this type of alkaloid therefore relate to pharmaceutical preparations or plant extracts, and sometimes to biological materials.

Several of the *Rauwolfia* alkaloids also occur in other plant genera, e.g., *Strychnos*, *Vinca*, *Catharanthus*, and *Mitragyna* alkaloids, and therefore also appear in other sections.

*13.2.1. Solvent systems*

*Rauwolfia* alkaloids are unstable in light, particularly in chloroform solution and on dry TLC plates[2,5,57,71,134,143]. Methanolic solutions are more stable[71]. With reserpine and rescinnamine, the 3-dehydro- and 3-iso (epimer) compounds are readily formed on TLC plates. Under certain conditions the lumi derivatives (tetradehydro compounds) are also formed. As anhydronium alkaloids they are much more polar and have strong fluorescence[134]. To compensate for the decomposition of reserpine after application on the TLC plate Frijns[143] introduced a correction factor in the quantitative analysis. The decomposition during development is negligible according to Frijns. However, after drying of the plates the 3-dehydro compounds are readily formed under the influence of light. This reaction can be used for detection and direct fluorodensitometric analysis[143]. Court[57] recommended developing the plates in the dark and the use of methanolic sample solutions. Reserpine and rescinnamine can also decompose in acidic or basic solution, yielding the hydrolysis products reserpic acid and methyl reserpate[71].

For the TLC analysis of *Rauwolfia* alkaloids a series of TLC systems have been described. Because of the weakly basic properties of many of the alkaloids neutral solvents have been used succesfully. Schlemmer and Link[4,5] described solvents system S1 for the separation of reserpine and rescinnamine and their decomposition products. According to Court[57], saturation of the chromatography tank is an important factor for the successful use of this TLC system. Prefractionation of plant extracts was also used to simplify the separation of complex alkaloid mixtures[57,189]. Phillipson and Shellard[63,65,76] studied the relationship between the stereochemistry of and the substituents in a number of heteroyohimbine alkaloids, including *Rauwolfia* alkaloids, and also their retention behaviour. TLC can be used to obtain information on the stereochemistry of these alkaloids, according to the authors.

Harris et al.[104] and Los and Court[111] found solvent system S10 to be suitable for the quantitative analysis of weakly basic alkaloids, including reserpine and rescinnamine. Owing to double fronting of this system other combinations of alcohols were tested. This led to the systems as described in Table 13.14[138]. Court et al.[138] found the initial spot diameter to be important for achieving satisfactory separations. The critical volume that could be applied in order to separate reserpine and rescinnamine in solvent system S1 depended on the solvent used for the sample (0.75 µl in methanol, 1 µl in chloroform and 1.5 µl in acetone for solvent S1, and 2 µl in chloroform and 3 µl in methanol for solvent S10, all relating to 1-5 µg of alkaloid). Extracts of plant material were best

**References p. 324**

prepared with benzene for weakly basic alkaloids and with diethyl ether, acetic acid or methanol for stronger bases.

Court and Habib[153] investigated about 300 solvent systems and found that no single solvent system could separate all the *Rauwolfia* alkaloids adequately for a quantitative determination. Solvent system S9 was found to be best for the difficult separations reserpine/rescinnamine and ajmalicine/reserpinine. The ten solvents described as the most appropriate by Court and co-workers[153,189] were systems S1 and S7-S15 in Table 13.14. The authors also discussed the relationship between structure and retention behaviour. For the quantitative assay of the alkaloids in *Rauwolfia* plant material at least five solvent systems were needed. To obtain a separation of alstonine and serpentine solvent system S4 can be used[59], as well as chloroform-methanol (85:15) in ammonia vapour[189] and *n*-butanol-chloroform-25% ammonia (50:50:2.5)[126], all in combination with silica gel.

For the difficult separation of reserpine and rescinnamine Hartmann and Schnabel[186] applied solvent system S5. Other TLC systems that have been reported to separate these two alkaloids are heptane - methyl ethyl ketone (1:1) in an ammonia atmosphere in combination with formamide-impregnated cellulose plates[6] and heptane - methyl ethyl ketone (2:1) saturated with water in combination with formamide-impregnated silica gel[11].

To separate reserpine from its decomposition products Pötter and Voigt[83] developed the silica gel plates with methanol - methyl ethyl ketone - heptane (1:1:1 and 8.4:33.6:58.0) in the same direction. A combination of TLC analysis and UV, IR and mass spectrometry in the identification of *Rauwolfia* alkaloids has been described by Habib and Court[170].

### *13.2.2. Detection*

Of non-specific detection methods, quenching of UV light on fluorescent plates is the most sensitive for *Rauwolfia* alkaloids. Of the modifications of Dragendorff's reagent, the Munier and Macheboeuf version is the most sensitive (R. Verpoorte, unpublished results). Iodoplatinate reagent has about the same sensitivity and the advantage that different colours are obtained for the various alkaloids (Table 13.16)[43,153,189]. Some detection reagents for *Rauwolfia* alkaloids are summarized in Table 13.12.

The various fluorescence colours of the alkaloids in UV light of 254 and 366 nm are useful in their identification (Table 13.16)[3,43,153,189].

Sams and Huffman[217] compared several methods to obtain optimal fluorescence for reserpine, suitable for quantitative analysis, and found that exposure to acetic acid vapour for 24 h yielded maximal fluorescence. The fluorescence was obtained immediately if the plates were sprayed with a 1% solution of *p*-toluenesulphonic acid in glacial acetic acid.

Court et al.[57,153,189] used several colour reactions for the detection of *Rauwolfia* alkaloids. The reagents used and the colours obtained for some alkaloids are summarized in Tables 13.13 and 13.14.

Teichert et al.[6] used concentrated nitric acid for the detection of sarpagine. Isaac et al.[77] described the TLC analysis of a series of β-carboline derivatives, including three *Rauwolfia* alkaloids. As detection reagents the following were used: xanthydrol reagent (no. 106), Ehrlich's reagent (no. 36e), Brentamine Fast Blue B reagent (no. 40a) ninhydrin reagent (no. 74c) and iron(III) chloride in sulphuric acid-perchloric acid reagent (no. 64).

For reserpine, TCBI reagent (no. 94)[179,183] (see Chapter 2, Table 2.3), π-acceptors[207] (see Chapter 2, Table 2.2) and iron(III) chloride glyoxylic acid reagent (no. 58)[167] have been used. Menn and McBain[66] described a detection method for cholinesterase inhibitors, including reserpine (no. 19) (see Chapter 2, p.15). Okumura et al.[180] used a flame-ionization detection method for alkaloids, including reserpine.

### *13.2.3. Quantitative analysis*

Schlemmer and Link[4,5] determined reserpine and rescinnamine spectrophotometrically in *Rauwolfia* preparations after separation on silica gel plates and elution with dioxane - 96% ethanol (1:1). The same method has also been used by several others[11,14,83]. Pötter and Voigt[83] modified the method by using chloroform-methanol (1:1) for the elution of the alkaloids. The same solvent was used by Bonati and Pesce[59] to elute serpentine and alstonine and their tetrahydro derivatives from silica gel. Other solvents have also been used to elute *Rauwolfia* alkaloids from silica gel plates: ethanol[208] for reserpine, chloroform-methanol (9:1) for reserpine and rescinnamine[79], chloroform - methanol (1:1) for ajmalicine[79], methanol for ajmaline, ajmalicine and yohimbine[148] and methanol - chloroform (9:1) for reserpine, rescinnamine, reserpiline and ajmalicine[120]. Seysen et al.[131] used different solvents for the elution of reserpine from silica gel depending on the quantitative method to be applied afterwards: chloroform - methanol (1:1, 9:1 and 1:9), dichloromethane - methanol (1:1), cyclohexane - methanol - dichloromethane (30:4:65), chloroform - water - potassium chloride and chloroform to which a drop of sodium hydroxide was added.

For the elution of reserpine from aluminium oxide chloroform has been used[116]. Rutkowska and Wojsa[21,60] eluted reserpine from silica gel - cellulose - starch (10:4:0.25) with a mixture of 40% acetic acid and 0.3% hydrogen peroxide (95:5) prior to its fluorimetric determination.

In a series of papers Court et al.[104,105,111,132,138,147,153] described the quantitative analysis of a number of *Rauwolfia* alkaloids in plant materials. The determinations were based on elution of the alkaloids from the sorbent before spectrophotometric analysis at an appropriate wavelength. The sorbent was collected from the plate by means of a microvacuum cleaner. In the same way the alkaloids could be transferred from the plate to potassium bromide micro-discs for IR spectroscopy[142]. In that case chloroform was used for the elution. For quantitative analysis ethanol and methanol were used for reserpine and ajmalicine and *n*-butanol - acetic acid - water (4:1:1) for rescinnamine, ajmaline and serpentine. These solvents gave better results than dioxane - ethanol (1:1). Because of the instability of the alkaloids the analysis has to be carried out in the dark or in subdued light. Habib and Court[147] described the quantitative analysis of *Rauwolfia* alkaloids by spectrophotometry after complexation with iodine. By this method interferences due to materials extracted from the adsorbent at shorter wavelengths is avoided. The method is applicable to all β-carboline alkaloids. The elution of the alkaloids from the sorbents was carried out in a separating funnel with chloroform - methanol (95:5) after addition of dilute ammonia or 0.9% sodium hydroxide solution.

Ragazzi and Veronese[50] used magnesium oxide as the sorbent, whereby quantitative analysis could be performed by dissolving both the adsorbent and the alkaloid in aqueous acid.

Manara[72] determined tritium-labelled reserpine by means of liquid scintillation counting. After isolation from biological material and purification by means of TLC the spots of reserpine were scraped off the plate and transferred into the counting vials containing naphthalene-dioxane scintillation fluid.

Direct quantitative analysis of *Rauwolfia* alkaloids has been applied in many instances. Hashmi et al.[110] developed a semi-quantitative method using circular TLC. Massa et al.[121] determined a number of alkaloids quantitatively by means of densitometry. Quantitative determinations based upon measurement of quenching of UV light (254 nm) on fluorescent plates or direct measurement of fluorescence of the alkaloids were found to be more sensitive than a determination based upon measurement at 400 nm after spraying with Dragendorff's reagent.

Hartmann and Schnabel[186] determined reserpine and rescinnamine by means of densitometry at 295 nm. A direct fluorimetric method was described by Kammerl and Mutschler (excitation wavelength 365 nm, emission measured at 495 nm)[157]. Heusser[37] determined reserpine and Hammer and Kaiser[144] both reserpine and rescinnamine after extraction from plant material by direct fluorodensitometry.

Tripp et al.[185] determined reserpine in human plasma by fluorodensitometry. The fluorescence was obtained by exposure to the acetic acid vapour (excitation wavelength 392 nm, emission determined at 540 nm). Sams and Huffman[217] modified this method; instead of exposure to acetic acid vapour for 24 h the plates were sprayed with 1% *p*-toluenesulphonic acid in glacial acetic acid. In this way the fluorescence colour was developed immediately. Frijns[134] developed the fluorescence by exposure of the alkaloid spots to light for at least 2 h. Reserpine and rescinnamine yielded the fluorescent 3-dehydro derivatives. The determination was performed by using an excitation wavelength of 365 nm and measuring the emission at 490 nm. Formation of the 3-dehydro derivatives was found to take place during the application of the sample, and a correction factor was therefore introduced to compensate for loss during the application. During the development of the plate no decomposition was observed.

Dombrowski et al.[184] determined 17-monochloroacetylajmaline and its metabolite ajmaline in plasma by direct fluorodensitometric analysis. Fluorescent spots were obtained after spraying with 6 *M* nitric acid and acetic acid and gently heating the plates. Emission was measured at 465 nm after excitation at 365 nm.

### *13.2.4. TAS technique and reaction chromatography*

Jolliffe and Shellard[123] applied the TAS technique to *Rauwolfia* preparations; Dragendorff's reagent revealed several spots. The conditions were oven temperature 275°C, distillation time 90 sec, 10-20 mg of sample mixed with 10 mg calcium hydroxide and silica gel of suitable moisture content as propellant.

Kaess and Mathis[43,122] described the reaction chromatography of some *Rauwolfia* alkaloids. Reserpine and rescinnamine could be oxidized to the lumi-compounds under the influence of heat. However, complete oxidation was obtained by first adding one drop of 1% potassium permanganate to the sample after application on the plate. This reaction permits the identification of the two alkaloids in a mixture with other alkaloids.

Reserpine, rescinnamine, methyl reserpate, yohimbine and serpentine contain ester groups that can be saponified with ethanolic potassium hydroxide (1 *N* in 90% ethanol) by heating in a sealed capillary at 100°C for 1 h. Reserpine and rescinnamine can be distinguished in this way by two-dimensional TLC; first reserpic acid and methyl reserpate are separated from trimethoxybenzoic acid and trimethoxycinnamic acid and then the two acids are separated [chloroform-cyclohexane-acetic acid (7:2:1)]. The plate is sprayed with 1% potassium permanganate solution; trimethoxycinnamic acid decolorizes the spray reagent and can be observed as a fluorescent spot, whereas trimethoxybenzoic acid does not react.

Ajmaline, yohimbine and sarpagine can be acetylated. The latter two alkaloids give the mono-O-acetyl compounds whereas ajmaline gives several products including, the diacetyl and the mono-O-acetyl compounds. A longer reaction time gives complete conversion to the diacetyl derivative of ajmaline.

**References p. 324**

Wilk and Brill[86] exposed TLC plates with applied alkaloids to iodine vapour for 18 h prior to development of the plates. For yohimbine a characteristic pattern of spots was observed.

## 13.3. HETEROYOHIMBINE AND OXINDOLE ALKALOIDS (*MITRAGYNA* AND *UNCARIA* ALKALOIDS)

### *13.3.1. Solvent systems*

In a series of papers, Phillipson and Shellard discussed the TLC behaviour of a number of heteroyohimbine and oxindole alkaloids, mainly originating from *Mitragyna* species. The chromatographic behaviour was explained in terms of substituent effects and configuration of the alkaloids[32,63,65,76,95,96].

It was found that in heteroyohimbine alkaloids methoxyl substitution in the aromatic ring lowers the $hR_F$ values. In alkaloids with a similar configuration the $hR_F$ values decrease in the order no substitution, 9-, 10- and 11-methoxy substitution. Concerning the stereochemistry of C-3 and C-20 of the alkaloids with an open E-ring (E-seco) and a closed E-ring, it was found that they could be arranged in order of decreasing $hR_F$ values as allo, normal, epiallo and pseudo, with the exception of the allo closed E-ring alkaloids with 19-$CH_3$ in the β-position[63,65,76]. For E-seco oxindole alkaloids[95] both the carbonyl function and $N_b$ were found to be involved in the adsorption process. Depending on the configuration steric hindrance can decrease the availability of these groups for hydrogen bonding with the silanol groups. The influence of hydroxy and methoxy substituents was also discussed.

The TLC behaviour of closed E-ring oxindole alkaloids was discussed[96] in terms of the availability of $N_b$ and the oxindole carbonyl function for hydrogen bond formation with the hydroxyl groups of the adsorbent. In Tables 13.18-13.21 the alkaloids and TLC systems used in the studies[63,76,95,96] are summarized.

Of the solvents mentioned[63] it was found that cyclohexane - chloroform - diethylamine (5:4:1) on silica gel plates could be used to distinguish between indole and oxindole alkaloids isolated from *Mitragyna* species (indole alkaloids have $hR_F$ values greater than 35 and oxindole alkaloids $hR_F$ values below 35). Chloroform in combination with aluminium oxide was found to have the widest applicability.

For the separation of the *Mitragyna* indole alkaloids chloroform - benzene (1:1) on aluminium oxide, diethyl ether on silica gel and chloroform - acetone (5:4) on aluminium oxide or silica gel were found to be most useful[63]. Shellard and Alam[93] used chloroform and chloroform - cyclohexane (7:3) with aluminium oxide plates, and diethyl ether and chloroform - acetone (5:4) with silica gel plates for the quantitative analysis of six oxindole alkaloids. None of the systems mentioned leads to a complete separation of all six oxindole alkaloids. Shellard et al.[68] described the TLC of 18 *Mitragyna* alkaloids on pH-gradient plates. Shellard and Lala[219] used four solvents for the identification of some *Mitragyna* alkaloids. The results are presented in Table 13.22.

Phillipson and Hemingway[178] described the use of chromatographic methods in combination with spectroscopic methods for the identification of alkaloids from herbarium samples of the genus *Uncaria* (Rubiaceae). About 60 heteroyohimbine, oxindole, roxburghine, simple β-carboline, pyridinoindoloquinolizidinone and gambirtannine types of indole alkaloids could be distinguished. The solvent systems used are listed in Table 13.23 and the detection methods in Table 13.24. Solvents S1-S5 were used for the separation of the tertiary alkaloids and solvents S6-S8 for the separation of the N-oxides.

### *13.3.2. Detection*

For the detection of the heteroyohimbine and oxindole alkaloids the usual alkaloid reagents can be used. For the detection of the oxindoles spraying with 0.2 *M* iron(III) chloride in 35% perchloric acid followed by heating at 60-65°C for 1 h or 120°C for 1.5 h has been used[95,97]. Concentrated nitric acid has been used for the detection of *Mitragyna* alkaloids, either as such or as the first step in the reaction of Vitali-Morin[94]. In the last case the plate is heated for 1 h at 130°C after spraying with concentrated nitric acid. Subsequently the plates are sprayed with an excess of freshly prepared 2.5% tetraethylammonium hydroxide solution in dimethylformamide. The detection methods used for *Uncaria* alkaloids by Phillipson and Hemingway[178] are summarized in Table 13.24.

### *13.3.3. Quantitative analysis*

Shellard and Alam discussed in a series of papers the quantitative analysis of *Mitragyna* alkaloids (rotundifoline, rhynchophylline, mitraphylline and their iso-compounds) by densitometric analysis[97,99] or by indirect spectrophotometric[93] or colorimetric[94,99] quantitative analysis.

In spectrophotometry, wavelengths of 223 and 242 nm were used for the oxindole alkaloids. The alkaloids were eluted from the sorbent with methanol; other solvents did not give complete recovery. Elimination of interfering substances extracted from the sorbent was achieved by extracting a similar area of sorbent with the same $hR_F$ value as the alkaloid to be determined from the same plate and using the solution as a reference. For mixtures of alkaloids that could not be separated completely, differential spectrophotometry was used to determine the alkaloids. Several independent TLC separations in combination with subtraction of the absorbance of the separated alkaloids from the absorbance of mixtures of alkaloids could also be used. In this way six oxindole alkaloids were analysed.

Vitali-Morin reaction has been used for colorimetric analysis[94]. The alkaloids were nitrated on the plate by spraying with concentrated nitric acid and heating at 130$^{o}$C for 1 h. Coloured spots were obtained, which were scraped off and extracted with dimethylformamide. The colour was produced by addition of a 2.5% solution of tetraethylammonium hydroxide in dimethylformamide. A reference solution was prepared by eluting the sorbent material with the same $hR_F$ value as the alkaloid to be analysed and treating it in the same way. When aluminium oxide was used as sorbent the results were unsatisfactory. However, better results were obtained if the TEAH reagent was sprayed on the plate prior to the extraction. Differential spectrophotometry could be used with unseparated compounds.

Densitometry has also been used for the quantitative analysis of *Mitragyna* alkaloids[97,99,106]. They were made visible by spraying with iron(III) chloride (0.2 *M*) in 35% perchloric acid and heating at 60-65$^{o}$C for 1 h or at 120$^{o}$C for 1.5 h. The latter procedure gave more stable colours. The determination was performed at 465 nm for the alkaloids tested.

By comparing the three methods mentioned above[99], it was found that in the analysis of mixtures of alkaloids isolated from plant material ultraviolet spectrophotometry was the most accurate and the densitometry the least accurate method. The advantage of the indirect methods is that the spot shape and the initial spot area do not influence the results as in the densitometric method. The complexity of the mixture of alkaloids is important for the choice of the method to be used; for mixtures of easily separated alkaloids all three methods can be used.

## 13.4. *VINCA* AND *CATHARANTHUS* ALKALOIDS

Because of the leukopenic and antineoplastic effects of some of the dimeric alkaloids isolated from *Vinca* and *Catharanthus* species (vincaleukoblastine, leurocristine, leurosine and leurosidine), a great deal of phytochemical research has been done on the alkaloids of these species. TLC in combination with selective spray reagents played an important role in this research.

Many of the alkaloids found in the plant genera mentioned also occur in other genera, viz., *Rauwolfia, Strychnos* and *Mitragyna*. The TLC analysis of these alkaloids has been dealt with in the previous sections.

*13.4.1. Solvent systems*

For the analysis of the alkaloids in *Vinca* species Cone et al.[26] described eleven solvent systems in combination with either aluminium oxide or silica gel (Table 13.25). Two-dimensional TLC was used for the identification of the alkaloids in complex mixtures. In Tables 26 and 27 the TLC systems suitable for particular separations are summarized.

Jakovljevic et al.[31] observed less tailing with aluminium oxide than with silica gel. The best results were obtained with aluminium oxide impregnated with 0.5 *M* lithium hydroxide and acetonitrile - absolute ethanol (95:5) or acetonitrile - benzene (3:7) as the solvent. Gröger and Stolle[44] used 0.5% potassium hydroxide-impregnated silica gel to reduce tailing; the solvent used was ethyl acetate or chloroform - acetone (9:1).

Farnsworth et al.[33] described a TLC evaluation of plant extracts. By means of three TLC systems and cerium(IV) ammonium sulphate (no. 12) and Dragendorff's spray reagents, a tentative identification of 63 alkaloids was achieved, even if reference alkaloids were not available in all instances. The alkaloids were divided into chromogenic classes, depending on their colour reactions (Table 13.30). This classification in combination with the three TLC systems permitted the identification (Table 13.28). Colour illustrations of spots typical of the chromogenic classes were included[33]. To obtain reproducible results, silica gel plates had to be used within 72 h after preparation. The mobile phases, ethyl acetate - absolute ethanol (3:1) and absolute methanol, should be dried over anhydrous sodium sulphate immediately prior to their use.

Farnsworth and Hilinski[46] applied chloroform - methanol (95:5) in combination with silica gel plates for the separation of the four dimeric alkaloids vincaleukoblastine, leurocristine, leurosine and leurosidine. However, better results were obtained by means of two-dimensional TLC using methanol as the second solvent.

Vachnadze et al.[156] used an identification scheme similar to that of Farnsworth et al.[33]. Thirty-eight alkaloids were identified via their colour reactions after spraying with 1% cerium(IV) sulphate in 85% phosphoric acid (no. 12) and their TLC behaviour on silica gel plates with benzene - ethyl acetate - methanol (2:3:1), benzene - ethyl acetate (2:3) and benzene - methanol (3:2) was discussed.

Kasymov et al.[82] described three TLC systems for the identification of some *Vinca* alkaloids (see Table 13.29).

### 13.4.2. *Detection*

For a survey of the possible spray reagents and their sensitivities, see Table 13.12 in Section 13.2 on *Rauwolfia* alkaloids (p.340).

Cerium(IV) ammonium sulphate reagent (no. 12) was described by Jakovljevic et al.[31] as highly specific for *Vinca* and *Catharanthus* alkaloids. They found that it gives the best results if the TLC plates are dried for 5 min at 100°C prior to spraying. This should be done within 30 min after development of the plate. If a 1:1 dilution of the reagent in water is used, repeated spraying may be necessary. The colours are stable if kept at 90-100°C and if moisture is excluded.

Farnsworth et al.[33] used this reagent in the identification of 63 *Vinca* and *Catharanthus* alkaloids. Eight alkaloids gave no colour and they were detected with Dragendorff's reagent, which was sprayed 24 h after the cerium(IV) ammonium sulphate reagent. No interference from the latter reagent was observed. Some alkaloids could be identified prior to spraying because of their fluorescence. The colours obtained with cerium(IV) ammonium sulphate reagent were observed at five time intervals (see Table 13.30). The first major colour was used for classification of the alkaloids, and subclasses were assigned on the basis of colour changes during the first 5 min after spraying. The characteristic colours of classes I and II and their subclasses are presented in coloured plates in the original publication[33]. The factors that influence the colours are the concentration of the alkaloid (particularly if the centre of the spot has a colour different to that of the edge), the $hR_F$ values (different spot sizes lead to different concentrations per unit area), residual solvent on the plate, heat and time between development of the plate and spraying.

Vachnadze et al.[156,176] used iron(III) chloride in perchloric acid in the analysis of *Vinca* alkaloids. Iron(III) chloride was used for some hydroxyindole alkaloids that did not give colours with cerium(IV) ammonium sulphate reagent. For the latter reagent Vachnadze et al.[156] found a relationship between the chromophoric group and the colours obtained. In this way α-methyleneindole, indoline, indole and hydroxyindole chromophores could be distinguished. Aromatic substituents influence the colours obtained (Table 13.31).

### 13.4.3. *Quantitative analysis*

Indirect quantitative analyses of *Vinca* and *Catharanthus* alkaloids have been performed spectrophotometrically[44,145,199,210] and by UV spectroscopy[101,133,205,206]

Karacsony et al.[40] used non-aqueous titration for the quantitative analysis of vincamine, after chromatographic separation. Mirkina and Shakirov[133] used chloroform - methanol (8:2) or chloroform - 1% tartaric acid in methanol (9:1) for the elution of vincamine from silica gel. Mudzhiri et al.[145] extracted akuammine from silica gel with 5% sulphuric acid, prior to spectrophotometric determination [with cerium(IV) ammonium sulphate at 470 nm]. Masoud et al.[101] determined the vincaleukoblastine content of *Catharanthus roseus* by UV spectroscopy at 214 nm, after two-dimensional TLC-separation and elution of the alkaloid with methanol. Sarin et al.[205] used hot chloroform - methanol (1:1) to elute ajmalicine from silica gel for its quantitative analysis in *Catharanthus roseus*.

The direct quantitative analysis of some *Vinca* alkaloids has been described by Massa et al.[121], who used densitometry after spraying with Dragendorff's reagent (400 nm) or measured the quenching of UV light on fluorescent plates (254 nm). Vachnadze[176] measured the spot areas after detection with cerium(IV) ammonium sulphate or with iron(III) chloride in perchloric acid in the quantitative analysis of 19 *Vinca* alkaloids.

The TLC analysis of *Vinca* and *Catharanthus* alkaloids is summarized in Table 13.22.

## 13.5. PHYSOSTIGMINE

### *13.5.1. Solvent systems*

Most of the chromatographic data available on physostigmine are found in a number of papers dealing with the identification of drugs (Table 13.1). For the assay of physostigmine in ophthalmic solutions, Berg[54] used TLC. Three solvents were found to be useful in the separation of physostigmine from its decomposition products, eserinol and rubreserine (Table 13.33). Two-dimensional TLC was performed with the combination of TLC systems S3 and S1 (Table 13.33). Rubreserine decomposed during the TLC separation if the mobile phase contained diethylamine; with dimethylamine the rate of decomposition was reduced to an acceptable degree. Güven and Unay[90] separated strychnine, brucine and physostigmine with methanol on alkaline silica gel plates. Paris and Paris[22] chromatographed physostigmine N-oxide by means of chloroform - methanol - diethylamine (90:10:5) on silica gel plates.

Smith[135] used chloroform - acetone (5:4) in combination with aluminium oxide plates. He observed that rubreserine reacted with dimethylamine and within 10 min a yellow compound was formed with an $hR_F$ value close to that of physostigmine. Solvent system S4 was used by Rogers and Smith[154] in the quantitative analysis of physostigmine.

### 13.5.2. *Detection*

Berg[54] found that the decomposition products of physostigmine did not give coloured spots with the Dragendorff's reagent. Physostigmine itself was detected with a sensitivity of about 0.1 - 1 μg with the different modifications of Dragendorff's reagent.

Iodine-containing reagents such as iodine vapour, iodine in chloroform and iodine solutions in potassium iodide can also be used to detect physostigmine. Iodoplatinate reagent gives a purple - violet colour with a sensitivity of about 0.1 μg and cerium(IV) ammonium sulphate gives (no. 12) a yellow-orange colour. Iron(III) chloride in perchloric acid (no. 62a) gives a faint brown colour only after heating. Güven and Unay[90] used a 10% aqueous solution of copper sulphate + 2% ammonia (5:1) (no. 27) as the spray reagent. Physostigmine gave a dark blue colour, which changed to red-brown after heating for 10 min at 110$^{o}$C.

The specific cholinesterase inhibitor detection method as described by Menn and McBain[66] (no. 19) is also suitable for physostigmine (see Chapter 2, p.15).

### 13.5.3. *Quantitative analysis*

Berg[54] found that neutral organic solvents and acidic polar solvents were unable to elute physostigmine completely from silica gel. With 0.1 *M* sodium hydroxide solution satisfactory results were obtained, permitting immediate spectrophotometric determination at 480 nm. For the separation of the alkaloid solvent system S3 was used.

Rogers and Smith[154] determined physostigmine indirectly after separation from its degradation products with solvent system S4. The alkaloid was eluted from aluminium oxide with methanolic hydrochloric acid. For correction for irrelevant absorbance in the UV spectrophotometric determination two methods were used: a differential method in which absorbance measurements were made at three wavelengths and a method in which orthogonal functions were applied to absorbance measurements at a set of nine wavelengths. The reproducibility of the elution method was found to be slightly better than the direct method described by Smith[135]. By the direct method physostigmine was determined by densitometry in the reflectance mode at 249 nm.

Massa et al.[121] determined physostigmine by means of densitometry after detection of the alkaloid with Dragendorff's reagent. Reflection was measured at 400 nm. Ebel et al.[128] assayed some alkaloids, including physostigmine, in eyedrops. The alkaloids were determined by means of spot area measurements. For physostigmine a 10% error was found, which therefore enables only semi-quantitative analyses to be performed.

### *13.5.4. TAS technique and reaction chromatography*

Jolliffe and Shellard[123] described the TAS technique for calaber beans. This resulted in one spot with a positive Dragendorff's reaction, identical with physostigmine. The conditions were oven temperature 275°C, distillation time 90 sec, sample mixed with equal amount of calcium hydroxide, and silica gel with a suitable moisture content as propellant. For TLC the solvent systems as described by Waldi[7] were used.

Stahl and Schmitt[171] also found physostigmine unchanged on the plate after TAS. The conditions were oven temperature 220°C, distillation time 2 min, and propellant 50 mg of Molecular sieve 4Å with 20% water. Wilk and Brill[86] described the reaction chromatography of physostigmine. After exposure of the TLC plate with the applied alkaloid to iodine vapour for 18 h, several spots characteristic of physostigmine were observed after development.

Kaess and Mathis[122] treated physostigmine with 0.1 *N* potassium hydroxide solution at 95°C. This led to the formation of a red oxidation product. In Table 13.34 the TLC analysis of physostigmine is summarized.

## 13.6. VARIOUS INDOLE ALKALOIDS

Of other types of indole alkaloids than those which have been dealt with above, little has been published with particular reference to their TLC separation.

McIsaac et al.[77] dealt with the chromatography of 73 β-carboline derivatives, of which five were terpenoid indole alkaloids. The $R_F$ values in three different systems were given, together with the colours observed in UV light and the colour reactions with seven spray reagents (reagents nos. 22, 33a, 36e, 40a, 64, 74c and 106).

The TLC of harman alkaloids was described by Lutomski et al.[73]. Of the twelve TLC systems tested, three were found to be most useful (S1-S3 in Table 13.36).

The quantitative analysis of harman alkaloids was reported by Messerschmidt[98,106]; the alkaloids were determined by means of fluorodensitometry using an excitation wavelength of 360 nm and measuring the emission at 378 nm. Poethke et al.[125] analysed harman alkaloids by eluting with methanol from the sorbent (silica gel), followed by UV spectroscopy. The alkaloids could be detected by means of their fluorescence (see Table 13.35) or by spraying with Dragendorff's reagent. The modification according to Thies and Reuther (no. 39e) was found to be the most sensitive modification of Dragendorff's reagent.

Some publications dealing with the TLC analysis of indole alkaloids, which have not been dealt with in any of the previous sections are summarized in Table 13.36.

REFERENCES


1 J. Kebrlé, H. Schmid, P. Waser and P. Karrer, *Helv. Chim. Acta,* 36 (1953) 102.
2 J. Bayer, *Pharmazie,* 13 (1958) 468.
3 F. Kaiser and A. Popelak, *Chem. Ber.,* 92 (1959) 278.
4 F. Schlemmer and E. Link, *Pharm. Ztg.,* 104 (1959) 646.
5 F. Schlemmer and F. Link, *Pharm. Ztg.,* 104 (1959) 1349.
6 K. Teichert, E. Mutschler and H. Rochelmeyer, *Deut. Apoth.-Ztg.,* 100 (1960) 477.
7 D. Waldi, K. Schnackerz and F. Munter, *J. Chromatogr.,* 6 (1961) 61.
8 N.R. Farnsworth, *Lloydia,* 24 (1961) 105.
9 G.H. Svoboda, *Lloydia,* 24 (1961) 173.
10 J. Bäumler and S. Rippstein, *Pharm. Acta Helv.,* 36 (1961) 382.
11 E. Ullman and H. Kassalitzky, *Arch. Pharm. (Weinheim),* 295 (1962) 37.
12 E. Vidic and J. Schütte, *Arch. Pharm. (Weinheim),* 295 (1962) 342.
13 J. Mokry, L. Dubravkova and P. Sefcovic, *Experientia,* 18 (1962) 564.
14 A. Linkkonen, *Farm. Aikakauslehti,* 71 (1962) 329.
15 N.R. Farnsworth, H.H.S. Fong, R.N. Blomster and F.J. Draus, *J. Pharm. Sci.,* 51 (1962) 217.
16 N.R. Farnsworth and K.L. Euler, *Lloydia,* 25 (1962) 186.
17 E. Caggiano and G.B. Marini-Bettolo, *Rend. Ist. Super Sanita,* 25 (1962) 375.
18 I. Sunshine, *Amer. J. Clin. Pathol.,* 40 (1963) 576.
19 C. Mathis and P. Duquénois, *Ann. Pharm. Fr.,* 21 (1963) 17.
20 W. Rusiecki and M. Henneberg, *Ann. Pharm. Fr.,* 21 (1963) 843.
21 U. Rutkowska and K. Wojsa, *Biul. Inst. Rosliuleczniczyck,* 9 (1963) 192; *C.A.,* 61 (1964) 15018f.
22 R.R. Paris and M. Paris, *Bull. Soc. Chim. Fr.,*(1963) 1597.
23 D. Schumann and H. Schmid, *Helv. Chim. Acta,* 46 (1963) 1996.
24 E. Papp and Z. Szabo, *Herba Hung.,* 2 (1963) 383.
25 M. Ikram, G.A. Miana and M. Islam, *J. Chromatogr.,* 11 (1963) 260.
26 N.J. Cone, R. Miller and N. Neuss, *J. Pharm. Sci.,* 52 (1963) 688.
27 W. Kamp, W.J.M. Onderberg and W.A. Seters, *Pharm. Weekbl.,* 98 (1963) 993.
28 L. Noguiera Prista, M.A. Ferreira and A.S. Roque, *Garcia Orta,* 12 (1964) 277; *C.A.,* 64 (1966) 3959f.
29 L. Noguiera Prista, M.A. Ferreira, A.S. Roque and A. Correia Alves, *Garcia Orta,* 12 (1964) 295; *C.A.,* 64 (1966) 3959f.
30 M. Hesse, W. v. Philipsborn, D. Schumann, G. Spiteller, M. Spiteller-Friedmann, W.I. Taylor, H. Schmid and P. Karrer, *Helv. Chim. Acta,* 47 (1964) 878.
31 I.M. Jakovljevic, L.D. Seay and R.W. Shaffer, *J. Pharm. Sci.,* 53 (1964) 553.
32 E.J. Shellard and J.D. Phillipson, *Kongr. Pharm. Wiss. Vortr. Originalmitt., 23, Münster, 1963,* (1964) 209; *C.A.,* 62 (1965) 6720d.
33 N.R. Farnsworth, R.N. Blomster, D. Damratoski, W. Meer and L.V. Cammarato, *Lloydia,* 27 (1964) 302.
34 I. Inagaki, S. Nishibe and T. Tokuhiro, *Nagoya Shiritsu Daigaku Yakugakubu Kenkyu Nempo,* 12 (1964) 38; *C.A.,* 64 (1966) 20168g.
35 V. Schwarz and M. Sarsunova, *Pharmazie,* 19 (1964) 267.
36 M. Sarsunova, J. Tölgyessy and M. Hradil, *Pharmazie,* 19 (1964) 336.
37 D. Heusser, *Planta Med.,* 12 (1964) 237.
38 E. Ragazzi, G. Veronese and C. Giacobazzi, in G.B. Marini-Bettolo Editor, *Thin-layer Chromatography,* Elsevier, Amsterdam, 1964, p. 150.
39 G. Grandolini, C. Galeffi, E. Montalvo, C.G. Casinovi and G.B. Marini-Bettolo, in G.B. Marini-Bettolo Editor, *Thin-layer Chromatography,* Elsevier, Amsterdam, 1964, p. 155.
40 E.M. Karacsony, I. Gyenes and C. Lorincz, *Acta Pharm. Hung.,* 35 (1965) 280; *C.A.,* 64 (1966) 4868d.
41 M. Quirin, J. Lévy and J. LeMen, *Ann. Pharm. Fr.,* 23 (1965) 93.
42 R. Paris, R. Rousselet, M. Paris and J. Fries, *Ann. Pharm. Fr.,* 23 (1965) 473.
43 A. Kaess and C. Mathis, *Ann. Pharm. Fr.,* 23 (1965) 739.
44 D. Gröger and K. Stolle, *Arch. Pharm. (Weinheim),* 298 (1965) 246.

45 E. Marozzi and G. Falzi, *Farmaco, Ed., Prat.*, 20 (1965) 302.
46 N.R. Farnsworth and I.M. Hilinski, *J. Chromatogr.*, 18 (1965) 184.
47 D. Giacopello, *J. Chromatogr.*, 19 (1965) 172.
48 L.K. Turner, *J. Forensic Sci. Soc.*, 5 (1965) 94.
49 M. Tomoda, *Kyoritsu Yakka Daigahu Kenkyu Nempo*, 10 (1965) 18; *C.A.*, 66 (1967) 98525h.
50 E. Ragazzi and G. Veronese, *Mikrochim. Acta*, (1965) 966.
51 W. Poethke and W. Kinze, *Pharm. Zentralhalle*, 104 (1965) 489.
52 A. Silva Santos, M. Conceicao Marques and A. Ralha, *Rev. Port. Pharm.*, 15 (1965) 363; *C.A.*, 64 (1966) 11533h.
53 E. Hultin, *Acta Chem. Scand.*, 20 (1966) 1588.
54 B.H. Berg, *Acta Pharm. Suecica*, 3 (1966) 209.
55 W.W. Fike, *Anal. Chem.*, 38 (1966) 1697.
56 A. Kaess and C. Mathis, *Ann. Pharm. Fr.*, 24 (1966) 753.
57 W.E. Court, *Can. J. Pharm. Sci.*, 1 (1966) 76.
58 M. Petkovic, *Farm. Glasnik*, 22 (1966) 229.
59 A. Bonati and E. Pesce, *Fitoterapia*, 37 (1966) 98.
60 U. Rutkowska and K. Wojsa, *Herba Pol.*, 12 (1966) 101; *C.A.*, 66 (1967) 79638h.
61 G.J. Dickes, *J. Ass. Public Anal.*, 4 (1966) 45.
62 A. Affonso, *J. Chromatogr.*, 21 (1966) 332.
63 J.D. Phillipson and E.J. Shellard, *J. Chromatogr.*, 24 (1966) 84.
64 I. Sunshine, W.W. Fike and H. Landesmann, *J. Forensic Sci.*, 11 (1966) 428.
65 J.D. Phillipson and E.J. Shellard, *J. Pharm. Pharmacol.*, S18 (1966) 5.
66 J.J. Menn and J.B. McBain, *Nature (London)*, 209 (1966) 1351.
67 F. Wartmann-Hafner, *Pharm. Acta Helv.*, 41 (1966) 406.
68 E.J. Shellard, M.Z. Alam and J. Armah, *Sci. Pharm. Proc. 25th, 1965*, 1 (1966) 305.
69 Chih-Chung Chen, Shu-Liang Chang and Jen-Lai Tai, *Yao Hsueh Hsueh Pao*, 13 (1966) 131; *C.A.*, 65 (1966) 8673c.
70 H. Guyot, J. Bachelier-Notter, M.J. Dupret and C. Evreux, *Ann. Med. Leg.*, 47 (1967) 250; *C.A.*, 68 (1968) 47933a.
71 E. Ullman and H. Kassalitzky, *Deut. Apoth.-Ztg.*, 107 (1967) 152.
72 L. Manara, *Eur. J. Pharmacol.*, 2 (1967) 136.
73 J. Lutomski, Z. Kowalewski, K. Drost and K. Schmidt, *Herba Pol.*, 13 (1967) 44; *C.A.*, 68 (1969) 6132v.
74 H.C. Hsiu, J.T. Huang, T.B. Shih, K.L. Yang, K.T. Wang and A.L.Lin, *J. Chin. Chem. Soc.*, 14 (1967) 161.
75 I. Zingales, *J. Chromatogr.*, 31 (1967) 405.
76 J.D. Phillipson and E.J. Shellard, *J. Chromatogr.*, 31 (1967) 427.
77 W.M. Mc Isaac, B.T. Ho, V. Estevez and D. Powers, *J. Chromatogr.*, 31 (1967) 446.
78 A. Noirfalise and G. Mees, *J. Chromatogr.*, 31 (1967) 594.
79 H. Wullen and E. Stainier, *J. Pharm. Belg.*, 22 (1967) 291.
80 A.H. Beckett, G.T. Tucker and A.C. Moffat, *J. Pharm. Pharmacol.*, 19 (1967) 273.
81 M.A. Elkicy, M. Karawya, S.K. Wahba and A.R. Kozman, *J. Pharm. Sci., U.A.R.*, 8 (1967) 201.
82 Sh.Z. Kasymov, Kh.N. Aripov, T.T. Shakirov and S.Yu. Yunusov, *Khim. Prir. Soedin.*, 3 (1967) 352.
83 H. Pötter and R. Voigt, *Pharmazie*, 22 (1967) 198.
84 B. Davidow, N.L. Petri and B. Quame, *Amer. J. Clin. Pathol.*, 50 (1968) 714.
85 P.E. Haywood and M.S. Moss, *Analyst (London)*, 93 (1968) 737.
86 M. Wilk and U. Brill, *Arch. Pharm. (Weinheim)*, 301 (1968) 283.
87 E. Röder, E. Mutschler and H. Rochelmeyer, *Arch. Pharm. (Weinheim)*, 301 (1968) 624.
88 H. Tomczyk, *Diss. Pharm. Pharmacol.*, 20 (1968) 63.
89 T. Kaniewska and B. Borkowski, *Diss. Pharm. Pharmacol.*, 20 (1968) 111.
90 K.C. Güven and O. Unay, *Eczacilik Bul.*, 10 (1968) 93; *C.A.*, 70 (1969) 14441b.
91 K. Lapina, *Farmatsiya (Moscow)*, 17 (1968) 54; *C.A.*, 70 (1969) 14434b.
92 G.B. Marini-Bettolo, F. Delle Monache, A. Gelabert de Brovetto and E. Corio, *J. Ass. Offic. Anal. Chem.*, 51 (1968) 185.
93 E.J. Shellard and M.Z. Alam, *J. Chromatogr.*, 32 (1968) 472.

94 E.J. Shellard and M.Z. Alam, *J. Chromatogr.*, 32 (1968) 489.
95 J.D. Phillipson and E.J. Shellard, *J. Chromatogr.*, 32 (1968) 692.
96 E.J. Shellard, J.D. Phillipson and D. Gupta, *J. Chromatogr.*, 32 (1968) 704.
97 E.J. Shellard and M.Z. Alam, *J. Chromatogr.*, 33 (1968) 347.
98 W. Messerschmidt, *J. Chromatogr.*, 33 (1968) 551.
99 E.J. Shellard and M.Z. Alam, *J. Chromatogr.*, 35 (1968) 72.
100 M. Debackere and L. Laruelle, *J. Chromatogr.*, 35 (1968) 234.
101 A.N. Masoud, N.R. Farnsworth, L.A. Sciuchetti, R.N. Blomster and W.A. Meer, *Lloydia*, 31 (1968) 202.
102 J.M.G.J. Frijns, *Pharm. Weekbl.*, 103 (1968) 929.
103 D. Gröger, *Pharmazie*, 23 (1968) 210.
104 M.J. Harris, A.F. Stewart and W.E. Court, *Planta Med.*, 16 (1968) 217.
105 W.E. Court, in E.J. Shellard (Editor) *Quantitative Paper- and Thin-layer Chromatography*, Academic Press, London, New York, 1968, p. 29.
106 E.J. Shellard, in E.J. Shellard Editor, *Quantitative Paper- and Thin-layer Chromatography*, Academic Press, London, New York, 1968, p. 51.
107 K. Gröningson and G. Schill, *Acta Pharm. Suecica*, 6 (1969) 447.
108 E. Stahl and E. Dumont, *J. Chromatogr. Sci.*, 7 (1969) 517.
109 M. Vanhaelen, *J. Pharm. Belg.*, 24 (1969) 87.
110 M.H. Hashmi, S. Parveen and N.A. Chughtai, *Mikrochim. Acta*, (1969) 449.
111 C. Los and W.E. Court, *Planta Med.*, 17 (1969) 164.
112 G. Szasz and G. Szasz, *Acta Pharm. Hung.*, 40 (1970) 38.
113 A. Wahlund and K. Gröningson, *Acta Pharm. Suecica*, 7 (1970) 615.
114 K. Gröningson, *Acta Pharm. Suecica*, 7 (1970) 635.
115 M. Petkovic, *Arh. Farm.*, 20 (1970) 173; *C.A.*, 73 (1970) 123461r.
116 N.A. Mirzazade, *Azerb. Med. Zh.*, 47 (1970) 26; *C.A.*, 73 (1970) 38588f.
117 M.L. Bastos, G.E. Kananen, R.M. Young, J.R. Monforte and I. Sunshine, *Clin. Chem.*, 16 (1970) 931.
118 M. Overgaard-Nielsen, *Dan. Tidsskr. Farm.*, 44 (1970) 7.
119 S. Gill, *Gdansk Tow. Nauk. Rozpr. Wydz.*, 3 (1970) 175; *C.A.*, 75 (1971) 64040u.
120 J.P. Devred, G. Parmentier and J. Parmentier, *Int. Symp. Chromatogr. Electrophor. Lect. Pap. 6th*, (1970) 417.
121 V. Massa, F. Gal and P. Susplugas, *Int. Symp. Chromatogr. Electrophor. Lect. Pap. 6th*, (1970) 470.
122 A. Kaess and C. Mathis, *Int. Symp. Chromatogr. Electrophor. Lect. Pap. 6th*, (1970) 525.
123 G.H. Jolliffe and E.J. Shellard, *J. Chromatogr.*, 48 (1970) 125.
124 J.D. Phillipson and N.G. Bisset, *J. Chromatogr.*, 48 (1970) 493.
125 W. Poethke, C. Schwartz and H. Gerlach, *Planta Med.*, 18 (1970) 303.
126 R. Verpoorte and F. Sandberg, *Acta Pharm. Suecica*, 8 (1971) 119.
127 E. Bennati, *Boll. Chim. Farm.*, 110 (1971) 664.
128 S. Ebel, W.D. Mikulla and K.H. Weisel, *Deut. Apoth.-Ztg.*, 111 (1971) 931.
129 V.P. Georgievskii, N.Y. Tsarenko, M.S. Schraiber and G.J. Khait, *Herba Pol.*, 17 (1971) 258.
130 G.S. Tadjer,and A. Lustig, *J. Chromatogr.*, 63 (1971) D44.
131 Seysen, Mottet, Crispin, Hennau, Gloesener, Devred, Vanhaelen, Louvet, Toth and R. Stainier, *J. Pharm. Belg.*, 26 (1971) 292.
132 M.S. Habib and W.E. Court, *J. Pharm. Pharmacol.*, 23 (1971) 230S.
133 R.A. Mirkina and T.T. Shakirov, *Khim. Prir. Soedin.*, 7 (1971) 65.
134 J.M.G.J. Frijns, *Pharm. Weekbl.*, 106 (1971) 605.
135 G. Smith, *Proc. Soc. Anal. Chem.*, 8 (1971) 66.
136 S. Zadeczky, D. Küttel and M. Takacsi, *Acta Pharm. Hung.*, 42 (1972) 7.
137 H. Sybirska and H. Gajdzinska, *Arch. Toxicol.*, 28 (1972) 296; *C.A.*, 77 (1972) 15111a.
138 W.E. Court, M.J. Harris and A.F. Stewart, *Can. J. Pharm. Sci.*, 7 (1972) 98.
139 E. Stahl, *Deut. Apoth.-Ztg.*, 112 (1972) 1154.
140 D.D. Datta, P.C. Bose and D. Ghosh, *East. Pharm.*, 15 (1972) 41.
141 J.K. Brown, L. Shapazian and G.D. Griffin, *J. Chromatogr.*, 64 (1972) 129.
142 W.E. Court, *J. Chromatogr.*, 73 (1972) 274.
143 G.E. Wright and T.Y. Tang, *J. Pharm. Sci.*, 61 (1972) 299.

144 F. Hammer and F. Kaiser, *Planta Med.*, 21 (1972) 5.
145 M.M. Mudzhiri, V.Yu. Vachnadze and K.S. Mudzhiri, *Soobshch. Akad. Nauk Gruz. SSR*, 67 (1972) 353; *C.A.*, 78 (1973) 47837d.
146 K.F. Ahrend and D. Tiess, *Zbl. Pharm.*, 111 (1972) 933.
147 M.S. Habib and W.E. Court, *Can. J. Pharm. Sci.*, 8 (1973) 81.
148 Z. Jung and M. Jungova, *Cesk. Farm.*, 22 (1973) 195.
149 E. Novakova and J. Vecerkova, *Cesk. Farm.*, 22 (1973) 347.
150 A. Fiebig, S. Kanafarska-Slotkowska and D. Choelkiewicz, *Farm. Pol.*, 29 (1973) 1087.
151 H.D. Crone and E.M. Smith, *J. Chromatogr.*, 77 (1973) 234.
152 R.A. van Welsum, *J. Chromatogr.*, 78 (1973) 237.
153 W.E. Court and M.S. Habib, *J. Chromatogr.*, 80 (1973) 101.
154 A.R. Rogers and G. Smith, *J. Chromatogr.*, 87 (1973) 125.
155 N.G. Bisset and J.D. Phillipson, *J. Pharm. Pharmacol.*, 25 (1973) 563.
156 V.Yu. Vachnadze, V.M. Malikov, Kh.T. Illyasova, K.S. Mudhiri and S.Yu. Yunusov, *Khim. Prir. Soedin.*, 9 (1973) 72.
157 E. Kammerl and E. Mutschler, *Pharm. Ztg.*, 118 (1973) 1905.
158 L. Angenot, *Thèse de Doctorat*, Liège, 1973.
159 K.F. Ahrend and D. Tiess, *Wiss. Z. Univ. Rostock, Math. Naturw. Reihe*, 22 (1973) 951.
160 A. Eichhorn and L. Kny, *Zbl. Pharm.*, 112 (1973) 567.
161 D.W. Chasar and G.B. Toth, *J. Chem. Educ.*, 51 (1974) 22.
162 P.D. Swaim, V.M. Loyola, H.D. Harlan and M.J. Carlo, *J. Chem. Educ.*, 51 (1974) 331.
163 A.C. Moffat and K.W. Smalldon, *J. Chromatogr.*, 90 (1974) 1.
164 A.C. Moffat and K.W. Smalldon, *J. Chromatogr.*, 90 (1974) 9.
165 K.G. Blass, R.J. Thibert and T.F. Draisey, *J. Chromatogr.*, 95 (1974) 75.
166 M. Malaiyandi, J.P. Barretta and M. Lanouette, *J. Chromatogr.*, 101 (1974) 155.
167 A. Puech, C. Duren and M. Jacob, *J. Pharm. Belg.*, 29 (1974) 126.
168 A.C. Moffat and B. Clare, *J. Pharm. Pharmacol.*, 26 (1974) 665.
169 R.J. Armstrong, *N.Z.J. Sci.*, 17 (1974) 15.
170 M.S. Habib and W.E. Court, *Planta Med.*, 25 (1974) 331.
171 E. Stahl and W. Schmitt, *Arch. Pharm (Weinheim)*, 308 (1975) 570.
172 A.K. Chowdhury and S.A. Chowdhury, *Bangladesh Pharm. J.*, 4 (1975) 11; *C.A.*, 83 (1975) 48262c.
173 J. Grabowska and E. Sell, *Farm. Pol.*, 31 (1975) 101.
174 Z. Legowska, K. Szymkovska and M. Zolmierowicz, *Farm. Pol.*, 31 (1975) 315; *C.A.*, 83 (1975) 125980q.
175 T.M. Holdstock and H.M. Stevens, *Forensic Sci.*, 6 (1975) 187.
176 V.Yu. Vachnadze, *Izv. Akad. Nauk Gruz. SSR, Ser. Khim.*, 1 (1975) 28.
177 M.S. Karawya, M.S. Hifnawy and H.S. Dahawi, *J. Ass. Offic. Anal. Chem.*, 58 (1975) 85.
178 J.D. Phillipson and S.R. Hemingway, *J. Chromatogr.*, 105 (1975) 163.
179 J.A. Vinson and J.E. Hooyman, *J. Chromatogr.*, 105 (1975) 415.
180 T. Okumura, T. Kadono and A. Isoo, *J. Chromatogr.*, 108 (1975) 329.
181 A.C. Moffat, *J. Chromatogr.*, 110 (1975) 341.
182 S.J. Stohs and G.A. Scratchley, *J. Chromatogr.*, 114 (1975) 329.
183 J.A. Vinson, J.E. Hooyman and C.E. Ward, *J. Forensic Sci.*, 20 (1975) 552.
184 L.J. Dombrowski, A.V.R. Crain, R.S. Browning and E.L. Pratt, *J. Pharm. Sci.*, 64 (1975) 643.
185 S.L. Tripp, E. Williams, W.E. Wagner Jr., and G. Lukas, *Life Sci.*, 16 (1975) 1167.
186 V. Hartmann and G. Schnabel, *Pharm. Ind.*, 37 (1975) 451.
187 J.R.B.J. Brouwers and P.E. Kamp, *Pharm. Weekbl.*, 110 (1975) 279.
188 A.M. Metwally, *Pharmazie*, 30 (1975) 92.
189 W.E. Court and P. Timmins, *Planta Med.*, 27 (1975) 319.
190 I.R. da S. Jardim and C.S. Tavares, *Rev. Bras. Farm.*, 56 (1975) 3; *C.A.*, 84 (1976) 35383p.
191 E. Vidic and E. Klug, *Z. Rechtsmed.*, 76 (1975) 283.

192 R.L. Munier and A.M. Drapier, *C.R. Acad. Sci., Ser. C*, 283 (1976) 719.
193 V.V. Drozhzhina, V.E. Chichiro and T.I. Bulenkov, *Farmatsiya (Moscow)*, 25 (1976) 44; *C.A.*, 86 (1977) 96046u.
194 L. Lepri, P.G. Desideri and M. Lepori, *J. Chromatogr.*, 116 (1976) 131.
195 F.W. Grant, *J. Chromatogr.*, 116 (1976) 230.
196 R. Verpoorte and A. Baerheim Svendsen, *J. Chromatogr.*, 124 (1976) 152.
197 L. Lepri, P.G. Desideri and M. Lepori, *J. Chromatogr.*, 123 (1976) 175.
198 A.N. Masoud, *J. Pharm. Sci.*, 65 (1976) 1585.
199 L.A. Sapunova and A.K. Prashchurovich, *Khim. Prir. Soedin.*, 12 (1976) 772; *C.A.*, 86 (1977) 117104d.
200 N.G. Bisset and J.D. Phillipson, *Lloydia*, 39 (1976) 263.
201 R. Verpoorte, *Thesis*, Leiden, 1976.
202 R.L. Munier and A.M. Drapier, *Chromatographia*, 10 (1977) 226.
203 R.L. Munier and A.M. Drapier, *Chromatographia*, 10 (1977) 290.
204 H. Kroeger, G. Bohn and G. Ruecker, *Deut. Apoth.-Ztg.*, 117 (1977) 1923.
205 J.P.S. Sarin, R.C. Nandi, R.S. Kapil and N.M. Khanna, *Indian J. Pharm.*, 39 (1977) 62.
206 Z.V. Robakidze, V.Yu. Vachnadze and K.S. Mudzhiri, *Izv. Akad. Nauk Gruz. SSR, Ser. Khim.*, 3 (1977) 223; *C.A.*, 88 (1978) 197701a.
207 G. Rücker and A. Taha, *J. Chromatogr.*, 132 (1977) 165.
208 J.E. Wallace, H.E. Hamilton, H. Skrdlant, L. Burkett and H. Schwertner, *J. Chromatogr.*, 138 (1977) 111.
209 A.N. Masoud, *J. Chromatogr.*, 141 (1977) D9.
210 L.A. Sapunova and A.K. Prashchuron, *Khim.-Farm. Zh.*, 11 (1977) 128; *C.A.*, 86 (1977) 117104d.
211 V.Yu. Vachnadze and K.S. Mudzhiri, *Khromatogr. Metody Farm.*, (1977) 156; *C.A.*, 90 (1979) 164704s.
212 L. Gagliardi, A.Amato, G. Ricciardi and S. Chiavarelli, *Riv. Tossicoli Sper. Clin.*, 7 (1977) 191; *C.A.*, 88 (1978) 32648a.
213 N.V. Rama Rao and S.N. Tandon, *Chromatographia*, 11 (1978) 227.
214 E. Marozzi, E. Cozza, A. Pariali, V. Gambaro, F. Lodi and E. Saligari, *Farmaco, Ed. Prat.*, 33 (1978) 195.
215 N.V. Rama Rao and S.N. Tandon, *J. Chromatogr. Sci.*, 16 (1978) 158.
216 R.A. de Zeeuw, F.J.W. Mansvelt and J.E. Greving, *J. Chromatogr.*, 148 (1978) 255.
217 R.A. Sams and R. Huffman, *J. Chromatogr.*, 161 (1978) 410.
218 W. Majak, R.E. McDiarmid and R.J. Bosse, *Phytochemistry*, 17 (1978) 301.
219 E.J. Shellard and P.K. Lala, *Planta Med.*, 33 (1978) 63.

TABLE 13.1

LITERATURE CITED IN CHAPTER 3 WHICH INCLUDES THE ANALYSIS OF INDOLE ALKALOIDS

| Alkaloid* | Ref.* | Alkaloid* | Ref.* |
|---|---|---|---|
| S,B,RSP,serp,serpin,ajmal,sarp, yoh,α-yoh,Ph | 7 | S | 130 |
| S,B,RSP,yoh,Ph | 12 | S,B,Ph | 136 |
| S,B | 35 | S,B,Ph | 146 |
| S,B,RSP,RSC,ajmal,sarp | 42 | S,B,yoh,Ph | 149 |
| RSP | 45 | S,B,Ph | 159 |
| S,B,RSP,ajmal,sarp,Ph,aspido-spermine,quebrachamine, uleine,harman | 47 | S,yoh | 163,164 |
| S,yoh | 55 | S,B | 168 |
| S,B,Ph | 61 | S,RSP | 169 |
| S,B,yoh | 64 | S | 181 |
| S,B,RSP,yoh,Ph | 74 | S | 191 |
| S,B,Ph | 78 | S,B | 192,202,203 |
| S | 85 | S,B,RSP,yoh,ajmal,ibogaine,Ph | 194,197 |
| | | S,B | 213,215 |
| | | yoh | 216 |

*Abbreviations:

| | | | |
|---|---|---|---|
| S | strychnine | RSC | rescinnamine |
| B | brucine | RSP | reserpine |
| α-C | α-colubrine | RSPine | reserpinine |
| β-C | β-colubrine | RSPline | reserpiline |
| V | vomicine | sarp | sarpagine |
| D | diaboline | serp | serpentine |
| PS | pseudostrychnine | serpin | serpentinine |
| WGA | Wieland Gumlich aldehyde | tetraphyl | tetraphylline |
| I | icajine | THalst | tetrahydroalstonine |
| N | novacine | yoh | yohimbine |
| PB | pseudobrucine | Lcrist | leurocristine |
| ret | retuline | Lsidine | leurosidine |
| holst | holstiine | Lsin | leurosine |
| S N-ox | strychnine N-oxide | VLB | vincaleukoblastine |
| ajmal | ajmaline | Vmin | vincamine |
| alst | alstonine | Ph | physostigmine |

TABLE 13.2

INDOLE ALKALOIDS (EXCLUDING ERGOT ALKALOIDS) IN DRUGS OF ABUSE ANALYSIS (SEE CHAPTER 12)

| Alkaloids | Ref. | Ref. in Chapter 12 |
|---|---|---|
| S,B,RSP | 10 | 8 |
| S | 18 | 22 |
| S,RSP | 84 | 102 |
| yoh,Ph | 117 | 141 |
| S | 141 | 203 |
| S | 152 | 240 |
| S | 165 | 289 |
| S,yoh,Ph | 198,209 | 356 |
| S | 204 | 369 |
| S | 212 | 386 |
| S | 214 | 391 |

TABLE 13.3

TLC SEPARATION OF SOME *STRYCHNOS* ALKALOIDS

TLC systems:
S1 Silica gel G, activated — Benzene - ethyl acetate - diethylamine (7:2:1)[124]
S2 Silica gel G, activated — Chloroform - diethylamine (9:1)[124]
S3 Silica gel $GF_{254}$, activated — Chloroform - cyclohexane - diethylamine (4:5:1)[92]
S4 Silica gel G, activated — Chloroform - cyclohexane - diethylamine (7:3:1)[124]
S5 Silica gel G, activated — Chloroform - acetone - diethylamine (5:4:1)[124]
S6 Silica gel G, activated — Diethyl ether - ethanol - diethylamine (95:5:10)[124]
S7 Silica gel G, activated — Diethyl ether - diethylamine (9:1)[124]

| Alkaloid | $hR_F$ values | | | | | | |
|---|---|---|---|---|---|---|---|
| | S1* | S2* | S3** | S4* | S5* | S6* | S7* |
| Strychnine | 42 | 70 | 38 | 52 | 40 | 82 | 24 |
| Brucine | 22 | 63 | 23 | 40 | 32 | 80 | 20 |
| α-Colubrine | 31 | 69 | | 51 | 37 | 81 | 19 |
| β-Colubrine | 30 | 69 | | 50 | 36 | 81 | 10 |
| 4-Hydroxystrychnine | 33 | 66 | | 49 | 39 | 80 | 21 |
| Pseudostrychnine | 50 | 70 | 37 | 54 | 53 | 82 | 46 |
| Pseudobrucine | 41 | 69 | 31 | 49 | 50 | 82 | 29 |
| Icajine | 60 | 81 | 50 | 68 | 60 | 91 | 32 |
| Novacine | 53 | 80 | 44 | 66 | 60 | 89 | 20 |
| Vomicine | 60 | 81 | 40 | 60 | 61 | 90 | 33 |
| Strychnine N-oxide | 0 | 2 | | 0 | 0 | 13 | 0 |
| Brucine N-oxide | 0 | 1 | | 0 | 0 | 11 | 0 |
| Diaboline | 29 | 52 | 42 | 30 | 36 | 72 | 22 |

*Estimated $hR_F$ values calculated from ref. 124.
**$hR_F$ values from ref. 92.

TABLE 13.4

TLC ANALYSIS OF SOME QUATERNARY ALKALOIDS, ALKALOID N-OXIDES AND TERTIARY ALKALOIDS[196]

TLC systems:
S1 Silica gel 60 $F_{254}$ pre-coated plates — Methanol - 0.2 *M* ammonium nitrate (3:2)
S2 Silica gel 60 $F_{254}$ pre-coated plates — Methanol - 2 *M* ammonia - 1 *M* ammonium nitrate (7:2:1)

| Compound | $hR_F$ values | | Compound | $hR_F$ values | |
|---|---|---|---|---|---|
| | S1 | S2 | | S1 | S2 |
| Dihydrotoxiferine | 27 | 24 | Caracurine V | 3 | 27 |
| C-Alkaloid H | 23 | 22 | Alcuronium | 33 | 43 |
| Caracurine V methoiodide | 16 | 14 | Tubocurarine | 20 | 40 |
| Bisnordihydrotoxiferine N-oxide | 24 | 74 | Macusine B | 60 | 57 |
| Bisnordihydrotoxiferine di-N-oxide | 29 | 79 | Melinonine A | 49 | 48 |
| Bisnor-C-alkaloid H N-oxide | 14 | 65 | Antirhine methochloride | 51 | 51 |
| Bisnor-C-alkaloid H di-N-oxide | 20 | 75 | Strychnine methochloride | 31 | 30 |
| Caracurine V N-oxide | 12 | 45 | Strychnine N-oxide | 33 | 72 |
| Caracurine V di-N-oxide | 17 | 72 | Strychnine | 19 | 63 |
| Bisnordihydrotoxiferine | 15 | 58 | Serpentine | 54 | 66 |
| Bisnor-C-alkaloid H | 5 | 50 | Alstonine | 57 | 66 |

TABLE 13.5

COLOURS OBTAINED BY SPRAYING WITH IODOPLATINATE REAGENT FOR SOME *STRYCHNOS* ALKALOIDS[56]

| Alkaloid | Colour |
|---|---|
| Strychnine | Blue-violet |
| Brucine | Blue |
| α-Colubrine | Blue-violet |
| β-Colubrine | Blue-violet |
| Vomicine | Pink-violet |
| Strychnine N-oxide | Blue-violet |

TABLE 13.6

DETECTION METHODS FOR *STRYCHNOS* ALKALOIDS

| Reagent | No. | Sensitivity (μg) | Background colour | Alkaloid colour | Ref. |
|---|---|---|---|---|---|
| Quenching 254 nm | | 0.01-0.1 | | | |
| Dragendorff modifications: | | | | | |
| Bregoff-Delwiche | 39a | 0.1-1 | Yellow | Orange | |
| Munier | 39b | 0.01-1 | Yellow | Orange | |
| + sodium nitrite | 39h | 0.01-1 | White | Orange-brown | |
| Munier and Macheboeuf | | 0.1-1 | Yellow | Orange | |
| + sodium nitrite | | 0.01-1 | White | Orange-brown | |
| Thies,Reuther and Vagujfalvi | 39f | | | | |
| + sulphuric acid | | 0.1-1 | White-yellow | Orange | |
| Iodine vapour | | 0.01-0.1 | Yellow | Brown | |
| Iodine in chloroform | 51b | 0.01-1 | Light yellow | Brown-orange | |
| Bouchardat | 52c | 0.1-1 | Dark yellow | Brown | |
| Potassium iodoplatinate | 56d | 0.01-1 | Pink-violet | Blue,violet,brown (see Table 13.5) | |
| Cerium(IV) sulphate in conc. sulphuric acid | 14a | 0.01-0.1 | Yellow | Various colours | |
| Cerium(IV) ammonium sulphate | 12 | 0.1-10 | Yellow | Various colours | |
| Cerium(IV) sulphate(1%) in sulphuric acid(10%) | 14b | 0.01-10 | Yellow | Various colours; some alkaloids do not colour, even after heating of the plate | |
| Cerium(IV) sulphate saturated in 65% nitric acid | 13 | 0.01-1 | Yellow | Various colours (Table 13.7) | |
| Cerium(IV) sulphate according Sonnenschein | 14c | 0.01-10 | Yellow | Various colours | |
| Iron(III) chloride(0.2 *M*) in perchloric acid(35%) | 62c | 0.1-10 | Light yellow | Various colours | 155 |
| Iron(III) chloride(0.05 *M*) in perchloric acid(35%) | 62b | 0.1-10 | Light yellow | Various colours (Table 13.7) | 3 |
| Copper sulphate(10%) in 2% ammonia | 27 | 1-10 | | Only some of the alkaloids give colours | 90 |
| Ammonium vanadate in nitric acid | 3 | 0.01-0.1 | | Various colours | 153 |
| Cinnamic aldehyde-hydrochloric acid | 21 | 0.1-1 | | Various colours | |
| TCBI reagent | 94 | | | Various colours | 179 |
| Iron(III) chloride in 50% $HNO_3$ | 61 | 0.1-10 | Yellow-white | Various colours | 54 |
| Sulphomolybdic acid | 90 | 0.1-10 | | Yellow-orange-brown | 153 |
| Phosphomolybdic acid | 79 | 0.1-10 | | Yellow-orange-brown | 153 |

*R. Verpoorte, unpublished results.

References p. 324

TABLE 13.7

COLOURS OBTAINED FOR SOME *STRYCHNOS* ALKALOIDS WITH CERIUM(IV) SULPHATE - SULPHURIC ACID (No. 14b) AND IRON(III) CHLORIDE - PERCHLORIC ACID (No. 62c) SPRAY REAGENTS

| Alkaloid | Cerium(IV) sulphate | | Iron(III) chloride-perchloric acid | | |
|---|---|---|---|---|---|
| | Immediate | 5 min/90°C | Immediate | 5 min/90°C | 30 min/90°C |
| Strychnine | | Pale violet | | Violet | Red |
| Brucine | Brown-orange | Orange-yellow | | Brownish-red | Yellow |
| α-Colubrine | Pale violet | Violet | | Pale violet | Greyish pink |
| β-Colubrine | Brown | Violet | | Grey | Greyish yellow |
| 4-Hydroxy-strychnine | Pale violet | Reddish purple | | Grey | Grey |
| 4-Hydroxy-3-methoxystrychnine | Pale yellow | Yellow | | Yellowish green | Chocolate brown |
| Diaboline | | Orange | | Light brown | Orange red |
| Strychnospermine | | Orange-brown | | Grey | Pink |
| Spermostrychnine | | Purple-orange | | Reddish purple | Greyish red |
| Novacine | | | | | Yellow |
| Icajine | | | | | Red |
| Vomicine | | | | | Grey |
| Angustine | | | Yellow | Yellow | Yellow-green |
| Macusine B | | | | Dark green-grey | Dark green-grey |
| Bisnor-dihydro-toxiferine | Violet, changing to yellow in the centre of the spot | | Blue-violet | Blue-violet | Blue-violet |
| Caracurine V | Violet, changing to yellow in the centre of the spot | | Blue-violet | Blue-violet | Blue-violet |
| Longicaudatine | Blue, changing to yellow in the centre of the spot | | Blue | Blue | |

TABLE 13.8

TLC ANALYSIS OF *STRYCHNOS* ALKALOIDS IN PLANT MATERIAL OR EXTRACTS

| Alkaloid* | Aim | Adsorbent | Solvent system | Ref. |
|---|---|---|---|---|
| S,B,α-C,β-C,V,PS,WGA, D,S N-ox,ret,holst | Separation | $SiO_2$ | $CHCl_3$-DEA(9:1)<br>MeOH<br>$CHCl_3$-MeOH(8:2,7:3)<br>Cyclohexane-DEA(9:1)<br>$CHCl_3$-$Me_2CO$-DEA(5:4:1) | |
| | | $Al_2O_3$ | $CHCl_3$<br>EtOAc-$CHCl_3$<br>EtOAc-benzene | 17 |
| S,B | Oxidation of B prior to TLC, to obtain separation of S and B | $SiO_2$ | BuOH-36%HCl(5:1) sat. with $H_2O$ | 20 |
| S,B | Separation of alkaloids on $Al_2O_3$ without binder | $Al_2O_3$ | Benzene-EtOH(9:1,8:2)<br>$Et_2O$-EtOH(97:3,95:5) | 35 |
| S,B | Indirect quantitative analysis | $Al_2O_3$ | $Et_2O$-EtOH(95:5) | 36 |
| S,B | Separation on MgO | 2% $MgSO_4$-impregnated $SiO_2$ | *n*-Hexane-$Me_2CO$(3:1)<br>$CHCl_3$-EtOAc(1:3) | 38 |
| S,B,α-C,β-C,V,D,S N-ox, ret,holst,dehydroB,PS, ethoxyS | Separation | $SiO_2$ | I. $CHCl_3$-MeOH(8:2)<br>II. *n*-BuOH-HCl(95:5) sat. with $H_2O$<br>Two-dimensional:I,II | 39 |
| S,B,α-C,β-C,V | Detection in *Strychnos nux vomica* leaves | 0.5 *M* NaOH-impregnated $SiO_2$ | I. $Me_2CO$<br>II. $CHCl_3$-MeOH-benzene(3:1:1)<br>I,II after each other in same direction | 41 |
| S,B | Detection in tinctures | $SiO_2$ | Benzene-$CHCl_3$(7:25)+5-10% DEA<br>$CHCl_3$+5-10% DEA<br>$CHCl_3$-$Me_2CO$(9:1)+5-10% DEA | |
| | | 0.5 *M* NaOH-impregnated $SiO_2$ | Benzene-$CHCl_3$(7:25),$CHCl_3$<br>$CHCl_3$-$Me_2CO$(9:1) | 42 |

TABLE 13.8 (continued)

| | | | | |
|---|---|---|---|---|
| S,B | Indirect quantitative analysis | $Al_2O_3$ | $Et_2O$-EtOH(9:1) | 51 |
| S,B,α-C,β-C,V | Reaction chromatography | $SiO_2$ | $CHCl_3$-$Me_2CO$-DEA(5:4:1)<br>Cyclohexane-$Me_2CO$-DEA(5:4:1) | 56 |
| S,B | Detection in tinctures | $SiO_2$ | Abs. EtOH-DEA(47.5:2.5) | 58 |
| S,B,α-C,β-C,PS | Detection in *Nux vomica* extracts and tinctures | $SiO_2$ | Benzene-EtOAc-DEA(7:2:1) | 67 |
| S,B | Determination in *Nux vomica* extracts | $Al_2O_3$ | Benzene-EtOAc(1:2) | 69 |
| S,B,α-C,β-C,V,PS,PB, S N-ox,D,WGA,ret,holst, dihydroB,N,I | Separation and identification of minor alkaloids of *Strychnos nux vomica* (Table 13.3) | $SiO_2$ | Cyclohexane-$CHCl_3$-DEA(5:4:1)<br>Benzene-EtOAc-DEA(7:2:1)<br>$CHCl_3$-MeOH(96:4)<br>Pyridine-EtOAc-$H_2O$(11.5:75:16.5), upper phase | 92 |
| S,B | Detection in *Nux vomica* | $SiO_2$ | Benzene-EtOAc-DEA(7:2:1) | 102 |
| S,B | Ion-pair chromatography | 1 *M* KSCN and 0.7 *M* $H_2SO_4$-impregnated cellulose | Cyclohexane-AmOH(1:1)<br>$CHCl_3$ | 107 |
| S,B | Detection in *Nux vomica* | $SiO_2$ | $Et_2O$-MeOH-17% $NH_4OH$(4:0.8:0.1) (2x) | 109 |
| S,B | Separation on mixed adsorbents | $SiO_2$-$Al_2O_3$-celite(1:1:1) | Ethanol-$CHCl_3$(9:1) | 119 |
| 30 *Strychnos* alkaloids | Correlations between $hR_F$ value and structure (Table 13.3) | $Al_2O_3$ or $SiO_2$ | $Et_2O$-DEA(9:1)<br>Benzene-EtOAc-DEA(7:2:1)<br>Cyclohexane-$CHCl_3$-DEA(3:7:1)<br>$CHCl_3$-DEA(9:1)<br>$CHCl_3$-$Me_2CO$-DEA(5:4:1)<br>$Et_2O$-EtOH-DEA(95:5:10)<br>$CHCl_3$-EtOH-DEA(95:5:10)<br>and the above mentioned solvents without DEA | 124 |
| S | Identification of seeds used in Indian and African jewellery | $SiO_2$ | MeOH-conc. $NH_4OH$(95:5) | 139 |
| S,B | Indirect quantitative analysis in homeopathic mother tincture | $SiO_2$ | Benzene-EtOAc-DEA(7:2:1) | 140 |

References p. 324

| | | | | |
|---|---|---|---|---|
| S,B,α-C,β-C,4-hydroxyS, D,4-hydroxy-3-methoxyS, spermostrychnine, strychnospermine | Identification in *Strychnos wallichiana* | $SiO_2$ | EtOAc-isoprOH-5.5% $NH_4OH$(45:35:20)<br>EtOAc-isoprOH-conc. $NH_4OH$(80:15:5),(100:2:1)<br>*n*-BuOH-0.1 *M* HCl-7.4% aq. $K_4Fe(CN)_6$(100:15:34) | 155 |
| S,B | Identification in *Nux vomica* preparations (also cotarnine and hyoscyamine) | | No details available | 172 |
| S | Identification | $SiO_2$ | Benzene-EtOH(95:5) | 174 |
| S,B | Indirect quantitative analysis in *Nux vomica* | $SiO_2$ | $Me_2CO$-benzene-$NH_4OH$(30:15:1) | 177 |
| S | Direct quantitative analysis (spot size) in *Nux vomica* (yohimbine as internal standard) | $SiO_2$ | $CHCl_3$-MeOH-$NH_4OH$(90:10:0.15) | 188 |
| Various Strychnos alkaloids (Table 13.4) | Separation of N-oxides and quaternary alkaloids | $SiO_2$ | MeOH-0.2 *M* $NH_4NO_3$(3:2)<br>MeOH-2 *M* $NH_4OH$-1 *M* $NH_4NO_3$(7:2:1) | 196 |

*For abbreviations, see footnote to Table 13.1.

TABLE 13.9

TLC ANALYSIS OF *STRYCHNOS* ALKALOIDS IN PHARMACEUTICAL PREPARATIONS AND AS PURE COMPOUNDS IN MIXTURES

| Alkaloid* | Other compounds | Aim | Adsorbent | Solvent system | Ref. |
|---|---|---|---|---|---|
| S | Atropine,scopolamine,veratrine,tetracaine | Separation | $SiO_2$ | $CHCl_3$-DEA(9:1)<br>$CHCl_3$-*n*-BuOH-MeOH-10% $NH_4OH$(12:9:9:1) | 27 |
| S,B | Opium alkaloids | Separation on MgO | 2% $MgSO_4$-impregnated MgO | *n*-hexane-$Me_2CO$(3:1)<br>$CHCl_3$-EtOAc(1:3) | 38 |
| S | Morphine,codeine,nicotine, chlorpromazine,chlorpromazine sulphoxide, tolazoline, imipramine | Decreased development time | $SiO_2$+Celite 545(1:1) | Methyl isobutyl ketone-AcOH-$H_2O$(20:10:5) | 48 |
| B | Atropine,aconitine,codeine | Separation on plaster of Paris | $CaSO_4$ | $CHCl_3$-AmOH-toluene-conc. HCl(50:1.5:1.5:0.25) | 62 |
| S,B | Quinine,quinidine,codeine, narcotine,papaverine,pilocarpine,scopolamine,veratrine | Separation with azeotropic mobile phases | $SiO_2$ | MeOH-benzene(39.1:60.9)<br>MeOH-$CHCl_3$-MeOAc(21.6:51.4:27.0)<br>MeOH-$CHCl_3$(23.0:47.0) | 87 |
| S,B | Physostigmine | Identification | Alkaline $SiO_2$ | MeOH | 90 |
| B | Atropine,scopolamine,belladonnine | Traverse pH-gradient TLC | pH gradient-impregnated $SiO_2$ | $CHCl_3$-EtOH(8:2) | 108 |
| S,B | Various alkaloids | Semi-quantitative circular TLC | $SiO_2$ | $CHCl_3$-EtOH(9:1) | 110 |
| S | Atropine, methylatropine, codeine,papaverine,xanthines,ephedrine,procaine, phenobarbital,chlorpromazine,melipramine,tripelennamine | Separation | $SiO_2$ | $Me_2CO$-cyclohexane-EtOAc(1:1:1) in $NH_3$ atmosphere | 112 |

| | | | | | |
|---|---|---|---|---|---|
| S | Papaverine, imipramine, desimipramine, secergan | Ion-pair chromatography | 0.5 g/g BuOH-impregnated acetylcellulose | 0.1 *M* NaCl | |
| | | | 0.5 g/g AmOH-impregnated acetylcellulose | 0.1 *M* NaCl, NaBr or $NaClO_4$ in 0.5 *M* $H_2SO_4$ | 113,114 |
| S | Atropine, homatropine, scopolamine, cocaine, morphine, meperidine | Separation on microslides | $SiO_2$ | $CHCl_3$-DEA(9:1)<br>$CHCl_3$-$Me_2CO$-DEA(5:4:1) | 118 |
| S | Asconerin, cardiamide, caffeine, neurosonine, opotonine | Separation | $SiO_2$ | I. $CHCl_3$-$Et_2O$-90% EtOH-DEA(35:5:10:15)<br>II. *n*-BuOH-MeOH-$Me_2CO$-$CHCl_3$(1:1:1:2:4)<br>Two dimensional: I, II | 150 |
| B | Caffeine, quinine | Separation | $SiO_2$ | MeOH | 161 |
| S,B | Caffeine, nicotine, quinine | Separation | $SiO_2$ | Benzene-EtOAc-DEA(10:10:3) | 162 |
| S,B | Cotarnine, hyoscyamine | Identification | | No details available | 172 |
| S,B | Reserpine, thebaine, codeine, aconitine, caffeine, quinine | Detection with FID | Sintered $SiO_2$<br>Sintered $Al_2O_3$ | $CHCl_3$-DEA(30:1)<br>Benzene-$CHCl_3$-DEA(36:8:1) | 180 |
| S | Codeine | Identification | $SiO_2$ | $CHCl_3$-$Me_2CO$-conc. $NH_4OH$(12:24:1) | 193 |

*For abbreviations, see footnote to Table 13.1.

TABLE 13.10

TLC ANALYSIS OF *STRYCHNOS* ALKALOIDS IN TOXICOLOGY

| Alkaloid* | Other compounds | Aim | Adsorbent | Solvent system | Ref. |
|---|---|---|---|---|---|
| S,B | | Oxidation B prior to TLC, to obtain separation of S and B | $SiO_2$ | BuOH-36% HCl(5:1) sat. with $H_2O$ | 20 |
| S | Opium alkaloids,nicotine, barbiturates,tranquillizers | Reduced development time by modified stationary phase | $SiO_2$+celite 545(1:1) | Methyl isobutyl ketone-AcOH-$H_2O$(20:10:5) | 48 |
| S | | Identification | | | 52 |
| S | Chlorpromazine,amidopyrine,mepromazine, nivaquine,hydroquinone | Detection in viscera | | No details available | 70 |
| S | Amphetamines,ephedrine, caffeine,nicotine | Stimulants and doping control in urine | $SiO_2$ | $CHCl_3$-$Me_2CO$-DEA(5:4:1)<br>MeOH-$Me_2CO$(1:1) | 80 |
| S | Amphetamine,caffeine | Doping control in biological specimens from horses | $SiO_2$ | MeOH | 81 |
| S | Securinine | Separation | $SiO_2$<br>$Al_2O_3$ | $CHCl_3$-$Me_2CO$-MeOH(7:2:1)<br>$CHCl_3$-benzene-EtOH(25:25:2) | 91 |
| S,B | Opium alkaloids,caffeine, cocaine,quinine,lobeline, sparteine | Identification doping in horse urine | $SiO_2$ | I. Hexane-$Me_2CO$-DEA(6:3:1)<br>II. $CHCl_3$-MeOH-DEA(95:5:0.05)<br>Two dimensional:I,II | 100 |
| S | | Detection of hashish and S | $SiO_2$ | Benzene-EtOH(95:5) | 174 |
| S | | S in heroin samples | $SiO_2$ | $CHCl_3$-MeOH-$Et_2O$-25% $NH_4OH$(75:25:5:1) | 187 |

TABLE 13.11

TLC SEPARATION OF SOME *RAUWOLFIA* ALKALOIDS

TLC systems:

| | | |
|---|---|---|
| S1 | Silica gel G, activated | Methanol-methyl ethyl ketone-heptane (8.4:33.6:58.0)[4] |
| S2 | Silica gel G, activated | Chloroform-acetone-diethylamine (5:4:1)[43] |
| S3 | Silica gel G, activated | Chloroform-cyclohexane-diethylamine (7:2:1)[43] |
| S4 | Silica gel G | Ethyl acetate-ethanol-dimethylformamide-diethylamine (30:60:5:5)[59] |
| S5 | Silica gel $F_{254}$ (fast flow-rate, ready-made plates) | Carbon tetrachloride-methanol (96:4)[186] |
| S6 | Silica gel 60 $F_{254}$ (ready-made plates) | Diisopropyl ether-methanol-ammonia (85:15:1)[186] |
| S7 | Silica gel G, activated | Xylene-isooctane-ethyl acetate-diethyl ether (15:45:5:40)[153,189] |
| S8 | Silica gel G, activated | *n*-Butanol-ethyl acetate-ethylene dichloride (10:30:60)[153,189] |
| S9 | Silica gel G, activated | Acetone-light petroleum (b.p. 40-60°C)-carbon tetrachloride-isooctane (35:30:20:15)[153,189] |
| S10 | Silica gel G, activated | Acetone-*n*-butanol-isooctane (33.6:8.4:58.0)[104,153,189] |
| S11 | Silica gel G, activated | Acetone-light petroleum (b.p. 40-60°C)-glacial acetic acid (45:45:10)[153,189] |
| S12 | Silica gel G, activated | Acetone-methanol-glacial acetic acid (70:25:5)[153,189] |
| S13 | Silica gel G, activated | *n*-Butanol-water-glacial acetic acid (4:1:1)[16,59,153,189] |
| S14 | Silica gel G, activated | Acetone-light petroleum (b.p. 40-60°C)-diethylamine (2:7:1)[153,189] |
| S15 | Silica gel G, activated | Acetone-methanol-diethylamine (7:2:1)[153,189] |
| S16 | Cellulose, impregnated with 20% formamide in acetone | Heptane-methyl ethyl ketone (1:1) in ammonia-saturated atmosphere[6] |

| Alkaloid | $hR_F$ values | | | | | | | | | | | | | | | |
|---|---|---|---|---|---|---|---|---|---|---|---|---|---|---|---|---|
| | S1* | S2 | S3 | S4 | S5 | S6 | S7* | S8* | S9* | S10* | S11* | S12* | S13* | S14* | S15* | S16 |
| Ajmalicine | 63 | 72 | 70 | | 78 | 57 | 20 | 74 | 57 | 75 | 18 | 60 | 70 | 89 | 96 | |
| Ajmaline | 5 | 52 | 38 | | 4 | 22 | 0 | 3 | 3 | 8 | 63 | 84 | 76 | 44 | 94 | 28 |
| Alstonine | 0 | | | 52 | | | 0 | 0 | 0 | 0 | 3 | 32 | 64 | 7 | 74 | |
| Aricine | 68 | | | | | | 33 | 82 | 72 | 82 | 18 | 78 | 69 | 91 | 96 | |
| Corynanthine | | 0 | 0 | | | | | | | | | | | | | |
| Deserpidine | 47 | | | | | | 0 | 63 | 39 | 55 | 58 | 91 | 78 | 52 | 96 | |
| Isoreserpiline | 61 | | | | | | 15 | 72 | 56 | 70 | 18 | 74 | 60 | 75 | 96 | |
| 12-Methoxyajmaline | 6 | | | | | | 0 | 4 | 3 | 10 | 58 | 85 | 76 | 50 | 95 | |
| Methyl deserpidate | 3 | | | | | | 0 | 3 | 3 | 3 | 3 | 29 | 58 | 29 | 93 | |
| Methyl reserpate | 3 | 31 | 11 | | 4 | 9 | 0 | 3 | 3 | 3 | 3 | 28 | 57 | 24 | 93 | |
| Norajmaline | 2 | | | | | | 0 | 2 | 2 | 3 | 42 | 72 | 67 | 16 | 90 | |
| Rauvomitine | 30 | | | | | | 0 | 24 | 15 | 15 | 20 | 42 | 65 | 67 | 94 | |
| Renoxidine | 0 | | | | | | 0 | 0 | 0 | 0 | 45 | 90 | 78 | 0 | 51 | |
| Rescinnamine | 40 | 72 | 60 | | 46 | 37 | 0 | 60 | 27 | 48 | 45 | 90 | 78 | 51 | 96 | 51 |
| Reserpiline | 30 | | | | | | 0 | 32 | 22 | 38 | 12 | 52 | 60 | 60 | 93 | |
| Reserpine | 45 | 72 | 60 | | 54 | 39 | 0 | 61 | 33 | 51 | 45 | 90 | 78 | 51 | 96 | 59 |
| 3-Dehydroreserpine | | | | | 2 | 2 | | | | | | | | | | |
| 3-Isoreserpine | | | | | 76 | 54 | | | | | | | | | | |
| Reserpic acid | | 0 | 0 | | | | | | | | | | | | | |
| Reserpinine | 66 | 72 | | | 84 | 65 | 29 | 82 | 71 | 82 | 18 | 78 | 69 | 91 | 93 | 89 |
| Sarpagine | 0 | 17 | 0 | | | | 0 | 0 | 0 | 3 | 4 | 4 | 66 | 5 | 81 | 3 |
| Serpentine | 0 | 27 | 13 | 43 | 0 | 0 | 0 | 0 | 0 | 0 | 3 | 29 | 62 | 6 | 70 | 6 |
| Serpentinine | | 59 | 44 | | | | | | | | | | | | | |
| Tetrahydroalstonine | 70 | | | | | | 45 | 92 | 76 | 85 | 18 | 83 | 70 | 93 | 96 | |
| Tetraphyllicine | 6 | | | | | | 0 | 4 | 4 | 6 | 15 | 33 | 64 | 69 | 91 | |
| Vomalidine | 44 | | | | | | 0 | 40 | 30 | 51 | 80 | 93 | 78 | 60 | 96 | |
| Yohimbine | 32 | | | | 17 | 32 | 0 | 15 | 23 | 35 | 3 | 40 | 52 | 66 | 94 | |
| α-Yohimbine | 47 | 63 | 50 | | 26 | 42 | 0 | 53 | 46 | 58 | 3 | 40 | 42 | 71 | 96 | |
| β-Yohimbine | | 55 | 26 | | | | | | | | | | | | | |
| ψ-Yohimbine | 3 | | | | | | 0 | 3 | 3 | 3 | 3 | 40 | 59 | 47 | 92 | |

*$hR_F$ values taken from ref. 189.

TABLE 13.12

DETECTION METHODS FOR *RAUWOLFIA* ALKALOIDS

| Reagent | No. | Sensitivity* | Background colour | Alkaloid colour | Ref. |
|---|---|---|---|---|---|
| Quenching, 254 nm | | 0.1-0.01 | | | |
| Fluorescence, 366 nm | | | | See Table 13.16 | |
| After drying | | 0.1-0.01 | | | |
| After spraying with acetic acid | | 0.1-0.01 | | | |
| After spraying with trichloroacetic acid sodium nitroprusside | | 0.1-0.01 | | | |
| Dragendorff's modifications: | | | | | |
| Bregoff-Delwiche | 39a | 0.1-1 | Yellow | Orange | |
| Munier | 39b | 0.1-1 | Yellow | Orange | |
| Munier + $NaNO_2$ | 39h | 0.1-1 | White | Orange-brown | |
| Munier + Macheboeuf | 39c | 0.1-1 | Yellow | Orange | |
| Munier + Macheboeuf + $NaNO_2$ | | 0.1-1 | White | Orange-brown | |
| Thies, Reuter and Vagujfalvi | 39f | 0.1-1 | Yellow | Orange | |
| + $NaNO_2$ | | 0.1-1 | White | Orange-brown | |
| + $H_2SO_4$ | | 0.1-1 | Yellow-white | Orange | |
| Iodine vapour | | 0.01-1 | Yellow | Brown | |
| Iodine in chloroform | 61b | 0.1 | Yellow-white | Brown-orange | |
| Bouchardat | 52c | 0.1-1 | Dark-yellow | Brown | |
| Potassium iodoplatinate | 56d | 0.1-1 | Pink-violet | Blue, violet, brown | |
| Cerium(IV) sulphate in conc. $H_2SO_4$ | 14a | 0.1-1 | Yellow | Various colours; some do not give colours | |
| Cerium(IV) ammonium sulphate | 12 | 0.01-1 | Yellow | Various colours; some do not give colours | |
| Cerium(IV) sulphate(1%) in 10% $H_2SO_4$ | 14b | 0.1-1 | Yellow | Various colours; some do not give colours | |
| Cerium(IV) sulphate, saturated in 65% $HNO_3$ | 13 | 0.01-1 | Yellow | Various colours; some do not give colours | |
| Cerium(IV) sulphate according to Sonnenschein | 14c | 0.1-10 | Yellow | Various colours; some do not give colours | |
| Iron(III) chloride in 50% $HNO_3$ | 61 | 0.1-10 | Yellow-white | Various colours | |
| Iron(III) chloride(0.2 *M*) in 35% $HClO_4$ | 62c | 0.1-1 | Yellow-white | Various colours | |
| Iron(III) chloride(0.05 *M*) in 35% $HClO_4$ | 62b | 0.1-1 | Yellow-white | Various colours | |
| Copper sulphate(10%) in 2% ammonia | 27 | 0.01-10 | Blue | Yellow-green; some do not give colours | 90 |
| Sulphomolybdic acid | 90 | 0.1-1 | | See Table 13.17 | 153 |
| Phosphomolybdic acid | 79 | 0.1-10 | | See Table 13.17 | 153 |
| Ammonium vanadate in $HNO_3$ | 3 | 0.01-1 | | See Table 13.17 | 153 |
| Cinnamic aldehyde/HCl vapours | 21 | 0.1-1 | | Various colours | |

*R. Verpoorte, unpublished results.

TABLE 13.13

COLOUR REACTIONS OF SOME *RAUWOLFIA* ALKALOIDS

| Alkaloid | Colour* | | | | | | Ref. |
|---|---|---|---|---|---|---|---|
| | Iodoplatinate | Silica gel $GF_{254}$: Fluorescence (254 nm)** | Silica gel G: Fluorescence (350 nm)** | 5% $FeCl_3$ in 35% $HClO_4$ | | | |
| | | | | A*** | B*** | C*** | |
| Ajmalicine | Pink | bl | Apple-gr | - | g-gr | g-gr | 43,153,189 |
| Ajmaline | Pink | br | v | rd | rd | rd | 43,153,189 |
| Alstonine | br | Intense bl | Intense bl | - | - | - | 189 |
| Aricine | Pink | br | or | - | g-br | g-gr | 153,189 |
| Deserpidine | Pink | bl | bl-gr | - | g-gr | gr-g | 153,189 |
| Lumirescinnamine | bl[a] | Intense bl | Intense bl | | | | 43 |
| Lumireserpine | bl[a] | Intense bl | Intense bl | | | | 43 |
| Iso-reserpiline | | | yel | br | bl | g-gr | 189 |
| 12-methoxyajmaline | | | Weak v | v | v | v | 189 |
| Methyl deserpidate | | | bl | - | g-gr | gr-g | 189 |
| Methyl reserpate | bl-v[a] | | bl | - | bl-gr | br-gr | 43,153,189 |
| Norajmaline | | | Weak v | or | or | or | 189 |
| Renoxidine | | | bl-gr | - | bl-gr | br-gr | 189 |
| Rauvomitine | | | Weak v | rd | rd | rd | 189 |
| Rescinnamine | Pink | Pale bl | bl-gr | - | bl-gr | br-gr | 43,153,189 |
| Reserpiline | br | br | yel | br | bl | g-gr | 43,153,189 |
| Reserpine | Pink | Pale bl | Apple-gr | - | bl-gr | br-gr | 43,153,189 |
| Reserpinine | Pink | Pale bl | Apple-gr | - | g-br | g-gr | 43,153,189 |
| Reserpic acid | bl-v[a] | | bl | | | | 43,189 |
| Sarpagine | br[a] | - | - | gr | pu-g | gr | 43,189 |
| Serpentine | br | Intense bl | Intense bl | - | - | - | 43,153,189 |
| Serpentinine | bl[a] | Intense bl | Intense bl | | | | 43 |
| Tetrahydroalstonine | | | bl-gr | - | g-gr | g-gr | 189 |
| Tetraphyllicine | Pink | br | v | rd | rd | rd | 153,189 |
| Vomalidine | | | Weak v | v | v | v | 189 |
| Yohimbine | Pink | bl-v | bl | - | g | gr-g | 153,189 |
| α-Yohimbine | Pink | bl-v | bl | - | g | gr-g | 43,153,189 |
| β-Yohimbine | bl-v[a] | | gr | | | | 43 |
| ψ-Yohimbine | | | bl | - | g | gr-g | 189 |

*rd=red;v=violet;g=grey;gr=green;bl=blue;or=orange;br=brown;pu=purple;yel=yellow.
**After drying for 1 h at 100°C.
***A=immediate,B=after heating with hair dryer,C=after heating at 110°C for 10 min[189].
[a]The colours for the iodoplatinate reagent and the fluorescence as described in ref. 43 are different from those in refs. 153 and 189. In the table the colours in refs. 153 and 189 are given, except those marked, which are taken from ref. 43, because they are not found in the other two references.

TABLE 13.14

CHROMOGENIC REACTIONS OF SOME *RAUWOLFIA* ALKALOIDS[43,153]

| Alkaloid | 1% cerium(IV) sulphate in 10% sulphuric acid (no.14b) | Sulphomolybdic acid (Fröhde's reagent) (no.90) | 5% iron(III) chloride in 50% nitric acid (no.61) | 0.5% phosphomolybdic acid in 50% nitric acid (no.79) | 1% ammonium vanadate in 50% nitric acid (no.3) |
|---|---|---|---|---|---|
| Ajmalicine | Grey | Green | Grey | Yellow | Green |
| Ajmaline | Crimson | Red | Deep red | Deep red | Deep red |
| Aricine | Brown | Deep green | Orange-brown | Pale green | Pale green |
| Deserpidine | Grey-brown | Green | Yellow-brown | White | Pale green |
| Rescinnamine | Green-brown | Yellow-green | Yellow-brown | Yellow-green | Yellow-green |
| Reserpiline | Violet | Pink | Brown | Yellow-green | - |
| Reserpine | Green-brown | Yellow-green | Yellow-green | Yellow-green | Yellow-green |
| Reserpinine | Yellow-brown | Green | Yellow | Yellow-green | Yellow-green |
| Serpentine | - | - | - | Buff | Buff |
| Tetraphyllicine | Crimson | Deep red | Pink | Deep red | Deep red |
| Yohimbine | Grey | Yellow-green | Yellow | Yellow-green | Grey |
| α-Yohimbine | Grey | Yellow | Yellow | Yellow-green | Grey |

TABLE 13.15

TLC ANALYSIS OF *RAUWOLFIA* ALKALOIDS IN PLANT MATERIALS

| Alkaloid* | Aim | Adsorbent | Solvent system | Ref. |
|---|---|---|---|---|
| RSP,RSC,serp,RSP decomp. | Indirect quantitative analysis(UV) (Table 13.14) | $SiO_2$ | Heptane-MeEtCO-MeOH(58.0:33.6:8.4)<br>MeOH-MeEtCO(4:1) | 4,5 |
| RSP,RSC,RSPine,serp,sarp, ajmal,yoh | Separation (Table 13.11) | $SiO_2$ | Heptane-MeEtCO-MeOH(58.0:33.6:8.4)<br>$CHCl_3$-EtOH(9:1) | |
| | | FMA-impregnated cellulose | Heptane-MeEtCO(1:1) in $NH_3$ atm. | 6 |
| RSP,RSC,ajmal,ajmalicine,yoh | Separation | $SiO_2$ | Heptane-MeEtCO-MeOH(58.0:33.6:8.4) | 14 |
| RSP | Indirect quantitative analysis in roots and pharmaceutical preparations | $SiO_2$+cellulose+starch (10:4:0.25) | MeEtCO-xylene-MeOH(10:10:2)<br>$Et_2O$-$CHCl_3$-EtOH-$NH_4OH$(28:8:2.5:1) | 21,60 |
| RSP,serp,serpin,ajmal, ajmalicine | Identification in plant extracts | $Al_2O_3$ | $CHCl_3$-$Me_2CO$(85:15)<br>Abs. EtOH<br>$CHCl_3$-$Me_2CO$-EtOH(90:5:5) | 25 |
| RSP | Determination RSP in *R. serpentina* | $SiO_2$ | $CHCl_3$-MeOH(93.5:6.5) | 37 |
| RSP,RSC,RSPine,serp,serpin, ajmal,ajmalicine,methoserpidine | Reaction chromatography (Tables 13.14, 13.16 and 13.17) | $SiO_2$ | $CHCl_3$-$Me_2CO$-DEA(5:4:1)<br>$CHCl_3$-cyclohexane-DEA(7:2:1) | 43 |
| RSP,RSC,RSPine,serp,ajmal, ajmalicine,aricine,deserpidine,yoh | Identification in plant materials | $SiO_2$ or $Al_2O_3$ | Heptane-MeEtCO-MeOH(6:3:1)<br>Heptane-MeEtCO-*n*-BuOH(6:3:1)<br>Heptane-MeEtCO-pyridine(70:15:15)<br>*n*-BuOH-$H_2O$-AcOH(4:1:1) | 57 |
| Serp,alst,THalst,ajmalicine | Indirect quantitative analysis(UV) (Table 13.14) | $SiO_2$<br>$Al_2O_3$ | EtOAc-EtOH-DMFA-DEA(30:60:5:5)<br>$CHCl_3$-benzene(8:2) | 59 |
| RSPine,isoRSPine,ajmal, THalst,tetraphyl | Relationship between structure and $hR_F$ values for *Mitragyna* alkaloids | | See section on *Mitragyna* alkaloids (p.316) | 63,65 |

TABLE 13.15 (continued)

| | | | | |
|---|---|---|---|---|
| RSPine,RSPline,isoRSPine,iso-RSPline,ajmalicine,THalst, aricine,tetraphyl,raunitidine, raumitorine,rauvanine,iso-raunitidine,epi-3-rauvanine | Effect of methoxy substitution and configuration on TLC behaviour of heteroyohimbine alkaloids | $SiO_2$<br><br><br><br>$Al_2O_3$ | Benzene-EtOAc(7:2)<br>$Et_2O$<br>$CHCl_3$-$Me_2CO$(5:4)<br>Benzene-EtOAc-DEA(7:2:1)<br>$CHCl_3$-benzene(1:1)<br>$CHCl_3$-benzene-DEA(1:1:0.01)<br>$Et_2O$<br>Cyclohexane-$CHCl_3$<br>Cyclohexane-$CHCl_3$-DEA(3:7:0.005)<br>Benzene-EtOAc(7:2) | 76 |
| corynanthine,yoh,raunitidine | TLC of 73 β-carbolines | $SiO_2$ | *n*-BuOH-AcOH-$H_2O$(4:1:1)<br>*n*-PrOH-$NH_4OH$(8:2)<br>$CHCl_3$-MeOH(9:1) | 77 |
| RSP,RSC,RSP acid,RSP acid methyl ester,isoRSP,3de-hydroRSP,serp,yoh,corynanthine,ajmalicine | Indirect quantitative analysis(UV) of RSP and decomposition products | $SiO_2$ | I. MeOH-MeEtCO-*n*-heptane(1:1:1)<br>II. MeOH-MeEtCO-*n*-heptane(8.4:33.6:58.0)<br>Subsequently I (6 cm) and II (16 cm) in same direction | 83,160 |
| rauvoxine,rauvoxinine | Among a number of oxindole alkaloids from *Mitragyna* | | See section on *Mitragyna* alkaloids (p.316) | 95 |
| RSP,RSC,serp | Indirect quantitative analysis(UV) (Table 13.14) | $SiO_2$ | *n*-BuOH-$Me_2CO$-isooctane(8.4:33.6:58)<br>MeOH-MeEtCO-*n*-heptane(8.4:33.6:58) | 104 |
| RSP,RSC,RSPine,RSPline, serp,ajmal,ajmalicine,yoh | Indirect quantitative analysis(UV) as in ref. 104 | $SiO_2$ | *n*-BuOH-$Me_2CO$-isooctane(8.4:33.6:58)<br>MeOH-MeEtCO-*n*-heptane(8.4:33.6:58)<br>*n*-BuOH-AcOH-$H_2O$(4:1:1) | 105,111, 132 |
| RSP,RSC,sarp,yoh,ajmal | Separation | $SiO_2$ | EtOAc-DEA(58:2) | 115 |
| Alst,serp,THalst,ajmalicine | Separation | $SiO_2$<br><br>$Al_2O_3$ | EtOAc-EtOH-DMFA-DEA(30:60:5:5)<br>*n*-BuOH-$CHCl_3$-25%$NH_4OH$(50:50:2.5)<br>$CHCl_3$-benzene(8:2)<br>$CHCl_3$-EtOAc(1:1) | 126 |
| RSP,RSC | Fluorodensitometric analysis in *Rauwolfia* extracts | 20% FMA-impregnated cellulose | Heptane-MeEtCO(2:1)+ 2% DEA | 129,144 |

References p. 324

| | | | | |
|---|---|---|---|---|
| RSP,RSC,RSPine,serp,ajmal, ajmalicine,aricine,yoh,deserpidine | Indirect quantitative analysis(UV) in *Rauwolfia* plant material | $SiO_2$ | *n*-BuOH-EtOH-$Me_2CO$-isooctane(1:1:4:4)<br>*n*-BuOH-EtOH-MeEtCO-isooctane(1:1:4:4)<br>*n*-BuOH-$Me_2CO$-isooctane(1:2:7)<br>*n*-PrOH-$Me_2CO$-isooctane(8.4:33.6:58) | 138 |
| RSP,RSC,RSPine,serp,ajmal, ajmalicine,aricine,yoh, deserpidine | Indirect quantitative analysis(UV) in *Rauwolfia* plant material | $SiO_2$ | Xylene-isooctane-EtOAc-$Et_2O$(15:45:5:40)<br>*n*-BuOH-EtOAc-ethylene dichloride (1:3:6)<br>$Me_2CO$-light petroleum (b.p. 40-60$^oC$)-$CCl_4$-isooctane(35:30:20:15)<br>$Me_2CO$-light petroleum (b.p. 40-60$^oC$)-AcOH (45:45:10)<br>$Me_2CO$-MeOH-AcOH(70:25:5)<br>$Me_2CO$-light petroleum (b.p. 40-60$^oC$)-$CCl_4$ (45:30:45) | 147 |
| Ajmal,ajmalicine,sarp,yoh | Indirect quantitative analysis in *Rauwolfia* | $SiO_2$ | $CHCl_3$-DEA(9:1)<br>$CHCl_3$-$Me_2CO$-DEA(4:5:1)<br>$CHCl_3$-benzene-DEA(2:7:1) | 148 |
| RSP,RSC,RSPine,RSPline, serp,ajmal,ajmalicine, aricine,yoh,α-yoh,deserpidine,tetraphyl | Indirect quantitative analysis(UV) in *Rauwolfia* plant material | | TLC systems S1 and S7-S15 in Table 13.14 | 153 |
| RSP,RSC,RSPine,RSP acid methyl ester,isoRSP, serp, ajmal,ajmalicine,yoh,corynanthine | Densitometric analysis in *Rauwolfia* extracts (Table 13.14) | $SiO_2$ | $CCl_4$-MeOH(96:4)<br>$isopr_2O$-MeOH-conc. $NH_4OH$(85:15:1)<br>THF-MeOH-AcOH(60:40:3) | 186 |
| 24 *Rauwolfia* alkaloids (Table 13.14) | TLC-behaviour of *Rauwolfia* alkaloids | | TLC systems S1 and S7-S15 in Table 13.14 | 189 |
| Alst,serp | Separation of N-oxides and quaternary alkaloids (Table 13.4) | $SiO_2$ | MeOH-0.2 *M* $NH_4NO_3$<br>MeOH-2 *M* $NH_4OH$-1 *M* $NH_4NO_3$(7:2:1) | 196 |

*For abbreviations, see footnote to Table 13.1.

TABLE 13.16

TLC ANALYSIS OF *RAUWOLFIA* ALKALOIDS IN PHARMACEUTICAL PREPARATIONS AND AS PURE COMPOUNDS IN MIXTURES

| Alkaloid* | Other compounds | Aim | Adsorbent | Solvent system | Ref. |
|---|---|---|---|---|---|
| RSP,RSC | Polyoxyethylene derivatives | Indirect quantitative analysis(UV) | FMA-impregnated $SiO_2$ | Heptane-MeEtCO(2:1) sat. with $H_2O$ | 11 |
| RSP | Chlordiazepoxide,meprobamate,promazine,chlorpromazine | Separation | $SiO_2$ | Benzene-dioxane-conc. $NH_4OH$(75:20:5)<br>$Me_2CO$-cyclohexane-EtOH(4:4:2)<br>Benzene-$Me_2CO$(4:1) | 49 |
| RSP,RSP acid, RSP acid methyl ester | Trimethoxybenzoic acid, papaverine,theophylline, theobromine,barbital, phenobarbital,polyoxyethylene derivatives | Indirect quantitative analysis (fluorimetric) | 2.5% $CaCl_2$-impregnated MgO | $Me_2CO$-hexane(1:1),(2:3),(2:3.5) | 50 |
| RSP | Homatropine,physostigmine, pilocarpine,solanine,solanidine,ephedrine,lobeline | Detection of cholinesterase inhibitors | Cellulose impregnated with 5% silicone 555 | $H_2O$-EtOH-$CHCl_3$(56:42:2) | 66 |
| RSP and decomposition products | | Decomposition of RSP in different solvents | $SiO_2$<br>FMA-impregnated $SiO_2$ | *n*-BuOH-MeEtCO-$H_2O$(2:1:1)<br>Heptane-MeEtCO(2:1) sat. with $H_2O$ | 71 |
| RSC | Various psychotropic drugs | Identification | $SiO_2$<br><br>0.1 *M* NaOH-impregnated $SiO_2$ | MeOH-12 *M* $NH_4OH$(100:1.5)<br>Benzene-EtOH-12 *M* $NH_4OH$(95:15:5)<br>Cyclohexane-benzene-DEA(75:20:15)<br>$Me_2CO$<br>$CHCl_3$-MeOH(9:1) | 75 |
| RSP,RSC,RSPline, ajmal,ajmalicine, yoh | Theophylline,chlorthalidone,hydroflumethiazide, isobutylhydrochlorthiazide,trichlormethiazide, hydrochlorthiazide,dihydralazine | Indirect quantitative analysis(UV or fluorimetric) | $SiO_2$<br>Cellulose impregnated with FMA-DMFA-$Me_2CO$ (20:15:65) | MeOH-MeEtCO-*n*-heptane(4.2:16.8:29)<br>*n*-Heptane-MeEtCO(2:1) sat. with FMA, chamber sat. with $NH_3$ | 79 |
| Ajmal | Various alkaloids | Semi-quantitative circular TLC | $SiO_2$ | $CHCl_3$-MeOH(9:1)<br>$CHCl_3$-EtOH(85:15) | 110 |

| | | | | | |
|---|---|---|---|---|---|
| RSP | | Indirect quantitative analysis | $Al_2O_3$ | Benzene-EtOH(9:1) | 116 |
| RSP,RSC,RSPline, ajmalicine | Isobutyl hydrochlorthiazide | Indirect quantitative analysis | $SiO_2$ | Cyclohexane-$CHCl_3$-MeOH-conc. $NH_4OH$ (50:60:10:0.5)<br>$CHCl_3$-cyclohexane-formaldehyde (10:70:30), followed by the same solvent in the ratio 5:70:30<br>$CHCl_3$-MeOH-cyclohexane-formaldehyde (15:3:70:30)<br>MeOH-$CHCl_3$-$NH_4OH$(10:5:0.1); $Et_2O$ to 100 ml | 120 |
| RSP | Benzthiazide, benziodarone | Indirect quantitative analysis | $SiO_2$ | Benzene-$Me_2CO$-MeOH(5:4:1) sat. with conc. $NH_4OH$<br>$CHCl_3$-MeOH(15:2) sat. with conc. $NH_4OH$<br>$CHCl_3$-cyclohexane-formaldehyde(1:3:7)<br>MeOH-cyclohexane-formaldehyde(5:70:30) | 131 |
| RSP,RSC and decomposition products | Chlorthalidone,dihydralazine,hydrochlorthiazide,furosemide,polythiazide | Fluorodensitometric analysis | $SiO_2$ | $CHCl_3$-MeOH(93:7) | 134 |
| RSP,isoRSP, 3,4dehydroRSP, lumiRSP | | Photodecomposition of RSP | $SiO_2$ | $CHCl_3$-MeOH(4:1)<br>*n*-Hexane-MeEtCO-MeOH(5:4:1) | 143 |
| RSP | Orphenadrine,meprobamate | Fluorodensitometric analysis | $SiO_2$ | $CHCl_3$-MeOH(8:2) | 157 |
| RSP,yoh | Ergot alkaloids,physostigmine,strychnine, brucine | New detection method | $SiO_2$ | *n*-BuOH-AcOH-$H_2O$(3:1:1)<br>$CHCl_3$-$Me_2CO$-DEA(5:4:1) | 167 |
| RSP | Binazine,hydrochlorthiazide,dihydralazine,glutethimide,theobromine, phenobarbital | Identification | $SiO_2$ | $Me_2CO$-MeOH-$CHCl_3$-benzene-isooctane-MeEtCO-25% $NH_4OH$(5:1.5:16:2:3:3:0.3)<br>*n*-BuOH-toluene-$H_2O$-AcOH(12:5:2:1.5)<br>Toluene-$Me_2CO$(8:5)<br>$Me_2CO$-toluene-dioxane-$H_2O$(4:4:7:1)<br>*n*-BuOH-$H_2O$-AcOH(17:5:2)<br>$Me_2CO$-$CHCl_3$-isooctane-toluene-25% $NH_4OH$(10:3:3:4:0.25) | 173 |

TABLE 13.16 (continued)

| Alkaloid* | Other compounds | Aim | Adsorbent | Solvent system | Ref. |
|---|---|---|---|---|---|
| RSP | Various thiazide diuretics and antihypertensive drugs | Separation | $SiO_2$ | MeEtCO-*n*-hexane(1:1,2:1 and 3:2)<br>$CHCl_3$-$Me_2CO$-triethanolamine(50:50:1.5) | 182 |
| RSP,RSC,ajmal | Thiabutazide,clopamide,dihydroergocristine methanesulphonate | Densitometric analysis | $SiO_2$ | $CCl_4$-MeOH(96:4)<br>isopr$_2$O-MeOH-$NH_4OH$(85:15:1)<br>THF-MeOH-AcOH(60:40:3) | 186 |
| Yoh | Strychnine | yoh as internal standard in quantitative analysis of strychnine | $SiO_2$ | $CHCl_3$-MeOH-$NH_4OH$(90:10:0.15) | 188 |
| RSP | Chlorthiazide,hydrochlorothiazide,spironolactone, triamterene | Indirect quantitative analysis(fluorimetric) | $SiO_2$ | $CHCl_3$-AcOH-MeOH(80:20:15) | 208 |

*For abbreviations, see footnote to Table 13.1.

References p. 324

TABLE 13.17

TLC ANALYSIS OF *RAUWOLFIA* ALKALOIDS IN BIOLOGICAL MATERIALS AND TOXICOLOGICAL ANALYSIS

| Alkaloid* | Aim | Adsorbent | Solvent system | Ref. |
|---|---|---|---|---|
| RSP | Quantitative analysis of tritium-labelled RSP | $SiO_2$ | $CHCl_3$-cyclohexane-DEA(4:5:1) | 72 |
| Ajmal | Identification in autopsy material | $SiO_2$ | $CHCl_3$-$Me_2CO$(9:1)<br>MeOH-$NH_4OH$(100:1) | 137 |
| Ajmal,17-monochloro-ajmal | Fluorodensitometric analysis in human plasma | $SiO_2$ | Benzene-AcOH-methanol(86:7:7) | 184 |
| RSP | Fluorodensitometric analysis in human plasma | $SiO_2$ | $CHCl_3$-$Me_2CO$(7:3) | 185 |
| RSP | Fluorodensitometric analysis in plasma | $SiO_2$ | $CHCl_3$-$Me_2CO$(7:3)<br>$CHCl_3$-MeOH(95:5)<br>Benzene-MeOH(8:2)<br>BuOH-$H_2O$-AcOH(4:1:1) | 217 |

*For abbreviations, see footnote to Table 13.1.

TABLE 13.18

TLC SYSTEMS USED TO STUDY THE CORRELATION BETWEEN THE STEREOCHEMISTRY OF SOME INDOLE AND OXINDOLE ALKALOIDS AND THEIR BEHAVIOUR IN TLC[63]

The alkaloids used in this study were reserpinine, tetrahydroalstonine, tetraphylline, ajmalicine, mitragynine, corynantheidine, speciogynine, dihydrocorynantheine, paynantheine, corynantheine, speciociliatine, mitraciliatine, isorhynchophylline, mitraphylline, isomitraphylline and speciophylline.

| System | Sorbent | Solvent* |
|---|---|---|
| S1 | Silica gel G, activated | Cyclohexane |
| S2 | Silica gel G, activated | Cyclohexane-chloroform(5:4) |
| S3 | Silica gel G, activated | Chloroform |
| S4 | Silica gel G, activated | Benzene-ethyl acetate(7:2) |
| S5 | Aluminium oxide G, activated | Chloroform-benzene(1:1) |
| S6 | Aluminium oxide G, activated | Cyclohexane-chloroform(3:7) |
| S7 | Silica gel G, activated | Diethyl ether |
| S8 | Aluminium oxide G, activated | Chloroform |
| S9 | Silica gel G, activated | Chloroform-acetone(5:4) |
| S10 | Silica gel G, activated | Methanol |
| S11 | Silica gel G, activated | Cyclohexane-diethylamine(9:1) |
| S12 | Silica gel G, activated | Cyclohexane-chloroform-diethylamine(5:4:1) |
| S13 | Silica gel G, activated | Chloroform-diethylamine(9:1) |
| S14 | Silica gel G, activated | Benzene-ethyl acetate-diethylamine(7:2:1) |
| S15 | Aluminium oxide G, activated | Chloroform-benzene-diethylamine(1:1:0.001) |
| S16 | Aluminium oxide G, activated | Cyclohexane-chloroform-diethylamine(3:7:0.005) |
| S17 | Silica gel G, activated | Diethyl ether-diethylamine(9:1) |
| S18 | Aluminium oxide G, activated | Chloroform-diethylamine(9:1) |
| S19 | Silica gel G, activated | Chloroform-acetone-diethylamine(5:4:1) |
| S20 | Silica gel G, activated | Methanol-diethylamine(9:1) |

*Solvent systems S1-S10 were also used in combination with 0.1 *M* sodium hydroxide-impregnated stationary phases.

TABLE 13.19

TLC SYSTEMS USED TO STUDY THE CORRELATION BETWEEN THE STEREOCHEMISTRY OF SOME HETERO-YOHIMBINE ALKALOIDS AND THEIR BEHAVIOUR IN TLC[76]

Alkaloids used in this study:

| Alkaloid | Type | Substituent in aromatic ring |
|---|---|---|
| E seco alkaloids (II): | | |
| Corynantheidine | Allo | H |
| Mitragynine | Allo | 9-$OCH_3$ |
| Corynantheine | Normal | H |
| Dihydrocorynantheine | Normal | H |
| Paynantheine | Normal | 9-$OCH_3$ |
| Speciogynine | Normal | 9-$OCH_3$ |
| Isocorynantheidine | Epiallo | H |
| Speciociliatine | Epiallo | 9-$OCH_3$ |
| Hirsutine | Pseudo | H |
| Mitraciliatine | Pseudo | 9-$OCH_3$ |
| Closed E ring alkaloids (III), C(19)-$CH_3$ $\alpha$ configurations: | | |
| Tetrahydroalstonine | Allo | H |
| Aricine | Allo | 10-$OCH_3$ |
| Reserpinine | Allo | 11-$OCH_3$ |
| Isoreserpiline | Allo | 10,11-di-$OCH_3$ |
| Ajmalicine | Normal | H |
| Tetraphylline | Normal | 11-$OCH_3$ |
| Akuammigine | Epiallo | H |
| Isoreserpinine | Epiallo | 11-$OCH_3$ |
| Reserpiline | Epiallo | 10,11-di-$OCH_3$ |
| Isoajmalicine | Pseudo | H |
| Mitrajavine | Pseudo | 9-$OCH_3$ |
| Closed E ring alkaloids (III), C(19)-$CH_3$ $\beta$ configurations: | | |
| Raunitícine | Allo | H |
| Raunitidine | Allo | 11-$OCH_3$ |
| Raumitorine | Normal | 10-$OCH_3$ |
| Rauvanine | Normal | 10,11-di-$OCH_3$ |
| Isoraunitidine | Epiallo | 11-$OCH_3$ |
| Epi-3-rauvanine | Pseudo | 10,11-di-$OCH_3$ |

| System | Sorbent | Solvent |
|---|---|---|
| S1 | Silica gel G, activated | Benzene-ethyl acetate(7:2) |
| S2 | Aluminium oxide G, activated | Chloroform-benzene(1:1) |
| S3 | Aluminium oxide G, activated | Chloroform-benzene-diethylamine(1:1:0.001) |
| S4 | Aluminium oxide G, activated | Benzene-ethyl acetate(7:2) |
| S5 | Aluminium oxide G, activated | Diethyl ether |
| S6 | Silica gel G, activated | Diethyl ether |
| S7 | Silica gel G, activated | Chloroform-acetone(5:4) |
| S8 | Aluminium oxide G, activated | Cyclohexane-chloroform(3:7) |
| S9 | Aluminium oxide G, activated | Cyclohexane-chloroform-diethylamine(3:7:0.005) |
| S10 | Silica gel G, activated | Benzene-ethyl acetate-diethylamine(7:2:1) |

TABLE 13.20

TLC SYSTEMS USED TO STUDY THE CORRELATION BETWEEN THE STEREOCHEMISTRY OF INDOLIZIDINE AND OXINDOLE ALKALOIDS AND THEIR BEHAVIOUR IN TLC[95]

Alkaloids used in this study:

| Alkaloid | Type | Substituent in aromatic ring |
|---|---|---|
| Isorhynchophylline | Normal | H |
| Rhynchophylline | Normal | H |
| Rotundifoline | Normal | 9-OH |
| Isorotundifoline | Normal | 9-OH |
| Rhynchociline | Normal | 9-$OCH_3$ |
| Ciliaphylline | Normal | 9-$OCH_3$ |
| Isospecionoxeine | Normal | 9-$OCH_3$ |
| Specionoxeine | Normal | 9-$OCH_3$ |
| Corynoxine | Allo | H |
| Corynoxine B | Allo | H |
| Mitragynine oxindole B | Allo | 9-$OCH_3$ |
| Speciofoline | - | 9-OH |

| System | Sorbent | Solvent |
|---|---|---|
| S1 | Silica gel G, activated | Diethyl ether |
| S2 | Silica gel G, activated | Benzene-ethyl acetate(7:2) |
| S3 | Silica gel G, activated | Cyclohexane-chloroform(3:7) |
| S4 | Silica gel G, activated | Chloroform |
| S5 | Silica gel G, activated | Chloroform-acetone(5:4) |
| S6 | Silica gel G, activated | Diethyl ether-ethanol(95:5) |
| S7 | Silica gel G, activated | Chloroform-ethanol(95:5) |
| S8 | Silica gel G, activated | Diethylether-diethylamine(9:1) |
| S9 | Silica gel G, activated | Benzene-ethyl acetate-diethylamine(7:2:1) |
| S10 | Silica gel G, activated | Cyclohexane-chloroform-diethylamine(3:7:1) |
| S11 | Silica gel G, activated | Chloroform-diethylamine(9:1) |
| S12 | Silica gel G, activated | Chloroform-acetone-diethylamine(5:4:1) |
| S13 | Silica gel G, activated | Diethyl ether-ethanol-diethylamine(95:5:10) |
| S14 | Silica gel G, activated | Chloroform-ethanol-diethylamine(95:5:10) |

TABLE 13.21

TLC SYSTEMS USED TO STUDY THE CORRELATION BETWEEN THE STEREOCHEMISTRY AND METHOXY SUBSTITUTION OF CLOSED E-RING OXINDOLES AND THEIR BEHAVIOUR IN TLC[96]

| System | Sorbent | Solvent |
|---|---|---|
| S1 | Aluminium oxide G, activated | Cyclohexane-chloroform(3:7) |
| S2 | Aluminium oxide G, activated | Cyclohexane-chloroform-diethylamine(3:7:0.005) |
| S3 | Aluminium oxide G, activated | Chloroform |
| S4 | Aluminium oxide G, activated | Diethyl ether-ethanol(95:5) |
| S5 | Aluminium oxide G, activated | Chloroform-diethylamine(9:1) |
| S6 | Silica gel G, activated | Benzene-ethyl acetate(7:2) |
| S7 | Silica gel G, activated | Diethyl ether |
| S8 | Silica gel G, activated | Diethyl ether-diethylamine(9:1) |
| S9 | Silica gel G, activated | Chloroform-acetone(5:4) |
| S10 | Silica gel G, activated | Diethyl ether-ethanol(95:5) |

Alkaloids used in this study:

| Alkaloid* | Type | C(19)-$CH_3$ | Substituent in aromatic ring | Oxindole ($CO/N_4$) |
|---|---|---|---|---|
| Isomitraphylline | Normal | α | H | Anti |
| Mitraphylline | Normal | α | H | Syn |
| Uncarine A (isoformosanine) | Normal | β | H | Anti |
| Uncarine B (formosanine) | Normal | β | H | Syn |
| Javaphylline | Normal | α | 9-$OCH_3$ | Anti |
| Isojavaphylline | Normal | α | 9-$OCH_3$ | Syn |
| Isopteropodine (uncarine C) | Allo | α | H | Anti |
| Pteropodine (uncarine E) | Allo | α | H | Syn |
| Speciophylline (uncarine D) | Epiallo | α | H | Syn |
| Uncarine F | Epiallo | α | H | Anti |
| Isocarapanaubine | Allo | α | 10,11-di$OCH_3$ | Anti |
| Carapanaubine | Allo | α | 10,11-di$OCH_3$ | Syn |
| Rauvoxinine | Epiallo | α | 10,11-di$OCH_3$ | Anti |
| Rauvoxine | Epiallo | α | 10,11-di$OCH_3$ | Syn |
| Majdine 1 | - | - | 10,11-di$OCH_3$ | - |
| Majdine 2 | - | - | 10,11-di$OCH_3$ | - |
| Majdine 3 | - | - | 10,11-di$OCH_3$ | - |
| Majdine 4 | - | - | 10,11-di$OCH_3$ | - |

TABLE 13.22

TLC ANALYSIS OF SOME *MITRAGYNA* ALKALOIDS[219]

TLC system:
  Silica gel G
Solvent systems:
  S1 Chloroform-acetone(5:4)
  S2 Chloroform-ethanol(6:1)
  S3 Ethyl acetate-isopropanol-ammonia(70:25:15)
  S4 Methanol

| Alkaloid | $hR_F$ values | | | |
|---|---|---|---|---|
| | S1 | S2 | S3 | S4 |
| Rotundifoline | 73 | 90 | 85 | 74 |
| Isorotundifoline | 58 | 85 | 82 | 62 |
| Rhynchophylline | 35 | 55 | 61 | 47 |
| Isorhynchophylline | 58 | 89 | 83 | 71 |
| Mitraphylline | 46 | 60 | 45 | 47 |
| Isomitraphylline | 57 | 67 | 48 | 57 |
| Hirsutine | 8 | 4 | 7 | 21 |
| Hirsuteine | 15 | 11 | 11 | 26 |
| Rhynchophylline N-oxide | 0 | 9 | 10 | 15 |
| Anti-rotundifoline N-oxide | 0 | 10 | 24 | 57 |

TABLE 13.23

TLC IDENTIFICATION OF *UNCARIA* ALKALOIDS[178]

TLC systems:

| | | |
|---|---|---|
| S1 | Silica gel G-GF$_{254}$(2:1), activated | Chloroform-acetone(5:4) |
| S2 | Silica gel G-GF$_{254}$(2:1), activated | Chloroform-ethanol(95:5) |
| S3 | Silica gel G-GF$_{254}$(2:1), activated | Diethyl ether-ethyl acetate(1:1) |
| S4 | Silica gel G-GF$_{254}$(2:1), activated | Ethyl acetate-isopropanol-conc. ammonia(100:2:1) |
| S5 | Silica gel G-GF$_{254}$(2:1), activated | Ethyl acetate-isopropanol-conc. ammonia(80:15:5) |
| S6 | Silica gel G-GF$_{254}$(2:1), activated | Ethyl acetate-isopropanol-conc. ammonia(60:35:5) |
| S7 | Silica gel G-GF$_{254}$(2:1), activated | Chloroform-methanol(6:1) |
| S8 | Silica gel G-GF$_{254}$(2:1), activated | Methanol-diethylamine(96:8) |

| Alkaloid | $hR_F$ values | | | | | | | |
|---|---|---|---|---|---|---|---|---|
| | S1 | S2 | S3 | S4 | S5 | S6 | S7 | S8 |
| Pentacyclic heteroyohimbines: | | | | | | | | |
| Ajmalicine (normal) | 72 | 60 | 62 | 71 | 88 | | | |
| Isoajmalicine (pseudo) | 17 | 27 | 7 | 23 | 79 | | | |
| Mitrajavine (pseudo) | 15 | 23 | 8 | 26 | 78 | | | |
| Tetrahydroalstonine (allo) | 83 | 72 | 77 | 81 | 90 | | | |
| Akuammigine (epiallo) | 50 | 45 | 34 | 50 | 85 | | | |
| 4-R akuammigine N-oxide | 0 | 0 | 0 | 0 | 24 | 38 | 50 | 70 |
| 19-epi-Ajmalicine (normal) | 78 | 63 | 68 | 75 | 88 | | | |
| 3-Iso-19-epi-ajmalicine (pseudo) | 20 | 39 | - | 27 | - | | | |
| Rauniticine (allo) | 43 | 25 | 35 | 60 | 88 | | | |
| Tetracyclic heteroyohimbines: | | | | | | | | |
| Dihydrocorynantheine (normal) | 69 | 44 | 60 | 76 | 86 | | | |
| Gambirine (normal) | 44 | 14 | 48 | 63 | 80 | | | |
| Speciogynine (normal) | 62 | 40 | 58 | 77 | 86 | | | |
| Hirsutine (pseudo) | 10 | 11 | 4 | 19 | 74 | | | |
| Hirsuteine (pseudo) | 16 | 15 | 8 | 25 | 76 | | | |
| Mitraciliatine (pseudo) | 10 | 10 | 4 | 18 | 74 | | | |
| Corynantheidine (allo) | 80 | 60 | 75 | 80 | 88 | | | |
| Mitragynine (allo) | 76 | 58 | 72 | 80 | 87 | | | |
| Isocorynantheidine (epiallo) | 33 | 21 | 27 | 56 | 82 | | | |
| Speciociliatine (epiallo) | 22 | 16 | 18 | 56 | 82 | | | |
| Pentacyclic oxindoles: | | | | | | | | |
| Isomitraphylline (normal A) | 68 | 45 | 45 | 60 | 74 | | | |
| Isomitraphylline N-oxide | 0 | 0 | 0 | 0 | 18 | 24 | 32 | 69 |
| Javaphylline (normal A) | 29 | 22 | 18 | 42 | 76 | | | |
| Mitraphylline (normal B) | 51 | 37 | 16 | 37 | 66 | | | |
| Mitraphylline N-oxide | 0 | 0 | 0 | 0 | 2 | 3 | 6 | 42 |
| Isopteropodine (allo A) | 72 | 47 | 56 | 69 | 77 | | | |
| Isopteropodine N-oxide | 0 | 0 | 0 | 0 | 24 | 35 | 44 | 71 |
| Pteropodine (allo B) | 68 | 47 | 50 | 65 | 77 | | | |
| Pteropodine N-oxide | 0 | 0 | 0 | 0 | 6 | 6 | 9 | 54 |
| Speciophylline (epiallo A) | 29 | 26 | 8 | 19 | 59 | | | |
| Speciophylline N-oxide | 0 | 0 | 0 | 0 | 2 | 3 | 5 | 53 |
| Uncarine F (epiallo B) | 60 | 38 | 38 | 52 | 77 | | | |
| Uncarine F N-oxide | 0 | 0 | 0 | 0 | 21 | 29 | 38 | 69 |
| Uncarine A, (normal A) | 70 | 45 | 49 | 65 | 76 | | | |
| Uncarine B (normal B) | 62 | 41 | 26 | 47 | 71 | | | |
| Tetracyclic oxindoles: | | | | | | | | |
| Isorhynchophylline (normal A) | 70 | 40 | 56 | 66 | 78 | | | |
| anti-Isorhynchophylline N-oxide | 0 | 0 | 0 | 0 | 21 | 38 | 42 | 75 |
| Isocorynoxeine (normal A) | 70 | 42 | 57 | 69 | 77 | | | |
| Rotundifoline (normal A) | 73 | 48 | 58 | 70 | 77 | | | |
| Rhynchociline (normal A) | 9 | 18 | 12 | 41 | 74 | | | |
| Rhynchophylline (normal B) | 35 | 22 | 11 | 29 | 66 | | | |

TABLE 13.23 (continued)

| Alkaloid | $hR_F$ values | | | | | | | |
|---|---|---|---|---|---|---|---|---|
| | S1 | S2 | S3 | S4 | S5 | S6 | S7 | S8 |
| Rhynchophylline N-oxide | 0 | 0 | 0 | 0 | 2 | 3 | 13 | 57 |
| Corynoxeine (normal B) | 43 | 22 | 15 | 35 | 65 | | | |
| Isorotundifoline (normal B) | 48 | 25 | 34 | 54 | 71 | | | |
| Ciliaphylline (normal B) | 27 | 11 | 19 | 27 | 63 | | | |
| Corynoxine (allo A) | 73 | 45 | 63 | 71 | 78 | | | |
| Corynoxine B (allo B) | 48 | 25 | 35 | 55 | 78 | | | |
| Speciofoline (epiallo A) | 67 | 42 | 46 | 62 | 75 | | | |
| Roxburghine C | 82 | 33 | 67 | 77 | 90 | | | |
| Roxburghine D | 53 | 33 | 45 | 66 | 89 | | | |
| Roxburghine E | 17 | 20 | 6 | 29 | 78 | | | |
| Harman | 29 | 11 | 16 | 37 | 72 | | | |
| Harmine | 17 | 6 | 7 | 26 | 67 | | | |
| Harmaline | 0 | 0 | 0 | 12 | 47 | | | |
| Angustine | 57 | 30 | 33 | 57 | 82 | | | |
| Angustidine | 44 | 20 | 17 | 39 | 74 | | | |
| Angustoline | 18 | 7 | 5 | 19 | 56 | | | |
| Gambirtannine | 87 | 85 | 81 | 85 | 90 | | | |
| Dihydrogambirtannine | 83 | 71 | 74 | 78 | 89 | | | |
| Oxogambirtannine | 83 | 79 | 70 | 78 | 87 | | | |

TABLE 13.24

IDENTIFICATION OF *UNCARIA* ALKALOIDS ON SILICA GEL LAYERS[178]

| Alkaloid | Colour* | | | | |
|---|---|---|---|---|---|
| | UV | | $FeCl_3/HClO_4$ (no. 62c) | Ehrlich's reagent (no. 36b) | 2% $Ce(SO_4)_2$/1 *M* $H_2SO_4$ |
| | 365 nm fl | 254 nm | | | |
| Pentacyclic heteroyohimbines: | | | | | |
| Ajmalicine | y/w | q | g→br | pu | - |
| Isoajmalicine | y/w | q | g→br | pu | - |
| Mitrajavine | or/y | q | g→br | pu | - |
| Tetrahydroalstonine | y/w | q | g→br | pu | - |
| Akuammigine | y/w | q | g→br | pu | - |
| 4-R akuammigine N-oxide | - | q | g→br | pu | - |
| 19-epi-Ajmalicine | y/w | q | g→br | pu | - |
| 3-Iso-19-epi-ajmalicine | y/w | q | g→br | pu | - |
| Rauniticine | y/w | q | g→br | pu | - |
| Tetracyclic heteroyohimbines: | | | | | |
| Dihydrocorynantheine | b/w | q | b/g→br | pu | - |
| Gambirine | - | q | g before heating→br | g | br |
| Speciogynine | y | q | g→br | pu | - |
| Hirsutine | b/w | q | g→br | pu | - |
| Hirsuteine | b/w | q | g→br | pu | - |
| Mitraciliatine | y | q | g→br | pu | - |
| Corynantheidine | b/w | q | b/g→br | pu | - |
| Mitragynine | y | q | g→br | pu | - |
| Isocorynantheidine | b/w | q | g→br | pu | - |
| Speciociliatine | y | q | g→br | pu | - |
| Pentacyclic oxindoles: | | | | | |
| Isomitraphylline | - | q | p | - | - |
| Isomitraphylline N-oxide | - | q | p | - | - |
| Javaphylline | - | q | p | - | - |
| Mitraphylline | - | q | p | - | - |
| Mitraphylline N-oxide | - | q | p | - | - |
| Isopteropodine | - | q | gr→p | - | - |
| Isopteropodine N-oxide | - | q | gr→p | - | - |
| Pteropodine | - | q | gr→p | - | - |
| Pteropodine N-oxide | - | q | gr→p | - | - |
| Speciophylline | - | q | gr→p | - | - |
| Speciophylline N-oxide | - | q | gr→p | - | - |
| Uncarine F | - | q | gr→p | - | - |
| Uncarine F N-oxide | - | q | gr→p | - | - |
| Uncarine A | - | q | p | - | - |
| Uncarine B | - | q | p | - | - |
| Tetracyclic oxindoles: | | | | | |
| Isorhynchophylline | - | q | b→p** | - | - |
| Anti-Isorhynchophylline N-oxide | - | q | b→p | - | - |
| Isocorynoxeine | - | q | b→p** | - | - |
| Rotundifoline | - | q | b→pu→br | p | Transient br |
| Rhynchociline | - | q | b→pu | p | - |
| Rhynchophylline | - | q | b→p** | - | - |
| Rhynchophylline N-oxide | - | q | b→p | - | - |
| Corynoxeine | - | q | b→p** | - | - |
| Isorotundifoline | - | q | b→pu→br | p | Transient br |
| Ciliaphylline | - | q | b→pu | p | - |
| Corynoxine | - | q | b→p** | - | - |
| Corynoxine B | - | q | b→p** | - | - |
| Speciofoline | - | q | b→pu→br | p | Transient br |

TABLE 13.24 (continued)

| Alkaloid | Colour* | | | | |
|---|---|---|---|---|---|
| | UV | | $FeCl_3/HClO_4$ (no. 62c) | Ehrlich's reagent (no. 36b) | 2% $Ce(SO_4)_2$/1 *M* $H_2SO_4$ |
| | 365 nm fl | 254 nm | | | |
| Roxburghine C | - | q | g→br | pu | r |
| Roxburghine D | - | q | g→br | pu | r |
| Roxburghine E | - | q | g→br | pu | or/r |
| Harman | b | b fl. | p/pu | - | - |
| Harmine | b | b fl. | pu→br | - | - |
| Harmaline | b | y/w fl. | pu→br | y | or |
| Angustine*** | y | y fl. | Transient gr | y | - |
| Angustidine*** | y | y fl. | Transient gr | y | - |
| Angustoline | y | y fl. | Transient gr | y | y |
| Gambirtannine | y/or | q | g | y | - |
| Dihydrogambirtannine | - | q | pu→g | p | - |
| Oxogambirtannine*** | y | y fl. | gr→br | y | br |

*Key to abbreviations: b = blue; br = brown; fl = fluorescence; g = grey; gr = green; or = orange; p = pink; pu = purple; q = fluorescence quenching; r = red; w = white; y = Yellow.

**The rate of change of colour of the unsubstituted tetracyclic oxindoles on prolonged heating is slower for rhynchophylline and isorhynchophylline than for the corynoxines or the corynoxeines.

***These alkaloids give yellow spots on spraying with Dragendorff's reagent, rather than orange.

TABLE 13.25

TLC ANALYSIS OF *VINCA* ALKALOIDS[26]

TLC systems:

| | Layer | Solvent |
|---|---|---|
| S1 | Aluminium oxide, activated | Chloroform-ethyl acetate(1:1) |
| S2 | Silica gel | Ethyl acetate-absolute ethanol(3:1) |
| S3 | Aluminium oxide, activated | Ethyl acetate-absolute ethanol(3:1) |
| S4 | Aluminium oxide, activated | Benzene |
| S5 | Aluminium oxide, activated | Chloroform |
| S6 | Silica gel | Chloroform |
| S7 | Aluminium oxide, activated | Benzene-chloroform(3:1) |
| S8 | Silica gel | Ethyl acetate-absolute ethanol(1:1) |
| S9 | Silica gel | Ethyl acetate |
| S10 | Aluminium oxide, activated | Ethyl acetate |
| S11 | Silica gel, impregnated with 0.5 *M* potassium hydroxide | Ethyl acetate-absolute ethanol(1:1) |

| Alkaloid | $hR_F$ values | | | | | | | | | Colour with cerium(IV) ammonium sulphate (no. 12) |
|---|---|---|---|---|---|---|---|---|---|---|
| | S1 | S2 | S3 | S4 | S5 | S6 | S7 | S8 | S9 | |
| Ajmalicine | 57 | 68 | 72 | 3 | 51 | 2 | 9 | 68 | 54 | Yellow |
| Carosidine | | 58 | | | | | | 59 | 10 | Yellow |
| Carosine | | 71 | | | | | | 65 | 24 | Purple-grey |
| Catharanthine | 77 | 59 | 74 | 12 | 77 | 3 | 37 | 58 | 38 | Green (fades quickly) |
| Catharine | 18 | 58 | 76 | | | | | 56 | 10 | Yellow |
| Catharosine | | 56 | | | | | | 58 | 8 | Purple |
| Isoleurosine | 35 | 22 | | 0 | 23 | 0 | 0 | | | Grey |
| Leurosine | 27 | 35 | | 0 | 20 | | | | | Grey |
| Lochnericine | 15 | 25 | 77 | | | | | 46 | 3 | Blue |
| Lochneridine | 0 | 0 | 21 | | | | | 4 | 0 | Blue-green |
| Lochnerine | 4 | 35 | 70 | | | | | 42 | 6 | Pale grey |
| Neoleurocristine | | 27 | | | | | | 43 | 3 | Blue |
| Neoleurosidine | | 6 | | | | | | 17 | 0 | Yellow-brown |
| Perivine | 5 | 30 | 48 | | | | | 39 | 11 | Lt. brown |
| Pleurosine | | 51 | 42 | | | | | 7 | 3 | Yellow |
| Serpentine | 3 | 0 | 11 | | | | | 0 | 0 | |
| Tetrahydroalstonine | 76 | 60 | 73 | 5 | 66 | 4 | 29 | 76 | 69 | Yellow-green |
| Vinblastine | 25 | 24 | 66 | 0 | 17 | 0 | 0 | 33 | 4 | Purple |
| Vincamicine | 3 | 9 | 42 | | | | | 20 | 0 | Blueish orange |
| Vincarodine | | 50 | | | | | | 50 | 10 | Blue (fades quickly) |
| Vindolicine | 24 | 46 | 73 | | | | | | | Blue |
| Vindolidine | | 15 | | | | | | 29 | 0 | Blue |
| Vindoline | 44 | | 68 | 0 | 53 | 3 | 6 | 57 | 20 | Crimson |
| Vindolinine | 55 | 37 | 70 | 0 | 51 | 0 | 9 | 44 | 13 | Orange |
| Virosine | 9 | 54 | 63 | | | | | 48 | 9 | Colourless |
| Sitsirikine | | 31 | | | | | | 45 | 9 | Yellow-green |

(From: The American Pharmaceutical Association).

TABLE 13.26

TLC SYSTEMS REQUIRED TO OBTAIN A SEPARATION BETWEEN SOME *VINCA* ALKALOIDS. NUMBERS REFER TO TLC SYSTEMS IN TABLE 13.25[26]

| Alkaloid | A | B | C | D | E | F | G | H | I | J | K | L | M | N | O | P | Q |
|---|---|---|---|---|---|---|---|---|---|---|---|---|---|---|---|---|---|
| Ajmalicine(A) | | 1 | 1 | 1 | 1 | 1 | 1 | 1 | 1 | 1 | 1 | 1,3 | 1 | 1 | 7 | 7 | 1 |
| Catharanthine(B) | 1 | | 1 | 1 | 1 | 1 | 1 | 1 | 1 | 1 | 2 | 1,3 | 1 | 1 | 1 | 1 | 1 |
| Catharine(C) | 1 | 1 | | 1 | | 3 | 1 | 1 | 1 | 1 | 1 | 4 | 1 | 3 | 1 | 1 | 6 |
| Isoleurosine(D) | 1 | 1 | 1 | | 5 | | 1 | | 1 | 1 | 1 | 3 | | | 7 | 7 | 3 |
| Leurosine(E) | 1 | 1 | | 5 | | | 1 | | 1 | 1 | 1 | 3 | | | 7 | 7 | 3 |
| Lochnericine(F) | 1 | 1 | 3 | | | | 1 | 3 | 6 | 1 | 1 | 3 | 1 | 3 | 1 | 1 | 3 |
| Lochneridine(G) | 1 | 1 | 1,3 | 1 | 1 | 1 | | 3 | 3 | 6 | 1 | 1,3 | 3 | 1 | 1 | 1 | 3 |
| Lochnerine(H) | 1 | 1 | 1,3 | | | 3 | 3 | | 6 | 3 | 1 | | 3 | | 1 | 1 | 3 |
| Perivine(I) | 1 | 1 | 1 | 1 | 1 | 6 | 6 | 6 | | 3 | 1 | 1 | 3 | 6 | 1 | 1 | 6 |
| Serpentine(I) | 1 | 1 | 1 | 1 | 1 | 1 | 1,6 | 3 | 3 | | 1 | 1 | 6 | 1 | 1 | 1 | 6 |
| Tetrahydroalstonine(K) | 1 | 2 | 1 | 1 | 1 | 1 | 1 | 1 | 1 | 1 | | 1 | 1 | 1 | 1 | 1 | 1 |
| VLB(L) | 1,3 | 1,3 | 4 | 3 | 3 | 3 | 3 | | 1 | 1 | 1 | | | 4 | | | 3 |
| Vincamicine(M) | 1 | 1 | 1 | | | 1 | 3 | 3 | 3 | 6 | 1 | | | 3 | 1 | 1 | 3 |
| Vindolicine(N) | 1 | 1 | 3 | | | 3 | 1 | | 6 | 1 | 1 | 4 | 3 | | 1 | 1 | |
| Vindoline(O) | 7 | 1 | 1 | 7 | 7 | 1 | 1 | 1 | 1 | 1 | 1 | | 1 | 1 | | 3 | 1 |
| Vindolinine(P) | 7 | 1 | 1 | 7 | 7 | 1 | 1 | 1 | 1 | 1 | 1 | | 1 | 1 | 3 | | 1 |
| Virosine(Q) | 1 | 1 | 6 | 3 | 3 | 3 | 3 | 3 | 6 | 6 | 1 | 3 | 3 | | 1 | 1 | |

(From: The American Pharmaceutical Association).

TABLE 13.27

SEPARATION OF CLOSELY RELATED DIMERIC *VINCA* ALKALOIDS

| Alkaloid | Suitable system for separation* |
|---|---|
| Leurosine<br>VLB<br>Isoleurosine | 2 or 8 |
| Leurosidine<br>Leurocristine<br>VLB or Leurosine | Develop first in S10, air dry for 10 min, then develop in S3. |
| Leurosidine Sulphate<br>Leurocristine Sulphate<br>VLB Sulphate or leurosine Sulphate | 11 |

*Sorbent systems as in Table 13.25.

(From: The American Pharmaceutical Association).

TABLE 13.28

TLC IDENTIFICATION OF *CATHARANTHUS* ALKALOIDS[33]

TLC systems:

| | | |
|---|---|---|
| S1 | Silica gel G, activated | Ethyl acetate-absolute ethanol(3:1) |
| S2 | Silica gel G, activated | *n*-Butanol-acetic acid-water(4:1:1) |
| S3 | Silica gel G, activated | Methanol |

| Alkaloid | Chromogenic class* | $hR_F$ values** | | |
|---|---|---|---|---|
| | | S1 | S2 | S3 |
| Vindoline | I-a | 55±3 | 32±2 | 61±3 |
| Vincoline | II-a | 64±1 | 38±2 | 61±1 |
| Vindolinine | | 39±5 | 40±2 | 46±4 |
| Akuammine | | 8±1 | 39±3 | 18±2 |
| Vincolidine | | 2±1 | 40±2 | 28±1 |
| Vindorosine | II-b | 57±3 | 33±2 | 60±5 |
| Vincaleukoblastine | II-c | 21±1 | 19±4 | 46±4 |
| Leurosidine | | 6±1 | 20±5 | 17±2 |
| Neoleurosidine | | 2±1 | 13±3 | 46±3 |
| Desacetylvincaleukoblastine | II-d | 8±1 | 15±3 | 16±2 |
| Vinaphamine | | 5±1 | 8±1 | 21±1 |
| Leurosivine | | 0±0 | 21±4 | 34±3 |
| Rovidine | II-e | 11±1 | 22±3 | 43±4 |
| Vincathicine | | 4±1 | 7±1 | 18±3 |
| Lochnerine | II-f | 23±3 | 46±3 | 54±3 |
| Ammorosine | | 22±2 | 52±3 | 56±3 |
| Isoleurosine | II-g | 46±4 | 24±3 | 66±3 |
| Leurosine | | 32±4 | 23±4 | 48±4 |
| Tetrahydroalstonine | III-a | 77±3 | 52±3 | 73±3 |
| Cavincine | | 72±4 | 52±3 | 71±4 |
| Ajmalicine | | 70±4 | 51±3 | 70±4 |
| Reserpine | | 69±3 | 62±2 | 74±4 |
| Vinosidine | | 66±3 | 61±2 | 75±4 |
| Cathindine | | 61±3 | 52±3 | 62±4 |
| Dihydrositsirikine | | 60±3 | 49±2 | 68±5 |
| Yohimbine | | 46±4 | 40±3 | 65±3 |
| Sitsirikine | | 27±4 | 46±1 | 54±2 |
| Cavincidine | | 8±1 | 47±2 | 34±3 |
| Carosidine | IV-a | 0±0 | 0±0 | 0±0 |
| Virosine | V-a | 48±4 | 42±2 | 66±4 |
| Vincarodine | | 44±3 | 41±2 | 65±3 |
| Neoleurocristine | V-b | 27±3 | 22±2 | 53±3 |
| Leurocristine | | 18±4 | 16±4 | 39±2 |
| Catharanthine | V-c | 58±6 | 51±3 | 57±4 |
| Perimivine | | 55±2 | 64±4 | 65±3 |
| Cathalanceine | V-d | 75±2 | 43±3 | 71±4 |
| Lanceine | | 69±3 | 35±2 | 70±3 |
| Lochnerivine | | 62±2 | 73±2 | 74±4 |
| Lochnerinine | V-e | 74±3 | 48±3 | 65±4 |
| Lochrovicine | | 70±3 | 50±2 | 58±2 |
| Lochnericine | | 31±3 | 34±3 | 60±5 |
| Lochrovine | | 11±2 | 65±2 | 24±1 |
| Akuammicine | | 10±2 | 48±3 | 18±2 |
| Lochneridine | | 1±1 | 53±3 | 28±3 |
| Lochrovidine | V-f | 66±2 | 35±2 | 58±1 |
| Pericyclivine | V-g | 57±2 | 54±4 | 55±3 |
| Vinaspine | | 29±3 | 56±4 | 52±2 |
| Ammocalline | VI-a | 15±1 | 45±2 | 24±3 |

**References p. 324**

TABLE 13.28 (continued)

| Alkaloid | Chromogenic class* | $hR_F$ values** | | |
|---|---|---|---|---|
| | | S1 | S2 | S3 |
| Carosine | VII-a | 62±2 | 41±3 | 66±3 |
| Catharicine | | 53±3 | 42±3 | 64±4 |
| Pericalline | VII-b | 16±2 | 53±3 | 22±4 |
| Catharine | VII-c | 46±3 | 34±4 | 70±6 |
| Vincamicine | | 3±2 | 9±4 | 55±4 |
| Vindolicine | VII-d | 35±3 | 5±2 | 59±5 |
| Vindolidine | | 12±1 | 0±0 | 60±2 |
| Mitraphylline | VIII-a | 47±4 | 36±2 | 61±4 |
| Maandrosine | | 36±2 | 39±1 | 70±5 |
| Perosine | | 24±3 | 45±4 | 47±3 |
| Perivine | | 23±2 | 46±3 | 48±3 |
| Perividine | | 22±1 | 51±2 | 62±4 |
| Alstonine | | 3±1 | 41±3 | 9±4 |
| Serpentine | | 0±0 | 38±4 | 5±2 |
| Pleurosine | | 0±0 | 23±4 | 0±0 |

*See Tables 13.30 and 13.31.
**Solvent systems as in Table 13.25.

(From: Journal of Natural Products (Lloydia).)

TABLE 13.29

TLC IDENTIFICATION OF SOME *VINCA* ALKALOIDS[82]

TLC systems:

| | | |
|---|---|---|
| S1 | Silica gel KSK G, activated | Ethyl acetate-methanol(9:1) |
| S2 | Silica gel KSK G, activated | Chloroform-methanol(9:1) |
| S3 | Silica gel KSK G, activated | Benzene-methanol(9:1) |

| Alkaloid | $hR_F$ values | | | Colours with 1% cerium(IV) ammonium sulphate in phosphoric acid (no. 12) |
|---|---|---|---|---|
| | S1 | S2 | S3 | |
| Akuammidine | 47 | 50 | 50 | Grey |
| Akuammine | 9 | 27 | 28 | Yellowish red |
| Akuammicine | 11 | 34 | 46 | Blue, fades |
| Vincanine | 10 | 45 | 45 | Green |
| Vincanidine | 9 | 30 | 35 | Brown |
| Vinervine | 10 | 25 | 30 | Violet, fading instantaneously |
| Vinervinine | 12 | 43 | 46 | Blue, fades after 15-20 min |
| Vincarine | 34 | 50 | 49 | Brick red, fades after 30 min |
| Vincamine | 43 | 69 | 63 | Yellow |
| Vincaridine | 34 | 33 | 41 | Red-yellow |
| Isoreserpiline | 75 | 76 | 66 | Violet |
| Kopsinine | 2 | 42 | 44 | Red |
| Pseudokopsinine | 45 | 65 | 55 | Red |
| Ervamine | 60 | 74 | 75 | Blue, changes to yellow after 1 h |
| Ervine | 70 | 79 | 65 | Dark yellow |
| Ervinidine | 74 | 71 | 67 | Violet, fades instantaneously |

(From: Plenum Publishing Co.).

TABLE 13.30

CRITERIA FOR CHROMOGENIC CLASSIFICATION OF *CATHARANTHUS* ALKALOIDS AND CHROMOGENIC COMPARISON[33]

| Chromogenic class | Colours following use of cerium(IV) ammonium sulphate reagent (no. 12) | | |
|---|---|---|---|
| | Immediate | Rapid changes | After 5 min |
| I-a | Crimson | Formation of yellow centre | Crimson with yellow centre |
| II-a | Orange-red | No change | Orange red |
| II-b | Orange-red | Formation of yellow centre | Yellow with orange-red edge |
| II-c | Orange | Quickly changes to lavender | Lavender |
| II-d | Orange | Slowly changes to lavender | Lavender |
| II-e | Orange | Slowly changes to yellow | Yellow |
| II-f | Orange | Quickly changes to orange with a grey edge | Grey |
| II-g | Orange | Grey centre quickly forms | Grey with light orange edge |
| III-a | Yellow-green | Changes to solid yellow almost immediately | Yellow |
| IV-a | Green | No change | Green |
| V-a | Blue | Transient. Colourless in a few seconds | Colourless |
| V-b | Blue | Blue fades slightly. Pale yellow centre may or may not form | Blue or blue with pale yellow centre |
| V-c | Blue with yellow centre | Yellow centre rapidly increases in size | Completely yellow or yellow with a slight blue edge |
| V-d | Blue with yellow centre | Yellow centre immediately changes to light orange. Centre rapidly enlarges | Light orange with small light blue edge |
| V-e | Blue with yellow centre | Yellow centre immediately changes to light orange. Centre does not increase in size | Blue with light orange centre |
| V-f | Blue-violet with yellow centre | Yellow centre immediately changes to light orange. Centre rapidly enlarges | Light orange with small blue-violet edge |
| V-g | Blue-green with yellow centre | Blue-green rapidly changes to grey | Grey |
| VI-a | Grey | No change | Grey |
| VII-a | Purple | Darkens slightly | Purple |
| VII-b | Purple | Lightens | Blue-grey |
| VII-c | Purple | Transient. Almost immediately changes to orange | Orange |
| VII-d | Purple with pink edge | No change | Purple with pink edge |
| VIII-a | No chromogenic response* | No change | No chromogenic response |

*Serpentine is a yellow alkaloid placed in category VIII-a which can be observed prior to use of the cerium(IV) ammonium sulphate reagent. No colour changes are apparent following treatment of serpentine with the reagent. At concentrations higher than 20 μg, most of the alkaloids in this class will give a faint yellow colour with the reagent.

TABLE 13.30 (continued)

| Alkaloid | Chromogenic class | Colours following use of the cerium(IV) ammonium sulphate reagent | | | | | |
|---|---|---|---|---|---|---|---|
| | | Immediate | Rapid changes | 5 min | 15 min | 60 min | 24 h |
| Vindoline | I-a | I-1 | I-2 | I-2 | I-2 | I-2 | I-3 |
| Vincoline | II-a | I-4 | - | I-4 | I-4 | I-4 | I-5 |
| Vindolinine | | I-4 | - | I-4 | I-4 | I-4 | I-5 |
| Akuammine | | I-4 | - | I-4 | I-4 | I-4 | I-5 |
| Vincolidine | | I-4 | - | I-4 | I-4 | I-4 | I-5* |
| Vindorosine | II-b | I-6 | - | I-6 | I-6 | I-7 | I-3 |
| Vincaleukoblastine | II-c | I-8 | I-9 | I-9 | I-9 | I-9 | var.** |
| Leurosidine | | I-8 | I-9 | I-9 | I-9 | I-9 | var. |
| Neoleurosidine | | I-8 | I-9 | I-9 | I-9 | I-9 | var. |
| Desacetyl VLB | II-d | I-8 | - | I-9 | I-9 | I-9 | var. |
| Vinaphamine | | I-8 | - | I-9 | I-9 | I-9 | var. |
| Leurosivine | | I-8 | | I-9 | I-9 | I-9 | var. |
| Rovidine | II-e | I-8 | I-10 | I-10 | I-10 | I-10 | var. |
| Vincathicine | | I-8 | I-10 | I-10 | I-10 | I-10 | var. |
| Lochnerine | II-f | I-8 | I-11 | I-11 | I-11 | I-11 | var. |
| Ammorosine | | I-8 | I-11 | I-11 | I-11 | I-11 | var. |
| Isoleurosine | II-g | I-8 | I-12 | I-13 | I-13 | I-13 | var. |
| Leurosine | | I-8 | I-12 | I-13 | I-13 | I-13 | var. |
| Tetrahydroalstonine | III-a | I-14 | I-15 | I-15 | I-15 | I-15 | I-15 |
| Cavincine | | I-14 | I-15 | I-15 | I-15 | I-15 | I-15 |
| Ajmalicine | | I-14 | I-15 | I-15 | I-15 | I-15 | I-15 |
| Reserpine | | I-14 | I-15 | I-15 | I-15 | I-15 | I-15 |
| Vinosidine | | I-14 | I-15 | I-15 | I-15 | I-15 | I-15 |
| Cathindine | | I-14 | I-15 | I-15 | I-15 | I-15 | I-15 |
| Dihydrositsirikine | | I-14 | I-15 | I-15 | I-15 | I-15 | I-15 |
| Yohimbine | | I-14 | I-15 | I-15 | I-15 | I-15 | I-15 |
| Sitsirikine | | I-14 | I-15 | I-15 | I-15 | I-15 | I-15 |
| Cavincidine | | I-14 | I-15 | I-15 | I-15 | I-15 | I-15 |
| Carosidine | IV-a | I-16 | - | I-16 | I-16 | I-16 | I-16 |
| Virosine | V-a | I-17 | NCR*** | NCR | NCR | NCR | NCR |
| Vincarodine | | I-17 | NCR | NCR | NCR | NCR | NCR |
| Neoleurocristine | V-b | I-18 | - | I-19 | I-19 | I-19 | var. |
| Leurocristine | | I-18 | - | I-18+ | I-18 | I-18 | var. |
| Catharanthine | V-c | I-20 | I-15 | I-15 | I-15 | I-15 | var. |
| Perimivine | | I-20 | - | I-15 | I-15 | I-15 | var. |
| Cathalanceine | V-d | I-20 | II-1 | II-2 | I-5++ | I-5 | I-5 |
| Lanceine | | I-20 | II-1 | II-2 | I-5++ | I-5 | I-5 |
| Lochnerivine | | I-20 | II-1 | II-2 | I-5++ | I-5 | I-5 |
| Lochnerinine | V-e | I-20 | II-1 | II-1 | II-1 | II-3 | II-4 |
| Lochrovicine | | I-20 | II-1 | II-1 | II-1 | II-2 | II-5 |
| Lochnericine | | I-20 | II-1 | II-1 | II-1 | II-3 | II-4 |
| Lochrovine | | I-20 | II-1 | II-1 | II-1 | II-2 | II-5 |
| Akuammicine | | I-20 | II-1 | II-1 | II-1 | II-2 | II-6 |
| Lochneridine | | I-20 | II-1 | II-1 | II-1 | II-2 | II-5 |
| Lochrovidine | V-f | II-7 | - | II-7 | II-7 | I-5 | I-5 |
| Piricyclivine | V-g | II-8 | - | II-10 | II-10 | II-10 | var. |
| Vinaspine | | II-8 | - | II-9 | II-10 | II-10 | var. |
| Ammocalline | VI-a | II-10 | - | II-10 | II-10 | II-10 | var. |
| Carosine | VII-a | II-11 | - | II-12 | II-12 | II-12 | II-12 |
| Catharicine | | II-13 | - | II-14 | II-14 | II-14 | II-14 |
| Pericalline | VII-b | II-15 | - | II-16 | II-17 | II-17 | II-18 |
| Catharine | VII-c | II-13 | I-8 | II-19 | II-19 | II-19 | I-3 |
| Vincamicine | | II-13 | I-8 | II-19 | II-19 | II-19 | I-14 |
| Vindolicine | VII-d | II-20 | - | II-20 | II-20 | II-20 | II-20 |
| Vindolidine | | II-20 | - | II-20 | II-20 | II-20 | II-20 |

TABLE 13.30 (continued)

| Alkaloid | Chromogenic class | Colours following use of the cerium(IV) ammonium sulphate reagent | | | | | |
|---|---|---|---|---|---|---|---|
| | | Immediate | Rapid changes | 5 min | 15 min | 60 min | 24 h |
| Mitraphylline | VIII-a | NCR | - | NCR | NCR | NCR | NCR |
| Maandrosine | | NCR | - | NCR | NCR | NCR | NCR |
| Perosine | | NCR | - | NCR | NCR | NCR | NCR |
| Perivine | | NCR | - | NCR | NCR | NCR | NCR |
| Perividine | | NCR | - | NCR | NCR | NCR | NCR |
| Alstonine | | NCR | - | NCR | NCR | NCR | NCR |
| Serpentine | | NCR+++ | - | NCR | NCR | NCR | NCR |
| Pleurosine | | NCR+++ | - | NCR | NCR | NCR | NCR |

*A blue or grey centre frequently forms after 24 h.
**Colour at 24 h varies from tan to grey, lavender or pale yellow.
***NCR = no chromogenic response to the reagent.
+Infrequently forms a pale yellow centre (I-19)
++Frequently with a slight pale blue edge.
+++Serpentine does not react with the reagent but is evident because of its natural yellow colour.

(From: Journal of Natural Products (Lloydia)).

TABLE 13.31

COLOUR REACTIONS OF SOME *VINCA* ALKALOIDS WITH A 1% SOLUTION OF CERIUM(IV) AMMONIUM SULPHATE IN 85% PHOSPHORIC ACID (NO. 12)[156]

Classification of colours:

| Colour on the scale | Colour | Colour on the scale | Colour |
|---|---|---|---|
| a-2 | Blackish | i-5 | Blueish green |
| a-3 | Greyish violet | i-1 | Isabella, pale yellow |
| a-6 | Pale greyish | j-3 | Pale sandy |
| b-1 | Dark blue | j-5 | Walnut |
| b-6 | Cream | k-1 | Smoky |
| c-4 | Dark ash | k-2 | Straw yellow |
| d-4 | Orange-pink | k-3 | Lemon yellow |
| d-6 | Ochre-yellow | l-5 | Carmine red |
| d-7 | Yellowish rust | l-7 | Golden yellow |
| e-2 | Yellow-orange | m-3 | Pale blue |
| e-4 | Leather brown | m-4 | Pinkish violet |
| e-5 | Pale lemon yellow | m-6 | Violet carmine |
| g-1 | Rusty | n-1 | Plum |
| g-3 | Pinkish lilac | n-5 | Flesh pink |
| g-5 | Lead grey | n-6 | Dark cream |
| h-2 | Dark violet | o-3 | Pale honey |
| h-3 | Ochre | o-4 | Orange red |
| i-1 | Grey-yellow | o-5 | Yellow |
| i-2 | Yellowish green | o-7 | Minium red |

| Alkaloid | Colour with a 1% solution of cerium(IV) ammonium sulphate in 85% phosphoric acid (no. 12) | | | | |
|---|---|---|---|---|---|
| | After spraying | After 5 min | After 15 min | After 1 h | After 24 h |
| α-Methyleneindoles: | | | | | |
| Vincanine (norfluorocurarine) | g-5 | Edge g-5<br>Centre i-1 | i-1 | i-1 | i-1 |
| Vincanidine | h-3 | h-3 | h-3 | o-5 | o-5 |
| Akuammicine | b-1 | b-1 interspersed with j-5 | Edge b-1<br>Centre j-5 | Edge b-1<br>Centre j-5 | h-3 |
| Vinervinine | b-1 | b-1 | Dark m-3 | Edge m-3<br>Centre j-1 | Light j-1 |
| Vinervine sulphate | a-3 | a-3 | a-3 | n-6 | - |
| Ervamine (*l*-vincadifformine | b-1 | b-1 | b-1 interspersed with j-5 | Edge g-5<br>Centre j-5 | e-5 |
| Ervinceine (*l*-methoxyvincadifformine | b-1 | Light b-1 | m-3 | k-1 | |
| Tabersonine | b-1 | h-2, edge n-1 | h-2, interspersed with e-2 | j-3 | Light j-3 |
| 16-Methoxytabersonine | m-3 | m-3 | m-3, interspersed with k-2 | k-2 | Light k-2 |
| Ervinidinine | b-1 | b-1 | n-1, interspersed with c-5 | k-2 | Light k-2 |
| Ervincinine | b-1 | m-3, interspersed with k-3 | Edge m-3<br>Centre k-3 | i-5 | i-5 |
| Ervinidine | b-1 | b-1, interspersed with e-2 | Edge m-3<br>Centre e-2 | Edge m-3<br>Centre e-2 | Pale e-2 |

TABLE 13.31 (continued)

| Alkaloid | Colour with a 1% solution of cerium(IV) ammonium sulphate in 85% phosphoric acid (no. 12) | | | | |
|---|---|---|---|---|---|
| | After spraying | After 5 min | After 15 min | After 1 h | After 24 h |
| Indoles: | | | | | |
| Kopsinine | d-4 | d-4 | d-4 | d-4 | d-4 |
| Kopsinilam | d-4 | d-4 | Dark d-4 | Dark d-4 | Dark d-4 |
| Kopsanone | Pale a-3 | Pale a-3 | Pale a-3 | Pale a-3 | Pale a-3 |
| Pseudokopsinine | l-5 | Dark d-4 | l-5 | l-5 | l-5 |
| Herbamine | l-5 | l-5 | l-5 | l-5 | b-6 |
| Majdinine | l-5 | l-5 | l-5 | n-5 | n-5 |
| Ajmaline | o-7 | l-5 | l-5 | n-5 | - |
| Vincamajine | l-5 | l-5 | l-5 | Edge n-5<br>Centre k-3 | k-3 |
| Akuammine | d-6 | d-7 | d-7 | d-7 | d-7 |
| Vincarine | o-4 | o-4 | o-4 | Pale o-4 | - |
| Herbadine | o-4 | n-5 | n-5 | n-5 | d-6 |
| Picrinine | g-1 | g-1 | Edge g-1<br>Centre e-2 | Edge e-4<br>Centre k-3 | m-4 |
| Vincaricine | l-7 | Edge d-6<br>Centre g-6 | Edge d-6<br>Centre g-6 | d-6 | d-6 |
| Vincarinine | l-7 | n-7 | d-6 | Pale d-6 | g-3 |
| 11-Hydroxypleio-carpamine | - | - | - | Pale b-6 | Pale b-6 |
| Akuammidine | a-3 | a-2 | c-4 | a-6 | a-6 |
| Tombosine | a-3 | c-4 | c-4 | a-6 | a-6 |
| Quebrachamine | m-4 | m-4, inter-spersed with i-1 | Edge m-4<br>Centre i-1 | Edge g-3<br>Centre i-2 | Pale n-1 |
| Vincamine | k-3 | k-3 | k-3 | k-3 | k-3 |
| Vincine | - | Pale k-3 | Pale k-3 | k-3 | Pale k-3 |
| Apovincamine | i-1 | i-1 | i-1 | i-1 | i-1 |
| Ervine | i-1 | i-1 | i-1 | i-1 | i-1 |
| Hydroxyindoles: | | | | | |
| Vinerine | - | g-5 | Edge g-5<br>Centre b-1 | n-1 | - |
| Vineridine | - | g-5 | Edge g-5<br>Centre b-1 | n-1 | - |
| Majdine | - | m-6 | n-1 | h-2 | - |
| Isomajdine | - | m-6 | n-1 | h-2 | - |

(From: Plenum Publishing Co.).

TABLE 13.32

TLC ANALYSIS OF *VINCA* AND *CATHARANTHUS* ALKALOIDS

| Alkaloid | Aim | Adsorbent | Solvent system | Ref. |
|---|---|---|---|---|
| Not specified | Separation of plant extracts | $SiO_2$ | $CHCl_3$-MeOH(95:5)<br>Two dimensional: I. $CHCl_3$-MeOH(95:5)<br>II. $CHCl_3$-MeOH(9:1) | 8 |
| VLB,Lcrist,Lsidine | Isolation from *Vinca* species | $Al_2O_3$ | EtOAc,followed by EtOAc-abs. EtOH(3:1) in the same direction | 9 |
| Vincadine,minovine,vincorine | Identification in *Vinca* species | $Al_2O_3$ | Benzene-$Me_2CO$(98.5:5) | 13 |
| Not specified | Phytochemical investigations of *Vinca* species | $SiO_2$ | $CHCl_3$-MeOH(95:5) | 15 |
| Vmin,isovmin | Separation | $SiO_2$ | Toluene-MeOH-$CHCl_3$(110:3:2) | 24 |
| 26 alkaloids (Table 13.25) | Analysis of *Vinca* alkaloids | | see Table 13.25, solvents S1-S11 | 26,28,29 |
| VLB,Lcrist,Lsin,Lsidine, vindoline,catharanthine, virosine,perivine,RSP,vinca-rodine,neoLsidine,ajmalicine, isoLsin,vincathicine | Assay methods for *Vinca* alkaloids | 0.5 *M* LiOH-impregnated $Al_2O_3$ | Acetonitrile-EtOH(95:5)<br>Acetonitrile-benzene(3:7) | 31 |
| 63 alkaloids (Table 13.28 and 13.30) | Identification | $SiO_2$ | EtOAc-abs. EtOH(3:1)<br>*n*-BuOH-AcOH-$H_2O$(4:1:1)<br>MeOH | 33 |
| Lsin | Constituents of *Vinca* | $SiO_2$ | EtOAc-abs. EtOH(3:1) | 34 |
| Vmin | Indirect quantitative analysis | | No details available | 40 |
| Vindoline,vindolinine,catha-ranthine,THalst,vindorosine, desacetylvindoline | Separation of monomers and dimers | 0.5% KOH-impregnated $SiO_2$ | EtOAc<br>$CHCl_3$-$Me_2CO$(9:1) | 44 |
| VLB,Lcrist,Lsin,Lsidine | Separation of dimeric alkaloids | $SiO_2$ | $CHCl_3$-MeOH(95:5)<br>Two dimensional: I. $CHCl_3$-MeOH(95:5)<br>II. MeOH | 46 |
| Serp,alst,Thalst,ajmalicine | Indirect quantitative analysis(UV) | $SiO_2$<br>$Al_2O_3$ | EtOAc-EtOH-DMFA-DEA(30:60:5:5)<br>$CHCl_3$-benzene(8:2) | 59 |

References p. 324

| | | | | |
|---|---|---|---|---|
| 16 *Vinca* alkaloids (Table 13.29) | Identification | $SiO_2$ | EtOAc-MeOH(9:1)<br>$CHCl_3$-MeOH(9:1)<br>Benzene-MeOH(9:1) | 82 |
| VLB | Assay of *Catharanthus roseus* for VLB | $SiO_2$ | I. $CHCl_3$-MeOH(95:5)<br>II. EtOAc-abs. EtOH(3:1)<br>Two-dimensional: I,II (both solvents developed twice) | 101 |
| Vmin | Indirect quantitative analysis(UV) in *Vinca* | $SiO_2$ | $CHCl_3$-MeOH(95:5), followed by $Me_2CO$-*n*-BuOH in the same direction | 133 |
| Akuammine | Indirect quantitative analysis (spectrophotometric) | $SiO_2$ | $CHCl_3$-EtOH(10:1)<br>EtOAc-EtOH(3:1) | 145 |
| 38 alkaloids (Table 13.31) | Qualitative characterization by means of TLC and spray reagents | $SiO_2$ | Benzene-EtOAc-MeOH(2:3:1)<br>Benzene-EtOAc(2:3)<br>Benzene-MeOH(3:2) | 156 |
| 19 alkaloids | Direct quantitative analysis (spot area) in *Vinca* species | $SiO_2$ | Benzene-EtOAc-MeOH(2:3:1)<br>Benzene-EtOAc(2:3)<br>Benzene-MeOH(3:2) | 176 |
| Minorine | Indirect quantitative analysis in *Vinca minor* | $SiO_2$ | EtOAc-$Me_2CO$(4:1) | 199 |
| Ajmalicine | Indirect quantitative analysis(UV) in *Catharanthus* | $SiO_2$ | $CHCl_3$-EtOAc(1:1) | 205 |
| Vmin | Indirect quantitative analysis(UV) in *Vinca minor* | $SiO_2$ | Benzene-EtOAc-MeOH(2:2:1,2:2:0.5) | 206 |
| Minorine | Indirect quantitative analysis | $SiO_2$ | EtOAc-$Me_2CO$(4:1) | 210 |
| 20 alkaloids | Identification in *Vinca* species | | No details available | 211 |

TABLE 13.33

TLC ANALYSIS OF PHYSOSTIGMINE

TLC systems:

| | | |
|---|---|---|
| S1 | Silica gel G, activated | Cyclohexane-chloroform-diethylamine(5:4:1)[54] |
| S2 | Silica gel G, activated | Chloroform-acetone-diethylamine(5:4:1)[54] |
| S3 | Silica gel G, activated | Chloroform-acetone-dimethylamine(33% in ethanol)(5:4:1)[54] |
| S4 | Aluminium oxide G (type E), activated | Chloroform-acetone(5:4)[135,154] |

| Compound | $hR_F$ values | | | |
|---|---|---|---|---|
| | S1 | S2 | S3 | S4 |
| Physostigmine | 35 | 61 | 74 | 61-67 |
| Rubreserine | 21 | 54 | 55 | 33-36 |
| Eserinol | 18 | 50 | 51 | 42-45 |

References p. 324

TABLE 13.34

TLC ANALYSIS OF PHYSOSTIGMINE

| Alkaloid | Other compounds | Aim | Adsorbent | Solvent system | Ref. |
|---|---|---|---|---|---|
| Ph N-oxide | N-Oxides of strychnine, atropine and hyoscyamine | Separation | $SiO_2$ | $CHCl_3$-MeOH-DEA(90:10:5) | 22 |
| Ph,rubreserine, eserinol | | Indirect quantitative analysis (Table 13.33) | $SiO_2$ | *n*-BuOH-$H_2O$-AcOH(4:1:1)<br>$CHCl_3$-MeOH-AcOH(75:20:5)<br>Cyclohexane-$CHCl_3$-DEA(5:4:1)<br>$CHCl_3$-$Me_2CO$-DEA(5:4:1)<br>$CHCl_3$-$Me_2CO$-DMA(33% in EtOH)(5:4:1) | 54 |
| Ph | RSP,pilocarpine,homatropine,solanine,solanidine,lobeline,ephedrine | Detection method for cholinesterase inhibitors | Cellulose impregnated with 5% silicone 555 | $H_2O$-EtOH-$CHCl_3$(56:42:2) | 66 |
| Ph | | Reaction chromatography | $SiO_2$ | Benzene-MeOH-$Me_2CO$-AcOH(70:20:5:5) | 86 |
| Ph | S,B | Identification | Alkaline $SiO_2$ | MeOH | 90 |
| Ph | | Reaction chromatography | $SiO_2$ | $CHCl_3$-$Me_2CO$-DEA(5:4:1) | 122 |
| Ph | Pilocarpine,homatropine, atropine | Direct semi-quantitative analysis (spot areas) | $SiO_2$ | $CHCl_3$-$Me_2CO$-DEA(5:4:1)<br>$CHCl_3$-DEA(9:1) | 128 |
| Ph,decomposition products | | Densitometric analysis (Table 13.33) | $Al_2O_3$ | $CHCl_3$-$Me_2CO$(5:4) | 135 |
| Ph | Atropine,ephedrine, tubocurarine | Separation of quaternary N-drugs | $SiO_2$ | 1 *M* HCl<br>1 *M* HCl-EtOH(1:1) | 151 |
| Ph,decomposition products | | Indirect quantitative analysis(UV) (Table 13.33) | $Al_2O_3$ | $CHCl_3$-$Me_2CO$(5:4) | 154 |
| Ph | | TAS technique, in calabar beans | $SiO_2$ | $Me_2CO$-$Et_2O$-conc. $NH_4OH$(50:50:3) | 171 |
| Ph | Atropine | Indirect quantitative analysis | $SiO_2$ | $CHCl_3$-$Me_2CO$-DEA(5:4:1) | 190 |

TABLE 13.35

TLC ANALYSIS VARIOUS INDOLE ALKALOIDS

| Alkaloid | Aim | Adsorbent | Solvent system | Ref. |
|---|---|---|---|---|
| Condylocarpine and akuammicine type of alkaloids, pleiocarpine | Chemical correlations between these alkaloids | $SiO_2$ | $CHCl_3$-MeOH(5:1)<br>Benzene-EtOAc-DEA(7:2:1)<br>$CHCl_3$-DEA(20:1)<br>$Me_2CO$-DEA(5:1) | 23 |
| Pleiocarpamine, normavacurine and related alkaloids | Structure elucidation | $SiO_2$ | $CHCl_3$-MeOH(5:1)<br>Benzene-EtOAc-DEA(7:2:1) | 30 |
| Harmaline,harman,harmine | Separation | $SiO_2$ | $CHCl_3$-MeOH-AcOH(60:35:5)<br>Two-dimensional: I. MeOAc-isoprOH-25% $NH_4OH$ (45:35:20)<br>II. BuOH-$H_2O$-AcOH(60:25:15) | 53 |
| Harman,harmine,harmalol, tetrahydroharman | Identification in plant material (Table 13.36) | $SiO_2$ | *n*-BuOH,MeEtCO<br>$CHCl_3$-MeOH(4:1) | 73 |
| β-Carbolines | Separation of 73 β-carboline derivatives | $SiO_2$ | *n*-BuOH-$H_2O$-AcOH(4:1:1)<br>*n*-prOH-$NH_4OH$(8:2)<br>$CHCl_3$-MeOH(9:1) | 77 |
| Tabersonine,β-yohimbine | Identification in *Amsonia* | $SiO_2$ | EtOAc-abs. EtOH(5:1)<br>$CHCl_3$-MeOH(95:5)<br>Two-dimensional: I. $CHCl_3$-MeOH(95:5)<br>II. MeOH | 88 |
| Harman alkaloids | Fluorodensitometric analysis | $SiO_2$ | Benzene-$CHCl_3$-abs. EtOH(4:1:1) | 98,106 |
| Harman alkaloids | Indirect quantitative analysis(UV) | $SiO_2$ | $CHCl_3$-MeOH(75:25) | 125 |
| Harman,harmine | Fluorodensitometric analysis in *Passiflora* | $SiO_2$ | $CHCl_3$-MeOH-$H_2O$(80:19:1) | 127 |
| Ibogaine,various drugs | Detection with TCBI spray reagent | $SiO_2$ | EtOAc-MeOH-$NH_4OH$(100:18:1.5)<br>MeOH-$NH_4OH$(100:1.5) | 183 |
| Gramine,dimethyltryptamine and related indoles | Fluorodensitometric analysis | $SiO_2$+Avicel (1:1) | PrOH-EtOAc-conc. $NH_4OH$-2-ethoxyethanol (60:15:3:5) | 218 |

TABLE 13.36

TLC ANALYSIS OF HARMAN ALKALOIDS[73]

TLC systems:

| | | |
|---|---|---|
| S1 | Silica gel G, activated | Methyl ethyl ketone |
| S2 | Silica gel G, activated | *n*-Butanol |
| S3 | Silica gel G, activated | Chloroform-methanol(4:1) |

| Alkaloid | $hR_F$ values | | | Fluorescence colour[125] |
|---|---|---|---|---|
| | S1 | S2 | S3 | |
| Harman | 74 | 57 | 63 | Blue |
| Harmine | 57 | 38 | 48 | Violet |
| Tetrahydroharman | 6 | 17 | 28 | |
| Harmalol | 1 | 4 | 18 | Green-blue |
| Harmol | | | | Green |
| Harmaline | | | | Blue-violet |

Chapter 14

# ERGOT ALKALOIDS

14.1. Solvent systems....375
14.1.1. Ergot alkaloids....375
14.1.2. LSD and derivatives....378
14.2. Detection....379
14.3. Quantitative analysis....380
14.4. TAS technique and reaction chromatography....382
References....383

The pharmaceutically important group of ergot alkaloids is derived from *d*-lysergic acid. Both the naturally occurring alkaloids, the semi-synthetic derivatives and the dihydro compounds are used in therapy. The semi-synthetic alkaloid *d*-lysergic acid diethylamide (LSD) has strong hallucinogenic effects and its abuse is widespread. The analysis of LSD has been subject of a number of publications, either in connection with the analysis of drugs of abuse in general or of hallucinogenic drugs (Tables 14.10 and 14.11).

The naturally occurring ergot alkaloids can be classified as clavine alkaloids and lysergic acid derivatives. The lysergic acid alkaloids can be divided into water-soluble alkaloids (including ergometrine) and water-insoluble or peptide alkaloids [the ergotamine group (ergotamine, ergosine) and the ergotoxine group (ergocristine, ergocryptine, ergocornine)]. The lysergic acid derivatives are readily epimerized into the stereoisomeric isocompounds having a different stereochemistry at C-8. The alkaloids derived from isolysergic acid, all having names ending with -inine, are not pharmacologically as active as the lysergic acid derivatives. In quality control of ergot alkaloids the separation of the stereoisomers is therefore of interest.

Lysergic acid derivatives are found not only in fungi but also in higher plants, known under the name "Morning Glory". They have been used for their hallucinogenic effects. Several publications deal with the analysis of such plants (Table 14.8).

## 14.1. SOLVENT SYSTEMS

### *14.1.1. Ergot alkaloids*

The first to describe the TLC analysis of ergot alkaloids was Rochelmeyer[1]. The solvent system was chloroform - ethanol (9:1 or 95:5) and the sorbent was

**References p. 383**

silica gel. The same solvents have been used in a series of investigations[3,4,7,29,38,41,45,49,57,63,108,136]; in others, similar solvents containing methanol instead of ethanol were used[11,18,23,39,41,50,64,69,72,80,81,86,88,91,101,102,122,129,132]. The solvent systems mentioned have been used in combination with alkalinized silica gel[19,28,31,56,91] or aluminium oxide plates[10,12,23,27,35,41,49,63,65,91,122]. Rochelmeyer and co-workers[2,7] separated the "classic" ergot alkaloids on formamide-impregnated cellulose plates (plates immersed in a 20% solution of formamide in acetone) with benzene - heptane - chloroform (6:5:3) followed by benzene - heptane (6:5). This combination also separates the alkaloids of the ergotoxine group. For the separation of clavine alkaloids ethyl acetate - ethanol - dimethylformamide (85:10:5) was used with silica gel plates[4,7]. The same TLC system, sometimes modified slightly, has been used in a series of investigations[6,9,10,20,91].

The separation of the clavine alkaloids has been described by various workers[4,6,7,9,10,11,27,48,57,58,80]. In Table 14.1 the results obtained with some of the solvent systems are summarized.

The separation of the ergometrine, ergotamine and ergotoxine groups of alkaloids is readily achieved, as can be seen from Tables 14.1 and 14.2. The alkaloids belonging to the ergotoxine group are difficult to separate. Zinser and Baumgärtel[15] used heptane - carbon tetrachloride - pyridine (1:3:2) (Table 14.1) on silica gel and developed the plates twice to separate ergocristine from ergocornine and ergocryptine. Hohmann and Rochelmeyer[16] separated the ergotoxine alkaloids and their dihydro derivatives with solvent system S5 (Table 14.1). McLaughlin et al.[20] tested 111 different solvent systems, but were not able to separate the three ergotoxine alkaloids in a single run. Ergocristine could be separated from the other two alkaloids with ethyl acetate - ethanol - dimethylformamide (13:0.1:1.9) on silica gel plates (Table 14.1). After elution from the plate ergocornine and ergocryptine could be separated on aluminium oxide with chloroform - diethyl ether - water (3:1:1). In many instances alkaloids had different $hR_F$ values when chromatographed in a mixture of alkaloids than when chromatographed alone. Agurell[23] modified solvent system S5 described by Hohmann and Rochelmeyer[16], and achieved a separation of ergotoxine alkaloids (S13 in Table 14.1). Prochazka et al.[32] used diethyl ether - ethyl acetate - 1 *M* ammonia (65:35:1) and silica gel plates, impregnated with formamide and alkalinized with ammonia, for the separation of the ergotoxine alkaloids.

Horak[47] separated the ergotoxine alkaloids by means of *n*-heptane - butyl acetate (1:1 or 65:35) saturated with formamide and ammonia in combination with silica gel plates. Gibson[55] used fluorobenzene - absolute ethanol - dimethylformamide (13:0.1:1.9) and Miscov et al.[89] used solvent system S23 (Table 14.1) to separate the ergotoxine alkaloids and their epimers. Reichelt and Kudrnac[95,100,114] described a number of solvent systems suitable for the separation of ergotoxine alkaloids (Table 14.1). They were used in combination with formamide- and ammonia-impregnated silica gel plates. Diisopropyl ether - tetrahydrofuran - toluene - diethylamine (70:15:15:0.1) was able to separate all components of the ergotoxine and ergotamine group and ergometrine, as well as their epimers. Sallam et al.[129] applied various solvent systems in combination with different types of silica gel. With chloroform - methanol - acetic acid (90:50:0.1) the ergotoxine alkaloids were separated, but their epimers not.

Hohmann and Rochelmeyer[16] separated the dihydro derivatives of ergot alkaloids on formamide-impregnated cellulose plates with ethyl acetate - *n*-heptane - diethylamine (5:6:0.02) (S5, Table 14.1). Reichelt and Kudrnac[100] used two solvents, S18 and S19 (Table 14.1), for the separation of the dihydro derivatives and Szepesy and co-workers[134,138] used acetone - 0.1 *M* ammonium carbonate - ethanol (32.5:67.5:1) on silica gel plates for the dihydro derivatives of the ergotoxine group alkaloids.

To obtain a complete resolution of all the alkaloids of the ergometrine, ergotamine and ergotoxine groups diisopropyl ether - tetrahydrofuran - toluene - diethylamine (70:15:15:0.1) and solvent systems S18, S19, S21 and S22 on formamide-impregnated silica gel plates gave good results[95,100,114]. The system described by Teichert et al.[2,7] is also suitable for this purpose.

With benzene - cyclohexane - diethylamine (5:2:0.01) on formamide-impregnated silica gel (S23) the ergotamine and ergotoxine alkaloids and their C-8 epimers (Table 14.1) can be separated; ergotamine did not move in this system[89].

In the purity control of ergot alkaloids the detection of their C-8 epimers is important, and a series of solvent systems has been proposed for this purpose[2-4,7,17,20,25,29,55,89,95,100,113-115]. The results of analyses performed with some of these TLC systems are summarized in Tables 14.1 and 14.2.

Reichelt studied the C-8 epimerization of ergot alkaloids in some organic solvents[113]. Blake[88] separated the lumi-compounds (10-hydroxy) of some ergot alkaloids (Table 14.3).

Ergot alkaloids found in higher plants (Morning Glory) were separated by means of a series of TLC systems[11,18,27,122]. The results obtained by Genest[27] are listed in Table 14.1.

Sarsunova and co-workers[63,105] and Sondack[103] separated some D- and L-lysergic acid diastereoisomers (ergometrine) with a number of solvents. TLC of the aci-compounds of some peptide alkaloids was performed by Sahli and Oesch[29] and others[15,28].

**References p. 383**

*14.1.2. LSD and derivatives*

TLC has been used extensively in the identification of LSD in drug seizures. In some instances the problem was to identify LSD and various other drugs of abuse and in others to detect and identify LSD and related compounds and other hallucinogenics.

Genest and Farmilo[19] used chloroform - methanol (9:1) in combination with base-impregnated silica gel plates for the analysis of LSD, some ergot alkaloids and other drugs of abuse. They also described a solvent system suitable for the TLC of products obtained by acidic hydrolysis of LSD and ergot alkaloids. Chloroform - methanol, often used for the separation of ergot alkaloids, is also one of the most widely used solvent systems for the analysis of LSD. It has been used in combination with silica gel plates and with silica gel plates impregnated with base. Phillips and Gardiner[56] tried the same solvent system in combination with different types of pre-coated plates. They preferred it to methanol and methanol - concentrated ammonia (100:1.5) on silica gel plates. The latter solvent system has been much used for the analysis of drugs of abuse, including LSD[43,74,77,85,96,110,124].

Cavallaro et al.[59] tested a number of TLC systems and found four of them (S1, S2, S3 and S4 in Table 14.2) to give best results in the analysis of LSD and related compounds.

Alliston et al.[70] used toluene - morpholine (9:1) on silica gel plates in the analysis of some erganes and tryptamines (Table 14.2). However, the solvent system could only be used in combination with a specific brand of pre-coated plates. Röder and Surborg[68] proposed azeotropic solvent systems for the analysis of opium alkaloids, LSD and mescalin.

Blake et al.[88] separated LSD and some ergot alkaloids and their lumi-derivatives (Table 14.3). The lumi-derivatives were prepared as an aid in the identification of the alkaloids. Fowler et al.[82] (Table 14.2) tested a number of TLC systems for the analysis of illicit ergot samples. For routine use acetone in combination with silica gel or aluminium oxide plates was found to be suitable. However, none of the systems was able to separate LSD unequivocally from the other ergot alkaloids tested. For this purpose a combination of two systems was needed. When neutral solvents were used the alkaloid spots were overspotted with sodium hydroxide solution, prior to development in the solvent system, to reduce tailing. A similar procedure was described by Phillips and Gardiner[56].

For the analysis of different alkyl-substituted amides of lysergic acid and isolysergic acid Bailey et al.[91] described a number of solvent systems combined with silica gel, aluminium oxide, 0.1 *M* sodium hydroxide-impregnated silica gel and different brands of pre-coated plates (Table 14.4). Aluminium oxide gave elongated spots. Pre-coated plates gave the best results if they were not activated before use.

Sperling[102] gave the $R_F$ values of some ergot alkaloids and lysergic acid derivatives in two solvent systems: chloroform - methanol (9:1) and chloroform saturated with ammonia - methanol (18:1), both on silica gel plates (Table 14.2).

Vinson et al.[110] studied the detection of drugs of abuse and gave the $R_F$ values of about 70 street drugs in two TLC systems commonly used in toxicological analysis (although in different proportions): ethyl acetate - methanol - ammonia (100:18:1.5) and methanol - ammonia (100:1.5). Güven and co-workers[128,133] described three TLC systems suitable for the analysis of LSD and some ergot alkaloids.

The analysis of LSD in biological materials has been dealt with in a number of investigations. However, because of the low doses used and the fact that most of the compound is metabolized into several other compounds, it is difficult to detect LSD in biological materials, such as urine[53]. In monkeys only 1% of unchanged LSD is found in the urine[126]. Extraction methods in the analysis of drugs of abuse are described in the section on opium alkaloids (p.239). The ion-exchange paper method and direct solvent extraction[53] were not sufficiently sensitive to detect LSD excreted in urine (5 μg/ml).

Kaistha and co-workers[83,109,126] used TLC for the identification of a series of drugs of abuse in urine, including LSD. Various TLC systems gave satisfactory results in the identification of LSD [see Table 12.13 (ref. 213) in Chapter 12, p.284].

Reviews of the analysis of LSD in drugs of abuse analysis have been published[37,76,126].

## 14.2. DETECTION

The ergot alkaloids can be detected by the common alkaloid spray reagents: Dragendorff's reagent and iodoplatinate reagent. When formamide-impregnated plates are used they have to be sprayed with sodium nitrite in hydrochloric acid to decompose the formamide, which interferes in the detection[64]. Holdstock and Stevens[107] described the recovery of alkaloids after spraying with iodoplatinate (see Chapter 2, p.14). However the recovery of ergotamine was poor.

The strong blue fluorescence of most ergot alkaloids in UV light is used in the detection of such alkaloids (sensitivity about 0.05 μg[19]). Most of the clavine alkaloids and the dihydro ergot alkaloids do not fluoresce, but the dihydro derivatives can be converted into yellow-green fluorescent compounds by irradiation with UV light (254 nm) for 20 min[20,86]. The fluorescence of the ergot alkaloids is intensified by addition of dimethylformamide to the mobile phase[111], impregnation of the stationary phase with formamide[123] or impregnation with a solution of paraffin in diethyl ether (1:9) after development[72,86].

With *p*-dimethylaminobenzaldehyde spray reagent (van Urk's reagent, Ehrlich's reagent) blue-purple spots are obtained for ergot alkaloids. A number of different modifications for the reagent has been described (nos. 36a-36h). Solutions of *p*-dimethylaminobenzaldehyde in cyclohexane[15,64,81], in 25% hydrochloric acid[3] to which a drop of iron(III) chloride solution is added for stabilization, in ethanol and concentrated hydrochloric acid[19], or in ethanol and concentrated sulphuric acid[30] have been used. The reagent in cylohexane requires, after spraying, exposure of the TLC plate to hydrogen chloride vapour to develop the blue colour. The sensitivity was 0.05 μg[15]. According to Genest[27], the blue colour can be stabilized by spraying with 1% sodium nitrite solution. Exposure to nitric acid vapour intensifies the colour[132].

Iron(III) chloride in perchloric acid (no. 62b) and iron(III) chloride in sulphuric acid (no. 63) have also been used as spray reagents. Both give characteristic colours, particularly for the clavine alkaloids (Table 14.5)[73,80].

Güven and co-workers[49,128,133] tried a series of spray reagents to detect ergot alkaloids. Spraying with 10% copper(II) sulphate - 2% ammonia (5:1) (no. 27) and heating at 105$^{o}$C for 15 min gave violet-brown spots. Spraying with 0.3% sodium 1,2-naphthoquinone-4-sulphonate in 50% ethanol (no. 88) gave red-purple spots against a light pink background after consecutive spraying with 10% hydrochloric acid and heating for 20 min at 110$^{o}$C. A similar treatment after spraying with 0.2% 2-nitrosonaphthol (no. 77) in ethanol gave blue-black spots against a yellow background. Other spray reagents described by these authors were 0.5% 2,6-dichloroquinone-4-chlorimide in methanol (brown spots), 0.3% congo red (pH 7) (red spots) and 0.5% quinone in dilute hydrochloric acid (blue spots). Rücker and Taha[125] used π-acceptors for the identification of alkaloids (see Table 2.2, p. 16). Similar reagents were also used by Heacock and Forrest[94] for the detection of some hallucinogenics, including LSD and lysergic acid amide (nos. 17, 41, 47, 92, 95a, 97a, 99 and 100b). The detection of LSD with TCBI reagent (no. 94) was described by Vinson et al.[110]. In the analysis of drugs of abuse several methods with consecutive spray reagents have been described (see Chapter 12, p. 235). LSD can also be detected by several of these methods[83,96,109,124].

## 14.3. QUANTITATIVE ANALYSIS

For the elution of the alkaloids after TLC separation a number of acidic solvents in connection with silica gel plates have been used: methanol - water - acetic acid (45:45:10)[3,20,78] and (2:2:1)[45,80]; methanol - water (4:6) containing 1% tartaric acid[15,45,80]; methanol - 1% aqueous tartaric acid (4:1)[29]; ethanol -

0.1 *M* sulphuric acid (5:15)[22,80]; methanol - sulphuric acid (95:5)[138] and 0.05 *M* sulphuric acid[11]. Röder et al.[45] tested several eluents based on methanol and organic acids and found methanol - water (4:6) containing 1% tartaric acid to be most effective. Keipert and Voigt[80] came to a similar conclusion; they also found chloroform - methanol (1:1) to be useful, and they preferred silica gel plates, because the results with aluminium oxide showed more spreading. Karacsony and Szarvady[12] preferred inactivated aluminium oxide because the alkaloids were otherwise too difficult to elute with their eluent chloroform - butanol (5:1). Georgievski[65] eluted the alkaloids from aluminium oxide with chloroform.

Basic elution solvents were used by Prochazka et al.[25] (acetone containing 1 g/l of ammonia), by Horak[48] [acetone - methanol - ammonia (47.5:47.5:5)] and by Reichelt[113] [acetone - methanol - 5% ammonia (50:50:1)]. After the elution the alkaloids were determined quantitatively either by spectrophotometry, using van Urk's reagent[3,20,22,25,45,48,80,113], fluorimetry[29] or UV spectrometry[15,17,32,33,136].

For the quantitative analysis of LSD methanol[19,30,59,101] and ethanol[44] have been used to elute the compound from the plates. According to Genest and Farmilo[19], the extraction must be carried out in the dark to prevent decomposition. After elution LSD is determined by fluorimetry[19,30,59], UV spectrometry[44,101], GLC[59] or spectrophotometry[101].

Several methods have been described for the direct densitometric determination of ergot alkaloids. Genest[27] determined the alkaloids present in Morning Glory by densitometry (at 570 nm) after detection of the alkaloids with *p*-dimethylaminobenzaldehyde, followed by spraying with 1% sodium nitrite solution to stabilize the colour. The densitometric determination was carried out 20 min after the first spray; the plates were kept in the dark. Vanhaelen and Vanhaelen-Fastré[81] also used van Urk's reagent for the densitometric determination (580 nm) and found the method to be more sensitive and specific than densitometry at the UV absorption maximum of the alkaloids (305 nm). After spraying the plates were stored for 2 h in the air and the light, prior to the densitometric analysis. Steinigen[77] and Bethke and Frei[112] used densitometry at the UV absorption maxima of the alkaloids.

Fluorodensitometry has been used in a series of investigations for the determination of ergot alkaloids. Massa et al.[61] determined ergometrine at 490 nm after excitation at 358 nm. Niwaguchi and Inoue[69] determined LSD at 410 nm after excitation at 330 nm. Frijns[72] described a densitofluorometric determination of ergometrine and ergotamine in pharmaceutical preparations. After development of the plates in the dark to prevent decomposition, they were impregnated with liquid paraffin in diethyl ether (1:9) to stabilize the fluorescence. This treatment gave a 10-fold enhancement of the fluorescence. The excitation wavelength used was 313 nm and the emission was measured at 420 nm. Dihydro alkaloids could be

**References p. 383**

analysed by a similar method, in which the plates, after development, were irradiated with UV light of 254 nm for 1 h, yielding fluorescent spots of the dihydro compounds[86]. Prosek et al.[116] determined the dihydro alkaloids of the ergotoxine group by using an excitation wavelength of 230 nm and measuring the emission at 250-320 nm. They found that the method gave better results than densitometry at the UV absorption maximum (280 nm). The same authors also described a quantitative determination for non-hydrogenated ergot alkaloids, using an excitation wavelength of 325 nm and measuring the emission at 445 nm[115,123]. By means of impregnation of the silica gel plates with formamide (30-40%) the best alkaloid separation was obtained and the fluorescence was very strong and fairly linear.

Eich and Schunack[111] determined ergot alkaloids in pharmaceutical preparations. To obtain optimum fluorescence of the alkaloids addition of dimethylformamide to the mobile phase was found to be important; the excitation wavelength used was 313 nm and the emission was measured at 445 nm. For the quantitative analysis of lysergic acid Rucman[119] used excitation at 360 nm and the emission was measured at 440 nm. For ergotamine Amin and Sepp[118] used an excitation wavelength of 365 nm and measured the emission at 450 nm, and developed the plates in the dark. Hatinguais et al.[137] determined the alkaloids in ergot by fluorodensitometry (excitation at 365 nm, emission at 450 nm). The plates were dried under nitrogen and used immediately for the densitometric analysis.

## 14.4. TAS TECHNIQUE AND REACTION CHROMATOGRAPHY

Jolliffe and Shellard[62] described the use of the TAS technique for a number of alkaloid-containing drugs, including ergot. They found some alkaloids in the micro-steam distillate, but the results were not consistent. Hiermann and Still[85] applied the TAS technique to the identification of LSD. They used methanol on a wad of asbestos as propellant and a temperature of 200°C for 60 sec.

The light sensitivity of ergot alkaloids and LSD has often been used in their identification (formation of lumi-compounds). Irradiation on the plate with UV light of 366 nm for 1 h[19] gave characteristic patterns of fluorescent spots for LSD and some ergot alkaloids. The patterns observed after irradiation depend, according to Andersen[54], on the time of exposure and the intensity of the UV light. Irradiation in chloroform solution gave more consistent spot patterns. The spots observed after irradiation at 366 nm of a number of ergot alkaloids and LSD were reported.

For the identification of LSD Niwaguchi and Inoue[69] irradiated the TLC plate with UV light of 254 nm; after development five additional spots were observed. Blake et al.[88] studied the optimum experimental conditions for conversion of ergot alkaloids and LSD into lumi-compounds (10-hydroxy). To obtain about equal amounts of the parent compound and the lumi-compound irradiation at 366 nm for 15 min in 10% acetic acid solution was required. The lumi-compounds do not fluoresce under UV light.

Radecka and Nigam[35] hydrogenated ergot alkaloids prior to TLC and GLC analysis. Ho and co-workers[75,92] converted LSD and bromo-LSD into dansyl derivatives after extraction from urine, to permit more sensitive TLC detection. For the identification of LSD and ergot alkaloids Genest and Farmilo[19] performed hydrolysis with 37% hydrochloric acid and obtained lysergic acid, amino acids and, in the case of LSD, diethylamine. The last compound is not detectable by TLC. The amino acids can be detected with ninhydrin reagent. Each alkaloid has its own characteristic pattern after hydrolysis.

Reimers[46] distinguished ergotamine from its dihydro derivative by evaporation of an ethanolic solution containing a small amount of acetic acid. Under these conditions ergotamine was partly epimerized, whereas the dihydro derivative was not affected. TLC of the residues after evaporation gave two spots for ergotamine and one for the dihydro derivative.

## REFERENCES


1 H. Rochelmeyer, *Pharm. Ztg.*, 103 (1958) 1269.
2 K. Teichert, E. Mutschler and H. Rochelmeyer, *Deut. Apoth.-Ztg.*, 100 (1960) 283.
3 M. Klavehn, H. Rochelmeyer and J. Seyfried, *Deut. Apoth.-Ztg.*, 101 (1961) 75.
4 M. Klavehn and H. Rochelmeyer, *Deut. Apoth.-Ztg.*, 101 (1961) 477.
5 D. Waldi, K. Schnackerz and F. Munter, *J. Chromatogr.*, 6 (1961) 61.
6 D. Gröger, V.E. Tyler, Jr., and J.E. Dusenberg, *Lloydia*, 24 (1961) 97.
7 K. Teichert, E. Mutschler and H. Rochelmeyer, *Z. Anal. Chem.*, 181 (1961) 325.
8 E. Vidic and J. Schütte, *Arch. Pharm. (Weinheim)*, 295 (1962) 342.
9 S. Agurell and E. Ramstad, *Lloydia*, 25 (1962) 67.
10 D. Gröger and D. Erge, *Pharmazie*, 18 (1963) 346.
11 W.A. Taber, L.C. Vining and R.A. Heacock, *Phytochemistry*, 2 (1963) 65.
12 E.M. Karacsony and B. Szarvady, *Planta Med.*, 11 (1963) 169.
13 Lien-Niang Li and Chi Chen Fang, *Yao Hsueh Hsueh Pao*, 10 (1963) 643; *C.A.*, 60 (1964) 7121c.
14 L. Wolf, B. Szarvady and E. Karacsony, *Acta Pharm. Hung.*, 34 (1964) 131; *C.A.*, 61 (1964) 6082c.
15 M. Zinzer and Ch. Baumgärtel, *Arch. Pharm. (Weinheim)*, 297 (1964) 158.
16 T. Hohmann and H. Rochelmeyer, *Arch. Pharm. (Weinheim)*, 297 (1964) 186.
17 V. Prochazka, F. Kavka, M. Prucha and J. Pitra, *Cesk. Farm.*, 13 (1964) 493.
18 T. Niwaguchi and T. Inoue, *J. Chromatogr.*, 43 (1969) 512.
19 K. Genest and C.G. Farmilo, *J. Pharm. Pharmacol.*, 16 (1964) 250.
20 J.L. McLaughlin, J.E. Goyan and A.G. Paul, *J. Pharm. Sci.*, 53 (1964) 306.
21 V. Schwarz and M. Sarsunova, *Pharmazie*, 19 (1964) 267.
22 Lien-Niang Li and Chi-Cheng Fang, *Yao Hsueh Hsueh Pao*, 11 (1964) 189; *C.A.*, 61 (1964) 2168b.
23 S. Agurell, *Acta Pharm. Suecica*, 2 (1965) 357.
24 R. Paris, R. Rousselet, M. Paris and M.J. Fries, *Ann. Pharm. Fr.*, 23 (1965) 473.

25 V. Prochazka, F. Kavka, M. Prucha and J. Pitra, *Cesk. Farm.*, 14 (1965) 154; *C.A.*, 66 (1967) 108274u.
26 M. Ram and K.L. Handa, *Indian J. Chem.*, 3 (1965) 42.
27 K. Genest, *J. Chromatogr.*, 19 (1965) 531.
28 W.N. French and A. Wehrli, *J. Pharm. Sci.*, 54 (1965) 1515.
29 M. Sahli and M. Oesch, *Pharm. Acta Helv.*, 40 (1965) 25.
30 A. Dihrberg and B. Newman, *Anal. Chem.*, 38 (1966) 1959.
31 J. Tyfczynska, *Diss. Pharm. Pharmacol.*, 18 (1966) 491.
32 V. Prochazka, F. Kavka, M. Prucha and J. Pitra, *Cesk. Farm.*, 15 (1966) 363.
33 L. Wichlinski and Z. Skibinski, *Farm. Pol.*, 22 (1966) 194; *C.A.*, 65 (1966) 10425g.
34 I. Zarebska and A. Ozarowski, *Farm. Pol.*, 22 (1966) 518; *C.A.*, 66 (1967) 40741m.
35 C. Radecka and I.C. Nigam, *J. Pharm. Sci.*, 55 (1966) 861.
36 I. Juhl and V. Waarst, *Arch. Pharm. Chem.*, 74 (1967) 887.
37 M. Lerner, *Bull. Narcotics*, 19 (1967) 39.
38 M. Petkovic, *Farm. Glas.*, 23 (1967) 445.
39 R.J. Martin and T.G. Alexander, *J. Ass. Offic. Anal. Chem.*, 50 (1967) 1362.
40 H.C. Hsiu, J.T. Huang, T.B. Shih, K.L. Yang, K.T. Wang and A.L. Lin, *J. Chin. Chem. Soc.*, 14 (1967) 161.
41 J.R. Bianchine, A. Niec and P.V.J. Macaraeeg, *J. Chromatogr.*, 31 (1967) 255.
42 A. Noirfalise and G. Mees, *J. Chromatogr.*, 31 (1967) 594.
43 E.G.C. Clarke, *J. Forensic Sci. Soc.*, 7 (1967) 46.
44 J.J. Thomas and L. Dryon, *J. Pharm. Belg.*, 22 (1967) 163.
45 K. Röder, E. Mutschler and H. Rochelmeyer, *Pharm. Acta Helv.*, 42 (1967) 407.
46 F. Reimers, *Arch. Pharm. Chem.*, 75 (1968) 1064.
47 P. Horak, *Cesk. Farm.*, 17 (1968) 37; *C.A.*, 68 (1968) 107904d.
48 P. Horak, *Cesk. Farm.*, 17 (1968) 89.
49 K.C. Güven and L. Eroglu, *Eczacilik Bul.*, 10 (1968) 53.
50 A. Noirfalise, *J. Pharm. Belg.*, 23 (1968) 387.
51 J.M.C.J. Frijns, *Pharm. Weekbl.*, 103 (1968) 929.
52 L. Wichlinski, *Acta Pol. Pharm.*, 26 (1969) 617.
53 S.J. Mulé, *J. Chromatogr.*, 39 (1969) 302.
54 D.L. Andersen, *J. Chromatogr.*, 41 (1969) 491.
55 M.S. Gibson, *J. Chromatogr.*, 44 (1969) 631.
56 G.F. Phillips and J. Gardiner, *J. Pharm. Pharmacol.*, 21 (1969) 793.
57 E. Röder, E. Mutschler and H. Rochelmeyer, *Z. Anal. Chem.*, 244 (1969) 46.
58 A. Kornhauser, M. Perpar and L. Gasperut, *Arch. Pharm. (Weinheim)*, 303 (1970) 882.
59 A. Cavallaro, G. Rossi and G. Elli, *Boll. Lab. Chim. Prov.*, 21 (1970) 303; *C.A.*, 74 (1971) 138709x.
60 M. Petkovic, *Farm. Glas.*, 26 (1970) 367.
61 V. Massa, F. Gal and P. Susplugas, *Int. Symp. Chromatogr. Electrophor. Lect. Pap. 6th*, (1970) 470.
62 G.H. Jolliffe and E.J. Shellard, *J. Chromatogr.*, 48 (1970) 125.
63 M. Sarsunova, M. Semonsky and A. Cerny, *J. Chromatogr.*, 50 (1970) 442.
64 M. Vanhaelen, *J. Pharm. Belg.*, 25 (1970) 175.
65 V.P. Georgievski, *Probl. Anal. Khim.*, 1 (1970) 94; *C.A.*, 74 (1971) 146428v.
66 G.F. Lozovaya, *Sud. Med. Ekspert.*, 13 (1970) 31; *C.A.*, 74 (1971) 74464d.
67 M. Sarsunova, M. Semonsky and V. Zikan, *Cesk. Farm.*, 20 (1971) 264.
68 E. Röder and K.H. Surborg, *Z. Anal. Chem.*, 256 (1971) 362.
69 T. Niwaguchi and T. Inoue, *J. Chromatogr.*, 59 (1971) 127.
70 G.V. Alliston, M.J. de Faubert Maunder and G.F. Phillips, *J. Pharm. Pharmacol.*, 23 (1971) 555.
71 G.F. Lozovaya, *Nauch. Tr. Irkutsk Gos. Med. Inst.*, 113 (1971) 113.
72 J.M.G.J. Frijns, *Pharm. Weekbl.*, 106 (1971) 865.
73 R. Voigt and P. Zier, *Pharmazie*, 26 (1971) 313.
74 M. Steinigen, *Pharm. Ztg.*, 116 (1971) 2072.
75 I.K. Ho, H.H. Loh and E.L. Way, *Proc. West. Pharmacol. Soc.*, 14 (1971) 183.
76 A. Sperling, *J. Chromatogr. Sci.*, 10 (1972) 268.
77 M. Steinigen, *Deut. Apoth.-Ztg.*, 112 (1972) 51.

78 Ya. Petrova, T. Tomova and L. Filipova, *Farmatsiya (Sofia)*, 22 (1972) 9; *C.A.*, 76 (1972) 158394k.
79 J.K. Brown, L. Shapazian and G.D. Griffin, *J. Chromatogr.*, 64 (1972) 129.
80 S. Keipert and R. Voigt, *J. Chromatogr.*, 64 (1972) 327.
81 M. Vanhaelen and R. Vanhaelen-Fastré, *J. Chromatogr.*, 72 (1972) 139.
82 R. Fowler, P.J. Gomm and D.A. Patterson, *J. Chromatogr.*, 72 (1972) 351.
83 K.K. Kaistha and J.H. Jaffe, *J. Pharm. Sci.*, 61 (1972) 679.
84 L.D. Vechkanova, A.I. Bankovski and A.N. Bankovskaya, *Khim. Prir. Soedin.*, 8 (1972) 483; *C.A.*, 78 (1973) 1631d.
85 A. Hiermann and F. Still, *Oesterr. Apoth. Ztg.*, 26 (1972) 337.
86 C.J.G.A. Bos and J.M.G.J. Frijns, *Pharm. Weekbl.*, 107 (1972) 111.
87 R.A. Egli, *Z. Anal. Chem.*, 259 (1972) 277.
88 E.T. Blake, P.J. Cashman and J.I. Thornton, *Anal. Chem.*, 45 (1973) 394.
89 V. Miscov, L. Rosca and E. Nichiforescu, *Farmacia (Bucharest)*, 21 (1973) 499; *C.A.*, 80 (1974) 100239e.
90 B. Pekic, S. Dordevic and K. Petrovic, *Farm. Glas.*, 29 (1973) 1; *C.A.*, 79 (1973) 50436y.
91 K. Bailey, D. Verner and D. Legault, *J. Ass. Offic. Anal. Chem.*, 56 (1973) 88.
92 H.H. Loh, I.K. Ho, T.M. Cho and W. Lipscomb, *J. Chromatogr.*, 76 (1973) 505.
93 R.A. van Welsum, *J. Chromatogr.*, 78 (1973) 237.
94 R.A. Heacock and J.R. Forrest, *J. Chromatogr.*, 78 (1973) 241.
95 J. Reichelt and S. Kudrnac, *J. Chromatogr.*, 87 (1973) 433.
96 N.H. Choulis, *J. Pharm. Sci.*, 62 (1973) 112.
97 K.F. Ahrend and D. Tiess, *Wiss. Z. Univ. Rostock, Math. Naturw. Reihe*, 22 (1973) 951.
98 M. Petkovic, *Acta Pharm. Jugoslav.*, 24 (1974) 23; *C.A.*, 81 (1974) 25833j.
99 D. Radulovic, Z. Glagojevic and D. Zivanov-Stakic, *Arh. Pharm.*, 24 (1974) 215; *C.A.*, 83 (1975) 103334e.
100 J. Reichelt and S. Kudrnac, *Cesk. Farm.*, 23 (1974) 13.
101 K. Khashalov, *Farmatsiya (Sofia)*, 24 (1974) 9; *C.A.*, 83 (1975) 37405a.
102 A.R. Sperling, *J. Chromatogr. Sci.*, 12 (1974) 265.
103 D.L. Sondack, *J. Pharm. Sci.*, 63 (1974) 1141.
104 J. Breiter, *Kontakte*, 3 (1974) 17.
105 M. Sarsunova and V. Hadrabova, *Cesk. Farm.*, 24 (1975) 81.
106 D. Smarzynska, A. Cegielska and A. Matuszewicz, *Farm. Pol.*, 31 (1975) 223.
107 T.M. Holdstock and H.M. Stevens, *Forensic Sci.*, 6 (1975) 187.
108 W. Debska and A. Owczarska, *Herba Pol.*, 21 (1975) 88; *C.A.*, 83 (1975) 136975t.
109 K.K. Kaistha, R. Tadrus and R. Janda, *J. Chromatogr.*, 107 (1975) 359.
110 J.A. Vinson, J.E. Hooyman and C.E. Ward, *J. Forensic Sci.*, 20 (1975) 552.
111 E. Eich and W. Schunack, *Planta Med.*, 27 (1975) 58.
112 H. Bethke and R.W. Frei, *Anal. Chem.*, 48 (1976) 50.
113 J. Reichelt, *Cesk. Farm.*, 25 (1976) 93.
114 J. Reichelt, *Cesk. Farm.*, 25 (1976) 213.
115 M. Prosek, E. Kucan, M. Katic and M. Bano, *Chromatographia*, 9 (1976) 273.
116 M. Prosek, E. Kucan, M. Katic and M. Bano, *Chromatographia*, 9 (1976) 325.
117 L. Lepri, P.G. Desideri and M. Lepori, *J. Chromatogr.*, 116 (1976) 131.
117a L. Lepri, P.G. Desideri and M. Lepori, *J. Chromatogr.*, 123 (1976) 175.
118 M. Amin and W. Sepp, *J. Chromatogr.*, 118 (1976) 225.
119 R. Rucman, *J. Chromatogr.*, 121 (1976) 353.
120 R. Aigner, H. Spitzy and R.W. Frei, *J. Chromatogr.*, 14 (1976) 381.
121 A.N. Masoud, *J. Pharm. Sci.*, 65 (1976) 1585.
122 J.M. Weber and T.S. Ma, *Mikrochim. Acta*, (1976) 217.
123 M. Prosek, E. Kucan, M. Katic and M. Bano, *Chromatographia*, 10 (1977) 147.
124 H. Kroeger, G. Bohn and G. Rücker, *Deut. Apoth.-Ztg.*, 117 (1977) 1923.
125 G. Rücker and A. Taha, *J. Chromatogr.*, 132 (1977) 165.
126 K.K. Kaistha, *J. Chromatogr.*, 141 (1977) 145.
127 A.N. Masoud, *J. Chromatogr.*, 141 (1977) D9.
128 K.C. Güven and T. Güneri, *Z. Anal. Chem.*, 283 (1977) 32.
129 L.A.R. Sallam, N. Naim and A.H. El-Refai, *Z. Anal. Chem.*, 284 (1977) 47.

130 E. Stahl, J. Brombeer and D. Eskes, *Arch. Kriminol.*, 162 (1978) 23.
131 M. Prosek, E. Kucan, M. Katic, M. Bano and A. Medja, *Chromatographia*, 11 (1978) 578.
132 E. Stahl and J. Brombeer, *Deut. Apoth.-Ztg.*, 118 (1978) 1527.
133 K.C. Güven and T. Altinkurt, *Eczacilik Bul.*, 20 (1978) 46.
134 L. Szepesy, I. Feher, G. Szepesi and M. Gazdag, *J. Chromatogr.*, 149 (1978) 271.
135 S.M. Hassan, *Pharmazie*, 33 (1978) 237.
136 P. Horvath, G. Szepesi and A. Kassai, *Planta Med.*, 33 (1978) 407.
137 P. Hatinguais, D. Beziat, P. Negol and R. Tarroux, *Trav. Soc. Pharm. Montpellie* 38 (1978) 329.
138 G. Szepesi, J. Molnar and S. Nyeredy, *Z. Anal. Chem.*, 294 (1979) 47.

TABLE 14.1

TLC SEPARATION OF ERGOT ALKALOIDS

TLC systems:

| System | Layer | Solvent |
|---|---|---|
| S1 | Silica gel, activated | Chloroform-ethanol(10:1)[4] |
| S2 | Silica gel G, activated | Ethyl acetate-ethanol-dimethylformamide(13:1:1)[9] |
| S3 | Silica gel G | Benzene-chloroform-ethanol(2:4:1)[15] |
| S4 | Silica gel G | Heptane-carbon tetrachloride-pyridine(1:3:2)[15], 2x |
| S5 | Cellulose, impregnated with 15% formamide in acetone | Ethyl acetate-*n*-heptane-diethylamine(5:6:0.02)[16] |
| S6 | Silica gel G, activated | Ethyl acetate-dimethylformamide-ethanol(13:9:0.1)[20] |
| S7 | Silica gel G, activated | Benzene-dimethylformamide(13:2)[20] |
| S8 | Aluminium oxide G, activated | Chloroform-diethyl ether-water(87.5:12.5:25), organic phase[20] |
| S9 | Silica gel G, activated | Chloroform-methanol(8:2)[23] |
| S10 | Silica gel G, activated | Chloroform-diethylamine(9:1)[23] |
| S11 | Silica gel G, activated | Chloroform-methanol-conc. ammonia(80:20:0.2)[23] |
| S12 | Aluminium oxide, activated | Chloroform-ethanol(96:4)[23] |
| S13 | Cellulose, impregnated with 15% formamide and 1% ammonia in acetone | Ethyl acetate-*n*-heptane-dimethylformamide(250:300:1)[23] |
| S14 | Aluminium oxide G | Chloroform-ethanol(96:4)[27] |
| S15 | Silica gel MN HR' | Acetone-piperidine(9:1)[27] |
| S16 | Silica gel MN HR' | Acetone-ethyl acetate-dimethylformamide(5:5:1)[27] |
| S17 | Silica gel G, activated | Chloroform-ethanol-acetone(6:4:4)[45] |
| S18 | Silica gel impregnated with 18 ml of formamide + 4 ml of 5% ammonia in 45 ml ethanol | Diisopropyl ether-tetrahydrofuran-diethylamine(80:20:0.2)[100] |
| S19 | Silica gel impregnated with 18 ml of formamide + 4 ml of 5% ammonia in 45 ml ethanol | Dibutyl ether-dichloromethane-diethylamine(60:40:0.2). Saturated with formamide[100] |
| S20 | Silica gel pre-coated (Silufol $UV_{254}$) | Chloroform-benzene-ethanol(40:20:10), saturated with 5% ammonia[114] |
| S21 | Silica gel pre-coated (Silufol $UV_{254}$), impregnated with 18 ml of formamide + 0.6 ml of 25% ammonia + acetone to 100 ml | Diisopropyl ether-toluene-ethanol-diethylamine(75:20:5:0.1)[114] |
| S22 | Silica gel pre-coated (Silufol $UV_{254}$), impregnated with 18 ml of formamide + 0.6 ml of 25% ammonia + acetone to 100 ml | Diisopropylether-toluene-tetrahydrofuran-*n*-heptane-diethylamine(50:15:15:20:0.1)[114] |
| S23 | Silica gel G, impregnated with formamide-ethanol(4:7.5) | Benzene-cyclohexane-diethylamine(5:2:0.01)[89] |
| S24 | Silica gel G $UV_{254}$, pre-coated | Acetone-0.1 *M* ammonium carbonate-ethanol(32.5:67.5:1)[138] |

TABLE 14.1 (continued)

| Alkaloid | $hR_F$ values | | | | | | | | | | | | | | | | | | | | | | | |
|---|---|---|---|---|---|---|---|---|---|---|---|---|---|---|---|---|---|---|---|---|---|---|---|---|
| | S1 | S2 | S3 | S4 | S5 | S6* | S7* | S8* | S9 | S10 | S11 | S12 | S13 | S14 | S15 | S16 | S17 | S18 | S19 | S20 | S21 | S22 | S23 | S24 |
| Ergometrine | 17 | 11 | 13 | 17 | | 17 | 8 | 0 | 27 | 2 | 40 | 21 | | 19 | 31 | 26 | 23 | 3 | 3 | 11 | 2 | 1 | 0 | 78 |
| Ergometrinine | 30 | | | | | 43 | 29 | 2 | 39 | 20 | 66 | 46 | | 50 | 52 | 53 | 28 | 14 | 10 | 30 | 7 | 5 | | 73 |
| Methylergometrine | | | | | | | | | | | | | | | | | | | | 16 | | | | |
| 1-Methylmethylergometrine | | | | | | | | | | | | | | | | | | | | 25 | | | | |
| Ergotamine | 51 | | 43 | 16 | 11 | 30 | 24 | 2 | 65 | 9 | 75 | 57 | | 48 | 47 | | 51 | 17 | 29 | 35 | 16 | 10 | 5 | 65 |
| Ergotaminine | 75 | | 73 | 48 | | 66 | 60 | 11 | 78 | 44 | 90 | 71 | | 72 | 63 | | 73 | 39 | 66 | 65 | 37 | 26 | 11 | 38 |
| Ergosine | 51 | | 43 | 21 | 17 | 34 | 27 | 2 | 63 | 10 | 73 | 58 | | | | | 51 | 28 | 33 | 35 | 23 | 14 | 22 | |
| Ergosinine | 75 | | | | | 70 | 65 | 11 | 76 | 45 | 89 | 66 | | | | | 73 | 56 | 72 | 65 | 49 | 36 | 31 | |
| Ergostine | | | | | | | | | 72 | 18 | 80 | 63 | | | | | | 35 | 52 | 48 | 31 | 20 | | |
| Ergostinine | | | | | | | | | 81 | 56 | 88 | 64 | | | | | | | | | | | | |
| Ergocornine | 69 | | 67 | 41 | 50 | 57 | 52 | 11 | 75 | 41 | 87 | 71 | 68 | | | | 68 | 69 | 78 | 57 | 57 | 40 | 58 | 60 |
| Ergocorninine | 81 | | 81 | | | 75 | 70 | 28 | 82 | 56 | 90 | 71 | 76 | | | | 77 | 80 | 94 | 71 | 94 | 74 | 65 | 33 |
| Ergocristine | 69 | | 67 | 36 | 41 | 53 | 47 | 11 | 75 | 37 | 87 | 72 | 62 | | | | 68 | 64 | 75 | 57 | 52 | 37 | 44 | 45 |
| Ergocristinine | 81 | | 81 | 62 | | 75 | 70 | 23 | 83 | 57 | 91 | 72 | 71 | | | | 77 | 80 | 94 | 71 | 91 | 69 | 52 | 16 |
| α-Ergocryptine | 69 | | 67 | 41 | 61 | 57 | 52 | 11 | 74 | 38 | 88 | 71 | 76 | | | | 68 | 77 | 83 | 57** | 72** | 51** | 71 | 53 |
| β-Ergocryptine | | | | | | | | | | | | | | | | | | | | 57 | 76 | 56 | | 50 |
| α-Ergocryptinine | 81 | | 81 | | | 75 | 73 | 33 | 81 | 57 | 90 | 72 | 81 | | | | 77 | 86 | 98 | 71*** | 98*** | 88*** | 79 | 25 |
| Dihydroergotamine | 11 | | 35 | | 9 | | | | | | | | | | | | | 13 | 15 | 32 | | | | |
| Acidihydroergotamine | | | 37 | | | | | | | | | | | | | | | | | | | | | |
| Dihydroergosine | | | | | | | | | | | | | | | | | | 22 | 17 | 32 | | | | |
| Dihydroergocornine | | | 53 | | 38 | | | | | | | | | | | | | 50 | 31 | 46 | | | | 63 |
| Dihydroergocristine | 14 | | 53 | | 30 | | | | | | | | | | | | | 43 | 36 | 46 | | | | 51 |
| Dihydroergocryptine | | | 53 | | 50 | | | | | | | | | | | | | 61 | 42 | 46 | | | | 58 |
| Dihydrolysergic acid | | | 0 | | | | | | | | | | | | | | | | | | | | | 55+ |
| Dihydrolysergic acid amide | | | 6 | | | | | | | | | | | | | | | 2 | 2 | | | | | |
| Lysenyl | | | | | | | | | | | | | | | | | | | | 43 | | | | |
| Agroclavine | 38 | 54 | | | | | | | 39 | 51 | 67 | 63 | | 68 | 60 | | 39 | | | | | | | |
| Chanoclavine I | 2 | 5 | | | | | | | 5 | 11 | 12 | 3 | | 2 | 35 | 5 | 5 | | | | | | | |
| Chanoclavine II | | | | | | | | | 6 | 7 | 10 | 3 | | | | | | | | | | | | |
| Costaclavine | | 16 | | | | | | | 5 | 66 | 24 | 56 | | | | | | | | | | | | |
| Elymoclavine | 13 | 23 | | | | | | | 17 | 11 | 34 | 29 | | 32 | 35 | 18 | 21 | | | | | | | |
| Festuclavine | | 36 | | | | | | | 23 | 48 | 51 | 60 | | | | | 34 | | | | | | | |
| Fumigaclavine A | | 74 | | | | | | | 57 | 60 | 78 | 71 | | | | | | | | | | | | |
| Fumigaclavine B | | 40 | | | | | | | 13 | 45 | 39 | 58 | | | | | | | | | | | | |
| Fumigaclavine C | | | | | | | | | 72 | 78 | 91 | 75 | | | | | | | | | | | | |

| Compound | 1 | 2 | 3 | 4 | 5 | 6 | 7 | 8 | 9 | 10 | 11 | 12 | 13 | 14 | 15 |
|---|---|---|---|---|---|---|---|---|---|---|---|---|---|---|---|
| Nor-agroclavine | | | | 16 | 30 | 34 | 23 | | | | | | | | |
| Penniclavine | 17 | 38 | | 21 | 7 | 35 | 5 | 5 | 35 | 30 | 29 | | | | |
| Isopenniclavine | 30 | 71 | | 41 | 10 | 56 | 8 | | | | 53 | | | | |
| Pyroclavine | | 51 | | 20 | 59 | 63 | 68 | | | | | | | | |
| Setoclavine | 42 | 68 | | 48 | 31 | 66 | 56 | | | | 49 | | | | |
| Isosetoclavine | 46 | 81 | | 54 | 38 | 68 | 60 | | | | 63 | | | | |
| α-Dihydrolysergol | | | | 8 | 9 | 22 | 18 | | | | | | | | |
| DL-4-Dimethylallyltryptophan | | | | 0 | 0 | 0 | 0 | | | | | | | | |
| 6-Methyl-$\Delta^{8,9}$-ergolene-8-carboxylic acid | | | | 0 | 0 | 0 | 0 | | | | | | | | |
| Lysergene | | 75 | | 58 | 44 | 74 | 63 | | | | | | | | |
| D-Lysergic acid | | | 0 | 0 | 0 | 0 | 0 | | | | | 0 | 0 | 0 | 0 |
| D-Lysergic acid amide | | | | 22 | 3 | 34 | 25 | 25 | 25 | 16 | | 5 | 3 | 13 | 4 |
| D-Isolysergic acid amide | | | | 51 | 22 | 68 | 39 | 44 | 48 | 46 | | 12 | 10 | 34 | 9 |
| D-Lysergic acid methylcarbinol amide | | | | 36 | 5 | 43 | 26 | | | | | | | | |
| Lysergine | | 64 | | 42 | 44 | 69 | 63 | | | | | | | | |
| Lysergol | | 27 | | 16 | 9 | 35 | 28 | 30 | 35 | 23 | | | | | |
| Isolysergol | | 57 | | 28 | 28 | 61 | 52 | | | | | | | | |
| D-Lysergyl-L-valine methyl ester | | | | 70 | 43 | 80 | 65 | | | | | | | | |
| LSD | | | | | | | | 68 | 63 | | | | | 44 | |
| Dihydroelymoclavine | | 12 | | | | | | | | | | | | | |
| Isolysergine | | 78 | | | | | | | | | | | | | |

*$hR_F$ from mixtures of these compounds.
**α-Ergocryptine.
***α- and β-ergocryptinine have the same $hR_F$ values.
+β-Dihydroergocryptine.

References p. 383

TABLE 14.2

TLC ANALYSIS OF LSD AND RELATED ALKALOIDS

TLC systems:

| | | |
|---|---|---|
| S1 | Silica gel H, activated | Chloroform-methanol-acetic acid(60:40:1)[59] |
| S2 | Silica gel H, activated | Acetone-chloroform-conc. ammonia(50:40:1.5)[59] |
| S3 | Aluminium oxide H, activated | Trichloroethane-acetone(1:1)[59] |
| S4 | Polyamide 11 $F_{250}$ | Chloroform-benzene-methanol-acetic acid (60:25:10:1)[59] |
| S5 | Silica gel $GF_{254}$ | Chloroform-methanol(9:1)[102] |
| S6 | Silica gel $GF_{254}$ | Chloroform saturated with ammonia-methanol(18:1)[102] |
| S7 | Silica gel, Eastman Chromagram 6060 sheets | Toluene-morpholine(9:1), unsaturated tank[70] |
| S8 | Silica gel $F_{254}$, pre-coated | Acetone[82] |
| S9 | Aluminium oxide $F_{254}$, pre-coated | Acetone[82] |
| S10 | Silica gel $F_{254}$, pre-coated | Acetone-chloroform(4:1)[82] |
| S11 | Silica gel $F_{254}$, pre-coated | Acetone-methanol(4:1)[82] |
| S12 | Silica gel $F_{254}$, pre-coated | Chloroform-methanol(4:1)[82] |
| S13 | Silica gel $F_{254}$, pre-coated | 1,1,1-Trichloroethane-methanol(9:1)[82] |
| S14 | Cellulose, sprayed with 5% sodium dihydrogen citrate and dried | *n*-Butanol-citric acid-water(870 ml + 4.8 g + 130 ml)[82] |

| Alkaloid | $hR_F$ values | | | | | | | | | | | | | |
|---|---|---|---|---|---|---|---|---|---|---|---|---|---|---|
| | S1 | S2 | S3 | S4 | S5 | S6 | S7 | S8 | S9 | S10 | S11 | S12 | S13 | S14 |
| Lysergic acid | 9 | 0 | 0 | 0 | 0 | 0 | 0 | 0 | 0 | 0 | 3 | 2 | 1 | 17 |
| Isolysergic acid | | | | | | | | | | | | | | |
| Lysergamide | 22 | 23 | 5 | 28 | 12 | 14 | | | 27 | | | | | |
| Isolysergamide | 22 | 53 | 14 | 28 | 40 | 33 | 12 | | | | | | | |
| LSD | 22 | 62 | 50 | 60 | 48 | 55 | 51 | 7 | 70 | 23 | 54 | 60 | 21 | 59 |
| Iso-LSD | 36 | 69 | 63 | 80 | 33 | 37 | 42 | 11 | 66 | 8 | 30 | 32 | 10 | 66 |
| 1-Acetyl lysergide | | | | | | | 61 | | | | | | | |
| Ergometrine | | | | | 10 | 13 | 11 | 10 | 26 | 7 | 44 | 36 | 9 | 31 |
| Ergometrinine | | | | | 34 | 31 | 22 | | | | | | | |
| Methylergometrine | | | | | 24 | 11 | 14 | 13 | 31 | 9 | 51 | 42 | 12 | 42 |
| Methysergide | | | | | 28 | | 33 | 13 | 33 | 9 | 51 | 54 | 21 | 47 |
| Ergotamine | | | | | 39 | 25 | 22 | 8 | 48 | 23 | 63 | 62 | 20 | 77 |
| Ergotaminine | | | | | 66 | 63 | 36 | | | | | | | |
| Dihydroergotamine | | | | | | | 15 | 19 | 40 | 12 | 54 | 60 | 20 | 79 |
| Ergosine | | | | | 41 | 20 | | 31 | 52 | 25 | 62 | 61 | 23 | 75 |
| Ergosinine | | | | | 60 | 58 | | | | | | | | |
| Ergocristine | | | | | 62 | 62 | 44 | 51 | 66 | 43 | 70 | 68 | 26 | 83 |
| Ergocristinine | | | | | 75 | 75 | | | | | | | | |
| Ergocryptine | | | | | 62 | 62 | | 51 | 68 | 43 | 69 | 67 | 27 | 83 |
| Ergocryptinine | | | | | 75 | 75 | | | | | | | | |
| Ergocornine | | | | | 62 | 62 | | 50 | 67 | 42 | 68 | 67 | 28 | 81 |
| Ergocorninine | | | | | 75 | 75 | | | | | | | | |
| Methylergometrinine | | | | | 36 | 29 | | | | | | | | |
| Ergostine | | | | | 48 | 40 | | | | | | | | |
| Ergostinine | | | | | 66 | 63 | | | | | | | | |
| Dihydroergocristine | | | | | | | | 35 | 64 | 27 | 64 | 65 | 24 | 83 |
| Psilocybin | | | | | | | 0 | | | | | | | |
| Psilocin | | | | | | | 46 | | | | | | | |
| N,N-Dimethyltryptamine | | | | | | | 54 | | | | | | | |
| N,N-Diethyltryptamine | | | | | | | 61 | | | | | | | |
| Ibogaine | | | | | | | 71 | | | | | | | |

TABLE 14.3[88]

TLC ANALYSIS OF PARENT ERGOT ALKALOIDS AND 10-HYDROXY DERIVATIVES

TLC system:
Silica gel G    Chloroform-methanol(9:1)

| Alkaloid | $hR_F$ value | Alkaloid | $hR_F$ value |
|---|---|---|---|
| Lysergic acid | 0 | Ergotaminine | 64 |
| 10-Hydroxylysergic acid | 0 | 10-Hydroxyergotaminine | 37 |
| LSD | 45 | Ergocryptine | 62 |
| 10-Hydroxy-LSD | 29 | 10-Hydroxyergocryptine | 29 |
| Ergometrine | 14 | Ergocristine | 74 |
| 10-Hydroxyergometrine | 3 | 10-Hydroxyergocristine | 66 |
| Methylergometrine | 18 | Dihydroergocristine | 48 |
| 10-Hydroxymethylergometrine | 5 | Irradiated dihydroergocristine | 48 |

TABLE 14.4

TLC ANALYSIS OF LSD AND RELATED DIALKYLAMIDES[91]

TLC systems:
S1 Silica gel G    1,1,1-Trichloroethane-methyl ethyl ketone-methanol(7:2:1)
S2 Silica gel G    Hexane-ethyl acetate-methanol(7:13:15)
S3 Silica gel G    Toluene-nitromethane-methanol(8:10:2)
S4 Silica gel G    Chloroform-methanol(9:1)
S5 Silica gel G    Benzene-dimethylformamide(9:1)
S6 Silica gel G    Benzene-dimethylformamide(13:2)
S7 Silica gel G    Ethyl acetate-ethanol-dimethylformamide(13:0.1:1.9)
S8 Silica gel G    Chloroform saturated with ammonia-methanol(18:1)
S9 Silica gel G    Chloroform-acetone(1:2)
These solvents were also tested with 0.1 *M* sodium hydroxide-impregnated silica gel and aluminium oxide plates, both pre-coated and home-made.

| Alkylamide derivative | $hR_F$ values | | | | | | | | |
|---|---|---|---|---|---|---|---|---|---|
| | S1 | S2 | S3 | S4 | S5 | S6 | S7 | S8 | S9 |
| Dimethyl | 10 | 33 | 13 | 29 | 14 | 23 | 20 | 39 | 10 |
| Isodimethyl | 4 | 15 | 4 | 10 | 9 | 15 | 14 | 25 | 5 |
| Diethyl | 25 | 51 | 25 | 40 | 33 | 45 | 40 | 50 | 25 |
| Isodiethyl | 8 | 25 | 14 | 12 | 17 | 29 | 25 | 31 | 9 |
| Methylpropyl | 24 | 50 | 24 | 39 | 30 | 41 | 38 | 49 | 22 |
| Isomethylpropyl | 6 | 24 | 13 | 10 | 17 | 28 | 23 | 30 | 8 |
| Ethylpropyl | 33 | 58 | 29 | 45 | 40 | 48 | 45 | 53 | 35 |
| Isoethylpropyl | 11 | 31 | 14 | 14 | 24 | 35 | 30 | 34 | 13 |
| Dipropyl | 39 | 62 | 33 | 50 | 48 | 52 | 50 | 55 | 44 |
| Isodipropyl | 20 | 42 | 15 | 20 | 33 | 40 | 36 | 38 | 23 |

TABLE 14.5

FLUORESCENCE COLOURS AND COLOURS OBTAINED WITH SOME SPRAY REAGENTS FOR ERGOT ALKALOIDS

| Alkaloid | Fluorescence | | *p*-Dimethylamino-benzaldehyde | 5 min after iron(III) chloride-sulphuric acid (no. 63)[73] | Iodo-platinate[5] |
|---|---|---|---|---|---|
| | 254 nm | 366 nm | | | |
| Agroclavine | - | - | Blue* | Wine red | |
| Chanoclavine | - | - | Blue* | Wine red | |
| Costaclavine | - | - | Blue* | Blue-violet | |
| Elymoclavine | - | - | Blue* | Wine red | |
| Festuclavine | - | - | Blue* | Wine red | |
| Fumigaclavine A | - | | Purple** | | |
| Fumigaclavine B | - | | Purple** | | |
| Fumigaclavine C | - | | | | |
| Nor-agroclavine | - | | | | |
| Penniclavine | + | + | Green* | | |
| Isopenniclavine | + | + | Green* | | |
| Pyroclavine | - | - | Blue* | Wine red | |
| Setoclavine | + | + | Green* | Yellow-green | |
| Isosetoclavine | + | + | Green* | Yellow-green | |
| Molliclavine | | - | Green* | | |
| Dihydrochanoclavine | | | | Blue | |
| Isodihydrochanoclavine | | | | Blue | |
| Dihydrolysergol | - | | | Wine red | |
| Iso-dihydrolysergol | | | | Wine red | |
| Lysergene | + | | Dark blue* | Yellow-green | |
| D-Lysergic acid | + | + | Purple*** | | |
| Iso-D-lysergic acid | + | + | | | |
| D-lysergic acid amide | + | + | Purple*** | | |
| Iso-D-lysergic acid amide | + | + | | | |
| Lysergine | + | | | | |
| Lysergol | + | | Blue* | Orange | |
| LSD | + | + | Purple*** | | |
| Iso-LSD | + | + | Purple*** | | |
| Ergometrine | + | + | Purple*** | Orange | White |
| Ergometrinine | + | + | Purple*** | | Violet-blue |
| Methylergometrine | + | + | Purple*** | | |
| 1-Methylmethylergo-metrine | + | + | Violet*** | | |
| Ergotamine | + | + | Purple*** | | Pink |
| Ergotaminine | + | + | Purple*** | | Pink |
| Ergosine | + | + | | | |
| Ergosinine | + | + | | | |
| Ergostine | + | + | | | |
| Ergostinine | + | + | | | |
| Ergocornine | + | + | | | |
| Ergocorninine | + | + | | | |
| Ergocristine | + | + | Purple*** | | Beige-light brown |
| Ergocristinine | + | + | | | Light-brown |
| Ergocryptine | + | + | | | |
| Ergocryptinine | + | + | | | |
| Dihydroergotamine | - | - | Violet*** | | Brownish |
| Dihydroergosine | - | - | | | |
| Dihydroergocornine | - | - | | | |
| Dihydroergocristine | - | - | | | Brownish |
| Dihydroergocryptine | - | - | | | |

*Ref. 1.
**Ref. 9, 4% *p*-DMAB in concentrated hydrochloric acid.
***Ref. 70, 5% *p*-DMAB in methanol-hydrochloric acid(1:1).

TABLE 14.6

LITERATURE CITED IN CHAPTER 3 WHICH INCLUDES THE ANALYSIS OF ERGOT ALKALOIDS

| Alkaloid* | Ref. | Alkaloid* | Ref. |
|---|---|---|---|
| Ergm,Ergmine,Ergta,Ergtaine,DiHErgta, Ergct,Ergctine,diHErgct | 5 | Ergta | 87 |
| | | Ergta | 97 |
| Ergta,Ergm | 8 | Ergta,Ergct,Ergm | 117,117a |
| Ergta,Ergm,Ergtox | 21 | LSD | 104 |
| Ergta | 24 | LSD | 121,127 |
| Ergta,Ergm | 40 | LSD | 124 |
| diHErgta | 42 | | |

*Abbreviations:

| | | | |
|---|---|---|---|
| Agcl | Agroclavine | Ergta | Ergotamine |
| Ccl | Costaclavine | Ergtaine | Ergotaminine |
| Chcl | Chanoclavine | MeErgm | Methylergometrine |
| Elcl | Elymoclavine | MeMeErgm | 1-Methylmethylergometrine = methysergide |
| Fcl | Festuclavine | | |
| Fucl | Fumigaclavine | Ergst | Ergostine |
| Mcl | Molliclavine | Ergstine | Ergostinine |
| Pcl | Penniclavine | Ergs | Ergosine |
| Pycl | Pyroclavine | Ergsine | Ergosinine |
| Scl | Setoclavine | Ergtox | Ergotoxine (= Ergct + Ergcp + Ergco) |
| Lysam | Lysergic acid amide = ergine | Ergct | Ergocristine |
| Lysac | Lysergic acid | Ergctine | Ergocristinine |
| LSD | β-Lysergide | Ergcp | Ergocryptine |
| Lysg | Lysergine | Ergcpine | Ergocryptinine |
| Lysgol | Lysergol | Ergco | Ergocornine |
| Ergm | Ergometrine = ergobasine = ergonovine | Ergcoine | Ergocorninine |
| Ergmine | Ergometrinine | diH | Dihydro- |

TABLE 14.7

TLC ANALYSIS OF ERGOT ALKALOIDS IN AND FROM FUNGI

| Alkaloid* | Aim | Adsorbent | Solvent system | Ref. |
|---|---|---|---|---|
| Ergta,Ergtox,Ergm, Ergmine,Lysac,Chcl, Elcl | Separation | $SiO_2$ | $CHCl_3$-EtOH(9:1) | 1 |
| Ergta,Ergs,Ergcr, Ergco,Ergcp and their C-8 isomers | Separation | FMA-impregnated cellulose | Developed first with benzene-heptane-$CHCl_3$ (6:5:3) followed by benzene-heptane(6:5) | 2,7 |
| Ergm,Ergta,Ergct and their C-8 isomers | Indirect quantitative analysis, colorimetric | $SiO_2$ | $CHCl_3$-EtOH(9:1) | 3 |
| Ergta,Ergs,Ergct, Ergco,Ergcp,Ergm, their C-8 isomers, Chcl,Elcl,Pcl, isoPcl,Agcl,Scl, isoScl, diHErgta, diHErgct | Separation (Table 14.1) | $SiO_2$ | $CHCl_3$-EtOH(20:1,10:1,5:1)<br>EtOAc-EtOH-DMFA(85:10:5) | 4,7 |
| Scl,Agcl,Pcl,ElCl, Chcl,Ergm | Identification in Paspalum ergot | 1% KOH-impregnated $SiO_2$ | EtOAc-EtOH-DMFA(13:1:1)<br>EtOH-EtOAc(8:2) | 6 |
| Agcl,Elcl,Pcl, isoPcl,Scl,isoScl, Chcl,lysergene, lysergol,isolysergol,Fcl,Pycl, lysergine,isolysergine,Ccl,Fucl A and B,Mcl,diHElcl, Lysac,isoLysac | Identification in Pennisetum ergot (Table 14.1) | $SiO_2$ | EtOAc-EtOH-DMFA(13:1:1)<br>$CHCl_3$-MeOH-FMA(14:4:1) | 9 |
| Chcl,Elcl,Pcl,Agcl, Scl,Fcl,Pycl,Ccl, Ergm,Lysam,iso-Lysam | Separation of clavine alkaloids | 1% KOH-impregnated $SiO_2$<br>$Al_2O_3$, acidic<br>$SiO_2$+$Al_2O_3$+$CaSO_4$ (52:60:18)<br>$SiO_2$+$Al_2O_3$+starch (65:35:5) | EtOAc-EtOH-DMFA(12:1:1)<br>EtOAc-dioxane-DMFA(5:5:1)<br>EtOAc-$Me_2CO$-DMFA(5:5:1)<br>$CHCl_3$-EtOH(96:4)<br>Cyclohexane-$CHCl_3$-MeOH (5:3:2)<br>$CHCl_3$-MeOH(8:2) | 10 |
| Ergta,Ergcp,Ergct, Ergco, their C-8 isomers,Ergm,Ergs, Lysac | Indirect quantitative analysis, colorimetric | $Al_2O_3$ | $CHCl_3$-EtOH(97:3)<br>$Et_2O$-EtOH(97:3) | 12 |
| Ergm,Ergta,Ergtox | Identification | $Al_2O_3$ | Benzene-abs. EtOH(10:0.5) | 13 |
| No details available | Examination of alkaloid content in Claviceps culture | | No details available | 14 |

TABLE 14.7 (continued)

| Alkaloid* | Aim | Adsorbent | Solvent system | Ref. |
|---|---|---|---|---|
| Ergm,Ergta,Ergs, Ergct,Ergcp,Ergco and their C-8 isomers | Separation prior to indirect quantitative analysis (colorimetric) (Table 14.1) | $SiO_2$<br>$Al_2O_3$ | Benzene-DMFA(13:2)<br>EtOAc-EtOH-DMFA(13:0.1:1.9)<br>$CHCl_3$-$Et_2O$-$H_2O$(3:1:1),(87.5:12.5:25), org. phase | 20 |
| Ergm,Ergta,Ergtox | Indirect quantitative analysis (colorimetric) | $Al_2O_3$ | Benzene-$CHCl_3$-abs. EtOH (7:3:0.5) | 22 |
| See Table 14.1 | Separation of ergot alkaloids, relation between $hR_F$ and structure | $SiO_2$<br><br><br><br><br>$Al_2O_3$<br>15% FMA, 1% $NH_4OH$-impregnated cellulose | I. $CHCl_3$-MeOH(8:2)<br>II. $CHCl_3$-DEA(9:1)<br>$CHCl_3$-MeOH-conc. $NH_4OH$ (80:20:0.2)<br>$CHCl_3$-MeOH-AcOH(4:3:3)<br>$CHCl_3$-EtOH(96:4)<br>EtOAc-*n*-heptane-DMFA (250:300:1)<br>Two-dimensional: I,II | 23 |
| Ergm,Ergmine, Ergta,Ergtaine | Indirect quantitative analysis (colorimetric) | $SiO_2$ | Data not available | 25 |
| Ergta,Ergm,Ergtox | Identification after column chromatographic separation | $SiO_2$ | $CHCl_3$-MeOH-1% tartaric acid(97:2:1) | 26 |
| Ergs,Ergct,Ergcp Ergco | Indirect quantitative analysis(UV) | FMA, $NH_4OH$-impregnated $SiO_2$ | Light petroleum-EtOAc-1 *M* $NH_4OH$(65:35:1)<br>Light petroleum-benzene-1 *M* $NH_4OH$(5:5:1)<br>Benzene-cyclohexane-1 *M* $NH_4OH$(5:5:1) | 32 |
| Data not available | Indirect quantitative analysis(UV) | $SiO_2$ | Benzene-$CHCl_3$-EtOH(2:4:1) | 33 |
| Ergta,Ergm,Ergco Ergct,Ergcp | Separation | $SiO_2$ | $CHCl_3$ sat. with conc. $NH_4OH$-$Me_2CO$(72:56)<br>$CHCl_3$-EtOH(9:1) | 38 |
| Ergm,Ergta,Ergs, Ergco,Ergcp,Ergct, their C-8 isomers, Agcl,Elcl | Indirect quantitative analysis (colorimetric) (Table 14.1) | $SiO_2$ | $CHCl_3$-EtOH-$Me_2CO$(6:4:4)<br>$CHCl_3$-EtOH(9:1) | 45 |
| Ergm,Ergta,Ergs, their C-8 isomers, Ergco,Ergct,Ergcp | Quantitative analysis | $SiO_2$ | *n*-Heptane-BuOAc(1:1,65:35), sat. with FMA and $NH_4OH$ | 47 |
| Ergm,Ergmine, Ergtox,isoLysac, lysg,Elcl,Lysac, Scl,isoScl,Fcl, Agcl,Pcl,Chcl, Ergta,Ergtaine, lysergen | Indirect quantitative analysis (colorimetric) | $SiO_2$ | Toluene-BuOH, sat. with 4 *M* HCl(6:4) | 48 |
| Ergta,Ergct, Ergco,Ergcp,Ergm | Separation | $SiO_2$, $Al_2O_3$ or 20% FMA-impregnated cellulose | $CHCl_3$-EtOH-$Me_2CO$(6:4:4)<br>$CHCl_3$-EtOH(9:1)<br>$CHCl_3$-MeOH(1:1)<br>Heptane-benzene-$CHCl_3$ (25:30:15)<br>Heptane-benzene(25:30)<br>$CHCl_3$-EtOH(96:4) | 49 |

TABLE 14.7 (continued)

| Alkaloid* | Aim | Adsorbent | Solvent system | Ref. |
|---|---|---|---|---|
| Ergm,Ergta,Ergct | Identification of drugs in Pharmacopeia | $SiO_2$ | Benzene-DMFA(13:2) | 51 |
| Ergct,α-Ergcp β-Ergcp,Ergco, Ergcpine,Ergctine, Ergcoine,Ergta,Ergs | Separation of ergotoxine alkaloids | $SiO_2$ | $C_6F_6$-DMFA-abs. EtOH (13:1.9:0.1) $C_6F_5H$-DMFA-abs. EtOH (13:1.9:0.1) | 55 |
| Agcl,diHErgco, diHErgct,diHErgta, Elcl,Lysam,Ergco, Ergct,Ergctine, Ergcp,Ergm,Ergs, Ergmine,Ergta, Ergtaine,Pcl | Separation with azeotropic solvents | $SiO_2$ | $CH_2Cl_2$-MeOH(92.7:7.3) $CHCl_3$-EtOH(92.0:8.0) $CHCl_3$-MeEtCO(17.0:83.0) $Me_2CO$-cyclohexane (67.5:32.5) | 57 |
| Lysac,Ergm,MeErgm, Ergta,Ergs,Ergct, Ergcp,Ergco,Agcl, Elcl,Lysergol | Separation of clavine alkaloids | $SiO_2$ | Benzene-$CHCl_3$-EtOH(4:4:1) *n*-BuOH sat. with $H_2O$ | 58 |
| Ergta,Ergcp,Ergm | Identification of drugs in Pharmacopeia | $SiO_2$ | $CHCl_3$-MeOH(95:5), developed twice | 64 |
| Ergm,Ergta,Ergcp, Ergct | Indirect quantitative analysis | $Al_2O_3$ | $CHCl_3$-MeOH(97:3) | 65 |
| Ergta,Ergtaine | Indirect quantitative analysis (colorimetric) | $SiO_2$ | $CHCl_3$-$Me_2CO$-DEA(5:4:1) | 78 |
| Ergta,Ergm,Chcl, Agcl | Indirect quantitative analysis (colorimetric) | $SiO_2$ | $CHCl_3$-MeOH(17:3),(4:1) | 80 |
| Ergct,Ergta,Ergm, Ergs,Ergco,Ergcp | Densitometric analysis | $SiO_2$ | $CHCl_3$-MeOH(98:2),(95:5), (93:7) | 81 |
| Ergm | Determination in ergot | | No details available | 84 |
| Ergs,Ergct,Ergcp, Ergco,Ergta,Ergm, Lysac,Ergctine, and their C-8 isomers | Identification in ergot (Table 14.1) | 0.1% NaOH-impregnated $SiO_2$ | *n*-Heptane-THF-toluene-$CHCl_3$ (5:4:1:5) *n*-Heptane-THF-toluene-EtOAc (5:4:1.5:4.5) *n*-Heptane-THF-toluene (2:4:5:,2:4:1,1:4:1) THF-toluene(4:1,3:2) | |
| | | FMA-impregnated $SiO_2$ | Benzene-cyclohexane-DEA (5:2:0.01) | 89 |
| Ergta | Determination in ergot | $SiO_2$ | $CHCl_3$-trichloroethylene-EtOH(9:2:1) | 90 |
| Ergta,Ergm,Ergs, Ergst,Ergco, Ergcp(α and β), Ergct and their C-8 isomers | Complete separation | FMA, $NH_4OH$-impregnated $SiO_2$ | $(Isopr)_2O$-THF-toluene-DEA (70:15:15:0.1) $(Isopr)_2O$-toluene-anh. EtOH-DEA(75:20:5:0.1) | 95 |

TABLE 14.7 (continued)

| Alkaloid* | Aim | Adsorbent | Solvent system | Ref. |
|---|---|---|---|---|
| Ergm,Ergcp,Ergct, Ergco and their C-8 isomers | Separation of the dextrorotatory from the levorotatory alkaloids | $SiO_2$ | Abs. EtOH,benzyl alcohol, $Me_2CO$,dioxane,MeOH-AcOH (58:2)<br>Dioxane-AcOH(58:2)<br>$Me_2CO$-DEA(58:2)<br>MeEtCO-DEA(58:2)<br>Dioxane-DEA(58:2)<br>Acetophenone-DEA(58:2)<br>$Me_2CO$-$CHCl_3$-DEA(30:27:3)<br>Dichloroethane-$Me_2CO$-DEA (37:20:3)<br>$Me_2CO$-tetralin-DEA(30:27:3)<br>$Me_2CO$-benzene-DEA(20:37:3)<br>$Me_2CO$-EtOAc-DEA(30:27:3) | 98 |
| Ergta,Ergtaine, Ergct | Separation by partition chromatography | pH 3-4 impregnated $SiO_2$ | $CHCl_3$-EtOH(9:1) | 108 |
| Ergta,Ergco,Ergct, Ergcp | Indirect quantitative analysis (colorimetric), stability in organic solvents | $SiO_2$ | MeOH-$Me_2CO$-25% $NH_4OH$ (50:50:2)<br>followed by $CHCl_3$-benzene-EtOH(75:15:10) | 113 |
| Ergta,Ergs,Ergct, Ergcp,Ergco, their C-8 isomers,Ergm, Lysac | Fluorodensitometric analysis | FMA-impregnated cellulose | EtOAc-*n*-heptane-DEA (5:6:0.005) | 115 |
| Lysac,isoLysac | Fluorodensitometric analysis | $SiO_2$ | EtOH-$H_2O$(8:2) | 119 |
| Ergta,Ergco,Ergs, Ergct,Ergcp, their C-8 isomers,Lysac | Fluorodensitometric analysis | FMA, $NH_4OH$-impregnated $SiO_2$ | $(Isopr)_2O$-THF-toluene-DEA (70:20:10:0.5) | 123,131 |
| Ergta,Ergm,Ergs, Ergco,Ergct,Ergcp, their C-8 isomers, Lysac,Agcl,Elcl, Chcl | Separation | $SiO_2$ or $Al_2O_3$ | $CHCl_3$-benzene-EtOH(40:10:3)<br>$CHCl_3$-MeOH(90:5)<br>$CHCl_3$-MeOH-AcOH(90:5:0.1) | 129 |
| Ergta,Ergs,Ergco, Ergct,Ergcp,Ergm | Separation | 0.1 *M* NaOH and FMA-impregnated $SiO_2$ | $CHCl_3$-$Et_2O$(3:10), developed twice | 135 |
| Ergta,Ergct | Indirect quantitative analysis(UV) in ergot | $SiO_2$ | $CHCl_3$-EtOH(9:1) | 136 |
| Ergta,Ergm,Ergs, Ergct,Ergco, Ergcp and their C-8 isomers | Fluorodensitometric analysis | FMA-impregnated cellulose | Hexane-EtOAc-12 *M* $NH_4OH$ (280:140:1) | 137 |
| Ergta,Ergm,Ergco, Ergct,Ergcp, their C-8 isomers, diHErgco,diHErgct, diHErgcp | Indirect quantitative analysis (Table 14.1) | $SiO_2$ | $Me_2CO$-0.1 *M* $(NH_4)_2CO_3$-EtOH(32.5:67.5:1) | 138 |

*For abbreviations, see footnote to Table 14.6.

TABLE 14.8

TLC ANALYSIS OF ERGOT ALKALOIDS IN HIGHER PLANTS

| Alkaloid* | Aim | Adsorbent | Solvent system | Ref. |
|---|---|---|---|---|
| Chcl,Elcl,Pcl, Ergm,Ergmine, Lysam,isoLysam, tryptophan | Identification in Morning Glory, indirect quantitative analysis | $SiO_2$ | $CHCl_3$-MeOH(17:3) | 11 |
| Lysam,isoLysam, Elcl,lysergol | Identification in Morning Glory seeds | $SiO_2$ | $Me_2CO$-piperidine(9:1)<br>$CHCl_3$-MeOH(4:1)<br>$Me_2CO$-EtOAc-DMFA(5:5:1) | 18 |
| Chcl,Elcl,Pcl, Agcl,Ergm, Ergmine Lysam, isoLysam,Lysergol, Ergta,Ergtaine, LSD | Densitometric analysis in Morning Glory seeds (Table 14.1) | $Al_2O_3$<br><br>$SiO_2$ | $CHCl_3$-EtOH(96:4)<br>$Me_2CO$-ethylpiperidine(9:1)<br>$Me_2CO$-piperidine(9:1)<br>$Me_2CO$-ethylpiperidine(9:1)<br>$Me_2CO$-EtOAc-DMFA(5:5:1) | 27 |
| Lysam,isoLysam, Ergm,Ergmine | Identification in the leaves of *Ipomoea* | $Al_2O_3$<br>$SiO_2$ | $CHCl_3$-EtOH(96:4)<br>$CHCl_3$-MeOH(9:1)<br>$Me_2CO$-piperidine(9:1)<br>$Me_2CO$ | 122 |

*For abbreviations, see footnote to Table 14.6.

TABLE 14.9

TLC ANALYSIS OF ERGOT ALKALOIDS IN PHARMACEUTICAL PREPARATIONS

| Alkaloid* | Other compounds | Aim | Adsorbent | Solvent system | Ref. |
|---|---|---|---|---|---|
| Ergta, Ergco, Ergcp, Ergct, their C-8 isomers, their diH-derivatives, Ergs, Ergm, acidiHErgct, Lysac, diHLysac, diHLysam | | Quality control (Table 14.1) | $SiO_2$ | Benzene-$CHCl_3$-EtOH(2:4:1)<br>Heptane-$CCl_4$-pyridine (1:3:2), developed twice | 15 |
| Ergco, Ergcp, Ergct, Ergta, their diH derivatives, Ergs | | Separation of dihydroderivatives (Table 14.1) | 15% FMA-impregnated cellulose | EtOAc-*n*-heptane-DEA (5:6:0.02) | 16 |
| Ergm, Ergmine, Ergta, Ergtaine | | Indirect quantitative analysis(UV) in solutions | $SiO_2$ | Benzene-prOH-1 *M* $NH_4OH$ (100:10:2) | 17 |
| Ergm, MeErgm, MeMeErgm, LSD Ergta, diHErgta, acidiHErgta, Ergmine, Ergtaine | Tropane alkaloids, barbiturates, caffeine, phenacetin, cyclizine, dimenhydrinate, diphenylpyraline | Identification and purity control | 0.1 *M* NaOH-impregnated $SiO_2$ | $CHCl_3$-MeOH(9:1)<br>$CHCl_3$-isoprOH-25% $NH_4OH$ (45:45:10) | 28 |
| Ergta, Ergtaine, their aci compounds, Ergm, Ergmine, MeErgm | Phenobarbital, homatropine MeBr, meprobamate | Indirect quantitative analysis (fluorimetric), quality control | $SiO_2$ | $CHCl_3$-EtOH(9:1) | 29 |
| Ergta, diHErgta, Ergm, MeErgm, Ergco, Ergct, Ergctine, diHErgtox | | Identification, purity control | 1% KOH-impregnated $SiO_2$ | $CHCl_3$-EtOH(9:1)<br>$CHCl_3$-EtOH-DEA(89:10:1) | 31 |
| No details available | | Identification | 15% FMA-impregnated talc | *n*-Heptane-THF-toluene (5:4:1) | 34 |
| Ergta | Tropane alkaloids, phenobarbital | Identification in tablets | $SiO_2$ | EtOH-$CHCl_3$-10% $NH_4OH$ (80:19:1) | 36 |

TABLE 14.9 (continued)

| Alkaloid* | Other compounds | Aim | Adsorbent | Solvent system | Ref. |
|---|---|---|---|---|---|
| Ergm,MeErgm, MeMeErgm | | Separation | $SiO_2$<br><br><br>$Al_2O_3$ | $CHCl_3$-MeOH(4:1,9:1)<br>$CHCl_3$-EtOH(4:1,9:1)<br>$CHCl_3$-DEA(9:1)<br>$CHCl_3$-benzene-AcOH (34:45:10)<br>$CHCl_3$-cyclohexane-DEA (70:30:0.5)<br>$CHCl_3$-DEA(9:1)<br>$CHCl_3$-EtOH(4:1,9:1,24:1) | 41 |
| Ergta, diHErgta, Ergtaine | | Reaction chromatography | $SiO_2$ | $CHCl_3$-EtOH(9:1) | 46 |
| Ergta,Ergco, Ergct,Ergcp, their C-8 epimers and diH derivatives,Ergm | | Identification | $SiO_2$ | $Me_2CO$-benzene-light petroleum(4:1:1) | 52 |
| Ergco | | Identification | $SiO_2$ | $Me_2CO$-triethanolamine(58:2)<br>Benzyl alcohol | 60 |
| MeErgm, MeErgmine and the same for the L-series | | Separation of the diastereoisomers | $SiO_2$ or $Al_2O_3$<br><br><br><br><br><br><br><br><br><br><br><br>$SiO_2$ | $Et_2O$-$H_2O$(95:5),$Et_2O$-EtOH-1 *M* imidazole(90:5:5), $CHCl_3$-$Me_2CO$(1:1)<br>$CHCl_3$-$Me_2CO$-0.1 *M* imidazole (50:45:5)<br>$CHCl_3$-$Me_2CO$-1 *M* imidazole (50:45:5)<br>$CHCl_3$-EtOH(9:1)<br>$CHCl_3$-EtOH-0.1 *M* imidazole (90:5:5)<br>$CHCl_3$-EtOH-1 *M* imidazole (90:5:5)<br>$Et_2O$-EtOH-$H_2O$(90:5:5)<br>$Et_2O$-EtOH-0.1 *M* imidazole (90:5:5)<br>$Et_2O$-EtOH-0.75 *M* imidazole (90:5:5)<br>$CHCl_3$-$Me_2CO$-$H_2O$(50:45:5)<br>$CHCl_3$-EtOH-$H_2O$(90:5:5) | 63 |
| Ergm,Ergta | Caffeine,tropane alkaloids, meclizine, cyclizine | Fluorodensitometric analysis | $SiO_2$ | $CHCl_3$-MeOH(75:25,9:1) | 72 |
| diHErgta, diHErgcp, diHErgct, diHErgco | | Fluorodensitometric analysis | $SiO_2$ | $CHCl_3$-MeOH(96:4) | 86 |
| Ergta,Ergm | Phenacetin, caffeine,codeine,phenobarbital, mecloxamine, aminopyrine | Separation | | No details available | 99 |

TABLE 14.9 (continued)

| Alkaloid* | Other compounds | Aim | Adsorbent | Solvent system | Ref. |
|---|---|---|---|---|---|
| Lysam, Ergta, Ergs, Ergct, Ergcp, Ergco, their C-8 epimers and dihydro derivatives, Lysac, Ergst, Ergm, Ergmine | | Identification and purity control (Table 14.1) | FMA, $NH_4OH$-impregnated $SiO_2$ | $(isopr)_2O$-THF-DEA (80:20:0.2)<br>$Bu_2O$-$CH_2Cl_2$-DEA(60:20:0.2) sat. with FMA | 100 |
| Ergm, Ergmine, D-Lysac D-2-propanolamide, L-Lysac L-2-propanolamide | | Separation of diastereoisomers of Ergm, purity control | $SiO_2$ | $Me_2CO$-MeOH-TrEA(25:4:1)<br>$CHCl_3$-$Me_2CO$-MeOH-TrEA (15:10:5:1)<br>$CHCl_3$-$Me_2CO$-MeOH-conc. $NH_4OH$(4:4:2:0.1,10:10:10:0.3)<br>$CHCl_3$-MeOH-conc. $NH_4OH$ (10:5:0.1)<br>prOH-$H_2O$-conc. $NH_4OH$ (88:12:1)<br>MeOH-*n*-BuOH(3:2)<br>EtOH-$H_2O$-AcOH(5:3:2) | 103 |
| MeErgm, MeErgmine, and their L-Lysac and L-isoLysac isomers, N-(D-6-methyl-8-ergolenyl)-N N'-diethylurea and the isoergolenyl compound | | Influence of *sec*.- and *tert*.-amines on the separation | $SiO_2$ | $Et_2O$-EtOH-aq. 0.5 *M* pyridine(90:5:5)<br>$CHCl_3$-EtOH-aq. 0.75 *M* pyridine(90:5:5) | 105 |
| Ergm, Ergta, Ergct, Ergtaine, Ergctine | | Fluorodensitometric analysis | $SiO_2$ | EtOAc-EtOH-DMFA-25% $NH_4OH$ (131:1:19:0.2) | 111 |
| diHErgta, MeErgm | | Densitometric analysis | $SiO_2$ | $CHCl_3$-EtOH-$NH_4OH$(70:30:2)<br>$CH_2Cl_2$-MeOH-$NH_4OH$ (80:15:0.1) | 112 |
| Lysam, Ergta, Ergs, Ergco, Ergct, Ergcp ($\alpha$ and $\beta$), their C-8 epimers and diH derivatives, Ergm, Ergmine, LSD, MeErgm, MeMeErgm, Lysac, Ergst | | Identification and purity control (Table 14.1) | $SiO_2$<br>FMA, $NH_4OH$-impregnated $SiO_2$ | $CHCl_3$-benzene-EtOH sat. with 5% $NH_4OH$(40:20:10)<br>$CHCl_3$-DEA(9:1)<br>$(isopr)_2O$-toluene-EtOH-DEA (75:20:5:0.1)<br>$(isopr)_2O$-toluene-THF-*n*-heptane-DEA(50:15:15:20:0.1)<br>$(isopr)_2O$-THF-DEA(90:10:0.2) | 114 |

TABLE 14.9 (continued)

| Alkaloid* | Other compounds | Aim | Adsorbent | Solvent system | Ref. |
|---|---|---|---|---|---|
| diHErgct, diHErgcp, diHErgco | | Fluorodensitometric analysis | 15% FMA-impregnated cellulose | EtOAc-*n*-heptane-DEA (4:6:0.2) | 116 |
| Ergta | Caffeine | Fluorodensitometric analysis in tablets | $SiO_2$ | $CHCl_3$-benzene(1:1) developed twice,followed by $CHCl_3$-EtOH(9:1),followed by $Me_2CO$-cyclohexane-MeOH (49:49:2) | 118 |
| Ergta, Ergtaine | Atropine, scopolamine, caffeine, barbiturates | Separation on silver-impregnated $SiO_2$ | $SiO_2$ | Not specified | 120 |
| diHErgco diHErgcp diHErgct | | Separation | $SiO_2$ | $Me_2CO$-0.1 *M* $(NH_4)_2CO_3$-EtOH (32.5:67.5:1) | 134 |

*For abbreviations, see footnote to Table 14.6.

TABLE 14.10

TLC ANALYSIS OF LSD AND ERGOT ALKALOIDS IN BODY FLUIDS

| Alkaloid* | Other compounds | Aim | Adsorbent | Solvent system | Ref. |
|---|---|---|---|---|---|
| LSD | Cannabis, mescaline | Identification | $SiO_2$ | $CHCl_3$-MeOH-AcOH (47.5:47.5:5)<br>$CHCl_3$-MeOH(9:1) | 50 |
| LSD | Narcotics, barbiturates, amphetamines, tranquillizers, psychomimetics | Identification | $SiO_2$ | EtOH-dioxane-benzene-$NH_4OH$ (5:40:50:5)<br>EtOAc-MeOH-$NH_4OH$(85:10:5)<br>MeOH-*n*-BuOH-benzene-$H_2O$ (60:15:10:15)<br>EtOH-pyridine-dioxane-$H_2O$ (50:20:25:5)<br>*tert*.-AmOH-*n*-$Bu_2O$-$H_2O$ (80:7:13) | 53 |
| Ergtox, Ergta | Atropine, barbiturates | Identification after poisoning | $SiO_2$ | $CHCl_3$-EtOH-25% $NH_4OH$ (20:5:1) | 66,71 |
| BromoLSD | Morphine, mescaline, amphetamines, catecholamines, $\Delta^9$-THC | TLC of dansyl derivatives on mini-plates | Polyamide | I. $H_2O$-HCOOH(100:1.5)<br>II. Benzene-AcOH(9:1)<br>III. $H_2O$+AcOH(50:1)<br>IV. $H_2O$+EtOH-*n*-BuOH+HCOOH (150:93:4:3)<br>Two-dimensional: I+IV,I+III or I+(III+I) | 75 |
| LSD | Narcotics, barbiturates, diazepoxides, amphetamines, mescaline, quinine | Drugs of abuse screening in urine (Table 12.13, p.284) | $SiO_2$ | EtOAc-cyclohexane-*p*-dioxane-MeOH-$H_2O$-$NH_4OH$(50:50:10:10:1.5:0.5,50:50:10:10:0.5:1.5)<br>EtOAc-cyclohexane-MeOH-$H_2O$-$NH_4OH$(70:15:8:0.5:2)<br>EtOAc-cyclohexane-MeOH-$NH_4OH$(70:15:10:5)<br>EtOAc-cyclohexane-$NH_4OH$ (50:40:0.1) | 83,109 |
| LSD | Morphine, naloxone, amphetamine | TLC of dansyl derivatives on mini-plates | Polyamide | Benzene-AcOH(9:1)<br>Benzene-HCOOH(200:3)<br>$H_2O$-EtOAc-EtOH(50:25:25)<br>Toluene-AcOH-DMFA(200:10:4)<br>Heptane-*n*-BuOH-AcOH (30:40:2.5)<br>AcOH-EtOH-$H_2O$-DMFA (25:20:50:20) | 92 |

*For abbreviations, see footnote to Table 14.6.

TABLE 14.11

TLC ANALYSIS OF LSD AND RELATED SUBSTANCES IN DRUG OF ABUSE SEIZURES

| Alkaloid* | Other compounds | Aim | Adsorbent | Solvent system | Ref. |
|---|---|---|---|---|---|
| LSD,Ergm, Ergta, Ergtaine, Ergmine, Ergct, Ergctine, diHErgta, Lysac, diHErgct, Ergtox | Barbiturates, opium alkaloids,heroin, amphetamines | Identification, reaction chromatography,indirect quantitative analysis (fluorimetric) | 0.1 *M* NaOH-impregnated $SiO_2$<br>$SiO_2$ | $CHCl_3$-MeOH(9:1)<br>$CHCl_3$-MeOH-28% $NH_4OH$(4:4:2) | 19 |
| LSD,Ergta, Ergm, diHErgta, MeErgm, MeMeErgm | Quinine,quinidine,indole, tryptamine | Identification, direct and indirect quantitative analysis (fluorimetric) | $SiO_2$<br>$Al_2O_3$ | 1,1,1-Trichloroethane-MeOH (9:1)<br>1,1,1-Trichloroethane-MeOH (98:2) | 30 |
| LSD | | Identification | $Al_2O_3$ | $CHCl_3$-EtOH(96:4) | 35 |
| LSD,isoLSD | | Identification | $SiO_2$<br>$SiO_2$-$Al_2O_3$ (1:1) | $CHCl_3$-MeOH(1:4)<br>$Me_2CO$ | 39 |
| LSD,MeMeErgm | Psilocin,Psilocybin,mescaline,tryptamines | Identification | $SiO_2$ | MeOH-$NH_4OH$(100:1.5)<br>$CHCl_3$-MeOH(9:1) | 43 |
| LSD,Ergta, Ergm | Psycholeptics and hallucinogens | Identification | $SiO_2$ | Benzene-light petroleum-$Me_2CO$-$NH_4OH$(35:35:35:1) | 44 |
| LSD | Cannabis, mescaline | Identification | $SiO_2$ | $CHCl_3$-MeOH-AcOH (47.5:47.5:5)<br>$CHCl_3$-MeOH(9:1) | 50 |
| LSD,Lysac, Ergta, Ergtaine, Ergmine, Ergct,Ergtox, Ergm,MeErgm, MeMeErgm | | Identification by UV degradation products | $SiO_2$ | $CHCl_3$-$Me_2CO$(1:4) | 54 |
| Lysac,Lysam, isoLysam, Ergm,MeErgm, isoLSD, Ergmine, MeMeErgm, Ergta,diH-Ergta,isoLSD, 1-Ac-LSD, Ergct,Ergtaine | | Identification | $SiO_2$<br>0.1 *M* NaOH-impregnated $SiO_2$ | MeOH,MeOH-28% $NH_4OH$ (100:1.5)<br>$CHCl_3$-MeOH(9:1) | 56 |

TABLE 14.11 (continued)

| Alkaloid* | Other compounds | Aim | Adsorbent | Solvent system | Ref. |
|---|---|---|---|---|---|
| LSD,Lysac, Lysam,iso-Lysam,isoLSD | | Identification, indirect quantitative analysis(GLC or fluorimetry (Table 14.2) | $SiO_2$ | $CHCl_3$-MeOH-AcOH(60:40:1)<br>$Me_2CO$-$CHCl_3$-conc. $NH_4OH$ (50:40:1.5)<br>*n*-BuOH-pyridine-$H_2O$(8:1:1)<br>*n*-BuOH-EtOAc-DEA(7:2:1)<br>$CH_2Cl_2$-dioxane-pyridine (5:4:1)<br>$CH_2Cl_2$-DEA(9:1)<br>Benzene-EtOAc-DEA(7:2:1)<br>Trichloroethane-MeOH-benzyl alcohol-citric acid solution(50:25:25:1)<br>Dioxane-DMFA-AcOH(60:40:2) | |
| | | | $Al_2O_3$ or $SiO_2$ | $CHCl_3$-MeOH(9:1)<br>*n*-BuOH-$H_2O$-AcOH(85:15:5) | |
| | | | $Al_2O_3$ | $CHCl_3$-MeOH-AcOH(85:10:5)<br>$CH_2Cl_2$-MeOH-$Me_2CO$(6:3:1)<br>Trichloroethane-MeOH-$Me_2CO$ (6:1:3)<br>Trichloroethane-$Me_2CO$(1:1)<br>$CHCl_3$-$Me_2CO$-AcOH(50:50:1) | |
| | | | Polyamide | $CHCl_3$-benzene-MeOH-AcOH (60:25:10:1) | 59 |
| LSD,isoLSD, N-D-6-methyl-8-isoergolenyl-N',N'-diethylcarbamide and its diastereoisomer | | Identification | $SiO_2$ | $CHCl_3$-EtOH-0.5% imidazole (90:5:5) | 67 |
| LSD | Opium alkaloids,mescaline | Separation with azeotropic solvents | $SiO_2$ | Benzene-MeEtCO-*n*-butylamine(75:15:10)<br>IsoprOH-1,2-dichloroethane-diisopropylamine (50:40:10)<br>1-Chlorobutane-EtOH-*n*-butylamine(80:10:10) | 68 |
| LSD | Quinine | Fluorodensitometric analysis | $SiO_2$ | MeOH-$CHCl_3$-*n*-hexane(1:4:2)<br>$CHCl_3$-MeOH(4:1)<br>$Me_2CO$-$CHCl_3$(4:1) | 69 |
| LSD,isoLSD, Lysac,Lysam, 1-Ac-LSD, Ergm,Ergmine, Ergta,Ergtaine,Ergct, MeErgm, diHErgta, MeMeErgm | Psilocybin, bufotenine, psilocin, bufotenine, ibogaine, N,N-diethyl- and N,N-dimethyltryptamine | Identification (Table 14.2) | $SiO_2$ | Morpholine-toluene(1:9)(unsat. tank)<br>$CHCl_3$-MeOH(9:1) | 70 |

TABLE 14.11 (continued)

| Alkaloid* | Other compounds | Aim | Adsorbent | Solvent system | Ref. |
|---|---|---|---|---|---|
| LSD | Opium alkaloids, narcotic analgesics, amphetamines, mescaline, psilocybin, psilocin, tryptamines | Fluorodensitometric analysis | $SiO_2$ | MeOH-25% $NH_4OH$(100:1.5)<br>$CH_2Cl_2$-MeOH(93:7) | 74,77 |
| LSD | Psilocybin tryptamines, amphetamines, mescaline, phencyclidine, strychnine | Screening of street drugs | $SiO_2$ | EtOAc-*n*-prOH-28% $NH_4OH$ (40:30:3) | 79 |
| LSD | Opium alkaloids, cocaine, mescaline, amphetamines | Identification, TAS technique | $SiO_2$ | MeOH-25% $NH_4OH$(100:1.5) | 85 |
| LSD, isoLSD, Lysac, Ergm, MeErgm, MeMeErgm, diHErgta, Ergta, diHErgct, Ergct, Ergcp, Ergs, Ergco, Lysam, Ergtoxine ethanesulphonate | | Identification in illicit ergot preparations (Table 14.2) | $SiO_2$ | $Me_2CO$, $Me_2CO$-$CHCl_3$(4:1), $Me_2CO$-MeOH(4:1)<br>$CHCl_3$, $CHCl_3$-$Me_2CO$(6:1), $CHCl_3$-MeOH(4:1,9:1), MeOH-conc. $NH_4OH$(100:1.5)<br>MeOH-acetate buffer(pH 4.5) (9:1), $CHCl_3$-cyclohexane-DEA(5:5:1)<br>1,1,1-Trichloroethane-MeOH (9:1,96:4,98:2,99:1)<br>Toluene-morpholine(9:1) | |
| | | | $Al_2O_3$ | $Me_2CO$<br>1,1,1-Trichloroethane-MeOH (9:1,96:4,98:2,99:1) | |
| | | | 5% $NaH_2$ citrate-impregnated cellulose | *n*-BuOH-citric acid-$H_2O$ (870 ml+4.8 g+130 ml) | 82 |
| LSD, isoLSD, Ergm, MeErgm, Ergtaine, Ergcp, Ergct, diHErgct and the 10-OH (lumi) derivatives | | Identification, reaction chromatography (Table 14.3) | $SiO_2$<br>$Al_2O_3$ | $CHCl_3$-MeOH(9:1)<br>$CHCl_3$-$Me_2CO$-$NH_4OH$(80:20:1) | 88 |
| LSD, diMeLysam, MePropLysam, EtPropLysam, diPropLysam and their iso-compounds | | Separation (Table 14.4) | $Al_2O_3$, $SiO_2$ or 0.1 *M* NaOH-impregnated $SiO_2$ | $CHCl_3$-MeOH(9:1)<br>1,1,1-Trichloroethane-MeEtCO-MeOH(7:2:1)<br>Hexane-EtOAc-MeOH(7:13:15)<br>Toluene-nitromethane-MeOH (8:10:2)<br>Benzene-DMFA(9:1,13:2)<br>EtOAc-EtOH-DMFA(13:0.1:1.9)<br>$CHCl_3$-$Et_2O$-$H_2O$(3:1:1), org. phase<br>$CHCl_3$- sat. with $NH_4OH$-MeOH(18:1)<br>$CHCl_3$-$Me_2CO$(1:2) | 91 |

TABLE 14.11 (continued)

| Alkaloid* | Other compounds | Aim | Adsorbent | Solvent system | Ref. |
|---|---|---|---|---|---|
| LSD | Opium alkaloids, cocaine, caffeine, mescaline, psilocybin, psilocin, amphetamines, barbiturates, tryptamines | Identification | $SiO_2$ | $CHCl_3$-$Et_2O$-MeOH-25% $NH_4OH$ (75:25:5:1) | 93 |
| LSD | Opium alkaloids, cocaine, amphetamines, barbiturates, marihuana | Separation and indirect quantitative analysis (UV) | $SiO_2$ | $CHCl_3$-$Me_2CO$(9:1)<br>EtOH-MeOH-conc. $NH_4OH$ (85:10:5)<br>MeOH-conc. $NH_4OH$(100:2)<br>$CHCl_3$-dioxane-EtOAc-conc. $NH_4OH$(25:60:10:5) | 96 |
| LSD | Morphine, hashish | Indirect quantitative analysis (UV) | $SiO_2$ | $CHCl_3$-MeOH(13:1,9:1), $Me_2CO$-MeOH(10:1)<br>$CH_2Cl_2$-MeOH(10:1), $Et_2O$-MeOH(3:1)<br>$CCl_4$-*n*-BuOH-MeOH(4:3:1)<br>Light-petroleum-$Et_2O$-MeOH (4:1:1) | 101 |
| LSD, isoLSD, Lysam, iso-Lysam, Lysac, Ergm, Ergmine, MeErgm, MeErgmine, MeMeErgm, Ergta, Ergos, Ergost, Ergosinine, Ergct, Ergco, Ergcp, Ergtaine, Ergostinine, Ergctine, Ergcpine, Ergcoine, diMeLSD, diMeisoLSD, MePropLSD, MePropisoLSD, EtPropLSD, EtPropisoLSD, diPropLSD, diPropisoLSD | | Identification (Table 14.2) | $SiO_2$ | $CHCl_3$-MeOH(9:1)<br>$CHCl_3$ (sat. with $NH_4OH$)-MeOH(18:1) | 102 |
| LSD | Parkopan | Identification | | No details available | 106 |

TABLE 14.11 (continued)

| Alkaloid* | Other compounds | Aim | Adsorbent | Solvent system | Ref. |
|---|---|---|---|---|---|
| LSD | Opium and tropane alkaloids, quinine, quinidine, nicotine, ibogaine, strychnine, mescaline, barbiturates, amphetamines, analgesics | Identification with selective spray reagent | $SiO_2$ | EtOAc-MeOH-$NH_4OH$ (100:18:1.5)<br>MeOH-$NH_4OH$(100:1.5) | 110 |
| LSD | Opium, tropane alkaloids, strychnine, quinine, nicotine, analgesics | Identification | $SiO_2$ | MeOH-conc. $NH_4OH$(100:1.5) | 124 |
| LSD, Ergta, Ergm, Ergco, Ergct, Ergcp | | Identification | $SiO_2$ | Benzene-$Me_2CO$-$Et_2O$-25% $NH_4OH$(4:6:1:0.3)<br>$CHCl_3$-benzene(5:4) sat. with FMA and mixed with 10 parts of MeOH<br>Benzene-*n*-heptane-$CHCl_3$-DEA (4:2:3:1) | 128, 133 |
| LSD | Psilocin, psilocybin | Identification in psilocybe impregnated with LSD | $SiO_2$ | $CHCl_3$-MeOH(9:1) | 130 |
| LSD | | Identification | $SiO_2$ | $CHCl_3$-MeOH(9:1) | 132 |

*For abbreviations, see footnote to Table 14.6.

Chapter 15

# *PSILOCYBE* ALKALOIDS

15.1. Solvent system ........................................................... 409
15.2. Detection ................................................................ 409
15.3. TAS technique ............................................................ 413
References .................................................................... 413

The separation and identification of psilocin and psilocybin have been dealt with in connection with the analysis of drugs of abuse. Both alkaloids are present in the mushroom *Psilocybe mexicana*, which is abused because of its hallucinogenic properties. In the body the alkaloid psilocybin is rapidly converted into psilocin by dephosphorylation[11].

## 15.1. SOLVENT SYSTEMS

The separation of the two alkaloids does not offer any problems, although most solvents described gave low $hR_F$ values for psilocybin. Table 15.1 lists some of the TLC systems used for the separation of the two alkaloids.

According to Steinigen[4,5], psilocybin is partly converted into psilocin in the solvent system methanol-ammonia (100:1.5), resulting in two spots for psilocybin. For the identification of the amphoteric psilocybin several workers have used polar acidic solvent systems in combination with silica gel plates[6,9,10,12,13].

## 15.2. DETECTION

For the detection of psilocin and psilocybin mostly iodoplatinate and *p*-dimethylaminobenzaldehyde reagent have been used. With iodoplatinate reagent dark purple colours are obtained[2], whereas with *p*-dimethylaminobenzaldehyde (no. 36i) blue and purple-blue colours are obtained for psilocin and psilocybin, respectively[8]. Stahl and co-workers[12,13] exposed the TLC plates to nitric acid vapour, after spraying with *p*-dimethylaminobenzaldehyde (no. 36c); the alkaloids are observed as blue-violet spots.

References p. 413

TABLE 15.1

TLC SEPARATION OF PSILOCIN (I) AND PSILOCYBIN (II) ON SILICA GEL PLATES

Saturated tanks, activated plates.

| Solvent system | $hR_F$ value | | Ref. |
|---|---|---|---|
| | I | II | |
| Ethyl acetate - methanol - ammonia (85:10:5) | | 57 | 2 |
| Methanol - 25% ammonia (100:1.5) | 34 | 5 | 1,4,5,6 |
| Propanol - 5% ammonia (50:35) | | | 6 |
| Butanol saturated with water - acetic acid (9:1) | | | 6 |
| Butanol - acetic acid - water (2:1:1) | | | 6 |
| Ethyl acetate - propanol - conc. ammonia (40:30:3) | 33 | 41 | 9,10 |
| Ethanol - acetic acid - water (6:3:1) | 60 | 31 | 9,10 |
| *n*-Butanol - acetic acid - water (65:13:22) | 30-35 | 15 | 12,13 |

TABLE 15.2

TLC ANALYSIS OF PSILOCIN AND PSILOCYBIN

| Alkaloid | Other compounds | Aim | Adsorbent | Solvent system | Ref. |
|---|---|---|---|---|---|
| Psilocin, psilocybin | LSD, methylsergide, tryptamines, mescaline | Identification of LSD | $SiO_2$ | $MeOH-NH_4OH$(100:1.5)<br>$CHCl_3-MeOH$(9:1) | 1 |
| Psilocybin | Opium alkaloids, LSD, quinine, marihuana, cocaine, benzodiazepines, chlorpromazine mescaline | Comparison of direct solvent extraction and ion-exchange paper in extraction of drugs of abuse from urine | $SiO_2$ | $EtOH$-dioxane-benzene-$NH_4OH$(5:40:50:5)<br>$EtOAc-MeOH-NH_4OH$(85:10:5)<br>$MeOH$-*n*-$BuOH$-benzene-$H_2O$(60:15:10:15)<br>$EtOH$-pyridine-dioxane-$H_2O$(50:20:25:5)<br>*tert.*-$AmOH$-*n*-$Bu_2O-H_2O$(80:7:13) | 2 |
| Psilocin, psilocybin | Ergot alkaloids, tryptamines, ibogaine | Identification in drug seizures (Chapter 14, Table 14.2, p. 390) | $SiO_2$ | Morpholine-toluene(1:9), unsat. tank<br>$CHCl_3-MeOH$(9:1) | 3 |
| Psilocin, psilocybin | Opium alkaloids, LSD, amphetamines, tryptamines, narcotic analgesics | Identification in drug seizures | $SiO_2$ | $MeOH$-25% $NH_4OH$(100:1.5) | 4,5 |
| Psilocin, psilocybin | | Review of analysis of halucinogenic drugs | $SiO_2$ | $PrOH$-5% $NH_4OH$(50:35)<br>$MeOH-NH_4OH$(100:1.5)<br>$BuOH$ sat. with $H_2O-AcOH$(9:1)<br>$BuOH-AcOH-H_2O$(2:1:1) | 6 |
| Psilocybin | Mescaline, strychnine, amphetamines, tryptamines, LSD, phencyclidine | Screening drug seizures | $SiO_2$ | $EtOAc-PrOH$-28% $NH_4OH$(40:30:3) | 7 |
| Psilocin, psilocybin | Opium alkaloids, LSD, amphetamines, tryptamines, barbiturates, ephedrine, caffeine, cocaine, mescaline strychnine | Identification in drug seizures | $SiO_2$ | $CHCl_3-Et_2O-MeOH$-25% $NH_4OH$(75:25:5:1) | 8 |

References p. 413

TABLE 15.2 (*continued*)

| | | | | | |
|---|---|---|---|---|---|
| Psilocin, psilocybin | Opium alkaloids, various alkaloids, barbiturates, amphetamines, various other compounds | Systematic identification of drugs of abuse (Chapter 12, Table 12.14, p.285) | $SiO_2$ | $CHCl_3$-$Et_2O$-MeOH-conc. $NH_4OH$(75:25:5:1)<br>EtOAc-PrOH-conc. $NH_4OH$(40:30:3)<br>MeOH-conc. $NH_4OH$(100:1.5)<br>EtOH-AcOH-$H_2O$(6:3:1) | 9,10 |
| Psilocin, psilocybin | | Review of analysis of drugs of abuse in urine | | | 11 |
| Psilocin, psilocybin | LSD | Identification in drug seizures using TAS technique | $SiO_2$ | *n*-BuOH-AcOH-$H_2O$(65:13:22) | 12,13 |

## 15.3. TAS TECHNIQUE

Stahl and co-workers[12,13] used the TAS technique in the analysis of psilocin and psilocybin. Equal amounts of plant material and lithium hydroxide were mixed and the alkaloids were volatilized by heating at 240°C for 120 sec; molecular sieve 4Å containing 20% water was the propellant. Psilocybin was converted into psilocin in this method.

## REFERENCES


1 E.G.C. Clarke, *J. Forensic Sci. Soc.*, 7 (1967) 46.
2 S.J. Mulé, *J. Chromatogr.*, 39 (1969) 302.
3 G.V. Alliston, M.J. de Faubert Maunder and G.F. Phillips, *J. Pharm. Pharmacol.*, 23 (1971) 555.
4 M. Steinigen, *Pharm. Ztg.*, 116 (1971) 2072.
5 M. Steinigen, *Deut. Apoth.-Ztg.*, 112 (1972) 51.
6 A. Sperling, *J. Chromatogr. Sci.*, 10 (1972) 268.
7 J.K. Brown, L. Shapazian and G.D. Griffin, *J. Chromatogr.*, 64 (1972) 129.
8 R.A. van Welsum, *J. Chromatogr.*, 78 (1973) 237.
9 A.N. Masoud, *J. Pharm. Sci.*, 65 (1976) 1585.
10 A.N. Masoud, *J. Chromatogr.*, 141 (1977) D9.
11 K.K. Kaistha, *J. Chromatogr.*, 141 (1977) 145.
12 E. Stahl, J. Brombeer and D. Eskes, *Arch. Kriminol.*, 162 (1978) 23.
13 E. Stahl and J. Brombeer, *Deut. Apoth.-Ztg.*, 118 (1978) 1527.

II.6. STEROIDAL ALKALOIDS

Chapter 16

# STEROIDAL ALKALOIDS

16.1. Solvent system....................415
16.2. Detection....................418
16.3. Quantitative analysis....................422
16.4. TAS technique and reaction chromatography....................422
References....................423

## 16.1. SOLVENT SYSTEMS

*Solanum* alkaloids occur in nature as glycosides (= glyco-alkaloids). The analysis of these alkaloids therefore concerns both glycosides and aglycones. Boll[2] described TLC system S1 (Table 16.1) for the separation of the aglycones and S2 and S3 for the separation of the glycosides present in *Solanum* species. Boll and Andersen[4] used chloroform-acetone on alkaline silica gel plates for the separation of the aglycones. Paquin and Lepage[7] separated the glycosides solanine and chaconine and their aglycone solanidine and obtained the best results with TLC system S4 (Table 16.1). Schreiber et al.[8] preferred solvents S5-S10 for the separation of a series of *Solanum* alkaloid aglycones and steroid sapogenins (Fig. 16.1). Solvent system S5 was found to be the best for an overall separation, whereas systems S6-S10 could be used for further differentiation of the groups.

Adam and Schreiber[10] used aluminium oxide plates and diethyl ether as the mobile phase to separate the glyco-alkaloids and Rönsch and Schreiber[20] used silver nitrate-impregnated plates with solvent systems S11, S12 and S13 for 5α-saturated and $\Delta^5$-unsaturated steroid alkaloids and steroid sapogenins (Fig. 2). Solvent system S12 was also used in continuous-flow TLC. Solvent system S13 in combination with aluminium oxide plates impregnated with 10% silver nitrate gave more diffuse spots than solvents S11 and S12 in combination with silica gel plates impregnated with 10% silver nitrate. However, system S13 could be used for the preparative separation of soladulcidine and solasodine.

Rozumek[24] separated some aglycones on silver nitrate silica gel plates and obtained a good separation with concentrations of 4.2 - 6.3% silver nitrate and dichloromethane - methanol (95:5) (developed three times). Gradients of silver nitrate were also used. By developing gradient plates from silver nitrate-containing to silver nitrate-free silica gel good separations were obtained, whereas development

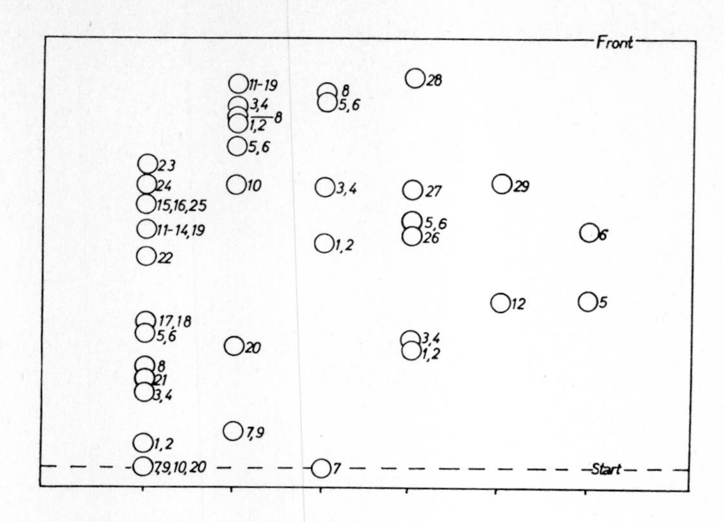


Fig. 16.1. Separation of *Solanum* alkaloids and steroid sapogenins[8].
1=solasodine
2=soladulcidine
3=tomatidenol
4=tomatidine
5=solanidine
6=demissidine
7=solanocapsine
8=solasodien
9=conessine
10=jervine
11=diosgenin
12=tigogenin
13=yamogenin
14=neotigogenin
15=smilagenin
16=sarsasapogenin
17=hecogenin
18=sisalagenin
19=9-dehydrohecogenin
20=digitogenin
21=gitogenin(?)
22=digalogenin(?)
23=cycloartenol
24=4α-methyl-5α-stigmasta-7,24(28)-dien-3-ol
25=β-sitosterol
26=5α-solasodan-3-one
27=5α-tomatidan-3-one
28=5α-solanidan-3-one
29=tigogenone
(Reproduced with kind permission of the authors).

in the opposite direction gave poorer results. Gradients perpendicular on the direction of development were also used, particularly to determine the optimum silver nitrate concentration. Continuous-flow TLC was found to give the best separations but, because of the more diffuse spots obtained in this way, the sensitivity was decreased.

Fayez and Saleh[25] added bromine to the mobile phase in order to separate the 5α-saturated and $\Delta^5$-unsaturated alkaloids on sodium acetate-impregnated silica gel plates. Solvent systems S3 and S14 were used for the identification of the alkaloidal glycosides and benzene-methanol (5:1) and chloroform-methanol (22:1.5) in combination with silica gel plates for the identification of the aglycones.

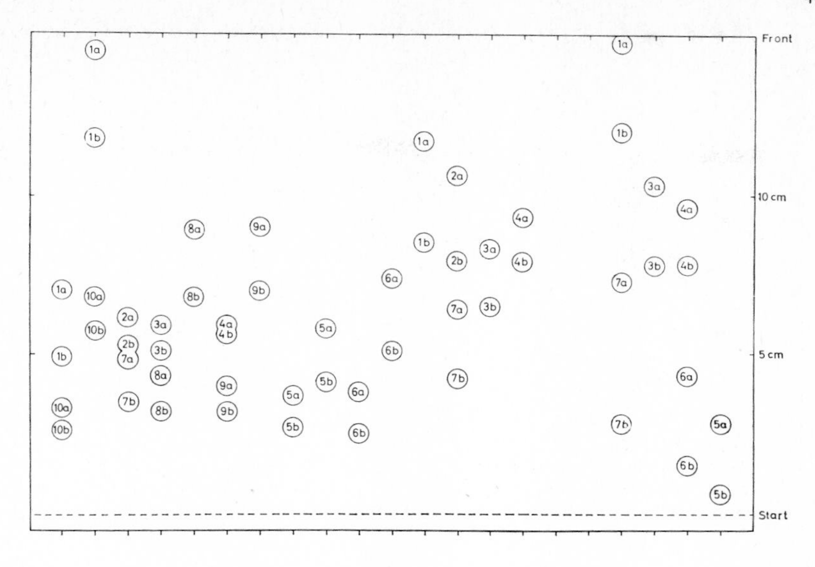


Fig. 16.2. Separation of 5α-saturated and $\Delta^5$-unsaturated steroid alkaloids on silver nitrate plates[20].

| | |
|---|---|
| 1a=demissidine | 6a=15α-hydroxytomatidine |
| 1b=solanidine | 6b=15α-hydroxytomatidenol |
| 2a=22-isodemissidine | 7a=15β-hydroxysoladulcidine |
| 2b=22-isosolanidine | 7b=15β-hydroxysolasodine |
| 3a=soladulcidine | 8a=(22*S*,25*R*)-22,26-epimino-5α-cholestan-3β,16β-diol |
| 3b=solasodine | 8b=(22*S*,25*R*)-22,26-epimino-cholest-5-en-3β,16β-diol |
| 4a=tomatidine | 9a=(22*S*,25*S*)-22,26-epimino-5α-cholestan-3β,16β-diol |
| 4b=tomatidenol | 9b=(22*S*,25*S*)-22,26-epimino-cholest-5-en-3β,16β-diol |
| 5a=15α-hydroxysoladulcidine | 10a=(22*R*,25*S*)-22,26-epimino-5α-cholestan-3β,16β-diol |
| 5b=15α-hydroxysolasodine | 10b=(22*R*,25*S*)-22,26-epimino-cholest-5-en-3β,16β-diol |

Hunter et al.[33] described the TLC analysis of 26 steroidal alkaloids with solvent systems S15-S22 on silica gel plates (Table 16.1). Both *Solanum* and *Veratrum* steroid alkaloids were analysed.

Zeitler[15] investigated the separation of *Veratrum* steroid alkaloids and found the solvent S23 in the proportions 18:2 or 14:6 to be suitable for a series of alkaloids (Table 16.2). Cyclohexane-absolute ethanol (17:3) in combination with silica gel plates was also used, but was found to be less suitable because elongated spots were formed. The plates were often developed several times in the solvents mentioned.

Labler and Cerny[6,12] used silica gel plates and diethyl ether or benzene, both saturated with concentrated ammonia, in an ammonia-saturated atmosphere to separate *Holarrhena* alkaloids. A number of steroidal alkaloids were separated (Figs. 16.3-16.7).

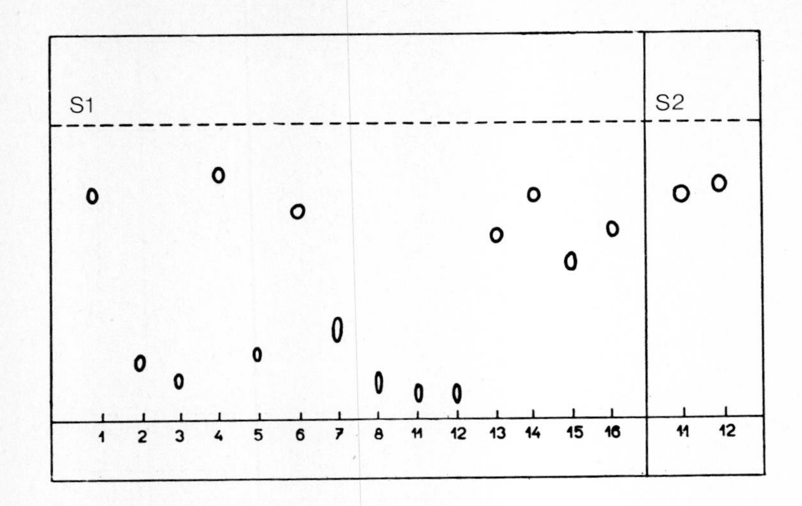


Fig. 16.3. Separation of cholestane and pregnane bases. TLC systems: S1=Silica gel G, benzene saturated with aqueous conc. ammonia; S2=silica gel G, diethyl ether saturated with aqueous conc. ammonia. Plates were developed in jars containing a dish with concentrated ammonia. Alkaloids (also for Figs. 16.4-16.7): 1=3α-dimethylaminocholest-5-ene; 2=3β-dimethylaminocholest-5-ene; 3=3β-dimethylamino-5α-cholestane; 4=3α-dimethylamino-5α-cholestane; 5=3α-dimethylamino-5β-cholestane; 6=3β-dimethylamino-5β-cholestane; 7=3α,20α-bis(dimethylamino)pregn-5-ene; 8=3β,20α-bis(dimethylamino)pregn-5-ene; 9=3α,20α-bis(dimethylamino)pregn-5-en-18-ol (tetramethylholarrhidine); 10=3β,20α-bis(dimethylamino)pregn-5-en-18-ol (tetramethylholarrhimine); 11=3β,20α-bis(dimethylamino)-5α-pregnane; 12=3β,20β-bis-(dimethylamino)-5α-pregnane; 13=3β-acetoxy-20α-dimethylamino-5α-pregnane; 14=3β-ace oxy-20β-dimethylamino-5α-pregnane; 15=20α-dimethylamino-5α-pregnane-3-one; 16=20β-c methylamino-5α-pregnan-3-one; 17=18-dimethylamino-5α-pregnan-3-one; 18=$\Delta^{N(20)}$-5α-co nanene; 19=$\Delta^{N(20)}$-5α-conanen-3β-ol; 20=N-desmethyl-5α-conanine; 21=5α-conanine; 22=3-keto-5α-conanine; 23=3β-hydroxy-N-desmethyl-5α-conanine; 24=3β-hydroxy-5α-conanine; 25=3α-dimethylamino-5α-conanine; 26=3β-dimethylamino-5α-conanine (dihydroconessine); 27=3α-dimethylaminoconan-5-ene (concuressine); 28=3β-dimethylaminoconan-5-ene (conessine); 29=3β-methylaminoconan-5-ene (isoconessimine); 30=3β-aminc conan-5-ene (conamine); 31=3β-dimethylamino-N-desmethylconan-5-ene (conessimine); 32=3β-methylamino-N-desmethylconan-5-ene (conimine).

## 16.2. DETECTION

For the detection of steroid alkaloids Dragendorff's and iodoplatinate spray reagents can be used. Because of the steroid skeleton, steroid detection reagents can also be used. Boll[2] used a 25% antimony trichloride solution in chloroform (Carr-Price reagent) (no. 70) for the detection of *Solanum* alkaloids. When heated at 110°C for 5 min the saturated compound tomatine gave a greyish violet colour, whereas the unsaturated compounds gave reddish violet colours. The colours obtained for some alkaloids with antimony trichloride reagent are summarized in Tabl 16.3[4]. The reagent can be used to distinguish saturated from unsaturated steroids.

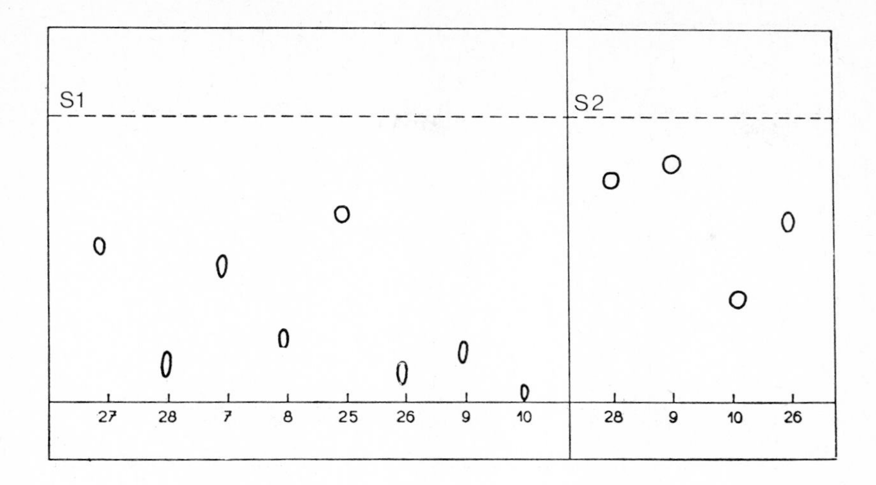


Fig. 16.4. Separation of fully methylated *Holarrhena* alkaloids. For TLC systems and alkaloids, see Fig. 16.3.

Schreiber et al.[8] used a saturated solution of cerium(IV) sulphate in 65% sulphuric acid (no. 14a) to detect steroidal compounds in *Solanum*. After heating for 15 min at 120$^{o}$C, brown to black-grey spots that showed fluorescence in UV light were observed; $\Delta^5$-unsaturated steroids could be observed without heating. According to the authors, the reagent could not be used with aluminium oxide plates or silver nitrate-impregnated plates. However, Rönsch and Schreiber found that it nevertheless could be used on silver nitrate plates without loss of sensitivity. Schreiber et al.[8] reported that *Solanum* alkaloids could be detected better with iodine (no. 51c). The non-nitrogen-containing triterpenes and steroids are less sensitive than steroidal alkaloids for this reagent. With iodine it is therefore possible to differentiate between both groups of compounds. Further, as a rather indifferent reagent, iodine can be removed after the detection has been performed, and the unchanged substances can be eluted from the plate. If silver-impregnated plates have been used they must first be sprayed with saturated potassium bromide or iodide solution and air-dried before iodine spray reagent can be used. Iodine spray reagent has also been used in other investigations[10,14,20].

According to Rönsch and Schreiber[20], the steroids are visible as white-grey spots on a darker background after spraying with 10% potassium bromide solution on silver-impregnated plates. For the detection of 3β-hydroxy-$\Delta^5$-compounds, Schreiber et al.[8] used Clarke's reagent (no. 44a), a solution of paraformaldehyde in phosphoric acid. About 10 min after spraying, the unsaturated alkaloids give a red-violet colour

**References p. 423**

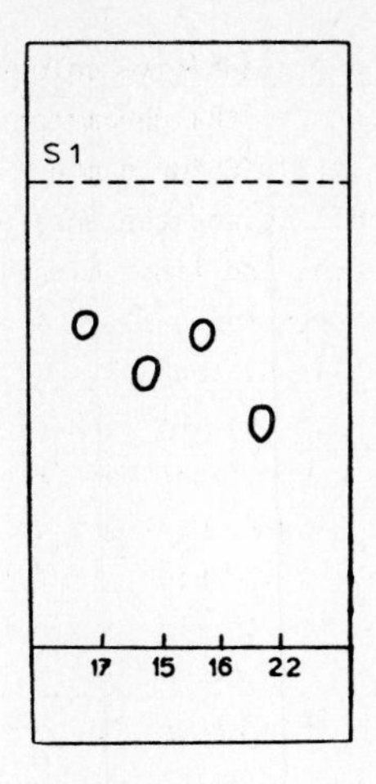


Fig. 16.5. Separation of pregnane bases substituted in positions 18 and 20. For TLC systems and alkaloids, see Fig. 16.3.

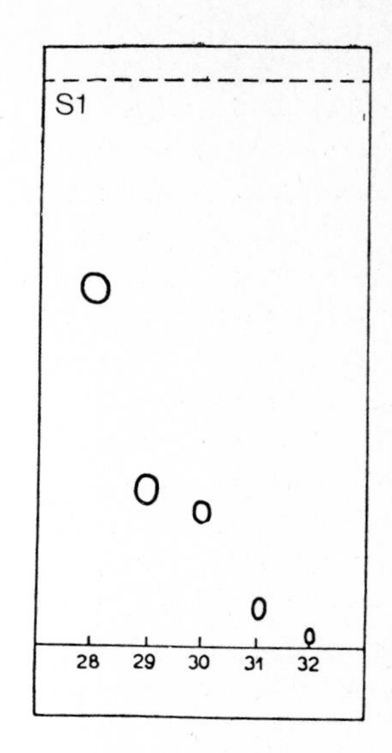


Fig. 16.7. Separation of conessine and related desmethyl bases. For TLC systems and alkaloids, see Fig. 16.3.

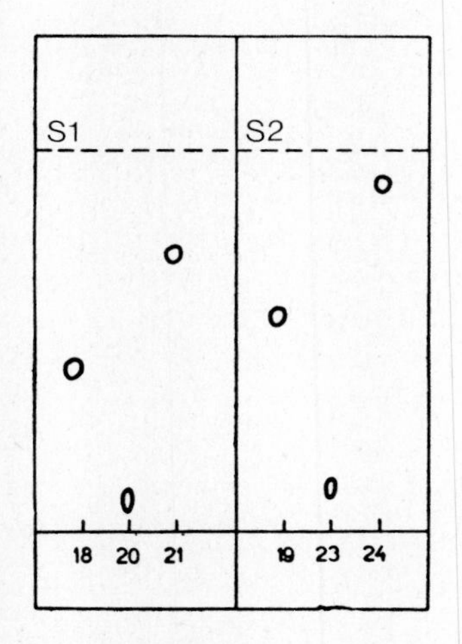


Fig. 16.6. Separation of pregnane bases containing pyrroline and pyrrolidine rings. For TLC systems and alkaloids, see Fig. 16.3.

(Reproduced with the kind permission of the authors and the Publishing House of the Czechoclovak Academy of Sciences).

and unsaturated sapogenins a blue colour. Saturated compounds give colour reactions if they are converted into 3-keto compounds prior to the chromatographic separation. Sachse and Bachmann[26] heated the plates at 110°C for 5 min after spraying with 0.1% paraformaldehyde in 85% phosphoric acid and the unsaturated compounds were observed as grey-violet spots in daylight and light blue fluorescent spots in UV light. Heating for longer or at a higher temperature led to brown-black spots in daylight and orange fluorescence in UV light.

Marquis reagent (formaldehyde in sulphuric acid) (no. 45) was used by Paquin and Lepage[7] for the detection of solanine, chaconine and solanidine. The purple-violet colour of the spots disappears very rapidly, however. Heftmann et al.[18] used 50% sulphuric acid for the detection of a number of steroids, including some alkaloids. The initial colours, colours after heating for 10 min on a hot-plate at 78°C and colours in UV light are summarized in Table 16.4.

Hunter et al.[33] used the 50% sulphuric acid as a spray reagent to detect a number of *Solanum* and *Veratrum* steroid alkaloids. After spraying, the plates were observed in daylight and UV light (366 nm) whilst lying on a hot-plate (80°C). The time of the appearance of a response was noted, in addition to the colour. The detection limit could be decreased by subsequently heating at 300°C until charring occurred. Table 16.5 summarizes the results.

Rozumek[24] found anisaldehyde in sulphuric acid containing 20% phosphomolybdic acid to be a very sensitive reagent (no. 5), particularly in combination with silver nitrate-impregnated plates (detection limit 0.2 μg). Fayez and Saleh[25] used 1% *p*-anisaldehyde in acetic acid containing 2% sulphuric acid (no. 4), and also Clarke's reagent (no. 44b) and sulphuric acid and antimony trichloride in acetic acid-chloroform (no. 7a). The sprayed plates were heated at 110°C for 2-3 min and the colours observed in daylight and UV light. The colours obtained for saturated and unsaturated compounds are listed in Table 16.6.

Borisov et al.[35] used 1% vanillin in concentrated sulphuric acid (no. 104) for the detection of *Solanum* glyco-alkaloids. Malaiyandi et al.[32] proposed ammonium metavanadate reagent (no. 103) for the detection of various alkaloids, including solanine, which gave a pink colour. Menn and McBain[19] developed a detection method for cholinesterase inhibitors, including solanine and solanidine (no. 19) (see Chapter 2, p.15). For the detection of *Veratrum* alkaloids, Zeitler[15] used 25% trichloroacetic acid in chloroform (no. 101). After heating for 10-15 min at 120°C the alkaloids showed fluorescence in UV light (366 nm): veratrine, cevadine and cevacine blue-green, veracevine and cevine yellow and the anhydro derivatives blue fluorescence.

Rücker and Taha[37] used π-acceptors for the detection of various alkaloids, including veratrine (see Chapter 2, Table 2.2, p.16). Jellema et al.[39] used optical brighteners for the detection of *Solanum* steroid alkaloids. Calcafluor M2R New (no. 10) was found to be ten times more sensitive than Dragendorff's reagent in the detection of the alkaloids. The alkaloids were observed as light blue spots against a blue background in longwave UV light (366 nm).

## 16.3. QUANTITATIVE ANALYSIS

Paquin and Lepage[7] studied the recovery of *Solanum* alkaloids from silica gel plates. With the solvents tested (1% sulphuric acid, 8% hydrochloric acid, 5% acetic acid, 95% ethanol and diethyl ether) recoveries of only 40-50% were found. Guseva et al.[14] used ethanol (or methanol)-acetic acid-water (4:1:5) to elute the *Solanum* glyco-alkaloids from the plates and determined them afterwards by spectrophotometry (Marquis reagent). Valovich and Nad[17] used chloroform to elute the less polar alkaloids from aluminium oxide plates. The glyco-alkaloid spots were collected and eluted by heating with 6% hydrochloric acid. To recover the alkaloids from silver nitrate-impregnated plates Rönsch and Schreiber[20] shook the collected silica gel for 2 h with 10% potassium cyanide solution and 2 *M* sodium hydroxide solution (8 and 2 ml, respectively, for one zone scraped off a 13x25 cm plate, prepared with 15 g of silica gel and 42 ml of 5% silver nitrate solution or 20 g of aluminium oxide and 30 ml of 5% silver nitrate solution).

Fayez and Saleh[25] carried out the direct quantitative analysis of *Solanum* alkaloids, after spraying with a 5% solution of phosphomolybdic acid in ethanol containing 4 ml of sulphuric acid per 100 ml. After heating at 110$^{o}$C for 10 min the spots were directly measured spectrophotometrically after removal of all sorbent except the spot to be measured.

Cadle et al.[38] determined glyco-alkaloids in potatoes by densitometry after spraying the TLC plates with a saturated solution of antimony(III) chloride in chloroform and subsequently heating at 150$^{o}$C for 4 min.

Massa et al.[27] reported the densitometric analysis (at 400 nm) of a series of alkaloids, including veratrine, after spraying with Dragendorff's reagent.

## 16.4. TAS TECHNIQUE AND REACTION CHROMATOGRAPHY

Jolliffe and Shellard[28] described the use of the TAS technique for the identification of alkaloid-containing crude drugs, including *Veratrum*. The following conditions were used: heating at 275$^{o}$C for 90 sec, 10-20 mg sample mixed with 10 mg of calcium hydroxide, with silica gel of suitable moisture content as propellant.

Reaction chromatography of veratrine was described by Wilk and Brill[22]. After spotting the alkaloid on the TLC plate, it is placed in iodine vapour for 18 h. When the plate is developed a characteristic pattern of spots was observed.

REFERENCES


1 S. Hermanek, V. Schwarz and Z. Cekan, *Collect. Czech. Chem. Commun.*, 26 (1961) 1669.
2 P.M. Boll, *Acta Chem. Scand.*, 16 (1962) 1819.
3 E. Vidic and J. Schütte, *Arch. Pharm. (Weinheim)*, 295 (1962) 342.
4 P.M. Boll and B. Andersen, *Planta Med.*, 10 (1962) 421.
5 R.R. Paris and M. Paris, *Bull. Soc. Chim. Fr.*, (1963) 1597.
6 L. Labler and V. Cerny, *Collect. Czech. Chem. Commun.*, 28 (1963) 2932.
7 R. Paquin and M. Lepage, *J. Chromatogr.*, 12 (1963) 57.
8 K. Schreiber, O. Aurich and G. Osske, *J. Chromatogr.*, 12 (1963) 63.
9 W. Kamp, W.J.M. Onderberg and W.A. van Seters, *Pharm. Weekbl.*, 98 (1963) 993.
10 G. Adam and K. Schreiber, *Z. Chem.*, 3 (1963) 100.
11 V. Schwarz and M. Sarsunova, *Pharmazie*, 19 (1964) 267.
12 L. Labler and V. Cerny, in G.B. Marini-Bettòlo (Editor), *Thin-Layer Chromatography*, Elsevier, Amsterdam, 1964, p. 144.
13 R. Paris, R. Rousselet, M. Paris and M.J. Fries, *Ann. Pharm. Fr.*, 23 (1965) 473.
14 A.R. Guseva, V.A. Paseshnichenko, M.G. Borikhina and R.K. Moiseev, *Biokhimiya*, 30 (1965) 260; *C.A.*, 63 (1965) 1656b.
15 H.J. Zeitler, *J. Chromatogr.*, 18 (1965) 180.
16 A. Haznagy, K. Szendrei and L. Toth, *Pharmazie*, 20 (1965) 651.
17 N.A. Valovich and L. Nad, *Med. Prom. SSSR*, 20 (1966) 44; *C.A.*, 64 (1966) 20168d.
18 E. Heftmann, S.T. Ko and R.D. Bennett, *J. Chromatogr.*, 21 (1966) 490.
19 J.J. Menn and J.B. McBain, *Nature (London)*, 209 (1966) 1351.
20 H. Rönsch and K. Schreiber, *J. Chromatogr.*, 30 (1967) 149.
21 A. Noirfalise and G. Mees, *J. Chromatogr.*, 31 (1967) 594.
22 M. Wilk and U. Brill, *Arch. Pharm. (Weinheim)*, 301 (1968) 282.
23 E. Röder, E. Mutschler and H. Rochelmeyer, *Arch. Pharm. (Weinheim)*, 301 (1968) 624.
24 K.E. Rozumek, *J. Chromatogr.*, 40 (1969) 97.
25 M.B.E. Fayez and A.A. Saleh, *Z. Anal. Chem.*, 246 (1969) 380.
26 J. Sachse and F. Bachmann, *Z. Lebensm. Unters.-Forsch.*, 141 (1969) 262.
27 V. Massa, F. Gal and P. Susplugas, *Int. Symp. Chromatogr. Electrophor. Lect. Pap. 6th*, (1970) 470.
28 G.H. Jolliffe and E.J. Shellard, *J. Chromatogr.*, 48 (1970) 125.
29 P. Than and P. Po, *Union Burma J. Sci. Technol.*, 3 (1970) 271; *C.A.*, 78 (1973) 102054y.
30 S. Zadeczky, D. Küttel and M. Takacsi, *Acta Pharm. Hung.*, 42 (1972) 7.
31 E. Novakova and J. Vecerkova, *Cesk. Farm.*, 22 (1973) 347.
32 M. Malaiyandi, J.P. Barette and M. Lanonette, *J. Chromatogr.*, 101 (1974) 155.
33 I.R. Hunter, M.K. Walden, J.R. Wagner and E. Heftmann, *J. Chromatogr.*, 118 (1976) 259.
34 E. Heftmann, Chromatography of Steroids. *Journal of Chromatography Library*, Vol. 8, Elsevier, Amsterdam, 1976, p. 117.
35 V.N. Borisov, L.A. Pikova, G.L. Zachepilova and A.I. Bankovskii, *Khim.-Farm. Zh.*, 10 (1976) 116; *C.A.*, 86 (1977) 127147j.
36 N.S. De Siqueira and A. Macan, *Trib. Farm.*, 44 (1976) 101; *C.A.*, 89 (1978) 39412y.
37 G. Rücker and A. Taha, *J. Chromatogr.*, 132 (1977) 165.
38 L.S. Cadle, D.A. Stelzig, K.L. Harper and R.J. Young, *J. Agr. Food Chem.*, 26 (1978) 1453.
39 R. Jellema, E.T. Elema and Th.M. Malingré, *J. Chromatogr.*, 176 (1979) 435.

TABLE 16.1

TLC SEPARATION OF *SOLANUM* ALKALOIDS (INCLUDING SOME *VERATRUM* STEROID ALKALOIDS[34])

TLC systems :

| System | Layer | Solvent |
|---|---|---|
| S1 | Silica gel G | Chloroform-methanol (19:1)[2] |
| S2 | Silica gel G | Ethyl acetate-pyridine-water (3:1:3), top layer[2] |
| S3 | Silica gel G | Chloroform-ethanol-1% ammonia (2:2:1), bottom layer[2] |
| S4 | Silica gel G | 95% Ethanol-acetic acid (3:1)[7] |
| S5 | Silica gel G, non-activated | Cyclohexane-ethyl acetate (1:1)[8] |
| S6 | Silica gel G, non-activated | Chloroform-methanol (6:4)[8] |
| S7 | Aluminium oxide G, activated | Cyclohexane-ethyl acetate (1:1)[8] |
| S8 | Aluminium oxide G, activated | *n*-Hexane-triethylamine (15:1)[8] |
| S9 | Silica gel G, non-activated | Light petroleum (b.p. 80-90°C)-benzene-ethyl acetate (85:5:10)[8] |
| S10 | Silica gel G, 10% silver nitrate-impregnated (spraying) and activated | Ethyl acetate-cyclohexane-96% ethanol (85:5:10)[8] |
| S11 | Silica gel G, 15% silver nitrate-impregnated and activated | Chloroform-methanol (9:1)[20] |
| S12 | Silica gel G, 10% silver nitrate-impregnated and activated | Chloroform-methanol (95:5)[20] |
| S13 | Aluminium oxide G, 10% silver nitrate-impregnated and activated | Chloroform-methanol (95:5)[20] |
| S14 | Silica gel G, activated | Benzene-methanol (10:7.5)[25] |
| S15 | Silica gel G, pre-coated | *n*-Hexane-ethyl acetate (1:1)[33] |
| S16 | Silica gel G, pre-coated | *n*-Hexane-ethanol (1:1)[33] |
| S17 | Silica gel G, pre-coated | Dichloromethane-methanol (23:2)[33] |
| S18 | Silica gel G, pre-coated | Dichloromethane-methanol (9:1)[33] |
| S19 | Silica gel G, pre-coated | Dichloromethane-acetone (4:1)[33] |
| S20 | Silica gel G, pre-coated | *n*-Hexane-acetone (1:1)[33] |
| S21 | Silica gel G, pre-coated | Dichloromethane-methanol-acetic acid (85:13:2)[33] |
| S22 | Silica gel G, pre-coated | Dichloromethane-methanol-ammonia (100:100:1)[33] |

| Alkaloid | $hR_F$ values | | | | | | | | | | | | |
|---|---|---|---|---|---|---|---|---|---|---|---|---|---|
| | S1 | S2 | S3 | S4 | S14 | S15 | S16 | S17 | S18 | S19 | S20 | S21 | S22 |
| $\Delta^{3,5}$-Tomatidiene | 81 | | | | | | | | | | | | |
| $\Delta^5$-Tomatidenol(3β) | 51 | | | | | 47 | 92 | 46 | 66 | 40 | 89 | 73 | 96 |
| Tomatidiene | 51 | | | | | 47 | 90 | 49 | 66 | 37 | 87 | 73 | 94 |
| Solasodine | 36 | | | | | 27 | 85 | 41 | 59 | 23 | 82 | 70 | 96 |
| 5α-Solasodanol(3β) | 36 | | | | | | | | | | | | |
| Solanidine | 32 | | | 62 | | 59 | 83 | 42 | 61 | 15 | 95 | 81 | 89 |
| 5α-Solanidanol(3β) | 32 | | | | | | | | | | | | |
| Solanine | | 22 | 26 | 22 | 20 | | | | | | | | |
| α-Chaconine | | 35 | 36 | 54 | 27 | | | | | | | | |
| Solasonine | | 24 | 33 | | 25 | | | | | | | | |
| Solamargine | | 45 | 55 | | 40 | | | | | | | | |
| Tomatine | | 29 | 29 | | 45 | | | | | | | | |
| Soladulcamarine | | 22 | 25 | | | | | | | | | | |
| α-Solamarine | | 32 | 33 | | | | | | | | | | |
| β-Solamarine | | 46 | 55 | | | | | | | | | | |
| γ-Solamarine | | 60 | 78 | | | | | | | | | | |
| Soladulcidine | | | | | | 26 | 87 | 40 | 53 | 23 | 85 | 71 | 97 |
| Veramine | | | | | | 10 | 75 | 27 | 35 | 9 | 60 | 68 | 93 |

TABLE 16.1. (*continued*)

| Alkaloid | $hR_F$ values | | | | | | | | | | | | |
|---|---|---|---|---|---|---|---|---|---|---|---|---|---|
| | S1 | S2 | S3 | S4 | S14 | S15 | S16 | S17 | S18 | S19 | S20 | S21 | S22 |
| 9α-Hydroxytomatidene | | | | | | 8 | 86 | 27 | 37 | 7 | 60 | 45 | 91 |
| 7α-Hydroxytomatidene | | | | | | 4 | 88 | 20 | 35 | 5 | 53 | 37 | 93 |
| 7α,11α-Dihydroxytomatidine | | | | | | 0 | 69 | 5 | 6 | 0 | 10 | 10 | 93 |
| 9α,11α-Dihydroxytomatidine | | | | | | 0 | 69 | 5 | 6 | 0 | 10 | 10 | 93 |
| Demissidine | | | | | | 58 | 85 | 21 | 33 | 19 | 87 | 57 | 97 |
| 5β-Solanidan-3-one | | | | | | 75 | 85 | 61 | 50 | 43 | 0 | 91 | 100 |
| Solanid-4-en-3-one | | | | | | 63 | 83 | 48 | 50 | 39 | 0 | 71 | 89 |
| Solanocapsine | | | | | | 0 | 2 | 0 | 0 | 0 | 0 | 5 | 12 |
| Tomatillidine | | | | | | 69 | 91 | 61 | 69 | 82 | 84 | 92 | 97 |
| Etioline | | | | | | 1 | 47 | 12 | 14 | 3 | 9 | 47 | 77 |
| Rubijervine | | | | | | 15 | 73 | 17 | 13 | 7 | 56 | 55 | 82 |
| Isorubijervine | | | | | | 19 | 81 | 27 | 33 | 11 | 60 | 43 | 91 |
| Veralobine | | | | | | 11 | 75 | 43 | 43 | 17 | 64 | 55 | 100 |
| Verarine | | | | | | 3 | 60 | 22 | 17 | 3 | 19 | 64 | 73 |
| Cyclopamine | | | | | | 3 | 46 | 25 | 21 | 5 | 17 | 57 | 81 |
| Jervine | | | | | | 0 | 35 | 26 | 19 | 3 | 14 | 53 | 83 |
| Veratramine | | | | | | 5 | 76 | 28 | 17 | 11 | 36 | 37 | 91 |
| Veramine | | | | | | 3 | 29 | 12 | 8 | 5 | 17 | 27 | 60 |
| Verazine | | | | | | 69 | 92 | 59 | 64 | 83 | 84 | 92 | 97 |
| Veralkamine | | | | | | 3 | 52 | 19 | 14 | 5 | 22 | 60 | 73 |

For separations with systems S5-S13, see Figs. 16.1 and 16.2.

TABLE 16.2

TLC SEPARATION OF *VERATRUM* ALKALOIDS[15]

TLC system:
S23 Silica gel $HF_{254}$, activated    Cyclohexane-diethylamine (9:1)

| Alkaloid | $hR_F$ values | Alkaloid | $hR_F$ values |
|---|---|---|---|
| Veratridine | 7 | Cevacine | 19 |
| Cevadine | 25 | Anhydrocevagenine triacetate | 24 |
| Veracevine | 0 | Anhydrocevadine triacetate | 45 |
| Cevagenine | 0 | Anhydroveracevine tetraacetate | 35 |
| Cevine | 3 | | |

TABLE 16.3

COLOUR REACTIONS OF NON-GLYCOSIDIC STEROID ALKALOIDS AND SAPOGENINS WITH ANTIMONY TRICHLORIDE REAGENT (no. 7c)[4]

| Compound | Colour with antimony trichloride | |
|---|---|---|
| | At 20°C | 10 min at 105°C |
| Solasodine | Red | Reddish violet |
| 5α-Solasodanol-(3β) | No colour | Greyish red |
| $\Delta^5$-Tomatidenol-(3β) | Red | Reddish violet |
| Tomatidine | No colour | Greyish red |
| $\Delta^{3,5}$- Tomatidiene | Red | Reddish violet |
| Diosgenin | Red | Reddish violet |
| Tigogenin | No colour | Yellow |
| Yamogenin | Red | Reddish violet |

TABLE 16.4

RESPONSES OF SOME STEROID ALKALOIDS TO 50% SULPHURIC ACID[18]

Abbreviations: BE = blue; BT = bright; DK = dark; GN = green; GY = grey; LC = lilac; LT = light; OE = olive; PE = pale; PK = pink; PU = purple; RE = rose; VY = very; YW = yellow; NIL = no response observed under experimental conditions.

| Alkaloid | Colour in daylight | | | Colour in UV |
|---|---|---|---|---|
| | Visible after (min) | Initial colour | Final colour* | (366 nm) |
| Holaphyllamine hydrochloride | 2.00 | PE-YW | YW-GY | VY-PE-BE |
| Holamine | 1.75 | PE-PK | LT-BR | DL-RD |
| Holaphylline | 1.50 | PE-OE | LT-BN | BE |
| Solasodine | 0.75 | PK | PU | BE |
| Tomatidine | 0.50 | PK | GN | BT-BE |
| $\Delta^5$-Tomatiden-3β-ol | 0.50 | PK | PU | BE |
| Solanidine | 0.50 | DK-PK | RD | BT-BE |

*After 10 min on a hot-plate at 78°C.

TABLE 16.5

RESPONSES OF STEROID ALKALOIDS TO 50% SULPHURIC ACID[33]

| Alkaloid | Daylight | | | | | UV light (366 nm) | | | | |
|---|---|---|---|---|---|---|---|---|---|---|
| | Starting | | Final colour* | MDA** (μg) | | Starting | | Final colour* | MDA** (μg) | |
| | Time (sec) | Colour* | | 80°C | 300°C | Time (sec) | Colour* | | 80°C | 300°C |
| Tomatidine | 130 | PK | BE-GN | 0.20 | 0.10 | 0 | BT-BE | GN | 0.04 | 0.04 |
| Tomatidenol | 86 | PK | RE | 0.25 | 0.10 | 76 | LT-BE | BT-BE | 0.05 | 0.05 |
| Soladulcidine | 300 | LT-GN | BE-GN | 1.00 | 0.25 | 210 | BE-GN | GN | 0.05 | 0.05 |
| Solasodine | 110 | PK | PU | 0.25 | 0.10 | 111 | LT-BE | BT-BE | 0.03 | 0.03 |
| Veramine | 100 | GY | GY | 0.85 | 0.10 | 23 | PK | TAN | 0.07 | 0.05 |
| 9α-Hydroxytomatidene | 250 | PK | GY | 0.90 | 0.25 | 131 | GY | TAN | 0.08 | 0.08 |
| 7α-Hydroxytomatidene | 80 | BE | BE-GY | 1.00 | 0.25 | 96 | PK-BE | GY-GN | 0.05 | 0.05 |
| 7α,11α-Dihydroxytomatidine | 270 | LT-BE | BE | 8.00 | 0.50 | 322 | PK-BE | PK | 0.30 | 0.30 |
| 9α,11α-Dihydroxytomatidine | NIL | NIL | NIL | 1.00 | 0.20 | 322 | PK-BE | PK | 0.20 | 0.10 |
| Demissidine | 132 | PK | PE-LC | 40.0 | 1.00 | 118 | BE | BE | 0.25 | 0.25 |
| Solanidine | 74 | RE | DK-RE | 0.25 | 0.10 | 0 | LT-BE | LT-BE | 0.10 | 0.10 |
| 5β-Solanidan-3-one | NIL | NIL | NIL | 14.0 | 0.25 | NIL | NIL | NIL | 2.00 | 0.05 |
| Solanid-4-en-3-one | 437 | LT-PK | NIL | 80.0 | 0.25 | 300 | BE-GY | GY-GN | 10.0 | 0.07 |
| Rubijervine | 60 | TAN | LC | 0.50 | 0.25 | 119 | DK-BE | PK | 0.06 | 0.04 |
| Isorubijervine | 53 | RE | LC | 0.90 | 0.35 | 89 | LT-BE | LT-BE | 0.08 | 0.06 |
| Veralobine | NIL | NIL | NIL | 80.0 | 0.25 | 360 | VY-LT-BE | VY-PE-BE | 2.00 | 0.07 |
| Verarine | 166 | YW | OE | 0.15 | 0.10 | 106 | YW | BT-YW | 0.02 | 0.02 |
| Cyclopamine | NIL | GY | GY | 0.25 | 0.10 | 0 | PK-BE | TAN | 0.05 | 0.10 |
| Jervine | 330 | LT-YW | GY | 0.50 | 0.30 | 151 | BE | BT-GN | 0.05 | 0.05 |
| Veratramine | 139 | YW | LT-OE | 0.40 | 0.01 | 59 | LT-BE | BT-YW | 0.02 | 0.02 |
| Veramarine | 70 | PK | OE | 1.20 | 0.07 | 64 | YW | BT-BE | 0.02 | 0.02 |
| Solanocapsine | NIL | NIL | NIL | 4.00 | 0.01 | NIL | NIL | NIL | 0.10 | 0.01 |
| Verazine | 120 | PK | LC | 2.00 | 0.90 | 86 | TAN | LT-LC | 0.40 | 0.07 |
| Tomatillidine | 80 | PK | LC | 0.20 | 0.10 | 92 | TAN | LT-LC | 0.10 | 0.05 |
| Etioline | 60 | PK | LC | 0.30 | 0.20 | 45 | LT-BE | BT-BE | 0.05 | 0.05 |
| Veralkamine | 118 | LC | PU-GY | 0.35 | 0.25 | 64 | TAN | TAN | 0.04 | 0.04 |

*For abbreviations, see Table 16.4.
**MDA = minimum detectable amount.

TABLE 16.6

DETECTION OF STEROID ALKALOIDS[25]

| Steroid alkaloids | 10% $H_2SO_4$ | | $SbCl_3$ (no. 7a) | | Clarke's (no. 44b) | | *p*-Anisaldehyde (no. 4) | |
|---|---|---|---|---|---|---|---|---|
| | Daylight | UV | Daylight | UV | Daylight | UV | Daylight | UV |
| 5α-Saturated | - | | - | Violet | - | Violet | Light blue | Blue |
| 5,6-Unsaturated | Pink | Violet | Pink | Yellow | Pink →dark blue | Yellow | Pink →dark blue | Yellow |

The change in colour occurs during heating of the plates.

TABLE 16.7

LITERATURE CITED IN CHAPTER 3 WHICH INCLUDES THE ANALYSIS OF STEROID ALKALOIDS

| Alkaloid | Ref. | Alkaloid | Ref. |
|---|---|---|---|
| V | 3 | V | 23 |
| ProtoV | 11 | Jerv,ProtoV A and B | 31 |
| Con | 13 | V | 31 |
| V | 21 | | |

Abbreviations:

| | | | |
|---|---|---|---|
| anhcevdtriOAc | Anhydrocevadine triacetate | sol | Solanine |
| anhcevgtriOAc | Anhydrocevagenine triacetate | solad | Solanidine |
| anhvctetraOAc | Anhydroveracevine tetraacetate | sold | Solasodine |
| cev | Cevine | soldanol | 5α-Solasodanol |
| cevc | Cevacine | solddiene | Solasodiene |
| cevg | Cevagenine | soldul | Soladulcamarine |
| cevd | Cevadine | soldulci | Soladulcidine |
| chac | Chaconine | solcap | Solacapsine |
| con | Conessine | soldone | 5α-Solasodan-3-one |
| dem | Demissidine | solm | Solamarine |
| hol | Holarrhenine | solmg | Solmargine |
| αOHsoldulci | 15α-Hydroxysoladulcidine | solol | 5α-Solanidanol |
| βOHsoldulci | 15β-Hydroxysoladulcidine | sols | Solasonine |
| αOHsold | 15α-Hydroxysolasodine | solsone | 5α-Solasodan-3-one |
| βOHsold | 15β-Hydroxysolasodine | Tdenol | $\Delta^5$-Tomatidenol |
| αOHtdenol | 15α-Hydroxytomatidenol | Tdiene | $\Delta^{3,5}$-Tomatidiene |
| αOHtomd | 15α-Hydroxytomatidine | tom | Tomatine |
| isodem | 22-Isodemissidine | tomd | Tomatidine |
| isosolad | 22-Isosolanidine | tomdone | 5α-Tomatidan-3-one |
| jerv | Jervine | V | Veratrine |
| protov | Protoveratrine | vtd | Veratridine |
| | | vc | Veracevine |

TABLE 16.8

TLC ANALYSIS OF *SOLANUM* ALKALOIDS IN AND FROM PLANT MATERIALS

| Alkaloid* | Aim | Adsorbent | Solvent system | Ref. |
|---|---|---|---|---|
| Tdenol,Tdiene,sold, sol,chac,sols,tom, α-, β- and γ-solm, soldul,tomd,soldanol, solol | Isolation from *Solanum* species (Table 16.1) | $SiO_2$ | EtOAc-pyridine-$H_2O$(3:1:3) (top layer)<br>$CHCl_3$-EtOH-1% $NH_4OH$(2:2:1) (bottom layer)<br>$CHCl_3$-MeOH(19:1) | 2 |
| sold,soldanol,tdenol, tomd,tdiene,tom,sols, solmg, α-, β- and γ-solm | Comparison of *Solanum* species | $SiO_2$<br>Alkaline $SiO_2$ | EtOAc-pyridine-$H_2O$(3:1:3) (top layer)<br>$CHCl_3$-MeOH(19:1)<br>$CHCl_3$-$Me_2CO$(3:1) | 4 |
| sol,chac,solad | Separation of *Solanum* alkaloids (Table 16.1) | $SiO_2$ | $CHCl_3$-AcOH-MeOH(85:2:13)<br>AcOH-95% EtOH(3:1, 1:1, 1:2, 1:3, 1:9)<br>Diisobutyl ketone-AcOH-$H_2O$ (8:5:1) | 7 |
| solad,dem,tdenol,tomd, sold,soldulci,solcap, soladona,solsone, tomdone,solddiene | Separation of *Solanum* alkaloids and steroid sapogenins (Table 16.1) | $SiO_2$<br><br><br>$Al_2O_3$<br><br>10% $AgNO_3$-impregnated $SiO_2$ | Cyclohexane-EtOAc(1:1)<br>$CHCl_3$-MeOH(6:4)<br>Light petroleum (b.p. 80-90°C)-benzene-EtOAc(85:5:10)<br>Cyclohexane-EtOAc(1:1)<br>*n*-Hexane-TrEA(15:1)<br>EtOAc-cyclohexane-96% EtOH(50:40:5) | 8 |
| sold,solad | Preparative separation | $Al_2O_3$ | $Et_2O$ | 10 |
| sols,solmg,sol,chac | Assay of steroid alkaloids in *Solanum* species, indirect quantitative analysis | $SiO_2$ | EtOAc-pyridine-$H_2O$(20:5:2) | 14 |
| sold,solddiene | Indirect quantitative analysis in *Solanum* species | $Al_2O_3$ | $CHCl_3$-MeOH(99:1) | 17 |

| | | | | |
|---|---|---|---|---|
| dem,solad,isodem, isosolad,soldulci,sold, tomd,tdenol,αOHsoldulci, αOHsold,αOHtomd,αOHtdenol, βOHsoldulci,βOHsold<br>3 stereoisomeric epimino-cholestan diols and epimino-cholest-5-en diols | Analytical and preparative separation of 5α-saturated and $\Delta^5$-unsaturated compounds (Table 16.1 and Fig. 16.2) | 5% $AgNO_3$-impregnated $SiO_2$<br>5% $AgNO_3$-impregnated $Al_2O_3$ | $CHCl_3$-MeOH(9:1, 95:5)<br>$CHCl_3$-$Et_2O$-AcOH(97:2.5:0.5)<br>$CHCl_3$-MeOH(95:5) | 20 |
| tdenol,sold,soldulci | Separation of main alkaloids in *Solanum* species | 6.3% $AgNO_3$-impregnated $SiO_2$ | $CH_2Cl_2$-MeOH(95:5) | 24 |
| sold,soldulci,tomd, tdenol,solad,dem, chac,solmg,sol,sols tom | Direct quantitative analysis and separation of saturated and unsaturated steroids (Table 16.1) | 4% NaOAc-impregnated $SiO_2$<br>$SiO_2$ | *n*-Hexane-EtOAc-EtOH(5:10:2) + 2 drops $Br_2$<br>Cyclohexane-EtOAc(1:3) + 2 drops $Br_2$<br>Benzene-MeOH(10:1) + 2 drops $Br_2$<br>*n*-Hexane-EtOAc(2:1) + 2 drops $Br_2$<br>Benzene-MeOH(10:7.5, 5:1)<br>$CHCl_3$-MeOH(22:1.5)<br>$CHCl_3$-EtOH-1% $NH_4OH$(2:2:1) | 25 |
| sol,solad | Detection of alkaloids in potatoes | $SiO_2$ | $CHCl_3$-MeOH-$H_2O$(5:5:1) | 26 |
| 26 steroid alkaloids (Tables 16.1 and 16.5) | Separation and identification | $SiO_2$ | *n*-Hexane-EtOAc(1:1)<br>*n*-Hexane-EtOH(1:1)<br>$CH_2Cl_2$-$Me_2CO$(4:1)<br>*n*-Hexane-$Me_2CO$(1:1)<br>$CH_2Cl_2$-MeOH-AcOH(85:13:2)<br>$CH_2Cl_2$-MeOH-$NH_4OH$(100:100:1) | 33 |
| sols,solmg,sold | Quantitative TLC analysis (no details available) | $SiO_2$ | $CHCl_3$-MeOH-25% $NH_4OH$(61:32:7) | 35 |
| sol | Identification in *Solanum* species | $SiO_2$ | Cyclohexane-$CHCl_3$-DEA(5:4:1) | 36 |
| sol,chac | Densitometric analysis in potatoes | $SiO_2$ | $CHCl_3$-95% EtOH-1% $NH_4OH$(2:2:1), org. phase | 38 |
| sol,tom,solad,sold, dem,tomd | Detection with optical brighteners | $SiO_2$ | *n*-BuOH-$H_2O$-HCOOH(4:5:1), upper phase<br>*n*-Hexane-$Me_2CO$(1:1) | 39 |

*For abbreviations, see footnote to Table 16.7.

References p. 423

TABLE 16.9

TLC ANALYSIS OF STEROID ALKALOIDS IN PLANT MATERIALS (EXCLUDING *SOLANUM*)

| Alkaloid* | Aim | Adsorbent | Solvent system | Ref. |
|---|---|---|---|---|
| 32 alkaloids (Figs. 16.3-16.7) | Separation of *Holarrhena* alkaloids | $SiO_2$ | Benzene or diethyl ether sat. with conc. ammonia in $NH_3$ atmosphere | 6,12 |
| vtd,V,cevd,vc,cev, cevg,cevc,anhcevgtriOAc anhcevdtriOAc,anhvcte-traOAc | Separation *Veratrum* alkaloids (Table 16.2) | $SiO_2$ | Cyclohexane-DEA(180:20 or 140:60)<br>Cyclohexane-abs. EtOH(170:30) | 15 |
| protoV A, protoV B | Control of aldulterants in *Primula* species | pH = 7.6-im-pregnated $SiO_2$ | $CHCl_3$-MeOH(1:1) | 16 |
| Not specified | TAS technique for *Veratrum* | $SiO_2$ | $CHCl_3$-$Me_2CO$-DEA(5:4:1)<br>$CHCl_3$-DEA(9:1)<br>Cyclohexane-$CHCl_3$-DEA(5:4:1)<br>Cyclohexane-DEA(9:1)<br>Benzene-EtOAc-DEA(7:2:1) | 28 |
| con | Identification in total alkaloids from *Holarrhena* | | No details available | 29 |

*For abbreviations, see footnote to Table 16.7.

TABLE 16.10

TLC ANALYSIS OF STEROID ALKALOIDS IN COMBINATION WITH OTHER COMPOUNDS

| Alkaloid* | Other compounds | Aim | Adsorbent | Solvent system | Ref. |
|---|---|---|---|---|---|
| sold,sold-diacetate | Various steroids | Separation | $Al_2O_3$ | Benzene-EtOH(98:2) | 1 |
| con,hol | Various isoquinoline and lupine alkaloids | Separation | $SiO_2$ | EtOAc-hexane-DEA(77.5:17.5:5)<br>Benzene-$CHCl_3$-DEA(20:75:5)<br>Hexane-dichloroethylene-DEA(20:75:5) | 5 |
| V | Tetracaine,atropine, scopolamine,strychnine | Separation | $SiO_2$ | $CHCl_3$-DEA(9:1)<br>$CCl_4$-*n*-BuOH-MeOH-10% $NH_4OH$(12:9:9:1) | 9 |
| sol,solad | Physostigmine,ephedrine,homatropine,lobeline,pilocarpine, reserpine | Detection of cholinesterase inhibitors | 5% silicone 555-impregnated cellulose | $H_2O$-EtOH-$CHCl_3$(56:42:2) | 19 |
| V | Various alkaloids | Reaction chromatography | $SiO_2$ | Benzene-MeOH-$Me_2CO$-AcOH(70:20:5:5) | 22 |
| V | Various alkaloids | Detection with $\pi$-acceptors | $SiO_2$ | $Me_2CO$-toluene-MeOH-$NH_4OH$(45:40:10:5) | 37 |

*For abbreviations, see footnote to Table 16.7.

# II.7. MISCELLANEOUS ALKALOIDS

## Chapter 17

## XANTHINE DERIVATIVES

17.1. Solvent systems........................................................435
17.2. Two-dimensional TLC.....................................................438
17.3. Detection...............................................................438
17.4. Quantitative analysis...................................................440
17.5. TAS technique and reaction chromatography...............................441
References....................................................................441

The emphasis in this chapter is on the naturally occurring xanthine derivatives caffeine, theobromine and theophylline. Because caffeine is often used in pharmaceutical analgesic and antipyretic preparations, most of the TLC analyses reported for xanthine derivatives concern the separation of such pharmaceutical mixtures (Table 17.6). A number of investigations dealing with the determination of drugs in urine or blood also take caffeine into account, because it is a common constituent of daily food (coffee, tea, etc.). Caffeine and its metabolites may therefore regularly be found by analysis of body fluids (Table 17.7). For optimal therapeutic use of theophylline its assay in plasma is important. A series of investigations have dealt with the determination of theophylline in plasma, etc. Reviews on the TLC analysis of xanthine derivatives have been published[5,71,174].

### 17.1. SOLVENT SYSTEMS

For the determination of caffeine in pharmaceutical preparations containing mixtures of analgesics and antipyretics such as salicylic acid derivatives, phenacetin and aminopyrine acidic solvent systems have often been used with silica gel plates. Examples of such solvent systems are benzene - diethyl ether - methanol - acetic acid, usually in the proportions (120:60:1:18)[2,48,64,65,70,77,91,97,106,115]. Schlemmer and Kammerl[102] found this system to be suitable for quantitative analysis. A related solvent is benzene - dioxane - acetic acid in various proportions[48,54,134,156]. Chloroform - ethanol in different ratios has also been used for pharmaceutical preparations containing caffeine[1,3,8,12,48,54,70,81,97,151,159]. Sarsunova and Schwarz[15,16,18] used aluminium oxide without a binder as the sorbent for the separation of some xanthine derivatives in pharmaceutical preparations.

References p. 441

Röder[68] preferred azeotropic solvent systems for similar purposes. For the separation of xanthine derivatives he used *n*-butylamine - ethanol - 1-chlorobutane (1:2:7) and diisopropylamine - 1,2-dichloroethane - isopropanol (1:4:5), both of which gave a complete separation of caffeine, theophylline and theobromine on silica gel plates. A number of useful TLC systems for the separation of xanthine derivatives are summarized in Table 17.1.

Chloroform - ethanol in various proportions in combination with silica gel plates is one of the most commonly used systems. In some instances this solvent system has been used in chromatography tanks saturated with ammonia[19,38,169]. Teichert et al.[3] used silica gel impregnated with a buffer (pH 6.8) and chloroform - ethanol (9:1). The influence of pH on the separation of the xanthines (caffeine, theophylline and theobromine) with chloroform - ethanol has been studied by Kraus and Dumont[71]. The influence of different concentrations of ammonia used for the pre-treatment of the chromatography chambers was studied, and also the influence of the treatment time. The best results were obtained with a treatment time of 20-30 min. With low concentrations of ammonia (2.5 and 5%) theobromine showed tailing and the authors suggested that this was due to a higher water content of the plate, rather than to differences in the ammonia concentration. It was also found that on suspension in water of silica gel samples equilibrated with different percentages of ammonia, all gave similar pH values (9.05-9.40). Pre-treatment with acids was also studied, as well as the influence of buffered solvent systems.

Dumont and Kraus[72] used pH-gradient plates to study the separation of the xanthine derivatives with a series of solvent systems: S1[3], S3[26], S4[50], S10[21], S4 in the proportions 5:5:1[14] and chloroform - diethyl ether (85:15)[4] gave the best results in the pH range 6.2-8.5. The three test compounds were well separated in solvent systems S3 and S1. The separation in system S10 was also satisfactory, but theophylline remained at the starting point.

Aigner et al.[157] used system S1 and methanol - acetone - triethanolamine (1:1:0.03) in combination with silver halide-impregnated silica gel plates. Stuchlik et al.[55] investigated the influence of pH and the concentration of sodium 2-naphthol-6,8-disulphonate in the mobile phase on the separation of ten xanthine derivatives on silica gel plates. The fluorescence of the sodium 2-naphthol-6,8-disulphonate is quenched by the xanthine derivatives and can be used for the detection.

Lüdy-Tenger[74] found basic systems to be preferable to acidic systems because of the tailing observed with caffeine in the latter. Good results were obtained with solvent systems containing ammonia-saturated chloroform (S8, S9). Kamp et al.[14] used carbon tetrachloride - chloroform - methanol (5:5:1) and Senanayake and Wyesekera[50] the same solvent in the proportions 8:5:1. The same solvent system has also been used by other workers[84,173]. Poey et al.[80] tested a number of TLC systems for the separation of caffeine, theophylline and theobromine. Systems S1 and S5 were found to give the best results (Table 17.1).

Rink and Gehl[32] separated some xanthine derivatives with isopropanol - chloroform - 10% aqueous piperazine (6:2:2) on silica gel plates impregnated with aqueous 10% piperazine. Ethyl acetate - methanol - ammonia (8:2:1) in combination with silica gel plates (S11) has been used for the analysis of xanthine derivatives in biological materials[78,173,176,178] and for the quantitative analysis of xanthine derivatives in cocoa[87]. Reichert[176] separated theobromine and theophylline in system S11. He also mentioned a number of xanthine derivatives that do not interfere in the determination of caffeine and theophylline when system S1 is used.

Heftmann and Schwimmer[78] described three solvents for the separation of eight xanthine derivatives. Chloroform - methanol (4:1) was found to be suitable for the more polar derivatives (xanthine, hypoxanthine, 1- and 7-methylxanthine), and system S11 for the separation of paraxanthine, caffeine, theobromine and theophylline. *n*-Butanol saturated with 2.8% ammonia on silica gel plates resolved all test compounds.

Amin and Sepp[159] used three consecutive solvents for the determination of ergotamine and caffeine in pharmaceutical preparations and in biological materials. The first solvent system was used to remove the plasma constituents, the second to detect ergotamine and the third for caffeine.

Acidic solvent systems have also been used for the separation of xanthine derivatives. Schunack et al.[47] and Schlemmer and Kammerl[102] found that acidic solvent systems gave more satisfactory results than basic solvents. They developed the silica gel plates respectively with chloroform - ethanol - formic acid (85:5:10) (developed twice) and benzene - diethyl ether - methanol - acetic acid (60:30:1:9). Baehler[9] used ethyl acetate - methanol - 12 *M* hydrochloric acid (18:2:0.05) and ethyl acetate - methanol - acetic acid (8:1:1) for the determination of caffeine, theobromine and theophylline. Warren[56] used butanol - water - acetic acid (4:1:1) in combination with pH 6.8 buffered silica gel plates for the determination of xanthine metabolites in blood. The plates were first developed in light petroleum to remove lipophylic compounds.

The separation of xanthine derivatives on cellulose plates has been described by Jalal and Collin[163] using the upper phase of butanol - hydrochloric acid - water (100:11:28) as the mobile phase. The separation of xanthine derivatives in mixtures with a number of other compounds on polyamide plates has been described by Hsiu et al.[39] (see Chapter 3, p.31) and Lyakina and Brutko[177].

**References p. 441**

## 17.2. TWO-DIMENSIONAL TLC

Two-dimensional TLC has been used to separate the components of pharmaceutical preparations[98,103,135]. For the separation of xanthine derivatives Schunack et al.[21] used solvents S10 and S7 and Lehmann and Martinod[37] used solvent S1 and ethyl acetate - methanol - acetic acid (8:1:1). These solvents could also be used to distinguish between natural and synthetic caffeine. Arnaud[152] and Milon and Antonioli[179] separated caffeine and its metabolites by two-dimensional TLC. As the first solvents were used chloroform - methanol (4:1) and chloroform - ethanol (9:1), respectively, and as the second solvent in both instances chloroform - acetone - *n*-butanol - 25% ammonia (3:3:4:1).

## 17.3. DETECTION

For the detection of xanthine derivatives quenching of UV light on fluorescent plates is widely used. With Dragendorff's reagent xanthine derivatives cannot be detected, except for those with an amino group in the side-chain (millophylline and bamifylline)[26]. Hauck[24] found, however, that xanthine derivatives could be detected with Dragendorff's reagent (Munier and Macheboeuf modification) when the plate was first sprayed with silver nitrate in 5% sulphuric acid and then with Dragendorff's reagent. The reverse sequence of the reagents could also be used[21,24]. For the colours obtained, see Table 17.2.

Pellerin et al.[97] sprayed first with 20% nitric acid and heated the plates at 100°C for 10 min before spraying with Dragendorff's reagent (Munier and Macheboeuf modification); caffeine gave a violet colour. Senanayake and Wyesekera[50] used a modified Dragendorff's reagent, to which they added a 4% solution of iodine in potassium iodide and ethanol (1:1). The reagent gave positive reactions with xanthine derivatives; however, the same colours as those obtained by spraying with iodine - potassium iodide in ethanol followed by hydrochloric acid in ethanol were observed.

Caffeine, theobromine and theophylline do not react with iodoplatinate reagent[39,125,176]. Several workers have reported positive reactions with iodoplatinate reagent for caffeine, but only after spraying with a series of other reagents[44,90,123]. With a solution of iodine in potassium iodide in ethanol used as spray reagent, followed by spraying with hydrochloric acid, different colours are observed for the various xanthine derivatives (Table 17.2)[3,6,14,21,28,71]. Senanayake and Wyesekera[50] found that sharper spots and stable colours (2-3 h) were obtained if the iodine solution and hydrochloric acid were combined to one spray reagent (no. 60e). Kamp et al.[14] used Bouchardat's reagent for caffeine, theobromine and theophylline (no. 60c).

With iodine - iron(III) chloride - tartaric acid in acetone - water (no. 53) various colours were obtained with xanthine derivatives (Table 17.2)[17,21,58,128]. Schmidt[114] sprayed consecutively with 2% iron(III) chloride and 0.1 *N* iodine solution, because reagent no. 53 easily decomposes as the iodine reacts with acetone to yield iodoacetone (tear-gas).

Sarsunova et al.[153] studied the effect of TLC detection with iodine on the determination of aromatic compounds. It was found that caffeine, theobromine and theophylline were not affected by the iodine detection and that iodine could therefore be used as a detection reagent without the risk of interference in the further analysis.

The murexide reaction is specific for xanthine derivatives. To accomplish this reaction on the TLC plate for the detection of xanthine derivatives various oxidants have been used. By spraying with 10% chloramine-T solution (no. 16), followed by 2 *M* hydrochloric acid and heating at 96-98$^{o}$C to remove excess of chlorine and successive exposure to ammonia vapour, caffeine and theophylline were observed as pink-red spots[2,77]. Bromine was used instead of chlorine by Netien et al.[27], Vanhaelen[57] and Egli[93]. Vanhaelen exposed the TLC plates to bromine vapour for 5 min and heated them at 120$^{o}$C for 10 min prior to exposure to ammonia vapour. Egli exposed the plates to bromine vapour for 10 min and ommitted heating: caffeine was seen as a faint pink spot and theophylline as a pink spot. On exposure to ammonia vapour the colours changed to pink-white and reddish, respectively. Netien et al. exposed the plates to bromine vapour for 3 h and subsequently heated them for 10 min. Poey et al.[80] studied other oxidants for the murexide reaction. Nitric acid, fuming nitric acid and potassium permanganate in sulphuric acid gave positive reactions only for caffeine. Hydrogen peroxide and bromine-water gave positive reactions only for caffeine and theophylline. Potassium perchlorate - hydrochloric acid (no. 82) gave positive reactions for caffeine, theophylline and theobromine. The red-violet colours of the spots obtained after spraying with the oxidant, heating at 110$^{o}$C for 30 min and exposure to ammonia vapour for 5 min were stable for at least 24 h. Caffeine could be detected at a level of about 2 μg and theobromine and theophylline at about 4 μg. A number of other xanthine derivatives could also be detected with this reagent. Conine and Paul[125] sprayed with iodoplatinate reagent after the murexide reaction.

In addition to the above-mentioned detection reagents a number of others have been used. Spraying with 1% mercury(I) nitrate (no. 69) gave grey-white spots on a dark background for theophylline and theobromine[16,58]. Baehler[9] used microsublimation to detect caffeine, theobromine and theophylline. The TLC plate was covered with a glass plate, which was cooled. The TLC plate was heated to about 250-260$^{o}$C and the xanthine spots were observed on the cooled glass plate (sensitivity about 5 μg).

**References p. 441**

A 1% cobalt nitrate solution in ethanol was used by Vanhaelen[57]. After heating at 110°C for 5 min blue-violet spots on a light blue background were observed. Schmidt[114] detected caffeine with Folin-Ciocalteu reagent (no. 43). Thielemann[106,110] used potassium hexacyanoferrate(III)(1%)-iron(III) chloride(2%) (1:1) (no. 80b) in the analysis of pharmaceutical preparations; caffeine gave an unstable brown colour. Okumura et al.[143] developed a flame-ionization detection method for alkaloids, including caffeine. Bohinc[30] reviewed detection methods in the TLC of xanthine derivatives.

## 17.4. QUANTITATIVE ANALYSIS

For the indirect determination of caffeine in pharmaceutical preparations a number of solvents have been described for eluting the compounds from the sorbent (silica gel):methanol[13], 95% ethanol - 0.2 *M* sodium hydroxide (1:1)[70], 1 *M* hydrochloric acid[103], magnesium oxide and water[122], ethanol[127,162] or 0.05 *M* sulphuric acid[162]. Szasz and Szasz[60] measured the blank values of silica gel $GF_{254}$ extracted with different solvents (acetic acid in water, sulphuric acid, ethanol and sodium hydroxide).

In analyses of xanthine derivatives in biological materials the following solvents have been used for the elution of these compounds from the sorbent: 0.1 *M* hydroxide[178] and ethanol[41] for silica gel plates and ethanol for aluminium oxide plates[112]. For determinations of xanthine derivatives in plant materials (including coffee and tea), chloroform[53,59,75] and 10% ammonia[163] have been used as eluents for silica gel plates. Arnaud[152] and Milon and Antonioli[179] described the determination of $^{14}$C-labelled caffeine and its metabolites in biological materials. After detection of the radioactive spots, they were scraped off and the activity was counted. Direct quantitative determinations of xanthine derivatives have been performed for analyses of both plant materials[50,76,84] and pharmaceutical preparations[172].

Densitometry has mostly been carried out at a wavelength of 272 nm, the absorbance maximum of xanthine derivatives. In this way caffeine[171,176] and theophylline[47,145,173] have been determined in biological materials. Reichert[176] discussed a number of compounds (including some metabolites) that did not interfere in the determination of caffeine and theophylline in serum, saliva or urine and described a TLC system suitable for the separation of theophylline and theobromine. Gupta et al.[173] used an internal standard in the analysis of theophylline in plasma.

Ebel and Herold[134,135,154] discussed the determination of caffeine in pharmaceutical preparations. They used phenacetin as an internal standard and two-dimensional TLC to separate some analgesic compounds. The quantitation of unseparated compounds was dealt with[154].

Theoretical aspects of direct densitometric quantitative analysis were discussed by Ebel and Hocke[150], with caffeine used as a model. The direct determination of caffeine in pharmaceutical preparations by densitometry at a wavelength of 272 nm has also been performed in other investigations[65,102,137,138,159]. According to Schlemmer and Kammerl[102], only with acidic solvent systems are satisfactory results obtained.

Malingré and Batterman[169] described the direct densitometric determination of caffeine in plant materials at a wavelength of 273 nm. Massa et al.[69,81] determined xanthine derivatives by measuring the quenching of UV light (254 nm) on fluorescent plates. The same method has been used in other investigations[86,87,91,175]. Malingré and Batterman[169] found this method to be less sensitive than densitometry at the absorption maximum of 273 nm.

## 17.5. TAS TECHNIQUE AND REACTION CHROMATOGRAPHY

Bican-Fister and Grdinic[94] used the TAS technique for the identification of drugs, including caffeine in mixtures with other drugs in tablets. The conditions used were heating at 250°C for 60 sec, using silica blue gel with 20% water content as propellant. Djurasinovic and Popovic[172] also used the TAS technique for the determination of caffeine in injections.

Kaess and Mathis[31] used the murexide reaction for the identification of caffeine on TLC plates. The reaction was performed prior to application on the plate.

Schmidt[104] reported the reaction chromatography of a series of drugs, using iodine as reagent.

## REFERENCES


1 H. Liebich, *Deut. Apoth.-Ztg.*, 99 (1959) 1246.
2 H. Gänshirt and A. Malzacher, *Arch. Pharm. (Weinheim)*, 293 (1960) 925.
3 K. Teichert, E. Mutschler and H. Rochelmeyer, *Deut. Apoth.-Ztg.*, 100 (1960) 283.
4 G. Machata, *Mikrochim. Acta*, (1960) 79.
5 E. Stahl, *Angew. Chem.*, 73 (1961) 646.
6 I.O. Cerri and G. Maffi, *Boll. Chim. Farm.*, 100 (1961) 951; *C.A.*, 57 (1962) 11467i.
7 J. Bäumler and S. Rippstein, *Pharm. Acta Helv.*, 36 (1961) 382.
8 T. Bican-Fister, *Acta Pharm. Jugosl.*, 12 (1962) 73; *C.A.*, 58 (1963) 13721d.
9 B. Baehler, *Helv. Chim. Acta*, 45 (1962) 309.
10 N.R. Farnsworth and K.L. Euler, *Lloydia*, 25 (1962) 186.
11 F. Tatsutoyo, T. Kido and H. Tanaka, *Yakuzaigaku*, 22 (1962) 269; *C.A.*, 59 (1963) 7319b.
12 G. Szasz, L. Khin and R. Budvari, *Acta Pharm. Hung.*, 33 (1963) 245.
13 H. Gänshirt, *Arch. Pharm. (Weinheim)*, 296 (1963) 129.
14 W. Kamp, W.J.M. Onderberg and W.A. Seters, *Pharm. Weekbl.*, 98 (1963) 993.
15 M. Sarsunova and V. Schwarz, *Pharmazie*, 18 (1963) 34.
16 M. Sarsunova and V. Schwarz, *Pharmazie*, 18 (1963) 207.

17 J. Zarnack and S. Pfeifer, *Pharmazie*, 19 (1964) 216.
18 V. Schwarz and M. Sarsunova, *Pharmazie*, 19 (1964) 267.
19 G. Szasz, M. Szasz-Zacsko and V. Polankay, *Acta Pharm. Hung.*, 36 (1965) 207.
20 R. Paris, R. Rousselet, M. Paris and M.J. Fries, *Ann. Pharm. Fr.*, 23 (1965) 473.
21 W. Schunack, E. Mutschler and H. Rochelmeyer, *Deut. Apoth.-Ztg.*, 105 (1965) 1551.
22 Chung Li, *Hua Hsueh Hsueh Pao*, 61 (1965) 518.
23 S. El. Gendi, W. Kisser and G. Machata, *Mikrochim. Acta*, (1965) 120.
24 G. Hauck, *Deut. Apoth.-Ztg.*, 105 (1965) 209.
25 W.W. Fike, *Anal. Chem.*, 38 (1966) 1697.
26 G.L. Szendey, *Arch. Pharm. (Weinheim)*, 299 (1966) 527.
27 G. Netien, N. Prum and A. Prum, *Bull. Trav. Soc. Pharm. Lyon*, 10 (1966) 39; *C.A.*, 67 (1967) 14870d.
28 T. Kaniewska and W. Zyzynski, *Diss. Pharm. Pharmacol.*, 18 (1966) 511; *C.A.*, 67 (1967) 14871e.
29 R. Adamski and T. Cieszynska, *Farm. Pol.*, 22 (1966) 731.
30 P. Bohinc, *Farm. Vestnik (Ljubljana)*, 17 (1966) 8; *C.A.*, 66 (1967) R22220m.
31 A. Kaess and C. Mathis, *Int. Symp. Chromatogr. Electrophor. Lect. Pap. 4th*, (1966) 525.
32 M. Rink and A. Gehl, *J. Chromatogr.*, 21 (1966) 143.
33 I. Sunshine, W.W. Fike and H. Landesman, *J. Forensic Sci.*, 11 (1966) 428.
34 F. Wartmann-Hafner, *Pharm. Acta Helv.*, 41 (1966) 406.
35 P. Schweda, *Anal. Chem.*, 39 (1967) 1019.
36 F. Reimers, *Arch. Pharm. Chem.*, 74 (1967) 531.
37 G. Lehmann and P. Martinod, *Arzneim.-Forsch.*, 17 (1967) 35.
38 M. Struhar and N. Thi Ban, *Farm. Obz.*, 36 (1967) 549; *C.A.*, 70 (1969) 105260h.
39 Hsing-Chien Hsiu, Jen-Tzaw Huang, Tsu-Bi Shih, Kun-Lin Yang, Kung Tsung Wang and Alice L. Lin, *J. Chin. Chem. Soc.*, 14 (1967) 161.
40 A. Noirfalise and G. Mees, *J. Chromatogr.*, 31 (1967) 594.
41 M.A. Elkiey, M. Karawya, S.K. Wahba and A.R. Kozman, *J. Pharm. Sci. UAR*, 8 (1967) 201.
42 V.M. Pechennikov, *Nauch. Tr. Aspir. Ordinatorov, 1-i Mosk. Med. Inst.*, (1967) 145; *C.A.*, 70 (1969) 99647w.
43 V.M. Pechennikov and A.Z. Knizhnik, *Nauch. Tr. Aspir. Ordinatorov, 1-i Mosk. Med. Inst.*, (1967) 146; *C.A.*, 70 (1969) 50490a.
44 B. Davidow, N.L. Petri and B. Quame, *Amer. J. Clin. Pathol.*, 50 (1968) 714.
45 F. Schmidt, *Arch. Pharm. (Weinheim)*, 301 (1968) 940.
46 V. Vukcevic-Kovacevic, *Arh. Farm.*, 18 (1968) 3; *C.A.*, 69 (1968) 99424a.
47 W. Schunack, E. Eich, E. Mutschler and H. Rochelmeyer, *Arzneim.-Forsch.*, 19 (1969) 1756.
48 M. Sarsunova, V. Schwarz, L. Krasnec and Kim Chi, *Chem. Zvesti*, 22 (1968) 118.
49 V.M. Pechennikov, A.Z. Knizhnik and P.L. Senov, *Farm. Zh. (Kiev)*, 23 (1968) 41; *C.A.*, 70 (1969) 22932c.
50 U.M. Senanayake and R.O.B. Wyesekera, *J. Chromatogr.*, 32 (1968) 75.
51 M. Debackere and L. Laruelle, *J. Chromatogr.*, 35 (1968) 234.
52 J.M.G.J. Frijns, *Pharm. Weekbl.*, 103 (1968) 929.
53 C.L. Franzke, K.S. Grunert, U. Hildebrandt and H. Griehl, *Pharmazie*, 23 (1968) 502.
54 M. Bachrata, J. Cerna and S. Szucsova, *Cesk. Farm.*, 18 (1969) 18.
55 M. Stuchlik, I. Csiba and L. Krasnec, *Cesk. Farm.*, 18 (1969) 91.
56 R.N. Warren, *J. Chromatogr.*, 40 (1969) 468.
57 M. Vanhaelen, *J. Pharm. Belg.*, 24 (1969) 87.
58 F. Schmidt, *Pharm. Ztg.*, 114 (1969) 1523.
59 C. Franzke, K.S. Grunert and H. Griehl, *Z. Lebensm.-Unters.-Forsch.*, 139 (1969) 85.
60 G. Szasz and G. Szasz, *Acta Pharm. Hung.*, 40 (1970) 38.
61 Z. Blagojevic and M. Skrlj, *Acta Pharm. Jugosl.*, 20 (1970) 133; *C.A.*, 74 (1971) 91242x.

62 V. Vukecevic-Kovacevic, *Bull. Sci. Cons. Acad. Sci. Arts RSF Yougosl. Sect. A*, 15 (1970) 238; *C.A.*, 73 (1970) 123554.
63 M.L. Bastos, G.E. Kananen, R.M. Young, J.R. Monforte and I. Sunshine, *Clin. Chem.*, 16 (1970) 931.
64 M. Overgaard-Nielsen, *Dan. Tidsskr. Farm.*, 44 (1970) 7.
65 W. Schlemmer, E. Kammerl and F.H. Klemm, *Deut. Apoth.-Ztg.*, 110 (1970) 833.
66 T. Avsic-Sernec, *Farm. Vestn. (Ljubljana)*, 21 (1970) 271; *C.A.*, 75 (1971) 40503w.
67 L.O. Kirichenko and F.Yu. Kagan, *Farm. Zh. (Kiev)*, 25 (1970) 42; *C.A.*, 73 (1970) 28951s.
68 E. Röder, *Int. Symp. Chromatogr. Electrophor. Lect. Pap. 6th, 1970*, (1971) 194.
69 V. Massa, *Int. Symp. Chromatogr. Electrophor. Lect. Pap. 6th, 1970*, (1971) 470.
70 F. Pellerin and D. Mancheron, *Int. Symp. Chromatogr. Electrophor. Lect. Pap. 6th, 1970*, (1971) 501.
71 L.J. Kraus and E. Dumont, *J. Chromatogr.*, 48 (1970) 96.
72 E. Dumont and L.J. Kraus, *J. Chromatogr.*, 48 (1970) 106.
73 A.S. Curry and D.A. Patterson, *J. Pharm. Pharmacol.*, 22 (1970) 198.
74 F. Lüdy-Tenger, *Pharm. Acta Helv.*, 45 (1970) 254.
75 W. Meissner and A. Walkowski, *Pr. Zakresu Towerozn. Chem. Wyzsza Szk. Ekon. Poznaniu, Zesz. Nauk. Ser. 1*, 40 (1970) 191; *C.A.*, 78 (1973) 56515z.
76 J. Washuettl, E. Bancher and P. Riederer, *Z. Lebensm.-Unters.-Forsch.*, 143 (1970) 253.
77 Van T. Lieu, *J. Chem. Educ.*, 48 (1971) 478.
78 E. Heftmann and S. Schwimmer, *J. Chromatogr.*, 59 (1971) 214.
79 S.J. Mulé, M.L. Bastos, D. Jukofsky and E. Saffer, *J. Chromatogr.*, 63 (1971) 289.
80 J. Poey, R. Denine, R. Merad-Boudia and N. Mrabet, *J. Eur. Toxicol.*, 3 (1971) 341.
81 V. Massa, F. Gal, P. Susplugas and G. Maestre, *Trav. Soc. Pharm. Montpellier*, 31 (1971) 167.
82 S. Zadeczky, D. Küttel and M. Takacsi, *Acta Pharm. Hung.*, 42 (1972) 7.
83 M.M. Baden, N.N. Valanju, S.K. Verma and S.N. Valanju, *Amer. J. Clin. Pathol.*, 57 (1972) 43.
84 A.G. Tchetche and J.A. Braun, *Ann. Univ. Abidjan Ser. C*, 8 (1972) 163.
85 H.J. Koenen, *Apothekersprakt. Pharm.-Tech. Assist.*, 18 (1972) 9.
86 L.R. Sellier and V. Torre, *Cienc. Ind. Farm.*, 4 (1972) 146; *C.A.*, 82 (1975) 160280u.
87 M.T. Cuzzoni and G. Gazzani, *Farmaco Ed. Prat.*, 27 (1972) 564.
88 E. Pawelczyck, Z. Plotkowiakowa and T. Malesza, *Farm. Pol.*, 28 (1972) 263; *C.A.*, 77 (1972) 39350d.
89 W.T. Fisher, A.D. Baitsholts and G.S. Grau, *J. Chromatogr. Sci.*, 10 (1972) 303.
90 J.E. Wallace, J.D. Biggs, J.H. Merritt, H.E. Hamilton and K. Blum, *J. Chromatogr.*, 71 (1972) 135.
91 G. Dertinger and H. Scholz, *Pharm. Ind.*, 34 (1972) 114.
92 Y. Imai, T. Kawakubo, I. Ohtake and M. Namakata, *Yakugaku Zasshi*, 92 (1972) 1074; *C.A.*, 78 (1973) 12279w.
93 R.A. Egli, *Z. Anal. Chem.*, 259 (1972) 277.
94 T. Bican-Fister and S. Grdinic, *Zentralbl. Pharm.*, 110 (1971) 1247.
95 K.F. Ahrend and D. Tiess, *Zentralbl. Pharm.*, 111 (1972) 933.
96 Z. Blagojevic, L. Glisovic and D. Zivanov-Stakic, *Arh. Farm.*, 23 (1973) 9; *C.A.*, 80 (1974) 63888t.
97 F. Pellerin, D. Dumitrescu-Mancheron and C. Chabrelle, *Bull. Soc. Chim. Fr.*, (1973) 123.
98 A. Fiebig, S. Kanafarska-Mlotkowska and D. Chodkiewicz, *Farm. Pol.*, 29 (1973) 1087.
99 R.A. van Welsum, *J. Chromatogr.*, 78 (1973) 237.
100 M.L. Bastos, D. Jukofsky and S.J. Mulé, *J. Chromatogr.*, 81 (1973) 93.

101 N.N. Valanju, M.M. Baden, S.N. Valanju, D. Mulligan and S.K. Verma, *J. Chromatogr.*, 81 (1973) 170.
102 W. Schlemmer and E. Kammerl, *J. Chromatogr.*, 82 (1973) 143.
103 A.M. Guyot-Herman and H. Robert, *J. Pharm. Belg.*, 28 (1973) 507.
104 F. Schmidt, *Krankenhaus-Apotheke*, 23 (1973) 10.
105 H. Thielemann, *Sci. Pharm.*, 41 (1973) 69.
106 H. Thielemann, *Sci. Pharm.*, 41 (1973) 71.
107 H.L. Wu and R.T. Wang, *Taiwan Ko Hsueh*, 27 (1973) 67; *C.A.*, 82 (1975) 7690b.
108 H.L. Wu and E.H. Chen, *Taiwan Yao Hsueh Tsa Chih*, 25 (1973) 32; *C.A.*, 84 (1976) 111747c.
109 K.F. Ahrend and D. Tiess, *Wiss. Z. Univ. Rostock, Math. Naturw. Reihe*, 22 (1973) 951.
110 H. Thielemann, *Z. Anal. Chem.*, 263 (1973) 336.
111 D. Radulovic, Z. Blagojevic and D. Zivanov-Stakic, *Arh. Farm.*, 24 (1974) 215; *C.A.*, 83 (1975) 103334e.
112 H. Sybirska and H. Gajkzinska, *Bromatol. Chem. Toksykol.*, 7 (1974) 189.
113 P.M. Kullberg and C.W. Gorodetzky, *Clin. Chem.*, 20 (1974) 177.
114 F. Schmidt, *Deut. Apoth.-Ztg.*, 114 (1974) 1593.
115 E. Curea and C. Opferman, *Farmacia (Bucharest)*, 22 (1974) 483; *C.A.*, 84 (1976) 111735x.
116 A. Wislocki, P. Martel, R. Ito, W.S. Dunn and C.D. McGuire, *Health Lab. Sci.*, 11 (1974) 13.
117 C. Klein, *Isr. Pharm. J.*, 17 (1974) 247; *C.A.*, 83 (1975) 10335f.
118 D.W. Chasar and G.B. Toth, *J. Chem. Educ.*, 51 (1974) 22.
119 P.D. Swaim, V.M. Loyola, H.D. Harlan and M.J. Carlo, *J. Chem. Educ.*, 51 (1974) 331.
120 J.H. Speaker, *J. Chromatogr. Sci.*, 12 (1974) 297.
121 A.C. Moffat and K.W. Smalldon, *J. Chromatogr.*, 90 (1974) 1, 9.
122 S.M. Khafagy, S.A. Metwally, A.N. Girgis and N. Rofael, *J. Drug. Res.*, 6 (1974) 75.
123 K.G. Blass, R.J. Thibert and T.F. Draisey, *J. Chromatogr.*, 95 (1974) 75.
124 A.C. Moffat and B. Clare, *J. Pharm. Pharmacol.*, 26 (1974) 665.
125 F. Conine and J. Paul, *Mikrochim. Acta*, (1974) 443.
126 R.J. Armstrong, *N. Z. J. Sci.*, 17 (1974) 15.
127 A.M. Soeterboek and M. van Tiel, *Pharm. Weekbl.*, 109 (1974) 962.
128 Z. Kubiuk, J. Porebski and T. Stozek, *Pol. J. Pharmacol. Pharm.*, 26 (1974) 581.
129 H. Thielemann, *Sci. Pharm.*, 42 (1974) 179.
130 A. Ito, N. Sekiyama, H. Ishii and Y. Kimura, *Tokyo Toritsu Eisei Kenkyusho Kenkyo Nempo*, 25 (1974) 265; *C.A.*, 82 (1975) 153881f.
131 H.L. Wu and T.C. Pan, *Taiwan Yao Hsueh Tsa Chih*, 26 (1974) 22; *C.A.*, 84 (1975) 111746b.
132 E. Curea and M. Martinovici, *Ann. Pharm. Fr.*, 33 (1975) 505.
133 D. Radulovic, Z. Blagojevic and D. Lilcic, *Arh. Farm.*, 25 (1975) 223; *C.A.*, 84 (1976) 140784t.
134 S. Ebel and G. Herold, *Chromatographia*, 8 (1975) 35.
135 S. Ebel and G. Herold, *Chromatographia*, 8 (1975) 569.
136 E. Curea and M. Martinovici-Fagarasan, *Clujul Med.*, 48 (1975) 253.
137 T. Inoue, M. Tatsuzawa, T. Ishii and Y. Inoue, *Eisei Shikenjo Hokoku*, 93 (1975) 31; *C.A.*, 85 (1976) 10490d.
138 T. Inoue, M. Tatsuzawa and S. Hashiba, *Eisei Shikenjo Hokoku*, 93 (1975) 83; *C.A.*, 85 (1976) 10489k.
139 F.E. Kagan, F.A. Mitchenko, L.O. Kirichenko and T.A. Koget, *Farm. Zh. (Kiev)*, 30 (1975) 75; *C.A.*, 83 (1975) 120940s.
140 A. Brantner, J. Vamos, E. Jeney and G. Szasz, *Gyogyszereszet*, 19 (1975) 10; *C.A.*, 83 (1975) 33099f.
141 A. Zobin and M. Gracza-Lukacs, *Gyogyszereszet*, 19 (1975) 455; *C.A.*, 84 (1976) 126802r.
142 K.K. Kaistha, R. Tadrus and R. Janda, *J. Chromatogr.*, 107 (1975) 359.
143 T. Okumura, T. Kadono and A. Isoo, *J. Chromatogr.*, 108 (1975) 329.

144 A.C. Moffat, *J. Chromatogr.*, 110 (1975) 341.
145 B. Wesley-Hadzija and A.M. Mattocks, *J. Chromatogr.*, 115 (1975) 501.
146 W.J. Serfontein, D. Botha and L.S. de Villiers, *J. Chromatogr.*, 115 (1975) 507.
147 P.A.F. Pranitis and A. Stolman, *J. Forensic Sci.*, 20 (1975) 726.
148 I. Hempel and H.D. Woitke, *Pharm. Prax.*, 10 (1975) 223.
149 E. Spratt, *Toxicol. Annu. 1974*, (1975) 229.
150 S. Ebel and J. Hocke, *Z. Anal. Chem.*, 277 (1975) 105.
151 M. Petkovic, *Arh. Farm.*, 25 (1976) 435; *C.A.*, 87 (1977) 90786j.
152 M.J. Arnaud, *Biochem. Med.*, 16 (1976) 67.
153 M. Sarsunova, B. Kakac and M.Ryska, *Cesk. Farm.*, 25 (1976) 156.
154 S. Ebel and G. Herold, *Chromatographia*, 9 (1976) 41.
155 M. Petkovic, *Farm. Glas*, 32 (1976) 363; *C.A.*, 86 (1977) 96075c.
156 G.K. Munshi and S.K. Das, *Indian Drugs Pharm. Ind.*, 11 (1976) 38; *C.A.*, 86 (1977) 60608q.
157 R. Aigner, H. Spitzy and R.W. Frei, *J. Chromatogr. Sci.*, 14 (1976) 381.
158 L. Lepri, P.G. Desideri and M. Lepori, *J. Chromatogr.*, 116 (1976) 131.
159 M. Amin and W. Sepp, *J. Chromatogr.*, 118 (1976) 225.
160 L. Lepri, P.G. Desideri and M. Lepori, *J. Chromatogr.*, 123 (1976) 175.
161 A.N. Masoud, *J. Pharm. Sci.*, 65 (1976) 1585.
162 F. Chrobok, *Krim. Forensische Wiss.*, 24 (1976) 157.
163 M.A.F. Jalal and H.A. Collin, *New. Phytol.*, 76 (1976) 277.
164 R. Cadorniga, M.A. Camacho and A. Dominguez-Gil, *Cienc. Ind. Farm.*, 9 (1977) 88; *C.A.*, 87 (1977) 172941v.
165 R. Codorniga, M.A. Camacho and A. Dominguez-Gil, *Cienc. Ind. Farm.*, 9 (1977) 114; *C.A.*, 87 (1977) 172946q.
166 K. Soviar, E. Dunckova and O. Moravkova, *Farm. Obz.*, 46 (1977) 441; *C.A.*, 90 (1978) 43868q.
167 S. Thunell, *J. Chromatogr.*, 130 (1977) 209.
168 A.N. Masoud, *J. Chromatogr.*, 141 (1977) D9.
169 T.M. Malingré and S. Batterman, *Pharm. Weekbl.*, 112 (1977) 1305.
170 L. Gagliardi, A. Amato, G. Ricciardi and S. Chiavarelli, *Riv. Tossicoli Sper. Clin.*, 7 (1977) 191; *C.A.*, 88 (1978) 32648a.
171 W. Feldheim, E.H. Reimerdes and G.R. Storm, *Z. Lebensm.-Unters.-Forsch.*, 165 (1977) 204.
172 S. Djurasinovic and R. Popovic, *Arh. Farm.*, 28 (1978) 163; *C.A.*, 90 (1979) 92480r.
173 R.N. Gupta, F. Eng and M. Stefanec, *Clin. Biochem.*, 11 (1978) 42.
174 E. Stahl and J. Brombeer, *Deut. Apoth.-Ztg.*, 118 (1978) 1527.
175 J. Sherma and M. Beim, *J. High Resolut. Chromatogr. Chromatogr. Commun.*, 1 (1978) 309.
176 M. Riechert, *J. Chromatogr.*, 146 (1978) 175.
177 M.N. Lyakina and L.I. Brutko, *Khim.-Farm. Zh.*, 12 (1978) 136; *C.A.*, 88 (1978) 126412w.
178 H. Rooseboom, H. Lingeman and G. Wiese, *Z. Anal. Chem.*, 292 (1978) 239.
179 H. Milon and J.A. Antonioli, *J. Chromatogr.*, 162 (1979) 223.

TABLE 17.1

TLC SEPARATION OF XANTHINE DERIVATIVES

TLC systems:

| | | |
|---|---|---|
| S1 | Silica gel G | Chloroform-ethanol(9:1)[80] |
| S2 | Silica gel, pre-coated | Chloroform-methanol(95:5)[173] |
| S3 | Silica gel $HF_{254}$ activated | Chloroform-acetone-methanol(1:1:1)[26] |
| S4 | Silica gel, pre-coated | Chloroform-carbon tetrachloride-methanol(8:5:1)[173] |
| S5 | Silica gel G | Chloroform-ethanol-diethylamine(88:10:2)[80] |
| S6 | Silica gel G | Chloroform-ethanol(9:1) in $NH_3$ atm.[19] |
| S7 | Silica gel $GF_{254}$ | Chloroform-ethanol-formic acid(88:10:2)[21] |
| S8 | Silica gel $GF_{254}$ | Chloroform sat. with 25% ammonia-acetone-*n*-propanol(12:6:2)[74] |
| S9 | Silica gel $GF_{254}$ | Chloroform sat. with 25% ammonia-acetone-isopropanol-*n*-propanol(16:2:1:1)[74] |
| S10 | Silica gel $GF_{254}$ | Benzene-acetone(3:7) in $NH_3$ atm.[21] |
| S11 | Silica gel, pre-coated | Ethyl acetate-methanol-ammonia(8:2:1)[173] |
| S12 | Silica gel G | Acetone-chloroform-*n*-butanol-25% ammonia(3:3:4:1)[17] |

| Alkaloid | $hR_F$ values | | | | | | | | | | | |
|---|---|---|---|---|---|---|---|---|---|---|---|---|
| | S1 | S2 | S3 | S4 | S5 | S6 | S7 | S8 | S9 | S10 | S11 | S12 |
| Caffeine | 90 | 44 | 50 | 37 | 80 | 89 | 57 | 82 | 87 | 70 | 85 | 78 |
| Theobromine | 60 | 22 | 43 | 15 | 50 | 51 | 34 | 55 | 54 | 31 | 68 | 47 |
| Theophylline | 75 | 23 | 49 | 23 | 60 | 21 | 45 | 25 | 27 | 6 | 38 | 26 |
| Inosine | 90 | | | | 90 | | | | | | | |
| Dyphylline | 35 | | 41 | | 65 | | 16 | | 12 | 17 | | |
| 7-(2'-Hydroxyethyl)theophylline | | | 44 | | | | 52 | | 43 | 65 | | 62 |
| 7-(2'-Hydroxypropyl)theophylline | | | 51 | | | | 37 | | 62 | 55 | | 69 |
| 3-Methylxanthine | | 6 | | 7 | | | | | | | 27 | |
| 1-Methyluric acid | | | | - | | | | | | | - | |
| 1,3-Dimethyluric acid | | 4 | | 4 | | | | | | | 21 | |
| 3-Isobutyl-1-methylxanthine | | 34 | | 38 | | | | | | | 51 | |
| Aminophylline | 65 | | | | 75 | | | | | | | |
| Methesculetol theophylline | 55 | | | | 90 | | | | | | | |
| Succiphylline | 80 | | | | 90 | | | | | | | |
| Xanturil | 65 | | | | 75 | | | | | | | |
| 6-Mercaptopurine | 60 | | | | 70 | | | | | | | |
| Dimenhydrinate | 70 | | | | 90 | | | | | | | |
| Uric acid | 0 | | | | 0 | | | | | | | |
| Bamifylline | | | 54 | | | | | | | | | |
| 8-Benzyltheophylline | | | 60 | | | | | | | | | |
| 8-Bromotheophylline | | | 42 | | | | | | | | | |
| 8-Chlorotheophylline | | | 41 | | | | | | | | | |
| 8-Nitrotheophylline | | | 45 | | | | | | | | | |
| Millophylline | | | 35 | | | | 3 | | | 81 | | |
| Etophylate | 20 | | 7 | | 25 | | | | | | | |
| 8-Bromocaffeine | | | 61 | | | | | | | | | |
| 8-Chlorocaffeine | | | 64 | | | | | | | | | |
| 1-(2'-Hydroxypropyl)theobromine | | | | | | | 39 | | | 50 | | |
| 1-(2',3'-Dihydroxypropyl)theobromine | | | | | | | 14 | | | 17 | | |
| 1-Hexyltheobromine | | | | | | | 72 | | | 85 | | |
| Fenethylline | | | | | | | 9 | | | 76 | | |
| 7-(2'-(1''-Methyl-2''-hydroxy-2''-phenylethylamino)ethyl)theophylline | | | | | | | 7 | | | 61 | | |
| 7-(2'-2''-hydroxy-2''-(3''',4'''-dihydroxyphenyl)ethyl amino)ethyltheophylline | | | | | | | 0 | | | 0 | | |

TABLE 17.2

DETECTION OF XANTHINE DERIVATIVES: COLOURS OBTAINED WITH SOME SPRAY REAGENTS[21]

| Alkaloid | Iodine in KI and ethanol, followed by hydrochloric acid in ethanol (no. 52d) | Iodine and $FeCl_3$ in acetone (no. 53) | Dragendorff's reagent, followed by $AgNO_3$ in $H_2SO_4$ (no. 86b) | Limit of detection (µg) |
|---|---|---|---|---|
| Caffeine | Red-brown | Red-brown | Red | <0.5 |
| Theobromine | Dark grey | Dark grey | Red | <0.5 |
| Theophylline | Blue grey | Dark brown | Red | <0.5 |
| Dyphylline | Light brown | Light brown | Dark brown | 5 |
| 7-(2'-Hydroxyethyl) theophylline | Dark brown | Dark brown | Light brown | <0.5 |
| 7-(2'-Hydroxypropyl) theophylline | Light brown | Light brown | Dark brown | 5 |
| Millophylline | Brown | Brown | Yellow | 1 |
| 1-(2'-Hydroxypropyl) theobromine | Grey brown | Grey | Brown | <0.5 |
| 1-(2',3'-Dihydroxy-propyl)theobromine | Light brown | Light brown | Light brown | 5 |
| 1-Hexyltheobromine | Dark blue | Dark grey | Yellow | <0.5 |
| Fenethylline | Brown | Brown | Yellow | 1 |
| 1,3-Dimethyl-7-[2'-(1''-methyl-2''-phenylethylamino)-ethyl]xanthine | Brown | Brown | Yellow | 1 |
| 1,3-Dimethyl-7-[2'-(2''-hydroxy-2''-3''',4'''-dihydroxy-phenyl)ethylamino] ethyl)xanthine | Light brown | Light brown | Light brown | 5 |

TABLE 17.3

LITERATURE CITED IN CHAPTER 3 WHICH INCLUDES THE ANALYSIS OF XANTHINE ALKALOIDS

| Alkaloid* | Ref. | Alkaloid* | Ref. |
|---|---|---|---|
| Caf,Tb,Tp | 17 | Caf,Tb,Tp | 109 |
| Caf,Tb | 18 | Caf,Tb,Tp | 114 |
| Caf | 20 | Caf | 121 |
| Caf,Tb,Tp | 23 | Caf | 124 |
| Caf | 25 | Caf,aminophylline | 126 |
| Caf | 33 | Caf | 144 |
| Caf | 39 | Caf,Tb,Tp,aminophylline | 158 |
| Caf | 40 | Caf,Tb,Tp,aminophylline | 160 |
| Caf,Tb,Tp | 82 | Caf | 161 |
| Caf,Tp | 93 | Caf | 168 |
| Caf | 95 | | |

*Abbreviations:

| | | | |
|---|---|---|---|
| Caf | Caffeine | 8-ClCaf | 8-Chlorocaffeine |
| Tb | Theobromine | 8-OmeCaf | 8-Methoxycaffeine |
| Tp | Theophylline | 8-OmeTb | 8-Methoxytheobromine |
| X | Xanthine | 3-MeX | 3-Methylxanthine |
| hypoX | Hypoxanthine | 1,3,7-triMeUrac | 1,3,7-Trimethyluric acid |
| paraX | Paraxanthine | 1,3-diMeUrac | 1,3-Dimethyluric acid |
| Dyp | Dyphylline | 1,7-diMeUrac | 1,7-Dimethyluric acid |
| OHEtTp | 7-(2'-Hydroxyethyl) theophylline | 3,7-diMeUrac | 3,7-Dimethyluric acid |
| | | 1-MeUrac | 1-Methyluric acid |
| Urac | Uric acid | 3-MeUrac | 3-Methyluric acid |
| Prox | Proxyphylline | 7-MeUrac | 7-Methyluric acid |
| 1-MeX | 1-Methylxanthine | 1,3,7-triMediH-Urac | 1,3,7-Trimethyldihydro uric acid |
| 7-MeX | 7-Methylxanthine | | |
| 8-MeCaf | 8-Methylcaffeine | | |

TABLE 17. 4

XANTHINE DERIVATIVES IN DRUGS OF ABUSE ANALYSIS (SEE CHAPTER 12)

| Alkaloid* | Ref. | Alkaloid* | Ref. |
|---|---|---|---|
| Caf | 24 | Caf | 116 |
| Tp | 35 | Caf | 120 |
| Caf | 44 | Caf | 123 |
| Caf | 51 | Caf | 125 |
| Caf,Tb,Tp | 63 | Caf | 142 |
| Caf | 73 | Caf | 146 |
| Caf | 79 | Caf | 149 |
| Caf | 83 | Caf | 161 |
| Caf | 89 | Caf | 167 |
| Caf | 90 | Caf | 168 |
| Caf | 99 | Caf | 170 |
| Caf | 100 | Caf | 174 |
| Caf | 101 | Caf | 147 |
| Caf | 113 | | |

*For abbreviations, see footnote to Table 17.3.

TABLE 17.5

TLC ANALYSIS OF XANTHINE DERIVATIVES IN PLANT MATERIALS, FOODS AND BEVERAGES

| Alkaloid* | Aim | Adsorbent | Solvent system | Ref. |
|---|---|---|---|---|
| Caf | Alkaloids screening | $SiO_2$ | $BuOH-H_2O-AcOH(4:1:1)$ | 10 |
| Caf | In Cola seeds | $SiO_2$ | $CHCl_3$,$CHCl_3$-cyclohexane, both with 5-10% DEA or in combination with 0.5 *M* NaOH or KOH-impregnated plates | 20 |
| Caf | In galenical plant preparations | $SiO_2$ | MeOH-EtOAc-HCl(7:1:2) | 27 |
| Caf,Tb | In Cola extracts | $SiO_2$ | $CHCl_3$-95% EtOH(9:1) | 29 |
| Caf,Tb | In Cola extracts and tinctures | $SiO_2$ | $CHCl_3$-MeOH(85:15) | 34 |
| Caf,Tb,Tp | In *Coffea*,purity test for Caf | $SiO_2$ | I. $CHCl_3$-EtOH(9:1)<br>II. EtOAc-MeOH-AcOH(8:1:1)<br>Two-dimensional: I,II | 37 |
| Caf,Tb | In Cola extracts | | No details available | 38 |
| Caf,Tb,Tp | In cocoa,direct quantitative analysis | $SiO_2$ | $CHCl_3$-$CCl_4$-MeOH(8:5:1) | 50 |
| Caf,Tb | In Cola extracts | $SiO_2$ | $CHCl_3$-MeOH(95:5) | 52 |
| Caf | In Cola extracts | $SiO_2$ | $CHCl_3$-MeOH(95:5) | 57 |
| Caf,Tb,Tp | In coffee,tea,maté,cola seeds and cocoa,indirect quantitative analysis(UV) | $SiO_2$ | $CHCl_3$-96% EtOH(9:1) | 53,59 |
| Caf | In coffee | $SiO_2$ | $BuOH-H_2O-AcOH(4:1:4)$ | 66 |
| Caf | In coffee | pH 6.8-impregnated $SiO_2$ | $CHCl_3$-EtOH(9:1) | 75 |
| Caf | In coffee beans,direct quantitative analysis (spot areas) | $SiO_2$ | $CHCl_3$-cyclohexane-AcOH(8:2:1) | 76 |
| Caf,Tb,Tp | Densitometric analysis | $SiO_2$ | $CHCl_3$-EtOH(95:5) | 81 |
| Caf | In *Coffea*,direct quantitative analysis (spot areas) | $SiO_2$ | $CHCl_3$-$CCl_4$-MeOH(8:5:1) | 84 |
| Caf | In tea | $SiO_2$ | Benzene-$Me_2CO$(3:7) | 85 |
| Caf,Tb,Tp, adenine | Densitometric analysis in cocoa | $SiO_2$ | EtOAc-MeOH-$NH_4OH$(8:2:1) | 87 |
| Caf | Indirect quantitative analysis (colorimetric) in tea,coffee beans and soft drinks | $SiO_2$ | Benzene-$Me_2CO$(3:7) in $NH_3$ atm. | 122 |
| Caf | In soft drinks | | No details available | 130 |
| Caf,Tb,Tp | In tea,coffee beans and cocoa,indirect quantitative analysis(UV) | Cellulose | $BuOH-H_2O-HCl(100:28:11)$ | 163 |
| Caf,Tb,Tp | Densitometric analysis in Cola seeds | $SiO_2$ | $CHCl_3$-EtOH(99:1) in $NH_3$ atm. | 169 |

*For abbreviations, see footnote to Table 17.3.

TABLE 17.6

TLC ANALYSIS OF XANTHINE DERIVATIVES IN PHARMACEUTICAL PREPARATIONS

| Alkaloid* | Other compounds** | Aim | Adsorbent | Solvent system | Ref. |
|---|---|---|---|---|---|
| Caf | C,Q,narcotine, Ampy,Antp,Ph | Identification in tablets | $SiO_2$ | $CHCl_3$-EtOH(99:1) | 1 |
| Caf,Tp | P,C | Identification | $SiO_2$ | Benzene-$Et_2O$-MeOH-AcOH (120:60:1:18)<br>Benzene-EtOH-AcOH(80:12:5)<br>$Et_2O$-$CHCl_3$-MeOH(50:50:1) | 2 |
| Caf,Tb,Tp | Antp | Identification | pH 6.8-impregnated $SiO_2$ | $CHCl_3$-96% EtOH(9:1) | 3 |
| Caf | Ampy,Ph,C,Phbarb, procaine,AcSal | Identification | $SiO_2$ | $Et_2O$,EtOAc<br>$CHCl_3$-EtOH(100:1) | 8 |
| Caf | Diph,Q,Ampy,Ph, Methampyrone, pyrabital, acetanilide | Identification | | No details available | 11 |
| Caf,Tp | C,P,ethylmorfine,homatropine methylbromide | Identification | $SiO_2$ | MeOH-$NH_4OH$(99:1)<br>$CHCl_3$-EtOH(9:1)<br>$CHCl_3$-$NH_4OH$(95:5)<br>MeOH-$CHCl_3$-$NH_4OH$(90:5:5)<br>MeOH-$Et_2O$-light petroleum (1:1:1)(1:4:4)<br>MeOH-AcOH(9:1) | 12 |
| Caf | Ph,Ampy,mandelic acid benzylester | Indirect quantitative analysis (UV) | $SiO_2$ | Cyclohexane-$Me_2CO$(4:5) | 13 |
| Caf | Q,P,Ph,C,Ampy, AcSal,Antp | Identification | $Al_2O_3$ | Benzene,benzene-EtOH(98:2, 95:5,9:1),$Et_2O$,$CHCl_3$, $CHCl_3$-EtOH(99:1)<br>$CHCl_3$-$n$-BuOH(98:2)<br>$CHCl_3$-$Me_2CO$(1:1) | 15 |
| Caf,Tb,Tp, aminophylline,MeCaf | Barbiturates | Identification | $Al_2O_3$ | Benzene-EtOH(98:2,95:5, 9:1,8:2,7:3)<br>$CHCl_3$-$n$-BuOH(98:2)<br>$CHCl_3$-$Me_2CO$(1:1) | 16 |
| OHEtTp,Tp | | Identification | $SiO_2$ | $CHCl_3$-isoprOH-$NH_4OH$(90:8:10) | 28 |
| Caf | Antp,C,acetanilide,AcSal,Ampy, Ph,Antp,Sal, Salam,Proph, aethallymal | Identification in tablets | $SiO_2$ | $CHCl_3$-$Me_2CO$(8:2)<br>EtOAc,93% EtOH,<br>$Et_2O$-benzene(7:3)<br>$Me_2CO$-5 $M$ $NH_4OH$(9:1)<br>Hexane-$CHCl_3$-AcOH(20:5:5) | 36 |
| Caf | Ampy,C,methampyrone,Phbarb | Identification | | No details available | 42 |
| Caf,Tb,Tp | Ampy,Antp,methampyrone,Phbu | Identification | $Al_2O_3$ | Benzene-EtOH(9:1)<br>Benzene-MeOH(9:1) | 43 |
| Tp,aminophylline | Salam,PhBu, sulfisomidin | Identification in suppositories | $SiO_2$ | Dioxane | 45 |
| Caf | Ampy,Antp,C, AcSal,Ph | Identification | $SiO_2$ | $Me_2CO$, 0.5 $M$ NaOH,$CHCl_3$,$H_2O$<br>BuOH-$H_2O$-AcOH(4:5:1)<br>BuOH-$H_2O$(1:1) | 46 |

TABLE 17.6 (continued)

| Alkaloid* | Other compounds** | Aim | Adsorbent | Solvent system | Ref. |
|---|---|---|---|---|---|
| Caf,Tb,Tp | Benzoic acid, AcSal,*p*-hydroxy-benzAc, Sal, MeSal,PhSal, methylparaben, propylparaben | Separation | $SiO_2$<br><br><br><br><br><br><br><br><br><br><br><br>$SiO_2$,$Al_2O_3$ (basic), $Al_2O_3$ (acidic) | $CHCl_3$-EtOH-AcOH(95:2.5:2.5)<br>$CHCl_3$-Cyclohexane-EtOH-AcOH (80:10:5:5)<br>Benzene-$Et_2O$-MeOH-AcOH (30:15:0.25:4.5)<br>Pentane-isopr$_2$O-AcOH (60:40:5)(90:25:4)<br>Benzene-dioxane-AcOH (70:26:4)<br>EtOH-$H_2O$-25% $NH_4OH$ (25:3:0.25)<br>Benzene-ethanol(8:2)<br>$CHCl_3$-EtOH(95:5)<br>$CHCl_3$-EtOH-AcOH(95:2.5:2.5)<br>$CHCl_3$-dioxane-AcOH(70:26:4) | 48 |
| Caf,Tb,Tp, Dyp | Ampy,Antp,Me-thampyrone, butadione | Identification | $SiO_2$ | EtOAc-$Me_2CO$-BuOH-10% $NH_4OH$ (5:4:3:1)<br>EtOAc-$Me_2CO$-25% $NH_4OH$(4:3:1)<br>EtOAc-$Me_2CO$(1:2)<br>$Me_2CO$-BuOH(1:2) | 49 |
| Caf | Ampy,Antp,benzAc Ph,Phbarb,EtM,C, Belladonna ex-tract | Identification | $SiO_2$ | $Me_2CO$-cylohexane(5:4)<br>$Me_2CO$-hexane(5:4)<br>$Me_2CO$-MeOH-DEA(1:1:0.03, 2:1:0.03)<br>$Me_2CO$-$CHCl_3$-DEA(2:1:0.03)<br>Benzene-dioxane-AcOH(5:4:1)<br>EtOH-$NH_4OH$(4:1)<br>$CHCl_3$-EtOH(4:1)<br>$CHCl_3$-benzene-AcOH(5:4:1)<br>EtOAc-MeOH-AcOH(8:1:1)<br>EtOAc-benzene(11:5) | 54 |
| Caf,Tp | Ampy,C,barbitur-ates | Identification in suppositories | $SiO_2$ | Benzene-$Et_2O$-AcOH(40:30:9)<br>Toluene-DEA(5:1) | 58 |
| Caf,Tb,Tp | C,P,S,AcSal,Ph, A,MeH,Eph,pro-caine,chlorpro-mazine,tripe-lennamine,meli-pramine | Identification | $SiO_2$ | $Me_2CO$-cyclohexane-EtOAc (1:1:1) in $NH_3$ atm. | 60 |
| Caf | AcSal,Ampy,Ph | Identification | $SiO_2$ | Cyclohexane-$CHCl_3$-DEA (5:4:1)<br>Cyclohexane-$CHCl_3$-AcOH (5:4:1) | 61 |
| Caf | AcSal,Ampy,C, Q,barbiturates, Ph | Identification | $SiO_2$ | $Me_2CO$,0.5 *M* NaOH | 62 |
| Caf | S,AcSal | Identification | $SiO_2$ | $CHCl_3$-$Me_2CO$-DEA(5:4:1)<br>Benzene-$Et_2O$-MeOH-AcOH (120:60:1:18) | 64 |
| Caf | | Densitometric analysis | $SiO_2$ | Benzene-$Et_2O$-AcOH-$H_2O$ (120:60:18:1) | 65 |
| Tp | Eph,dimedrol | Indirect quanti-tative analysis | $Al_2O_3$ | Benzene-EtOH(9:1) | 67 |

References p. 441

TABLE 17.6 (continued)

| Alkaloid* | Other compounds** | Aim | Adsorbent | Solvent system | Ref. |
|---|---|---|---|---|---|
| Caf, 8-ClTp | Diph,meclozine, chlorphenoxamine | Separation with azeotropic solvents | $SiO_2$ | Diisopropylamine-1,2-dichloroethane(1:9)<br>Diisopropylamine-isoprOH (1:9) | 68 |
| Caf | Acetanilide,Q, Meacetanilide, Ph,Ampy,Antp | Identification and indirect quantitative analysis(UV) | $SiO_2$ | Benzene-$Et_2O$-MeOH-AcOH (120:120:1:18)<br>$CHCl_3$-96% EtOH(99:1)<br>Cyclohexane-$CHCl_3$-pyridine (20:60:5) | 70, 97 |
| Caf | AcSal,Ph | Identification | $SiO_2$ | Benzene-$Et_2O$-MeOH-AcOH (120:60:1:18) | 77 |
| Caf,Tb,Tp | | Densitometric analysis | $SiO_2$ | $CHCl_3$-EtOH(95:5) | 81 |
| Caf,Tb,Tp | C,EtM,Phbarb,Ph, Ampy,AcSal (see Chapter 3) | Identification and indirect quantitative analysis (UV) | $SiO_2$ | 70% EtOH<br>$Me_2CO$-cyclohexane-EtOAc (1:1:1)<br>$CHCl_3$-$Me_2CO$(1:1)<br>$CHCl_3$-MeOH-cyclohexane (7:3:1)<br>All solvents in $NH_3$ atm. | 82 |
| Tp,Dyp | Eph,Phbarb,prednisolone,prednisone,P,ethyl *p*-aminobenzoate | Quantitative analysis | $SiO_2$ | $CH_2Cl_2$-MeOH-AcOH(90:10:3) | 86 |
| Caf | benzac,P,Ampy,C Phbarb | Identification on microslides | $SiO_2$<br>$Al_2O_3$ | Benzene-$Me_2CO$(3:7) in $NH_3$ atm.<br>Benzene-$Me_2CO$(7:3) | 88 |
| Caf | AcSal,Ph | Densitometric analysis | $SiO_2$ | Benzene-$Et_2O$-AcOH-MeOH (60:30:9:0.5) | 91 |
| Caf | Ampy,C,Ph,barbiturates | TAS technique for identification in tablets | $SiO_2$ | $Et_2O$ | 94 |
| Caf,Tb,Tp | Ampy,AcSal,adenosine,Eph,P, RSP,Phbarb,Ph, nikethamide | Identification on microslides | | No details available | 96 |
| Caf | nikethamide,S | Identification | $SiO_2$ | I. $CHCl_3$-$Et_2O$-90% EtOH-DEA (35:5:10:1.5)<br>II. $CHCl_3$-$Me_2CO$-MeOH-BuOH (4:2:1:1)<br>Two-dimensional: I,II | 98 |
| Caf, OHEtTp | No,C,Phbarb, Diph,PhEph, thiamine | Densitometric analysis | $SiO_2$ | PrOH-25% $NH_4OH$(88:2)<br>Cyclohexane-$CHCl_3$-DEA (90:18:12)<br>Benzene-$Et_2O$-MeOH-AcOH (60:30:1:9)<br>PrOH-EtOAc-$H_2O$-BuOH(4:3:3:2) | 102 |
| Caf | AcSal,khelline, Q,pholcodine, promethazine | Identification and indirect quantitative analysis(UV) | $SiO_2$ | Two-dimensional:<br>I. BuOH-$H_2O$-AcOH(4:1:5), org. phase<br>II. tartaric acid in $H_2O$ (1.5 g/l) | 103 |

TABLE 17.6 (continued)

| Alkaloid* | Other compounds** | Aim | Adsorbent | Solvent system | Ref. |
|---|---|---|---|---|---|
| Caf | Ampy,barbiturates | Identification | $SiO_2$ | Cyclohexane-$Me_2CO$(4:5) | 105 |
| Caf | AcSal,Sal,Salam | Identification | $SiO_2$ | Benzene-$Et_2O$-MeOH-AcOH (120:60:1:18) | 106 |
| Caf | Ampy,acetaminophen | Identification | | No details available | 107 |
| Caf | Ampy,acetaminophen,Ph | Identification in folk medicines | | No details available | 108 |
| Caf | Ph,Antp,Ampy, mandelic acid benzyl ester | Identification | $SiO_2$ | $Me_2CO$-cyclohexane(5:4)<br>Benzene-$Me_2CO$(8:2) | 110 |
| Caf | Ampy,Phbarb,Ph, C,ergotamine, mecloxamine | Identification | | No details available | 111 |
| Caf | AcSal,Salam,Ph, Ampy,P | Identification | $SiO_2$ | $Et_2O$-benzene-MeOH-AcOH (20:10:5:2)<br>Toluene-$Me_2CO$-EtOH-25% $NH_4OH$ (45:45:7:3) | 115 |
| Caf | AcSal,Ampy,C,Ph | Identification and quantitative analysis | | No details available | 117 |
| Caf | | Indirect quantitative analysis (colorimetric) | $SiO_2$ | Benzene-$Me_2CO$(3:7) in $NH_3$ atm. | 122 |
| Caf | Ph,Salam,propylphenazone | Indirect quantitative analysis (UV) | $SiO_2$ | $Et_2O$ | 127 |
| Caf | Ampy,phenprobamate,ascorbic acid | Identification | $SiO_2$ | 95% EtOH<br>Cyclohexane-$Me_2CO$(4:5)<br>$CHCl_3$-$Me_2CO$-95% EtOH (70:30:3)<br>Cyclohexane-$CHCl_3$-AcOH (4:5:1)<br>95% EtOH-AcOH(9:1) | 128 |
| Caf | Ampy,Ph,acetaminophen,methyltestosterone | Identification in folk medicines | | No details available | 131 |
| Caf | C,Phbarb,promethazine | Identification | $SiO_2$ | Hexane-95% EtOH-DEA(25:10:5) | 132<br>136 |
| Tp | Adrenaline, benzac,nikethamide,adenosine | Identification | | No details available | 133 |
| Caf | Ph | Densitometric analysis | $SiO_2$ | Benzene-dioxane-AcOH(60:20:2) | 134 |
| Caf | AcSal,Ph,acetaminophen | Densitometric analysis | $SiO_2$ | Two-dimensional:<br>I. *n*-Hexane-dioxane-HCOOH (45:40:2)<br>II. $Me_2CO$-$CHCl_3$-HCOOH (30:30:1) | 135 |
| Caf | Ph,Q,Ampy | Densitometric analysis | | No details available | 137 |

TABLE 17.6 (continued)

| Alkaloid* | Other compounds** | Aim | Adsorbent | Solvent system | Ref. |
|---|---|---|---|---|---|
| Caf | Benzac | Densitometric analysis | | No details available | 138 |
| Caf,Tb | Platyphylline,P, Phbarb,salsoline | Identification | $Al_2O_3$ | Benzene-MeOH(9:1) | 139 |
| Caf,Tb,Tp | C,EtM,P,A,Acsal, Ph,MeHomatropine barbiturates, methampyrone | Chromatographic behaviour with single-component mobile phases | $SiO_2$ | MeOH,EtOH,PrOH,BuOH,AmOH, $Me_2CO$,MeEtCO,$MeCOCH_2CH_2Me_2$, EtOAc,BuOAc,AmOAc,$Et_2O$, $isopr_2O$,$Bu_2O$,$CH_2Cl_2$,$CHCl_3$, $CCl_4$,benzene,cyclohexane, light petroleum | 140 |
| Caf,Tp | Eph,lidocaine | Identification | $SiO_2$ | EtOH(96%) in $NH_3$ atm. | 141 |
| Tp,OHEtTp | P,Phbarb,MeA | Identification in suppositories | $SiO_2$ | MeOH-$Me_2CO$-conc. HCl (90:10:4)<br>$Me_2CO$-$CHCl_3$-*n*-BuOH-conc. $NH_4OH$(3:3:4:1) | 148 |
| Caf | | Densitometric analysis | $SiO_2$ | $Me_2CO$-$CHCl_3$(5:1) | 150 |
| Caf | C,AcSal,Ampy,Ph, barbiturates | Identification | $SiO_2$ | $CHCl_3$-$Et_2O$,$CHCl_3$-EtOH, benzene-$Me_2CO$-$Et_2O$-$H_2O$, $CHCl_3$-$Et_2O$-$H_2O$<br>No further details available | 151 |
| Caf | Ph | Direct quantitative analysis | $SiO_2$ | $CHCl_3$-HCOOH(50:2) | 154 |
| Caf | Ampy,C,Ph, Phbarb,AcSal | Identification | $SiO_2$ | No further details available | 155 |
| Caf,Tp, aminophylline | AcSal,Benzac, acetaminophen, methaqualone, Ph,chlorpheniramine,pyrithyldione | Identification | $SiO_2$ | *n*-BuOH-$H_2O$-$NH_4OH$(90:8:2)<br>Benzene-dioxane-AcOH (90:2.5:4) | 156 |
| Caf,Tp,Tb, X | Ergotamine, ergotaminine, A,Scop,barbiturates | Separation on $Ag^+$-impregnated $SiO_2$ | $Ag^+$-impregnated $SiO_2$ | $CHCl_3$-EtOH(9:1)<br>MeOH-$Me_2CO$-triethanolamine (1:1:0.03) | 157 |
| Caf | Ergotamine | Densitometric analysis | $SiO_2$ | Consecutively:<br>I. $CHCl_3$-benzene(1:1) (2 x 15 cm)<br>II. $CHCl_3$-EtOH(9:1) (1 x 20 cm)<br>III. $Me_2CO$-cyclohexane-MeOH (49:49:2)(1 x 20 cm) | 159 |
| Caf | AcSal,Ph | Identification | $SiO_2$ | Benzene-$Me_2CO$(3:7)<br>BuOAc-$Me_2CO$-BuOH-$NH_4OH$ (5:4:3:1) | 164 |
| Caf | AcSal,Ampy,Par, Phbu,dextropropoxyphene | Identification | | No details available | 165 |
| Caf,Tb,Tp, OHEtTp aminophylline | Ampy,Eph,P,barbiturates, mebrophenhydamine,radobelin | Identification | $SiO_2$ | No further details available | 166 |

TABLE 17.6 (continued)

| Alkaloid* | Other compounds** | Aim | Adsorbent | Solvent system | Ref. |
|---|---|---|---|---|---|
| Caf | Benzac | Direct quantitative analysis (planimetric), TAS technique | $SiO_2$ | Benzene-$Me_2CO$(3:7),sat. with 25% $NH_4OH$ | 172 |
| Caf | AcSal,Ph | Densitometric analysis | $KC_{18}$,reversed-phase | MeOH-0.5 *M* NaCl(1:1) | 175 |
| Caf,Tb | Ampy,AcSal,Ph, Phbarb,methampyrone | Identification | Polyamide | $CHCl_3$-96% EtOH-AcOH(96:4:1)<br>Cyclohexane-$CHCl_3$-MeEtCO (6:1:2)<br>EtOAc-cyclohexane-MeOH-25% $NH_4OH$ (70:15:10:5)<br>Cyclohexane-$CHCl_3$-dioxane (5:3:1)<br>Hexane-BuOH-AcOH(10:20:1) | 177 |

*For abbreviations, see footnote to Table 17.3.

**Abbreviations:

| | | | |
|---|---|---|---|
| A | Atropine | Par | Paracetamol |
| AcSal | Acetylsalicylic acid | Ph | Phenacetine |
| Ampy | Aminopyrine | Phbarb | Phenobarbital |
| Antp | Antipyrine | Phbu | Phenylbutazone |
| B | Brucine | PhEph | Phenylephrine |
| Benzac | Benzoic acid | PhSal | Phenylsalicylic acid |
| C | Codeine | Proph | Propyphenazone |
| Diph | Diphenhydramine | Q | Quinine |
| Eph | Ephedrine | RSP | Reserpine |
| EtM | Ethylmorphine | S | Strychnine |
| MeA | Methylatropine | Sal | Salicylic acid |
| MeSal | Methylsalicylic acid | Salam | Salicylamide |
| No | Noscapine | Scop | Scopolamine |
| P | Papaverine | | |

TABLE 17.7

TLC ANALYSIS OF XANTHINE DERIVATIVES IN BIOLOGICAL MATERIALS (URINE, BLOOD, ETC.)

See also analysis of drugs of abuse (TABLE 17.4).

| Alkaloid* | Other compounds** | Aim | Adsorbent | Solvent system | Ref. |
|---|---|---|---|---|---|
| Caf,Tb, Tp | Barbiturates, Ampy,Antp, Sal,Ph | Identification | $SiO_2$ | $CHCl_3$-$Et_2CO$(85:15) | 4 |
| Caf | Various opium alkaloids and related compounds,A,S,B, Q,nicotine, Antp,RSP,Ampy, Proph,benoxinate,methylphenidate,cocaine,procaine | Identification | $SiO_2$ | MeOH-$Me_2CO$-triethanolamine(1:1:0.03) | 7 |
| Caf | Amphetamine,S | Detection in horse urine, indirect quantitative analysis (UV) | 0.1 *M* NaOH-impregnated $SiO_2$ | MeOH | 41 |
| Tp,Dyp, Prox | | Direct quantitative analysis in blood (UV) | $SiO_2$ | $CHCl_3$-EtOH-HCOOH(85:5:10) (2 x) | 47 |
| Caf,Tb, parax | | Caffeine metabolites in blood | pH 6.8-impregnated $SiO_2$ | BuOH-$H_2O$-AcOH(4:1:1)<br>BuOH-$H_2O$-HCOOH(33:7:1) | 56 |
| Caf,Tb, Tp,etophylate,aminophylline,succiphylline, veinartan, Dyp,dimenhydrinate xanturil, Urac, 6-mercaptopurine, adenyl, adenosine triphosphate | Bromodiphenhydramine | Identification by means of murexide reaction (Table 17.1) | $SiO_2$ | $CHCl_3$-EtOH(9:1)<br>$CHCl_3$-EtOH-DEA(88:10:2)<br>Benzene-$Me_2CO$(3:7)<br>EtOAc-MeOH-AcOH(8:1:1)<br>$Me_2CO$-$CHCl_3$-*n*-BuOH-$NH_4OH$ (3:4:4:1)<br>$CHCl_3$-EtOH-HCOOH(88:10:2)<br>Cyclohexane-$CHCl_3$-DEA (5:4:1)<br>Benzene-EtOAc-DEA(7:2:1) | 80 |
| Caf,Tb | Ampy | Detection in horse urine | $SiO_2$ | $CHCl_3$-MeOH(9:1) | 92 |
| Caf | Ampy,glutethimide,Q,hydroxyzine | Detection in cadaveric material | $SiO_2$<br>$Al_2O_3$ | MeOH-$NH_4OH$(100:1)<br>$CHCl_3$-$Et_2O$(1:1) | 112 |
| Tp,Caf | Eph,diazepam, Phbarb,C,Diph | Densitometric analysis of Tp in plasma | $SiO_2$ | $CHCl_3$-MeOH(9:1) | 145 |

TABLE 17.7 (continued)

| Alkaloid* | Other compounds** | Aim | Adsorbent | Solvent system | Ref. |
|---|---|---|---|---|---|
| Caf,Tp, Tb,paraX, 1-MeX, 3-MeX, 7-MeX, 1,3,7-triMeUrac, 1,3-diMe-Urac,1,7-diMeUrac, 3,7-diMe-Urac, 1-MeUrac, 3-MeUrac, 7-MeUrac, 1,3,7-tri-MediHUrac, triMeal-lantoin | N-methylurea N,N-dimethylurea | Direct quantitative analysis of metabolites of $^{14}C$-labelled Caffeine in urine | $SiO_2$ | I. $CHCl_3$-MeOH(4:1)<br>II. $CHCl_3$-$Me_2CO$-*n*-BuOH-conc. $NH_4OH$<br>Two dimensional: I,II | 152 |
| Caf | Ergotamine | Densitometric analysis in blood | $SiO_2$ | Consecutively:<br>I. $CHCl_3$-benzene(1:1) (2 x 15 cm)<br>II. $CHCl_3$-EtOH(9:1) (1 x 20 cm)<br>III. $Me_2CO$-cylohexane-MeOH (49:49:2)(1 x 20 cm) | 159 |
| Caf | Phbarb,P | Indirect quantitative analysis(UV) | $SiO_2$ | MeOH | 162 |
| Caf | | Densitometric analysis in blood | $SiO_2$ | $CHCl_3$-$Me_2CO$(9:1) | 171 |
| Caf,Tp, Tb,1,3-diMeUrac, hypoX, 1-MeUrac, 3-MeUrac, 3-iso-butyl-1-Me-X | Amobarb,dextromethorphan,Eph, hydroxyzine, Phbarb,salbutamol,terbutalin,glyceryl guaiacolate | Densitometric analysis of theophylline in plasma (Table 17.1) | $SiO_2$ | $CHCl_3$-MeOH(95:5)<br>$CCl_4$-$CHCl_3$-MeOH(8:5:1)<br>EtOAc-MeOH-$NH_4OH$(8:2:1) | 173 |
| Caf,Tp, Tb,1-MeX, 3-MeX,7-MeX, 1,7-diMeX, 3-MeUrac, 1-MeUrac, 1,3-diMe-Urac,Urac, X,hypoX | Bilirubin, haemoglobin | Densitometric analysis in serum, saliva and urine | $SiO_2$ | $CHCl_3$-MeOH(9:1)<br>EtOAc-MeOH-25% $NH_4OH$(8:2:1) | 176 |
| Caf,Tb, Tp,3-MeX, Urac | | Indirect quantitative analysis (UV) of theophylline in plasma | $SiO_2$ | EtOAc-MeOH-$NH_4OH$(80:20:15) | 178 |

TABLE 17.7 (continued)

| Alkaloid* | Other compounds** | Aim | Adsorbent | Solvent system | Ref. |
|---|---|---|---|---|---|
| Caf | | Indirect quantitative analysis of $^{14}C$-labelled caffeine in plasma | $SiO_2$ | I. $CHCl_3$-EtOH(9:1)<br>II. $Me_2CO$-*n*-BuOH-$CHCl_3$-25% $NH_4OH$(3:4:3:1)<br>Two dimensional: I,II | 179 |

*For abbreviations, see footnote to Table 17.3.
**For abbreviations, see second footnote to Table 17.6.

TABLE 17.8

TLC ANALYSIS OF PURE XANTHINE DERIVATIVES

| Alkaloid* | Aim | Adsorbent | Solvent system | Ref. |
|---|---|---|---|---|
| Caf,Tb,Tp | Separation | $SiO_2$ | $CCl_4$-$CHCl_3$-MeOH(5:5:1) | 14 |
| Caf,Tb,Tp,OHEtTp, Prox | Separation (Table 17.1) | $SiO_2$ | $Me_2CO$-$CHCl_3$-*n*-BuOH-25% $NH_4OH$(3:3:4:1) | 17 |
| Caf,Tb,Tp | Separation (Table 17.1) | $SiO_2$ | $CHCl_3$-EtOH(9:1) in $NH_3$ atm. | 19 |
| Caf,Tb,Tp and various derivatives (Table 17.1) | Identification and indirect quantitative analysis(UV) | $SiO_2$ | I. Benzene-$Me_2CO$(3:7)<br>II. $CHCl_3$-EtOH-HCOOH (88:10:2)<br>III. $CH_2Cl_2$-MeOH(92:8)<br>Two-dimensional: I,II | 21 |
| Caf,Tb,Tp | Separation | $Al_2O_3$ | BuOH-isoAmOH-HCl-$H_2O$ (8:2:1:20) | 22 |
| Caf,Tb,Tp,various derivatives (Table 17.1) | Separation | $SiO_2$ | $CHCl_3$-$Me_2CO$-MeOH(1:1:1) | 26 |
| Caf,Tb,Tp,X, hypoX,Urac,guanine,adenine | Separation | 10% piperazine-impregnated $SiO_2$ | IsoprOH-$CHCl_3$-10% aq. piperazine(6:2:2) | 32 |
| Caf,Tb,Tp | To distinguish between natural and synthetic caffeine | $SiO_2$ | I. $CHCl_3$-EtOH(9:1)<br>II. EtOAc-MeOH-AcOH(8:1:1)<br>Two-dimensional: I,II | 37 |
| Caf,Tb,Tp,Dyp, OHEtTp,1-(2,3-dihydroxypropyl)Tb, 8-MeCaf,8-ClCaf, 8-OMeCaf,8-OMeTb | Separation | $SiO_2$ | Buffer(pH 3,7 or 10) with 0.025,0.050 or 0.075 *M* sodium 2-naphthol-6,8-disulphonate | 55 |
| Caf,Tb,Tp,Dyp, OHEtTp,aminophylline,bamifylline, 1-hexylTb,etophylate, OHpropTb, xanthinol niacinate | Separation with azeotropic solvents | $SiO_2$ | *n*-Butylamine-EtOH-1-BuCl (1:2:7)<br>Diisopropylamine-1,2-dichloroethane-isoprOH (1:4:5) | 68 |
| Caf,Tb,Tp | Influence of pH on TLC behaviour | $SiO_2$ | $CHCl_3$-96% EtOH(99:1) in 2.5,5.0,10.0 or 20.0% $NH_4OH$ atm.,12 or 25% HCl atm. 10% HCOOH atm.,or $H_2O$ atm. | |
| | | $SiO_2$ | Buffer of pH 3.6,3.8,4.15, 9.9,10.05 or 11.4 | 71 |
| Caf,Tb,Tp | Separation on pH-gradient plates | $SiO_2$ | $CHCl_3$-$Et_2O$(85:15)<br>$CCl_4$-$CHCl_3$-MeOH(5:5:1)<br>$CCl_4$-$CHCl_3$-MeOH(8:5:1)<br>$CHCl_3$-$Me_2CO$-MeOH(1:1:1)<br>Benzene-$Me_2CO$(3:7)<br>$CHCl_3$-EtOH(9:1) | 72 |
| Caf,Tb,Tp,Prox, Etofyllin,Dyp | Separation (Table 17.1) | $SiO_2$ | $CHCl_3$ sat. with 25% $NH_4OH$-$Me_2CO$-*n*-prOH(12:6:2)<br>$CHCl_3$-sat.with 25% $NH_4OH$-$Me_2CO$-isoprOH-*n*-prOH (16:2:1:1) | 74 |

TABLE 17.8 (continued)

| Alkaloid* | Aim | Adsorbent | Solvent system | Ref. |
|---|---|---|---|---|
| Caf,Tb,Tp,paraX,X, hypoX,7-MeX,1-MeX, aminophylline | Separation | $SiO_2$ | $CHCl_3$-MeOH(4:1)<br>EtOAc-MeOH-28% $NH_4OH$(8:2:1)<br>*n*-BuOH sat. with 2.8% $NH_4OH$ | 78 |
| Caf | Separation from Q and B on microslides | $SiO_2$ | MeOH,EtOH,$Me_2CO$,$CHCl_3$,benzene | 118 |
| Caf | Separation from nicotine,Q,S and B on plates prepared by spraying adsorbent | $SiO_2$ | Benzene-EtOAc-DEA(10:10:3) | 119 |
| Caf,Tb,Tp | Separation | $SiO_2$ | $Me_2CO$-$CHCl_3$-BuOH-25% $NH_4OH$ (3:3:4:1)<br>$CHCl_3$-$Et_2O$(9:1)<br>$CHCl_3$-EtOH(9:1) | 129 |
| Caf | Densitometric analysis | $SiO_2$ | $Me_2CO$-$CHCl_3$(5:1) | 150 |
| X,Caf,Tb,Tp | Separation on $Ag^+$-impregnated $SiO_2$ | $SiO_2$ or $Ag^+$-impregnated $SiO_2$ | $CHCl_3$-EtOH(9:1)<br>MeOH-$Me_2CO$-triethanolamine (1:1:0.03) | 157 |

*For abbreviations, see footnote to Table 17.3.

Chapter 18

# DITERPENE ALKALOIDS

18.1. Solvent system ........................................................ 461
18.2. Detection ........................................................ 462
18.3. Quantitative analysis ........................................................ 462
18.4. TAS technique and reaction chromatography ........................................................ 462
References ........................................................ 463

## 18.1. SOLVENT SYSTEMS

Most work on the TLC of diterpene alkaloids concerns the alkaloid aconitine and its analysis in plant materials and in toxicology. For the separation of aconitine and related alkaloids from plant materials a number of different TLC systems have been used with neutral, acidic and basic mobile phases in combination with silica gel.

Denoël and Van Cotthem[3] used *n*-butanol-acetic acid (100:14 or 100:10) saturated with water to separate aconitine and pseudoaconitine. Fischer and Weixlbaumer[6] separated aconitine, benzoylaconine and aconine with isopropanol-methanol-23% ammonia (36:24:1), prior to their quantitative analysis (Table 18.1). Wartmann-Hafner[10] tested several solvent systems for the separation of the alkaloids present in aconitine-containing extracts and tinctures. Cyclohexane-diethylamine (9:1) and methanol were found to be most suited. Frijns[13] used two-dimensional TLC to analyse aconitine-containing preparations. The first solvent, chloroform-diethylamine (9:1), separated the alkaloids from the ballast compounds, and the second, cyclohexane-chloroform-diethylamine (5:4:1), separated the alkaloids.

Ragazzi et al.[4] used magnesium oxide impregnated with 2.5% calcium chloride for the analysis of a number of alkaloids, including aconitine. *n*-Hexane-methanol (4:1) and ethyl acetate or ethyl acetate-acetone (4:1) were used as solvent systems. Hiermann[22] also used magnesium oxide as the stationary phase to separate aconitine, benzoylaconine and aconine with light petroleum (b.p. 60-80$^{o}$C)-acetone-methanol (25:8:1) (Table 18.1).

Yuan-Lung Chu et al.[7] found basic aluminium oxide (activity IV) in combination with light petroleum-diethyl ether (1:10) to be suitable for the separation of aconite alkaloids. Strzelecka[11] preferred aluminium oxide for the separation of the alkaloids present in a *Delphinium* species. Elatine and methyllycaconitine could be separated and identified with cyclohexane-chloroform-methanol (10:6:1).

**References p. 463**

Golkiewicz et al.[12] investigated the alkaloids of a *Consolida* species by means of TLC. The results of their studies are summarized in Table 18.2. Soczewinski and Golkiewicz[15] studied the correlation of the adsorption affinity of organic substances with their acid - base properties. Among the compounds studied were a number of *Consolida* alkaloids. It was found that for silica gel plates developed with acetone or propanol, the adsorption affinity was determined mainly by the basicity of the tertiary nitrogen atom.

## 18.2. DETECTION

The most commonly used detection reagent for aconite alkaloids is Dragendorff's reagent. However, iodoplatinate reagent, iodine vapour[14], 0.1 *N* iodine solution[6] and cobalt thiocyanate[22] (no. 26b) have also been used for this type of alkaloid. With iodoplatinate reagent aconitine gives a red-brown colour[1] and with cobalt thiocyanate blue spots against a pink background[22]. Okumura et al.[23] used flame-ionization detection for a number of alkaloids, including aconitine.

## 18.3. QUANTITATIVE ANALYSIS

Indirect quantitative analysis of aconitine in aconite preparations has been described by Fischer and Weixlbaumer[6]. After TLC separation the alkaloid was eluted from the plate with chloroform-isopropanol (2:1). After evaporation of the solvent the residue was dissolved in 1% hydrochloric acid and the absorbance measured at 234 nm. Before use the silica gel plates were washed with chloroform-isopropanol (2:1). Because of difficulties in the elution of the alkaloid from the silica gel, Hiermann[22] used magnesium oxide as the sorbent. After TLC separation the spots of the alkaloids were scraped off the plate and the alkaloid plus stationary phase dissolved in 2% hydrochloric acid. The absorbance was measured at 234 nm. Massa et al.[17] described a densitometric determination of alkaloids at 400 nm after detection with Dragendorff's reagent. Aconitine was also determined.

## 18.4. TAS TECHNIQUE AND REACTION CHROMATOGRAPHY

Identification of aconite by means of the TAS technique was described by Jolliffe and Shellard[18]. A 10 - 20-mg amount of plant material was mixed with 10 mg of calcium hydroxide and heated at 275°C for 90 sec. Silica gel with a suitable moisture content was used as the propellant.

Reaction chromatography of aconitine was used by Kaess and Mathis[8]. By saponification of aconitine (heating for 1 h at 100°C in 0.1 *M* potassium hydroxide in

95% ethanol) complete conversion into aconine was obtained. If the saponification was performed in the cold the intermediate, benzoylaconine, could also be observed on the plates.

In Chapter 3, refs. 1, 2, 5, 19, 20 and 21 (from this chapter) include the analysis of aconitine.

REFERENCES

1 D. Waldi, K. Schnackerz and F. Munter, *J. Chromatogr.*, 6 (1961) 61.
2 E. Vidic and J. Schütte, *Arch. Pharm. (Weinheim)*, 295 (1962) 342.
3 A. Denoël and B. Van Cotthem, *J. Pharm. Belg.*, 18 (1963) 346.
4 E. Ragazzi, G. Veronese and C. Giacobazzi, in G.B. Marini-Bettõlo (Editor), *Thin-Layer Chromatography*, Elsevier, Amsterdam, 1964, p. 149.
5 R. Paris, R. Rousselet, M. Paris and M.J. Fries, *Ann. Pharm. Fr.*, 23 (1965) 473.
6 R. Fischer and H. Weixlbaumer, *Pharm. Zentralhalle*, 104 (1965) 298.
7 Yuan-Lung Chu, Chih-Chen Lu and Jen-Hung Chu, *Yao Hsueh Hsueh Pao*, 12 (1965) 381; *C.A.*, 64 (1966) 7966g.
8 A. Kaess and C. Mathis, *Int. Symp. Chromatogr. Electrophor. Lect. Pap. 4th*, (1966) 525.
9 A. Affonso, *J. Chromatogr.*, 21 (1966) 332.
10 F. Wartmann-Hafner, *Pharm. Acta Helv.*, 41 (1966) 406.
11 H. Strzelecka, *Diss. Pharm. Pharmacol.*, 19 (1967) 81.
12 W. Golkiewicz, M. Przyborowska and E. Soczewinski, *Diss. Pharm. Pharmacol.*, 20 (1968) 635.
13 J.M.G.J. Frijns, *Pharm. Weekbl.*, 103 (1968) 929.
14 M. Struhar, V. Springer and J. Pekarek, *Cesk. Farm.*, 18 (1969) 62.
15 E. Soczewinski and W. Golkiewicz, *Chem. Anal. (Warsaw)*, 14 (1969) 465.
16 M. Vanhaelen, *J. Pharm. Belg.*, 24 (1969) 87.
17 V. Massa, F. Gal and P. Susplugas, *Int. Symp. Chromatogr. Electrophor. Lect. Pap. 6th*, (1970) 470.
18 G.H. Jolliffe and E.J. Shellard, *J. Chromatogr.*, 48 (1970) 125.
19 G.S. Tadjer, *J. Chromatogr.*, 63 (1971) D44.
20 E. Novakova and J. Vecerkova, *Cesk. Farm.*, 22 (1973) 347.
21 K.F. Ahrend and D. Tiess, *Wiss. Z. Univ. Rostock, Math. Naturw. Reihe*, 22 (1973) 951.
22 A. Hiermann, *Zentralbl. Pharm. Pharmakother. Laboratoriumdiagn.*, 113 (1974) 1247.
23 T. Okumura, T. Kadono and A. Isoo, *J. Chromatogr.*, 108 (1975) 329.
24 G.K. Munshi, V. Mudgal and R. Chanda, *J. Inst. Chem. (India)*, 48 (1976) 297; *C.A.*, 87 (1977) 11684y.
25 P.N. Varma amd S.K. Talwar, *Indian J. Pharm.*, 39 (1977) 104.

TABLE 18.1

TLC ANALYSIS OF ACONITE ALKALOIDS

TLC systems:

S1 Magnesium oxide, activated — Light petroleum (b.p. 60-80$^{o}$C)-acetone-methanol (25:8:1)[22]

S2 Silica gel G — Isopropanol-methanol-23% ammonia (36:24:1)[6]

S3 Silica gel G — Chloroform-acetone-diethylamine (5:4:1)[8]

| Alkaloid | $hR_F$ values | | |
|---|---|---|---|
| | S1 | S2 | S3 |
| Aconitine | 75 | 76 | 70 |
| Benzoylaconine | 47 | 46 | 65 |
| Aconine | 10 | 14 | 0 |

References p. 463

TABLE 18.2

TLC ANALYSIS OF *CONSOLIDA REGALIS* ALKALOIDS[12]

0.35 mm silica gel G

| Alkaloid | $pK_a$ | $hR_F$ values | | | | | | | |
|---|---|---|---|---|---|---|---|---|---|
| | | Ethyl acetate | *n*-Propanol | Acetone | Methanol | Chloroform-methanol 45:5 | Benzene-diethylamine (48:2) | Cyclohexane-*n*-propanol-diethylamine (42:4:2) | Cyclohexane-*n*-propanol-diethylamine (46:2:2) |
| Elatine | 5.33 | 48 | 55 | 89 | 92 | 92 | 42 | 46 | - |
| Delcosine (F) | 5.50 | 7 (7)* | 30 (29) | 35 | 78 | 45 (44) | 9 | 22 | (12) |
| Delsoline (D) | 5.60 | 14 (17) | 53 (48) | 76 | 84 | 79 (76) | 27 | 50 | (35) |
| Eldeline | 5.76 | 42 | 43 | 78 | 90 | 57 | 23 | 52 | - |
| Condelphine | 6.45 | 9 | 23 | - | 68 | - | - | - | - |
| Lycoctonine (G) | 7.50 | 7 (7) | 11 (10) | 12 | 34 | 17 (22) | 9 | 28 | (18) |
| Methyllycaconitine | - | 12 | 20 | 51 | 65 | 50 | 26 | 40 | - |
| Anthranoillycoctonine | - | 29 | 34 | 56 | 66 | 62 | 23 | 48 | - |
| Aconitine | 7.23 | 25 | 29 | 49 | 57 | 26 | 41 | 59 | - |

*$hR_F$ values in parentheses refer to alkaloid extract.

TABLE 18.3

TLC ANALYSIS OF DITERPENE ALKALOIDS IN PLANT MATERIALS

| Alkaloid* | Aim | Adsorbent | Solvent system | Ref. |
|---|---|---|---|---|
| Ac,PseudoAc | Separation | $SiO_2$ | *n*-BuOH-AcOH(100:14), sat. with $H_2O$<br>*n*-BuOH-AcOH(100:10), sat. with $H_2O$ | 3 |
| Ac,Acn,BAcn | Indirect quantitative analysis (UV) (Table 18.1) | $SiO_2$ | IsoprOH-MeOH-23% $NH_4OH$(36:24:1)<br>IsoprOH-MeOH(2:1) | 6 |
| Ac,MesAc,HypAc, delphline,delcosine,methyllycaconitine,delsemine | Separation | $Al_2O_3$, basic | Light petroleum-$Et_2O$(1:10) | 7 |
| A,Acn,BAcn | Reaction chromatography (Table 18.1) | $SiO_2$ | $CHCl_3$-$Me_2CO$-DEA(5:4:1) | 8 |
| Ac | Identification in plant material | $SiO_2$ | Cyclohexane-DEA(9:1)<br>MeOH | 10 |
| Methyllycaconitine, elatine | Identification in *Delphinium* species | $Al_2O_3$ | Cyclohexane-$CHCl_3$-MeOH(10:10:1) or (10:6:1) | 11 |
| Ac,elatine,delcosine,delsoline, eldeline,condelphine,lycoctonine, methyllycaconitine, anthranoillycoctonine | Identification in *Consolida* species (Table 18.2) | $SiO_2$ | EtOAc, PrOH, $Me_2CO$, MeOH<br>I. $CHCl_3$-MeOH(45:5)<br>II. Benzene-DEA(48:2)<br>Cyclohexane-PrOH-DEA(42:4:2) or (46:2:2)<br>Two-dimensional: I,II | 12 |
| Ac | Identification | $SiO_2$ | I. $CHCl_3$-DEA(9:1)<br>II. Benzene-$CHCl_3$-DEA(5:4:1)<br>Two-dimensional: I,II | 13 |
| Ac | Identification and quantitative analysis (PC) | $SiO_2$ | IsoprOH-MeOH-25% $NH_4OH$(36:24:1) | 14 |
| Elatine,delsoline, eldeline,condelphine,licoctonine | Correlation between $pK_a$ and retention | $SiO_2$ | $Me_2CO$, PrOH | 15 |
| Ac | Identification | $SiO_2$ | Benzene-$Et_2O$-MeOH(14:4:4) | 16 |

References p. 463

| | | | | |
|---|---|---|---|---|
| Ac | TAS technique | $SiO_2$ | $CHCl_3$-$Me_2CO$-DEA(5:4:1)<br>$CHCl_3$-DEA(9:1)<br>Cyclohexane-$CHCl_3$-DEA(5:4:1)<br>Cyclohexane-DEA(9:1)<br>Benzene-EtOAc-DEA(7:2:1) | 18 |
| Ac,Acn,BAcn | Indirect quantitative analysis (UV) (Table 18.1) | MgO | Light petroleum (b.p. 60-80$^{o}$C)-$Me_2CO$-MeOH(25:8:1) | 22 |
| Ac | Identification in homeopathic drugs | $SiO_2$ | Benzene-EtOAc-DEA(7:2:1) | 24 |
| Ac | Identification in tinctures | $SiO_2$ | MeOH-$NH_4OH$(10:1.5) | 25 |

*Abbreviations: Ac = aconitine; Acn = aconine; BAcn = benzoylaconine; MesAc = mesaconitine; HypAc = hypaconitine.

TABLE 18.4

TLC ANALYSIS OF DITERPENE ALKALOIDS IN COMBINATION WITH OTHER COMPOUNDS

| Alkaloid | Other compound | Aim | Adsorbent | Solvent system | Ref. |
|---|---|---|---|---|---|
| Aconitine | Cinchona alkaloids, hydrastine | Separation of MgO plates | 2.5% $CaCl_2$-impregnated MgO | *n*-hexane-$Me_2CO$(4:1)<br>EtOAc<br>EtOAc-$Me_2CO$(4:1) | 4 |
| Aconitine | Atropine,codeine, brucine | Separation on $CaSO_4$ | $CaSO_4$ | $CHCl_3$-AmOH-toluene-conc. HCl (50:1.5:1.5:0.25) | 9 |

Chapter 19

# COLCHICINE AND RELATED ALKALOIDS

19.1. Solvent systems.......................................................469
19.2. Detection.............................................................469
19.3. Quantitative analysis..................................................470
19.4. TAS technique..........................................................470
References..................................................................470

## 19.1. SOLVENT SYSTEMS

Santavy and co-workers[2,7,13] studied the PC and TLC analysis of alkaloids related to colchicine and draw conclusions about the structure and TLC behaviour. Most of the investigations carried out in this field concern the identification of colchicine in plant materials, extracts or tinctures. Eitel et al.[16] described the analysis of biological materials for colchicine. Some of the TLC systems used in these investigations are summarized in the Tables 19.1 and 19.2. Because of its neutral character, colchicine can be analysed in neutral solvent systems in combination with silica gel plates. The TLC of colchicine in connection with drug identification schemes has been dealt with in Chapter 3[1,3,9,10,18,19,22,23,25,27].

A review of the chromatography of methylenedioxyphenyl compounds has been given by Fishbein and Falk[14]. This included the analysis of some colchicine-type alkaloids.

## 19.2. DETECTION

For the detection of colchicine and related alkaloids Dragendorff's reagent and iodoplatinate reagent can be used. Some of the colours obtained with the latter reagent are summarized in Table 19.1. According to Haag-Berrurier and Mathis[20], antimony trichloride in chloroform (no. 7b) is more sensitive than iodoplatinate reagent, but it is less selective. Potesilova et al.[7] found that the very sensitive Oberlin-Zeisel reaction could be used in TLC for detection of alkaloids with a tropolone-ring. Alkaloids with a methoxy group attached to the tropolone-ring react only after previous treatment with hydrogen chloride vapour for 30 min. Then the plates could be sprayed with iron(III) chloride solution (no. 57).

Phenolic tropolone alkaloids can be seen as yellow spots on TLC plates in daylight[7]; a yellow colour is also observed with iodoplatinate reagent, but for other tropolone alkaloids the colour is brown (Table 19.1).

References p. 470

Detection of colchicine with Dragendorff's reagent is less sensitive than detection by quenching of UV light on fluorescent plates[6]. Grant[26] used aqueous cobalt thiocyanate (no. 26a) for the detection of various alkaloids, including colchicine (see Chapter 2, p.15). Kaniewska and Borkowski[11] used chromotropic acid (no. 20) for the detection of some alkaloids, including colchicine; copper(II) sulphate in ammonia (no. 27) has also been used for the same purpose[4], and iron(III) chloride in hydrochloric acid (no. 59) has been used for the detection of phenolic alkaloids[7,8,20].

## 19.3. QUANTITATIVE ANALYSIS

Indirect determination of colchicine has been used in several studies. To elute the alkaloid from the sorbent Bonati and Bacchini[5] used water, whereas Potesilova et al.[7] and Dusinsky et al.[8] used ethanol. The alkaloids were determined by means of UV spectrometry. To avoid photodecomposition the quantitative analysis should be carried out in the dark[7].

## 19.4. TAS TECHNIQUE

The TAS technique for the identification of *Colchicum* plant material has been described by Jolliffe and Shellard[17] with an oven temperature of 275°C and a distillation time of 90 sec. The sample was mixed with calcium hydroxide and silica gel with a suitable moisture content was used as the propellant.

## REFERENCES

1 D. Waldi, K. Schnackerz and F. Munter, *J. Chromatogr.*, 6 (1961) 61.
2 O.A. Neumüller, H.J. Kuhn, G.O. Schenck and F. Santavy, *Justus Liebigs Ann. Chem.*, 674 (1964) 122.
3 V. Schwarz and M. Sarsunova, *Pharmazie*, 19 (1964) 267.
4 K.C. Güven and O. Pekin, *Eczacilik Bul.*, 8 (1966) 163; *C.A.*, 65 (1966) 19930a.
5 A. Bonati and M. Bacchini, *Fitoterapia*, 37 (1966) 24.
6 F. Wartmann-Hafner, *Pharm. Acta Helv.*, 41 (1966) 406.
7 H. Potesilova, J. Hrbek, Jr. and F. Santavy, *Collect. Czech. Chem. Commun.*, 32 (1967) 141.
8 G. Dusinsky, F. Machovicova and M. Tyllova, *Farm. Obr.*, 36 (1967) 397; *C.A.*, 70 (1969) 22933d.
9 H.C. Hsiu, J.T. Huang, T.B. Shih, K.L. Yang, K.T. Wang and A.L. Lin, *J. Chin. Chem. Soc.*, 14 (1967) 161.
10 A. Noirfalise and G. Mees, *J. Chromatogr.*, 31 (1961) 61.
11 T. Kaniewska and B. Borkowski, *Diss. Pharm. Pharmacol.*, 20 (1968) 111.
12 J.M.G.J. Frijns, *Pharm. Weekbl.*, 103 (1968) 929.
13 H. Potesilova, J. Santavy, A. El-Hamidi and F. Santavy, *Collect. Czech. Chem. Commun.*, 34 (1969) 3540.
14 L. Fishbein and H.L. Falk, *J. Chromatogr. Chromatogr. Rev.*, 11 (1969) 1.
15 M. Vanhaelen, *J. Pharm. Belg.*, 24 (1969) 87.

16 N.H. Eitel, S.L. Wallace and B. Omokoku, *Biochem. Med.*, 4 (1970) 181.
17 G.H. Jolliffe and E.J. Shellard, *J. Chromatogr.*, 48 (1970) 125.
18 R.A. Egli, *Z. Anal. Chem.*, 259 (1972) 277.
19 K.F. Ahrend and D. Tiess, *Zbl. Pharm.*, 111 (1972) 933.
20 M. Haag-Berrurier and M.C. Mathis, *Ann. Pharm. Fr.*, 31 (1973) 457.
21 M. Kasim and H. Lange, *Arch. Exp. Veterinaermed.*, 27 (1973) 601.
22 K.F. Ahrend and D. Tiess, *Wiss. Z. Univ. Rostock, Math. Naturw. Reihe*, 22 (1973) 951.
23 R.J. Armstrong, *N.Z.J.Sci.*, 17 (1974) 15.
24 L.I. Churadze, P.A. Yavich, P.Z. Beridze and Ch.A. Chikhlidze, *Farmatsiya (Moscow)*, 25 (1976) 41; *C.A.*, 86 (1977) 34308x.
25 L. Lepri, P.G. Desideri and M. Lepori, *J. Chromatogr.*, 116 (1976) 131.
26 F.W. Grant, *J. Chromatogr.*, 116 (1976) 230.
27 L. Lepri, P.G. Desideri and M. Lepori, *J. Chromatogr.*, 123 (1976) 175.

TABLE 19.1

TLC ANALYSIS OF COLCHICINE AND RELATED ALKALOIDS[7,13]

TLC systems:
- S1 Silica gel G — Benzene-ethyl acetate-diethylamine(5:4:1) + 8% methanol
- S2 Silica gel G — Benzene-ethyl acetate-diethylamine(5:4:1)
- S3 Silica gel G — Chloroform-acetone-diethylamine(7:2:1)
- S4 Silica gel G — Chloroform-acetone-diethylamine(7:2:1) + 8% methanol
- S5 Silica gel G — Benzene-ethyl acetate-diethylamine(7:2:1)

| Alkaloid | $hR_F$ values | | | | | Colour | | | | | | |
|---|---|---|---|---|---|---|---|---|---|---|---|---|
| | S1 | S2 | S3 | S4 | S5 | Day light | UV light(366 nm) | HCl vapour | $FeCl_3$ (no. 57) | Iodoplatinate | $SbCl_3$ (no. 7b) Immediately | After heating ($100^oC$) |
| Neutral and phenolic alkaloids: | | | | | | | | | | | | |
| Colchicine | 56 | | 61 | | | | yel-br | yel | br | Red br-beige | yel | yel |
| N-formyldesacetylcolchicine | 44 | | 54 | | | | beige | yel | br | Red br-beige | yel | yel |
| Cornigerine | 58 | | 63 | | | | yel | yel | br | br | yel | or-yel |
| 2-Demethylcolchicine | 22 | | 27 | | | yel | Dark viol | yel | br | yel | yel | yel |
| 2-Acetyl-2-demethylcolchicine | 53 | | 56 | | | | Pink | yel | br | | | |
| 2-Ethyl-2-demethylcolchicine | 53 | | 58 | | | | Light br | yel | br | | | |
| 3-Demethylcolchicine | 32 | | 36 | | | yel | Dark viol | yel | br | yel | yel | yel |
| 3-Acetyl-3-demethylcolchicine | 51 | | 57 | | | | Pink | yel | br | | | |
| 3-Ethyl-3-demethylcolchicine | 49 | | 68 | | | | Pink | yel | br | | | |
| 3-Propyl-3-demethylcolchicine | 54 | | 67 | | | | Pink | yel | br | | | |
| Colchiceine | 27 | | 32 | 44 | | yel | br-yel | yel | br | | | |
| N-Formyldesacetylcolchiceine | 18 | | 26 | 95 | | yel | Light blue | yel | br | | | |
| O-Acetylcolchiceine | 37 | | 54 | 95 | | | Pink | yel | br | | | |
| O-Benzoylcolchiceine | 96 | | 96 | 96 | | | br-viol | yel | br | | | |
| N-Benzoyl-N-desacetylcolchiceine | 0 | | 0 | 48 | | yel | Blue-green | yel | br | | | |
| Isocolchicine | 38 | | 42 | | | | Pink | yel | br | | | |
| 6-Hydroxycolchicine | 20 | | | | 0 | | | | | br-beige | | |

| | | | | | | | | | | | | | |
|---|---|---|---|---|---|---|---|---|---|---|---|---|---|
| Alkaloids with basic character: | | | | | | | | | | | | | |
| Desacetylcolchicine | 35 | | | | | | Blue | yel | br | | | | |
| Desacetylcolchiceine | 0 | | 0 | 52 | | yel | Blue-green | yel | br | | | | |
| Desacetylisocolchicine | 22 | | | | | | Light blue | yel | br | | | | |
| Demecolcine | | 68 | 70 | | 56 | | Bronze | yel | br | br-beige | or-br | yel | |
| Demecolceine | | 9 | 18 | | | yel | Dark viol | yel | br | | | | |
| 2-Demethyldemecolcine | | 30 | 49 | | 22 | yel | Dark viol | yel | br | yel | or-br | yel | |
| 3-Demethyldemecolcine | | 51 | 56 | | 31 | yel | Dark viol | yel | br | yel | or-br | yel | |
| 3-Ethyl-3-demethyldemecolcine | | 50 | 62 | | | | Light blue | yel | br | | | | |
| 3,N-Diacetyl-3-demethyl-demecolcine | | 55 | 67 | | | | Blue | yel | br | | | | |
| N-Methyldemecolcine | | 82 | 87 | | 78 | | Light viol | yel | br | yel | yel | yel | |
| N-Formyldemecolcine | 64 | 53 | 55 | | 42 | | viol | yel | br | | | | |
| N-Acetyldemecolcine | | 62 | 65 | | | | yel-br | yel | br | | | | |
| N-Acetylisodemecolcine | | 48 | 54 | | | | viol | yel | br | | | | |
| N-Propionyldemecolcine | | 70 | 73 | | | | viol | yel | br | | | | |
| N-Benzoyldemecolcine | | 38 | 44 | | | | viol | yel | br | | | | |
| Kreysigine | 86 | | | | 84 | | | | | or-yel | | | |
| Kreysiginine | | | | | 52 | | | | | Green | | | |
| Autumnaline | | | | | 56 | | | | | br-yel | | | |
| O-Methylandrocymbine | | | | | 64 | | | | | viol | | | |
| Alkaloid glycoside: | | | | | | | | | | | | | |
| Colchicoside | 7 | | | | | | Light blue | yel | br | | | | |
| Lumi-derivatives: | | | | | | | | | | | | | |
| β-Lumicolchicine | 82 | | | | | | Grey | | | Pink | Beige-yel | Red-br-viol | red |
| γ-Lumicolchicine | 70 | | | | | | Grey | | | Black-grey beige | Beige-yel | Red-br-viol | red |
| β-Lumidemecolcine | | 79 | | | 74 | | Light grey | | | br-beige | Light beige | Red-br | |
| γ-Lumidemecolcine | | 65 | | | | | Light pink | | | br-beige | Light beige | Red-br | |
| N-Acetyllumidemecolcine | | 83 | | | | | Light blue | | | | Light beige | Red-br | |
| β-Lumicornigerine | 83 | | | | 50 | | | | | br-beige | | | |
| γ-Lumicornigerine | 73 | | | | 18 | | | | | br-beige | | | |
| γ-Lumi-2-demethyldemecolcine | 30 | | | | 0 | | | | | Beige | | | |
| β-Lumi-3-demethylcolchicine | 64 | | | | | | | | | Beige | | | |

br=brown; or=orange; yel=yellow; viol=violet.

TABLE 19.2

TLC SEPARATION OF COLCHICINE AND RELATED ALKALOIDS[2]

TLC system:
Silica gel G, activated    Ethyl acetate-ethanol (8:2)

| Alkaloid | $hR_F$ values | Alkaloid | $hR_F$ values |
|---|---|---|---|
| Colchicine | 15 | β-Lumicolchicine | 48 |
| Demecolcine | 10 | γ-Lumicolchicine | 32 |
| N-Acetyldemecolcine | 9 | Lumidemecolcine | 10 |
| α-Lumicolchicine | 27 | N-Acetyllumidemecolcine | 42 |

References p. 470

TABLE 19.3

TLC ANALYSIS OF COLCHICINE AND RELATED ALKALOIDS

| Alkaloid | Aim | Adsorbent | Solvent system | Ref. |
|---|---|---|---|---|
| Colchicine and derivatives (Table 19.2) | Identification of photodecomposition product demecolcine | $SiO_2$ | EtOH-EtOAc(2:8) | 2 |
| Colchicine | Identification | $SiO_2$ | MeOH | 4 |
| Colchicine, colchicoside | Indirect quantitative analysis in seeds and extracts | $SiO_2$ | $CHCl_3$-DEA(9:1)<br>BuOH-$H_2O$-AcOH(4:1:1) | 5 |
| Colchicine | Identification in plant material and extracts | $SiO_2$ | $CHCl_3$-MeOH(85:15)<br>Benzene-EtOAc-DEA(7:2:1) | 6 |
| Colchicine and related alkaloids (Table 19.1) | Detection in plant material | $SiO_2$ | Benzene-EtOAc-DEA(7:2:1, 5:4:1 and 5:4:1) + 8% MeOH<br>$CHCl_3$-$Me_2CO$-DEA(7:2:1 and 7:2:1) + 8% MeOH | 7 |
| Colchicine, demecolcine | Indirect quantitative analysis | $SiO_2$ | $CHCl_3$-MeOH(95:5) | 8 |
| Colchicine | Identification in plant material and extracts | $SiO_2$ | $CHCl_3$-DEA(9:1) | 12 |
| Colchicine and related alkaloids (Table 19.1) | Detection in plant material, indirect quantitative analysis (UV) | $SiO_2$ | Benzene-EtOAc-DEA(5:4:1) + 8% MeOH<br>Benzene-EtOAc-DEA(7:2:1) plate treated with 1*M* $NH_4OH$ | 13 |
| Colchicine | Identification in plant material and extracts | $SiO_2$ | $CHCl_3$-MeOH-FMA(7:2:1) | 15 |
| Colchicine and derivatives | Determination in biological material | $SiO_2$ | $CHCl_3$-$Me_2CO$-DEA(5:4:1) | 16 |
| Not specified | TAS technique for *Colchicum* | $SiO_2$ | $CHCl_3$-$Me_2CO$-DEA(5:4:1)<br>$CHCl_3$-DEA(9:1)<br>Cyclohexane-$CHCl_3$-DEA(5:4:1)<br>Cyclohexane-DEA(9:1)<br>Benzene-EtOAc-DEA(7:2:1) | 17 |

TABLE 19.3 (*continued*)

| Alkaloid | Aim | Adsorbent | Solvent system | Ref |
|---|---|---|---|---|
| Colchicine, colchiceine, demecolcine, β-lumicolchicine | Identification in plant and seeds extracts | $SiO_2$ | I. Benzene-EtOAc-DEA(5:4:1)<br>II. Benzene-EtOAc-DEA(5:4:1) + 8% MeOH<br>Subsequently I 10 cm, II 15 cm | 20 |
| Colchicine | Semi-quantitative, toxicological analysis | $SiO_2$ | EtOH-benzene(100:15) | 21 |
| Colchicine<br>Colchamine | Indirect quantitative analysis | | No details available | 24 |

Chapter 20

# IMIDAZOLE ALKALOIDS

20.1. Solvent systems .......... 477
20.2. Detection .......... 477
20.3. Quantitative analysis .......... 478
20.4. TAS technique and reaction chromatography .......... 478
References .......... 479

TLC analysis of pilocarpine has mostly been performed in connection with the analysis of pharmaceutical ophthalmic preparations. However, pilocarpine has also been included in some studies on general alkaloid identification by TLC; in Chapter 3, refs. 1, 2, 4, 6, 17, 19 and 20 (from this chapter) include the analysis of pilocarpine.

## 20.1. SOLVENT SYSTEMS

For the separation of some tropane alkaloids, physostigmine and pilocarpine in ophthalmic preparations several solvents have been described (Bradley et al.[7]). Dijkhuis[11] used solvents S1 and S2 (Table 20.1) to detect contamination of eye-drops with atropine, homatropine, scopolamine and pilocarpine. Massa et al.[15] separated pilocarpine from isopilocarpine and pilocarpic acid in the solvent systems butanol-water-acetic acid (10:3:2) and chloroform-benzene-methanol-acetic acid (6:3:3:1), both with silica gel plates. Both systems were used for the analysis of ophthalmic preparations containing pilocarpine. The former system was also used for the determination of pilocarpine in plant materials.

Ebel et al.[16] used the common TLC systems chloroform-acetone-diethylamine (5:4:1) and chloroform-diethylamine (9:1) for the semi-quantitative analysis of pilocarpine and homatropine in eyedrops on silica gel plates. The former solvent system in the proportions 10:8:2 was used by Vanhaelen[14] for the identification of pilocarpine in plant material. Röder et al.[9] described some azeotropic solvent mixtures for the separation of alkaloids, including pilocarpine.

## 20.2. DETECTION

Dragendorff's reagent in its different modifications has mostly been used. Dijkhuis[11] sprayed with Dragendorff's reagent (Munier and Macheboeuf modification), followed by 5% sodium nitrite solution. The sensitivity for pilocarpine increased

References p. 479

from 1 to 0.025 μg, whereas the sensitivity of quenching of UV light (254 nm) was found to be 300 μg on fluorescent plates. With iodoplatinate reagent pilocarpine gives a light brown colour[1]. Schmidt[20] also used iodine-iron(III) chloride followed by iodine (no. 60) and mercury(I) nitrate (no. 69) as detection reagents for pilocarpine. Rücker and Taha[22] detected a series of alkaloids, including pilocarpine, by means of π-acceptors (see Chapter 2, Table 2.2, p.16). Menn and McBain[5] developed a detection method for cholinesterase inhibitors (no. 19), including pilocarpine (see Chapter 2, p.15).

## 20.3. QUANTITATIVE ANALYSIS

Direct semi-quantitative analysis by means of circular TLC has been described for a series of alkaloids, including pilocarpine, by Hashmi et al.[10]. Ebel et al.[16] also developed a semi-quantitative method for the determination of pilocarpine and homatropine in eyedrops. After spraying with Dragendorff's reagent (Munier and Macheboeuf modification) the spot areas were compared. Massa et al.[12] used densitometry after spraying with Dragendorff's reagent (reflection mode, 400 nm). A similar method was described for the determination of pilocarpine, isopilocarpine and pilocarpic acid in pharmaceutical preparations[15].

## 20. 4. TAS TECHNIQUE AND REACTION CHROMATOGRAPHY

The TAS technique for the determination of pilocarpine in plant materials has been described by Jolliffe and Shellard[13]. They used a temperature of 275°C and a distillation time of 90 sec. The sample (10-20 mg) was mixed with an about equal amount of calcium hydroxide, and silica gel of suitable moisture content was used as the propellant. Stahl and Schmitt[21] analysed samples of 10 mg of *Jaborandi folium* by heating at 250°C for 120 sec. As the propellant they used 50 mg of molecular sieve 4Å with 20% water.

Kaess and Mathis[3] applied reaction chromatography to pilocarpine. The alkaloid was converted into the salt of pilocarpic acid by treatment with 0.1 *M* potassium hydroxide solution. Wilk and Brill[8] described the reaction chromatography of a number of compounds, including pilocarpine, by using iodine vapour as reagent. After application of the alkaloids on the TLC plate, they were exposed to iodine vapour for 18 h. After development of the plates each alkaloid gave a characteristic pattern of spots. Schmidt[18] also used iodine as a reagent in the identification of a series of drugs, including pilocarpine.

REFERENCES

1 D. Waldi, K. Schnackerz and F. Munter, *J. Chromatogr.*, 6 (1961) 61.
2 E. Vidic and J. Schütte, *Arch. Pharm. (Weinheim)*, 295 (1962) 342.
3 A. Kaess and C. Mathis, *Int. Symp. Chromatogr. Electrophor. Lect. Pap. 4th*, (1966) 525.
4 G.J. Dickes, *J. Ass. Public Anal.*, 4 (1966) 45.
5 J.J. Menn and J.B.McBain, *Nature (London)*, 209 (1966) 1351.
6 Hsing-Chien Hsiu, Jen-Tzaw Huang, Tsu-Bi Shih, Kun-Lin Yang, Kung Tsung Wang and A.L. Lin, *J. Chin. Chem. Soc.*, 14 (1967) 161.
7 T.J. Bradley, M.S. Parker and M. Barnes, *J. Hosp. Pharm.*, 24 (1967) 17.
8 M. Wilk and U. Brill, *Arch. Pharm. (Weinheim)*, 301 (1968) 282.
9 E. Röder, E. Mutschler and H. Rochelmeyer, *Arch. Pharm. (Weinheim)*, 301 (1968) 624.
10 M.H. Hashmi, S. Parveen and N.A. Chughtai, *Mikrochim. Acta*, (1969) 449.
11 I.C. Dijkhuis, *Pharm. Weekbl.*, 104 (1969) 1317.
12 V. Massa, F. Gal and P. Susplugas, *Int. Symp. Chromatogr. Electrophor. Lect. Pap. 6th*, (1970) 470.
13 G.H. Jolliffe and E.J. Shellard, *J. Chromatogr.*, 48 (1970) 125.
14 M. Vanhaelen, *J. Pharm. Belg.*, 25 (1970) 175.
15 V. Massa, F. Gal, P. Susplugas and G. Maestre, *Trav. Soc. Pharm. Montpellier*, 30 (1970) 267.
16 S. Ebel, W.D. Mikulla and K.H. Weisel, *Deut. Apoth.-Ztg.*, 111 (1971) 931.
17 S. Zadeczky, D. Küttel and M. Takacsi, *Acta Pharm. Hung.*, 42 (1972) 7.
18 F. Schmidt, *Krankenhaus-Apotheke*, 23 (1973) 10.
19 E. Novakova and J. Vecerkova, *Cesk. Farm.*, 22 (1973) 347.
20 F. Schmidt, *Deut. Apoth.-Ztg.*, 114 (1974) 1593.
21 E. Stahl and W. Schmitt, *Arch. Pharm. (Weinheim)*, 308 (1975) 570.
22 G. Rücker and A. Taha, *J. Chromatogr.*, 132 (1977) 165.

TABLE 20.1

TLC ANALYSIS OF PILOCARPINE[11]

TLC systems:

S1 Silica gel $F_{254}$ pre-coated plates Benzene-acetone-diethyl ether-5% ammonia (40:60:10:3.4)

S2 Silica gel $F_{254}$ pre-coated plates Cyclohexane-chloroform-diethylamine (3:7:1)

| Alkaloid | $hR_F$ values | |
|---|---|---|
| | S1 | S2 |
| Pilocarpine | 40 | 55 |
| Atropine | 10 | 58 |
| Homatropine | 10 | 65 |
| Scopolamine | 52 | 75 |
| Cocaine | 100 | 100 |
| Physostigmine | 65 | 95 |

TABLE 20.2

TLC ANALYSIS OF PILOCARPINE

| Alkaloid* | Other compounds | Aim | Adsorbent | Solvent system | Ref. |
|---|---|---|---|---|---|
| pil | | Reaction chromatography | $SiO_2$ | $CHCl_3$-$Me_2CO$-DEA(5:4:1) | 3 |
| pil | Physostigmine, ephedrine, homatropine, lobeline, reserpine, solanine, solanidine | Detection of cholinesterase inhibitors | 5% Silicone 555-impregnated cellulose | $H_2O$-EtOH-$CHCl_3$(56:42:2) | 5 |
| pil | Atropine, homatropine, cocaine, physostigmine | Identification in ophthalmic solutions | | No details available | 7 |
| pil | Various alkaloids | Reaction chromatography | $SiO_2$ | Benzene-MeOH-$Me_2CO$-AcOH(70:20:5:5) | 8 |
| pil | Various alkaloids | Separation with azeotropic solvents | $SiO_2$ | MeOH-benzene(39.1:60.9)<br>MeOH-$CHCl_3$-MeAc(21.6:51.4:27.0)<br>MeOH-$CHCl_3$(23.0:47.0) | 9 |
| pil | Various alkaloids | Semi-quantitative circular TLC | $SiO_2$ | $CHCl_3$-MeOH(9:1)<br>$CHCl_3$-EtOH(85:15) | 10 |
| pil | Atropine, homatropine, scopolamine | Identification as impurities in ophthalmic solutions | $SiO_2$ | Benzene-$Me_2CO$-$Et_2O$-5% $NH_4OH$(40:60:10:3.4)<br>Cyclohexane-$CHCl_3$-DEA(3:7:1) | 11 |
| pil | Various alkaloids | Densitometric analysis | | Not specified | 12 |
| pil | | TAS technique for *Pilocarpus* | $SiO_2$ | $CHCl_3$-$Me_2CO$-DEA(5:4:1)<br>$CHCl_3$-DEA(9:1)<br>Cyclohexane-$CHCl_3$-DEA(5:4:1)<br>Cyclohexane-DEA(9:1)<br>Benzene-EtOAc-DEA(7:2:1) | 13 |
| pil | | Identification in *Pilocarpus* | $SiO_2$ | $CHCl_3$-$Me_2CO$-DEA(10:8:2) | 14 |
| pil, isopil, pilac | | Densitometric analysis | $SiO_2$ | BuOH-$H_2O$-AcOH(10:3:2)<br>$CHCl_3$-benzene-MeOH-AcOH(6:3:3:1) | 15 |

TABLE 20.2 (*continued*)

| Alkaloid* | Other compounds | Aim | Adsorbent | Solvent system | Ref. |
|---|---|---|---|---|---|
| pil | Homatropine | Semi-quantitative analysis in ophthalmic solutions | $SiO_2$ | $CHCl_3$-$Me_2CO$-DEA(5:4:1)<br>$CHCl_3$-DEA(9:1) | 16 |
| pil | Various compounds | Reaction chromatography | $SiO_2$ | Toluene-MeOH-$Me_2CO$-AcOH(70:20:5:5) | 18 |
| pil | | TAS technique for *Pilocarpus* | $SiO_2$ | $Me_2CO$-$Et_2O$-conc. $NH_4OH$(50:50:3) | 21 |
| pil | Various alkaloids | Detection method | $SiO_2$ | $Me_2CO$-toluene-MeOH-$NH_4OH$(45:40:10:5) | 22 |

*pil=pilocarpine; pilac=pilocarpic acid; isopil=isopilocarpine.

Chapter 21

# QUATERNARY AMMONIUM COMPOUNDS

21.1. Solvent system....483
21.1.1. Normal-phase chromatography....483
21.1.2. Partition chromatography....484
21.1.3. Ion-pair chromatography....484
References....485

Quaternary alkaloids are highly polar compounds and their TLC analysis presents some special problems. Therefore, a separate chapter is devoted to this type of compound. The individual alkaloids are also dealt with in the apropriate alkaloid chapter, but in this chapter a series of TLC systems suitable for the analysis of quaternary alkaloids and some synthetic or semi-synthetic quaternary compounds are discussed.

## 21.1. SOLVENT SYSTEMS

To overcome the problems connected with the TLC analysis of quaternary ammonium compounds, three approaches are possible:

1. Normal-phase chromatography with highly polar solvent systems;
2. partition chromatography;
3. ion-pair chromatography.

### *21.1.1. Normal-phase chromatography*

Several workers[1,3,8,14,19] found aluminium oxide to be suitable for the separation of quaternary ammonium compounds. According to Waldi[1], the analysis of quaternary compounds was possible on aluminium oxide using acidic solvent systems. The compounds tested did not move in basic solvents or on silica gel plates. Similar results were reported by Taylor[3] for choline derivatives.

MacLean and Jewers[19] used basic aluminium oxide with acetone-water (85:15 or 9:1) to separate salts of monoquaternary compounds. The cations of the quaternary compounds could best be separated on basic aluminium oxide with chloroform-methanol-conc. ammonia (6:3:1) or on acidic aluminium oxide with chloroform-methanol (4:1 or 85:15) (developed twice).

For the analysis of some quaternary isoquinoline alkaloids, methanol-water-25% ammonia (8:1:1)[15] in combination with silica gel plates has been used. Hsiu

References p. 485

et al.[13] used polyamide plates for a series of alkaloids including a number of quaternary alkaloids (see Chapter 3, Table 3.7, p.31), whereas Klöppel et al.[18] made use of silica gel as the sorbent in connection with separation studies on 62 quaternary ammonium compounds. Acetone-acetic acid-25% hydrochloric acid (10:85:5) or methanol-acetic acid-25% hydrochloric acid (10:85:5) was used as mobile phase (Table 21.1). Stevens and Moffat[21] also used an acidic solvent system [methanol-0.2 *M* hydrochloric acid (8:2)] in combination with silica gel plates (Table 21.2), as did Giebelmann et al.[26,27] (Table 21.3).

*21.1.2. Partition chromatography*

Giacopello[5] preferred to analyse a series of alkaloids on impregnated cellulose plates (Chapter 3, Table 3.9, p.33).

Calderwood and Fish[10,17] separated a series of *Fagara* alkaloids by means of TLC on cellulose (Chapter 11, Table 11.1, p.188).

*21.1.3. Ion-pair chromatography*

Manthey and Amundson[6] reported that TLC separations could be improved by using alcoholic solution of inorganic salts in combination with silica gel plates. Similar solvent systems were used by Jane[22] in the HPLC analysis of drugs of abuse. De Zeeuw et al.[23] and Verpoorte and Baerheim Svendsen[25] used silica gel plates in combination with alcoholic salt solutions in the analysis of some quaternary ammonium compounds (Table 21.4) and quaternary alkaloids and alkaloid N-oxides (Table 21.5); see also Chapter 13, Table 13.4, p.330. Variation of the molarity of the salt solutions or of the ratio of salt solution to methanol can be used to obtain optimal $hR_F$ values. Instead of alcoholic solutions acetone solutions can also be used (R. Verpoorte, unpublished results).

Gröningson and Schill[16] described the ion-pair chromatography of some tertiary and quaternary alkaloids using reversed-phase systems. The stationary phase was cellulose impregnated with an organic salt solution, and the mobile phase was pentanol or pentanol-chloroform (1:1).

Gordon[7] separated quaternary ammonium salts on silica gel with neutral solvents. For a particular cation different $hR_F$ values were obtained, depending on the counter ion used. The influence of the anion on the mobility of the cations was discussed.

## REFERENCES


1 D. Waldi, *Naturwissenschaften*, 50 (1963) 614.
2 H. Bayzer, *Experientia*, 20 (1964) 233.
3 E.H. Taylor, *Lloydia*, 27 (1964) 96.
4 M.R. Gasco and G. Gatti, *Boll. Chim. Farm.*, 104 (1965) 639.
5 D. Giacopello, *J. Chromatogr.*, 19 (1965) 172.
6 J.A. Manthey and M.E. Amundson, *J. Chromatogr.*, 19 (1965) 522.
7 J.E. Gordon, *J. Chromatogr.*, 20 (1965) 38.
8 G. Sullivan and L.R. Brady, *Lloydia*, 28 (1965) 68.
9 E. Hultin, *Acta Chem. Scand.*, 20 (1966) 1588.
10 J.M. Calderwood and F. Fish, *J. Pharm. Pharmacol.*, 18 (1966) 119S.
11 Ch. Wollmann, S. Nagel and E. Scheibe, *Pharmazie*, 21 (1966) 665.
12 J.L. Kiger and J.G. Kiger, *Ann. Pharm. Fr.*, 25 (1967) 601.
13 H.C. Hsiu, J.T. Huang, T.B. Shih, K.L. Yang, K.T. Wang and A.L. Lin, *J. Chin. Chem. Soc.*, 14 (1967) 161.
14 A. Fiori and M. Marigo, *J. Chromatogr.*, 31 (1967) 171.
15 M.T. Wa, J.L. Beal and R.W. Doskotch, *Lloydia*, 30 (1967) 245.
16 K. Gröningson and G. Schill, *Acta Pharm. Suecica*, 6 (1969) 447.
17 J.M. Calderwood and F. Fish, *J. Pharm. Pharmacol.*, 21 (1969) 126S.
18 A. Klöppel, D. Post, G. Schneider and H. Schütz, *Z. Anal. Chem.*, 252 (1970) 279.
19 W.F.H. MacLean and K. Jewers, *J. Chromatogr.*, 74 (1972) 297.
20 H.D. Crone and E.M. Smith, *J. Chromatogr.*, 77 (1973) 234.
21 H.M. Stevens and A.C. Moffat, *J. Forensic Sci. Soc.*, 14 (1974) 141.
22 I. Jane, *J. Chromatogr.*, 111 (1975) 227.
23 R.A. de Zeeuw, P.E.W. van der Laan, J.E. Greving and F.J.W. Mansvelt, *Anal. Lett.*, 9 (1976) 831.
24 R. Kinget and A. Michoel, *J. Chromatogr.*, 120 (1976) 234.
25 R. Verpoorte and A. Baerheim Svendsen, *J. Chromatogr.*, 124 (1976) 152.
26 R. Giebelmann, S. Nagel, Ch. Brunstein and E. Scheibe, *Zentralbl. Pharm.*, 115 (1976) 339.
27 R. Giebelmann, *Zentralbl. Pharm.*, 116 (1977) 1011.
28 J.P. Franke, J. Wijsbeek, J.E. Greving and R.A. de Zeeuw, *Arch. Toxicol.*, 42 (1979) 115.

TABLE 21.1

TLC SEPARATION OF SOME QUATERNARY AMMONIUM COMPOUNDS[18]

TLC systems:
S1 Silica gel G Acetone-acetic acid-25% hydrochloric acid (10:85:5)
S2 Silica gel G Methanol-acetic acid-25% hydrochloric acid (10:85:5)

| Compound | $hR_F$ values | |
|---|---|---|
| | S1 | S2 |
| Carbachol | 18 | |
| Choline | 60 | |
| Succinylcholine | 2 | |
| Acetylcholine | 27 | |
| Muscarine | 7 | |
| Butylscopolamine | | 36 |
| Methylscopolamine | | 23 |
| Methylatropine | | 19 |
| Berberine | 55 | |
| Dimethyltubocurarine | 9 | |
| Benzalkonium | 56 | |
| Neostigmine | 18 | |
| Gallamine | 1 | |
| Thiazinamium | 17 | |
| Acriflavine | 16 | |
| Serpentine | 54 | |
| Hydrastinine | 22 | |
| Sanguinarine | 56 | |
| Tubocurarine | 14 | |
| Pyridostigmine | 27 | |
| Thiamine | 9 | |

TABLE 21.2

TLC SEPARATION OF SOME QUATERNARY AMMONIUM COMPOUNDS[21]

TLC systems:

| | | |
|---|---|---|
| S1 | Cellulose Cel 300-25UV$_{254}$, pre-coated | Tetrahydrofuran-[1 g of ammonium formate and 5 ml of formic acid (98%) in 100 ml of water] (21:9) |
| S2 | Silica gel G25, pre-coated | Methanol-0.2 *M* hydrochloric acid (8:2) |

| Compound | $hR_F$ values | |
|---|---|---|
| | S1 | S2 |
| Methylatropine | 95 | 35 |
| Azamethonium | 40 | 10 |
| Bretylium | 94 | 40 |
| Cetrimide | 100 | 50 |
| Decamethonium | 56 | 16 |
| Gallamine | 34 | 5 |
| Guanethidine | 56 | 50 |
| Hexamethonium | 36 | 10 |
| Paraquat | 22 | 10 |
| Suxamethonium | 35 | 10 |
| Suxethonium | 40 | 23 |
| Tubocurarine | 85 | 40 |
| Acetylcholine | 70 | 60 |
| Choline | 60 | 60 |
| Pancuronium | 80 | Not tried |

TABLE 21.3

TLC SEPARATION OF SOME QUATERNARY AMMONIUM COMPOUNDS[26]

Silica gel G, 0.5 mm, activated.
Solvent systems:

S1 Acetone-1 *M* hydrochloric acid (1:1)
S2 96% Ethanol-1 *M* hydrochloric acid (1:1)
S3 Dioxane-1 *M* hydrochloric acid (1:1)
S4 Methylal-1 *M* hydrochloric acid (1:1)
S5 Dimethylformamide-1 *M* hydrochloric acid (1:1)
S6 Methyl ethyl ketone-1 *M* hydrochloric acid, lower phase
S7 Tetrahydrofuran-acetone-1 *M* hydrochloric acid (1:1:2)
S8 Dimethylformamide-tetrahydrofuran-acetone-2 *M* hydrochloric acid (1:1:1:1)
S9 Acetone-acetic acid-25% hydrochloric acid (10:85:5)
S10 Methanol-acetic acid-25% hydrochloric acid (10:85:5)
S11 Pyridine-acetic acid-water-methanol (5:10:10:75)

| Compound | $hR_F$ values | | | | | | | | | | |
|---|---|---|---|---|---|---|---|---|---|---|---|
| | S1 | S2 | S3 | S4 | S5 | S6 | S7 | S8 | S9 | S10 | S11 |
| Choline | 70 | 54 | 65 | 67 | 80 | 73 | 72 | 75 | | | |
| Acetylcholine | 67 | 45 | 58 | 61 | 77 | 66 | 65 | 72 | | | |
| Methacholine | 63 | 45 | 56 | 57 | 73 | 61 | 61 | 71 | | | |
| Carbachol | 69 | 52 | 65 | 63 | 79 | 75 | 71 | 77 | | | |
| Tetrylammonium | 53 | 29 | 43 | 42 | 60 | 53 | 49 | 58 | | | |
| Bromocholine | 66 | 50 | 57 | 59 | 73 | 70 | 65 | 71 | | | |
| Neostigmine | 62 | 45 | 55 | 52 | 72 | 60 | 60 | 72 | | | |
| Pyridostigmine | 59 | 40 | 45 | 48 | 67 | 58 | 58 | 65 | | | |
| Benzalkonium | 73 | 74 | | | | | | | 95 | 96 | 46 |
| Tetramethylammonium | 72 | 49 | | | | | | | 19 | 25 | 11 |
| Pralidoxime | 78 | 58 | | | | | | | 40 | 49 | 24 |
| Benzilylcholine | 79 | 69 | | | | | | | 60 | 61/70 | 24 |
| Butylscopolamine | 80 | 72 | | | | | | | 66 | 71 | 42 |
| Alkonium | 72 | 71 | | | | | | | 93 | 87 | 43 |
| Acriflavine | 84 | 75 | | | | | | | 90 | 65/87 | 62/74 |
| Methylrosanilinium | 71 | 57 | | | | | | | 30 | 89 | 45/75 |
| Pyrvinium | 66 | 55 | | | | | | | 43 | 92 | 50/77 |
| Benzoylcholine | 71 | 65 | | | | | | | 50 | 53 | 17 |
| Butyrylcholine | 65/74 | 37/59 | | | | | | | 2/38 | 9/45 | 17 |
| Butyrylthiocholine | 48/63/74 | 18/32/61 | | | | | | | 3/48 | 7/53 | 20 |
| Methylatropine | 76 | 59 | | | | | | | 35 | 39 | 21 |
| Tetra-*n*-butylammonium | 79 | 70 | | | | | | | 90 | 92 | 70 |
| Pentamethonium | 40 | 12 | 35 | 35 | 63 | 41 | 42 | 37 | | | |
| Hexamethonium | 40 | 13 | 37 | 27 | 63 | 39 | 42 | 37 | | | |
| Decamethonium | 47 | 20 | 44 | 30 | 65 | 43 | 44 | 47 | | | |
| Oxamethonium | 35 | 10 | 31 | 24 | 65 | 34 | 33 | 35 | | | |
| Methyloxamethonium | 37 | 11 | 33 | 24 | 70 | 36 | 37 | 35 | | | |
| Azamethonium | 61 | 30 | 60 | 55 | 82 | 67 | 62 | 58 | | | |
| Suxamethonium | 47 | 20 | 46 | 38 | 68 | 49 | 49 | 49 | | | |
| Tubocurarine | 73 | 59 | | | | | | | 39 | 39 | 14 |
| Hydroxytriethonium | 59 | 34 | | | | | | | 9 | 9 | 4 |
| Gallamine | 22 | 5 | 22 | 14 | 56 | 17 | 24 | 23 | | | |

TABLE 21.4

TLC SEPARATION OF SOME QUATERNARY AMMONIUM COMPOUNDS[23]

TLC systems:

S1 Silica gel 60-$F_{254}$, pre-coated — 0.5 *M* NaBr in methanol, in unsaturated tank
S2 Silica gel 60-$F_{254}$, pre-coated — 0.5 *M* NaI in chloroform-methanol (2:8), in unsaturated tank
S3 Silica gel 60-$F_{254}$, pre-coated, impregnated with 0.5 *M* NaBr — $CHCl_3$-MeOH (7:3), in saturated tank

| Compound | $hR_F$ values | | |
|---|---|---|---|
| | S1 | S2 | S3 |
| Benzalkonium | 84 | 94 | 74 |
| Butylscopolamine | 69 | 88 | 50 |
| Propantheline | 59 | 83 | 67 |
| Poldine | 55 | 82 | 58 |
| Thiazinamium | 47 | 78 | 59 |
| Methylscopolamine | 41 | 68 | 33 |
| Thiamine | 35 | 55 | 17 |
| Thiazinamium sulphoxide | 29 | 53 | 43 |
| Stercuronium | 24 | 64 | 48 |
| Dimethyltubocurarine | 17 | 59 | 53 |
| Gallamine | 7 | 35 | 4 |
| Paraquat | 2 | 8 | 1 |

References p. 485

TABLE 21.5

INFLUENCE OF DIFFERENT SALTS, SALT CONCENTRATION AND DIFFERENT SOLVENT RATIOS ON THE $hR_F$ VALUES OF SOME QUATERNARY AND TERTIARY ALKALOIDS AND ALKALOID N-OXIDES AS OBSERVED IN TLC ON SILICA GEL[25]

| Salt | Molarity of salt solution | Ratio of methanol to salt solution | Compound | | | | | | |
|---|---|---|---|---|---|---|---|---|---|
| | | | Alstonine | Dihydro-toxiferine | Alcuronium | Melinonine A iodide | Strychnine N-oxide | Strychnine | Cara-curine V |
| $NH_4Cl$ | 0.2 | 3:2 | 57 | 27 | 37 | 51 | 30 | 28 | 4 |
| $NH_4NO_3$ | 0.2 | 3:2 | 61 | 26 | 34 | 49 | 40 | 25 | 3 |
| NaCl | 0.2 | 3:2 | 59 | 24 | 37 | 48 | 49 | 29 | 4 |
| $MgCl_2$ | 0.2 | 3:2 | 63 | 34 | 44 | 54 | 33 | 38 | 36 |
| $CaCl_2$ | 0.2 | 3:2 | 64 | 33 | 44 | 55 | 33 | 40 | 38 |
| $CaCl_2$ | 0.1 | 3:2 | 57 | 26 | 34 | 48 | 28 | 26 | 20 |
| $NH_4NO_3$ * | 0.2 | 3:2 | 47 | 9 | 14 | 30 | 46 | 16 | 0 |
| $NH_4NO_3$ ** | 0.2 | 3:2 | 59 | 10 | 14 | 32 | 49 | 27 | 9 |
| $NH_4OOCCH_3$ | 0.2 | 3:2 | 54 | 14 | 24 | 40 | 34 | 21 | 2 |
| $NH_4OH$ | 0.2 | 3:2 | 10 | 1 | 0 | 3 | 60 | 30 | 3 |
| $(NH_4)_2CO_3$ | 0.2 | 3:2 | 38 | 5 | 7 | 25 | 45 | 16 | 1 |
| $NH_4HCO_3$ | 0.2 | 3:2 | 36 | 4 | 4 | 21 | 42 | 13 | 0 |
| 0.1 *M* $NH_4OOCCH_3$-0.1 *M* $CH_3COOH$ | 0.2 | 3:2 | 54 | 11 | 14 | 43 | 33 | 39 | 4 |
| $NH_4Cl$ | 0.2 | 2:3 | 55 | 24 | 32 | 48 | 30 | 28 | 0 |
| $NH_4Cl$ | 0.2 | 1:4 | 26 | 10 | 16 | 35 | 21 | 25 | 0 |
| $NH_4Cl$ | 0.2 | 4:1 | 51 | 13 | 23 | 38 | 23 | 17 | 5 |
| $NH_4Cl$ | 1 | 3:2 | 64 | 45 | 53 | 59 | 32 | 50 | 44 |
| $NH_4NO_3$ | 1 | 3:2 | 65 | 47 | 54 | 60 | 32 | 50 | 45 |

* Before development the plate was exposed to ammonia vapour for 30 min.
** The plate was first developed with ethyl acetate-isopropanol-25% ammonia (9:7:1) and then allowed to dry in the open air for 30 min, before development in the solvent system mentioned.

References p. 485

TABLE 21.6

INFLUENCE OF AMMONIA CONCENTRATION AND SOLVENT RATIO ON THE $hR_F$ VALUES OF SOME QUATERNARY AND TERTIARY ALKALOIDS AND ALKALOID N-OXIDES IN METHANOL-AQUEOUS AMMONIA-AMMONIUM NITRATE SOLUTIONS, AS OBSERVED IN TLC ON SILICA GEL[25]

| Solution | Total molarity of aqueous solution | Ratio of methanol to salt solution to ammonia solution | Compound | | | | | | |
|---|---|---|---|---|---|---|---|---|---|
| | | | Alstonine | Dihydro-toxiferine | Alcuronium | Melionine A iodide | Strychnine N-oxide | Strychnine | Cara-curine V |
| 1 *M* $NH_4NO_3$-2 *M* $NH_4OH$ | 1.67 | 7:1:2 | 68 | 29 | 37 | 44 | 66 | 58 | 28 |
| 1 *M* $NH_4Cl$-2 *M* $NH_4OH$ | 1.67 | 7:1:2 | 68 | 24 | 40 | 47 | 66 | 57 | 29 |
| 1 *M* NaCl-2 *M* $NH_4OH$ | 1.67 | 7:1:2 | 69 | 23 | 40 | 49 | 67 | 58 | 28 |
| 0.1 *M* $NH_4NO_3$-2 *M* $NH_4OH$ | 1.37 | 7:1:2 | 52 | 0 | 1 | 11 | 68 | 60 | 31 |
| 1 *M* $NH_4NO_3$-0.1 *M* $NH_4OH$ | 0.4 | 7:1:2 | 56 | 11 | 19 | 36 | 47 | 22 | 3 |
| 1 *M* $NH_4NO_3$-1 *M* $NH_4OH$ | 1 | 7:1:2 | 66 | 28 | 37 | 47 | 63 | 48 | 17 |
| 1 *M* $NH_4NO_3$-1 *M* $NH_4OH$ | 1 | 5:2:3 | 75 | 51 | 64 | 66 | 76 | 56 | 22 |
| 1 *M* $NH_4NO_3$-1 *M* $NH_4OH$ | 1 | 5:1:4 | 75 | 40 | 58 | 61 | 75 | 59 | 26 |
| 1 *M* $NH_4NO_3$-1 *M* $NH_4OH$ | 1 | 5:4:1 | 73 | 53 | 66 | 67 | 69 | 32 | 12 |
| 1 *M* $NH_4NO_3$-1 *M* $NH_4OH$ | 1 | 27:1:2 | 28 | 1 | 3 | 11 | 32 | 24 | 3 |
| 1 *M* $NH_4NO_3$-1 *M* $NH_4OH$ | 1 | 2:3:5 | 81 | 76 | 79 | 79 | 89 | 65 | 7 |
| 1 *M* $NH_4NO_3$-1 *M* $NH_4OH$ | 1 | 3:3:4 | 84 | 73 | 83 | 83 | 89 | 63 | 14 |

TABLE 21.7

TLC ANALYSIS QUATERNARY AMMONIUM COMPOUNDS

| Compound | Aim | Adsorbent | Solvent system | Ref. |
|---|---|---|---|---|
| Berberine, choline, neostigmine, padisal | Separation of quaternary compounds | $Al_2O_3$ | Cyclohexane-$CHCl_3$-AcOH(45:45:10)<br>Cyclohexane-$CHCl_3$-EtOH-AcOH(4:3:2:1) | 1 |
| Choline,acetylcholine,chlorcholine, tetramethylammonium | Separation of quaternary compounds | Cellulose | *n*-BuOH-$H_2O$-AcOH(4:5:1)<br>$CHCl_3$-MeOH-$H_2O$(75:22:3) | 2 |
| Choline,acetylcholine,carbamylcholine, succinylmonocholine,succinyldicholine, β-methylcholine,methacholine,bethanechol | Separation of choline and derivatives | $Al_2O_3$ | *n*-BuOH-$H_2O$-90% HCOOH(12:7:1), upper phase<br>*n*-BuOH-$H_2O$-AcOH(4:1:1)<br>*n*-BuOH-$H_2O$-AcOH(4:5:1), upper phase | 3 |
| Tubocurarine | Separation of curarimetics | Carboxymethylcellulose | 0.075 *M* aq. NaCl | 4 |
| Tembetarine,magnoflorine,berberine, pseudoberberine,palmatine,berberrubine, chelerythrine,sanguinarine and various tertiary alkaloids (Chapter 3, Table 3.9 3.10, p. 33) | Separation of alkaloids on cellulose | 0.5 *M* KCl impregnated cellulose | *n*-BuOH-36% HCl(98:2) sat. with $H_2O$ | |
| | | 0.2 *M* $KH_2PO_4$-impregnated cellulose | *n*-BuOH sat with $H_2O$<br>*sec*.-BuOH sat with $H_2O$ | |
| | | 0.5 *M* $KH_2PO_4$-impregnated cellulose | IsoprOH-$H_2O$(3:1) | |
| | | cellulose | *n*-BuOH-$H_2O$-AcOH(10:3:1) | 5 |
| Quaternary ammonium salts | Separation of quaternary ammonium salts | $SiO_2$ | $CHCl_3$-EtOH(24:1, 9:1)<br>$Me_2CO$-N-methylacetamide | 7 |
| Betaine,choline,muscarine | Separation | $Al_2O_3$ | MeOH-$CCl_4$-AcOH(28:12:1) | 8 |
| Quaternary alkaloids(tubocurare), betaine,choline | Separation | $SiO_2$ | EtOH-HCOOH-$H_2O$-FMA(5:1:1:3) | 9 |
| Tembetarine,N-methylcanadine,1-hydroxy-2,9,10-trimethoxy-N,N-dimethylaporphine | Separation of *Fagara* alkaloids | Cellulose | *n*-BuOH-$H_2O$-AcOH(10:3:1)<br>*tert*.-AmOH-isoAmOH-HCOOH-$H_2O$(1:1:1:5) | 10 |

| | | | | |
|---|---|---|---|---|
| Tubocurarine,decamethonium,suxamethonium,gallamine,mephenesine | Identification of muscle relaxant agents | $SiO_2$ | $Me_2CO$-2 *M* HCl(1:1)<br>$Me_2CO$-1 *M* HCl(1:1)<br>$Me_2CO$-0.5 *M* HCl(1:1)<br>$Me_2CO$-0.1 *M* HCl(1:1)<br>0.5 *M* HCl<br>MeOH-1 *M* HCl(1:1)<br>96% EtOH-1 *M* HCl(1:1)<br>Dioxane-$Me_2CO$-1 *M* HCl(0.5:0.5:1)<br>Dioxane-1 *M* HCl(1:1)<br>IsoprOH-1 *M* HCl(1:1) | 11 |
| 36 quaternary ammonium compounds | Identification | $SiO_2$ | $CHCl_3$-isoprOH-conc. $NH_4OH$(5:5:1) | 12 |
| Jatrorrhizine,magnoflorine,menisperine,phellodendrine,tubocurarine, various tertiary alkaloids (Chapter 3, Table 3.7, p.31) | Separation of alkaloids on polyamide plates | Polyamide | Dioxane-cyclohexane-DEA(10:20:0.5)<br>$CHCl_3$-cyclohexane-DEA(10:20:0.5)<br>MeEtCO-cyclohexane-DEA(20:30:0.5)<br>EtOH-$CHCl_3$-AcOH(20:200:0.5)<br>$H_2O$-EtOH-pyridine(10:0.5:0.3)<br>Cyclohexane-EtOAc-prOH-DMA(30:2.5:0.9:0.1)<br>$H_2O$-EtOH-DMA(88:12:0.1) | 13 |
| Acetylcholine,decamethonium,choline, gallamine,*d*-tubocurarine,succinyl chloride | Detection of curarimetics in biological materials | Acidic $Al_2O_3$<br>$Al_2O_3$ | $CHCl_3$-MeOH(8:2)<br>MeOH-$H_2O$-AcOH(92:5:3) | 14 |
| Palmatine,magnoflorine,columbamine, jatrorrhizine | Identification of alkaloids in *Stephania* species (Chapter 11, Table 11.2, p.188) | $SiO_2$ | MeOH-$H_2O$-$NH_4OH$(8:1:1) | 15 |
| Methylatropine,methylscopolamine, butylscopolamine,atropine,scopolamine, papaverine,strychnine,brucine | Ion-pair chromatography | Cellulose impregnated with 0.7 *M* $H_2SO_4$ + 0.7 *M* NaCl (or NaBr, KSCN, $NaClO_4$) | AmOH<br>AmOH-$CHCl_3$(1:1) | 16 |
| Candicine,coryneine,tembetarine,magnoflorine,N-methylcorydine,N-methylisocorydine,laurifoline,xanthoplanine,palmatine,berberine,chelerythrine,nitidine | TLC and electrophoresis of quaternary *Fagara* alkaloids (Chapter 11, Table 11.1, p.188) | Cellulose | 0.1 *M* HCl<br>*n*-BuOH sat. with 2 *M* HCl<br>*n*-BuOH-pyridine-$H_2O$(6:4:3) | 17 |

References p. 485

TABLE 21.7 (*continued*)

| Compound | Aim | Adsorbent | Solvent system | Ref. |
|---|---|---|---|---|
| Butylscopolamine,methylatropine,methylscopolamine,berberine,serpentine,hydrastinine,tubocurarine,sanguinarine and 55 other quaternary ammonium compounds | Ion-pair extraction of quaternary ammonium compounds, and their identification (Table 21.1) | $SiO_2$ | $Me_2CO$-AcOH-25% HCl(10:85:5)<br>MeOH-AcOH-25% HCl(10:85:5)<br>MeOH-$H_2O$-AcOH-pyridine(75:10:10:5) | 18 |
| Columbamine,palmatine,coptisine,jatrorrhizine, 14 quaternary ammonium compounds and some salts thereof | Separation | Basic $Al_2O_3$<br>Acidic $Al_2O_3$ | $Me_2CO$-$H_2O$(85:15, 9:1)<br>$CHCl_3$-MeOH-conc. $NH_4OH$(6:3:1)<br>$CHCl_3$-MeOH(4:1, 85:15) | 19 |
| Tubocurarine,ephedrine,atropine, physostigmine, 15 quaternary ammonium compounds and 4 tertiary nitrogen compounds | Identification on microslides | $SiO_2$ | 1 *M* HCl<br>EtOH-1 *M* HCl(1:1) | 20 |
| Methylatropine,azamethonium,bretylium, cetrimide,decamethonium,gallamine,guanethidine,hexamethonium,paraquat,suxamethonium,suxethonium,tubocurarine, acetylcholine,choline,pancuronium | Identification in urine and blood (Table 21.2) | Cellulose<br>$SiO_2$ | THF-(1 g $NH_4$ formate-5 ml HCOOH/100 ml $H_2O$) (21:9)<br>MeOH-0.2 *M* HCl(8:2) | 21 |
| Dimethyltubocurarine,butylscopolamine, methylscopolamine and 9 quaternary nitrogen compounds | Ion-pair chromatography of quaternary compounds (Table 21.4) | $SiO_2$<br>NaBr-impregnated $SiO_2$ | 0.5 *M* NaBr in MeOH<br>0.5 *M* NaI in $CHCl_3$-MeOH(1:4)<br>$CHCl_3$-MeOH(7:3) | 23 |
| Pancuronium and hydrolysis products | Separation | $SiO_2$ | *n*-BuOH-pyridine-AcOH-20% $NH_4Cl$(60:40:12:48), organic phase | 24 |
| Dihydrotoxiferine,C-alkaloid H,toxiferine,alcuronium,tubocurarine,macusine B,melinonine A, antirhine methochloride,strychnine methochloride, various tertiary alkaloids | Separation of quaternary alkaloids and alkaloid N-oxides (Tables 21.5 and 21.6 and Chapter 13, Table 13.4, p.330) | $SiO_2$ | MeOH-0.2 *M* $NH_4NO_3$(3:2)<br>MeOH-2 *M* $NH_4OH$-1 *M* $NH_4NO_3$(7:2:1) | 25 |

| | | | | |
|---|---|---|---|---|
| Butylscopolamine,methylatropine tubocurarine, 29 other quaternary ammonium compounds | Identification (Table 21.3) | $SiO_2$ | $Me_2CO$-1 *M* HCl(1:1)<br>96% EtOH-1 *M* HCl(1:1)<br>Dioxane-1 *M* HCl(1:1)<br>Methylal-1 *M* HCl(1:1)<br>DMF-1 *M* HCl(1:1)<br>MeEtCO-1 *M* HCl, lower phase<br>THF-$Me_2CO$-1 *M* HCl(1:1:2)<br>DMFA-THF-$Me_2CO$-2 *M* HCl(1:1:1:1)<br>$Me_2CO$-AcOH-25% HCl(10:85:5)<br>MeOH-AcOH-25% HCl(10:85:5)<br>Pyridine-AcOH-$H_2O$-MeOH(5:10:10:75) | 26 |
| Methylatropine,atropine,butylscopolamine,scopolamine | Two-dimensional separation (Chapter 7, Table 7.2, p.101) | $SiO_2$ | I. MeOH<br>II. MeOH-1 *M* HCl(1:1)<br>III. MeOH-AcOH-$H_2O$(16:1:3)<br>IV. *n*-BuOH-AcOH-$H_2O$(4:1:5)<br>Two-dimensional: I,II, I,III or I,IV | 27 |
| Choline,methylatropine,neostigmine, oxyphenonium,propantheline,succinylcholine,thiazinanium,decamethonium, hexamethonium,pancuronium,pentolonium, tubocurarine | Isolation and determination | $SiO_2$ | I. MeOH<br>II. MeOH-$H_2O$(4:1)<br>III. MeOH-$H_2O$-$NH_4OH$(100:21:3) + 4 g $NH_4$ acetate<br>I, II and III consecutively | 28 |

APPENDIX

DETECTION METHODS AND SPRAY REAGENTS

1. *Ammoniacal silver nitrate*
   *Preparation.* To 5 ml of 0.59 *M* silver nitrate solution is added 6 *M* ammonia dropwise, until a clear solution is obtained (to be prepared freshly).
   *Procedure.* After spraying the chromatogram is heated at 90°C for 5 min.
2. *Ammonium selenite-sulphuric acid (Lafon's reagent)*
   *Preparation.* 5% ammonium selenite in concentrated sulphuric acid.
3. *Ammonium vanadate-nitric acid*
   *Preparation.* 1% ammonium vanadate in 50% nitric acid.
4. *Anisaldehyde-sulphuric acid*
   *Preparation.* 0.5 ml of anisaldehyde is dissolved in 50 ml of acetic acid and 1 ml of concentrated sulphuric acid is added.
   *Procedure.* After spraying the chromatogram is heated at 120°C for 15 min.
5. *Anisaldehyde-sulphuric acid with phosphomolybdic acid*
   *Preparation.* 0.5 ml of anisaldehyde is dissolved in acetic acid-methanol (10:85). To the solution 5 ml of concentrates sulphuric acid and 20 g of phosphomolybdic acid are added.
   *Procedure.* After spraying the chromatogram is heated at 110°C for 5-10 min.
6. *Antimony pentachloride*
   *Preparation.* 20% antimony pentachloride in chloroform.
7. *Antimony trichloride*
   *(a)Preparation.* 20 g of antimony trichloride are dissolved in a mixture of 20 ml of acetic acid and 60 ml of chloroform.
   *Procedure.* After spraying the chromatogram is heated at 110°C for 2-3 min and is observed in daylight and UV light.
   *(b)Preparation.* Saturated solution of antimony trichloride in chloroform.
   *Procedure.* Colours are observed immediately after spraying and after heating the chromatogram at 100°C for 10 min. Observations are made in daylight and UV light (366 nm).
   *(c)Preparation.* 25% antimony trichloride in chloroform.
   *Procedure.* After spraying the chromatogram is heated at 110°C for 5 min.
8. *Bromocresol green*
   *Preparation.* A 0.1% alcoholic solution of bromocresol green, 0.5 *M* sodium phosphate buffer (pH 5.5) and water (2:2:1) are mixed.

9. *Bromophenol blue*

*Preparation.* 0.1% bromophenol blue in ethanol.

10. *Calcafluor M2R New*

*Preparation.* 0.02% Calcafluor M2R New in methanol.

*Procedure.* The chromatogram is dipped in the reagent and after drying is observed in UV light (366 nm).

11. *Cerium(IV) ammonium nitrate-hydroxylamine*

*Preparation.* (I) 5% cerium(IV) ammonium sulphate in acetone.
(II) 5% hydroxylammonium chloride in 80% acetone.

*Procedure.* The chromatogram is subsequently sprayed with I and II and dried with hot air. Heating at 110$^{o}$C for 5 min can increase the sensitivity.

12. *Cerium(IV) ammonium sulphate*

*Preparation.* 1 g of cerium(IV) ammonium sulphate is dissolved in 99 g of 85% phosphoric acid by heating the mixture on a hot-plate for 5-10 min. The spray reagent can be used as such or after dilution with an equal volume of water, in which case, however, more than one spraying may be necessary.

13. *Cerium(IV) sulphate-nitric acid*

*Preparation.* 0.3% cerium(IV) sulphate in 65% nitric acid.

*Procedure.* The chromatogram is observed immediately after spraying, after heating with hot air and after heating at 110$^{o}$C for 15 min.

14. *Cerium(IV) sulphate-sulphuric acid*

*(a)Preparation.* A saturated solution of cerium(IV) sulphate in 65% sulphuric acid.

*Procedure.* The chromatogram is observed in daylight and UV light immediately after spraying, after heating with hot air and after heating at 110$^{o}$C for 15 min.

*(b)Preparation.* 1% cerium(IV) sulphate in 10% sulphuric acid.

*Procedure.* The chromatogram is observed immediately after spraying, after heating with hot air and after heating at 120$^{o}$C for 15 min.

*(c)(According to Sonnenschein):*

*Preparation.* 0.1 g of cerium(IV) sulphate is suspended in 4 ml of water, 1 g of trichloroacetic acid is added and the mixture is boiled. Concentrated sulphuric acid is added dropwise until a clear solution is obtained.

*Procedure.* The chromatogram is observed immediately after spraying, after heating with hot air and after heating at 110$^{o}$C for 15 min.

15. *Chloramine-T*

*Preparation.* A saturated solution of chloramine-T in 10% acetic acid.

*Procedure.* After spraying the chromatogram is observed in UV light.

16. *Chloramine-T (Murexide reaction)*

*Preparation.* 10% chloramine-T in water.

*Procedure.* After spraying the chromatogram with the reagent it is further sprayed with 1 *M* hydrochloric acid and heated at 96-98$^{o}$C until no chlorine can be smelled. Subsequently the chromatogram is exposed to ammonia vapour.

17. *Chloranil*

*Preparation.* 1% chloranil in acetonitrile.

18. *1-Chloro-2,4-dinitrobenzene-bromothymol blue*

*Preparation.* (I) 0.5% 1-chloro-2,4-dinitrobenzene in ethanol.
(II) 0.05% bromothymol blue in ethanol.

*Procedure.* The chromatogram is subsequently sprayed with I and II.

19. *Cholinesterase inhibition*

*Preparation.* (I) Human blood plasma.
(II) A 0.6% solution of bromothymol blue in 0.1 *M* sodium hydroxide is mixed with 1% aqueous acetylcholine chloride solution (1:16).

*Procedure.* After separation of the compounds on a cellulose plate impregnated with 5% Dow Corning Silicone 555 the chromatogram is sprayed with I. After 30 min at room temperature the chromatogram is sprayed with II. Substances with an acetylcholinesterase-inhibiting effect are observed as blue spots against a yellow background.

20. *Chromotropic acid (Gaebel's reagent)*

*Preparation.* 10g of sodium 1,8-dihydroxynaphthalene-3,6-disulphonate (sodium salt of chromotropic acid) are dissolved in 100 ml of water. The spray reagent is prepared by mixing 1 part of this solution with 5 parts of 65% sulphuric acid (can be stored only for 1 day).

*Procedure.* The chromatogram is observed immediately after spraying and after heating at 105$^{o}$C for 30 min.

21. *Cinnamaldehyde*

*Preparation.* 1% cinnamaldehyde in methanol.

*Procedure.* After spraying the chromatogram is dried and exposed to hydrochloric acid vapour.

22. *Citric acid-acetic anhydride*

*Preparation.* 2% citric acid in acetic anhydride.

23. *CNTNF reagent*

*Preparation.* 1% 9-dicyanomethylene-2,4,7-trinitrofluorene in acetonitrile.

24. *Cobalt nitrate*

*Preparation.* 1% cobalt nitrate in ethanol.

*Procedure.* After spraying the chromatogram is heated at 110$^{o}$C for 5 min.

25. *Cobalt rhodanide*

*Preparation.* 3 g of ammonium rhodanide and 1 g of cobalt nitrate are dissolved in 20 ml of water.

26. *Cobalt thiocyanate*

*(a)Preparation.* 10 g of ammonium thiocyanate and 10 g of cobalt nitrate hexahydrate are dissolved in 100 ml of water.

*(b)Preparation.* 3 g of ammonium thiocyanate and 1 g cobalt chloride are dissolved in 20 ml of water.

27. *Copper(II) sulphate-ammonia*

*Preparation.* 10% copper(II) sulphate solution is mixed with 2% ammonia solution (5:1).

*Procedure.* After spraying the chromatogram is heated at 105°C for 15 min.

28. *Dansyl chloride*

*Preparation.* 0.05% 1-dimethylaminonaphthalene-5-sulphonyl chloride in acetone.

*Procedure.* After spraying the chromatogram is observed in UV light.

29. *DDQ reagent*

*Preparation.* 0.2% 2,3-dichloro-5,6-dicyanoquinone in acetonitrile.

30. *Denigès reagent*

*Preparation.* (I) 3% hydrogen peroxide.
(II) 10% ammonia.
(III) 2% copper(II) sulphate.

*Procedure.* After spraying the chromatogram with I and drying, it is sprayed with II and III.

31. *o-Dianisidine in dilute hydrochloric acid*

*Preparation.* A 0.5% *o*-dianisidine solution in dilute hydrochloric acid is mixed with an equal volume of an aqueous 10% sodium nitrite solution.

32. *Diazotized benzidine (Wachtmeister's reagent)*

*Preparation.* (I) 5 g of benzidine and 14 ml of concentrated hydrochloric acid are dissolved in water and diluted to 1000 ml.
(II) 10% aqueous sodium nitrite. The spray reagent is prepared by mixing equal volumes of I and II, with stirring, at 0°C. The spray reagent can be stored for only a few hours.

33. *Diazotized sulphanilic acid (Pauly's reagent)*

*(a)Preparation.* (I) 0.3% solution of sulphanilic acid in 8% hydrochloric acid.
(II) 5% aqueous sodium nitrite. The spray reagent is prepared freshly by mixing 25 ml of I with 2.5 ml of II.

*Procedure.* After spraying the chromatogram it is further sprayed with 20% sodium carbonate solution.

*(b)Preparation.* (I) 5 g of sulphanilic acid and 53 ml of concentrated hydrochloric acid are dissolved in 1 l of water.
(II) 0.5% aqueous sodium nitrite. The spray reagent is prepared freshly by mixing equal parts of I and II.

*34. 2,6-Dibromo-p-benzoquinone-4-chlorimide (Gibb's reagent)*

*Preparation.* (I) 10% aqueous sodium acetate.

(II) 1% 2,6-dibromo-*p*-benzoquinone-4-chlorimide in ethanol.

*Procedure.* The chromatogram is subsequently sprayed with I and II.

*35. 2,6-Dichloro-p-benzoquinone-4-chlorimide (Gibb's reagent)*

*(a)Preparation.* 0.25% 2,6-dichloro-*p*-benzoquinone-4-chlorimide in methanol.

*(b)Preparation.* 2% 2,6-dichloro-*p*-benzoquinone-4-chlorimide in methanol.

*Procedure.* The spray reagent cannot be used in connection with chromatograms developed in diethylamine containing solvent systems. The chromatogram is observed 15 min after spraying and after heating on a hot-plate.

*36. p-Dimethylaminobenzaldehyde*

*(a)Preparation.* 0.5% *p*-Dimethylaminobenzaldehyde in cyclohexane.

*Procedure.* After spraying the chromatogram is exposed to hydrochloric acid vapour.

*(b)Preparation.* 1% *p*-dimethylaminobenzaldehyde in 96% ethanol.

*Procedure.* After spraying the chromatogram is exposed to hydrochloric acid vapour.

*(c)Preparation.* 1 g of *p*-dimethylaminobenzaldehyde is dissolved in a mixture of 50 ml of 36% hydrochloric acid and 50 ml of ethanol.

*Procedure.* After spraying the chromatogram is heated at 50°C for 2-3 min. Exposure to nitric acid vapour intensifies the colours.

*(d)Preparation.* 0.5 g of *p*-dimethylaminobenzaldehyde is dissolved in a mixture of 5 ml of 37% hydrochloric acid and 95 ml of ethanol.

*(e)Preparation.* A 2% solution of *p*-dimethylaminobenzaldehyde in 95% ethanol is mixed with 6 *M* hydrochloric acid (8:2).

*Procedure.* After spraying the chromatogram is heated at 110°C for 1 min.

*(f)Preparation.* 200 mg of *p*-dimethylaminobenzaldehyde are dissolved in 100 ml of 25% hydrochloric acid to which one drop of a 10% iron(III) chloride solution has been added.

*Procedure.* After spraying the chromatogram is heated with hot air.

*(g)Preparation.* 0.8 g of *p*-dimethylaminobenzaldehyde is dissolved in a mixture of 10 ml of concentrated sulphuric acid and 90 ml of 95% ethanol.

*(h)Preparation.* 125 g of *p*-dimethylaminobenzaldehyde are dissolved in a mixture of 65 ml of concentrated sulphuric acid and 35 ml of water. To the solution 0.05 ml of 5% iron(III) chloride solution is added. The reagent must be used within 7 days.

*(i)Preparation.* 0.5 g of *p*-dimethylaminobenzaldehyde is dissolved in a mixture of 53 ml of concentrated sulphuric acid and 50 ml of water. To this solution 0.5 ml of 10.5% iron(III) chloride solution is added.

37. *3,5-Dinitrobenzoyl chloride*

*Preparation.* 5% 3,5-dinitrobenzoyl chloride in ethyl acetate.

38. *DNFB reagent*

*Preparation.* 0.2% 2,4-dinitrofluorobenzene in acetonitrile.

39a. *Dragendorff's reagent (Bregoff-Delwiche)*

*Preparation.* (I) 8 g of bismuth subnitrate is dissolved in 20 ml of 25% nitric acid.

(II) 20 g of potassium iodide is dissolved in a mixture of 1 ml of 25% hydrochloric acid and 5 ml of water. I and II are mixed and water is added until an orange-red colour is obtained (the volume is about 95 ml). The solution is filtered and diluted to 100 ml with water. The spray reagent is prepared by subsequently mixing 20 ml of water, 5 ml of 6 *M* hydrochloric acid, 2 ml of stock solution and 6 ml of 6 *M* sodium hydroxide solution. If bismuth hydroxide is not completely dissolved by shaking, 6 *M* hydrochloric acid is added until a clear solution is obtained. The stock solution is storable for longer periods; the spray reagent is stable for 10 days if stored cool.

39b. *Dragendorff's reagent (Munier)*

*Preparation.* (I) 1.7 g of bismuth subnitrate is dissolved in 100 ml of 20% tartaric acid solution.

(II) 16 g of potassium iodide are dissolved in 40 ml of water.
I and II are mixed. The spray reagent is prepared by mixing 50 ml of the stock solution, 100 g of tartaric acid and 500 ml of water.

39c. *Dragendorff's reagent (Munier and Macheboeuf)*

*Preparation.* (I) 2.5 g of bismuth subnitrate are dissolved in a mixture of 20 ml of water and 5 ml of acetic acid.

(II) 4 g of potassium iodide are dissolved in 10 ml of water.
I and II are mixed. This stock solution can be stored for at least 6 months. The spray reagent is prepared by mixing 5 ml of the stock solution with 10 ml of acetic acid and diluting to 100 ml with water.

39d. *Dragendorff's reagent (Robles)*

*Preparation.* 5g of potassium iodide are dissolved in a mixture of 12.5 ml of water and 3 ml of 4 *M* hydrochloric acid. 1 g of bismuth nitrate is dissolved in this solution by boiling. After cooling, 12.5 ml of water are added. To this solution 5 g of potassium iodide and 5 g of citric acid are added. The solution is further diluted with 75 ml of 0.1 *M* hydrochloric acid.

*39e. Dragendorff's reagent (Thies and Reuter)*

*Preparation.* 2.6 g of bismuth subcarbonate and 7.0 g of sodium iodide (dried over concentrated sulphuric acid for 24 h) are boiled with 25 ml of acetic acid for a few minutes, left overnight and filtered off (sodium acetate), then 20 ml of the solution are diluted with 80 ml of ethyl acetate. The stock solution can be stored in sealed containers. The spray reagent is prepared by mixing 20 ml of the stock solution, 50 ml of acetic acid and 120 ml ethyl acetate. To this solution 10 ml of water are added.

*39f. Dragendorff's reagent (Thies and Reuter, Vagujfalvi modification)*

*Preparation.* The spray reagent is prepared by mixing 20 ml of stock solution (see 39e), 50 ml of acetic acid and 120 ml of ethyl acetate.

*Procedure.* After spraying the chromatogram with the reagent, it is further sprayed with 10% sulphuric acid.

*39g. Dragendorff's reagent (Trabert)*

*Preparation.* 1.3 g of bismuth subcarbonate is added to 12.0 ml of boiled water. and 15.0 g of potassium iodide are added. After addition of 3.8 ml of 2 *M* sulphuric acid the mixture is filtered (potassium sulphate) and diluted to 100 ml with water. To this stock solution 0.1 g of sodium hydrogen sulphite is added for conservation. The spray reagent is prepared by mixing 4 ml of stock solution, 30 ml of methanol and 20 ml of diethyl ether, filtering and adding of 0.3 ml of acetic acid and 2 ml of water. The spray reagent is storable for at least 1 week if kept in the dark in well sealed containers.

*39h. Dragendorff's reagent-sodium nitrite*

*Preparation.* (I) Dragendorff's reagent (Munier) (no. 39b).
(II) 10% aqueous sodium nitrite.

*Procedure.* The chromatogram is subsequently sprayed with I and II.

*40. Fast Blue B salt*

*(a)Preparation.* 0.5% aqueous Fast Blue B salt solution is mixed with an equal volume carbonate-hydrogen carbonate buffer (pH 9.2).

*(b)Preparation.* 30 ml of Fast Blue B salt are dissolved in 0.1 *M* sodium hydroxide solution. The reagent must be prepared freshly.

*41. Fluoranil*

*Preparation.* 1% fluoranil in acetonitrile.

*42. Fluorescamine*

*Preparation.* 0.02% fluorescamine in anhydrous acetone.

*Procedure.* After spraying the chromatogram is observed in UV light.

*43. Folin-Ciocalteau reagent*

*Preparation.* 10g of sodium tungstate and 2.5 g of sodium molybdate are dissolved in 70 ml of water and 5 ml of 85% phosphoric acid and 10 ml of concentrated hydrochloric acid are added. The mixture is refluxed for 10 h and then 15 g of lithium sulphate, 5 ml of water and one drop of bromine are added. The mixture is boiled for 15 min, cooled and diluted to 100 ml with water. The reagent should not be green. The spray reagent is obtained by dilution with 3 volumes of water.

*Procedure.* Spray first with 20% aqueous sodium carbonate solution, dry the chromatogram, then spray with the reagent.

44. *Formaldehyde-phosphoric acid (Clarke's reagent)*

*(a) Preparation.* 0.03 g of paraformaldehyde is dissolved in 100 ml of 85% phosphoric acid (stable for 1 week).

*Procedure.* The colours develop about 10 min after spraying.

*(b) Preparation.* 0.1 g of paraformaldehyde is dissolved in 100 ml of 85% phosphoric acid.

*Procedure.* After spraying the chromatogram is heated at 110°C for 5 min. The chromatogram is observed in daylight and UV light.

45. *Formaldehyde in sulphuric acid (Marquis reagent)*

*Preparation.* 2 ml of 40% formaldehyde in water are mixed with 100 ml of 55% sulphuric acid.

*Procedure.* After spraying the chromatogram is heated at 105-110°C for 15-30 min.

46. *Formaldehyde in sulphuric acid and iron(III) chloride (reagent of D'Aloy and Valdiguié)*

*Preparation.* To a solution of formaldehyde in sulphuric acid iron(III) chloride is added (no further details available).

47. *HNS reagent*

*Preparation.* Saturated solution of 2,2',4,4',6,6'-hexanitrostilbene in acetonitrile.

48. *Hydrogen peroxide-acetic anhydride*

*Preparation.* (I) 30% hydrogen peroxide solution containing 2-4 mg/ml of sodium pyrophosphate.

(II) Acetic anhydride-light petroleum (b.p. 80-100°C)-benzene (1:4:5).

(III) 1 g of *p*-dimethylaminobenzaldehyde is dissolved in ethanol diethylene glycol monoethyl ester-hydrochloric acid (70:30:1.5).

*Procedure.* Spray with I, heat at 90-100°C for 15 min, after cooling spray with II and heat at 90-100°C for 15 min. Observe the alkaloids as brown spots, fluorescent in UV light. Spray with III, heat for 5-15 min and observe the alkaloids as blue spots.

49. *Hydrogen peroxide-potassium hexacyanoferrate(III)*

*Preparation.* (I) 3% aqueous hydrogen peroxide solution.

(II) 5% aqueous potassium hexacyanoferrate(III) solution.

*Procedure.* Spray with I, heat at 100°C for 10 min, spray the chromatogram with II and heat at 100°C for 10 min.

*50. Hydroxylamine-iron(III) chloride*

*Preparation.* (I) Mix 1 volume of 14% hydroxylammonium chloride solution in methanol and 4 volumes of 3.5 *M* aqueous potassium hydroxide solution and filter the solution.

(II) 2% iron(III) chloride in 100 ml of 10% hydrochloric acid.

*Procedure.* Spray with I, leave the chromatogram for 10 min and spray with II.

*51a. Iodine in carbon tetrachloride*

*Preparation.* 0.5% solution of iodine in carbon tetrachloride.

*51b. Iodine in chloroform*

*Preparation.* 0.5% solution of iodine in chloroform.

*51c. Iodine in hexane.*

*Preparation.* Saturated solution of iodine in *n*-hexane.

*51d. Iodine in methanol*

*Preparation.* 1% solution of iodine in methanol.

*52. Iodine-potassium iodide*

*(a)Preparation.* 5g of iodine are dissolved in a solution of 20 g of potassium iodide in 100 ml of water. Prior to use the solution is diluted with water (1:50).

*(b)Preparation.* 1 g of iodine is dissolved in a solution of 10 g of potassium iodide in 50 ml water, to which 2 ml of acetic acid has been added. The spray reagent is prepared by dilution to 100 ml with water.

*(c)Preparation.* 2 g of iodine are dissolved in a solution of 4 g of potassium iodide in 100 ml of water. To this solution an equal volume of 1 *M* hydrochloric acid is added.

*(d)Preparation.* 3 g of iodine and 1 g of potassium iodide are dissolved in 100 ml of ethanol.

*Procedure.* After spraying with this reagent the chromatogram is sprayed with 12.5% hydrochloric acid in ethanol.

*(e)Preparation.* 2 g of iodine and 2 g of potassium iodide are dissolved in 50 ml of 95% ethanol and 50 ml of 25% hydrochloric acid are added.

*(f)Preparation.* 2 g of iodine are dissolved in 50 ml of 95% ethanol and a solution of 2 g of potassium iodide in 16.2 ml water is added. When the solution is clear 33.8 ml of concentrated hydrochloric acid are added.

*53. Iodine-iron(III) chloride*

*Preparation.* 2 g of iodine and 5 g of iron(III) chloride are dissolved in 100 ml of acetone-20% aqueous tartaric acid (45:55).

54. *Iodine-pyrrole*

*Procedure.* The chromatogram is placed in iodine vapour for 10 min, followed by exposure to pyrrole vapour for 1 h.

55. *Iodopalladate*

*Preparation.* 100 ml of 8% aqueous potassium iodide solution are mixed with 1.5 ml of 10% aqueous palladium chloride solution and the solution is filtered.

56. *Iodoplatinate reagent*

*(a) Preparation.* The spray reagent is prepared freshly by mixing 3 ml of 10% aqueous hexachloroplatinic acid solution with 97 ml of water and 100 ml of 6% aqueous potassium iodide solution.

*(b) Preparation.* The spray reagent is prepared by mixing 5 ml of 5% aqueous hexachloroplatinic acid solution and 45 ml of 10% aqueous potassium iodide solution and diluting to 100 ml with water. The reagent must be prepared freshly.

*(c) Preparation.* The reagent is prepared by mixing 5 ml of 0.1% aqueous hexachloroplatinic acid solution, 0.3 ml of 1 *M* potassium iodide, 0.5 ml of 2 *M* hydrochloric acid and 25 ml of acetone. The reagent must be prepared freshly.

*Procedure.* The chromatogram is dipped in the reagent.

*(d) Preparation.* The spray reagent is prepared by mixing equal parts of 0.3% aqueous hexachloroplatinate solution [1 g of platinum(IV) chloride dissolved in 333 ml of 1 *M* hydrochloric acid] and 6% aqueous potassium iodide solution. The reagent must be prepared freshly.

*Acidified iodoplatinate reagent*

*(e) Preparation.* The spray reagent is prepared by mixing 5 ml of 5% aqueous hexachloroplatinate, 45 ml of 10% aqueous potassium iodide and diluting with water to 100 ml. Immediately before spraying 10 ml of concentrated hydrochloric acid are added to 100 ml of the reagent. The reagent must be prepared freshly.

*(f) Preparation.* The spray reagent is prepared by mixing 5 ml of 5% aqueous hexachloroplatinic acid, 45 ml of 10% aqueous potassium iodide, 50 ml of water and 100 ml of 2 *M* hydrochloric acid. The reagent must be prepared freshly.

57. *Iron(III) chloride*

*Preparation.* Saturated solution of iron(III) chloride in water.

*Procedure.* The chromatogram is exposed to hydrochloric acid vapour for 30 min and subsequently sprayed with the reagent.

58. *Iron(III) chloride-glyoxylic acid reagent*

*Preparation.* Dissolve 0.5 g of glyoxylic acid and 0.05 g of iron(III) chloride in a mixture of 60 ml of acetic acid, 10 ml of water and 30 ml of concentrated sulphuric acid.

*Procedure.* The chromatogram is observed immediately after spraying and after heating at 110°C for 5-10 min.

*59. Iron(III) chloride-hydrochloric acid*

*Preparation.* 1% iron(III) chloride in 0.5 *M* hydrochloric acid.

*60. Iron(III) chloride-iodine*

*Preparation.* (I) 2% aqueous iron(III) chloride solution.
(II) 0.1 *M* aqueous iodine solution.

*Procedure.* The chromatogram is subsequently sprayed with I and II.

*61. Iron(III) chloride-nitric acid*

*Preparation.* 5% iron(III) chloride in 50% nitric acid.

*Procedure.* The chromatogram is observed immediately after spraying and after heating at 105°C for 5 min.

*62. Iron(III) chloride-perchloric acid*

*(a)Preparation.* 1 ml of 0.5 *M* iron(III) chloride solution is mixed with 50 ml of 35% perchloric acid.

*Procedure.* The chromatogram is observed immediately after spraying, after heating with hot air and after heating at 110°C for 30 min.

*(b)Preparation.* 5 ml of perchloric acid and 2 ml of 0.05 *M* iron(III) chloride are diluted to 100 ml with water.

*(c)Preparation.* 0.2 *M* iron(III) chloride in 35% perchloric acid.

*Procedure.* The chromatogram is observed immediately after spraying, after heating with hot air and after heating at 110°C for 30 min.

*63. Iron(III) chloride-sulphuric acid*

*Preparation.* 0.05% iron(III) chloride in 50% sulphuric acid.

*Procedure.* The chromatogram is observed 5 min after spraying.

*64. Iron(III) chloride-sulphuric acid-perchloric acid*

*Preparation.* 98 ml of 30% sulphuric acid, 2 ml of 5% iron(III) chloride solution and 100 ml of 36% perchloric acid are mixed.

*Procedure.* After spraying the chromatogram is heated.

*65. Isatin*

*Preparation.* 0.2% of isatin is dissolved in acetone containing 4% acetic acid.

*Procedure.* After spraying the chromatogram is heated at 105-110°C for 5-10 min. Nornicotine colours blue.

*66. König's reaction*

*(a)Preparation.* 3% 4-chloroaniline in methanol.

*Procedure.* The chromatogram is exposed to cyanogen bromide vapour for 5 min; after 5 min in the open air the chromatogram is sprayed with the reagent. To obtain a complete reaction the procedure is repeated three times.

*(b) Preparation.* 2% aniline in ethanol.

*Procedure.* After spraying the chromatogram with the reagent, it is exposed to cyanogen bromide vapour.

*(c) Preparation.* 1% benzidine in ethanol.

*Procedure.* After spraying the chromatogram with the reagent, it is exposed to cyanogen bromide vapour.

*(d) Preparation.* 2% *p*-aminobenzoic acid in methanol is mixed with an equal volume 0.1 *M* phosphate buffer (pH 7).

*Procedure.* After spraying the chromatogram with the reagent, it is exposed to cyanogen bromide vapour.

67. *Labat's reagent*

*Preparation.* 10 ml of 5% gallic acid solution in ethanol are mixed with 200 ml of concentrated sulphuric acid.

68. *Mercury(II) acetate-anisaldehyde-phosphomolybdic acid*

*Preparation.* (I) Saturated solution of mercury(II) acetate in methanol.
(II) 0.5 ml of anisaldehyde is dissolved in 50 ml of acetic acid to which 1 ml of concentrated sulphuric acid has been added (to be prepared freshly).
(III) 10% phosphomolybdic acid in methanol.

*Procedure.* Spray the chromatogram with I and dry it at 80°C, spray with a mixture of II and III (9:1) and heat at 120°C for 15 min.

69. *Mercury(I) nitrate*

*Preparation.* 1% mercury(I) nitrate in water.

70. *Mercury(I) nitrate in nitric acid*

*Preparation.* 4% mercury(I) nitrate in 3% nitric acid.

*Procedure.* After spraying the chromatogram is heated at 110°C for 15 min.

71. *Methyl orange*

*Preparation.* 0.1% methyl orange in ethanol.

72. *Millon's reagent*

*Preparation.* 5g of mercury are dissolved in 10 g of fuming nitric acid, then 10 ml water are added.

73. *Murexide reaction*

*Preparation.* Mix 20 ml of concentrated hydrochloric acid and 80 ml of 10% aqueous potassium perchlorate (to be prepared freshly).

*Procedure.* After spraying the chromatogram is heated at 110°C for 30 min and is subsequently placed in ammonia vapour for 5 min.

74. *Ninhydrin*

*(a) Preparation.* 0.2% ninhydrin in *n*-butanol-2 *M* acetic acid (95:5)

*Procedure.* After spraying the chromatogram is heated at 110°C for 5 min.

*(b) Preparation.* 0.1 g of ninhydrin is dissolved in 20 ml of methanol and 2 ml of sulphuric acid and 5 ml of acetic acid are added.

*(c)Preparation.* 1% ninhydrin in butanol to which a few drops of pyridine have been added.

*Procedure.* After spraying the chromatogram is heated.

75. *Ninhydrin-phenylacetaldehyde*

*(a)Preparation.* 0.8 g of ninhydrin and 1.8 ml of phenylacetaldehyde are dissolved in 100 ml of ethanol.

*(b)Preparation.* (I) 0.4% ninhydrin in pH 9 phosphate buffer.

*Procedure.* The chromatogram is subsequently sprayed with reagent I and II. The chromatogram is observed 10 min after spraying the second reagent. The reaction can be speeded up by heating the chromatogram a few minutes at 100$^{\circ}$C.

76. *p-Nitraniline*

*Preparation.* (I) 2.5 g of *p*-nitraniline are dissolved in 1 *M* hydrochloric acid and the solution is diluted with 500 ml of ethanol. (II) 5% aqueous sodium nitrite solution. The spray reagent is prepared by mixing 100 ml of I and 20 ml of II and cooling to 5-10$^{\circ}$C for 10 min. The reagent should be freshly prepared and used when cold.

77. *Nitrosonaphthol*

*Preparation.* 0.2% 2-nitrosonaphthol in ethanol.

*Procedure.* After spraying with the reagent the chromatogram is sprayed with 10% hydrochloric acid and the chromatogram is heated at 110$^{\circ}$C for 20 min.

78. *Phenothiazine-bromine-ammonia*

*Preparation.* 0.05% phenothiazine in ethanol.

*Procedure.* The chromatogram is sprayed with the reagent, dried for 10 min and placed in bromine vapour for 10 min.10 min after removal from the bromine chamber the colours are observed. Then the plate is placed in ammonia vapour for 10 min and again the colours are observed. Instead of exposure to bromine, exposure to iodine for 1 h followed by drying for 20 min and exposure to ammonia vapour for 20 min can be applied.

79. *Phosphomolybdic acid*

*Preparation.* 0.5 phosphomolybdic acid in 50% nitric acid.

80. *Potassium hexacyanoferrate(III)-iron(III) chloride*

*Preparation.* A 5% aqueous solution of potassium hexacyanoferrate(III) is mixed with an equal volume of a 1% iron(III) chloride solution.

81. *Potassium hexacyanoferrate(III)-iron(III) chloride*

*(a)Preparation.* 1% aqueous potassium hexacyanoferrate(III) solution is mixed with an equal volume of a 2% aqueous iron(III) chloride solution (to be prepared freshly just before spraying).

*Procedure.* After spraying the chromatogram with the reagent a further spray with 2 *M* hydrochloric acid may intensify the colours.

*(b)Preparation.* 57 mg of potassium hexacyanoferrate(III) and 7.8 mg of potassium hexacyanoferrate(II) are dissolved in 100 ml water.

*Procedure.* After spraying the chromatogram is observed in UV light.

82. *Potassium perchlorate-hydrochloric acid*

*Preparation.* The reagent is prepared by mixing 20 ml of concentrated hydrochloric acid and 80 ml of 10% aqueous potassium perchlorate solution. The reagent must be prepared freshly.

*Procedure.* The chromatogram is sprayed with the reagent, heated at 110°C for 30 min and placed in ammonia vapour for 5 min.

83. *Potassium permanganate (alkaline)*

*Preparation.* 1% aqueous potassium permanganate solution is mixed with an equal volume of a 5% aqueous sodium carbonate solution.

84. *Quinone*

*Preparation.* 0.5% quinone in dilute hydrochloric acid.

85. *Rhodamine B*

*Preparation.* 0.25% rhodamine B in ethanol.

*Procedure.* Observe the colours in daylight and UV light.

86. *Silver nitrate-Dragendorff's reagent*

*(a)Preparation.* (I) 2% aqueous silver nitrate solution.
(II) 5% sulphuric acid.
(III) Dragendorff's reagent (no. 39c).

*Procedure.* The chromatogram is subsequently sprayed with reagents I, II and III.

*(b)Preparation.* (I) 1% silver nitrate solution in 5% sulphuric acid.
(II) Dragendorff's reagent (no. 39c).

*Procedure.* The chromatogram is subsequently sprayed with reagents I and II.

*(c)Preparation.* (I) Saturated solution of silver sulphate in 10% sulphuric acid.
(II) Dragendorff's reagent (no. 39c).

*Procedure.* The chromatogram is subsequently sprayed with reagents I and II.

87. *Sodium iodate*

*Preparation.* 5% sodium iodate in 1% ammonia.

*Procedure.* Observe the colours in UV light (365 nm).

88. *Sodium 1,2-naphthoquinone-4-sulphonate*

*Preparation.* 0.3% sodium 1,2-naphthoquinone-4-sulphonate in 50% ethanol.

*Procedure.* After spraying the chromatogram with the reagent, it is further sprayed with 10% hydrochloric acid and heated at 110°C for 20 min.

89. *Sodium nitrite*

*Preparation.* 2% sodium nitrite in water.

*Procedure.* After spraying the chromatogram is exposed to ammonia vapour.

90. *Sulphomolybdic acid in sulphuric acid (Fröhde's reagent)*

*Preparation.* 1% ammonium molybdate in concentrated sulphuric acid.

*Procedure.* The chromatogram is observed immediately after spraying with the reagent and after heating at 105°C for 5 min.

91. *Sulphovanadic acid in sulphuric acid (Mandelin's reagent)*

*Preparation.* 1% ammonium vanadate in concentrated sulphuric acid.

92. *TACOT reagent*

*Preparation.* 0.01% solution of tetranitro-2,3:5,6-dibenzo-1,3a,4,6a-tetraazapentalene in acetonitrile.

93. *Tannin*

*Preparation.* 1 g of tannin is dissolved in 1 ml of ethanol, and then further diluted to 10 ml with ethanol.

94. *TCBI reagent*

*Preparation.* 0.1 g of N,2,6-trichloro-*p*-benzoquinone imine is dissolved in chloroform-dimethyl sulphoxide (9:1) saturated with sodium hydrogen carbonate (stable for 4 months if stored in a brown bottle at about 4°C).

*Procedure.* After spraying the chromatogram with the reagent, it is heated at 110°C for 1-2 min.

95. *TCNE reagent*

*(a) Preparation.* 1% tetracyanoethylene in acetonitrile.

*(b) Preparation.* 0.2% tetracyanoethylene in acetonitrile.

96. *TCNQ reagent*

*Preparation.* 0.2% 7,7,8,8-tetracyanoquinodimethane in acetonitrile.

97. *TetNF reagent*

*(a) Preparation.* 1% 2,4,5,7-tetranitro-9-fluorenone in acetonitrile.

*(b) Preparation.* 0.2% 2,4,5,7-tetranitro-9-fluorenone in acetonitrile.

*Procedure.* The chromatogram is observed immediately after spraying and after heating.

98. *Tetraphenylborate-fisetin (or quercetin)*

*Preparation.* (I) 1% sodium tetraphenylborate in water-saturated methyl ethyl ketone.

(II) 1 ml of 0.1% fisetin (or quercetin) in methanol is mixed with 6 ml of ethanol.

*Procedure.* Spray the chromatogram subsequently with I and II.

99. *TNB reagent*

*Preparation.* 1% 1,3,5-trinitrobenzene in acetonitrile.

100. *2,4,7-Trinitrofluorenone (TNF reagent)*

*(a) Preparation.* 0.5% 2,4,7-trinitrofluorenone in benzene.

*Procedure.* After spraying the chromatogram is heated at 110°C for 10 min.

*(b)Preparation.* 0.2% 2,4,7-trinitrofluorenone in acetonitrile.

*Procedure.* The chromatogram is observed immediately after spraying and after heating.

*(c)Preparation.* 1% 2,4,7-trinitrofluorenone in acetonitrile.

*101. Trichloroacetic acid*

*Preparation.* 25% trichloroacetic acid in chloroform.

*Procedure.* After spraying the chromatogram is heated at 120°C for 10-15 min, and it is observed in daylight and UV light.

*102. Trichloroacetic acid-sodium nitroprusside*

*Preparation.* 3% sodium nitroprusside in 50% aqueous trichloroacetic acid.

*Procedure.* After spraying the chromatogram is observed in UV light.

*103. Vanadium pentoxide*

*Preparation.* 1.62 g of anhydrous ammonium metavanadate is dissolved in 125 ml of concentrated sulphuric acid at 150°C. The deep red solution is cooled and 125 ml of ice-cold water are added, yielding an orange solution of 0.5% vanadium pentoxide in 50% sulphuric acid. The spray reagent is prepared by diluting 10-fold.

*Procedure.* After spraying the chromatogram is heated at 50°C for 3 min and at 110°C for 3-7 min.

*104. Vanillin-sulphuric acid*

*Preparation.* 1% vanillin in concentrated sulphuric acid.

*Procedure.* After spraying the chromatogram is heated at 100°C for 1 min.

*105. Vitali-Morin reaction*

*Preparation.* (I) Concentrated nitric acid.

(II) 2.5% tetraethylammonium in dimethylformamide.

*Procedure.* The chromatogram is sprayed with I; in the case of silica gel the plate is subsequently heated at 130°C for 1 h, and in the case of aluminium oxide at 140-150°C for 1 h. Then the chromatogram is sprayed with an excess of freshly prepared reagent II.

*106. Xanthydrol*

*Preparation.* 1% xanthydrol in 95% ethanol to which a few drops of concentrated hydrochloric acid have been added.

# INDEXES

## *How to find your way*

The various subjects discussed in the chapters are presented in the Subject Index, as well as the botanical names of plants mentioned in the text. Solvent systems, detection, quantitative analysis (including densitometry), TAS-technique and reaction chromatography are not included in the Subject Index. These subjects are dealt with separately for each group of alkaloids and may be found in the list of contents of each chapter.

All compounds with a pertinent reference to $hR_F$ values or detection methods, either in the text or in a table, are listed in the Compound Index. For the various groups of alkaloids, tables summarizing all the available literature are presented at the end of each chapter. Compounds listed in these tables are not included in the Compound Index. Hence, to find all data relating to a particular compound, it may be necessary to consult both the Compound Index and the appropriate tables.

## SUBJECT INDEX

Abbreviations for solvents VII
Aconite alkaloids 461-467
  analysis in plant material 466, 467
Activation of adsorbents 3, 4
Alkaloid N-oxides 61, 63, 484, 490, 491
Aluminium oxide 4
Amaryllidaceae alkaloids 184, 219
Amphetamines 236, 242-244
Aporphine alkaloids 180, 181, 207-209, 212
*Areca* alkaloids 64
Artefact formation 53-56
*Atropa* alkaloids 91-112
Barbiturates 242-244
Basic modifiers in solvent systems 3, 5
Belladonna alkaloids 91-112
Benzophenanthridine alkaloids 174, 175, 195-197, 213
Benzylisoquinoline alkaloids 175-177, 198-200, 211
Bisbenzylisoquinoline alkaloids 177-179, 200-206
Butylated cocaine derivatives 113, 114, 120
Cactus alkaloids 159-169
  analysis in biological materials 167, 168
  analysis in drug seizures 169
  analysis in plant material 165
Caffeine metabolites 438, 440, 456-458
Caffeine in pharmaceutical preparations 435, 437, 440, 441, 450-455
Caffeine in plant materials 440, 441, 449

β-Carbolinium derivatives 313, 323
*Catharanthus* alkaloids 318-321, 359-369
*Chelidonium* alkaloids 174, 175
Chemisorption 3
*Cinchona* alkaloids 131-157
  analysis in biological material 157
  analysis in drug seizures 156
  analysis in food and beverages 157
  analysis in pharmaceutical preparations 154, 155
  analysis in plant material 150-153
Classification of solvents 5-8
Clavine alkaloids 376, 379, 380, 389, 390, 392
Coca leaves 114, 116, 125
Cocaine metabolites 113-115, 121-123, 127, 128, 240
Colchicine and related alkaloids 469-476
*Conium* alkaloids 61, 63, 64
*Consolida* alkaloids 462, 465
*Corydalis* alkaloids 189
Curarimetics 178, 179, 204-206, 483-495
*Cytisus* alkaloids 62, 64
*Delphinium* alkaloids 461
Detection methods and spray reagents 497-512
Dihydro ergot alkaloids 376, 377, 379, 381, 382, 388, 390-392
Diterpene alkaloids 461-467
  analysis in plant materials 466, 467
Double spot formation 3, 9, 160
Dragendorff's reagent 11-14
Dragendorff's reagent false positive reaction 12-14
Drugs of abuse 113-129, 159, 169, 233-306
  analysis in biological materials 291-299
  analysis in drug seizures 302-306
Ergometrine alkaloids 376, 377, 388-392
Ergot alkaloids 54, 375-408
  analysis in biological materials 403
  analysis in fungi 394-397
  analysis in higher plants 398
  analysis in pharmaceutical preparations 399-402
Ergotamine alkaloids 376, 377, 388-392
Ergotoxine alkaloids 376, 377, 388-392
Extraction from biological fluids 239-245
Extraction methods 51-53
*Fagara* alkaloids 180, 188, 484
False positive reaction with Dragendorff's reagent 12-14
*Fumaria* alkaloids 173
*Genista* alkaloids 62
Glyco-alkaloids 415 (see *Solanum* alkaloids)
Harman alkaloids 323, 373
Heteroyohimbine alkaloids 316-318, 350-358 (see also *Strychnos*, *Rauwolfia*, *Vinca* and *Catharanthus* alkaloids)
*Holarrhena* alkaloids 417, 419, 420
*Hydrastis* alkaloids 174, 175-177, 198-200
*Hyoscyamus* alkaloids 91-112
Indole alkaloids 307-414
Imidazole alkaloids 477-482
Ipecacuanha alkaloids 182-184, 215-218
Isolation methods 51-53
Isoquinoline alkaloids 159-220
Jaborandi folium 478
Leguminosae alkaloids 62, 63, 70, 71
*Lobelia* alkaloids 61, 64, 66
Local anaesthetics 114, 115, 121

LSD and derivatives 378-408
analysis in drug seizures 404-408
Lumi-ergot alkaloids 377, 391
*Lupine* alkaloids 62, 63
Lysergic acid derivatives 377-379, 388-392
*Mitragyna* alkaloids 316-318, 350-354
Morphinane alkaloids 213
Morphine metabolites 288, 300, 301
Nicotine metabolites 79
N-oxides 61, 63, 484, 490, 491
Opium alkaloids 221-306
analysis in opium 265-269
analysis in pharmaceutical preparations 272-280
analysis in plant material 265-269
quality control 270, 271
Oxindole alkaloids 316-318, 350-358
Papaveraceae alkaloids 173, 180, 181, 184, 211-214 (see also opium alkaloids)
Pavine alkaloids 213, 214
*Peumus boldus* alkaloids 180, 181, 207
Peyote cactus alkaloids 159-169
Phenylethylamines 159-169
Phthalide alkaloids 175-177, 198-200, 211, 212
Physostigmine decomposition products 321, 322
Pilocarpine and related alkaloids 477-482
analysis in plant material 481
analysis in pharmaceutical preparations 477, 478
*Piper* alkaloids 64
Piperidine alkaloids 61-77
analysis in plant material 73-75
analysis in pharmaceutical preparations 76, 77
Potassium iodoplatinate reagent 14
Primary amines 12, 15, 159-169, 236
Protoberberine alkaloids 172-175, 188-194, 212, 213
Protopine alkaloids 172-175, 188-194, 213
Pseudotropine alkaloids 113-129
analysis in biological materials 127, 128
analysis in drug seizures 129
analysis in pharmaceutical preparations 126
analysis in plant material 120, 125
Psilocybe alkaloids 409-413
*Psilocybe mexicana* 409
*Punica* alkaloids 64
Pyridine alkaloids 61-77, 79-90
analysis in plant material 73-75, 85
analysis in pharmaceutical preparations 76, 77
Pyrrolizidine alkaloids 61-77
analysis in pharmaceutical preparations 76, 77
analysis in plant material 73-75
Pyrrolidine alkaloids 61-77
analysis in pharmaceutical preparations 76, 77
analysis in plant material 73-75
*Rauwolfia* alkaloids 310-316, 339-349
analysis in biological materials 349
analysis in plant material 343-345
analysis in pharmaceutical preparations 346-348
Rhoeadine alkaloids 181, 182, 209, 210, 214
Quinoline alkaloids 131-157
Quinolizidine alkaloids 61-77
analysis in plant material 73-75
analysis in pharmaceutical preparations 76, 77

Quaternary alkaloids 159, 178, 179, 188, 204-206, 483-495
Quaternary ammonium compounds 483-495
Salicylic acid derivatives 435
Sample application 9
Saturation chromatography tank 5
Secondary amines 12, 15, 159-169
*Sedum* alkaloids 61, 66
Silica gel 3, 4
*Skytanthus* alkaloids 61
Solanaceae alkaloids 91-112
*Solanum* alkaloids 415-419, 421, 422, 424, 425, 430, 431
Solvent selectivity 5-8
Steroidal alkaloids 415-433
  analysis in plant material 430-431
Steroid sapogenins 415, 416, 421, 426
*Strychnos* alkaloids 307-310, 330-338
  analysis in pharmaceutical preparations 336, 337
  analysis in plant material 333-335
Terpenoid indole alkaloids 307-373
*Thalictrum* alkaloids 172, 177, 178, 201
Theophylline in biological material 440, 456-458
Tobacco alkaloids 79-90
  analysis in biological materials 89-90
  analysis in plant material 85
  in drugs of abuse analysis 86-88
Tropane alkaloids 91-129
Tropine alkaloids 91-112
  in drugs of abuse analysis 104
  analysis in biological material 112
  analysis in pharmaceutical preparations 108-110
  analysis in plant material 105-107
Tropolone alkaloids 469
Tryptamines 378, 390
*Uncaria* alkaloids 316-318, 355-358
*Veratrum* alkaloids 417, 421, 422, 424, 425
*Vinca* alkaloids 318-321, 359-369
Xanthine derivatives 435-460
  analysis in biological materials 440, 456, 458
  analysis in food and beverages 449
  analysis in pharmaceutical preparations 450-455
  analysis in plant material 440
Xanthine metabolites 437, 438, 440

COMPOUND INDEX

Acetaminophen 282
Acetophenazine 17, 283
Acetophenone 66
3β-Acetoxy-20α-dimethylamino-5α-pregnane 418
3β-Acetoxy-20β-dimethylamino-5α-pregnane 418
7β-Acetoxy-1-methoxymethyl-1,2-dehydro-8α-pyrrolizidine 67
Acetylcholine 486-488
Acetylcodeine 257, 259, 262
O-Acetylcolchiceine 472
N-Acetyldemecolcine 473, 474
2-Acetyl-2-demethylcolchicine 472
3-Acetyl-3-demethylcolchicine 472
Acetyldihydrocodeine 258
Acetyldihydrocodeinone 23, 25, 258
Acetylethylmorphine 257, 259
N-Acetylisodemecolcine 473
Acetyllobeline 66
N-Acetyllumidemecolcine 473, 474
1-Acetyl lysergide 390
N-Acetylmescaline 160, 163, 164
Acetylmethadol 284, 289
O-Acetylmorphine 233, 257, 259, 262, 287, 289
O-Acetylnandinine 24
Acetylpholcodine 257, 259
Acetyl salicylic acid 36, 285
Acidihydroergotamine 388
Aconine 461, 464
Aconitine 27, 29, 30, 37, 461, 462, 464, 465
Acriflavine 486, 488
Adenocarpine 70
Adlumidine 198, 211
Adlumine 198, 212
Agroclavine 388, 392
Ajmalicine 312-314, 321, 339, 341, 342, 350, 351, 355, 357, 359-361, 364
Ajmaline 20, 33, 37, 40-45, 313-315, 339, 341, 342, 367
Akagerine 54
Akuammicine 361, 362, 364, 366
Akuammidine 362, 367
Akuammigine 351, 355, 357
4-R-akuammigine N-oxide 355, 357
Akuammine 321, 361, 362, 364, 367
Alborine 213
Alcuronium 330, 490, 491
C-Alkaloid H 330
Alkonium 488
Allocryptopine 34, 189, 190, 195, 213
Allosedamine 66
Alphaprodine 258, 289
Alpigenine 209, 214
Alpinine 209, 214
Alpinone 213
Alstonine 312, 313, 330, 339, 341, 362, 365, 490, 491
Alvodine 287
3β-Aminoconan-5-ene 420
Aminophenazone 39
Aminophylline 40-45, 446
Aminopyrine 25, 435
Amitriptyline 39, 122
Ammocalline 361, 364
Ammorosine 361, 364
Amobarbital 17, 282, 285
Amphetamine 17, 39, 122, 234, 237, 238, 242, 283, 284, 286

Amurensine 214
Amurensinine 214
Amurine 212, 213
Amuroline 212
Amuronine 212
Amylocaine 23
Anabasine 79, 80, 81, 83
Anacrotine 69
Anagyrine 70, 71
Anapawine 219
Anatabine 83
7-Angelylheliotridine 67
7-Angelylretronecine 67
Angustifoline 70
Angustidine 356, 358
Angustine 57, 332, 356, 358
Angustoline 356, 358
Anhalamine 163
Anhalidine 163
Anhalinine 163
Anhalonidine 163
Anhalonine 163, 164
Anhydrocevadine triacetate 425
Anhydrocevagenine triacetate 425
Anhydroplatynecine 67
Anhydroveracevine tetraacetate 425
Anileridine 17, 287
Anthranoillycoctonine 465
Anti-isorhynchophylline N-oxide 355, 357
Antirhine methochloride 330
Anti-rotundifoline N-oxide 354
Antistine 122
Apoatropine 20, 35, 91, 100-102
Apomorphine 17, 23, 24, 257, 259, 283
Aposcopolamine 100-102
Apovincamine 367
Arecoline 21, 40, 42-44, 61, 63, 69, 79
Argemonine 213
Aricine 339, 341, 342, 351
Armepavine 33, 198, 211
Aspidospermine 21, 33
Atropine 16, 17, 20, 23-25, 27, 30-32, 35-37, 40, 42-44, 55, 91-95, 100-103, 122, 283, 285, 477, 480
Atropine methyl nitrate 16, 30, 101
Atropine N-oxide 101, 102
Autumnaline 473
Azamethonium 487, 488
Bamifylline 25, 438, 446
Bebeerine 200
Belladonnine 24, 100
Benzalkonium chloride 36, 486, 488, 489
Benzilylcholine 488
Benzocaine 17, 23, 121, 286
Benzoylaconine 461, 463, 464
Benzoylcholine 488
O-Benzoylcolchiceine 472
N-Benzoyldemecolcine 473
N-Benzoyl-N-desacetylcolchiceine 472
Benzoylecgonine 46, 47, 113-117, 120-129
Benzoylnorecgonine 113, 117, 121
8-Benzyltheophylline 446
Berbamine 200-202
Berberine 27, 34, 40-45, 54, 56, 172-174, 188-190, 195, 198, 212, 486
Berberrubine 34, 189, 190
Bicuculine 198, 211
Bisnor-C-alkaloid H 330, 332
Bisnor-C-alkaloid H N-oxide 330
Bisnor-C-alkaloid H di-N-oxide 330
Bisnorargemonine 213
Bisnordihydrotoxiferine 330, 332
Bisnordihydrotoxiferine N-oxide 330
Bisnordihydrotoxiferine di-N-oxide 330

$3\beta,20\alpha$-bis-(Dimethylamino)-$5\alpha$-pregnane 418
$3\beta,20\beta$-bis-(Dimethylamino)-$5\alpha$-pregnane 418
$3\alpha,20\alpha$-bis-(Dimethylamino)-pregn-5-ene 418, 419
$3\beta,20\alpha$-bis-(Dimethylamino)-pregn-5-ene 418, 419
$3\alpha,20\alpha$-bis-(Dimethylamino)-pregn-5-en-18-ol 419
$3\beta,20\alpha$-bis-(Dimethylamino)-pregn-5-en-18-ol 419
Boldine 20, 40-45, 180, 181, 207
Bracteoline 212
Bretylium 487
8-Bromocaffeine 446
Bromocholine 488
Bromodiphenhydramine 282
8-Bromotheophylline 446
Brucine 16, 20, 23, 30-33, 35-37, 40-47, 307-310, 321, 329-338
Brucine N-oxide 330
Brunsdonnine 219
Bufotenine 17
Bulbocapnine 20, 207, 212
Bulphanamine 219
Butabarbital 282
Butacaine 121
Butaperazine 27
Butylscopolamine 23, 30, 92, 101, 486, 488, 489
Butyrylcholine 488
Butyrylthiocholine 488
Caffeine 23, 31, 32, 36, 37, 40-45, 122, 236, 283, 285, 440-441, 446, 447
Calycotomine 70
Canadine 189, 198
Candicine 164, 188
Cannabidiol 286
Cannabinol 286
Capnoidine 198, 211
Caracurine V 330, 332, 490, 491
Caracurine V di-N-oxide 330
Caracurine V methoiodide (toxiferine) 330
Caracurine V N-oxide 330
Caranine 219
Carapanaubine 353
Carbachol 486, 488
Carbetidine 258
Carbinoxamine 25, 282
Carbromal 282
Carnegine 164
Carosidine 359, 361, 364
Carosine 359, 362, 364
Cathalanceine 361, 364
Catharanthine 359-361, 364
Catharicine 362, 364
Catharine 359, 360, 362
Catharosine 359
Cathindine 361, 364
Cavincine 361, 364
Cavindicine 361, 364
Cephaeline 20, 36, 182-184, 215, 231
Cepharanthine 200
Cetrimide 487
Cevacine 421, 425
Cevadine 43, 44, 425
Cevagenine 425
Cevine 421, 425
Chaconine 415, 421, 424
Chanoclavine 388, 392
Chelerythrine 34, 188, 195, 213
Chelidonine 195, 213
Chelilutine 195, 213
Chelirubine 195, 213
Chlorcyclizine 282
Chlordiazepoxide 17, 27, 283, 284, 290
Chlormezanone 283
8-Chlorocaffeine 446

Chloroquine 17, 25, 283
8-Chlorotheophylline 446
Chlorothiazide 17
Chlorpheniramine 282, 284
Chlorpromazine 17, 27, 122, 281, 283, 284, 290
Chlorprothixene 283
Choline 486, 487, 488
Chondocurarine 204
Ciliaphylline 352, 356, 357
Cinchonidine 32, 36, 37, 40-45, 131-157
Cinchonine 20, 31-33, 36, 37, 40-45, 131-157
Cinchoninone 133, 141, 149
Cinnamylcocaine 114, 116, 120
Citisine 23
Cocaine 16, 17, 21, 23-25, 27, 30-32, 35, 37, 39, 40, 42-44, 46, 47, 113-117, 120-129, 236, 244, 281, 282, 284, 285, 287, 290, 480
Codamine 33, 211
Codeine 16, 17, 20, 23, 25, 27, 30-33, 35-37, 39, 46, 47, 122, 213, 223-226, 228, 229-232, 237, 257-260, 262, 281, 283-285, 287-290
Codeine N-oxide 258
Colchiceine 472
Colchicine 20, 23, 31, 35, 40-45, 469, 470, 472, 474
Colchocoside 473
α-Colubrine 310, 330-332
β-Colubrine 310, 330-332
Columbamine 172, 173, 188-190
Conamine 420
$\Delta^{N(20)}$-5α-Conanene 420
$\Delta^{N(20)}$-5α-Conanen-3β-ol 420
5α-Conanine 420
Concuressine 419
Condelphine 465
γ-Coneceine 64
Conessimine 420
Conessine 37, 416, 419, 420
Coniine 61, 64, 79
Conimine 420
Coptisine 189, 190, 212
Corlumidine 198, 212
Corlumine 198, 211
Cornigerine 472
Corydaline 189, 190, 213
Corydalmine 189
Corydine 207, 212
Corynantheidine 350, 351, 355, 357
Corynantheine 350, 351
Corynanthine 339
Coryneine 188
Corynoxeine 356, 357
Corynoxine 352, 356, 357
Corynoxine B 352, 356, 357
Corypalline 164
Corytuberine 207, 212
Costaclavine 388, 392
Cotarnine 21, 24, 198, 211, 257, 259
Cotinine 80, 83
Crinamidine 219
Crinamine 219
Crinidine 219
Crinine 219
Crispatine 68
Criwelline 219
Cryptopine 190, 213, 257, 262
Cupreidine 133, 144, 149
Cupreine 20, 133, 141, 149
Cyclanoline 31
Cyclazocine 284
Cycleanine 200
Cyclizine 238
Cycloartenol 416
Cyclopamine 425, 427
Cytisine 63, 70, 71
Dauricine 200
Decamethonium 487, 488

Dehydrocorydaline 189
2-Dehydroemetine 215
9-Dehydrohecogenine 416
2-Dehydroisoemetine 215
$\Delta^5$-Dehydrolupanic acid 70
$\Delta^5$-Dehydrolupanine 70
$\Delta^{11}$-Dehydrolupanine 70
3-Dehydroreserpine 339
Delcosine 465
Delsoline 465
Demecolceine 473
Demecolcine 473, 474
2-Demethylcolchicine 472
3-Demethylcolchicine 472
2-Demethyldemecolcine 473
3-Demethyldemecolcine 473
Demethyleneberberine 189, 190
Demissidine 416, 417, 425, 427
Desacetylcolchiceine 473
Desacetylcolchicine 473
Desacetylisocolchicine 473
Desacetylvincaleukoblastine 361, 364
Deserpidine 339, 341, 342
Desimipramine 39
Desipramine 27
N-Desmethyl-5α-conanine 420
Desoxyretronecine 67
Dextromethorphane 17, 238, 258
Dextromoramide 23, 39, 258
Diaboline 330, 332
3,N-Diacetyl-3-demethyldemecolcine 473
Diacetyllobelanidine 66
Diacetylretrorsine 69
Diallylbarbituric acid 282
Diazepam 27, 283, 284, 290
Dicentrine 33
Dicyclomine 17, 283
Diethylmorphine 257
N,N-Diethylnicotinamide 23
Diethylpropion 17
N,N-Diethyltryptamine 390
Digalogenine 416
Digitogenine 416
Dihydrochanoclavine 392
Dihydrocinchonidine 131-157
Dihydrocinchonidine N-oxide 141
Dihydrocinchonine 131-157
Dihydrocodeine 20, 25, 30, 31, 122, 257, 258, 283
Dihydrocodeinone 20, 25, 27, 30, 35, 258, 283, 287, 289, 290
Dihydroconessine 419
Dihydrocorynantheine 350, 351, 355, 357
Dihydrocupreidine 141, 149
Dihydrocupreine 141, 149
Dihydroelymoclavine 389
Dihydroepicinchonidine 131-134, 141, 143-147, 149
Dihydroepicinchonine 131-134, 141, 143-147, 149
Dihydroepiquinidine 131-134, 141, 143-147, 149
Dihydroepiquinine 131-134, 141, 143-147, 149
Dihydroergocornine 35, 388, 392
Dihydroergocristine 20, 35, 388, 390-392
Dihydroergocryptine 35, 388, 392
Dihydroergosine 388, 392
Dihydroergotamine 20, 23, 35, 388, 390, 392
Dihydrogambirtannine 356, 358
Dihydrohydroxycodeinone(oxycodone) 30, 35, 39, 224, 257, 289, 290
Dihydrohydroxymorphinone 289, 290
Dihydrolysergic acid 388
Dihydrolysergic acid amide 388
α-Dihydrolysergol 389
Dihydromorphine 287, 289

Dihydromorphinone 20, 23, 30, 35, 235, 258, 281, 284, 287, 289
7α,11α-Dihydroxytomatidine 425, 427
9α,11α-Dihydroxytomatidine 425, 427
1-(2',3'-Dihydroxypropyl)theobromine 446, 447
Dihydroquinidine 131-157
Dihydroquinidine N-oxide 141
Dihydroquinine 131-157
Dihydrositsirikine 361, 364
Dihydrotoxiferine 330, 490, 491
Dimenhydrinate 282, 446
Dimethindene 282
3,4-Dimethoxy-β-phenethylamine 163
DL-4-Dimethylallyltryptophan 389
3α-Dimethylamino-5β-cholestane 418
3α-Dimethylamino-5α-cholestane 418
3β-Dimethylamino-5α-cholestane 418
3β-Dimethylamino-5β-cholestane 418
3α-Dimethylaminocholest-5-ene 418
3β-Dimethylaminocholest-5-ene 418
3α-Dimethylaminoconan-5-ene 419
3β-Dimethylaminoconan-5-ene 419, 420
3α-Dimethylamino-5α-conanine 419
3β-Dimethylamino-5α-conanine 419
3β-Dimethylamino-N-desmethylconan-5-ene 420
18-Dimethylamino-5α-pregnan-3-one 420
20α-Dimethylamino-5α-pregnan-3-one 418, 420
20β-Dimethylamino-5α-pregnan-3-one 418, 420
1,3-Dimethyl-7-[2'-(2''-hydroxy-2''-3''',4'''-dihydroxyphenyl)ethyl-amino)ethyl]xanthine 447
1,3-Dimethyl-7-[2'-(1''-methyl-2''-phenylethylamino)-ethyl]xanthine 447
N,N-Dimethyl-β-phenethylamine 164
Dimethyltryptamine 17, 390
Dimethyltubocurarine 486, 489
1,3-Dimethyluric acid 446
Diosgenine 416, 426
Diphenhydramine 17, 282, 284
Diphenylhydantoin 17, 282
Diphenylpyraline 282
α,β'-Dipyridyl 83
Dixyrazine 27
Dolichotheline 164
Domesticine 207, 212
Doxylamine 17, 25
Dyphylline 446, 447
Ecgonine 23, 46, 47, 113, 116, 120-123
Echimidine 68
Echinatine 68
Echiumine 68
Ectylurea 283
Elatine 461, 465
Eldeline 465
Elwesine 219
Elymoclavine 388, 392
Emetamine 215
Emetine 21, 30, 32, 36, 37, 40-44, 182-184, 215, 231
Ephedrine 16, 17, 20, 23, 25, 27, 31, 32, 36, 39, 54, 164, 321, 283-285
19-epi-Ajmalicine 355, 357
Epialpinine 214
Epicinchonidine 131-134, 141, 143-147, 149
Epicinchonine 131-134, 141, 143-147, 149
Epi-13-hydroxy-α-isolupanine 70
Epi-13-hydroxylupanine 70
Epilupinine 70
(22R,25S)-22,26-Epimino-5α-cholestan-3β,16β-diol 417
(22S,25R)-22,26-Epimino-5α-cholestan-3β,16β-diol 417

(22S,25S)-22,26-Epimino-5α-cholestan-3β,16β-diol 417
(22R,25S)-22,26-Epimino-cholest-5-en-3β,16β-diol 417
(22S,25R)-22,26-Epimino-cholest-5-en-3β,16β-diol 417
(22S,25S)-22,26-Epimino-cholest-5-en-3β,16β-diol 417
Epinephrine 17
Epipapaverrubine G 214
Epiquinidine 131-134, 141, 143-147, 149
Epiquinine 131-134, 141, 143-147, 149
Epi-3-rauvanine 351
Ergocornine 376, 388, 390, 392
Ergocorninine 388, 390, 392
Ergocristine 20, 40-45, 376, 388, 390-392
Ergocristinine 20, 388, 390, 392
Ergocryptine 376, 388, 390-392
Ergocryptinine 388, 390, 392
Ergometrine 20, 31, 32, 35, 40-45, 376, 377, 381, 388, 390-392
Ergometrinine 388, 390, 392
Ergosine 388, 390, 392
Ergosinine 388, 390, 392
Ergostine 388, 390, 392
Ergostinine 388, 390, 392
Ergotamine 16, 20, 31, 32, 35, 37, 40-45, 376, 377, 379, 381-383, 388, 390, 392, 437
Ergotaminine 388, 390-392
Ergotoxine 32, 376, 377 (see also ergocornine, ergocristine and ergocryptine)
Ervamine 362, 366
Ervinceine 366
Ervincinine 366
Ervine 362, 367
Ervinidine 362, 366
Ervinidinine 366
Eserinol 321, 370
Ethinamate 17, 282
2-Ethyl-2-demethylcolchicine 472
3-Ethyl-3-demethylcolchicine 472
3-Ethyl-3-demethyldemecolcine 473
Ethylmorphine 25, 27, 30-32, 36, 39, 40-45, 224, 225, 230, 257, 258, 259, 283, 287, 289
Etioline 425, 427
Etophyllate 446
Eucatropine 40, 42-44
Europine 68
Fagarine II 34
Fangchinoline 202
Fenethylline 446, 447
Festuclavine 388, 392
Flexamine 219
Flexine 219
Flexinine 219
Flufenazine 27
N-Formyldemecolcine 473
N-Formyldesacetylcolchiceine 472
N-Formyldesacetylcolchicine 472
Fulvine 68
Fumigaclavine A 388, 392
B 388, 392
C 388, 392
Gallamine 486-489
Galanthamine 219
Galanthine 219
Galugenine 214
Gambirine 355, 357
Gambirtannine 356, 358
Gentianine 57
Gitogenine 416
Glaucamine 209, 214
Glaucine 201, 203, 212
Glaudine 214
Glaziovine 212
Glutethimide 17, 244, 282, 286, 290
Gnoscopine 212

Grantianine 68
Guanethidine 487
Haemanthamine 219
Haemanthidine 219
Harmaline 356, 358, 373
Harmalol 373
Harman 33, 323, 356, 358, 373
Harmine 356, 358, 373
Harmol 373
Hayatidine 200
Hayatine 200
Hayatinine 200
Hecogenine 416
Heleurine 67
Heliosupine 68
Heliotridane 67
Heliotridine 67, 69
Heliotrine 68, 69
Heptabarbital 17, 282
Herbadine 367
Herbamine 367
Hernandesine 201
Heroin 17, 23, 24, 25, 27, 30, 31, 33, 121, 228, 233-235, 238, 239, 257-260, 262, 281, 283, 285, 287, 289
Hexamethonium 487, 488
Hexobarbital 282
Hexobendine 25
1-Hexyltheobromine 446, 447
Hippawine 219
Hippeastrine 219
Hirsuteine 354, 355, 357
Hirsutine 351, 354, 355, 357
Histamine 164
Histidine 164
Holamine 426
Holaphyllamine 426
Holaphylline 426
Holocaine 121
Homatropine 16, 20, 24, 30, 31, 36, 40, 42-44, 91-94, 100-103, 477, 480
Homatropine methylbromide 16, 36, 100, 103
Homatropine N-oxide 101
Homochelidonine 195, 213
Homolinearisine 212
Homolycorine 219
Hordenine 20, 163, 164
Hunnefoline 190, 213
Hunnemanine 34, 190, 213
Hydrochlorthiazide 17
Hydrocotarnine 211
13-Hydrolupanine 70
Hydrastine 24, 40-45, 176, 177, 198, 211
Hydrastinine 21, 175-177, 198, 486
10-Hydroxycodeine 257
6-Hydroxycolchicine 472
3β-Hydroxy-5α-conanine 420
3β-Hydroxy-N-desmethyl-5α-conanine 420
7-(2'-2''-Hydroxy-2''-(3''',4'''-dihydroxyphenyl)ethylamino)ethyl-theophylline 446
2'-Hydroxy-5,9-dimethyl-2-cyclopropyl-methyl-6,7-benzomorphan 289
2'-Hydroxy-5,9-dimethyl-2-(3,3-dimethyl-allyl)-6,7-benzomorphan 289
*dl*-2'-Hydroxy-5,9-dimethyl-2-phenethyl-6,7-benzomorphan 289
10-Hydroxyergocristine 391
10-Hydroxyergocryptine 391
10-Hydroxyergometrine 391
10-Hydroxyergotaminine 391
7-(2'-Hydroxyethyl)theophylline 446, 447
Hydroxyheliotridane 67
13-Hydroxy-α-isolupanine 70
10-Hydroxy-LSD 391
4-Hydroxylupanine 70
10-Hydroxylysergic acid 391

β-Hydroxymescaline 164
7β-Hydroxy-1-methoxymethyl-1,2-dehydro-8α-pyrrolizidine 67
4-Hydroxy-3-methoxystrychnine 332
7α-Hydroxy-1-methyl-1,2-dehydro-8α-pyrrolizidine 67
7β-Hydroxy-1-methylene-8α-pyrrolizidine 67
1-Hydroxymethyl-1,2-epoxy-8α-pyrrolizidine 67
7β-Hydroxy-1-methylene-8β-pyrrolizidine 67
3-Hydroxy-N-methylmorphinan 258, 289
*l*-3-Hydroxymorphinan 289
11-Hydroxypleiocarpamine 367
1-(2'-Hydroxypropyl)theobromine 446, 447
7-(2'-Hydroxypropyl)theophylline 446, 447
Hydroxysparteine 70
15α-Hydroxysoladulcidine 417
15β-Hydroxysoladulcidine 417
15α-Hydroxysolasodine 417
15β-Hydroxysolasodine 417
4-Hydroxystrychnine 330, 332
7α-Hydroxytomatidene 425
9α-Hydroxytomatidene 425, 427
15α-Hydroxytomatidenol 417
7α-Hydroxytomatidine 427
15α-Hydroxytomatidine 417
Hydroxytriethonium 488
*l*-2'-Hydroxy-2,5,9-trimethyl-6,7-benzomorphan 289
Hydroxyzine 27
*l*-Hyoscyamine (see also atropine) 93
Hypoxanthine 437
Ibogaine 17, 40-45, 390
Icajine 308, 330, 332
Imipramine 27, 39, 122, 283, 284
Indicine 67
Indole 69
Inosine 446
Integerrimine 68
Iproniazid 284
Isoajmalicine 351, 355, 357
Isoboldine 212
3-Isobutyl-1-methylxanthine 446
Isocarapanaubine 353
Isochondodendrine 200
Isocolchicine 472
Isoconessimine 420
Isocorydine 201, 207, 212
Isocorynantheidine 350, 355, 357
Isocorynoxeine 355, 357
Isocorypalmine 212
22-Isodemissidine 417
Isodihydrolysergol 392
3-Iso-19-epi-ajmalicine 355, 357
Isofangchinoline 202
Isojavaphylline 353
22-Isolanidine 417
Isoleurosine 359-361, 364
Iso-LSD 389-392
α-Isolupanine 70
Isolysergene 392
D-Isolysergic acid 389
D-Isolysergic acid amide 389, 390, 392
Isolysergic acid dimethylamide 391
Isolysergic acid dipropylamide 391
Isolysergic acid ethylpropylamide 391
Isolysergic acid methylpropylamide 391
Isolysergine 389, 392
Isolysergol 389
Isomajdine 367
Isomitraphylline 350, 353-355, 357
Isomitraphylline N-oxide 355, 357
Isoniazid 284
Isopenniclavine 389, 392
Isophoramine 70

Isopilocarpine 477, 478
Isopteropodine 353, 355, 357
Isopteropodine N-oxide 355, 357
Isoraunitidine 351
Isoreserpiline 339, 341, 351, 362
3-Isoreserpine 339
Isoreserpinine 351
Isoretronecanol 67
Isorhoeadine 209, 214
Isorhoeagenine 214
Isorhynchophylline 350, 352, 354, 355, 357
Isorotundifoline 352, 354, 356, 357
Isorubijervine 425, 427
Isosetoclavine 389, 392
α-Iso-sparteine 33, 70
Isothebaine 207, 212
Isotetrandrine 200, 202
Isotrilobine 33, 200
Isovincoside-lactam 57
Jacobine 68
Jacoline 68
Jaconine 68
Jacozine 68
Jatrorrhizine 31, 172, 173, 188, 189 190
Javaphylline 353, 355, 357
Jervine 36, 416, 425, 427
Ketobemidone 25, 258, 289
3-Keto-5α-conanine 420
8-Ketosparteine 70
Kopsanone 367
Kopsinilam 367
Kopsinine 362, 367
Krelagine 219
Krepowine 219
Kreysigine 473
Kreysiginine 473
Lamprolobine 70
Lanceine 361, 364
Lasiocarpine 68, 69
Latericine 198, 211
Latifoline 68
Laudanidine 211, 227, 257
Laudanine 211, 224, 226, 231, 257
Laudaninine 226
Laudanosine 33, 198, 201, 202, 211, 257
Laurifoline 188
Leurocristine (vincristine) 318, 319, 360, 361, 364
Leurosidine 318, 319, 360, 361, 364
Leurosine 318, 319, 359, 360, 361, 364
Leurosivine 361, 364
Levo dromoran 287
Levomepromazine 27
Levorphanol 281
Lidocaine 17, 121, 285
Limacine 202
Lobelanidine 32, 66, 76
Lobelanine 32, 64, 66
Lobeline 21, 30, 32, 40-45, 62, 64, 66
Lochnericine 359-361, 364
Lochneridine 359-361, 364
Lochnerine 359-361, 364
Lochnerinine 361, 364
Lochnerivine 361, 364
Lochrovicine 361, 364
Lochrovidine 361, 364
Lochrovine 361, 364
Lophophorine 160, 163, 164
LSD 17, 234, 238, 284, 290, 378-383, 389-392
α-Lumicolchicine 474
β-Lumicolchicine 473, 474
γ-Lumicolchicine 473, 474
β-Lumicornigerine 473
γ-Lumicornigerine 473
Lumidemecolcine 474

β-Lumidemecolcine 473
γ-Lumidemecolcine 473
β-Lumi-3-demethylcolchicine 473
γ-Lumi-2-demethyldemecolcine 473
Lumirescinnamine 341
Lumireserpine 341
Lupanic acid 70
Lupanine 33, 70, 71
Lupinine 63, 70
Lycoctonine 465
Lycorenine 219
Lycorine 219
Lysenyl 388
Lysergene 389
Lysergic acid 238, 382, 389-392
Lysergic acid amide 389, 390
*d*-Lysergic acid diethylamide 17, 234, 238, 284, 290, 378-383, 389-392
Lysergic acid dimethylamide 391
Lysergic acid dipropylamide 391
Lysergic acid ethylpropylamide 391
D-Lysergic acid methylcarbinol amide 389
Lysergic acid methylpropylamide 391
Lysergide 285
Lysergine 389
Lysergol 389, 392
D-Lysergyl-L-valine methyl ester 389
Maandrosine 362, 365
Macusine B 330, 332
Magnoflorine 31, 33, 172, 173, 188-190, 212
Magnoline 200
Majdinine 367
Majdine 367
Majdine 1 353
Majdine 2 353
Majdine 3 353
Majdine 4 353
Mandelic acid 93, 101
Masonine 219
Matrine 70
Mecambridine 190, 213
Mecambrine 207, 212
Mecambroline 207, 212
Meclizine 282
Meconine 198
Melinonine A iodide 330, 490, 491
Menisperine 31
Mepazine 283
Meperidine 17, 28, 35, 122, 258, 281, 283-285, 287, 289, 290
Mephentermine 17
Mephenytoin 17
Meprobamate 232, 283, 286
Mepyramine 25
6-Mercaptopurine 446
Mescaline 17, 159-169, 281, 284, 285, 287, 290
Metanephrine 164
Methacholine 488
Methadone 17, 23, 27, 122, 238, 258, 281, 283-285, 287, 289, 290
Methamphetamine 17, 122, 283, 284, 286
Methapyrilene 17, 25, 282, 284, 285
Methaqualone 27, 284, 285
Methesculetol theophylline 446
12-Methoxyajmaline 339, 341
1-Methoxymethyl-1,2-dehydro-8α-pyrrolizidine 67
1-Methoxymethyl-1,2-epoxypyrrolizidine 67
*l*-3-Methoxy-N-methylmorphinan 289
*l*-3-Methoxymorphinan 289
4-Methoxy-β-phenethylamine 163
16-Methoxytabersonine 366
7-Methoxy-1,2,3,4-tetrahydroisoquinoline 164
8-Methoxy-1,2,3,4-tetrahydroisoquinoline 164
3β-Methylaminoconan-5-ene 420

3β-Methylamino-N-desmethylconan-5-ene 420
Methylamphetamine 39
O-Methylandrocymbine 473
O-Methylanhalonidine 163
Methylatropine 486-488
N-Methylcoclaurine 200
Methylcodeine 257
N-Methylconiine 64
N-Methylcorydine 188
N-Methylcytisine 67, 68
N-Methyldemecolcine 473
Methyldeserpidate 339, 341
N-Methyl-14-O-desmethylepiporphyroxin 214
Methyldihydromorphinone 289
N-Methyl-3,4-dimethoxy-β-phenethylamine 163
Methyldopa 17
Methylecgonidine 114, 116
Methylecgonine 46, 47, 121
Methylenedioxyamphetamine 237
1-Methylenepyrrolizidine 67
6-Methyl-$\Delta^{8,9}$-ergolene-8-carboxylic acid 389
Methylergometrine 388, 390-392
Methylergometrinine 390
7-(2'-(1''-Methyl-2''-hydroxy-2''-phenylethylamino)ethyl)theophylline 446
N-Methyl-5-hydroxy-1,2,3,4-tetrahydroisoquinoline 164
N-Methyl-6-hydroxy-1,2,3,4-tetrahydroisoquinoline 164
N-Methyl-7-hydroxy-1,2,3,4-tetrahydroisoquinoline 164
N-Methyl-8-hydroxy-1,2,3,4-tetrahydroisoquinoline 164
N-Methylisocorydine 188
N-Methyllaurotetanine 207
Methyllycaconitine 461, 465
N-Methylmescaline 163, 164
N-Methyl-4-methoxy-β-phenethylamine 163
1-Methylmethylergometrine 388, 390, 392
N-Methyloridine 212
Methyloxamethonium 488
N-Methylparavallarine 55
N-Methyl-β-phenethylamine 163, 164
Methylphenidate 17, 283-285
O-Methylpsychotrine 182, 215
Methylreserpate 315, 339, 341
Methylrosanilinium 488
Methylscopolamine 92, 486, 489
4-Methyl-5-stigmasta-7,24(28)-dien-3-ol 416
O-Methylthalmetine 201
N-Methyltyramine 163, 164
1-Methyluric acid 446
1-Methylxanthine 437
3-Methylxanthine 446
7-Methylxanthine 437
Methyprylon 282
Metoclopramide 25
Metopon 287
Millophylline 438, 446, 447
Mitraciliatine 350, 351, 355, 357
Mitragynine 350, 351, 355, 357
Mitragynine oxindole B 352
Mitrajavine 351, 355, 357
Mitraphylline 317, 350, 353-355, 357, 362, 365
Mitraphylline N-oxide 355, 357
Mogadon 28
Molliclavine 392
Monoacetyllobelanidine 66
17-Monochloroacetylajmaline 315
Monocrotaline 68, 69
Morphine 16, 20, 23, 24, 28, 30-33, 35, 37, 39, 42-44, 46, 47, 120, 122, 213, 223, 226-232, 234, 235,

237-240, 257-260, 262, 281, 283-285, 287-290
Morphine 3-ethereal sulphate 288
Morphine glucuronide 288
Morphine N-methyl iodide 288
Morphine N-oxide 288
Multiflorine 33
Muramine 190, 213
Muscarine 486
Myosmine 83
Nadaurine 213
Nalorphine 239, 258, 281, 288-290
Naloxone 234, 284, 290
Nandinine 34
Narceine 20, 23, 24, 30, 33, 37, 40-45, 212, 223, 227, 228, 230, 231, 257-259, 262
Narcissamine 219
Narcissidine 219
Narcotoline 212, 227, 257
Nartazine 219
Narwedine 219
Neflexine 219
Neoleurocristine 359, 361, 364
Neoleurosidine 359, 361, 364
Neopine 257
Neostigmine 486, 488
Neotigogenine 416
Nerinine 219
Nerispine 219
Nicotine 17, 23, 28, 30, 79-81, 83, 122, 281, 283, 285, 288
*m*-Nicotine 83
Nicotine di-N-oxide 83
Nicotine N-oxide 83
Nicotone 83
Nicotyrine 79, 83
Nisentil 287
Nitidine 188, 195
8-Nitrotheophylline 446
Nor-agroclavine 389, 392
Norajmaline 339, 341
2-N-Norberbamine 202
Norchelidonine 195, 213
Norcodeine 288, 289
Norcotinine 83
Norecgonine 121
Norepinephrine 164
Norisocorydine 207
Normeperidine 56, 289
Normethadone 258
Normethanephrine 164
Normorphine 239, 281, 288, 289
Nornarceine 212
Nornicotine 79-81, 83
2-N-Norobamegine 202
Nortryptiline 17, 39, 283
Noscapine 21, 23, 24, 30-33, 35-37, 40-47, 212, 223, 224, 226-232, 257-259, 262, 283, 287
Novacine 308, 330, 332
Nuceferine 212
Nuciferoline 212
Numorphan 287
Obamegine 201, 202
Ocoteine 33
Octopamine 164
Opianic acid 198
Opipramol 25, 28
Orensine 70
Oreodine 214
Oreogenine 214
Oridine 212
Orientalinone 212
Otosenine 68
Oxamethonium 488
Oxanamide 283
Oxazepam 17
Oxogambirtannine 356, 358
17-Oxolupanine 70
Oxomemazine 25
Oxomuramine 213

Oxo-powelline 219
17-Oxosparteine 67, 70
Oxyacanthine 200
Oxycodone 258
Oxylupanine 71
Oxysanguinarine 195, 213
Palaudine 211
Palmatine 34, 56, 172, 173, 188-190, 213
Pancuronium 487
Papaveraldine 211, 257
Papaverine 16, 21, 23, 24, 28, 30-33, 35, 36, 37, 40-45, 200, 211, 223, 224, 226, 227, 229-232, 257-259, 262, 283, 285, 287
Papaverinol 257
Papaverrubines A-G 214
Paraquat 487, 489
Paraxanthine 437
Parkamine 219
Paynantheine 350, 351
Pellotine 160, 163
Penareine 219
Penduline 202
Penniclavine 389, 392
Pentabarbital 282
Pentamethonium 488
Pentazocine 17, 122, 284, 285
Pericalline 362, 364
Periciazine 28
Pericyclivine 361, 364
Perimivine 361, 364
Perividine 362, 365
Perivine 359, 360, 362, 364
Peronin 287
Perosine 362, 365
Perphenazine 17, 28, 283
Phaeantine 200, 202
Phellodendrine 31
Phenacetin 282, 435, 440
Phenazocine 281, 287
Phenazone 39
Phencyclidine 285
Phendimetrazine 17
β-Phenethylamine 163, 164
Pheniramine 282
Phenmetrazine 25, 284
Phenobarbital 17, 282, 285
Phentermine 17
Phenylephrine 17, 164, 283
Phenylpropanolamine 283
Phenyltoloxamine 282
Phenytoin 286
Pholcodine 232, 257-259
Pholedrine 25
Physostigmine 20, 23, 30, 31, 33, 35, 36, 40-45, 285, 321-323, 330, 331, 477, 480
Physostigmine N-oxide 321
Picrinine 367
Pilocarpic acid 477, 478
Pilocarpine 16, 20, 30, 31, 35, 36, 477-482
Pilocereine 163
Piminodine 289
Pipradrol 284
Platynecine 67, 69
Platyphylline 68
Pleurosine 359, 362
Pluviine 219
Poldine 489
Porphyroxine 257
Powellamine 219
Powelline 219
Pralidoxime 488
Procaine 17, 23, 25, 28, 121, 122, 282, 285, 287
Prochlorperazine 283
Proline 164
Promazine 17, 28, 283, 284
Promethazine 17, 28, 282
Promoton 28

Pronuciferine 212
Propantheline 489
N-Propionyldemecolcine 473
Propoxyphene 122, 281, 282, 284, 285, 289, 290
3-Propyl-3-demethylcolchicine 472
Prostigmine 40, 42-44
Prothipendyl 25
Protoemetine 215
Protopine 34, 189, 190, 195, 213, 257
Protoveratrine A 32, 36, 43, 44
Pseudoaconitine 461
Pseudoberberine 34
Pseudobrucine 330
Pseudococaine 46, 47, 124-129
Pseudokopsinine 362, 367
Pseudomorphine 213, 227, 260, 288
Pseudostrychnine 54, 330
Pseudotropine 101
Psicain new 21
Psilocin 285, 390, 409, 410, 413
Psilocybin 239, 285, 290, 390, 409, 410, 413
Psychotrine 182, 215
Pteropodine 353, 355, 357
Pteropodine N-oxide 355, 357
Pycnamine 200, 202
Pyrathiazine 282
Pyridostigmine 486, 488
Pyrilamine 282
Pyroclavine 389, 392
Pyrrobutamine 282
Pyrvinium 488
Quebrachamine 33, 367
Quinacrine 283
Quinidine 17, 20, 23, 25, 30, 32, 33, 35-37, 40-45, 131-157, 283
Quinidinone 133, 141, 149
Quinine 16, 17, 20, 23, 25, 28, 30, 31, 32, 33, 35-37, 39, 40-45, 122, 131-157, 235, 236, 283-285, 287, 290
Raumitorine 351
Rauniticine 351, 355, 357
Raunitidine 351
Rauvanine 351
Rauvomitine 339, 341
Rauvoxine 353
Rauvoxinine 353
Rauwolscine 20, 339, 341, 342
Renoxidine 339, 341
Repanduline 33, 202
Rescinnamine 37, 311-315, 339, 341, 342
Reserpic acid 339, 341
Reserpiline 313, 339, 341, 342, 351
Reserpine 16, 17, 20, 31, 33, 35, 37, 40-45, 283, 310-315, 339-349, 361, 364
Reserpinine 312, 339, 341, 342, 350, 351
Retamine 70
Reticuline 211, 226, 227, 257
Retronecanol 67, 69
Retronecine 67, 69
Retronecine trachelanthate 68
Retronecine viridiflorate 68
Retrorsine 68, 69
Retusamine 68
Retusine 68
Rhodamine B 28
Rhoeadine 209, 214
Rhoeagenine 209, 214
Rhombifoline 70
Rhynchociline 352, 355, 357
Rhynchophylline 317, 352, 354, 355, 357
Rhynchophylline N-oxide 354, 356, 357
Ridelliine 68
Rinderine 68
Roemerine 207, 212
Rosmarinine 68, 69

Rotoxamine 282
Rotundifoline 317, 352, 354, 355, 357
Rovidine 361, 364
Roxburghine C 356, 358
Roxburghine D 356, 358
Roxburghine E 356, 358
Rubijervine 425, 427
Rubreserine 321, 370
Salicylic acid 282
Salsolidine 164
Salsoline 164
Salutaridine 213
Sanguinarine 34, 174, 175, 195, 213, 486
Santiagine 70
Sarothamnine 70
Sarpagine 20, 33, 37, 313, 315, 339, 341
Sarracine 68
Sarsasapogenine 416
Sceleratine 68
Scopine 102
Scopolamine 16, 20, 23-25, 28, 30-32, 35-37, 40, 42-44, 91-95, 100-103, 283, 285, 477, 480
Scopolamine N-oxide 101, 102
Scopoline 21, 40, 42-45, 95, 100, 101
Scoulerine 213
Secobarbital 17, 282, 285
Sedamine 66
Sedinine 66
Sedridine 66
Senecionine 68, 69
Seneciphylline 68
Senkirkine 68
Serpentine 20, 312,315, 330, 339, 342, 353, 360, 362, 365, 486
Serpentinine 20, 339, 341
Setoclavine 389, 392
Sinactine 213
Sisalagenine 416
β-Sitosterol 416
Sitsirikine 359, 361, 364
Smilagenine 416
Soladulcamarine 424
Soladulcidine 415-417, 424, 427
α-Solamarine 424
β-Solamarine 424
γ-Solamarine 424
Solamargine 424
5α-Solanidanol (3β) 424
5α-Solanidan-3-one 416
5β-Solanidan-3-one 425-427
Solanid-4-en-3-one 425, 427
Solanidine 415-417, 421, 424, 425, 427
Solanine 415, 421, 424
Solanocapsine 416, 425, 427
5α-Solasodanol (3β) 424
5α-Solasodanol (3β) 426
5α-Solasodan-3-one 416
Solasodien 416
Solasodine 415-417, 424, 426, 427
Solasonine 424
Sparteine 21, 23, 30, 35, 37, 40, 41-45, 63, 70, 71, 79
Speciociliatine 351, 355, 357
Speciofoline 352, 356, 357
Speciogynine 350, 351, 355, 357
Speciophylline 350, 353, 355, 357
Speciophylline N-oxide 355, 357
Spectabiline 68
Spermidine 42, 43
Spermine 42, 43
Spermostrychnine 332
Sphaerocarpine 70
Stercuronium 489
Strigosine 69

Strychnine 16, 17, 20, 23, 25, 28, 30-33, 35-37, 40-47, 55, 238, 283, 285, 307-310, 321, 329-338, 490, 491
Strychnine methochloride 330
Strychnine N-oxide 330, 331, 490, 491
Strychnospermine 332
Stylopine 190, 212
Succinylcholine 486
Succiphylline 446
Sulfamerazine 17
Sulfathiazole 17
Supinidine 67
Supinine 68, 69
Suxaethonium 487
Suxamethonium 487, 488
Synephrine 164
Tabersonine 366
Talbutal 282
Tazettine 219
Tembetarine 33, 188
Tenuipine 33
Tetra-*n*-butylammonium 488
Tetracaine 28, 122, 287
Tetracycline 284
Tetrahydroalstonine 339, 341, 350, 351, 355, 357, 359-361, 364
Tetrahydroberberine 33, 190, 212
$\Delta^9$-Tetrahydrocannabinol 17, 286
Tetrahydrocheilanthifoline 189, 190
Tetrahydrocolumbamine 189, 212
Tetrahydrocoptisine 189
Tetrahydrocorysamine 190, 212
Tetrahydroharman 373
Tetrahydropalmatine 189, 190, 212
Tetrahydropseudoberberine 33
Tetrahydrothalifendine 201
Tetramethylammonium 488
Tetramethylholarrhidine 419
Tetramethylholarrhimine 419
Tetrandrine 202
Tetraphyllicine 339, 341, 342
Tetraphylline 350, 351
Tetrylammonium 488
Thalicarpine 178, 201
Thalictricavine 189, 190, 213
Thalidasine 201
Thalidesine 201
Thaliphendine 188-190
Thaliphendlerine 201
Thalmelatine 201
Thalmetine 201
Thalmineline 201
Thebaine 21, 23, 30, 33, 35, 37, 46, 47, 213, 223, 224, 226, 228-230, 257, 258-260, 262, 287
Thenyldiamine 282
Theobromine 32, 36, 40-45, 435-440, 446, 447
Theophylline 36, 40-45, 435-440, 446, 447
Thermopsine 70
Thiamine 486, 489
Thiamylal 282
Thiazinanium 486, 489
Thiazinanium sulphoxide 489
Thiopental 286
Thiopropazate 17, 283
Thioproperazine 25, 28
Thioridazine 283, 284
Thorazine 122
Tigogenine 416, 426
Tigogenon 416
5$\alpha$-Tomatidan-3-on 416
Tomatidenol 416, 417, 427
$\Delta^5$-Tomatidenol (3$\beta$) 424, 426
Tomatidiene 424
$\Delta^{3,5}$-Tomatidiene 424, 426
Tomatidine 416, 417, 426, 427

Tomatillidine 425, 427
Tomatine 418, 424
Tombosine 367
Trifluoperazine 17, 28, 283, 284
Triflupromazine 28
Trilobine 33
Trimeprazine 17, 283
Trimethoprim 25
Tripelennamine 25, 281, 282
Tripolidine 282
Trithioridazine 28
Tropacocaine 21, 24, 46, 47, 116, 120
Tropic acid 93, 95, 101
Tropine 16, 24, 40, 42-45, 91, 95, 100-102
Tropine N-oxide 101
Tropinon 95, 101
Tubocurarine 40-44, 178, 179, 204, 330, 486-488
Tubocurine 31, 204, 205
Tubotaiwine 54
Tyramine 163, 164
Uleine 33
Uncarine A 353, 355, 357
Uncarine B 353, 355, 357
Uncarine F 353, 355, 357
Uncarine F N-oxide 355, 357
Undulatine 219
Uric acid 446
Veracevine 421, 425
Veralkamine 425, 427
Veralobine 425, 427
Veramarine 427
Veramine 424, 425, 427
Verarine 425, 427
Veratramine 425, 427
Veratridine 30, 425
Veratrine 16, 23, 421, 422
Verazine 425, 427
Veronamine 201
Vinaphamine 361, 364
Vinaspine 361, 364
Vincaleukoblastine (vinblastine) 318, 319, 321, 359-361, 364
Vincamajine 367
Vincamicine 359, 360, 362, 364
Vincamine 320, 321, 362, 367
Vincanidine 362, 366
Vincanine 362, 366
Vincaricine 367
Vincaridine 362
Vincarine 362, 367
Vincarinine 367
Vincarodine 359, 361, 364
Vincathicine 361, 364
Vincine 367
Vincolidine 361, 364
Vincoline 361, 364
Vincoside-lactam 57
Vindolicine 359, 369, 362, 364
Vindolidine 359, 362, 364
Vindoline 359-361, 364
Vindolinine 359-361, 364
Vindorosine 361, 364
Vineridine 367
Vinerine 367
Vinervine 362, 366
Vinervinine 362, 366
Vinosidine 361, 364
Virosine 359-361, 364
Vomalidine 339, 341
Vomicine 308, 310, 330-332
Wieland-Gumlich aldehyde 54
Xanthine 437, 446
Xanthoplanine 188
Xanturil 446
Yamogenine 416, 426
Yohimbine 20, 30, 31, 37, 39, 40-45, 285, 313, 315, 316, 339, 341, 342, 361, 364
β- Yohimbine 339, 341
ψ-Yohimbine 339, 341